A02181993001

D1716602

HANDBOOK OF PLANT VIRUS DISEASES

SUTIC, DRAGOLJUB D

SB736.S836 1999

NW

HANDBOOK OF PLANT VIRUS DISEASES

Dragoljub D. Šutić, Ph.D.
Professor
Department of Plant Protection
Faculty of Agriculture
University of Belgrade
Belgrade, Yugoslavia

Richard E. Ford, Ph.D.
Professor
Department of Crop Sciences
Plant Pathology Faculty
College of Agricultural, Consumer and Environmental Sciences
University of Illinois at Urbana–Champaign
Urbana, Illinois

Mališa T. Tošić, Ph.D.
Professor
Department of Plant Protection
Faculty of Agriculture
University of Belgrade
Belgrade, Yugoslavia

CRC Press
Boca Raton   London   New York   Washington, D.C.

Acquiring Editor: John Sulzycki
Project Editor: Carol Whitehead
Marketing Manager: Barbara Glunn, Jane Lewis, Arline Massey, Jane Stark
Cover design: Dawn Boyd
PrePress: Gary Bennett, Kevin Luong
Manufacturing: Carol Slatter

Library of Congress Cataloging-in-Publication Data

Šutić, Dragoljub D.
    Handbook of plant virus diseases / by Dragoljub D. Šutić, Richard E. Ford, Mališa T. Tošić.
       p.   cm.

    Includes bibliographical references (p.   ) and index.
    ISBN 0-8493-2302-9
    1. Virus diseases of plants--Diagnosis--Handbooks, manuals, etc. 2. Plant viruses--Identification--Handbooks, manuals, etc. I. Ford, Richard Earl, 1933- . II. Tošić, Mališa T. III. Title.
SB736.S836 1998
632'.8--dc21
                                                                                                                                                        98-53248
                                                                                                                                                         CIP

    This book contains information obtained from authentic and highly regarded sources. Reprinted material is quoted with permission, and sources are indicated. A wide variety of references are listed. Reasonable efforts have been made to publish reliable data and information, but the author and the publisher cannot assume responsibility for the validity of all materials or for the consequences of their use.

    Neither this book nor any part may be reproduced or transmitted in any form or by any means, electronic or mechanical, including photocopying, microfilming, and recording, or by any information storage or retrieval system, without prior permission in writing from the publisher.

    The consent of CRC Press LLC does not extend to copying for general distribution, for promotion, for creating new works, or for resale. Specific permission must be obtained in writing from CRC Press LLC for such copying.

    Direct all inquiries to CRC Press LLC, 2000 Corporate Blvd., N.W., Boca Raton, Florida 33431.

    Trademark Notice: Product or corporate names may be trademarks or registered trademarks, and are only used for identification and explanation, without intent to infringe.

© 1999 by CRC Press LLC

No claim to original U.S. Government works
International Standard Book Number 0-8493-2302-9
Library of Congress Card Number 98-53248
Printed in the United States of America 1 2 3 4 5 6 7 8 9 0
Printed on acid-free paper

Contents

Preface vii

The Authors ix

Acknowledgments xiii

Introduction xv

Symbols of Host Names xix

Symbols of Symptoms and Non-Host Names xxi

Classification and Nomenclature: Plant Virus Families and Groups xxiii

Chapter 1   Virus Diseases of Food Graminoid Plants 1
    Virus Diseases of Maize (*Zea Mays*) 1
    Virus Diseases of Wheat (*Triticum aestivum*) 24
    Virus Diseases of Barley (*Hordeum vulgare*) 38
    Virus Diseases of Oat (*Avena sativa*) 48
    Virus Diseases of Rye (*Secale cereale*) 53
    Virus Diseases of Rice (*Oryza sativa*) 54
    Sorghum Virus Diseases (*Sorghum bicolor*) 67
    References 71

Chapter 2   Virus Diseases of Forage Feed Plants 85
    Virus Diseases of Graminoids 85
        Virus Diseases of Bromegrass (*Bromus inermis*) 85
        Virus Diseases of Cocksfootgrass (*Dactylis glomerata*) 86
        Virus Diseases of Couch Grass (*Agropyron repens*) 90
        Virus Diseases of False Oatgrass (*Arrhenatherum elatius*) 90
        Virus Diseases of Fescuegrass (*Festuca spp.*) 91
        Virus Diseases of Forage Sorghum (*Sorghum bicolor*) 92
        Virus Diseases of Foxtail Millet (*Setaria spp.*) 92
        Virus Diseases of Molinia (*Molinia coerulea*) 93
        Virus Diseases of Panic Grasses (*Panicum spp.*) 93
        Virus Diseases of Ryegrass (*Lolium perenne* and *Lolium multiflorum*) 94
        Virus Diseases of Sweet Vernal Grass (*Anthoxanthum odoratum*) 96
        Virus Diseases of Timothygrass (*Phleum pratense*) 96
    Virus Diseases of Weed Grasses 97
        Virus Diseases of Johnsongrass (*Sorghum halepense*) 97

Virus Diseases of Forage Legume Plants..99
    Virus Diseases of Alfalfa (*Medicago sativa*) 99
    Virus Diseases of Red Clover (*Trifolium pratense*) 106
    Virus Diseases of Alsike Clover (*Trifolium hybridum*)....................................114
    Virus Diseases of Crimson Clover (*Trifolium incarnatum*)..............................114
    Virus Diseases of Subterranean Clover (*Trifolium subterraneum*).................115
    Virus Diseases of White Clover (*Trifolium repens*)..115
    Virus Diseases of Birdsfoot Trefoil (*Lotus corniculatus*)................................117
    Virus Diseases of Vetch (*Vicia Spp.*)...117
    Other Forage Legumes: Hosts for Some Viruses 117
References 118

Chapter 3   Vegetable Virus Diseases 125

Virus Diseases of Solanaceous Vegetable Plants 125
    Virus Diseases of Pepper (*Capsicum annuum*) 126
    Virus Diseases of Tomato (*Lycopersicon lycopersicum*) 134
    Virus Diseases of Eggplant (*Solanum melongena*) 148
Virus Diseases of Cruciferous Vegetables 150
    Virus Diseases of Cabbage (*Brassica oleracea var. Capitata*), Cauliflower
    (*B. oleracea var. Botrytis*), and Brussels Sprouts (*B. oleracea var. Gemmifera*)...150
    Virus Diseases of Radish (*Raphanus sativus*)...153
    Virus Diseases of Turnip (*Brassica rapa*)...155
    Cruciferous Vegetables: Natural Alternative Host for Some Viruses 157
Virus Diseases of Lily (*Allium spp.*) Vegetables..157
Virus Diseases of Cucurbitaceous Vegetable Plants 162
    Virus Diseases of Cucumber (*Cucumis sativus*) 162
    Virus Diseases of Melon (*Cucumis melo*), Watermelon (*Citrullus vulgaris*),
    Squash (*Cucurbita maxima*), and Courgette (*Cucurbita pepo*) 167
Virus Diseases of Legume Vegetables..173
    Virus Diseases of Bean (*Phaseolus vulgaris*) 174
    Virus Diseases of Pea (*Pisum sativum*) 186
    Virus Diseases of Broad Bean (*Vicia faba*) 194
    Virus Diseases of Cowpea (*Vigna spp.*) 199
Virus Diseases of Umbelliferous Vegetables..203
    Virus Diseases of Carrot (*Daucus carota*) 203
    Virus Diseases of Celery (*Apium graveolens*) 206
    Virus Diseases of Parsnip (*Pastinaca sativa*)..207
    Virus Diseases of Parsley (*Petroselinum crispum*) 208
Virus Diseases of Compositeae Vegetables 208
    Virus Diseases of Artichoke (*Cynara scolymus*)...208
    Virus Diseases of Endive (*Chicorium endivia*)...211
    Virus Diseases of Lettuce (*Lactuca sativa*)...211
Virus Diseases of Chenopodaceae Vegetables..216
    Virus Diseases of Spinach (*Spinacia oleracea*) 216
References 219

Chapter 4   Viral Disease of Industrial Plants 229

Virus Diseases of Peanut (*Arachis hypogaea*) 229

    Virus Diseases of Potato (*Solanum tuberosum*) 240
    Virus Diseases of Tobacco (*Nicotiana tabacum*) 259
    Virus Diseases of Sugar Beet (*Beta vulgaris*)..276
    Virus Diseases of Sugarcane (*Saccharum officinarum*) 292
    Virus Diseases of Sunflower (*Helianthus annuus*)..297
    Virus Diseases of Soybean (*Glycine max.*)...299
    Virus Diseases of Hop (*Humulus lupulus*)...305
    References 310

Chapter 5   **Virus Diseases of Fruit Trees**..**321**
    Virus Diseases of Pome Fruits 330
        Virus Diseases of Apple (*Malus sylvestris*)..330
        Virus Diseases of Pear (*Pyrus communis*) and Those Assumed Caused
        by Viruses..341
        Virus Diseases of Quince (*Cydonia oblonga*) and Those of Quince
        Assumed Caused by Viruses...346
    Virus Diseases of Stone Fruit Trees 347
        Plum Virus Diseases 348
        Virus Diseases of Peach (*Prunus persica*) 360
        Virus Diseases of Apricot (*Prunus armeniaca*) 370
        Virus Diseases of Sweet Cherry (*Prunus avium*) 374
        Virus Diseases of Sour Cherry (*Prunus cerasus*) 386
        Virus Diseases of Almond (*Prunus communis*) 388
    Virus Diseases of Shell Fruits 389
        Virus Diseases of Walnut (*Juglans regia*) 389
        Virus Diseases of Hazelnut (*Corylus avelana*) 391
        Virus Diseases of Chestnut (*Castanea sativa*) 392
    Virus Diseases of Some Tropical And Subtropical Fruit Crops 392
        Virus Diseases of Avocado (*Persea americana*) 392
        Virus Diseases of Banana (*Musa paradisiaca*) 394
        Virus Diseases of Citrus (*Citrus spp.*)...396
        Virus Diseases of Fig (*Ficus carica*)...407
        Virus Diseases of Olive (*Olea europea*)..408
        Virus Diseases of Palm (*Palmaceae*) 409
        Virus Diseases of Papaya (*Caprica papaya*) 410
    References 414

Chapter 6   **Virus Diseases of Small Fruits** **433**
    Virus Diseases of Strawberry (*Fragaria chiloensis*)...434
    Virus Diseases of Raspberry (*Rubus Idaeus* and *R. occidentalis*).....................................443
    Virus Diseases of Black Raspberry (*Rubus occidentalis*) 452
    Virus Diseases of Ribes spp. 457
        Virus Diseases of Gooseberry (*Ribes sp.*)...457
        Virus Diseases of Black Currant (*Ribes sp.*)...458
        Virus Diseases of Red Currant (*Ribes sp.*)...460
    Virus Diseases of Blueberry (*Vaccinium sp.*)...462
    References 467

Chapter 7 Virus and Virus-Like Diseases of Grapevine (*Vitis spp.*)...................................477
    References 496

Glossary...503

Index 535

Preface

Physical and biological technologies have taken a quantum leap in aiding research into the elucidation of plant viruses and thus the diseases caused by them. In fact, the genome of numerous plant viruses has been carefully sequenced. Almost concomitantly, the host plant genomes have been diagrammed in such detail that now one can deduce which nucleotide sequences (genes) of a virus interact with its host genes. We are on the threshold of completing all components of the gene-for-gene hypothesis. Individual viral genes have been discovered by mutational analyses that dictate the specific symptom formation.

One carefully done series of experiments by Tošić et al. (1990. *Plant Dis.* 74: 80:549) have verified that viral strains can be determined by a differential set of sorghum varieties with equal accuracy as their detection by serological or biochemical nucleotide or amino acid sequences. The tactics employed by plant viruses and biochemical changes caused within host plants after infection have made the disease process more easily understood. The PCR (polymerase chain reaction) machines and biochemical gel technologies have revolutionized our science.

Early in the 20th century, viral diseases were diagnosed in plants primarily by the symptoms developed during the infection process. Understanding how symptoms arise and change during the growth of the host plant allows even more accuracy in viral disease diagnosis.

However, even with all these newly discovered facts, the bottom line of plant pathological research is to elucidate how to prevent, eradicate, and/or control the movement of phytopathogenic viruses. A very recent new book *Plant Virus Disease Control* by Hadidi et al. (1998. APS Press, St. Paul, MN) is a welcome treatise on control because host genetic resistance is usually the best answer that plant pathologists have been able to offer. Since most diagnosticians or integrated pest management (IPM) scouts do not have many of the highly sophisticated biotechnologies readily available, the authors of this book perceived a need for a handbook primarily for diagnosis by symptoms that also lists some of the major specific characteristics of each virus. With the aid of more than a dozen recent source books, this handbook provides an authoritative, descriptive, and symptomatic signature of each virus disease. The study of inclusions in various portions of cells infected by viruses is a useful adjunct for aiding diagnosis. Readers are referred to the *CRC Handbook of Viruses Infecting Legumes* (1991) by Edwardson and Christie as that authority.

We chose to highlight food and feed plants because of their requisites to life. The American Phytopathological Society's APS Press (St. Paul, MN) has published two dozen excellent compendia on numerous crop disease subjects during the last 20 years. They contain excellent color photographs of many of the virus-infected plants that are the subject of this book; therefore, they serve as excellent companion books to this one.

Ornamental crops are known to be quite susceptible and broadly infected with viruses. In addition to the APS Press *Compendia of Ornamental Diseases*—Roses 1983, Rhododendron and Azalea 1986, Chrysanthemum 1997, Flowering Potted Plants 1995, and Foliage Plants 1987, a book solely about ornamental viruses has been published in French (Albouy, J. and J.-C. Devergne. 1998. *Maladies à Virus des Plantes Ornamentales.* INRA Editions, Versailles. 492 pp.). These authors cover the subject essentially as we have done for the crops. The other book about urban plants, also in French, covers virological aspects of nonagricultural plants (Angers (France Nov. 1996). 1997. *La Plante dans la Ville.* Ed Riviere, L. M. INRA Editions, Versailles. 352 pp.).

These all serve as good companion sources to this handbook. Virus diseases in forest crops are least well published and we made no attempt to summarize and describe them in this handbook.

J. H. Hill (1991. Standardization of Terminology for description of viruses: Problems and Opportunities. *Advances in Culture Collections* 1:11-22) traced the history of how plant virus nomenclature and classification have evolved over the last several decades. Much progress has been made as university research scientists have gathered regularly every 3 years at the International Congresses of Virology to share research results and organize all viruses into a logical, understandable system. We use the 1991 Classification and Nomenclature for this book.

Taxonomic authorities on plant viruses have been the Reports of the ICTV (International Committee for the Taxonomy of Viruses) plus the Commonwealth Mycological Institute and Association of Applied Biologists (CMI/AAB) *Descriptions of Plant Viruses*, edited by B.D. Harrison and A.F. Murant. The abbreviations or sigla used follow closely those listed in the two-volume set of *Atlas of Plant Viruses*, edited by R.I.B. Francki, R.G. Milne, and T. Hatta (1985). Having perceived redundancies as we reviewed the literature, the need occurred to us to suggest that some of the sigla—especially when two or more identical acronyms refer to quite separate and unrelated viruses—should be revised for more logic and consistency. We have used such a new revised acronym system in this book.

This handbook is written for anyone interested in plant pathology as a science and/or plant diseases as models of complex biological interactions of two independent or codependent organisms. Practical, applied scientists will benefit from this treatise. Also, highly specialized basic scientists can gain an appreciation of some of the elementary facts about classic diseases as well as new ones caused by viruses they may be using for basic research. Specialists who will especially benefit are professionals whose business depends on accurate diagnosis of causal agents of disease, plant clinic diagnosticians, agricultural inspectors, integrated pest management (IPM) scouts, plant science teachers, and extension specialists or private consultants/practitioners. Students seeking higher education in the plant sciences, especially pest management and in particular plant pathology, will be interested.

The authors have tried to make this handbook easily readable. We take sole credit for any errors or oversights and will appreciate these being called to our attention.

Some sections are more detailed than others, which reflects generally the relatively earlier discovery of the disease and also its relative economic importance reflected in a greater volume of accumulated literature.

The Authors

Dragoljub Šutić, Professor Emeritus, received a B.Sc. in Agronomy from the Faculty of Agriculture in 1949, a B.Sc. in Biology from the Faculty of Natural Sciences in 1954, and in 1955 earned his Ph.D. in Plant Pathology from the University of Belgrade, Beograd, Yugoslavia. He then was awarded a 1-year postdoctoral leave to specialize in Plant Virology in the Virology Center at the Plant Pathology Station, Versailles, France (1957/58). He also made study trips and visits to universities and research institutes in the U.S., France, Italy, Germany, Holland, Belgium, Great Britain, Norway, Sweden, Hungary, Czechoslovakia, Bulgaria, and Greece.

At the Faculty of Agriculture, University of Belgrade, he was appointed Teaching and Research Assistant in 1949, Assistant Professor in 1959, Associate Professor in 1961, and Professor in 1969, respectively, of Plant Pathology. He taught courses in Plant Pathology, Advanced Mycology, Plant Bacteriology, Anatomy and Physiology of Diseased Plants, and Plant Virus Diseases. Although retired in 1985, Dr. Šutić maintains an active research and publishing program. He served as major professor for 10 Ph.D. graduate students and as a member of over 30 M.S. and Ph.D. defense committees.

Dr. Šutić's scientific professional memberships include the Serbian Plant Protection Society, Yugoslav Plant Pathology Society, Yugoslav Microbiological Society, and Yugoslav Plant Protection Association. He was appointed Head of the Plant Pathology Division, after which he became a member of the Council and of many committees in the Faculty of Agriculture. He has also served in the following capacities: a member of the councils of the Institute for Plant Protection and Environment, Beograd, Yugoslavia, and Laboratory for Electron Microscopy, University of Belgrade; President and Vice President of Plant Protection Association, Belgrade; President of the Serbian Microbiological Society; President and Vice President of the Yugoslavian Microbiological Society; Vice President of the Yugoslavian Plant Pathology Society; Member of the Council of Mediterranean Phytopathological Union (Firenze, Italy); Editorial Board member of the *Journal of Phytopathology* (Blackwell Wissenschafts-Verlag GesmbH, Berlin, Germany); member of the International Committee on the Taxonomy of Viruses; and member of the International Committee for Fruit Tree Virus Diseases. He participated in 16 international congresses and as chairman of many sections.

Dr. Šutić was named a correspondent member in 1976 of the French Agricultural Academy (Paris, France).

Dr. Šutić is the author of over 160 scientific papers and book chapters, four books, and co-author of three books.

He received the Honorary Plaque of the Faculty of Agriculture, University of Belgrade, in 1970; Silver Wreath of the Republic of Serbia in 1979; Nolit's Award for the book *Plant Virus Diseases* in 1983; Honorary Charter member of the Union of Plant Protection Societies of Yugoslavia in 1986; and Honorary Charter Member of the Plant Protection Association, Belgrade, in 1986. He received an Honorary Plaque in 1986 and the Golden Medal in 1996 of the Fruit Tree Research Institute, Čačak.

Dr. Šutić's primary research interests are plum pox potyvirus (Sharka), graminaceous viruses, vegetable viruses, sugar beet viruses, and host-plant resistance to viruses. He led in developing successful international cooperation on these research areas in addition to other subjects.

He was principal investigator on four international research projects: (1) Sharka or plum pox virus diseases; (2) selection of superior indigenous ecotypes and genotypes among *Prunus* spp. plus investigation of resistance of selections and cultivars to the Sharka (plum pox) virus; (3) seed transmission of forage legume viruses (all three counterpart projects U.S.–Yugoslavia); and (4) Sharka, la maladie à virus des arbres fruitiers (collaborative project France–Yugoslavia).

Richard E. Ford, Professor Emeritus, received his B.S. in Botany in 1956 while working as a research technician on the oak wilt disease at Iowa State University, Ames. He taught general botany and introductory plant physiology laboratory for one semester. Both M.S. and Ph.D. (specialty Virology) in Plant Pathology were earned in 1959 and 1961, respectively, at Cornell University, Ithaca, New York. He accepted a Research Plant Pathologist position in 1961 with the U.S. Department of Agriculture, jointly as an Assistant Professor at Oregon State University in Corvallis, to research legume viruses especially those infecting peas. The discovery of maize dwarf mosaic potyvirus in 1965 caused concern in the U.S. corn belt, whereupon Dr. Ford established a plant virology research laboratory as Associate Professor at Iowa State University. He also taught general virology, plant virology, and physiology of disease courses and was promoted to Professor in 1969. He became Professor and Head, Department of Plant Pathology, University of Illinois, Urbana, in 1972. He maintained active virology research programs and taught several courses during his 20 years of administration. Dr. Ford's program attracted five postdoctoral researchers and five sabbatic professors who researched maize dwarf mosaic and soybean mosaic potyviruses. As major professor, he guided 13 M.S. and 12 Ph.D. candidates. He strongly believed the department head should participate actively in the education of each graduate student; therefore, he served on the examining committee and critically edited the theses of 89 M.S. and 116 Ph.D. candidates at the University of Illinois.

Dr. Ford served on and chaired numerous committees in university, professional scientific societies, administrative searches, grower and seed organizations, international organizations, and for regional and national research. He focused on and actively helped design local, regional, national, and international integrated pest management (IPM) programs that evolved during his career. He enjoyed administrative reviews of programs in 10 Departments of Plant Pathology, teaching in IPM workshops, chairing 22 scientific paper sessions, presenting 80 invited talks, participating in 25 international scientific congresses, and lecturing in 24 countries.

Dr. Ford was a sabbatical Visiting Professor in Canada at the Agriculture Canada Research Station, Vancouver (1985), at the University of Wisconsin (1993), and in Australia at the Queensland University for Technology, Brisbane (1994).

Dr. Ford has been a member of 12 scientific societies, including two as charter member, and an officer of the American Phytopathological Society (APS) (President 1982–1983); North Central Division of APS (President 1978–1979); International Working Group on Legume Viruses (Executive Secretary 1990–1993); Intersociety Consortium for Plant Protection (Chairman 1984); the Association of USA Department Heads of Plant Pathology (Chairman 1976–1978); and Consortium for International Crop Protection (member since 1978, Chairman, Board of Directors 1990–1992 and 1994–1998, and Executive Director 1999–present), among others.

Dr. Ford evaluated over 1200 grant proposals for state, regional, national, and international competitions. He served six years on the editorial board of *Phytopathology* and currently serves on the Editorial Board of the *Journal of Phytopathology* (Germany). He wrote 150 peer-reviewed journal articles and abstracts, 20 book chapters, 2 books, over 35 popular articles, obituaries, and editorials, and 10 biennial research reports for the Department of Plant Pathology, Illinois Agricultural Research Station.

Dr. Ford, an active leader in community service, social and fraternal organizations, listed in 19 different *Who's Who* biographies, is a member of eight honorary societies, including Phi Kappa Phi and Sigma Xi (President 1982, Illinois Chapter). His awards include Fulbright Distinguished Professor; Fellow APS; Fellow AAAS; Fellow National Academy of Sciences, India; and Distinguished Service, North Central Division of APS, among others.

Dr. Ford's research interests center around the potyviruses infecting corn (maize) and legumes. His lifetime research has resulted in identifying natural host resistance or mediating it against viruses, discovering new viruses, elucidating virus–virus interactions, studying the physiological effects of viral infection in plants, and exploring natural and experimental host ranges of potyviruses. Cooperative research with Drs. Shukla (Australia) and Tošić (Yugoslavia) culminated in a major revision of the nomenclature and taxonomy of the maize–sugarcane potyvirus family.

Mališa Tošić, Professor of Plant Pathology, received his B.Sc. in Agronomy from the Faculty of Agriculture, University of Belgrade, Beograd-Zemun, Yugoslavia in 1959. He earned his M.Sc. in Plant Pathology and Ph.D. in Plant Virology from the same university in 1964 and 1971, respectively. He did a Postdoctoral virology specialization for one year in the Department of Botany and Plant Pathology, Iowa State University, Ames, Iowa, under the guidance of Professor R.E. Ford, researching potyviruses that infect maize, sugarcane, sorghum, and Johnsongrass. He also has made study trips and visits to several universities in the U.S. (University of Illinois, Urbana, Illinois; Iowa State University, Ames, Iowa; The Ohio State University, Experiment  Station, Wooster, Ohio); Italy (Istituto di Fitovirologia Applicata, Torino); The Netherlands (Virology Department, Agricultural University, Wageningen); Germany; France; Hungary; Bulgaria; Russia; Great Britain; and Austria. He was a guest at national annual scientific meetings of the American Phytopathological Society (1971 and 1973) and the Indian Phytopathological Society (1997).

Dr. Tošić was elected to the position of Teaching and Research Assistant at the Faculty of Agriculture, University of Belgrade in 1959, then promoted to Assistant Professor in 1971, Associate Professor in 1977, and Professor of Plant Pathology in 1982. He has taught Plant Mycology and now teaches Plant Virus Diseases and Diseases of Field and Vegetable Crops. He has been major professor for nine M.S. and six Ph.D. candidates in Plant Pathology.

He is a member of The Serbian Plant Protection Society, having served as Vice President for one term. He was also elected Chairman of The Plant Pathology Division at the Faculty of Agriculture, University of Belgrade, for two terms. He is the Yugoslavian representative in the Mediterranean Phytopathological Union (Firenze, Italy), and a member of the Editorial Board, *Journal of Phytopathology* (Blackwell Wissenschafts-Verlag GesmbH, Berlin, Germany).

M. Tošić is the author of over 130 scientific papers and book chapters, co-author of three books, and author of one book. His current research interests are potyviruses infecting maize, sugarcane, sorghum, and Johnsongrass, and also virus diseases of cereals, sugar beet, vegetables, etc. He contributed significantly to plant viruses serology (N-terminus serology) and the application of differential cvs. in the identification of plant viruses, especially potyviruses infecting maize, sugarcane, sorghum, and Johnsongrass.

Acknowledgments

Our special appreciation and thanks goes to Mrs. Vesna Vidas for her expert translation of the original draft of the text from Cyrillic into English at the Faculty of Agriculture, University of Belgrade, Beograd, Serbia (Yugoslavia).

Our sincere thanks to Mrs. Sandy Waterstradt who with her staff typed and computerized the entire edited copy in addition to their regular office duties in the Department of Plant Pathology, University of Illinois, Urbana-Champaign; and to Sharon Gocking, assisted by Carol Matthews, who faithfully and patiently incorporated numerous corrections as we proceeded through several draft editions; and finally, to Mrs. I. Radha for help in typing the peanut virus section (DVRR).

The times our many scientific friends and colleagues offered expert advice, agreed to read sections, and showed support as we developed this handbook are very much appreciated.

Introduction

This handbook presents data and ideas from numerous publications in a concise form. The authors acknowledge and thank each scientist who published his/her work, that humankind can benefit from such dedicated efforts. This handbook attempts to draw together and to present a uniform overview of the essential information of many plant diseases caused by viruses.

The objectives of this book are to

- **PRESENT** basic information about viral-caused and viral-like diseases of cultivated crops that may aid in their diagnoses and possible control
- **SUMMARIZE** the current knowledge about each virus-induced disease in many economically important cultivated crops
- **SERVE** as an introductory text for students of plant pathology, teachers of plant pathology, plant virology, horticulture, agronomy, and general agriculture
- **PROVIDE** useful information for plant clinics, diagnostic laboratory services, and biologists interested in the complex interactions between plants and their viral pathogens

This book is organized into the following sections to aid users in locating the description of a disease of interest. Each previous similar book is arranged in a unique order, as determined by their authors. This book is no exception. It has certain advantages and disadvantages. Cultivated plants are arranged in groups according to their final destinations and uses after harvest. Some of the diseases caused by virus infection and the resultant epidemiology are characteristic within these groups. The control measures also tend to have aspects in common within these groupings. In-depth coverage of epidemiology is not attempted, although some mention is made for a few viruses where its understanding leads to better control tactics. A recent report resulting from a symposium/congress (Nelson and Orum, 1998. Local and regional spatial analysis of plant virus epidemics with geographic information systems (GIS) and geostatistics. Sultan Qaboos Univ. (Oman). *J. Sci. Res. Agric. Sci.* 3:85–93) shows how growers and producers must focus on the epidemic levels of viruses in numerous field production systems for field, vegetable, fruit, and forage crops. In the same journal (Alabouvette, 1998. Management of diseases induced by soil-borne pathogens, solarization and biological control, p. 65–76) the author discussed the impossibility of eradicating, plus the difficulty of controlling, soil-borne pathogens (viruses) and their vectors. Drastic chemical measures to eradicate them have failed. Thus, the key is to attempt—wherein possible—to reduce density levels of pathogens and stimulate beneficial organisms in soils.

1. The *Graminoid Food and Feed Plants* section contains descriptions and pictures of viral diseases, including maize dwarf mosaic, wheat streak, barley yellow dwarf, and other viruses.
2. The *Forage Feed Plants* section contains descriptions and pictures of viral diseases, including alfalfa mosaic, red clover vein mosaic, *Phleum* mottle, and others.
3. The *Vegetable Food Plants* section contains descriptions and pictures of viral diseases, including tomato mosaic, cucumber mosaic, bean yellow mosaic, cowpea aphid-borne mosaic, and many others.

4. The *Industrial Food and Feed Plants* section contains descriptions and pictures of viral diseases, including potato X and Y, tobacco mosaic, tobacco etch, beet mosaic, sugarcane mosaic, sunflower mosaic, soybean mosaic, and peanut stunt, among others.
5. The *Fruit Tree Food Plants* section contains descriptions and pictures of viral diseases, including apple mosaic, apple chlorotic leafspot, plum pox, prunus necrotic ringspot, peach rosette mosaic, walnut line pattern, cherry leaf roll, cherry ringspot, citrus variegation, banana bunchy top, chestnut mosaic, and papaya ringspot, among many other viruses.
6. The *Small Fruit Food Plants* section contains descriptions and pictures of viral diseases, including strawberry crinkle, strawberry latent ringspot, raspberry vein chlorosis, and blueberry leaf mottle, among others.
7. The *Grapevine Food Plant* section contains descriptions and pictures of viral diseases, including grapevine fanleaf, grapevine chrome mosaic, arabis mosaic, and others.

Geographic distribution, host range, symptoms, and pathogenesis are listed for each virus, followed by a reference section at the end of each chapter. Within each host group of plants, a distinction is made between those diseases with adequate proof of viral pathogenicity by listing them under a heading of "diseases caused by a virus" as contrasted with a heading of "virus-like diseases," which suggests to the reader that adequate data are not available to fulfill Koch's Postulates for etiology (proof that the virus isolated can cause disease).

A second heading lists plants as alternative hosts for some viruses that naturally infect a broad range of plant genera and species. This differentiates such diseases from those considered as major economic factors in food/feed production.

A third heading is labeled "diseases *assumed* to be caused by viruses." They have symptoms reminiscent or suggestive of those caused by viruses but yet without knowledge of transmission characteristics (mechanical or biological agents) and presumed viral particles as measured by physical-chemical means or by electron microscopy or isolated following purification from infected plant sap. Proof of virus involvement is often based on symptomology and graft transmissions.

The authors wanted to include life cycle diagrams and even a cartoon of the genome of each virus, wherein known, but they opted to stay with the central focus of providing information adequate to aid in initial diagnosis of viral-caused diseases. Several "atlases" of viruses contain such basic informations about chemical constitution and genome organizations. (Ackermann et al., 1998. *Virus Life in Diagrams*. CRC Press LLC, 240 pp; Mazzone, 1998. *CRC Handbook of Viruses: Mass-Molecular Weight Values and Related Properties*. CRC Press LLC, 208 pp; Francki et al., 1985. *Atlas of Plant Viruses*, Vol. I, 222 pp. and II, 284 pp. CRC Press LLC; Edwardson and Christie, 1991. *Handbook of Viruses Infecting Legumes*. CRC Press LLC, 413 pp; among others).

The authors have attempted to avoid the unnecessary repetition of long viral names by using abbreviations (acronyms) for each virus and virus-like disease. Viral acronyms are used commonly among scientists to eliminate the repeated use of the often lengthy common names. Acronyms, consisting of abbreviations of a susceptible host (locations optional)/symptoms and the word "virus" to represent diseases caused by viruses in plants, are the standard in scientific literature and are accepted generally worldwide as designating their identity. Confusion can exist when several hosts that start with the same alphabetical letter are referred to in relation to common symptom types. The creation and assignment of acronyms is still going through a process of standardization by the International Committee for Virus Taxonomy. The authors deemed it appropriate to develop a consistent system for (1) assigning acronyms to each new virus described and named and (2) revising currently used acronyms to more accurately describe the virus and avoid any potential confusion that now exists. The principle currently in use for naming a virus is first to name the host on which the virus was originally discovered, sometimes to indicate a specific country or area, then to describe the key symptoms, and finally to add the word "virus" once it has been proven as the causal agent. The symptom should be diagnostic and should differentiate that particular virus from any other virus that infects the same host. In order to enable the reader to use the acronyms, the authors have begun to develop and use a system that they believe will help avoid confusion. For each first capital letter in the acronym that is identical for other hosts (e.g., alfalfa, almond, apple, apricot, arabis, artichoke, etc.), the use of a second lower-case letter is proposed, usually the last letter of the word or one that phonetically suggests the host name. Therefore, A = alfalfa, Ad = almond, Ap = apple, At = apricot, Ar = arabis, Ak = artichoke; and papaya = Pa, parsley = Py, passion fruit = Pf, pea = Pe, peach = Ph, peanut = Pt, pepper = Pp, potato = P, plum = Pl, etc. For such hosts, it may be quite confusing—even to the specialists—as to which one is intended if one reads only the acronym without seeing the full word. Likewise for symptoms M = mosaic, Mt = mottle, Sk = streak, St = striate, Sp = stripe, Sn = stunt, etc., the acronyms used are consistent, have logic, and it is recommended that they be adopted by virologists responsible for naming the viruses. Such a standardized approach will alleviate confusion and misunderstandings in this matter. The authors see the need for and propose a second lower-case letter to aid in such distinctions.

A complete index of all viruses mentioned in the book is provided at the end of the book with both full names, acronyms, and—parenthetically—the current acronym most commonly used in the literature.

This handbook proposes and uses the following procedure for creating acronyms.

1. a. The first letter is a capital letter that represents the first letter of the common name of the host on which the symptoms were observed and from which the virus was isolated.
   b. A lower-case letter follows, often the last letter of the host name, or if already assigned to a previously used acronym, then a letter that is phonetically suggestive of the plant.
   c. Should the host name consist of two (or more) words, the first letter of the second word should be the lower-case letter.
2. d. The second capital letter should be the first letter of a location (state, region, or country).
   e. If no location is indicated, the second capital letter (or first after location) should be the letter from the first word of symptom description; for example, Wisconsin pea streak virus should be PeWSkV; European plum line pattern virus should be PlELmPtV; etc.
   f. The second letter of symptom description should be the last letter, lower-case, in the word or the next most phonetically logical letter to alleviate confusion.
3. g. The third capital letter should be from the second word of the symptom description.
   h. The final capital letter designates the pathogen (i.e., V = virus, Vd = viroid, P = phytoplasma, S = spiroplasma, or D = disease, which indicates that the causal agent has not yet been proven or is of uncertain identity)
4. A number or a letter following V, Vd, D, etc. indicates a strain or an unrelated virus, often labeled that way because the resultant symptoms of these viruses lacked distinctive markings from others already described.

The authors take full responsibility for any omissions or errors. Due to the occurrence of many diseases and voluminous literature available, the authors were unable to cite all authors and publications relating to each disease. Nonetheless, we appreciate receiving any corrections or suggestions for a potential new edition of this book.

SYMBOLS OF HOST NAMES

Symbol	Host	Symbol	Host	Symbol	Host
A	Alfalfa	Cy	Cymbidium	Po	Pangola
Ab	Abaca	E	Eggplant	Pp	Pepper
Ad	Almond	EC	Epirus cherry	Pr	Pear
Ag	Agropyron			Ps	prune, (Prunus)
Ak	Artichoke	F	Fig	Pt	Peanut
Am	Arrhenatherum	Fe	Fescue (Festuca)	Py	Parsley
An	Abutilon	Fx	Foxtail		
Ap	Apple			Q	Quince
Ar	Arabis	g	Grass		
At	Apricot	G	Grapevine	R	Rice
Av	Avocado	Gb	Gooseberry	Ra	Radish
Ax	Anthoxanthum	Gc	Garlic	Rb	Raspberry
		Gt	Groundnut	RCu	Red currant
B	Bean	Gu	Guinea	RCl	Red clover
Ba	Banana			Rg	Ryegrass
Bb	Blackberry	H	Hop	Ri	Ribgrass
BB	Broad Bean	Hn	Henbane	Ro	Rose
Bc	Broccoli	Hr	Horseradish	Rs	Rubus
BC	Black currant	Ht	Hazelnut		
Bd	Bidens			S	Spinach
Be	Blackeye	Jg	Johnsongrass	Sa	Satsuma
Bg	Blackgram			Sb	Strawberry
Blb	Blueberry	L	Lettuce	Sc	Sugarcane
Bm	Bramble	Lg	Legume	SCh	Sour cherry
Br	Brome	Lk	Leek	Sf	Sunflower
BRb	Black raspberry	Lo	Lolium	Sh	Shallot
Bt	Beet	Lu	Lucerne	Sl	Sowthistle
By	Barley			Sn	Sanguinolenta
		M	Maize	Sq	Squash
C	Cucumber	Ma	Malva	Sr	Sorghum
Ca	Carnation	Mb	Mung bean	Src	Subterranean clover
Cb	Cranberry	Mn	Melon	Stc	Sweetclover
Cc	Coconut	Mo	Molinia	Su	Southern
Ccy	Chickory	Ms	Melilotus	SVg	Sweet vernal grass
Ce	Celery			Sw	Sowbane
Cf	Cauliflower	O	Onion	Sy	Soybean
Cg	Cabbage	Og	Oatgrass		
Ch	Cherry	Ol	Olive	T	Tobacco
Cic	Crimson clover	Ou	Ourmia	Tb	Thimbleberry
Ck	Cocksfoot	Ot	Oat	Tm	Tomato
Cl	Clover			Tr	Tolare
Cm	Chrysanthemum	P	Potato	Tu	Turnip
Cn	Constricta	Pa	Papaya	Ty	Timothy
Co	Carrot	Pe	Pea		
Cp	Cowpea	Pf	Passion fruit	W	Wheat
Cr	Cereal	Pg	Panic grass (Panicum)	Wb	Wineberry
Crn	Corn	Ph	Peach	WCl	White clover
Cs	Citrus	Pl	Plum	Wm	Watermelon
Ct	Chestnut	Pm	Phleum	Wt	Walnut
Cu	Currant	Pn	Petunia		
Cv	Cassava	Pnp	Parsnip	Z	Zucchini

SYMBOLS OF SYMPTOMS AND NON-HOST NAMES

Symbol	Symptom/Non-host Name	Symbol	Symptom/Non-host Name	Symbol	Symptom/Non-host Name
A	American	Dh	Death	Ln	Line
Aa	Australian	Dk	Dieback	Lr	Leafroll
Ac	Aucuba	Dl	Dapple	Ls	Leafspot
Ad	Asteroid	Dn	Deformation	Lt	Latent
An	Andean	Dp	Drop		
Ap	Aphid	Ds	Disease	M	Mosaic
As	Assistor	Dx	Dixie	Md	Mild
Ay	Aspermy			Mg	Mottling
		e	eye	Mp	Mop
b	Borne	E	European	Ms	Measles
B	Black	Eg	Edge	Mt	Mottle
Ba	Banding	El	Early	My	Mycoplasma
Bc	Bacilliform	En	Enation		
Bd	Bud	Ep	Epirus	N	Necrosis
Bg	Bulgarian	Es	Eyespot	Nc	Necrotic
Bh	Blotch	Et	Etch	Nn	Northern
Bi	Big	Ex	Exocortis	Nt	Net
Bk	Bark				
Bl	Blue	F	Fan	O	Ochre
Br	Browning	Fa	Failure	Ol	Oil
Bs	Blister	Fj	Fiji	Ou	Ourmia
Bt	Bract	Fk	Fleck		
Bu	Bushy	Fl	False	P	Pox
By	Bunchy	Fn	Fino	Pc	Platycarpa
		Fr	Freckle	Pd	Pseudo
C	Chlorotic	Ft	Fruit	Pg	Pitting
Ca	Calico			Ph	Phloem
Ccg	Cadang-cadang	G	Golden	Pi	Pit
Cg	Concave gum	Gg	Grooving	Pk	Pucker
Ch	Chrome	Gl	Gall	Po	Pod
Ck	Crak	Gm	Gummy	Ps	Psorosis
Cl	Clearing	Gn	Green	Pt	Pattern
Cm	Common	Gr	Grassy	Py	Phytoplasma
Cn	Canker				
Cp	Clump	HB	Hoja Blanca	R	Ring
Cr	Crinkle			Ra	Rayado
Cs	Chlorosis	I	Italian	Rb	Rhabdo
Ct	Cristacortis	Ic	Isometric	Rd	Ragged
Cs	Chlorotic spot	Id	India(n)	Re	Red
Cc	Cryptic	Im	Impietrature	Rg	Rugose
Cu	Curl(y)	In	Infectious	Rh	Rough
Cx	Cachexia-xyloporosis	Iv	Interveinal	Rl	Roll
Cy	Corky			Ro	Rosette
"C:"	C type (of anything)	L	Leaf	Rp	Rasp
		Lc	Leaf curl	Rs	Ringspot
D	Dwarf	Ld	Limited	Rt	Rattle
De	Decline	Ll	Little	Rv	Reversion

xxi

SYMBOLS OF SYMPTOMS AND NON-HOST NAMES (CONTINUED)

Symbol	Symptom/Non-host Name	Symbol	Symptom/Non-host Name	Symbol	Symptom/Non-host Name
Ry	Rusty	Spe	Speckle	Tp	Top
		Spl	Split	Tr	Tulare
s	Spot	Spo	Spot(ted)	Tt	Transient
S	Stain	Spr	Spur	Tw	Twig
Sb	Stubby	Spy	Spy	Ty	Transitory
Sbc	Sunblotch	Sr	Sterile	Tz	Tristeza
Sc	Scar	Ss	Shoestring		
Sd	Seed	St	Striate	U	Utah
Sdb	Seed-borne	Stm	Stem		
Se	Severe	Stn	Stony	V	Vein
Ser	Serrano (Mx)	Str	Star	Vg	Variegation
Sg	Seedling	Sty	Sooty		
Sh	Serch	Su	Southern	W	Wilt
Si	Sinaloa (Mx)	SV	Satellite virus	Wd	Wood(y)
Sk	Streak	Sy	Scaly	Wn	Winter
Skn	Skin			Wr	Wart
Sl	Spindle	T	True	Ws	Western
Sm	Summer	Tb	Tuber	Wt	White
Sn	Stunt	Td	Twisted		
So	Soil-borne	Tg	Tungro	Y	Yellow
Sp	Stripe	Tl	Tillering	Yg	Yellowing
Spa	Spiroplasma	Tn	Thin	Ys	Yellows

Potato viruses have been labeled with letters rather than with symptom descriptions or with numbers as other viruses. We have maintained those designations as commonly published, whereby the capital letter symbol follows the word virus. Therefore, the acronym for the potato viruses has a letter that follows capital V (i.e., potato viruses Y or A are designated by the acronyms PVY or PVA, respectively). The strains of such viruses are also designated by superscript letters attached after the acronym. The acronyms for other viruses maintain the custom that the numbers follow capital V for virus or Vd for viroid.

Classification and Nomenclature: Plant Virus Families and Groups

Family or Group	Morphology	Nucleic Acid Type	Configuration	Type Host	Type Virus
Alfalfa mosaic[1]	Bacilliform	ssRNA	3 + strands	Alfalfa	Alfalfa mosaic virus
Bromovirus	Icosahedral	ssRNA	3 + strands	Bromegrass	Brome mosaic virus
*Bunyaviridae**	Spherical	ssRNA	3 − strands	Tomato	Tomato spotted wilt virus
Capillovirus	Rod	ssRNA	1 + strand	Apple	Apple stem growing virus
Carlavirus	Rod	ssRNA	1 + strand	Carnation	Carnation latent virus
Carmovirus	Isometric	ssRNA	1 + strand	Carnation	Carnation mottle virus
Caulimovirus[2]	Isometric	dsDNA	circular	Cauliflower	Cauliflower yellows virus
Closterovirus	Rod	ssRNA	1 + strand	Sugar beet	Sugar beet yellows virus
Commelina yellow mottle[1]	Bacilliform	dsDNA	1 circular	Commelina	Commelina yellow mottle virus
Comovirus	Isometric	ssRNA	1 + strand	Cowpea	Cowpea mosaic virus
Cryptovirus	Isometric	dsRNA	2 segments	White clover	White clover cryptic virus I & II
Cucumovirus	Isometric	ssRNA	3 + strands	Cucumber	Cucumber mosaic virus
Dianthovirus	Isometric	ssRNA	2 + strands	Carnation	Carnation ringspot virus
Fabavirus	Isometric	ssRNA	2 + strands	Broad bean	Broad bean wilt virus
Furovirus	Rod	ssRNA	2 + strands	Wheat	Soil-borne wheat mosaic virus
Geminivirus I	Isometric	ssDNA	1 circular	Maize	Maize streak virus
II	Isometric	ssDNA	1 circular	Sugar beet	Beet curly top virus
III[3]	Isometric	ssDNA	2 circular	Bean	Bean golden mosaic
Hordeivirus	Helical	ssRNA	3 + strands	Barley	Barley stripe mosaic virus
Ilarvirus	Isometric	ssRNA	3 + strands	Tobacco	Tobacco streak virus
Luteovirus	Isometric	ssRNA	1 + strand	Barley	Barley yellow dwarf virus
Maize chlorotic dwarf[1]	Isometric	ssRNA	1 + strand	Maize	Maize chlorotic dwarf virus
Marafivirus	Isometric	ssRNA	1 + strand	Maize	Maize rayado fino virus
Necrovirus	Isometric	ssRNA	1 + strand	Tobacco	Tobacco necrosis virus
Nepovirus	Isometric	ssRNA	2 + strands	Tobacco	Tobacco ringspot virus
Parsnip yellow fleck[1]	Isometric	ssRNA	1 + strand	Parsnip	Parsnip yellow fleck virus
Pea enation mosaic[1]	Isometric	ssRNA	2 + strands	Pea	Pea enation mosaic virus
Potexvirus	Rod	ssRNA	1 + strand	Potato	Potato virus X
Potyvirus	Rod	ssRNA	1 + strand	Potato	Potato virus Y
Reoviridae[4]	Icosahedral	dsRNA	10–12 segments	*Agallia constricta*	Wound tumor virus
*Rhabdoviridae**,[5] A	Bacilliform	ssRNA	1 − strand	Lettuce	Lettuce necrotic yellow virus
B	Bacilliform	ssRNA	1 − strand	Potato	Potato yellow dwarf virus
Sobemovirus	Icosahedral	ssRNA	1 + strand	Bean	Southern bean mosaic virus
Tenuivirus	Rod	ssRNA	4 −? strands	Rice	Rice stripe virus
Tobamovirus‡	Rod	ssRNA	1 + strand	Tobacco	Tobacco mosaic virus
Tobravirus	Rod	ssRNA	2 + strands	Tobacco	Tobacco rattle virus
Tombusvirus	Isometric	ssRNA	1 + strand	Tomato	Tomato bushy stunt virus
Tymovirus	Icosahedral	ssRNA	1 + strand	Turnip	Turnip yellow mosaic virus

1. Nonitalic type indicates distinct viruses that have not yet been assigned a family name.
2. First discovery of a dsDNA virus infectious in plants.
3. First discovery of a ssDNA virus infectious in plants.
4. The only plant virus discovered first in a leafhopper; rice is considered a type plant host for this family.
5. A = subgroup transmitted by aphids; B = subgroup transmitted by leafhoppers.
* Groups in which a glycoprotein envelope surrounds the virion.
‡ First virus shown to have infectious RNA, and to be a ribonucleoprotein, purified to paracrystallinity, seen by electron microscopy, composed of identical subunits, reassembled from its parts, and the first plant gene for virus resistance cloned was the N gene against TMV.

After Francki, R.I.B., et al. 1991. Classification and Nomenclature of Viruses; 5th Report of the International Committee on Taxonomy of Viruses. *Archives of Virology*, Suppl. 2. Springer-Verlag, NY, 450 pp.

1 Virus Diseases of Food Graminoid Plants

VIRUS DISEASES OF MAIZE (*Zea Mays*)

Maize is grown worldwide in temperate zones and in tropical and subtropical areas. It ranks third in the world, immediately following wheat and rice in its distribution.

Maize viruses occur in all areas where the crop grows. Damsteegt (1981) reported on the natural occurrence of more than 40 distinct viruses and strains in maize crops. Some of these viruses cause occasionally serious economic damages in maize to limit production, whereas some are endemic, as the basis of the extent of damage caused. Control measures of eradication and prevention of infection of those viruses economically most damaging, as well as measures aimed at reducing crop losses, are indispensable to high yields. The *Compendium of Corn Diseases* (Shurtleff, 1986; White, 1999). provides an excellent companion treatment of all diseases of maize.

MAIZE DWARF MOSAIC POTYVIRUS (MDMV)

Properties of Viral Particles

Viral particles are filamentous with an average size of 750 × 13 nm (Williams and Alexander, 1965). Particles of some isolates vary in length from 550 to 900 nm (average 741 nm), from 650 to 850 nm (average 755 nm), etc. (Tošić, 1965).

The particles contain single-stranded RNA of a molecular weight of ca. 3.32×10^6 Da (Berger et al., 1989), composing approximately 5% of the particle weight (Jones and Tolin, 1972). The protein coat is a single species polypeptide with a native configuration molecular weight of 30.7 kDa (von Baumgarten and Ford, 1981; Jensen et al., 1986). Viral particle sedimentation coefficient is 176 ± 5S (Tošić and Ford, 1974). Pin-wheel inclusion bodies are induced by MDMV infection which serve as a rapid diagnostic aid (Figure 1).

Stability in Sap

MDMV has a thermal inactivation point of 55°C, a dilution endpoint of 10^{-5}, and an *in vitro* longevity at 2 to 3°C of 192 hours (Williams and Alexander, 1965). Other MDMV isolates described in the U.S. have thermal inactivation points that range between 56 and 58°C, *in vitro* longevity up to 72 hours, and a dilution endpoint between 1×10^{-3} and 20×10^{-3} (Tošić and Ford, 1974).

Serology

The virus features moderate immunogenic activity when rabbits are injected with viral particles (Shepherd, 1965; Tošić and Ford, 1974). Antisera have been applied successfully in enzyme-linked immunosorbent assays (Jarjees and Uyemoto, 1984), immunosorbent electron microscopy (Derrick, 1975), and electro-blot immunoassays (Shukla et al., 1989, 1989a).

Strains

MDMV consists of several isolates, infective in maize, sugarcane, and sorghum. MDMV was identified as a sugarcane mosaic virus (ScMV) strain (Shepherd, 1965). The relationships among these strains are now being elucidated with newer research technologies.

Comparative serological analyses of 17 ScMV and MDMV strains (from the U.S. and Australia) in electro-blot immunoassay, with cross-absorbed polyclonal antibodies directed toward surface-located, virus-specific N-termini of coat proteins have shown that these strains belong to

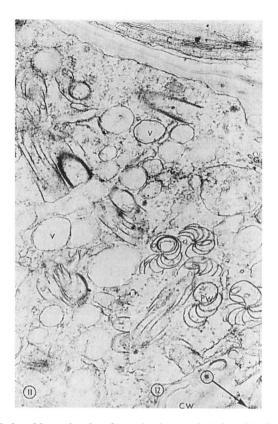

FIGURE 1 Inclusions induced by maize dwarf mosaic virus strain A in cells of corn.

the particular potyvirus species for which the following names have been proposed: maize dwarf mosaic virus (MDMV), sugarcane mosaic virus (ScMV), Johnsongrass mosaic virus (JgMV), and sorghum mosaic virus (SrMV) (Shukla et al., 1989a). The uniqueness of these species has also been confirmed by comparative investigation of symptomatological reactions of differential (sorghum and oats) plants, inoculated with isolates infective for maize, sugarcane, and sorghum (Tošić et al., 1990).

Based on such investigations, it now has been concluded that strains MDMV-A, -D, -E, and -F—which differ clearly from other strains infective in maize, sugarcane, and sorghum—belong to MDMV.

Geographic Distribution

MDMV is spread worldwide wherever maize and sorghum are grown. It has been described in North and South America, Europe, Africa, and Asia, whereas there are no reports on its distribution in Australia (Ford et al., 1989).

Maize dwarf mosaic virus was first observed in North America (Janson and Ellett, 1963; Williams and Alexander, 1965) where it has been intensively investigated. In Europe, this disease was first described as the maize mosaic virus (Panjan, 1960; Tošić, 1962) and prior to that as the sorghum red stripe and maize mosaic virus (Grancini, 1957; Lovisolo, 1957; Lovisolo and Aćimović, 1961), which was later named European maize mosaic virus in order to distinguish it from other similar diseases (Šutić, 1983).

Economic Importance

MDMV infections have a detrimental effect on plant growth, flowering, grain yields, and quality of the commercial crop, and of seed maize in particular. Such detrimental effects have been described by numerous authors. Paliwal et al. (1968), for example, determined that the virus postponed flowering for 5 days, reduced plant height by some 50%, ear and whole plant weight by approximately 74%, and dry matter content by 36 to 57%. Teyssandier et al. (1983) described maize dwarf mosaic virus as the economically most important disease in Argentina, causing yield reductions of up to 90%. Investigations carried out in South Carolina showed that in 100% incidence of plant infection, yields were reduced 62%. Each 1% increase in incidence of infected plants reduced yields by 63 kg ha^{-1} (Kingsland, 1980). Investigations performed with 15 hybrids in Hungary showed that the number of infected plants ranging from 3 to 14% caused 14 to 24% yield reductions (Sum et al., 1979). Some observations in the former Yugoslavia showed that MDMV reduced plant growth by 10 to 19% and caused a 42% decrease in yield (Šutić, 1983). Conditions that favored disease development in some regions resulted in maize seed yield reductions of more than 70% and in some cases a complete loss (Tošić et al., 1990a).

MDMV substantially decreased grain oil content up to 31% without changing the ratio among individual fatty acids (Pešić et al., 1977). Investigations have also shown that the virus reduces only slightly the maize leaf protein content, which is highly important for the use of green matter in animal nutrition. Tu et al. (1968) determined that photosynthesis was reduced in infected maize plants by 13% more than the reduction in numbers of chloroplasts and by 7% more than the reduction in chloroplast content.

Other research shows that MDMV increases the susceptibility of maize to some harmful pathogenic fungi. Thus, previous infection with MDMV increased the susceptibility of mosaic-affected plants to the fungus *Gibberella zeae* by 2.5 times (Tošić et al., 1977). In some hybrids, mosaic-affected plants are more susceptible to *Ustilago maydis* by some 21% (Ivanović, 1979) and to *Helminthosporium turcicum* by some 13% (Panić et al., 1978). The increase in susceptibility of mosaic-affected plants was also observed in infections by other fungi: *Puccinia sorghi*, *Setosphaeria turcica*, and *Cochliobolus heterostrophus* (Tošić, 1988), *Fusarium moniliforme* (Futrell and Scott, 1969), *Helminthosporium pediculatum*, and others (Tu and Ford, 1971).

Host Range

All maize dwarf mosaic virus plant hosts known to date are monocotyledonous in the Gramineae family. Host susceptibility has frequently been tested with different virus isolates so that it is difficult to compare the results obtained in these investigations. The most complete reviews of host susceptibility to different isolates were given by Ford and Tošić (1972), Rozenkranz (1983, 1987), and Tošić and Ford (1972). Maize, cultivated sorghum (*Sorghum bicolor*) and common millet (*Panicum miliaceum*) are the most frequent hosts among annual plants. Other annual natural hosts include *Chloris virgata*, *Digitaria sanguinalis*, *Echinochloa* (*Panicum*) *crusgalli*, *Eragrostis cilianensis*, *Eriochloa gracilis*, *Leptochloa filiformis*, *Paspalum dilatatum*, *Setaria faveri*, *S. glauca*, *S. italica*, *S. verticillata*, *Sorghum bicolor* (*S. sorghum*), *S. saccharatum*, and *S. sudanense*.

The perennial natural host group comprises *Erianthus maximus*, *Saccharum edule*, *S. officinarum*, *S. robustum*, *Sorghum halepense*, and *Tripsacum dactyloides*.

Experimentally infected hosts include more than 200 plant species of the graminoid family in the genera: *Aegilops*, *Agrostis*, *Alopecurus*, *Amphicarpum*, *Andropogon*, *Aristida*, *Botriochloa*, *Bouteloua*, *Brachiaria*, *Bromus*, *Cenchrus*, *Chasmanthium*, *Chloris*, *Ctenium*, *Dactyloctenium*, *Danthonia*, *Digitaria*, *Echinochloa*, *Ehrharta*, *Eleusine*, *Elyenorus*, *Eragrostis*, *Eremopoa*, *Eremopyrum*, *Erianthus*, *Euchlaena*, *Glyceria*, *Hackelochloa*, *Heteropogon*, *Lagurus*, *Leersia*, *Leptochloa*, *Lolium*, *Manisuris*, *Miscanthus*, *Muhlenbergia*, *Oryza*, *Panicum*, *Paspalum*, *Pennisetum*, *Phacelurus*, *Phalaris*, *Phleum*, *Polypogon*, *Saccharum*, *Sacciolepis*, *Schizachyrium*, *Setaria*, *Sorghastrum*, *Sorghum*, *Sporobolus*, *Stipa*, *Tricholaena*, *Tridens*, *Triplasis*, *Uniola*, *Urochloa*, *Zea*, *Zizania*, and *Zizaniopsis* (Rozenkranz, 1983, 1987; Tošić and Ford, 1972). None of the 70 investigated dicotyledonous herbaceous plants is susceptible to the maize dwarf mosaic virus (Ford and Tošić, 1972).

Maize and sorghum are suitable diagnostic hosts for MDMV. Seedlings of various sorghum cvs. may also be easily infected and display symptoms of mosaic or necrosis.

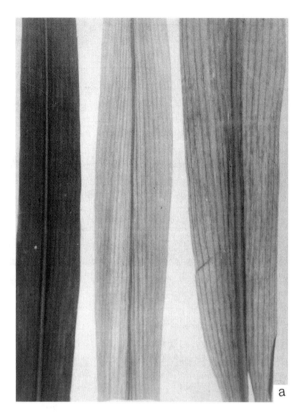

FIGURE 2 Maize dwarf mosaic virus strain A infected *Sorghum bicolor*.

Perennial sorghum, Johnsongrass (*Sorghum halepense*) (Figure 2), currently is the most suitable plant for the multiplication and long-lasting maintenance of the virus and its individual strains.

A highly suitable assay plant for this virus is sorghum cv. Atlas whose leaves display symptoms of roundish necrotic lesions within 1 week following inoculation, whereas the typical mosaic appears on the noninoculated leaves (Tošić and Ford, 1972; Tošić et al., 1990). Other sorghum cvs. (SA 8735, Rio, Martin, NM 31, R 430, OKY 8, Tamaran, Aunis, Trudex, and TX 2786) (Tošić et al., 1990), as well as sweet corn cultivars, are also used for assays. According to the latest data, MDMV does not infect oat plants and thus differs from the highly similar Johnsongrass mosaic virus (JgMV) (Tošić et al., 1990).

Based on specific reactions in sorghum cvs., in addition to differences in serological and biochemical properties, a mutual differentiation of the maize dwarf mosaic virus (MDMV), Johnsongrass mosaic virus (JgMV), sugarcane mosaic virus (ScMV), and sorghum mosaic virus (SrMV) was also performed. MDMV is characterized as infective in all sorghum cvs., but not infective in oats as is JgMV. In most sorghum cvs., MDMV causes a mosaic symptom pattern and in the Atlas, Rio, NM 31, R 430, and Aunis cvs., and less often also necrosis of new leaves. Thus, MDMV differs from the ScMV and SrMV strains that cause different symptoms in sorghum cvs. A specific uniqueness of MDMV is its ability to infect Johnsongrass, which is not susceptible to ScMV and SrMV (Ford et al., 1989; Tošić et al., 1990) (Figure 3).

Incidence of Infection

MDMV belongs to the group of viruses regularly occurring in maize. Natural infection rates range from a trace to 100%, depending on genotype susceptibility, inoculum presence, vector activity, and environmental conditions for plant and vector development.

Virus Diseases of Food Graminoid Plants

FIGURE 3   Maize dwarf mosaic virus strain A infected *Sorghum halepense*.

Disease incidence in some maize fields of southern Ohio varied, for example, from a trace to 50% (Janson and Ellett, 1963). In some cases in Argentina, 32% infected plants (on average 13%) was recorded (Yossen et al., 1983). Mechanical inoculation of plants in Venezuela resulted in infection rates ranging from 85% in the North Star hybrid to 25% in the most resistant hybrid Ballringer (Anzola et al., 1980). Natural infection of maize plants recorded in former Yugoslavia ranged from 5 to 30% (Šutić et al., 1969). The majority of maize hybrids are considered susceptible to MDMV with the possible infection rates ranging from 25 to 50% and even 80%. The number of infected plants in sweet corn and some inbred lines often reach 100% under ideal epidemiological conditions (Tošić, 1988).

Symptoms

The primary symptoms of disease in maize are mosaic with various leaflet decolorations, stunting, sterility, and premature plant death (Figure 4).

Mosaic symptoms appear first at the base of the youngest leaf, still in the sheath. Chlorotic streaks first observed at the bases of these leaves, gradually spread along the leaflet parallel with the veins and similarly affect all the leaves formed thereafter. These chlorotic streaks multiply simultaneously with the growth of the leaves, forming a streaky mosaic, or merging into chlorotic stripes and lines on one or both sides of the main leaf vein; chlorotic (yellow) color often dominates in apical leaves, with green color remaining only in form of streaks, stripes, and lines, mainly close to the veins (Figure 5). Chlorotic, yellow plant apices appear at the beginning of the vegetation period and in years favorable for early infection.

FIGURE 4 Maize dwarf mosaic virus strain A natural infection in volunteer sweet corn (Iowa).

Mosaic and yellowish color are frequently accompanied by the appearance of reddish-violet color. This type of change is particularly characteristic for parent lines in seed maize crops. Reddish-violet elongated spots, stripes whose formation usually starts from leaf tips or margins, appear on older leaves. The occurrence of this color on diseased leaves may lead to the wrong conclusion that it is a consequence of a nutrient deficiency.

A characteristic and important symptom is leaf necrosis, which can also cause plant death. It occurs especially on plants in maize seed production fields. Leaf tissue necrosis and plant withering may in some cases be mistaken for symptoms of soil water deficiency.

Mosaic-affected inbred line plants in seed production fields are particularly subject to early death. Massive scale young plant dwarfing, prior to tasseling and silking, prevents maize seed production and such fields are ploughed under and used for other purposes.

MDMV is also characterized by stunting of plants, which is more pronounced in case of earlier infection as a result of disorders in plant growth and development. Stunting is characterized by insufficient elongation of internodes following infection. All internodes along the entire stalk are shortened and often show rosette-like forms at the apices. Such pathological changes are particularly pronounced in seed maize crops whose parents seldom reach the height of 0.5 m. MDMV was named to accurately reflect these plant symptoms.

Due to structural and physiological disorders, sterile plants also occur; whereas in cases of early infections, ears are either not formed at all or remain incompletely fertilized (Figure 6). Disorders in ear fertilization result from increased pollen sterility or diachronic plant tasseling and silking. In growth and pollination, infected plants lag one or two weeks behind the healthy ones. Fertilized mosaic-affected plants express a delayed maturity such that at grain harvest many such plants are in transition between the milk and the waxy stage or are in the waxy stage.

FIGURE 5 Maize dwarf mosaic virus strain A infected *Zea mays*.

Leaf mosaic is the main symptom of maize dwarf mosaic virus infection in Johnsongrass, plus chlorotic streaks that often merge together (Figure 7). Chlorotic streaks and stripes on older leaves turn a reddish-violet color (these symptoms are described in more detail in the chapter on Johnsongrass viruses).

MDMV symptoms in cultivated sorghum are similar to those in maize. On plants of susceptible genotypes, such as the majority of broom corn cultivars, red stripes and leaf necrosis, accompanied by early plant dwarfing, occur in particular (these symptoms are described in more detail in the chapter on cultivated sorghum viruses).

Pathogenesis

In nature, maize dwarf mosaic virus is preserved in perennial and annual hosts. Research investigations have shown that Johnsongrass (*Sorghum halepense*), as a perennial weed plant, is the most important host for long-term virus preservation and transmission (Šutić and Tošić, 1966). Sehgal (1966) maintains that this plant is the general natural reservoir of the dwarf mosaic virus also in the state of Missouri (U.S.). Hollings and Brunt (1981) cite data proving that Johnsongrass is the most important perennial host of the maize dwarf mosaic virus strain ScMV (MDMV was earlier considered an ScMV strain) that infects maize crops in Kansas (Sill, 1966) and other states of the U.S. As with sorghum, other natural hosts also represent a source of infection of lesser or greater importance.

The results of investigations regarding virus persistence in maize seed seem contradictory. Thus, for example, according to some authors, MDMV does not retain infectivity and is not transmitted with maize seed (Bancroft et al., 1966; Sehgal, 1966; Signoret, 1974). Yet, others have shown that

FIGURE 6 Poor kernel fill in maize infected by maize dwarf mosaic virus strain A; healthy (left).

FIGURE 7 Maize dwarf mosaic virus strain A natural infection in Johnsongrass (Iowa).

MDMV transmits through maize seed in low percentages. Thus, Shepherd and Holdeman (1965) succeeded in transmitting an ScMV strain with maize seed in 0.4% of the cases. Of 29,735 maize seeds assayed, Williams et al. (1968) observed only two infected plants. Kerlan et al. (1974) determined virus transmission in seed of the OH43 line of 0.01 to 0.1% of the cases. While testing 11,833 plantlets of the maize line L21 originating from seed produced on infected plants, Tošić and Šutić (1977) observed only one infected plantlet (i.e., in 0.008% of the cases). MDMV was transmitted with the seed of an inbred maize line to 0.2% of the plants (Hill et al., 1974) and with sweet corn seed only to 0.005% of the plants (Mikel et al., 1984). These data show that maize dwarf mosaic virus is rarely transmitted with maize seed, and then only in very low percentages. Such inconsistency might be due to the heterogeneity of the assayed maize isolates which, as recently established, may belong to the MDMV, or to the ScMV, JgMV, and SrMV species.

MDMV is transmitted either with the sap of infected plants (mechanically) or with the aid of vectors. Mechanical transmission is especially useful for experimental work. Thus far, mechanical virus transmission apparently has no significant influence on natural virus spreading.

Some 20 aphid species transmit MDMV in a nonpersistent manner, some of which are well-known according to their distribution: *Acyrthosiphon pisum, Aphis craccivora, Macrosiphum (Sitobion) avenae, Myzus persicae, Rhopalosiphon maidis, R. padi, Schizaphis graminum*, and others. *Macrosiphum festucae* has recently been discovered as a new vector of this virus in cultivated sorghum (Mijavec, 1989).

The extent of spread of MDMV depends on the source of infection, vector activity, and distribution of susceptible plants. For example, infection distribution of MDMV-infected maize in some localities depends directly on the number of infected Johnsongrass plants. *Schizaphis graminum* and *Myzus persicae* aphids become infective by feeding on a diseased sorghum for 1 minute; and likewise they can transmit MDMV with 1 minute of probing. *Rhopalosiphon maidis* may become infective in this way in 2 minutes, which is usually also sufficient for virus transmission to healthy plants. By transmitting the virus from Johnsongrass, *S. graminum* infected up to 46%, *M. persicae* up to 43%, and *R. maidis* up to 26% of maize plants (Tošić and Šutić, 1968).

With respect to the source of infection and modalities of virus transmission, natural maize infection occurs in the following primary cycle:

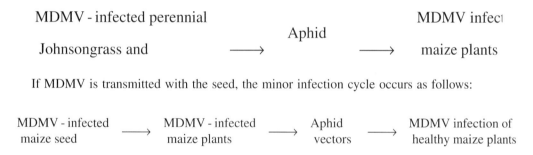

If MDMV is transmitted with the seed, the minor infection cycle occurs as follows:

MDMV - infected maize seed $\longrightarrow$ MDMV - infected maize plants $\longrightarrow$ Aphid vectors $\longrightarrow$ MDMV infection of healthy maize plants

Dynamics of virus epidemiology differ, depending on the conditions in individual countries. In Yugoslavia, for example, numerous initial infections were observed at the beginning of June, which increased markedly by mid-July, up to 85% for the cultivar Kansas, 77% for the Novi Sad golden dent corn, and 61% for the Krajina flint corn of the total number of diseased plants at the end of the vegetation period. Substantially fewer plants were observed on which disease symptoms appeared first in August. Isolated and rare new infections occurred in early September (Tošić and Šutić, 1974).

Control

Protection of maize against MDMV is best ensured with the use of resistant genotypes. The most important preventive measures include the elimination of weed host plants, primarily Johnsongrass, foxtail (*Setaria* spp.), barnyard grass (*E. crusgalli*), and others, to reduce the availability of initial inoculum. Use only healthy seed for sowing to avoid the danger of initial infection sources. Rogueing to eliminate the first infected plants in a crop is especially useful in maize seed production.

Maize and sorghum production, if possible, should be spatially separated because these susceptible plants are capable of becoming sources of mutual infection and should also be isolated from plants susceptible to plant aphids (barley, etc.).

Damage resulting from MDMV can be somewhat alleviated with earlier sowing, also ensuring earlier growth of plants which are, thus, less susceptible to the virus at the time of massive aphid flight. Although lacking significant experimental results, reducing aphid populations is usually recommended. For example, the number of MDMV-infected plants was 15.4% after treatment with mineral oil, 17% after treatment with a mineral oil plus aphicide, 28% after treatment with aphicides alone, and 23% in the check (nontreated) plots (Ferro et al., 1980). If possible to predict aphid flights, the first pesticide treatment should be applied immediately prior to a massive aphid flight.

Testing many cvs., inbred lines, and hybrids in former Yugoslavia for plant susceptibility to MDMV showed significant differences (Tošić and Mišović, 1967; Tošić et al., 1978; 1979). Under conditions of natural infection, the following percentages of infected plants were recorded in cvs.: Krajina flint corn = 2, Novi Sad golden dent = 18, and Kansas 1859 = 26. In another locality, the percentages of infection were L-106 = 4, L-202 = 7, L-4 = 11, L-206 = 16, L-232 = 96, L-214 = 99, and some others = 100 (most susceptible to infection).

The fact that hybrids are more resistant to MDMV as compared with their parents is highly important for maize production. Of 13 hybrids tested in former Yugoslavia, 8 showed greater resistance than that of their individual parents. The resistance of the remaining five hybrids was assessed as intermediate (average of the sum of the resistance for both parents). Similar results with evidence of greater resistance of hybrids were obtained in Romania. The majority of 39 Romanian and foreign single cross (69%) and 55 double cross (75%) hybrids was found MDMV resistant. Double cross hybrids obtained by crossing resistant with susceptible single crosses were only slightly infected under natural conditions (Pop et al., 1966). Some self-pollinated maize lines (I11-A. and Pa 32), tested in the U.S., displayed high resistance to a maize dwarf mosaic virus isolate and this resistance was transmitted to their hybrids. However, I11-A. was susceptible to another isolate, suggesting that different virus isolates are not dependent on the same genetic control in this line (Wernham and MacKenzie, 1967).

Maize dwarf mosaic virus ranks among numerous other potyviruses for which genetic resistance in maize plants has been discovered (Scheifele and Wernham, 1969). Contemporary research of genetic resistance and resistance carrier genes (Mikel et al., 1984a; Rosenkranz and Scott, 1984), as well as of the transmission of this resistance to the offspring (Roane et al., 1983), open new prospects for breeding work aimed at the creation of new genetic materials resistant to this virus.

The new biotechnological approaches to prevent viral replication in maize will soon pay dividends. The coat protein gene of the MDMV has been cloned, sequenced, and now inserted into maize germplasm. These strategies, coupled with time-proven plant breeding technologies, will provide even greater advantages in preventing MDMV infection in areas of high risk (Anonymous, 1992).

Maize Chlorotic Dwarf Waikavirus (MCDV) [Type]

Properties of Viral Particles

Viral particles are isometric, 30 nm in diameter, molecular weight 8.8×10^6 Da, and sedimentation coefficient $83 \pm 6S$. The nucleocapsid core consists of single-stranded RNA of 3200 kDa, meaning some 36% of the virion weight. Particle density in CsCl amounts to 1.51 g cm^{-3} (Matthews, 1981).

According to some authors, two to three protein species of molecular weight ranging from 18 to 30 kDa comprise the viral particle structure (Gingery et al., 1978; Gingery, 1981), suggesting the existence of a multipartite RNA genome.

Stability in Sap

There is no available information (Gingery et al., 1978).

Serology

Results of serological analyses indicate that viral particles are immunogenically active. A number of serological methods have been applied successfully in virus research: immunofluorescence assay, microimmune agar gel double diffusion, enzyme-linked immunosorbance, serologically specific electron microscopy, etc.

Serological methods are those most frequently applied in virus research and disease diagnosis, since other methods currently are less suitable. MCDV is not mechanically transmitted and its confirmation with biotests using vectors requires more time.

Strains

The type MCDV strain was collected in Ohio (loc. cit. Gingery et al., 1978). Some Johnsongrass isolates of this virus differed from the typical strain according to plant stunting, discoloration, and leaf tearing symptom intensity, although no serological differences have been observed.

Geographic Distribution

MCDV was first discovered in the U.S., where it appeared on an epiphytotic scale as early as 1960, causing serious maize stunting (Bradfute et al., 1972; 1972a; Pirone et al., 1972; Rosenkranz, 1969). This virus, listed as a problem in maize production, has spread into regions north of the Gulf of Mexico, in states along the Ohio River Valley, in Pennsylvania, and in areas from the Atlantic coast toward eastern Texas (Gingery et al., 1978; Gingery, 1981).

The distribution of MCDV is conditioned by the abundant presence of Johnsongrass, in which both the virus and its leafhopper vector *Graminella nigrifrons* overwinters.

Economic Importance

MCDV can reduce the yields of susceptible maize hybrids up to 91% and, together with MDMV, ranks among the economically most important viruses in continental U.S. (Gordon et al., 1981). As a maize stunting agent, based on its spreading epidemiology and the extent of damages caused, this virus is much more important than other maize stunting agents, such as maize stunt spiroplasma (MSnSp) and maize bushy stunt phytoplasma (MBuSnPy). Investigations in the U.S. during 1973 to 1975 proved that MCDV was present in 91% of samples with corn-stunt-like symptoms gathered in 16 states (Gordon and Nault, 1977), whereas MSnSpa and MBuSnPy were identified only in 2% of the cases. These findings are particularly important in view of the application of methods of maize protection against stunting agents.

Host Range

MCDV has a comparatively narrow range of hosts belonging only to the Gramineae family. Only 18 species of gramineous hosts, both cultivated and weeds, have been recorded as susceptible.

Maize is economically the most important natural virus host. Other natural hosts include: Johnsongrass (*Sorghum halepense*), sorghum (*Sorghum bicolor*), sudan grass (*S. vulgare* cv. *Sudanense*), millet (*Panicum miliaceum*, *Pennisetum glaucum*), and wheat (*Triticum aestivum*), including annual weed grasses, crabgrass (*Digitaria sanguinalis*), and foxtails (*Setaria* spp.) (Gingery et al., 1978). Maize is a differential virus host displaying striking (evident) disease symptoms. Wheat is a natural, symptomless host. *Zea perennis* and *Tripsacum* spp. are resistant and potential sources of resistance genes (Gingery et al., 1978).

Symptoms

MCDV causes a variety of symptoms on naturally infected maize plants in the field. Various symptoms include plant stunting, leaf discoloration, and leaf tearing.

Plant stunting is the most frequent form of disease when plants lagging behind in growth display reduced vitality and yielding potential. MCDV was named based on the characteristic dwarfing symptom.

Leaf discoloration manifests itself as leaf yellowing, leaf reddening, and marginal chlorosis. Initial leaf discoloration symptoms are mild chlorosis usually appearing between the smallest leaf veins, followed by the development of chlorotic leaf stripe symptoms. A specific initial leaf symptom is also vein clearing, usually not occurring on plants infected with the MDMV. Chlorosis also appears along leaf margins and, therefore, this symptom has also been described as marginal leaf chlorosis. General leaf yellowing and reddening are final chlorotic changes.

MCDV is particularly damaging because it causes tearing of infected leaves, a symptom characteristic as a pathological change in leaf growth-habit and morphology.

Symptoms occurring in plants inoculated in greenhouses are mild and not completely characteristic of those recorded in natural infections. Chlorotic mottle at the base of maize whorl leaves usually appears under such conditions.

The stunting and leaf discoloration symptoms may be similar to those caused by other pathogens (e.g., MDMV and MSnSpa), leading to potentially incorrect disease diagnosis. Therefore, the use of serological and other methods is recommended to obtain correct diagnoses.

Pathogenesis

MCDV overwinters in infected Johnsongrass, a perennial virus host. During the winter, the virus persists in rhizomes and therefrom it translocates into new shoots developing at the beginning of the season. Following feeding on infected shoots, leafhoppers (*Graminella nigrifrons*) transmit the virus into healthy plants, thus spreading the infection in nature. Infected Johnsongrass serves as a permanent primary source of infection. Winter wheat presumably can also serve as a primary source of infection, although no proof has been reported (Gingery et al., 1978; Gingery, 1981).

All infected perennial and annual plants represent sources of secondary infection taking place during the season.

Graminella nigrifrons is the only known vector for natural spread of MCDV, whereas *Deltacephalus sonorus* and *Exitiana exitiosa* have been used as experimental vectors for MCDV. Vector transmission has not been proved for *Dalbulus maidis*, *D. eliminatus*, *Baldulus tripsaci*, *Stirellus bicolor*, or *Macrosteles frascifrons* (Nault et al., 1973; Nault and Bradfute, 1977). *G. nigrifrons* is capable of transmitting MCDV semipersistently from infected to healthy plants in two hours. During semipersistent transmission, after the acquisition of viral particles, the vector maintains its infective ability for a few days; whereas for persistent transmission, the vector maintains inoculative ability during its lifetime (Gingery et al., 1978; Gingery, 1981). Both males and females actively transmit MCDV particles, but those particles acquired by nymphs are not transmitted following molt, nor is it transovarially transmitted to new generations (Gingery et al., 1978).

No evidence exists for transmission of this virus either mechanically or through the seed of infected plants.

Control

MCDV field spread requires a combination of the presence of the source of infection, an adequate population of the vector leafhopper *Graminella nigrifrons*, susceptible maize cultivars, and other hosts. The primary recommendation is to eliminate the perennial virus host—Johnsongrass—taking care to also destroy the rhizomes that represent a reservoir for the viruses. Although slightly useful, this method unfortunately is difficult to apply in most maize production areas. Adequate insecticides can also reduce the vector population; for example, a systemic insecticide Carbofuran decreased such a population, simultaneously also reducing MCDV disease occurrence in the maize crop (All et al., 1976; 1977; Pitre, 1968). The application of such chemicals is limited by the price of the insecticide and its potential harmfulness to human health. Early maize sowing, prior to intensive vector activity, is also useful to reduce damages due to this virus.

The most important protection method is to grow tolerant maize hybrids, although some hybrids yielded satisfactorily in spite of the infection with MCDV (Findley et al., 1977). Maize lines and hybrids that have displayed a tolerant reaction should be grown in regions in which MCDV

MAIZE ROUGH DWARF FIJIVIRUS (MRgDV)

Properties of Viral Particles

MRgDV particles are isometric with double capsids and double-stranded RNA. The diameter of viral particles stained with uranyl acetate is 65 nm, and of those stained with phosphotungstic acid 70 to 75 nm (Lesemann, 1972; Milne et al., 1973). Characteristic A-spikes are located on the outer coat of the virus with internal B-spikes situated below each A-spike (Milne et al., 1973). MRgDV contains RNA-dependent polymerase in subviral particles (SVp), but not in the outer envelope (Ikegami and Francki, 1976).

The MRgDV capsid contains seven polypeptides and its genome is composed of ten RNA segments, six of which are probably monocistronic (Boccardo and Milne, 1975). Molecular weights of the polypeptides vary, corresponding to the coding capacity of each genome segment, ranging from 64 to 139×10^3. Spikeless viral particles (following chemical treatment) have a diameter of 50 to 55 nm and contain three polypeptides of 123, 126, and 139 kDa (Matthews, 1981). The molecular weight of the MRgDV genome is 18.9 kDa

The sedimentation coefficient of SVp is 400S, with a 260/280 ratio of 1.5 (loc. cit. Lovisolo, 1971).

Based on the morphological structure of the virion, genome RNA composition, and (*Fulgoroidea*) insect vectors, MRgDV belongs to the subgroup of two reoviruses characterized by a genome composed of ten segments, of which eight are the same size.

Stability in Sap

Virus thermal inactivation point is 55 to 60°C. Partially purified virus preparations remain serologically active even after 37 days at room temperature, as well as during 10 minutes at 70°C, but they become inactive at 80°C (Wetter et al., 1969). Longevity *in vitro* of viral particles at –20°C is indefinite. The optimum for preservation is pH 6.5 (Shikata, 1981).

Serology

MRgDV has moderate immunogenic activity in agar gel diffusion and slide precipitation tests. MRgDV particles from either infected plants or planthoppers display the same antigenic activity (Lovisolo, 1971).

External capsomers in viral particles are labile with slight antigenic activity, whereas SVps are stable and highly antigenic (Luisoni et al., 1975). The virus displays a close serological relationship with the rice black-streaked dwarf virus (RBSkDV), the Pangola stunt virus (PoSnV) (Milne and Luisoni, 1977), and the cereal tillering disease virus (CrTlDsV) (Matthews, 1981; Shikata, 1981; Signoret, 1988).

Strains

Harpaz and Klein (1969) described the "dwarfing" and the "non-dwarfing" strains of the virus differing based on symptoms in infected plants and their relations with the vectors. Insufficient information is available on the strains of this virus.

Geographic Distribution and Economic Importance

Maize rough dwarf disease was first described in 1949 in northern Italy in introduced susceptible maize cultivars (Fenaroli, 1949). Later, it was observed mainly in European countries. Outside Europe, it has been described only in Israel. MRgDV occurs in European countries of Italy, Switzerland, Spain, France, and Czechoslovakia (Lovisolo, 1971; Harpaz, 1972). MRgDV is widespread in those regions in which the vector planthoppers *Laodelphax striatellus* are found.

The maize rough dwarf disease is economically damaging because it causes severe dwarfing and nearly complete sterility in maize, particularly in early infections. A limited number of adventitious roots appear on infected plants. The extent of damages depends on the conditions for disease development and the susceptibility of the cultivars. Its economic importance worldwide is comparatively limited due to its small area of distribution. Based on economic importance in Europe, MRgDV is classified as minor (Signoret, 1988).

Host Range

Although first described in maize in 1949, the viral nature of MRgDV was confirmed 10 years later (Fenaroli, 1949). In addition to maize, other natural virus hosts are *Digitaria sanguinalis* and *Echinochloa crusgalli*. Other natural hosts among Gramineae are *Lolium multiflorum* and *L. perenne*. Sixteen experimental hosts belong to the grass family, including wheat, barley, oats, etc. (Lovisolo, 1971).

Suitable diagnostic maize plants include Lagune Elite, Nevé Jaar 22, and Wisconsin 641 AA. If young seedlings of these plants (in the coleoptile stage) are infected with the aid of the planthopper *Laodelphax striatellus*, dwarf plants develop a dark green color with numerous enations along leaf veins. Rice, only slightly susceptible to MRgDV, is highly susceptible to RBSkDV, and thus can be used for diagnostic distinction of these viruses. Sugarcane is resistant to MRgDV, is highly susceptible to RBSkDV, and thus can also be used for diagnostic distinction of these viruses. Sugarcane is resistant to MRDV and therefore can be used as a filter plant for distinguishing it from the Fiji disease virus (FjDsV), to which this host is highly susceptible. Virus hosts for local lesion reaction are unknown (Lovisolo, 1971).

Suitable hosts for MRgDV multiplication and preservation include maize, barley, wheat, and oats (Lovisolo, 1971).

Symptoms

Characteristic leaf galls, vein thickening, and vein enations appear on the lower leaf surface, along leaf veins, on leaf sheaths, and on roots of maize plants. Symptoms of similar appearance may also be caused by the toxic action of the saliva of planthoppers feeding on maize leaves. In such cases, the elimination of planthoppers results in symptom disappearance.

The virus causes atrophy of individual kernels, ear tips in particular, and of the whole ear. Adventious roots are shorter with fewer secondary hairs, whereas longitudinal fissures appear on thicker roots.

Early infections cause pronounced internode shortening, dwarfing, and plants obviously lag behind in growth. These symptoms become increasingly less pronounced with later infections until older plants when infected show no signs of dwarfing; thus, only galls along leaf veins remain as visible symptoms of disease. The rough leaf surface caused by galls and dwarfing resulted in the name maize rough dwarf disease.

Symptoms of twisted leaves and broken leaf edges sometimes occur in experimentally infected maize, wheat, barley, and oats. A characteristic cytopathological symptom is the presence of numerous viral particles and their aggregates (viroplasm) in neoplastic phloem cells and mesophyll cells close to the phloem.

Physiological disorders are characterized by abnormal asparagine accumulation in infected plants (Harpaz and Applebaum, 1961). Asparagine and glutamine, directly or indirectly, involve the biosynthesis of purines, pyrimidines, nucleosides, and nucleotides. Larger amounts of amides in viral-diseased plants versus in plants infected with other pathogens can be explained by the involvement of these compounds in the structure of the virus nucleic acid.

Pathogenesis

MRgDV persists in naturally infected plants belonging to the family of grasses. Its transmission mechanically or through seed has not been recorded.

The Delphacidae planthopper *Laodelphax striatellus* is the only vector species known to naturally transmit MRgDV persistently. Acquisition by the planthopper requires 1 day; incubation requires 10 to 15 days, following which the planthopper can inoculate a plant in 5 hours (Harpaz et al., 1965). Few of the offspring of viruliferous females remain infective because the majority die during embryonic development. Viral particles and aggregates (viroplasm) were found in the cells of many *L. striatellus*, proof that the virus multiplies in this vector. Under certain experimental conditions, leafhoppers *Delphacodes propinqua* and planthoppers *Javesella pellucida* and *Sogatella vibix* have the ability to transmit MRgDV.

Since planthopper is the only known vector, obviously virus distribution and spread depend on the conditions influencing the development and activity of this insect.

Control

Several viral control measures are recommended, based on epidemiologically important conditions. Naturally infected host plants should be destroyed. Timely planting ensures more mature (more resistant) plants during the most intense planthopper (i.e. virus vector) activity. Planthoppers should also be controlled with chemicals in borders of weed plants belonging to the grasses family some 20 to 30 days prior to maize seeding in maize production fields.

MAIZE MOSAIC RHABDOVIRUS (MMV)

Properties of Viral Particles

MMV particles are bacilliform; in negatively stained preparations, their bullet shape is characteristic.

Particle sizes are in plant cross-sections 242×48 nm and in leaf dip preparations 225×90 nm (Herold, 1972). Particle dimensions of one maize isolate were 140×63 nm (loc. cit. Cornuet, 1987). Peters (1981) classified this virus species as a definitive member of the plant rhabdovirus group with particle size of 300×75 nm.

MMV particles consist of an outer envelope and the inner nucleocapsid. The outer envelope of lipoprotein nature consists of lipid membranes with embedded protein components. This envelope serves to preserve the nucleocapsid, whereas its decomposition results in the loss of the viral particle regularity. Synthesis of the outer envelope is not coded by the viral genome, but is a product of the host cell whose components surround viral particles.

The nucleocapsid core consists of a single-stranded noninfective negative sense (–)RNA. The nucleic core is surrounded by protein capsids whose synthesis is coded by the viral genome. MMV nucleocapsids contain the replicase (transcriptase) enzyme as its structural component, which transcribes (–)RNA into (+)RNA, ensuring synthesis of viral proteins. The synthesized positive sense (+)RNA is the template from which negative sense (–)RNA strands are formed with the aid of the same enzyme. Therefore, (+)RNA genetically guides the synthesis of viral components by the host cell and is, thus, the infective component.

Stability in Sap

Stability of MMV in sap is comparatively short-lived. In the crude sap of infected leaves, diluted 1/10 in 0.01 M phosphate buffer pH 7.5, as well as in partially purified preparations viral particles retain infectivity at 4°C after 24 and 48 hours (Herold, 1972). Sap infectivity is controlled by infection of the leafhopper *Peregrinus maidis*, the only active virus vector in nature.

Serology

MMV belongs to the group of rhabdoviruses representing an antigenic complex composed of a number of different weakly antigenic proteins (Herold, 1972). Antisera of 1/4 titre were obtained by injecting guinea pig with a partially purified virus from the plant sap. Serological analyses used for identification may aid in the investigation of possible virus localization in host plants and vector.

Strains

Strains of MMV have not been described to date because isolates from maize seem identical. However, differentiation of isolates with the aid of various hosts and antigen complexes of these viruses can be expected. The structure and mode of leafhopper transmission of MMV is similar to viruses such as potato yellow dwarf and wheat striate mosaic viruses, which both multiply in host plants and in insect vectors. Therefore, MMV is suitable for the investigation of mutual relations in the development of animal and plant viruses.

Geographic Distribution and Economic Importance

MMV, which ranks among the most widespread viruses of Gramineae plants, has been described in Hawaii, Cuba, Trinidad, Tanzania, Puerto Rico, Mauritius, Surinam, and Venezuela (Herold, 1972), and the disease was recorded in the Caribbean and in India (Francki et al., 1981). Since this mosaic disease predominates in tropical countries, Šutić (1983) labeled it the "tropical mosaic" in order to better differentiate it from other types of mosaic described in maize. MMV distribution depends on the movement and environmental conditions for the growth and development of the natural vector *Peregrinus maidis*.

MMV ranks among the rhabdoviruses that cause plant diseases of most economic importance. It is extremely damaging due to its widespread distribution and the severe stunting of maize. Serious losses were registered in the maize crop in Venezuela (Francki et al., 1981). Lastra (1977) reported that MMV is capable of infecting up to 60% of maize plants in favorable (mild) weather conditions during September to January.

Host Range

MMV infects a comparatively narrow range of host plants, limited only to the Gramineae family, with maize as its primary natural host (Herold, 1972). Other natural hosts are *Rottboellia exaltata* and *Setaria vulpiseta*, and experimental hosts include *Euchlena mexicana* and *Sorghum*.

Susceptible maize cultivars (such as Sicarigua and the *Theobromina blance* line) are diagnostic.

Symptoms

Symptoms in maize and other hosts include yellow spots, leaf stripes, and plant stunting. Light-green and yellow spots first appearing on the leaves of infected maize plants, gradually develop into a form of mosaic mottling, broken with stripes between and along the tiny leaf veins. Parts of infected leaves may turn completely yellow and necrotic. Symptoms characteristic for the tropical mosaic differ substantially from those of the European maize mosaic type.

Similar symptoms may also occur on maize stalks, leaf sheaths, and ears. In early infections, all internodes are shortened, whereas later infections cause internode shortening only above the infection point. Due to morphogenic changes, plants lag behind in growth and their apices frequently become twisted.

Pathogenesis

The virus is preserved in natural hosts. No transmission of MMV in the seed from infected plants has been obtained to date.

This virus has not been transmitted mechanically with the sap of infected plants in nature. The primary vector is the leafhopper *Peregrinus maidis*. All nymphs and adults, both males and females, actively transmit MMV in a persistent manner. Only 1 day or even less is sufficient for the leafhopper to acquire MMV, which retains infectivity from 11 days to 7 weeks following incubation. Vector efficiency of leafhoppers seems to increase when the latent virus that infects *P. maidis* is absent or present only at low titers (Herold, 1972).

Since *P. maidis* is the only known vector, disease pathogenesis depends directly on environmental conditions that influence leafhopper development.

Control

Preventive measures recommended for virus control include early destruction of infected plants as sources of infection, spatial isolation of susceptible plants in production areas, and measures aimed at the reduction of the *P. maidis* populations, such as chemical control. However, planting resistant maize cultivars has priority over all other control methods.

Maize Streak Geminivirus (MSkV)

Properties of Viral Particles

MSkV particles are isometric, occurring mostly in pairs. The geminate particles are of five-angular shape and 30×20 nm, and single particles are of six-angular shape and 20×20 nm (Bock et al., 1974). The ratio between geminate and single particles varies (8:1; 3:1) and depends on the method of preparation. Since most viral particles are joined in pairs, the group was named gemini (gemini – "twins").

MSkV nucleocapsids contain single-stranded, closed circular DNA molecules. DNA molecular weight is 710 kDa and protein molecular weight is 28 kDa. The gemini particle sedimentation coefficient is 70S according to Cornuet (1987); but according to Bock et al. (1974), it is 76S for geminate particles and 54S for single particles.

Stability in Sap

The biophysical constraints determined for one virus strain isolated from maize are: thermal inactivation point not below 60°C, dilution endpoint greater than 1/1000, and longevity at room temperature not less than 24 hours (Bock et al., 1974). Particle infectivity in crude plant sap was assayed by transmission with the aid of vectors since this virus cannot be mechanically transmitted.

Serology

MSkV has moderate to pronounced immunogenic activity (Bock et al., 1974) and antisera are successful in gel diffusion and tube tests, respectively.

Strains

Several virus strains (isolates) have been described based on host susceptibility and disease symptoms. McLean (1947) described two: one as strongly virulent (severe) (A) and one as mild (B) in maize. MSkV isolates from sugarcane (Uba-virus) did not induce protection in maize against the A and B strains. MSkV isolates from wild grasses *Sporobolus*, *Eleusine*, and *Paspalum* will infect maize, but not sugarcane. MSkV seems to have considerable adaption to its hosts; thus, the differentiation of isolates or strains of the MSkV are expected.

Assays of MSkV antisera have revealed certain serological differences among isolates obtained from different hosts. It was shown, for example, that the sugarcane isolate is related closely, and the guinea grass (*Panicum maximum*) distantly, to the maize isolate (Bock et al., 1974).

Geographic Distribution

MSkV is widely distributed in individual areas. It has been recorded in African countries, on Mauritius and Madagascar, and observed in India and also probably in other southern Asian countries. It has not been recorded yet in European countries.

Host Range

Hosts of MSkV include only the grasses (Gramineae) family, the most important of which are maize, sugarcane, and millet (*Eleusine coracana*). In Africa, MSkV has been recorded in the Andropogoneae (*Cymbopogon, Imperate, Saccharum, Rottboeillia*), Eragrosteae (*Dactyloctenium, Diphachne, Eleusine, Eragrostis, Leptochloa, Setaria*), Sporoboleae (*Sporobolus*), Zoysieae (*Tragus*), Maydeae (*Zea, Euchlaena*), Hordeae (*Triticum*), and Avenae (*Avena*) families (loc. cit. Bock et al., 1974). Maize is diagnostic only for the strains that infect it.

Symptoms

General disease symptoms are chlorotic streaks visible in interveinal and narrow stripes along the leaf veins (Figure 8). Initial symptoms are round white spots at the base of the leaflets, which then gradually elongate and appear as broken chlorotic streaks. Streaks further elongate, continue to fuse in both length and width, resulting in the final disease symptom (i.e., narrow chlorotic stripes that cover the leaf surface).

FIGURE 8 Typical maize streak. (Courtesy V. D. Damsteegt. Reproduced with permission from APS.)

On susceptible maize, sugarcane, and millet (*Eleusine coracana*), streaks take on a highly specific appearance, for which the disease was named.

Pathogenesis

Many susceptible plant species grow permanently in regions where this virus is widespread, which provides ample MSkV inoculum to infect economically important crops.

Leafhoppers only in the *Cicadulina* genus are natural virus vectors. Five species have been described as vectors of MSkV: *C. mbila*, *C. storeyi*, *C. bipunctella zeae*, *C. latens*, and *C. parazeae*. By feeding on infected plants, leafhoppers acquire the virus within 1 hour, with a latent period in leafhoppers of 6 to 12 hours (at 30°C), whereafter MSkV inoculation can occur within 5 minutes. All nymphal instars maintain active vectorship which is maintained during molting. MSkV is not transmitted transovarially to next generations. Once infected, vectors remain viruliferous for life, although data have been recorded on virus multiplication in the vector body (loc. cit. Bock, 1974).

Conditions optimal for MSkV multiplication in the leafhopper are also important epidemiological factors.

The virus is transmitted neither in the seed nor mechanically with the sap of infected plants.

Control

For prevention of infection, growing of susceptible plant species from isolates and elimination or reduction of reservoirs of MSkV infections are essential. Reduction of leafhopper populations, both in production and other nearby fields where these insects may reside, is advisable. The possibility of virus spread during the season can be reduced by growing various mixtures of plant species.

New solutions for disease control suggest the need to breed and introduce virus-tolerant or -resistant maize hybrids. Genes regulating resistance of maize to MSkV have been discovered (Gorter, 1959; Rose, 1978). Maize cultivars featuring resistance and tolerance are currently used in breeding programs in South Africa, Rhodesia, Kenya, and Nigeria (Goodman, 1981).

FIGURE 9   Maize chlorotic mottle. (Courtesy J. Castillo L. Reproduced with permission from APS.)

Other Maize Virus Diseases

Maize White Line Mosaic Virus (MWtLnMV)

MWtLnMV is isometric, 35-nm diameter. Some MWtLnMV isolates are also in association with a satellite-like virus of approximately 17 nm in diameter. MWtLnMV particles have a single-stranded RNA of approximate 1.25×10^6 MW and a single protein structural unit of 32 or 35.3 kDa. The satellite-like virus also features a single-stranded RNA of 0.4×10^6 MW and a protein of 24.7 kDa.

Widespread in the U.S., France, and Italy, MWtLnMV causes a stunt and severe leaf mosaic in plants whose cobs do not fill out and contain only a few kernels. Yield reductions range up to 45%. It has a narrow host range only among monocotyledonous plants.

MWtLnMV is not transmitted mechanically and transmission through seed is less than 0.01%. It may be transmitted by soil-borne fungi (de Zoeten and Reddick, 1984).

Maize Chlorotic Mottle Machlomovirus (MCMtV)

MCMtV is isometric and approximately 30 nm in diameter. Viral particles probably contain single-stranded RNA of 1400 to 1600 kDa, and a single species protein of 23.9 to 27.0 kDa.

The MCMtV, widespread in Peru, Argentina, Mexico, and the U.S., causes symptoms of chlorotic leaf mottle (Figure 9) necrosis, and plant stunting. Infected plants scarcely form ears and witness an early death. Yields from infected plants under natural conditions are reduced by 10 to 15%, or by 59% under experimental conditions. Virus hosts belong only to the Gramineae family. The virus is transmitted neither mechanically, nor through seed. Chrysomelid beetles are natural vectors of MCMtV (Gordon et al., 1984).

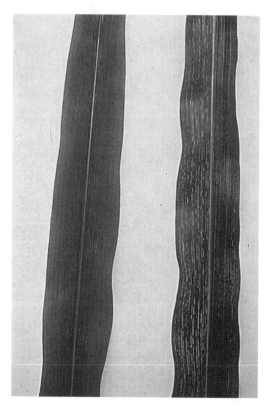

FIGURE 10 Rayado Fino; healthy (left). (Courtesy R. Gámez. Reproduced with permission from APS.)

Maize Stripe Tenuivirus (MSpV)

MSpV particles are filamentous, of undetermined length, and 3 nm in diameter. They contain five species of single-stranded RNA, ranging from 0.52 to 3.01×10^6 MW and a single species of protein 32.7×10^3 MW.

MSpV has been observed in maize in tropical countries of the U.S. (Florida), Venezuela, Peru, Guadaloupe, Nigeria, Sao Tome, Kenya, Botswana, Mauritius and Reunion, Australia, Costa Rica, and possibly the Philippines. Its distribution depends on reservoir hosts and the presence of the vector *Peregrinus maidis* planthopper. MSpV infects naturally only a few species of Gramineae, including maize, sorghum, and itchgrass (*Rottboellia exaltata*). Leaf stripes and stunting of plants, accompanied by the production of little grain, are the most pronounced symptoms in maize.

MSpV cannot be transmitted mechanically or through seed. The planthopper *Peregrinus maidis*, the only known vector, transmits the virus persistently and transovarially (Gingery, 1985).

Maize Rayado Fino Marafivirus (MRaFnV)

MRaFnV particles are isometric, approximately 31 nm in diameter, and contain single-stranded RNA of 2000 kDa and a single-peptide protein of 25 kDa.

Wide distribution occurs in the tropical areas of Central America, Mexico, Peru, Uruguay, Brazil, Colombia, in the southern U.S., and probably in the Caribbean islands. Virus hosts are Gramineae plants. Symptoms in infected maize include leaf spots and striping, chlorosis, formation of holes in the leaves, wilting and death of plants and in susceptible cultivars, reduced size of roots, and ears bearing little grain (Figure 10).

MRaFnV is transmitted neither mechanically nor through maize seed. Leafhoppers *Dalbulus maidis* are natural vectors transmitting the virus persistently (Gámez, 1980).

Maize Leaf Fleck Virus (MLFkV)

MLFkV was first observed in 1950 in California (Stoner, 1952), and later in Bulgaria (Atanasoff, 1965). Its symptoms are irregular roundish spots between veins near older leaf tips. Spots enlarge, later appearing on lower leaf surfaces. Initially, yellowish spots gradually appear as greasy blotches (a drop of oil spilled on paper) with a necrotic center, which is the diagnostic symptom. Leaf margins and tips are affected first, then entire leaflets become dry. MLFkV is of minor economic importance (Gordon et al., 1981).

MLFkV is neither mechanically, seed, nor soil borne (Stoner, 1952), but is transmitted persistently by aphids *Rhopalosiphum prunifoliae*, *R. maidis*, and *Myzus persicae*. *R. prunifoliae* fed on infected maize acquires the virus within 10 minutes and, when fed on a healthy plant, can inoculate it in 5 minutes. After feeding on infected maize, *R. prunifoliae* and *M. persicae* remain viruliferous for life. Susceptible plant hosts include maize (some cultivar used in production), sweet corn, popcorn, and *Phalaris tuberosa* var. *stenoptera*.

Atanasoff (1965; 1966) found MLFkV widespread in Bulgaria, causing mottling of kernels and probably male sterility in maize. He also reported MLFkV as seed-borne and probably also the cause of leafspot mosaic in wheat and other Gramineae plants.

Similar maize disease symptoms also observed in Yugoslavia have not been experimentally reproduced in healthy plants to date. Based on the scarcity of available data, the infectivity of this disease should be further investigated.

MAIZE, NATURAL ALTERNATIVE HOST FOR SOME VIRUSES

Sugarcane Mosaic Potyvirus (ScMV)

Maize mosaic as a symptom type was first described by Brandes (loc. cit. Gordon et al., 1981) in 1920. More recent investigations showed that ScMV also causes mosaic in other Gramineae plants.

ScMV has been described on all continents, consisting of several strains: A, B, D, E, and F in the U.S.; BC, SC, and Sabi in Australia; and the former B strain of maize dwarf mosaic virus (MDMV-B = ScMV-MDB) in the U.S. and Europe (Teakle et al., 1989). Until recently, ScMV was considered a unique virus species also comprising maize dwarf mosaic (MDMV), sorghum mosaic (SrMV), and Johnsongrass mosaic viruses (JgMV). Based on the latest chemical, serological, and host range symptomatology data, these viruses have been reclassified into separate species (Shukla et al., 1989). Now, ScMV strains have been placed in a new taxonomic division.

Infected maize plants display symptoms of mosaic, mottling, chlorosis, stunting, and decreased fertility.

Important factors for sugarcane mosaic virus on maize epidemiology are sugarcane production fields, as a perennial source of infection regularly infecting maize and other hosts, in close proximity. ScMV is transmitted nonpersistently by many aphids, such as *Dactynotus ambrosiae*, *Hysteroneura setariae*, *Rhopalosiphum maidis*, *Toxoptera graminis*, etc. (Teakle et al., 1989). ScMV transmission through maize seed was reported (von Wechmar and Chauhan, 1984).

Johnsongrass Mosaic Potyvirus (JgMV)

JgMV has recently been separated from the ScMV group (Shukla et al., 1989). First described in Australia as a MDMV strain (Taylor and Peres, 1968) and later as a strain of ScMV infective in Johnsongrass, it now stands alone (Teakle and Grylls, 1973). JgMV was described in the U.S. as a strain of MDMV infective in oats (MDMV-O) (McDaniel and Gordon, 1985).

Symptoms in JgMV-infected plants mimic those caused by MDMV. In addition to mosaic and yellowing, JgMV-diseased maize plants also display reddening, accompanied by necrosis. Infected maize plants express a delay in maturity plus a substantial reduction in yields.

FIGURE 11 Mottling caused by sorghum mosaic virus, formerly sugarcane mosaic virus strain H. (Courtesy R. Toler. Reproduced with permission from APS.)

Johnsongrass is the primary perennial source of inoculum in the epidemiology of the JgMV disease in maize. *Aphis craccivora*, *A. gossypii*, *Myzus persicae*, and *Rhopalosiphum maidis* transmit JgMV nonpersistently. Limited research has failed to show seed transmission of JgMV in maize seed.

Sorghum Mosaic Potyvirus (SrMV)

SrMV was recently separated from the ScMV group (Shukla et al., 1989). Research has shown that the H and I strains do not belong to ScMV, but represent a separate virus that was named after cultivated sorghum—its natural host.

Symptoms of SrMV on maize also mimic those caused by MDMV, ScMV, and JgMV (Figure 11). The classification of SrMV as a separate species and the elaboration of methods for its identification now will aid in delineating its distribution and better understanding the disease it causes in maize as its secondary host.

Barley Yellow Dwarf Luteovirus (ByYDV)

Maize naturally infected with ByYDV was reported in the U.S. in 1977 (Stoner, 1977). More recently, it has been reported as naturally infecting maize in European countries (France, Italy, and former Yugoslavia), and in Morocco (Gordon et al., 1983), seemingly becoming increasingly important as a maize pathogen. Since ByYDV is widespread in barley, oats, wheat, and other grasses and since it is naturally transmitted by an omnipresent aphid population, one would expect a more common presence of ByYDV in maize than reported.

Symptoms on infected maize differ, depending on both plant genotype and external environmental factors, usually more pronounced in nature than in greenhouses (Refatti et al., 1987). Stoner (1977) reported visible purpling and reddening on lower leaf surfaces, followed by rapid leaf necrosis, but only slight stunting of infected plants. Signoret (1983) reported infected plants as generally stunted, with yellow spots and stripes appearing mainly on younger leaves and markedly reduced yields. Refatti et al. (1987) reported severe damage and total crop losses caused by ByYDV in maize inbred lines. Gordon et al. (1981) recorded less damage caused by ByYDV in maize than in infected barley and oats plants.

ByYDV is transmitted in nature by aphids in a persistent manner. The most active aphid vectors are *Rhopalosiphum padi*, *Metopolophium dirhodum*, and *Sitobion avenae* (Osler et al., 1987). All cereals are harvested before the maize crop, which remains in the field available to aphid attacks. In France, large *R. padi* populations develop in maize fields in late September, when often more than 1000 aphids per plant occur (Henry, 1987). Those infected maize plants represent alternative hosts and reservoirs rich in ByYDV, which then transmit during senescence of maize to winter cereal and other grass crops. Brown et al. (1984) reported that economic losses resulted in winter grain crops in eastern Washington when plants were infected immediately following emergence by the virus originating from irrigated, ByYDV-infected maize plants.

Maize inbred lines and hybrids differ according to their susceptibility to ByYDV. For example, some inbred lines used for commercial hybrids are susceptible to ByYDV. Symptomless maize lines and hybrids are tolerant, but not resistant to the virus (Refatti et al., 1987). Lorenzoni et al. (1987) investigated the ByYDV tolerance of more than 1200 inbred lines in the field in Italy. Some lines were tolerant and others highly susceptible; one group of lines showed changeable, inconsistent symptoms. It is important to stress that some lines used in hybrid breeding were tolerant, a characteristic transmitted with the dominant gene that enables the creation of tolerant maize hybrids with the use of only one tolerant parent.

In view of the endemic nature of ByYDV and of maize as an alternative host, the following recommendations may help reduce losses in maize crops: delay sowing of susceptible crops; control virus vectors with the aid of systemic insecticides; and grow tolerant maize lines and hybrids.

Cereal Chlorotic Mottle Virus (CrCMtV)

CrCMtV is in the Rhabdovirus group, with bacilliform particles of 240×65 nm. It may contain single-stranded, negative-sense RNA and five polypeptides of differing molecular weights possibly similar to the proteins (G, N, NS, M1, and M2) based on other Rhabdovirus research (Greber, 1982; Peters, 1981). Although serologically different, CrCMtV resembles the American wheat striate mosaic virus (WAStMV), which has a Cicadellid leafhopper vector common to both viruses and the viruses have nearly identical host ranges.

CrCMtV has been described only in eastern Australia (Greber, 1982a). Plants in the Gramineae family are the only known virus hosts. Important natural virus hosts include maize, barley (*Hordeum vulgare*), wheat (*Triticum aestivum*), and several weed grasses. CrCMtV causes fine chlorotic striations, plant stunting, and tassel sterility (Greber, 1979; 1981). It usually infects 0 to 10% and in some cases even up to 60% of plants. Infected biennial and perennial plants of the Gramineae family represent important virus reservoirs. CrCMtV is transmitted neither mechanically, nor through maize seed. Its natural vectors are Cicadellid leafhoppers.

Wheat Streak Mosaic Virus (WSkMV)

WSkMV was first discovered in the sweet corn crop in Idaho in 1957 (Finley, 1957) and later in various maize cultivars in several states of the U.S. (Gordon et al., 1981). The intensity of its occurrence is more or less pronounced, depending on the presence of the sources of infection, the vectors, and maize genotype susceptibility.

WSkMV infects wheat, maize, and other gramineous plants in nature, but not Johnsongrass or Atlas sorghum (Gordon and Nault, 1977; Knoke et al., 1974). WSkMV causes mild mosaic and mottle pattern of greens and yellows in maize leaves, in many ways similar to MDMV and ScMV symptoms on maize. WSkMV is of minor economic importance in maize.

The virus is transmitted with the sap of infected plants, but its primary natural vector is the Eriophid mite, *Aceria tulipae* (Williams et al., 1967) and *A. tošićhella* (Tošić, 1971). WSkMV-infected plants of cereal, hay, and pasture crops represent field reservoir sources of infection.

Brome Mosaic Virus (BrMV)

BrMV was first discovered in maize in South Dakota in 1967 (Moline, 1973; Stoner et al., 1967), and has been studied in some detail (Ford et al., 1970). Although generally spread among a large number of gramineous plants, no information is available on its distribution in maize in other countries. BrMV is of minor economic importance in maize production.

Cucumber Mosaic Virus (CMV)

CMV was first discovered on maize as its natural host in 1934 in Florida (Wellman, 1934). Later, natural infections of maize with CMV were reported in California (Stoner, 1949) and Kansas (Sill, 1966). Via inoculation with a typical CMV strain and its gladiolus isolate, Bridgmon (1951) experimentally induced disease symptoms in maize plants. Natural occurrence of maize infection with CMV in Europe has been reported only in former Yugoslavia (Panjan, 1966).

Symptoms of CMV infection in maize are irregular mosaic stripes on the leaves of infected maize plants. CMV-infected seedlings exhibit delayed growth and die. Infected leaves of older plants when reduced in growth cause the plants to appear stunted.

Based on distribution and intensity, CMV is of minor importance in maize production. However, as a maize pathogen, it should receive due attention in view of the ubiquitous universal distribution and nearly unlimited host range (more than 700 plant species), plus the fact that it is transmitted nonpersistently by numerous aphid species that are naturally very active vectors.

VIRUS DISEASES OF WHEAT (*Triticum aestivum*)

Acreage of wheat sown ranks first among cereals and, based on production, it holds a leading position in human nutrition in the world. Numerous, heterogenous pathogen groups cause yield reductions and are capable of completely destroying wheat crops; therefore, they are very important among production factors. The *Compendium of Wheat Diseases* (Wiese 1987) provides a thorough treatment of all important diseases of wheat.

Viruses are important among wheat pathogens because, not only do they represent limiting production factors, but they are difficult or nearly impossible to control directly. Many of these viruses occur widespread wherever wheat is grown and have general world importance. However, some viruses are distributed only in narrower geographic areas, which results from the presence of certain virus vectors and environmental conditions necessary for their development. Specific geographical identification is occasionally designated in the names of some viruses.

Control or prevention of virus infection has become an indispensable factor for modern wheat production, both nationally and generally worldwide.

Wheat Striate Mosaic Rhabdovirus (WStMV)

WStMV is a name under which two virus species have been described. One virus, discovered in the U.S. and Canada (Lee, 1964; Slykhuis, 1953), was designated as the American WStMV; and the other, in European countries, as the European WStMV (Slykhuis and Watson, 1958). These two viruses resemble each other in symptoms only, but differ markedly in other characteristics by which they were described as separate virus species.

Wheat American Striate Mosaic Rhabdovirus (WAStMV)

Properties of Viral Particles

WAStMV particles are bacilliform of 250×75 nm dimensions, with a sedimentation coefficient of 875S and a buoyant density in sucrose of 1.22 (g cc^{-1}) (Sinha et al., 1976).

Viral particles are comprised of 68% protein, 25% lipid, 5% RNA, and 2% carbohydrate (Sinha et al., 1976). The protein component includes proteins of four different sizes: L (145×10^3), G (92×10^3), N (59×10^3), and M (25×10^3) MW (Trefzger-Stevens and Lee, 1977).

Viral RNA is single stranded, of 2.2×10^6 Da MW, with a sedimentation coefficient of 31.5S (Sinha et al., 1976). The presence of different proteins in the viral particle implies the existence of several genome RNA types.

Stability in Sap

This trait was assessed by injecting an extract from viruliferous leafhoppers into virus-free insects. The TIP is 55°C; the DEP is 10^{-5}; and the LIV at both 4 and 10°C is 3 days (Lee and Bell, 1963).

Serology

Like other Rhabdoviruses, WAStMV is weakly immunogenic. Antisera were obtained that reacted with WAStMV antigens both in infected plants and in viruliferous vectors (Sinha, 1968), indicating that this virus multiplies equally well in the plant host and in the insect vector.

The large size of the viral particles makes tube precipitin or ring precipitin tests most suitable for serological analyses. Gel diffusion tests substantially reduce the titre of antiserum preparations to 1/32 or less (Jackson et al., 1981). Antisera with different titres were obtained by intramuscular injection of partially purified preparations: 1/160 (ring precipitin test), 1/40 (tube precipitin test), and 1/5 (gel diffusion test). The highest antiserum titre (1/320) in the ring precipitin test was obtained by a weekly injection, each of 4 mg, of purified virus for 8 weeks (Sinha and Behki, 1972). Jackson et al. (1981) stressed the potential serological affinity between cereal chlorotic mottle virus (CrCMtV) and WStMV, as well as between oat striate mosaic virus (OtStMV) and WStMV.

Strains

No strains have been reported. New data might become available from serological analyses of purified virus isolates and research on variations in several countries.

Geographic Distribution and Economic Importance

WAStMV has been described in North America. First discovered in 1953 in South Dakota (Slykhuis, 1953), it was found later in the north central regions of the U.S., as far south as Illinois. The incidence observed in fields surveyed in Canada sometimes was less than 1% and occasionally as high as 25% in the U.S. (Slykhuis, 1962; Timian, 1960).

WAStMV causes economic damage in highly susceptible cultivars and in locations with optimal environmental conditions for plant infection. In severe epiphytotics, the growth and seed set of spring wheat is markedly delayed, seed set is reduced 30 to 60%, and seeds are shriveled. Plants infected early in the season die (Slykhuis and Sherwood, 1964). Due to a restricted, localized distribution, WAStMV is of limited economic importance in world production.

Host Range

Virus hosts comprise at least 20 plant species belonging only to the Gramineae family. Hosts susceptible to WAStMV include *Triticum aestivum* (wheat), *Avena sativa* (oats), *Secale cereale* (rye), *Zea mays* (maize), and several native grasses (loc. cit. Jackson et al., 1981). All durum wheat cvs. are highly susceptible; some winter wheat cultivars are susceptible; and for barley and oats cultivars, the majority is moderately susceptible to WAStMV.

Recommended diagnostic plants are wheat (*Triticum durum* cv. Ramsey), oats (*Avena sativa* cv. Victory), barley (*Hordeum vulgare* cv. Vantage), maize cv. Gaspe flint, *Eragrostis cilianensis* cv. Lutati, and *Panicum capillare* (Sinha and Behki, 1972).

Symptoms

Leaf striation is the general symptom in wheat, accompanied by weak chlorosis that becomes more intensive and necrotic as leaves age. Plants express a delay in growth and stunting. Grain crinkling and sterility of heads can occur in highly susceptible cultivars.

It is rather difficult to distinguish wheat striate mosaic symptoms from other wheat mosaic diseases early in disease development. Chlorotic and parallel streaks characterize wheat striate mosaic, but insufficiently different to distinguish from similar striate mosaic disease symptoms described in Europe and Australia and winter wheat mosaic symptoms observed in Russia.

The European and the American wheat striate mosaics can usually be distinguished by the species of Cicadellid leafhoppers that transmit the virus. None of the WAStM viruses could be transmitted simultaneously by Cicadellid leafhoppers of the Delphacideae and Jassidae families. This led Sinha and Behki (1972) to conclude that these two mosaic diseases are caused by different wheat viruses. The WAStMV is closely related to the Russian winter wheat mosaic, based on bacilliform particle size (260 × 60 nm), and transmitted by the Cicadellid leafhopper vectors (*Psamotettix striatus*) of the Jassidae family.

Pathogenesis

WAStMV is preserved naturally over seasons in infected biennial cereals and in some native grasses.

Two Cicadellid leafhoppers (Jassidae)—*Endria inimica* and *Elymana virescens*—serve as natural vectors of the virus. *E. inimica* is of basic importance and an efficient vector capable of acquiring and transmitting the virus in all development stages. The minimal acquisition period can be shorter than 1 minute (Slykhuis, 1963). The acquisition period depends on the temperatures necessary for insect preservation; for example, 4 days at 32°C and 24 days at 16°C. Virus transmission can occur at 10°C (Slykhuis and Sherwood, 1964). Infected leafhoppers remain viruliferous for long periods, some throughout their lives (Slykhuis, 1963; Slykhuis and Sherwood, 1964). WAStMV multiplies during the latent period in the insect up to 1000× concentrations (Sinha and Chiykowski, 1967; Paliwal, 1968).

The virus is transmitted mechanically, but probably not, or seldom, transmitted with the seed (Sinha and Behki, 1972).

Control

Cicadellid leafhoppers play the most important role in preservation and spread of WAStMV; thus, control of leafhoppers represents a most important method of plant protection against infection by this virus.

The incidence of infected plants in nature can be reduced by rogueing infected plants (where possible) and by spatial isolation of susceptible plants in production fields. Growing resistant and tolerant plants is the best method to reduce yield losses due to infection with WAStMV.

Wheat European Striate Mosaic Rhabdovirus (WEStMV)

A bacilliform WEStMV causing the European wheat striate mosaic type disease was discovered in Europe. First described in England, and later in Germany and Spain, it can also occur in Denmark, Poland, and Romania (Brčak, 1979). Natural virus hosts include wheat (*Triticum aestivum*), oat (*Avena sativa*), barley (*Hordeum vulgare*), *Bromus mollis*, *Lolium multiflorum*, *L. perenne*, and *Phleum pratense*. Several experimental hosts exist among the *Dactylis*, *Eragrostis*, and *Poa* genera.

Chlorotic striated spots occur on the leaves of WEStMV-infected wheat plants as the main disease symptom. With age, leaves become more or less chlorotic with white and dark brown necrotic spots, causing complete leaflet desiccation in susceptible wheat cultivars. Plants infected early in the seedling stage sometimes die 2 to 3 months following infection, whereas later infections cause plant decline shortly before earing, a delay in growth, partially fertilized spikes, and incompletely formed grains.

The leaf serration symptom is characteristic of this disease on infected barley plants.

White and yellow striated spots appear on the leaves and stalks of diseased oat plants. WEStMV-infected plants become orange to light red, wilt, and die. This disease can reduce the yield by as much as 10%.

Active virus vectors in nature are Cicadellid leafhoppers (Delphacidae): *Javesella pellucida*, *J. obscurella*, and *J. dubia*. They transmit the virus in a persistent manner. Virus multiplication in Cicadellid leafhoppers requires 10 to 14 days before transmission can occur. In *J. pellucida*, this virus is transovarially transmitted to the next generation (Brčak, 1979; Serjeant, 1967). It is transmitted neither mechanically nor through the seed.

WHEAT SOIL-BORNE MOSAIC TOBAMOVIRUS (WSoMV)

Properties of Viral Particles

The dimensions of the rigid, rod-shaped viral particles, which vary with individual strains, range between 100–160 × 20 nm and 280–300 × 20 nm (Proeseler, 1986). Some particles 300-nm long occur as dimers composed of two 160-nm long rods. Particles of 300 nm are infective, but not those of 160 nm (Gumpf, 1971). Sedimentation coefficients (Nebraska isolate) are 172S and 211S for 160 and 300 nm rods, respectively.

WSoMV particles contain two single-stranded RNA classes: that from rods 160-nm long contain only 24.3S RNA, and that from 300-nm long contain both 24.3S RNA and 33.4S RNA (Gumpf, 1971), with molecular weights of 1.06×10^6 and 2.0×10^6, respectively. RNA comprises 5% of the viral particle, with no information on its protein composition (Brakke, 1971).

Stability in Sap

Strains of WSoMV differ slightly, but the thermal inactivation point is 60 to 65°C, dilution endpoint is 10^{-2} to 10^{-3} and *in vitro* longevity at 18 to 21°C is 48 to 96 hours. The virus can remain infective for several years in *Polymyxa graminis* spores in dry soils (Cornuet, 1987), and more than 11 years in dry plant leaves (Brakke, 1971).

Serology

WSoMV expresses strong immunogenic activity by injecting the purified virus into rabbits (Brakke, 1971). Serological analyses were performed with agar diffusion (Gumpf, 1971; Rao and Brakke, 1969) and ELISA (Proeseler, 1986) tests.

Strains

Green, yellow, and one rosetting strains of WSoMV have been described. Isolates from Japan and the U.S. are related serologically, although those from Japan also infected tobacco and *Zea mays* (loc. cit. Brakke, 1971).

Geographic Distribution and Economic Importance

The WSoMV has been recorded in the U.S., Europe, and Japan, but the relationship among their isolates is unclear (Ikata and Kawai, 1940; McKinney, 1925; 1953; Saito et al., 1962; Sill, 1958). In Europe, the virus was described in wheat, both in Italy and in the U.S.S.R. (Agarkov, 1956; Canova, 1964; 1966). The virus has not been positively identified in Yugoslavia although its sporadic occurrence in some localities is suspected.

WSoMV belongs to the group of diseases considered economically most damaging. In mild epidemics, wheat yields were reduced 10 to 20%; while in severe cases, this reduction was more than 50%. This disease has jeopardized wheat production in some regions of the U.S.

Host Range

In nature, WSoMV infects wheat, barley, and rye. Other virus hosts include monocotyledonous plants—the genera *Agropyron, Agrostis, Aira, Bromus, Dactylis, Festuca, Holcus, Lolium, Panicum, Phleum, Poa, Secale*, and *Triticum*. Experimental hosts among dicotyledonous plants are *Chenopodium amaranticolor* and *C. quinoa*. *Triticum spelta* is recommended as an assay plant; it displays characteristic symptoms of natural infections in November.

Symptoms

Disease symptoms vary, depending on plant susceptibility and temperature conditions for their development. Although mosaic disease symptoms on winter wheat, barley, and rye may appear already in the fall, they are not well expressed at that time. Resumed development of infected plants in the spring yields chlorotic discoloration restricted to leaf veins, then finally symptoms of excessive tillering, pronounced stunting, and rosetting. Roots of infected plants are also stunted and inflorescences develop poorly with few small grains. See Figures 12 and 13.

Pathogenesis

WSoMV is transmitted by the fungus *Polymyxa graminis*. Viral infectivity is preserved in the soil for 6 years in its resting spores. Resting spores form groups (sori) of 6 to 10, and sometimes as many as 100 members. Zoospores released from resting spores function to infect epidermal cells and root hairs. Following incubation, zoospores develop into zoosporangia (in 8 to 9 days) and can affect the entire root system (Rao, 1968). WSoMV particles are firmly bound inside zoospores and resting spores, which makes it difficult to isolate and inactivate the virus (Rao and Brakke, 1969).

The virus can be mechanically transmitted under low temperatures with infective sap (inoculum) prepared in pH 6.2 buffer; then the plants are placed in the dark following inoculation. No information exists on virus transmission through the seed.

Control

Implementing virus control is difficult since *Polymyxa graminis* assures WSoMV a site for its persistence in resting spores for which successful control measures are complicated, particularly in case of large acreages. Protection of wheat in the U.S. is most successful by growing resistant or tolerant cultivars.

WHEAT STREAK MOSAIC POTYVIRUS (WSkMV)

Properties of Viral Particles

WSkMV is a filamentous particle of 700 × 15 nm. The possible infectivity of shorter particles has not been proved (Brakke, 1971a). The WSkMV sedimentation coefficient is 165S. WSkMV contains a single-stranded (+)RNA, consisting of a single molecule of 2800 kDa, with a sedimentation constant of 40S (Brakke, 1971a).

Stability in Sap

WSkMV stability varies, depending on the mode of preparations and the storage temperatures. The TIP is 52°C, the DEP is from 10^{-3} to 2×10^{-3} (Tošić, 1971), and under certain conditions, WSkMV retained infectivity in dry leaves for 16 years (Brakke, 1971a).

Serology

WSkMV is moderately immunogenic by injecting rabbits with the purified virus (Brakke, 1971a). Satisfactory antiserum titres should be prepared from the purified virus due to low concentrations of viral particles in the crude or partially purified sap.

Virus Diseases of Food Graminoid Plants

FIGURE 12 Winter wheat infected by wheat soil-borne mosaic virus in a variety nursery with susceptible wheat hill-plots surrounding resistant varieties (center of picture).

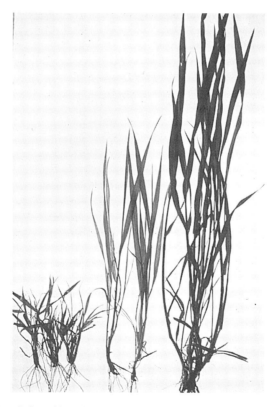

FIGURE 13 Winter wheat infected by wheat soil-borne mosaic virus; varieties Bison (left), susceptible sage (center) and Resistant Abe (right).

TABLE 1
Reduced Growth and Yield of Some Wheat Cultivars Infected with WSkMV

Wheat Feature	Wheat Cultivars (reduction in %)		
	San Pastore	Bankut 1205	Bezostaja 1
Number of secondary stalks	0.5	20.8	0.0
Stalk length comprising the spike	15.6	45.9	25.3
Number of grains per spike	25.2	52.0	35.9
Yield, grams per plant	58.9	63.8	53.2
1000 grain weight, grams	13.7	22.6	15.6

Strains
Some WSkMV isolates have been identified as mild strains based on their virulence. The virus is not related serologically with any known wheat virus.

Geographic Distribution and Economic Importance
WSkMV occurs worldwide, in Africa (Egypt), Asia, North America (the U.S. and Canada), Europe (Romania, former U.S.S.R., former Yugoslavia), and in the Middle East. It is one of the most important viruses in the central Great Plains of the U.S. In some countries (former Yugoslavia), it is the most frequently observed and isolated virus (Šutić, 1983; Šutić and Tošić, 1964a).

WSkMV ranks as economically damaging, especially for winter wheat when early infections of WSkMV in the fall immediately follow plant emergence. Characteristic damage includes reduced plant growth, secondary stalk decline, and reduced yield and grain weight (see Table 1).

How much WSkMV damages plant growth and yielding potential is directly proportional to the number of infected plants in the production fields.

Host Range
All wheat cvs. are natural and experimental hosts. Natural hosts also include oats (*Avena sativa*, *A. strigosa*), barley (*Hordeum vulgare*, *H. distichon*), some maize cvs., and spontaneous flora plants (*Cynodon dactylon*, *Eragrostis cilianensis*, *Lolium multiflorum*, *Panicum capillare*, *P. crusgalli*, *P. sanguinale*, *Setaria glauca*, *S. viridis*, and *Cenchrus pauciflorus*).

Experimental hosts are numerous, such as *Agropyron crossa*, *A. ovata*, *Avena barbata*, *A. fatua*, *Eragrostis trichoides*, *Lolium temulentum*, *Panicum miliaceum*, *Phalaris canariensis*, *Poa compressa*, *Secale cereale*, *Setaria italica*, *Triticum compactum*, *T. dicoccum*, *T. durum*, *T. monococcum*, and *Zea mays* (some cvs.).

Wheat cvs. with systemic infections are used as assay plants. No suitable local lesion host is known.

Symptoms
Primary symptoms of WSkMV are changes in leaves and in plant growth. Initial symptoms in early spring appear as chlorotic spots and streaks on the leaves and, simultaneously with plant growth and disease development, these streaks gradually cover the entire leaf, starting from the tips and margins of the leaves. Leaves become chlorotic, gradually twist, and dry shortly prior to plant earing.

Plants infected with WSkMV early in the season are reduced to half their normal size (Figure 14). and usually are sterile (the number of sterile spikelets increases by 26 to 58%, depending on the wheat cv.), often forcing the producers to plough fields with severe epidemics. Mites developing and feeding on wheat as their host also play a key role in dramatic symptom occurrence.

Important physiopathological changes occur in infected plants (Tošić, 1971a). For example, chlorophyll content in infected San Pastore plants was reduced by 35% in the early growth stage and by 17% at the time of wheat earing. In the Bankut 1205 plants, this reduction was 57% and

Virus Diseases of Food Graminoid Plants

FIGURE 14 Wheat streak mosaic virus-infected hybrid wheat; healthy (left and right).

37%, respectively; and in Bezostaja 1, 25% and 42%, respectively. Reduced chlorophyll content directly reflects less plant photosynthetic activity, which causes delayed growth. The activity of the respiratory enzyme peroxidase also increases at initiation of infection, which mirrors the initial increase of respiration as a general infected plant defensive mechanism.

Panić and Tošić (1969) observed that WSkMV influences differently the number of ectodesmata in the external wall of epidermal cells. Characteristically, during incubation and following the appearance of the first symptoms, the number of ectodesmata increased by 44% in susceptible wheat cultivar Bankut 1205, but decreased by 62% in a more resistant San Pastore cultivar. Further studies are needed on the possible influence of ectodesmata as related to plant susceptibility to WSkMV infections.

Pathogenesis

WSkMV persists in infected wheat plants throughout the year and is transmitted from infected to healthy plants by the eriophyid mite. The incidence of infected plants depends on the severity of the source of infection and on vector activity. Following harvest, these vectors transmit WSkMV to volunteer wheat plants, where it remains infectious until the end of the season; then, from them, the virus is next transmitted onto young winter wheat plants. Infected winter crops represent sources of infection for the next season, which completes the infection cycle.

Eriophyid mites *Aceria tulipae* are natural virus vectors. The eriophyid mite *Aceria tošichella* has been described in former Yugoslavia as a new vector (Tošić, 1971). Mites are capable of ingesting WSkMV into their bodies only in the nymphal stages, but then transmitting it both as nymphs and adults. Mites acquire WSkMV by feeding on diseased plants for 15 minutes and they remain infective for up to 9 days. The virus is not transmitted transovarially to the next generations (loc. cit. Brakke, 1971a).

As polyphagous organisms mites are naturally widespread, their colonization of spontaneous plants of the grasses family (*Avena* spp., *Hordeum* spp., *Panicum* spp., *Bromus* spp. and others) can serve as WSkMV hosts. By moving from infected wheat plants, mites resume feeding and infect these plants, thus broadening the reservoir base and increasing sources of infection of wheat. Mites (less than 0.25 mm) are transported by winds over great distances, thus also contributing to virus transmission. This explains the spread of WSkMV in the direction of strong prevailing winds observed in nature. Mites, which tolerate low winter temperatures comparatively well, overwinter in winter wheat and other host plants and continue their development and vector role in the spring.

WSkMV can be transmitted mechanically with the sap of infected plants, but this mode of transmission is of little importance in its natural spread.

Control

Several methods are recommended for prevention of WSkMV spread in wheat production. The control of weed plants, both as virus and mite hosts, markedly reduces the sources of infection, especially in border zones of wheat production in small plots.

As with weeds, destruction of volunteer wheat plants is also recommended because these stray plants help maintain a bridge between crops for the development of viruses and mites. Chemical control of mites in wheat and weed plants is effective, but should be applied only in cases of severe infection. Improved cultural operations ensure increased plant vitality, which can alleviate losses due to this disease.

Resistant and tolerant wheat cultivars should be planted in WSkMV high-risk areas. Bankut 1205, San Pastore, and Bezostaja 1 are susceptible cultivars. While testing many cultivars, Bellingham et al. (1957) were unable to find even one with satisfactory resistance to WSkMV, and the authors rated the cultivars with yield reductions of 7 to 20% as tolerant. An important recommendation regards breeding of cultivars by chromosome substitution from *Agropyron repens* (Proeseler, 1986a).

Wheat Spindle Streak Mosaic Potyvirus (WSlSkMV)

Properties of Viral Particles

WSlSkMV particles are filamentous and among the longest known soil-borne viruses. Viral particle sizes vary based on method of preparations; that is, 190 to 1975 × 12.8 nm in leaf dip preparations (Slykhuis and Polak, 1971), and over 3000 × 18 nm in thin-sectioned wheat leaves (Hooper and Wiese, 1972; Langenberg and Schroeder, 1973). Purification tends to fragment filamentous forms, resulting in shorter particles. The chemical composition of viral particles is insufficiently known, but based on the main characteristics, WSlSkMV belongs to subgroup 2 of Potyviruses (Signoret, 1988a).

WSlSkMV is similar to the wheat yellow mosaic virus observed in Japan in that it is serologically related and cross protection has also been shown (Usugi and Saito, 1979). WSlSkMV also resembles the European barley yellow mosaic virus since they belong to similar Potyvirus groups and both are transmitted by soil fungi (Signoret, 1988a).

Stability in Sap

WSlSkMV particles are highly unstable in the sap extracted from infected plants. In the sap extracted in 0.1 M Na_2SO_3 or 0.1 M phosphate buffer at pH 7, the virus remains infective for 3 to 4 days at 5°C, or for 1 day at 20°C, and its thermal inactivation point is 50°C (Slykhuis, 1976).

Serology

No data are available.

Strains

Insufficient information exists to assign virus strains.

FIGURE 15 Wheat yellow mosaic virus caused by wheat spindle streak mosaic virus. (Courtesy M. V. Wiese. Reproduced with permission from APS.)

Geographic Distribution and Economic Importance

WSlSkMV has been discovered in North America (Ontario, Canada, and in some northern states of the U.S.), India, and France. Nothing appears to prevent its spread to other European countries as well (Signoret et al., 1977).

WSlSkMV may cause pronounced reductions in wheat yield, depending on the level of soil infestation of, and other external factors influencing, the soil fungus *Polymyxa graminis* vector development. Signoret (1988a) observed reductions of wheat yields of 2 to 18%, and Slykhuis (1976) observed reductions in contaminated soils of 25 to 50%.

Host Range

Triticum eastivum in Canada and the U.S. and *T. durum* in southern France are the only reported natural hosts of WSlSkMV (Slykhuis, 1976). Numerous other monocotyledonous and dicotyledonous plants tested by experimental inoculations of WSlSkMV failed to cause symptoms of disease.

Symptoms

Mosaic, the primary symptom of disease, appears on infected winter wheat plants. Mosaic symptoms are first characterized as chlorotic and later as necrotic spindle-shaped streaks on the leaves (see Figure 15). Typical symptoms develop on lower leaf surfaces during cool spring periods. Infected wheat plants are slightly stunted, with reduced tillering and somewhat reduced yields.

Pathogenesis

WSlSkMV, preserved in long-lived resting spores of *Polymyxa graminis* in the soil, can yield infectivity for more than 8 years (Slykhuis, 1976). Zoospores, released from resting spores, infect other plants, thus spreading the WSlSkMV infection in the fields. As a rule, winter wheat is infected in the fall, although disease symptoms appear the next spring. The incubation period at 5 to 15°C lasts 90 to 30 days; thus, infection intensity is less severe with later sowing (Signoret, 1988a). On the other hand, infection intensity increases from one year to the next when wheat is grown in monoculture.

Although WSlSkMV can be transmitted mechanically, this does not seem to greatly influence its natural spread. No information exists on WSlSkMV transmission through wheat seed.

Control

We recommended use of noninfected fields for crop production because infection of wheat by WSlSkMV occurs in the soil. Delayed sowing in the autumn can be useful where less contaminated soils and soils with suspected (unconfirmed) infections must be used. Chemical soil treatment can reduce the populations of the virus vector *Polymyxa graminis*, although more theoretical than practicable (applicable). As reported in the U.S., Canada, and France, in practice, growing of more tolerant wheat cultivars is the most reliable method to reduce losses caused by WSlSkMV.

OTHER VIRUS DISEASES OF WHEAT

Wheat Winter Mosaic Rhabdovirus (WWnMV)

The bacilliform WWnMV particles (260 × 60 nm) were observed in both infected plants (Razvyazkina and Polyakova, 1967) and viruliferous vectors (Razvyazkina and Polyakova, 1970).

The identical shape and size of WWnMV particles with those of the wheat dwarf virus (WDV) described in Russia and other eastern European countries suggest that these two viruses belong to the same species. This assumption becomes even more feasible because both viruses have the common vector—leafhopper *Psammotettix alienus*.

WWnMV was first discovered in Russia (Zazhurilo and Sitnikova, 1939) and later in Poland (Hoppe, 1974), and its presence is also suspected more broadly in eastern Europe, reducing wheat yields 10 to 15% and in some cases as much as 40 to 80% (loc. cit. Šutić, 1983).

Wheat, oats, barley, and rye are natural hosts of WWnMV. Other hosts include *Agropyron cristatum*, *Avena byzantina*, *A. fatua*, *Calamagrostis epigeios*, *Setaria glauca*, and *S. viridis*. Plants resistant to WWnMV infection include *Agropyron repens*, *Bromus inermis*, *Sorghum sudanense*, and *Zea mays* (loc. cit. Šutić, 1983).

WWnMV causes severe damage in winter wheat, while spring wheat is less affected. Light-green striped spots appear between parallel leaf veins, particularly on upper leaves, while lower leaves are often symptomless. The occurrence of general chlorosis usually masks mosaic symptoms. Other characteristic symptoms include reduced growth and excessive tillering, which results in a dwarfed rosetting appearance accompanied by phloem necrosis.

WWnMV is transmitted neither with the sap nor through seed. The leafhopper vector *Psammotettix striatus* acquires WWnMV in the nymph and adult stages and transmits it in a persistent manner (Sukhov and Sukhova, 1940; Zazhurilo and Sitnikova, 1939). The leafhopper *P. alienus* is a vector of WWnMV in Poland. The acquisition period by the vector requires 13 to 27 days (21 days average). The incubation period for WWnMV in wheat plants is 7 to 29 days (14 days average). *Psammotettix striatus* transovarially transmits WWnMV particles in a high percentage (72%) (Shaskolskaya, 1962).

Geographical distribution of WWnMV is restricted and probably depends on the movement of the Cicadellid leafhopper. Control by chemical and cultural operations that destroy or prevent occurrence of Cicadellid leafhoppers are important tactics.

Wheat Dwarf Disease (Presumed Virus) (WDDs)

WDDs has been described in Russia, Sweden, and the eastern European countries of Bulgaria, former Czechoslovakia, and Romania (Vacke, 1961). A similar soil-borne and mechanically transmitted wheat dwarf virus disease has been recorded in former Yugoslavia. This mechanically transmitted entity is probably related to some other virus naturally present in these infected samples (Šutić and Tošić, 1966; Tošić, 1962a).

Winter wheat represents the primary host of WDDs and other cereals are potential hosts. Clearly visible chlorotic blotches on infected leaves in the spring gradually form necrotic zones with age and the plants die. WDDs infection delays growth and their tillers, which form compact rosettes, lodge. WDDs-stunted plants are either spikeless or have short sterile spikes which causes yield reductions. The overall damage depends directly on the number of infected plants. The leafhopper *Psammotettix alienus* transmits the virus in a persistent manner (Vacke, 1961).

Losses caused by WDDs can be attenuated primarily by growing tolerant cultivars and also by the eradication of its vectors.

Wheat Chlorotic Streak Rhabdovirus (WCSkV)

WCSkV with particles are 355×55 nm *in situ* and 355×62 nm *in vitro* (Signoret et al., 1972). First discovered in the *Triticum durum* crop in southern France, WCSkV was found later also in the northern and central parts of France (Signoret et al., 1976). Cicadellid leafhoppers *Laodelphax striatellus* transmit the virus in a persistent manner. WCSkV is identical to the barley yellow striate mosaic virus (ByYStMV) described in Italy, based on particle shape and the same vector.

Triticum spp. are natural virus hosts of WCSkV, including *T. durum*, *T. compactum*, and *T. aestivum*, which is the most susceptible (Signoret et al., 1977a). Symptoms on WCSkV-infected plants are chlorotic streaks mostly on emerging apical leaves, followed by head mosaic and yellowing. Glumes remain short and underdeveloped and grain set is reduced by 80%. Grains that reach maturity are shriveled, with a reduced weight of 50% (Signoret et al., 1976).

The *Laodelphax striatellus* vectors acquire WCSkV in 10 minutes, enter a latent period, then remain capable of transmitting it for a long period (Signoret et al., 1977a). The virus is not transmitted through the seed, and its natural grass reservoirs are infected wheat and *Agropyrum* plants (Cornuet, 1987).

Wheat Yellow Leaf Closterovirus – Subgroup I (WYLV)

WYLV is elongated and flexuous, 1600 to 1850×12 nm (Inouye, 1976). It has been reported only in Japan (Inouye et al., 1973).

Hosts of WYLV are plants of the Gramineae family, including wheat (*Triticum aestivum*), barley (*Hordeum vulgare*), rye (*Secale cereale*), oats (*Avena sativa*), *Agropyron kamoji*, and *Lolium multiflorum* (Inouye et al., 1973).

Initial symptoms on new leaves of WYLV-infected wheat plants are diffused chlorotic flecks, followed later by leaf tip yellowing and rapid plant death. Infected seedlings either die or remain stunted, forming few tillers. On oat leaves, orange to red discolorations appear and plants die (Inouye, 1976).

The WYLV vector, aphid *Rhopalosiphon maidis*, transmits the virus in a semipersistent manner. It is transmitted neither through the seed nor mechanically with the sap of infected plants (Inouye, 1976). *Agropyron*, a common native weed, may be an important reservoir of WYLV.

WHEAT: NATURAL ALTERNATIVE HOST FOR SOME VIRUSES

Agropyron Mosaic Virus (AgMV)

Wheat is the natural host of AgMV, economically important, and found in the U.S., Canada, New Zealand, former U.S.S.R., Finland, Germany, and Bulgaria (Slykhuis, 1973; Šutić, 1983).

Natural hosts of AgMV are switch grass *Agropyron repens*, on which it was first described, and *Triticum* spp. It causes systemic infection in several gramineous plants and local infection in the experimental dicotyledonous host *Chenopodium quinoa*. The mild (green) virus strain causes weakly discernible symptoms on wheat leaves, and the severe (yellow) strain causes pronounced stunting in wheat (Slykhuis, 1973). Severe strains cause serious stunt damage and reduced yields in infected plants. Damage caused by AgMV is comparable to WSkMV in Bulgaria (Kovaćevski et al., 1977).

AgMV is preserved in natural weed sources, such as infected wheat and switch grass. Its vectors are *Eriophyid* mites *Phytocoptes (Abacarus) hystrix*. The virus can also be soil-borne with the aid of infected switch grass rhizomes. It is transmitted with the sap of infected plants. No information is available on seed transmission of AgMV.

The elimination of switch grass as the primary natural reservoir for AgMV may enable a substantial reduction of yield losses. Chemical control of mite vectors, both in wheat and in switch grass or other weed plant populations, is useful in areas characterized by a high disease incidence.

Barley Stripe Mosaic Virus (BySpMV)

Wheat is a natural host for BySpMV as well as other cereals; for example, barley (the main host), oats, maize, and numerous grasses. Transmission of BySpMV occurs in the pollen and seed of the majority of gramineous host plants, which contributes to its extensive distribution worldwide.

Symptom intensity on wheat leaves ranges from a mild stripe mosaic to necrosis, followed by complete dwarfing. A number of virus strains, including the most common mild strain and the moderate strains, have been designated, based on virulence of symptoms on wheat and other host plants (Atabekov and Novikov, 1989).

In natural infections, BySpMV does not cause substantial reductions in commercial wheat yields, regardless of its transmission through the seed cycle. However, under experimental conditions, wheat crops from the seed of infected plants show reduced yields and poor seed quality (McNeal et al., 1976).

BySpMV is readily seed transmitted and pollen transmitted from infected plants in nature (Bennett, 1969). Insufficient data exists on the seed transmission of BySpMV in naturally infected wheat plants. However, the virus was transmitted to 71% of seedlings with the seed from mechanically inoculated/infected wheat plants and to 87% of seedlings grown from mechanically inoculated barley seed (Hagborg, 1954). This virus can survive 20 years in wheat seed (McNeal et al., 1976) and several years in barley seed (Scott, 1961). In contrast to wheat, barley seed seems less favorable for virus preservation (Jackson and Lane, 1981). Although transmitted mechanically with the sap of infected plants, this method seemingly has little significance for wheat in nature. Virus vectors are still unknown.

The use of healthy, virus-free seed is the primary method for successful virus control. Therefore, particular attention should be dedicated to the production of healthy seed and seed phytosanitary control.

Barley Yellow Dwarf Virus (ByYDV)

Wheat is susceptible to ByYDV, an economically important host of this virus among cereals. Omnipresent in wheat, ByYDV may be the most widely spread and frequent wheat pathogen in the world. It infects several wheat (*Triticum* spp.) species, bread wheat (*T. aestivum*) being the most important among them, but also including durum wheat (*T. turgidum* var. *durum*) and Triticale (×*Triticosecale*).

ByYDV incidence in wheat in individual countries and from one year to another varies, depending on the environmental conditions that favor vector occurrence and development, genotype susceptibility, and cultural operations in wheat production. Reported incidences of ByYDV ranged from 5 to 10%, considered economically insignificant, to 50% economically important losses, and even 80 and 100%, economic disaster (total loss).

Yield reductions in wheat correspond directly to the number of infected plants. ByYDV infections of winter wheat in the fall cause significantly more damage than infections in the spring. Some field trials showed that reductions in wheat yield following autumn infections were 63%, in contrast with yield reductions of 41% following spring infection (Conti et al., 1990). Gildow (1990) measured 100% plant infection 4 weeks after sowing, with a concomitant yield decrease of 87% under optimal weather conditions; whereas plant inoculations 6 weeks after sowing resulted in only 30% yield reductions. Research under controlled conditions in irrigated winter wheat inoculated with ByYDV yielded 50% less than noninfected controls; also, percentage wheat yield losses were about half of the percentage of infected plants in the field (Gildow, 1990).

Indirect detrimental effects such as reduction of cold hardiness and predisposition to scab and Septoria diseases are an important aspect of the overall damages caused by ByYDV, particularly in winter wheat (loc. cit. Haber, 1990).

The restriction of the development of ByYDV in the phloem of infected plants causes characteristic symptoms such as yellowing, reddening, and brittleness of leaves. Leaf discoloration begins from leaf tips and leaves develop reduced in size. Infected plants express delayed maturity, growth, and thus, dwarfing. ByYDV infections of early winter wheat cause plants to attain only one-third to one-half of their normal height, wherein earing seldom occurs, grain size is reduced, and yields are negligible.

Vectors of ByYDV are aphids, which transmit the virus in a persistent manner. Vectors include *Rhopalosiphon padi, R. maidis, Metopolophium (Acyrthosiphon) dirhodum, Macrosiphum (Sitobion) avenae, Schizaphis graminum*, and others.

As the season progresses, vectors transmit ByYDV from naturally infected plants to winter cereals where it overwinters. Infected plants are sources of infection for all cereal species, including maize and weed grasses. Winter-sown cereals and perennial grasses represent permanent reservoirs of ByYDV inoculum in the continuum of maintenance of the virus development cycle. More severe plant infections occur in production under irrigation when plants remain vigorous and green during longer periods of time, which is particularly important for vector feeding and activity. ByYDV is neither transmitted with the sap from infected plants nor through the seed of infected plants.

Based both on the development cycle of ByYDV and the activity of its vectors, the following recommendations are suggested to help reduce yield losses. Winter cereals should be sown at the latest time possible and spring cereals at the earliest time possible. In areas with a history of severe infections, crop rotations including introduction of non-host crops are recommended. Early seedling infections can be reduced with the in-furrow application of granular systemic insecticides at the time of sowing (Gildow, 1990).

Tolerant or resistant wheat cultivars should be used in production. Among Ecuadorean wheats (*Triticum aestivum*), for example, Tungurahuo cv. is more resistant as compared to Altar and Chimborazo cvs. (Fox and Tola, 1990). The Yugoslav wheat cultivar NS 879 displayed a high resistance level, with the final incidence of infection caused by ByYDV less than 5% at season's end (Tandon et al., 1990). Conti et al. (1990) reported an offspring resistant to ByYDV was obtained by crossings between wheat and its wild relatives such as *Elymus (Agropyron)* spp., but also stressing the need for more effort to insert this resistance into useful cultivars.

Breeding of resistant cultivars represents an integral part of research programs as long-term solutions for ByYDV control in cereals (e.g., in Australia and New Zealand). These programs include conventional breeding methods, production of synthetic resistance genes, and transfer of ByYDV resistance from *Agropyron* spp. into wheat (Johnstone et al., 1990).

Barley Yellow Striate Mosaic Virus (ByYStMV)

ByYStMV in the Rhabdovirus group, subgroup A, is identical to the wheat chlorotic streak virus (WCSkV) (Conti, 1988). Detailed information on this virus is found in the chapters on other wheat virus diseases — WCSkV and barley virus disease.

TABLE 2
Effects of BrMV on the Growth and Yield of Two Wheat Cultivars

Properties	Wheat cvs. (reduction in %)	
	San Pastore	Bezostaja 1
Number of secondary tillers per plant	48.6	8.0
Stalk length with spike comprised	17.6	13.0
Number of kernels per spike	46.6	31.3
Yield per plant	66.4	47.7
1000 grain weight	9.8	4.4

Brome Mosaic Virus (BrMV)

BrMV was isolated first from awnless brome (*Bromus inermis*) plants and described as the brome grass mosaic virus. Wheat is a natural host of this virus. BrMV is widespread in African, North American, and European countries.

Natural hosts of BrMV include several cultivated gramineous plants (wheat, barley, oats, maize), plus several grass species. Dicotyledonous plants are experimental virus hosts and *Chenopodium* species in which the virus causes local lesions are suitable indicators of virus infections. BrMV causes elongated mosaic spots and stripes on wheat leaves. Plants infected early in the season express delayed growth and have significantly shortened spikes. The number of secondary tillers per plant in infected susceptible cultivars may be reduced 50%. Similarity of symptoms makes it difficult to distinguish this virus from the wheat streak mosaic virus.

In Europe, BrMV has minor effects on crops. Tošić (1971a) observed in experimental infections that BrMV reduced both tillering, plant growth, and yields (Table 2).

The pathological effect of BrMV on individual plants is serious, but the extent of overall crop damage depends on the number of infected plants in the field. BrMV remains infective and is both preserved and transmitted from year to year in perennial hosts (e.g., *Bromus inermis* and others). Several animal vectors and plant species are involved in the spread of BrMV, a trait that differs substantially from all other virus species. These vectors include nematodes, beetles, mites, and the rust fungus—*Puccinia graminis* f. sp. *tritici* (Cooper, 1988). The wide variety of vectors implies a nonspecific, atypical relationship between the vectors and the virus. The virus is transmitted mechanically with the sap of infected plants, which may be important for the natural spread of BrMV.

Control of BrMV is difficult to ensure because possibly not all modes of its transmission and spread are sufficiently known. However, spatial isolation of wheat production fields from other monocotyledonous virus hosts is a useful practice. Vector control in cases in which virus vectors are known is also recommended. All plant residues, as well as native plants in which the virus survives from the time of harvest to the next sowing, should be destroyed. Adequate crop rotation contributes to the decrease of the presence of nematode virus vectors in the soil. Data pertaining to genotypes of wheat resistant to BrMV are available (Panarin and Dubonosov, 1975; Panarin and Zabavina, 1977); also for barley (Cooper, 1988).

VIRUS DISEASES OF BARLEY (*Hordeum vulgare*)

See the *Compendium of Barley Diseases* (Mathre, 1997) for a complete treatment of all diseases of barley.

BARLEY STRIPE MOSAIC HORDEIVIRUS (BySpMV)

Properties of Viral Particles

BySpMV consists of a multi-component structure containing genetic information necessary for infection distributed into two to five types of specific viral particles, depending on the strain. Each particle contains its own genome (RNA) component. Two or three of these genome components are necessary to cause infection. Functionally, it is a tripartite genome structure (Jackson and Brakke, 1973; Lane, 1974). Significant detail is available describing genome properties for each segment (Atabekov and Novikov, 1989).

The dimensions of the viral particles vary from 108 to 175 × 22 nm, depending on the method of viral particle preparation and measurement techniques (loc. cit. Jackson and Lane, 1981). For example, some virus isolates in the former Yugoslavia measured 80 to 230 × 30 nm, with the most frequent length as 130 to 170 nm (average 153 nm) (Šutić and Tošić, 1964). The sedimentation coefficient, which varies for individual strains, is 185S. BySpMV has a molecular weight of 26,000 kDa (Atabekov and Novikov, 1989).

BySpMV particles contain 96% protein and 3.7 to 4% RNA (Atabekov and Novikov, 1989; Brakke, 1979). The coat protein is 21 kDa. The molecular weights of the single-stranded RNAs are 1400, 1170, 1040, and 930 kDa (Cornuet, 1987).

Stability in Sap

Jackson and Lane (1981) summarized the results of the investigation of biophysical virus properties in infected plant sap that included these results: TIP ranges from 65 to 68°C, DEP from 5×10^{-2} to 1×10^{-4}, and LIV is about one month.

Serology

BySpMV is highly immunogenic. Serological analyses enable speedy and easy confirmation of the presence of BySpMV, both in the sap of infected plants and in the seed. All BySpMV strains are serologically related (Carroll, 1980), plus it is distantly related to the lychnis ringspot virus (Gibbs et al., 1963).

Strains

Strains of BySpMV were designated based on characteristic disease symptoms on barley, wheat, and oat plants. They are yellow leaf, white leaf, dwarf, and necrotic strains, which based on their virulence in these hosts are latent, mild, moderate, and severe strains, respectively. Some strains differ based on their ability to infect oats and *Chenopodium amaranticolor* (McKinney and Greeley, 1965), and on their ability to infect some barley or wheat seed (Jackson and Lane, 1981). Still other strains differ according to the different numbers of RNA components necessary for infection. For example, two RNA components are required for the virulence of the Type and the Russian strains, whereas the Norwich strain needs three RNA components (Jackson and Brakke, 1973; Lane, 1974). Because a wide variation of characteristics of BySpMV strains exist, only well-known and previously identified virus strains should be used for trials.

Geographic Distribution and Economic Importance

BySpMV is spread worldwide, primarily due to its transmission through seed in a high percentage of cases. BySpMV is known in North America, Asia, Australia, England, France, Germany, the former U.S.S.R., Denmark, Romania, and former Yugoslavia (Šutić, 1983).

BySpMV is an economically damaging disease because it reduces plant vigor, and causes floret sterility and shriveling of seed in infected plants. These changes result in barley yield reductions, the extent dependant on the mode and time of infection, cultivar susceptibility, and general conditions of the barley crop development.

FIGURE 16 Barley stripe mosaic virus on barley; some systemically infected leaves are nearly white (top). (Courtesy V. Pederson. Reproduced by permission from APS.)

Reduced barley yields in naturally infected fields prior to the application of control measures in Montana and North Dakota amounted to between 25 and 30% (Eslick, 1953; Timian and Sisler, 1955). Barley yields where infection occurred in the seedling stage were reduced by 60% (Hagborg, 1954); and in inoculated plants, reductions were 90% (McKinney, 1953). Spaar and Schumann (1977) reported barley yields reduced by 60%, depending on cultivar susceptibility and time of infection. In field trials, barley yield reductions are directly proportional to the fraction of infected seed sown (Jackson and Lane, 1981).

Host Range

Gramineous plants are natural hosts, whereas experimentally, BySpMV also may infect the plants of a few dicotyledonous species. In nature, BySpMV occurs primarily on barley, wheat, oat, maize, and numerous grasses. According to the findings of several authors summarized by Jackson and Lane (1981), 213 gramineous species (23 featuring latent infection) belonging to the 10 plant tribes Agrostideae, Andropogoneae, Aveneae, Chlorideae, Festuceae, Hordeae, Oryzeae, Paniceae, Phalarideae, and Tripsaceae are susceptible to BySpMV.

Susceptible (with systemic symptoms of infection) dicotyledonous plants include *Spinacia oleracea* (Chenopodiaceae), *Commelina communis* (Commelinaceae), *Nicotiana tabacum* (Solonaceae), and those without symptoms—that is, with latent infection are *Gomphrena globosa* (Amaranthaceae) and *Primula malacoides* (Primulaceae). Local lesions appear on *Beta vulgaris*, *Chenopodium album*, *C. amaranticolor*, *C. ambrosiodes*, *C. botrys*, *C. giganteum*, *C. hybridum*, and *C. quinoa* (Jackson and Lane, 1981).

Useful virus sources for experimental work are barley (*Hordeum vulgare*) and wheat (*T. aestivum*), and suitable assay plants with local lesions *C. amaranticolor* and *C. quinoa* (Atabekov and Novikov, 1989).

Symptoms

BySpMV (Figure 16) is characterized by several symptoms based on the differing stages of disease in the acute stage immediately following infection and consists of striped mosaic and yellow or whitish chlorosis that appears first near the base of the youngest leaves (Figure 17). During this stage, leaf tissue necrosis and growth arrestment occur, especially in susceptible cultivars (Šutić, 1983).

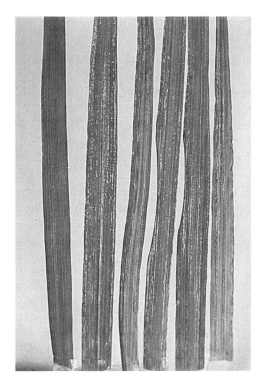

FIGURE 17 Barley stripe mosaic virus in barley.

In the chronic systemic infection stage, chlorotic stripes broken by yellow and whitish areas form between the veins of developed leaves. Once these spots and areas merge, the leaves take on a generally chlorotic appearance characterizing weakened physiological ability. Due to growth arrestment, plants and individual plant parts remain dwarfed, the extent of dwarfing dependant on cultivar susceptibility and strain virulence. Specific changes in the development of floral parts that occur include: individual florets become sterile; anthers fail to open, unfavorably influencing pollen formation and release; and the resultant seeds are shriveled. Deviations in floral development directly influence—decrease—plant yield.

Pathogenesis
BySpMV is transmitted with the seed and pollen of infected plants, enabling it to retain infectivity in host plants from generation to generation. Although not fully known, the mechanism of seed infection is assumed to involve BySpMV penetration into the apical floral meristem and therefrom into the anthers and pistil, then into the developing seed (Bennett, 1969).

Percentage transmission of BySpMV through the seed depends on barley cultivars, virus strains, environmental factors, and stage of plant development at the time of infection. Among 219 Campana barley seed lots tested in Montana (U.S.), 55% were infected. Seed infection in individual samples within one lot ranged from 5 to 50% (Afanasiev, 1956). The percentage seed transmission recorded for experimentally infected Domaći Ozimi barley seed in former Yugoslavia ranged from 37% in one, to 55% in another (Šutić and Tošić, 1966a). BySpMV transmission through barley seed can be as high as 90%, and rarely 100% has been recorded.

In nature, BySpMV is transmitted with the pollen, most certainly the reason for occurrence of seed infection in high percentages and its spread to greater distances. The presence of BySpMV in the seed embryo has been established; thus, one can assess percentage seed infection by mechanical transmission of the embryo sap into test plants (Hamilton, 1965).

BySpMV is easily transmitted mechanically, but this mode does not seem to be of primary importance for its spread in nature. No vectors are known.

Control

Buying and sowing healthy seed is the main method for partial or complete elimination of infection and for the reduction of losses caused by BySpMV in production fields. Virus elimination will enable the production of economically important BySpMV-free cultivars and the use of their seed for commercial crop production. In the breeding and selection process, priority should be given to either virus-free cultivars or those in which reinfection rarely occurs under natural conditions.

Control, to date, of BySpMV through the use of resistant cultivars does not seem feasible on a worldwide scale (Jackson and Lane, 1981). Breeding of resistant cultivars is difficult due to different reactions of barley genotypes to the virus and to the absence of resistance genes to all viral strains. Anticipated positive results from some current breeding projects are awaited.

BARLEY YELLOW DWARF LUTEOVIRUS (BYYDV)

Properties of Viral Particles

The type isolate of ByYDV in the Luteovirus group was described as MAV (Matthews, 1982). Isometric viral particles vary from 20 to 30 nm, depending on the isolate and its mode of preparation (loc. cit. Rochow and Duffus, 1981). The MAV isolate sedimentation coefficient is 115 to 120S (Rochow, 1970). The nonspecific vector isolate contains a single RNA molecule of 1800 kDa (Mehrad et al., 1979). The RNA molecular weight of the RPV and the RMV vector-specific isolates is 1900 kDa (Gildow et al., 1983). The virion contains a single major protein of 23.5 kDa for an MAV isolate and 24.5 kDa for a PAV isolate (Scalla and Rochow, 1977).

Stability in Sap

The TIP in sap for MAV and RPV isolates of ByYDV is 65 and 70°C, respectively (Heagy and Rochow, 1965), and the DEP is 1×10^{-3} (loc. cit. Šutić, 1983). Concentrations of ByYDV in plants are low because it is restricted only to phloem tissue. The virus is stable in the plant sap, which is particularly suitable for experimental work.

Serology

ByYDV is readily immunogenic and, because of low virus concentration in expressed plant sap, serological tests should be done with partially purified virus preparations (Rochow, 1970). Antisera of different isolates have enabled the establishment of the serological relationships among these isolates through cross and comparative investigations. All isolates classified into two serologically differing groups include group 1 comprising RPV and RMV and group 2 including MAV and PAV isolates. Isolate SAG displayed some antigenic relationship with the PAV isolate. Some ByYDV isolates also reveal differing serological relatedness with other Luteoviruses.

Strains

Research has shown that many isolates differ according to virulence, cytopathological, and serological features and their relationship with aphid species as virus vectors. Some differences are large enough to represent possible variants of distinct virus species.

Based on their virulence in oat plants (cv. Coast Black) and their ability to be transmitted by aphids (*Macrosiphum avenae*, *Rhopalosiphum maidis*, *R. padi*, and *Schizaphis graminum*), Rochow (1970) described four isolates: RMV, RPV, MAV, and PAV. The SAG vector-specific isolate was described by Gill (1969). All these isolates are the subject of comparative investigations of their biological variability.

Geographic Distribution and Economic Importance

ByYDV occurs worldwide except in tropical regions. It is probably the most widespread and common cereal virus, also known as the cereal yellow dwarf virus. ByYDV is reported in North

FIGURE 18 Yellowing of barley leaves infected with barley yellow dwarf virus. (Courtesy B. Cunfer. Reproduced with permission from APS.)

America, Europe, Asia, Australia, and New Zealand (Slykhuis, 1958; 1962a) and in Great Britain, France, Italy, Bulgaria, Hungary, former Yugoslavia, and other European countries (Šutić, 1983).

ByYDV ranks among the most economically damaging of viruses because it is so widespread and causes significant yield losses to barley, wheat, and other cereals. The losses are particularly large in early-sown winter barley and other winter cereal crops that incur early seedling infections in the fall. Experimental infections reduced barley yields in the Victoria province (Australia) by 9 to 79%, whereas infections of older plants resulted in 6 to 9% yield losses (Smith, 1980). Yield losses in winter barley ranged from 30 to 40% in field trials in Rothamstead under conditions of natural infection (Dowson et al., 1986). The incidence of infected plants, which can range from 10 to 90%, is the main factor influencing reductions in barley yield. Information on reductions in wheat yields caused by ByYDV can be found in the chapter on wheat virus diseases—ByYDV on wheat.

Host Range

Hosts of ByYDV are only monocotyledonous members of the Gramineae family, although a few members of the Cyparaceae family may also be experimentally infected (Lapierre and Maroquin, 1988). Some 100 gramineous plants are susceptible to ByYDV (Rochow and Duffus, 1981).

Cultivated barley, oats, wheat, triticale, rice, rye, maize, and sorghum are economically important virus hosts. Other important virus hosts are numerous widely distributed lawn, pasture, and field border grasses, some species belonging to *Bromus*, *Cynodon*, *Dactylis*, *Digitaria*, *Echinochloa*, *Eragrostis*, *Erharta*, *Festuca*, *Lolium*, *Paspalum*, *Phalaris*, *Phleum*, *Pennisetum*, *Poa*, and others. ByYDV has not been observed in or experimentally transmitted into dicotyledonous plants.

Suitable diagnostic and assay plants for the investigation of ByYDV are winter oats (*Avena byzantina*, cvs. Coast Black and California Red) and spring oats (*A. sativa*, cvs. Clintland and Blenda), which react by leaf reddening (Rochow, 1970).

Symptoms

In barley, wheat, oats, and rye, ByYDV causes disease symptoms of leaf chlorosis and plant dwarfing. Leaf yellowing in barley is characterized by a golden yellow color, also known as "yellow dwarfing" (Figure 18). Symptom intensity depends on the stage of development during which the plant was infected. Leaves of infected seedlings turn yellow within 10 days. Yellowing first occurs on the leaf margins, gradually spreading toward the leaf base, and finally encompassing the entire leaflet. Infected leaves are erect, rather rigid, and firm. Initial phloem cell necrosis also spreads to other neighboring tissues. Plants show delayed growth, have a rosette-like appearance, and fail to develop grain heads.

Infected plants of some cereals feature poor tillering, although in a few barley cultivars tillering may be increased. Rooting of infected plants is reduced, causing disorders in water and nutrient absorption. Similar symptoms sometimes may be caused by abiotic factors (excessive soil moisture, drought, nitrogen deficiency, and low temperatures) difficult to distinguish from those of the infective yellow dwarf disease. Groups of infected plants are observed in the field in form of "yellow plaques" varying in size.

Other details on the symptoms in infected wheat and oat plants are found in the chapters on barley yellow dwarf virus on wheat and barley yellow dwarf virus on oats.

Pathogenesis

ByYDV is preserved naturally in wild and cultivated perennial grasses, which provide long-term sources of infection. These "green bridges" ensure for ByYDV a continuous infective inoculum in nature. Volunteer plants that grow in fields following harvest play an important role in the infection of winter cereals (winter barley, winter wheat, etc.); that is, during the same year, ByYDV is transmitted from volunteer plants into winter crops, from which it is further transmitted the next year.

Vectors of ByYDV in nature include more than 20 aphid species that transmit the virus in a persistent (circulative) manner. Infected females do not transmit ByYDV transovarially to their progeny. The most abundant and important vectors *Macrosiphum* (*Sitobion*) *avenae*, *Rhopalosiphum maidis*, *R. padi*, and *Schizaphis graminum*. The time required for aphid colonization and initiation of feed colonization, which also marks the beginning of virus acquisition, depends on aphid species. *S. avenae* required 30 minutes for this process in oats (cv. Clinton 64), although 65% of the aphids tested started feeding successfully only after 90 minutes (Scheller and Shukle, 1986). *R. padi* needed 60 minutes and *S. graminum* 200 minutes to penetrate the phloem of oat plants (cv. California Red) and to begin feeding (Montllor and Gildow, 1986).

The importance of the role of individual vectors in virus transmission depends on their distribution, development cycles, and living/feeding habits with available hosts. For example, aphids of the major species, *Rhopalosiphum padi*, in Victoria transmitted the virus 23 to 61% of the time, whereas this percentage for *M. miscanthi avenae* and *R. maidis* was 20 to 44% and 8%, respectively (Smith, 1980). In England, virulent isolates are transmitted most frequently by *R. padi* and mild isolates most frequently by *M. avenae* and *Metopolophium dirhodum* (Plumb, 1974).

Rochow (1969) studied the role of vectors in the transmission of four isolates featuring different virulences into oats (cv. Coast Black) and concluded that the RMV isolate is regularly transmitted by *R. maidis* and seldom by *R. padi*, *M. avenae*, and *S. graminum* aphids. He also observed that the RPV (mildly virulent) isolate is regularly transmitted by *R. padi*, sometimes by *Schizaphis graminum*, and rarely by *R. maidis* and *M. avenae* aphids; also, the MAV (moderately virulent) isolate is regularly transmitted by *M. avenae* and seldom by *R. padi*, *R. maidis*, and *R. graminum*; and finally, the PAV (severely virulent) isolate is regularly transmitted by *R. padi* and *M. avenae*, sometimes by *S. graminum*, and seldom by *R. maidis* aphids.

The virus is transmitted neither with the sap of infected plants, nor through seed. It was experimentally transmitted from one barley plant to another with the aid of *Cuscuta campestris* (Timian, 1964). The fact that aphids are the primary vectors in nature helps explain its worldwide distribution and its ranking as the most widespread cereal and wild grass virus.

Control

ByYDV control measures include cultural methods in plant production, control of aphids with either insecticides or biological methods, as well as searching for and breeding of tolerant or resistant plant genotypes.

The main cultural measures consist of the adjustment of the time of winter and spring crop sowing in order to avoid peak aphid populations when the majority of plant infections occur. Therefore, fall sowing as late as possible and spring sowing as early as possible are recommended.

Another important method is aphid control with the application of systemic insecticides. In-furrow distribution of granular aphicides at sowing is recommended for the first chemical treatment, followed by one to three sprayings with systemic aphicides in autumn and spring. As a rule, spraying of autumn plants is more efficient than spraying in the spring. The time of spraying should be adjusted to weather conditions and aphid population densities.

Resistant or tolerant cultivars should be planted for cereal production. The level of tolerance in four barley cultivars recently tested was ranked in the order Post > Perry > Harrison > Darra (Gildow, 1990).

Contemporary research activities focus on the creation of cereal cultivars resistant to ByYDV by incorporation of Yd2 genes for resistance from Ethiopian barley. The barley cultivar Vixen provisionally recommended for production in regions in which ByYDV was obtained this way (Barker, 1990). Other positive research results along these lines are expected.

An excellent book summarizes in significant detail all aspects of the history and research on ByYDV (D'Arcy and Burnett, 1995).

BARLEY YELLOW MOSAIC POTYVIRUS (ByYMV)

Properties of Viral Particles

ByYMV particles are filamentous and of two sizes: 275×13 nm and 550×13 nm (Inouye and Saito, 1975); whereas those studied in England were most frequently 650×11 nm (Hill, 1980). The virus contains two RNA species of 2800 and 1400 kDa, and two proteins of 35 and 29 kDa (Huth, 1988).

The TIP and LIV are unknown, and the DEP is 10^{-2}. ByYMV is moderately immunogenic by intravenous and intramuscular injection of rabbits with the partially purified virus (Inouye and Saito, 1975). Two types of particles were discovered in infected barley with the aid of serological analyses. Both types are morphologically identical and soil transmitted, but only one is easily transmitted with infected plant sap (Huth, 1988).

Geographic Distribution and Economic Importance

ByYMV, a rather new barley pathogen, was first described in Japan (Inouye and Saito, 1975). The European discoveries were first in England (Hill, 1980), and then in Germany (Huth, 1980), France, and Belgium. It has been reported in China. ByYMV is assumed to be much more widespread than reported.

Since ByYMV is soil borne, it is difficult to control. Where found, it is, *inter alia*, a most damaging pathogen, reducing barley yields 10 to 90%, depending on interactions among availability of inoculum, soil type, climatic conditions, and cultivar susceptibility (Huth, 1988). Most pronounced losses occur in areas with low winter and spring temperatures.

Host Range

ByYMV, a member of a rare virus group, is narrowly specialized to infect only plants of the *Hordeum* genus (*H. sativum* and *H. spontaneum*). Natural infections were discovered only in winter barley (Huth, 1988). All attempts to inoculate and infect experimentally *Triticum aestivum*, *Avena sativa*, *Oryza sativa*, *Chenopodium amaranticolor*, and *Nicotiana tabacum* plants remained unsuccessful (Inouye and Saito, 1975). At the end of April, ByYMV was discovered in key winter barley cultivars (Igri, Maris Otter, and Sonja) in England. Most winter barley cultivars in Germany are susceptible.

Symptoms

Small yellow spots and stripes appear first on the youngest leaves of infected plants. Symptoms usually occur on the third leaf, which starts rolling and causes an erect habit and a pointed appearance. Symptom intensity depends on temperature conditions during disease development. At low temperatures (below 5°C), disease symptoms are sharp; but at temperatures above 15 to 18°C, plants develop normally and leaves are edged with light green stripes; otherwise without symptoms of disease, ByYMV-infected plants are only half the size of healthy ones, with stunted roots and fewer fertile tillers. The disease can be observed following sowing in December and yellow "patches" in the fields are particularly noticeable in March/April.

Pathogenesis

ByYMV remains infective in and probably is transmitted by the fungus *Polymyxa graminis*. Viral particles that persist in the resting spores are carried by the released zoospores, which move actively in the soil. Following their germination, ByYMV is carried into the plant during infection of the roots of healthy plants. Therefore, pathogenesis depends on the vector *P. graminis*, whose activity is primarily conditioned by soil moisture and acidity. ByYMV can retain infectivity in the resting spores for several years, and longer in air-dried soils.

The virus can be transmitted mechanically with infective sap. Other methods of transmission are unknown.

Control

The nature of the persistence in and the transmissibility by a soil-borne fungus causes difficulty for control of ByYMV. No adequate measures exist for protection of barley in contaminated soils. Therefore, the primary recommendation is to grow barley in noninfested soils. Damages on soils lightly colonized by *Polymyxa graminis* may be further reduced by adequate crop rotation and cultural practices favoring soil aeration and drying. Susceptible and high-quality barley cultivars can be grown on heavy clay soils that are unfavorable for *P. graminis* colonization, thus also for ByYMV disease development.

However, the most effective way to reduce yield losses is to grow cultivars tolerant or resistant to ByYMV. Nine European cultivars and several barley lines resist ByYMV infection (Huth, 1982; 1984; Takahashi et al., 1973).

BARLEY YELLOW STRIATE MOSAIC RHABDOVIRUS (BYYSTMV)

Properties of Viral Particles

ByYStMV particles, bacilliform with both ends rounded, are 300 to 320 × 45 nm (Appiano and Conti, 1974; Conti and Plumb, 1977). The virus contains a single-stranded RNA, a protein (currently not analyzed), and a lipid envelope. Based on its morphology and composition in infected cells, ByYStMV is a plant Rhabdovirus placed into subgroup A.

Stability in Sap

Virus stability in barley sap, tested by injecting into *Laodelphax striatellus* leafhoppers, has a TIP of from 50 to 60°C, a DEP of 10^{-2} to 10^{-3}, and LIV of 2 to 4 days at 5°C and 1 to 2 days at 22°C (Conti, 1980).

Serology

ByYStMV is moderately immunogenic by intramuscular injection of rabbits (loc. cit. Milne and Conti, 1986; Milne et al., 1986), depending on the temperature.

Strains

ByYStMV is identical to wheat chlorotic streak virus from France and closely related to maize sterile stunt virus from Australia. The Australian virus is probably a strain (MS) of ByYStMV with intermediate serological properties between ByYStMV and the northern cereal mosaic virus (Milne and Conti, 1986).

Geographic Distribution and Economic Importance

ByYStMV has been reported only in Italy (Conti, 1972; 1980) and southern France (Signoret et al., 1972; 1976; 1977a). In Italy, it was observed in barley plants used for the testing of a field leafhopper collection (Conti, 1969; 1972). Since ByYStMV is transmitted by leafhoppers, its occurrence can be expected in other southern European countries.

ByYStMV ranks among the least damaging of viruses due to its occurrence in limited geographical areas and in low incidence (i.e., rates of infection recorded were 5 to 8% in winter wheat crops).

Host Range

Hosts of ByYStMV include only monocots, Gramineae family. Natural virus hosts in Europe include wheat, *Agropyron repens*, oats, and rye; also, natural infections in Morocco occur in maize (loc. cit. Milne and Conti, 1986).

Experimental inoculations and infections were successful on 26 species of Gramineae in the following genera: *Lagurus*, *Avena*, *Trisetum*, *Briza*, *Bromus*, *Dactylis*, *Poa*, *Hordeum*, *Lolium*, *Triticum*, *Secale*, *Phalaris*, *Sorghum*, *Zea*, *Panicum*, and *Setaria*. No dicotyledonous plants tested have been found susceptible to ByYStMV.

Symptoms

Yellow striate mosaic is the most obvious symptom in the spring and disappears later in the season. Conti (1980) classified disease symptoms based on infected plant reactions into four groups: (1) pronounced stunting and changeable striate mosaic symptoms on leaves (Agrostideae, Avenae, Hordeae); (2) chlorotic striations (Andropogonaceae, Maydeae, and Paniceae); (3) leaf striating (Festuceae); and (4) plant apical yellowing (Phalarideae).

A number of plant species are recommended as diagnostic plants with characteristic symptoms, including *Avena sativa* cv. Alba (striate yellow mosaic, stunt, and reddening of lower leaves), *Bromus* spp. (dwarfing and leaf striating), *Sorghum halepense*, and *S. vulgare* (chlorotic striations followed by necrosis along leaf margins, plus reduced growth), and others.

Pathogenesis

Vectors of ByYStMV are *Laodelphax striatellus* planthoppers transmitting it in a persistent manner in nature, and *Javesella pellucida* planthoppers transmitting ByYStMV under experimental conditions (loc. cit. Milne and Conti, 1986). *L. striatellus* acquires the virus after feeding for at least 1 hour, which then becomes able to transmit ByYStMV after a latent period of 9 to 29 days; it maintains infectivity for life. The virus is transmitted transovarially and the larvae become viruliferous 2 weeks after hatching (Conti, 1980).

The virus is transmitted neither mechanically nor through plant seed, meaning that the Cicadellid planthopper vector activity plays a decisive role in pathogenesis.

Control

The lack of data on virus control is probably because of its limited area of distribution. Judged by pathogenesis, chemical control of the planthopper vector is useful to reduce losses in areas of greater virus incidence.

OTHER VIRUS DISEASES OF BARLEY

Cereal Tillering Disease Reovirus (CrTlDsV)

CrTlDsV, a member of the Reovirus group, subgroup 2, is known only in northern Europe (Milne and Luisoni, 1977a). It was recognized for the first time in 1971 in barley and oats in Sweden (Lindsten et al., 1973).

Virus hosts are gramineous plants. Naturally infected barley displays excessive tillering, severe dwarfing, and malformed leaves with serrated margins. CrTlDsV is serologically related to maize rough dwarf virus (MRgDV) (Lapierre, 1988). Virus vectors are *Laodelphax striatellus* planthoppers. Transovarial virus transmission is unknown (Lapierre, 1988).

Cereal Chlorotic Mottle Rhabdovirus (CrCMtV)

CrCMtV is a member of the Rhabdovirus group and is distributed in Queensland, Australia, and New South Wales (Greber, 1976; 1977; 1977a). Natural virus hosts include barley, maize, sweet corn, *Digitaria ciliaris, Dinebra retroflexa, Echinochloa colona, Eragrostis cilianensis, Eleusine coracana,* and *E. indica* (Jackson et al., 1981). Characteristic disease symptoms are stunting and chlorotic leaf striping. Natural virus vectors are Jassid leafhoppers *Nesoclutha pallida* (Greber, 1979).

Cereal Northern Mosaic Virus (CrNnMV)

CrNnMV discovered in 1944 on barley (*Hordeum vulgare*) and millet (*Setaria italica*) in Japan causes light green stripes and spots on leaves and leaf sheaths. Some plants remain sterile and form many tillers (50 to 60). Vectors that transmit CrNnMV in a persistent manner include *Laodelphax striatellus, Delphacodes albifascia, Ukanodes sapporonus,* and *Muellerianella fairmairei* (Ito and Fukushi, 1944; Jackson et al., 1981).

BARLEY: NATURAL ALTERNATIVE HOST FOR SOME VIRUSES

Barley is also the host of some other viruses described in oats and wheat sections. These viruses are oat pseudo-rosette virus (OtPdRsV), wheat striate mosaic virus (WStMV), wheat streak mosaic virus (WSkMV) (Figure 19), wheat soil-borne mosaic virus (WSoMV), and wheat winter mosaic virus (WWnMV).

Details on these viruses are described on primary hosts.

VIRUS DISEASES OF OAT (*Avena sativa*)

Several oat viruses have been described in the U.S. and in southern and eastern Europe where oat is grown on substantial acreages. Some of these viruses are transmitted by Cicadellid planthoppers (oat blue dwarf, sterile dwarf, and pseudo-rosette viruses), and others are soil-borne (oat mosaic) or sap-transmissible (oat necrotic mottle) viruses. Oat is the natural host of several viruses whose primary hosts are graminoid plants.

OAT BLUE DWARF VIRUS (OtBlDV) (MEMBER OF THE MAIZE RAYADO FINO VIRUS GROUP)

Properties of Viral Particles

OtBlDV particles are isometric, 28 to 30 nm in diameter, and have a sedimentation coefficient of 119S. Virus RNA is single stranded and apparently homologous (2130 kDa) (Banttari and Zeyen, 1973).

The DEP in infected barley extract, tested with the aid of leafhopper vectors, is 1/256 to 1/512, and the LIV is 10 minutes at 60°C and 16 days at 24°C (Banttari and Zeyen, 1973). OtBlDV is weakly immunogenic.

Geographic Distribution

Dwarfing caused by OtBlDV was discovered in North America and Europe two decades ago, yet American and European isolates have not been characterized.

Host Range

OtBlDV is rare because it is capable of infecting both monocotyledonous and dicotyledonous plants. Its sometimes symptomless hosts belong to seven different plant families. Suitable diagnostic plants

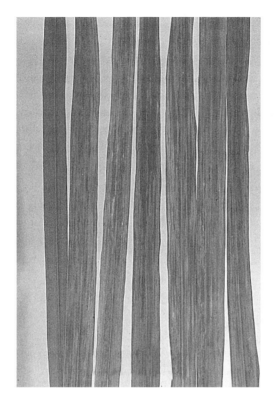

FIGURE 19 Symptoms of wheat streak mosaic virus-infected barley.

include oats, common flax (*Linum usitatissimum*), common chickweed (*Stellaria media*), and others (Banttari and Zeyen, 1973).

Symptoms

Several diagnostically important symptoms appear on OtBlDV-infected oat plants (Banttari and Zeyen, 1973). Easily distinguishable among these symptoms are a bluish-green coloration of infected leaves with necrosis under field conditions, enations on leaf veins and stem, and floret sterility. Infected oat plants are dwarfed and tend toward excessive tillering. Enations along leaf veins on common flax and chickweed leaves are characteristic symptoms. Infected flax leaves are malformed and crinkled.

OtBlDV causes economic damage through dwarfing of plants and floret sterility in oats and other grasses.

Pathogenesis

Cicadellid leafhoppers transmit OtBlDV in a persistent manner. *Macrosteles fascifrons* in North America and *M. laevis* in Europe are the vectors that acquire the OtBlDV from infected plants in 15 minutes and transmit it to healthy plants after 7 days of incubation. Individual insects may remain viruliferous more than 2 months (Banttari and Zeyen, 1970).

The virus is transmitted neither with infected plant sap nor through oat seed. Vectors play a dominant role in virus spread and plant infection in view of the modes of transmission.

Control

Although no particular virus controls have been designed, chemical control of vectors and planting more tolerant or resistant oat cultivars may help prevent infection or reduce, and thus alleviate, losses caused by OtBlDV in the field.

Oat Sterile Dwarf Reovirus (OtSrDV)

Properties of Viral Particles

OtSrDV particles are isometric and 65 to 70 nm in diameter. The double-stranded viral RNA is divided into ten segments, of total molecular weight 18,400 kDa. The composition of the double-shell protein is unknown (Boccardo and Milne, 1980).

The inner protein shell is immunogenic, but little is known about the immunogenic property of the outer virus shell. The mild and the normal virus strains were described in Sweden (Lindsten, 1973). OtSrDV is serologically related to *Lolium* enation virus and *Arrhenatherum* blue dwarf virus, which suggests that they are related strains of one virus (Boccardo and Milne, 1980). Its features place it in the Reovirus group, subgroup 2.

Geographic Distribution

OtSrDV seems to spread only in the European countries of former Czechoslovakia (losses of 70 to 100% were recorded), Poland, Great Britain, Germany, Finland, Norway, and Sweden (Pruša, 1958; Pruša et al., 1959; Lindsten, 1959; 1961; 1961a; Brčak, 1979).

Host Range

Oat, wheat, barley, rye, and maize cultivars are natural hosts for OtSrDV. Other susceptible plants in nature belong to the genera *Arrhenatherum*, *Cynosurus*, *Lolium*, *Phalaris*, and *Poa* (Boccardo and Milne, 1980). OtSrDV is latent in *Bromus mollis*, *B. tectorum*, *Triticum arvensis*, and *Festuca pratensis* plants (Spaar and Schumann, 1977).

Symptoms

Symptom appearance and severity vary, depending on host species (Boccardo and Milne, 1980). The first disease symptoms occur on oats as dark green and, later on, yellow striated spots with leaf stunting. Small enations appear on lower leaf surfaces. Later, entire leaves turn yellow, red, and partially also violet-red. Flowers are mostly sterile and dwarfed plants have a rosette-like appearance.

OtSrDV-infected *Lolium perenne* and *L. multiflorum* that form enations on the nodes of the flowering stem and spike are highly susceptible and usually die. Barley and wheat are less susceptible hosts of OtSrDV (Boccardo and Milne, 1980).

Pathogenesis

Sources of virus infections are annual and perennial virus hosts. The virus overwinters in its perennial hosts (*Lolium perenne* and others), from which it is transmitted year to year. Overwintering may occur in diapausing vector nymphs.

Planthoppers (Delphacidae) are vectors of OtSrDV that transmit it in a circulative (persistent) manner; four species are important: *Javesella pellucida* (the most important one), *J. discolor*, *J. dubia*, and *Dicranotropis hamata* (Boccardo and Milne, 1980). The shortest acquisition period is 30 to 60 minutes; the latent period is 3 to 4 weeks and then vectors remain viruliferous throughout life. OtSrDV is also transmitted transovarially by *J. pellucida* at a low percentage (0.2%) (Vacke, 1966).

OtSrDV is transmitted neither with plant sap nor through the seed; thus, vectors are decisive in virus spread and persistence in nature.

Control

Due to plant sterility, OtSrDV ranks among highly damaging plant viruses that should by all means be controlled in areas of greater incidence. The following preventive measures are recommended: spatial isolation in susceptible plant production, cultural operations aimed at elimination and destruction of spontaneous virus hosts, and chemical control of vectors immediately prior to the occurrence of dense populations.

OAT PSEUDO-ROSETTE VIRUS (OtPdRoV)

Properties of Viral Particles

Donchenko first reported the pseudo-rosette disease of oats in Siberia (Sukhov and Vovk, 1938), the etiology of which is not yet known. Spaar and Schumann (1977) reported bacilliform particles of $167 \pm 20 \times 57$ nm. Smith (1972) reports two types of viral particles: bacilliform of 60-nm diameter, and elongated, rod-shaped 500 to 600×40 nm. Information is lacking on biophysical and biochemical properties of OtPdRoV.

Geographic Distribution

Geographic distribution of OtPdRoV is limited, that is, described only in Siberia (Brysgalowa, 1945; Sukhov and Vovk, 1938; Sukhov and Sukhova, 1940) and Japan (Smith, 1972). This distribution and slow spread likely result from the specific relationship between the virus and its vector and specific or optimal conditions for vector development.

Host Range

All *Avena sativa*, *A. strigosa*, and *A. byzantina* varieties are susceptible to OtPdRoV. Natural hosts include wheat, barley, rye, wheat × *Agropyron* hybrids, millet, sorghum, and maize (Damsteegt, 1981; Jackson, 1981), *Agropyron repens*, *Bromus inermis*, *Oryza sativa*, *Panicum crusgalli*, *Setaria viridis*, and other grasses.

Symptoms

Characteristic symptoms that appear on infected oat plants include shoot proliferation with the formation of many tillers (50 to 60), giving the plant a tufted appearance. Sometimes in early summer, leaves become reddish, then brown (darker) and more rigid. Inflorescences emerge only partially from the sheaths. A proliferation of branches may form on inflorescences to give a tufted appearance or, vice versa, branching may be reduced. Significant changes in the formation of anthers and ovules cause partial or complete sterility. Similar symptoms are also observed on barley, rye, and maize.

Pathogenesis

OtPdRoV overwinters in its vectors and infected perennial grasses (*Agropyron repens*, *Bromus inermis*, *Echinochloa crusgalli*, and *Setaria viridis*). *S. viridis* is particularly important because it attracts Cicadellid planthopper vectors.

Virus vectors are *Laodelphax striatellus* planthoppers, which transmit OtPdRoV in a circulative (persistent) manner (Sukhov and Vovk, 1938a). Upon feeding on infected plants, these vectors can acquire the virus within 6 hours, optimally 2 to 3 days. The minimal latent period lasts 6 days, after which healthy plants can be inoculated in 5 to 10 minutes during vector feeding (Sukhov and Vovk, 1938a). Vectors play a dominant role in the spread of OtPdRoV because it is neither sap, nor seed transmissible.

Control

Cultural operations and chemical control measures aimed at the elimination of the weed sources of infection and at reducing the planthopper vector populations are highly recommended to prevent disease spread.

OTHER VIRUS DISEASES OF OAT

Oat Mosaic Virus (OtMV)

Particles of OtMV are rod shaped, of 600 to 750×12 to 14 nm. The TIP is 44 to 46°C; DEP, depending on the mode of inoculation, is 1×10^{-2} to 5×10^{-3}; and LIV at 20°C is 24 and 48 hours (Hebert and Panizo, 1975).

OtMV is widespread in the U.S., where it was first described. It has been discovered in England (MacFarlane et al., 1968) and is assumed to cause oat mosaic in New Zealand (Slykhuis, 1962a). OtMV causes extensive damage, reducing yields by 25 to 50%. The majority of infected plants of susceptible cultivars remain sterile.

Only the *Avena* genus, specifically *A. sativa* (the natural host) and *A. byzantina* (experimental host), is susceptible to OtMV. Oat cultivars Nysel and Peniarth are suitable assay plants. The first symptoms of leaf mottling are observed in the spring; then later, the warmer weather partially masks the disease. The leaf spot mosaic virus strain causes mainly indistinct yellow spots on the leaves and streaks near the leaf base. The leaf tip mosaic virus strain causes mosaic mottling on the tips of the youngest leaves and mosaic of upper leaf sheaths.

The virus is soil borne by the fungus *Polymyxa graminis*. It retains infectivity in the resting spores of this fungus on plant roots, wherefrom it is transmitted by the release of zoospores, which initiate new infections. The virus is mechanically transmitted with difficulty and transmission through the seed is unknown.

OtMV is difficult to control because it is preserved in the soil. Due to early infections, autumn sowing is less desirable than spring sowing. Selection of noncontaminated soils for and the use of tolerant cultivars are useful plant protection methods.

Oat Soil-Borne Stripe Virus (OtSoSpV)

OtSoSpV consists of rigid particles 140 and 300 nm long (Hariri and Lapierre, 1988). Discovered in United Kingdom and France, it is identical to the new oat golden stripe virus (OtGSpV) in the U.K. (MacFarlane and Plumb, 1968; 1978).

OtSoSpV has been recorded only in winter oats with symptoms clearly discernible in April, wherefrom brilliant yellow spots and tissue necrosis become evident on apical leaves in June and July. Mechanically inoculated *Nicotiana clevelandii* and *Chenopodium amaranticolor* plants react to infection by producing local lesions, whereas *N. debneyi* plants react to infection sites with chlorotic lesions (Hariri and Lapierre, 1988). The soil fungus *Polymyxa graminis* probably is the virus vector.

An important oat crop protection method is production on noncontaminated soils. Crin Noir oat cv. has displayed tolerance to OtSoSpV infection.

Oat Striate Mosaic Virus (OtStMV)

OtStMV particles are bacilliform or bullet shaped and 210×75 nm in sectioned plant cells (loc. cit. Jackson et al., 1981) and belong to the Rhabdovirus group. OtStMV was described in 1975 in Illinois (Jedlinski, 1976). Oat and wheat are its main hosts. Symptoms on infected leaves include striated mosaic and necrosis. Plants are stunted due to internode shortening.

The leafhopper *Graminella nigrifrons* is efficient at transmitting OtStMV in a persistent manner. The vectors acquire the virus during feeding for 1 to 2 days on infected oats; then it requires an incubation period of 24 days at 25°C before transmission. The latent period in inoculated oats is 27 to 35 days at 24°C (Jackson et al., 1981). The vectors acquire the virus both as nymphs and as adults. No evidence exists for transovarial transmission of OtStMV.

Oat Necrotic Mottle Virus (OtNcMtV)

OtNcMtV viral particles are filamentous of approximately 720×11 nm dimensions, but no information exists about their chemical composition (Gill, 1976). The TIP is 50°C; the DEP 1×10^{-3} and LIV is from 3 to 7 days at 23°C to 6 to 8 weeks at 4°C (Gill, 1976). OtNMtV has good immunogenic activity. No strains have been described.

OtNcMtV has been described only in Canada. It infects only plants of the Gramineae family. Susceptible hosts are oat cultivars and both wild and cultivated grasses in the *Bromus*, *Lolium*, and *Poa* genera.

Initial symptoms on oat leaves are chlorotic lines that gradually turn into dark leaf mottling with necrotic lines and zones. The virus is transmitted mechanically with the sap of infected plants, but other modes of transmission have not been discovered.

OAT: NATURAL ALTERNATIVE HOST FOR SOME VIRUSES

Oat is the natural and/or experimental host of several viruses primarily occurring in cereals. These viruses are generally spread and predominate in those countries in which oat is grown on substantial acreages. Losses caused by these viruses are due to genotype susceptibility, vector distribution, and activity plus ecological conditions favorable for vector development. The following viruses belong to this group.

Barley Yellow Dwarf Virus (ByYDV)

ByYDV, similar to its presence in other cereal crops, is generally spread throughout areas of oat cultivation. Economic damage occurs in those countries in which oats are grown on substantial acreages. In Sweden, for example, this virus causes yield reductions in spring oat (*Avena sativa*), particularly in years when sowing is delayed (Rydén, 1990). Similarly, as in the case of barley, one or two sprayings with adequate insecticides for aphid vector control alleviates yield losses of oats due to ByYDV. For example, two sprays in 1983 increased oat yields by 16 and 38% and by 18 and 19% in 1986 when compared with control plants (Rydén, 1990). Attempts at hybridization have reduced damages with the use of the resultant resistant plants. For example, in the cross *A. sterilis* × *A. sativa*, the resistance of the first parent was transmitted to the progeny which contributes to the improvement of oat yield among other features, thus increasingly favorable results can be visualized (Qualset, 1990).

Some other virus diseases on oats include barley yellow striate mosaic virus (ByYStMV), cereal tillering disease virus (CrTlDsV) (Lindsten, 1973; Lindsten et al., 1973), maize rough dwarf virus (MRhDV), cereal northern mosaic virus (CrNnMV), and wheat winter mosaic virus (WWnMV).

Detailed information on the properties of these viruses is described on primary hosts.

VIRUS DISEASES OF RYE (*Secale cereale*)

Rye is grown worldwide. Based on acreages sown, Europe and the former U.S.S.R. are the largest world producers of this crop, which is also used for human nutrition, although somewhat less extensively than wheat, rice, and maize. Rye products also play an important role in animal nutrition.

Although rye is susceptible to many viruses, none has been described on rye as its main host. Thus, rye is predominantly an alternative host for viruses whose main hosts include cereals, rice, maize, and other weed grass species. Detailed information on the features of these viruses is provided in their descriptions on main hosts.

Based on damage caused, viruses clearly are important in rye production. Infected plants represent sources of infection for other adjacent cereal crops, particularly in cases where insect vectors aid in the movement and spread of infective viruses in nature. In view of their potential importance, only viruses occurring in rye as their natural or experimental host will be cited here.

MAIZE VIRUSES

Maize Streak Virus (MSkV).

Wheat Viruses
Wheat American striate mosaic virus (WAStMV), wheat soil-borne mosaic virus (WSoMV), wheat dwarf virus (WDV), wheat streak mosaic virus (WSkMV), wheat winter mosaic virus (WWnMV), and wheat yellow leaf virus (WYLV).

Barley Viruses

Barley stripe mosaic virus (BySpMV), barley yellow dwarf virus (ByYDV), and barley yellow striate mosaic virus (ByYStMV).

Oat Viruses

Oat blue dwarf virus (OtBlDV), oat pseudo-rosette virus (OtPdRoV), and oat sterile dwarf virus (OtSrDV).

Common Cereal Viruses

Cereal chlorotic mottle virus (CrCMtV), cereal tillering disease virus (CrTlDsV), and northern cereal mosaic virus (CrNnMV).

Rice Viruses

Rice dwarf virus (RDV), rice gall dwarf virus (RGlDV), rice hoja blanca virus (RHBV), and rice ragged stunt virus (RRdSnV).

Weed Grass Viruses

Agropyron mosaic virus (AgMV), brome mosaic virus (BrMV), cocksfoot mild mosaic virus (CkMdMV), cocksfoot mottle virus (CkMtV), lolium mottle virus (LmMtV), and phleum mottle virus (PmMtV).

Other Plant Viruses

Cucumber mosaic virus (CMV), unidentified virus, dark green rye and oat stunting agent (Agarkov, 1969) and unidentified virus, rye stripe, and wheat mosaic agent (Hoppe, 1969).

VIRUS DISEASES OF RICE (*Oryza sativa*)

Rice is grown worldwide. Based on acreages sown, it ranks second immediately following wheat; but based on tonnes of production, it ranks first in the world. It plays a leading role in human nutrition and sustaining a burgeoning population primarily in subtropical and tropical regions of Asia, Africa, and America, in which its importance can be compared with that of the wheat crop in temperate parts of the world. An excellent companion book for this section is the *Compendium of Rice Diseases* (Webster and Gunnell, 1992), which covers all pathogens and diseases of rice.

Rice is the primary (main) and the secondary host for many viruses that damage quality of product and quantity of yield plus representing limiting production factors under certain conditions.

Eleven virus species have been described on rice as their main host. The most numerous among these viruses is the Reovirus group (four species), then the rice stripe virus group (three species), and finally, viruses belonging to different groups (four species). Rice is also an important secondary (alternative) host for some other cereal and gramineous plant viruses.

The largest number of rice viruses has been described in Asian, African, and American countries where the rice crop is grown most intensively and extensively. Only ByYDV (syn. rice giallume disease) and maybe rice dwarf virus have been recorded in southern European countries.

RICE DWARF REOVIRUS (RDV)

Properties of Viral Particles

RDV particles are isometric, 70 nm in diameter. The double-stranded viral RNA consists of 12 differing-length segments and constitutes 11% of the viral particle molecular weight. No data are available on the protein composition of viral particles (Iida et al., 1972). The sedimentation coefficient of viral particles containing RNA is 510S and undefined for the RNA-free particle type (Iida et al., 1972).

Stability in Sap

Since RDV is transmitted in a persistent manner, stability properties were determined via assays in leafhopper vector extracts. The RDV TIP is 40 to 45°C; the DEP is 1×10^{-4} to 1×10^{-5}; and LIV is 48 to 72 hours at 4°C; whereas at –40°C, RDV remains infective in infected rice leaves and vectors for one year (Iida et al., 1972).

Serology

RDV is immunogenically very active by intramuscular injections of rabbits. These antisera react with antigens extracted both from infected plants and from insect vectors (Kimura, 1962).

Strains

No virus strains have been described.

Geographic Distribution and Economic Importance

This virus is widespread in Japan and Korea (Iida et al., 1972). Although reported in Bulgaria (Atanasov and Dodov, 1961), its relatedness or identity with the Asian virus has not been proved.

RDV-infected plants have shortened internodes and substantial growth delay. Early season infections are most dwarfed and they form few flowers or remain flowerless. Such severe changes cause significant damage to rice production.

Host Range

Rice is both the natural host and the plant most susceptible to this virus. Other susceptible plants include *Alopecurus fulvus*, *Panicum miliaceum*, *Echinochloa crusgalli*, and *Poa pratensis*. Rye, wheat, and oats are slightly susceptible and maize is resistant to RDV.

Symptoms

Primary symptoms of RDV disease are spots appearing along leaf veins. Yellowish and whitish spots or shorter and longer streaks on leaves appear early in the season. Whitish spots occur on older leaves. (See Figure 20.) A general appearance of stunted plants with pronounced delay of growth, plus secondary tillers simultaneously formed on shortened internodes, cause infected plants to take on a rosette-like shape. Growth of roots is delayed and they develop horizontally along the surface. Inflorescences seldom form on RDV-infected plants.

Pathogenesis

The leafhoppers *Nephotettix cincticeps*, *N. apicalis*, and *Inazuma dorsalis* transmit the virus in a persistent manner. *N. cincticeps*, the main natural vector, acquires the virus by feeding on infected rice plants for 1 day. Young nymphs are most efficient vectors, capable of acquiring the virus in 1 minute; vectors transmit RDV to young rice seedlings by feeding for 30 minutes. The virus is transovarially transmitted at high percentages to the next generations of leafhoppers (Iida et al., 1972).

The virus is neither transmitted mechanically nor through rice seed from infected plants.

Control

Since RDV probably overwinters more effectively in Cicadellid leafhoppers than in its plant hosts, control of Cicadellid leafhoppers, particularly in their overwintering habitats, represents one of the most important plant protection measures. Resistant rice cultivars should be given priority in rice production where possible.

RICE BLACK-STREAKED DWARF REOVIRUS (RBSkDV)

Properties of Viral Particles

Two types of RBSkD viral particles with respective diameters of about 75 to 80 and 50 to 55 nm occur in infected plant cells and vectors (Shikata, 1974). The double-stranded virus RNA represents

FIGURE 20 Chlorotic specks and streaks caused by rice dwarf virus. (Reproduced with permission from APS.)

a genome divided into ten segments. This virus is possibly a member of the Reovirus group, subgroup 2 (Shikata, 1981).

Stability in Sap

Biophysical features of RBSkDV as determined in the infected rice and insect vector extracts differ somewhat. The TIP in rice sap is 60°C. LIV at 4°C in rice sap is 7 days, but 6 days in the insect extract. The DEP in rice sap is 1×10^{-5}, but 1×10^{-6} in the vector extract. RBSkD viral particles remain infective in frozen rice leaves for nearly a year (Shikata, 1974).

Serology

RBSkDV is immunologically active. Antisera obtained by intramuscular injection of rabbits with 60-nm viral particles are successful in ring precipitin tests and in agar gel diffusion tests (Luisoni et al., 1973), and a serological relatedness with the maize rough dwarf virus has been reported.

Strains

No data on specific viral strains are available.

Geographic Distribution and Economic Importance

RBSkDV is widespread in Japan. RBSkDV causes severe rice stunting. Inflorescences are rarely, or not, formed on infected plants, which ranks it among the most economically damaging of rice viruses, also occurring in maize, wheat, and barley crops in Japan.

Host Range

Some 25 hosts of RBSkDV belong exclusively to the Gramineae family. Rice, maize, wheat, oats, and barley are of particular importance. Rice is a suitable diagnostic plant.

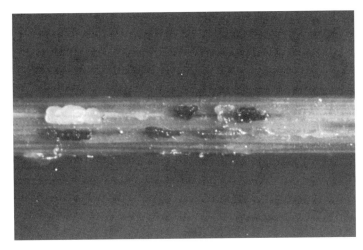

FIGURE 21 Galls on a rice culm caused by black-streaked dwarf virus. (Courtesy A. Shinkai. Reproduced with permission from APS.)

Symptoms

General symptoms observed on RBSkDV infected plants include severe plant stunting and waxy swellings along the veins on the lower leaf surface (Figure 21). These swellings represent a neoplastic proliferation of infected cells. Darkening and twisting of young leaves are also typical. Disease symptoms are severely pronounced in highly susceptible rice, japonica-type cultivars, and milder in less susceptible, indica-type cultivars (Shikata, 1974).

Pathogenesis

RBSkDV is transmitted naturally only by Cicadellid planthoppers. Neither transmission mechanically with infected plant sap, nor through seed has been demonstrated.

Laodelphax striatellus, *Unkanodes sapporona*, and *U. albifacia* are virus vectors (Shikata, 1974). Virus acquisition from infected rice plants requires 30 minutes; the incubation period in vectors requires 7 to 35 days; and transmission to healthy plants requires at least 5 minutes. The virus is not transovarially transmitted (Shikata, 1974).

Control

Chemical control of vectors may help to reduce the extent of infection and virus spread in rice fields. In severely infested areas, priority should be given to sowing resistant rice types.

RICE GALL DWARF REOVIRUS (RGlDV)

Properties of Viral Particles

RGlDV particles are polyhedral, 65 nm in diameter. Virus RNA is double-stranded and contains 12 segments with a total molecular weight of 16,900 kDa (Hibi et al., 1984). Viral particles contain seven different molecular weight proteins (Omura et al., 1985).

RGlDV is considered a member of the Phytoreovirus group (Milne and Lovisolo, 1977; Fenner and Gibbs, 1983) since leafhoppers, but not planthoppers, are vectors and since transovarial passage occurs. Morphological transmission and host range similarities of RGlDV are close to the RDV. One substantial difference is that RGlDV is found located in parenchymal phloem cells, whereas RDV occurs in different cells of infected plants. Another difference is that RGlDV and RDV are not related serologically and they differ based on RNA and protein particle molecular weights (Omura and Inouye, 1985). No information exists on virus strains.

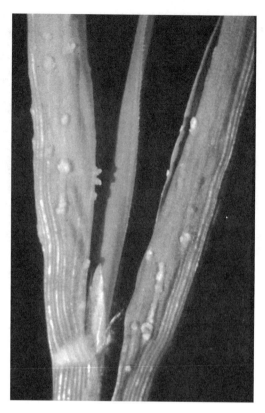

FIGURE 22 Galls caused by rice gall dwarf virus, first light green and somewhat translucent, then turn white. (Reproduced with permission from APS.)

Geographic Distribution and Economic Importance

RGlDV has been described recently in Thailand, Malaysia, and China (loc. cit. Omura and Inouye, 1985). Early season infections by RGlDV cause plant dwarfing, flowering delay, and grain shriveling, which result in substantial reductions in rice yields.

Host Range and Symptoms

Rice is the natural virus host. Other hosts infected experimentally with the aid of vectors include wild rice (*Oryza nifipagon*), barley, wheat, rye, oats, Italian ryegrass, and *Alopecurus aequalis* (loc. cit. Omura and Inouye, 1985). *Oryza sativa* cv. Taichung Native 1 is the diagnostic plant for RGlDV.

Whitish galls, after which this disease was named, appear on lower leaf surfaces and on the outer side of leaf sheaths (Figure 22). RGlDV infection delays plant growth. Infected plants gradually turn dark green with age. Symptoms of disease in experimental hosts are similar to those in rice.

Pathogenesis

The virus retains infectivity in infected rice plants and in Cicadellid leafhopper vectors in nature. Vectors of RGlDV include *Nephotettix nigropictus*, *N. cincticeps*, *N. malayanus*, *N. virescens*, and *Recilia dorsalis*. *N. nigropictus* is the most effective of all vectors, transmitting RGlDV in 2 to 95% of the cases (Inouye and Omura, 1982; Morinaka et al., 1982). RGlDV particles are transmitted transovarially from generation to generation with an efficiency up to 100%.

The vectors transmit RGlDV to plants in a persistent manner. The virus is transmitted neither mechanically nor by seed. Chemical control of vectors in rice fields reduces incidence of infection and decreases yield losses caused by RGlDV.

Rice Ragged Stunt Reovirus (RRdSnV)

Properties of Viral Particles

RRdSnV particles are polyhedral, 50 nm in diameter; with outer spikes, the diameter is 65 nm (Milne et al., 1982). The double-stranded RNA consists of eight segments with a total weight of 11,600 kDa (Boccardo and Milne, 1980a). No information characterizing virus proteins is available. This subgroup 2 Reovirus (Senboku et al., 1978) features pronounced and complete immunogenic activity. No one has published information on possible strains.

Geographic Distribution and Economic Importance

RRdSnV, described only recently, occurs in China, India, Indonesia, Japan, Malaysia, the Philippines, Sri Lanka, Taiwan, and Thailand (Milne et al., 1982). RRdSnV ranks among the viruses that cause economically important damage in rice production. In Indonesia, rice ragged stunt occurs on 34 to 76% of plants in some rice cultivars, causing yield losses of 17 to 47%. Nearly total (80 to 100%) losses caused by RRdSnV were recorded in India (Milne et al., 1982).

Host Range and Symptoms

Natural hosts of RRdSnV are limited to rice: *Oryza sativa*, *O. latifolia*, and *O. nivara*. Hibino (1979) experimentally infected maize, barley, rye, and possibly oats; however, the results are contradictory. *O. sativa* is the virus diagnostic plant.

Based on published data, RRdSnV particles occur in both cytoplasm of parenchymal cells and sieve tubes of the phloem tissue. RRdSnV causes hypertrophy in parenchymal cells, resulting in enation formation on lower leaf surfaces and leaf sheaths (Hibino et al., 1979; Milne, 1980). Initially, these enations are small, whitish protuberances, later turning brown and resulting in the appearance of ragged leaves (Figure 23). Infected young rice plants are dwarfed with increased branching on the internodes. Dark green at the beginning, infected plants remain green after heading. RRdSnV is also characterized by delayed flowering, incomplete panicle development, and shriveled grain, causing reductions in rice yields. Most japonica- and indica-type rice cultivars are susceptible to RRdSnV, although each is resistant under field conditions and in greenhouses (Milne et al., 1982). Symptoms on experimentally infected plants are similar to those occurring in naturally infected rice.

Pathogenesis

In nature, RRdSnV retains infectivity in rice plants and in the rice brown planthopper *Nilaparvata lugens*, the only known vectors (Milne et al., 1982). *N. lugens* requires at least 3 hours for acquisition of RRdSnV, an average latent period of 9 days (2 to 33 days), and a minimal inoculation period of 1 hour. Males and females transmit RRdSnV and nymphs are more active than adults. The vector remains viruliferous for 1 to 4 weeks and some for life. This virus is neither transmitted transovarially (Milne et al., 1982), mechanically, nor through seed from infected plants.

The mode of RRdSnV transmission and spread indicates the need for chemical control of vectors to reduce disease incidence in the field and allow higher yields. Priority should be to select resistant rice cultivars, particularly in areas of high rice ragged stunt incidence.

Rice Stripe Tenuivirus (RSpV) [Type]

Properties of Viral Particles

RSpV particles are filamentous, of indefinite length, and 8 nm in diameter. The RSpV particles sediment into three components: the middle (M), the bottom (B), and the nB component. Single-stranded viral RNA consists of four segments of 1.9, 1.4, 1.0, and 0.9 ($\times 10^{-6}$) MW (Toriyama, 1983), which accounts for 12% of the total viral particle weight. Only nB particle components are infective. The viral capsid contains a single protein species of 32 kDa. Sedimentation coefficients of the three sediments are 65S (M), 80S (B), and 98S (nB) (Toriyama, 1983).

FIGURE 23 Serrated leaf margins caused by rice ragged stunt virus. (Reproduced with permission from APS.)

Stability in Sap

These biological constants vary, depending on the host origin of the extracts tested (Toriyama, 1983). The DEP in vector extracts is 1×10^{-4} to 1×10^{-5} and that in infected plant extracts is from 1×10^{-3} to 1×10^{-4}. The TIP is 50 to 55°C; the LIV in vector extracts is 4 days at 4°C; and in infected plant extracts is 8 to 12 months at –20°C.

Serology

RSpV has high immunogenic activity. Antisera were successfully obtained by inoculating rabbits (Toriyama, 1983).

Strains

Individual virus variants occur based on symptom characteristics and virus transmissibility by vectors, but no specific strains have been described.

Geographic Distribution and Economic Importance

Rice stripe disease, first described in Japan, was later found in Korea, China, Taiwan, and the former U.S.S.R. (Toriyama, 1983). RSpV is economically damaging and capable of causing substantial yield reductions, especially if young rice plants become infected. Damages, due to severe pathological changes that occur in infected plants, are primarily delayed emergence and ear maturity.

Host Range and Symptoms

The virus has been reported in 37 species in the Gramineae family among the natural hosts rice, maize, wheat, oats, foxtail, millet, and a number of wild grasses. Rice is the diagnostic plant of choice for RSpV.

Chlorotic stripes appear first on infected rice leaves, followed by chlorosis of the entire plant. Simultaneously with symptom development, chlorotic stripes gradually turn brown and become necrotic. Panicle formation is either poor or completely absent in infected plants, resulting in substantial yield reductions. Similar leaf striping, discoloration, and stunting are also observed on maize and wheat leaves.

Pathogenesis

Natural virus hosts represent sources of infection from which the virus is transmitted and spread onto susceptible field plants with the aid of vectors. Vectors that transmit RSpV in a persistent manner include *Laodelphax striatellus* brown planthoppers, *Unkanodes sapporona*, *U. albifascia*, and *Terthoron albovizzatum* (Toriyama, 1983). *L. striatellus*, the most active vector, has the shortest acquisition feeding period of 15 minutes, with an optimum of 1 day. RSpV has an incubation period in the vector of 5 to 21 days and the minimum feeding time for inoculation of plants is 1 hour when only one half of the viruliferous vectors are capable of transmitting adequate RSpV to infect rice seedlings. RSpV particles are transmitted transovarially to some 90% of progeny insects (Toriyama, 1983).

Mechanical inoculation seldom results in infection and RSpV also is not transmitted through seed from infected plants.

Control

Infections and damages caused by RSpV can be reduced by chemical control of the vectors. The use of tolerant and resistant rice cultivars is also recommended. Japanese upland- and indica-type rice varieties are generally tolerant and/or resistant to this virus (Toriyama, 1983).

RICE HOJA BLANCA TENUIVIRUS (RHBV)

Properties of Viral Particles

The RHBV filamentous particles are of indefinite length and 3 to 4 nm in diameter. Viral RNA accounts for about 12% of the total particle weight; a single coat protein species of RHBV is 34 kDa (Morales and Niessen, 1985). According to the same authors, the sedimentation coefficient, depending on virus conformation, is from 63 to 97S. RHBV has pronounced immunogenic activity. No proof exists for virus strains, whereas different infected plant reactions result from environmental variability; however, isolates from *Echinochloa colona* may be a strain.

Geographic Distribution and Economic Importance

RHBV is widespread in North and South America only, in Brazil, Belize, Colombia, Costa Rica, Cuba, Dominican Republic, El Salvador, Ecuador, Guatemala, Guyana, Honduras, Mexico, Nicaragua, Panama, Peru, Puerto Rico, Surinam, the U.S., and Venezuela (Everett and Lamey, 1969).

RHBV ranks among viruses causing economically important rice yield reductions. Seed set in infected plant panicles is either absent or poor.

Host Range and Symptoms

Rice is the primary and natural virus host. Other Gramineae experimentally infected with viruliferous planthoppers include *Avena sativa*, *Digitaria horizontalis*, *Hordeum vulgare*, *Leptochloa filiformis*, and *Secale cereale* (Morales and Niessen, 1985). *Oryza sativa* is the virus diagnostic plant.

The hoja blanca disease was named based on the elongated chlorotic and whitish stripes on infected rice leaves (see Figure 24). In early infections, these stripes become necrotic and the plants die; RHBV causes stunt, resulting in partial or complete sterility. Infections that occur later in the season influence plant yields to a lesser degree.

Pathogenesis

In nature, RHBV remains infective both in infected rice plants and vectors. Rice planthoppers *Sogatodes orizicola* transmit the virus in a persistent manner, and *S. cubanus* is a virus vector only

FIGURE 24 Chlorosis of rice leaves caused by rice hoja blanca. (Reproduced with permission from APS.)

under experimental conditions (Morales and Niessen, 1985). The minimal acquisition period is about 1 hour, whereas the minimal period for the vector to inoculate a plant is 3 to 24 hours (Galvez, 1967; McMillan et al., 1962). Nymphs plus female and male adults can transmit RHBV. Females can pass the RHBV transovarially in a serial manner to ten subsequent generations without access to a source of RBHV (Galvez, 1967). RHBV is transmitted neither mechanically nor through rice seed from infected plants.

Control

Seed damage and yield loss caused by RHBV can be alleviated with chemical control of vectors. Resistant and tolerant rice cultivars should be given priority in rice production. Cultivar Bluebonnet 50 is susceptible and Columbia 1 is highly tolerant to RHBV (Morales and Niessen, 1985).

RICE GRASSY STUNT TENUIVIRUS (RGRSNV)

Properties of Viral Particles

RGrSnV particles are filamentous, occasionally circular, of 950 to 1350 × 6 to 8 nm. Viral RNA consists of four segments of 1450, 1300, 1200, and 1150 kDa (Toriyama, 1985). It contains a single protein species of 31 kDa, although Toriyama (1985) reported two proteins of 34 and 31.5 kDa.

The DEP of RGrSnV in rice leaf extracts is 1×10^{-2} to 1×10^{-3} and in vector extracts is 1×10^{-3} to 1×10^{-6}. LIV in leaf extracts is less than 12 hours at 24°C and more than 6 days at 4°C. RGrSnV is strongly immunogenic in rabbits (Hibino, 1986). Several virus strains denoted as "wilted stunt," "grassy stunt B," "grassy stunt Y," strain 1 ("ordinary"), and strain 2 have been described, based on variations in symptoms and susceptibility of rice plants.

Geographic Distribution and Economic Importance

The rice grassy stunt disease has been reported in several Asian countries: India, Indonesia, Japan, Malaysia, the Philippines, Sri Lanka, Taiwan, and Thailand (Hibino, 1986).

RGrSnV causes economic damage in tropical and subtropical Asia. Plants infected early fail to form, or form only a small number of panicles bearing unfilled grains. Yields of infected plants in greenhouses varied, depending on the cultivars and the time of their infection. Reductions in yield of plants infected during the first month ranged from 70 to 100%; and when infected during the second month, reductions were 0 to 80% (Palmer et al., 1978; Iwasaki et al., 1980).

Host Range and Symptoms

Rice is the only natural host of RGrSnV. The rice cvs. Taichung Native 1, Shan-san-sa-san, and Reicho serve as diagnostic plants.

Symptoms in RGrSnV-infected rice plants vary, depending on cultivars and time of infection (Hibino, 1986). The most severe symptoms occur on seedlings infected early in the season. Pronounced stunting of plants is accompanied by excessive tillering and erect leaves. Also, spots differing in shape and size occur on infected leaves. Usually, the disease partially or completely prevents panicle formation. Infections that occur later in the season cause yellowing and the formation of brown panicles bearing abortive kernels.

Pathogenesis

Infected rice plants and insect vectors serve as natural reservoirs for RGrSnV. Planthoppers *Nilaparvata lugens*, *N. bakeri*, and *N. muiri* transmit RGrSnV in a persistent manner (Hibino, 1986). Vector efficiency depends primarily on the development stage of the insects, virus strains, and environmental conditions. The minimum acquisition time for *N. lugens* is 1 hour; its latent period is 3 to 28 days (average, 8 days); and the minimum time for successful inoculation is 9 minutes (Hibino, 1986). The insects retain the ability to transmit RGrSnV their entire life, but do not pass RGrSnV transovarially to their progeny.

RGrSnV is not mechanically transmitted, nor is it transmitted through rice seed from infected plants.

Control

Chemical control of vectors is inadequate to effect economic reward, especially in tropical areas in which rice is grown year-round and also because of migrations of vectors great distances. In Indonesia, the Philippines, and Vietnam, the disease caused by RGrSnV is controlled successfully by growing resistant cultivars derived from a gene isolated from *Oryza nivara* plants (Ling et al., 1970; Khush and Ling, 1974). However, plant resistance effectiveness may be endangered by the occurrence of strains of RGrSnV with new virulence genes in selected growing areas.

OTHER VIRUS DISEASES OF RICE

Rice Necrosis Mosaic Virus (RNMV)

RNMV is a Potyvirus (group) with particles of 275 to 550 × 13 to 14 nm (Hollings and Brunt, 1981). No information is available on chemical composition. The TIP in plant sap is 60 to 65°C; the DEP is 5×10^{-3} to 1×10^{-4}; and LIV at room temperature is 7 to 14 days (Inouye and Fujii, 1977). RNMV is immunogenically active, but no strains have been described.

RNMV is widespread in Japan. Rice is its only host known to date. Yellow flecks and streaks clearly visible on lower leaf surfaces, and usually not on new leaves, occur in infected plants (Figure 25). Necrotic flecks on stalk bases and leaf sheaths are diagnostic. Infected plants are stunted, with a reduced number of tillers (Inouye and Fujii, 1977).

The soil fungus *Polymyxa graminis* is the vector of RNMV. The virus, harbored in the resting spores of *P. graminis*, is transmitted via zoospores. Zoospores infect root hairs of rice seedlings

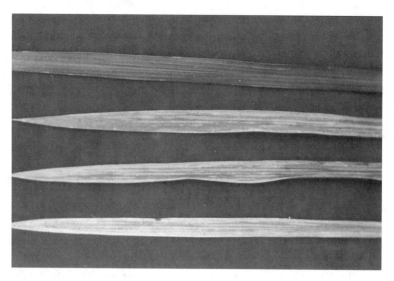

FIGURE 25 Rice necrosis mosaic virus on three leaves; healthy (top). (Courtesy T. Inouye). (Reproduced with permission from APS.)

and effect inoculation, thus infection by RNMV. RNMV is transmitted also through the seed of infected plants in 2.6 to 5.3% of the cases (Fujikawa et al., 1972). Most Japanese paddy rice cultivars are susceptible to infection by RNMV; however, Kanto No. 52 and several other cultivars show resistance (Fujii, 1975).

Rice Transitory Yellowing Rhabdovirus (RTyYgV)

RTyYgV has bullet-shaped viral particles of 96 × 120 nm in leaf dip preparations and 94 × 193 nm in leaf sections (Shikata, 1972). RTyYgV expresses strong immunogenic activity, but the chemical composition is unknown. No strains of RTyYgV have been described.

RTyYgV was described in Taiwan. Rice is its only reported natural host. Primary symptoms in RTyYgV-infected plants include yellowing, stunting, and a reduced number of tillers (Figure 26). Japonica-type rice cultivars are more susceptible with more severe symptoms than the indica types (Shikata, 1972).

Rice green leafhoppers *Nephotettix apicalis* and *N. cincticeps* serve as vectors to transmit the virus in a persistent manner. Following incubation, the virus persists in these vectors for 3 to 29 days (*N. apicalis*) or 21 to 30 days (*N. cincticeps*). RTyYgV is transmitted neither transovarially (Hsieh et al., 1970) mechanically, nor through the seed.

Rice Tungro Waikavirus (RTgV)

RTgV has an isometric shape and a diameter of 30 to 33 nm. Few studies have been made on basic characteristics, such as chemical composition of RNA and protein, immunogenic activity, and transmission through seed (Gálvez, 1971). Two virus strains, severe (S) and mild (M), have been designated, based on symptoms on the differential plant cvs. FK 135 and Achech.

RTgV is widespread in the Philippines, Malaysia, Indonesia, Pakistan, Thailand, and India. Rice tungro disease ranks among the economically most important virus diseases in South Asia (Gálvez, 1971).

Oryza spp. serve as the main hosts of RTgV, which may also infect other plants in the Gramineae family. Symptoms of disease occur in infected *Eleusine indica* plants, whereas *Echinochloa colonum* and *E. crusgalli* are symptomless hosts. General symptoms on RTgV-infected plants include mosaic, mottling, and yellowing of leaves and stunting of plants.

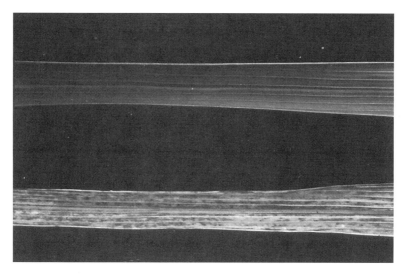

FIGURE 26 Rice transitory yellowing virus; healthy (top). (Courtesy E. Shikata. Reproduced with permission from APS.)

Vectors of RTgV are *Nephotettix impicticeps* green leafhoppers, *N. apicalis*, *N. impicticeps* × *N. apicalis*, and *Recilia dorsalis*. Adult insects and all five larval stages transmit RTgV in a semipersistent manner. Following the acquisition access period, all active vector insect stages lose their ability to transmit RTgV after 5 days, and in 50% of the cases already after 24 hours (Gálvez, 1971). RTgV is not mechanically transmitted.

Rice Yellow Mottle Sobemovirus (RYMtV)

RYMtV is a possible member of the Southern Bean mosaic virus group (Matthews, 1981). RYMtV particles are isometric and 25 nm in diameter. Viral RNA is single stranded, of 1400 kDa. No data are published on protein composition (Bakker, 1975). RYMtV has strong immunogenic activity. The TIP of RYMV in plant sap is 65°C; the DEP is 1×10^{-6}; and LIV is 99 days (at 20°C) or 260 days (at 4°C). No strains have been described for RYMtV (Bakker, 1975).

RYMtV is restricted to the Kisuma area in western Kenya. RYMtV infects plants only in the Gramineae family, Oryzeae and Eragrostideae in particular (Bakker, 1975). *Oryza sativa* is the diagnostic for RYMtV. Primary disease symptoms in infected rice are yellow streaks distributed along leaf veins, delayed growth, a reduced number of tillers, and failure to form panicles.

Vectors of RYMtV are Chrysomelid beetles Criocerinae, Cryptocephalinae, Galerucinae (*Sesselia pusilla*), Halticineae (*Chetocnema* spp.), and Hispinae (*Trichispa sericea*), plus the long-horned grasshopper *Conocephalus merumontanus* (Bakker, 1975). The relationship between RYMtV and its vector differs with various hosts. *S. pussila* and *T. sericea* may acquire the virus in 15 minutes and require 15 minutes to inoculate plants. *S. pussilla* remains viruliferous for 2 to 3 days, but *T. sericea* for only 1 day (Bakker, 1974). RYMtV is not transmitted through seed, yet is easily transmitted mechanically with the sap of infected plants—which is a key difference from other rice viruses.

RICE: NATURAL ALTERNATIVE HOST FOR SOME VIRUSES

Barley Yellow Dwarf Luteovirus (ByYDV)

Rice is economically quite important as a host of ByYDV. First observed in rice in Italy in 1955 (Corbetta, 1967), ByYDV in 1970 became a serious problem in rice production. Initially based on symptoms, ByYDV was described as the rice yellows virus (rice "giallume" virus), but later it was established as a nonspecific isolate of ByYDV (Osler, 1984; Belli et al., 1986). A number of plants with rice yellows symptoms similar to the Italian yellow disease were noted in former Yugoslavia (near Kočani), but the virus has yet to be identified (Šutić, 1983).

Rice yellows due to ByYDV infection may cause significant rice yield reductions, depending on the level of susceptibility of the rice cultivar, the time of infection, and the presence of sources of infection and virus vectors. Severe infections of ByYDV cause early plant death and yield reductions ranging from 40 to 100%. Losses due to yellows diseases are greater when infection occurs earlier in the season. For example, inoculation in the stage of the first leaf development reduced rice yields by 85% (cv. S. Andrea) to 98% (cv. Balilla); whereas in later inoculations in the tiller initiation stage, these reductions were 33% (cv. Roma) to 65% (cv. Balilla) (Moletti et al., 1990). The greatest damages due to this disease can be anticipated in regions where susceptible rice cultivars are planted continuously over longer periods of time.

Primary symptoms on infected plants occur in late May or early June. The greatest percentages of infected plants are recorded in late June and early July. Plant leaves gradually become yellow, originating at leaf tips and margins. Main leaf veins are green initially, whereas whole leaves later become yellow, yellow-orange, and dark green. Dark-colored necrosis sometimes also occurs on infected leaves. ByYDV-infected plants express delayed growth, with panicles underdeveloped or not formed. The disease is striking in the appearance of yellow patches in fields. As the disease develops, these zones gradually merge together. ByYDV infections are easily noticed by the yellow color and dwarfing of plants.

In nature, ByYDV infects many wild and cultivated annual and perennial plants belonging to the Gramineae family. These plants represent sources of ByYDV for infection of rice during the season and serve as overseason reservoirs. The perennial weed *Leersia oryzoides*, common in rice fields, plus a suitable aphid vector is significant for the development of the RYV isolate cycle (Osler et al., 1980). *L. oryzoides* easily multiplies and spreads via both seeds and stolons (Conti, 1983); thus, it occurs regularly in rice fields. Among the aphid vectors are *Rhopalosiphum padi* and *Sitobion avenae*, which transmit ByYDV in a persistent manner. The susceptibility of the majority of cultivated rice cultivars plays an important role in the pathogenesis.

Several measures are recommended to prevent spread of infection: eliminate weeds from rice fields, border zones, and irrigation ditches; use chemicals to control aphid vectors, taking care not to destroy the pathogens that provide biological control of vectors (the *Aphelinus* genus) and avoid contamination of irrigation water; grow resistant cvs. such as Arborio, Cristallo, Carnaroli and Navile, and avoid the use of susceptible cvs. such as Balilla and Bahia on contaminated soils (Osler et al., 1977). Resistant cvs. and breeding lines are available in Italy with the introduction of a genotype of the old Italian cv. Vialone Nero, the only known source of rice yellows resistance (Moletti and Osler, 1978; Moletti et al., 1979).

Maize Dwarf Mosaic Virus (MDMV)

Some rice cvs. are symptomless hosts of MDMV (Rosenkranz, 1981). Other data show rice susceptible to MDMV-A and ScMV-MDB (=MDMV-B) strains (Tošić and Ford, 1972). See the chapter on virus diseases of maize for detailed information on MDMV.

Oat Pseudo-rosette Virus (OtPdRoV)

Some authors rank rice among natural hosts of OtPdRoV (Spaar and Schumann, 1977a). Details of OtPdRoV are in the chapter on virus diseases of oat.

Ryegrass Mosaic Virus (RgMV)

Information pertaining to RgMV is contradictory (Slykhuis, 1972). Mulligan (1966) reported on the British strain of RgMV. Details on RgMV are found in the chapter on virus diseases of graminoid forage plants.

SORGHUM VIRUS DISEASES (*Sorghum bicolor*)

Sorghum, a crop grown worldwide and particularly important in tropical and subtropical areas, is becoming increasingly important in temperate zones. The largest acreages of sorghum are in Asia, Africa, and North and Central America. Sorghum is used directly for human food and in animal feed. Additionally, it provides valuable raw materials for several processing industries. An important companion book is the *Compendium of Sorghum Diseases* (Frederiksen, 1986).

Among other plants in the family Poaceae, sorghum is susceptible to many viruses. Seldom the primary host, it is most often the alternate host for many viruses described on other graminoid plants. As an alternate host, sorghum represents an important source of infection from which many viruses are transmitted and spread into other species of Poaceae in nature.

Viruses occur in the sorghum crops wherever grown. Many among them cause significant changes in infected plants, including stunting, death, and yield losses, which represent limiting production factors under certain conditions.

Sorghum Mosaic Potyvirus (SrMV)

Properties of Viral Particles

SrMV particles are flexuous filaments of 750 nm, which contain single-stranded RNA and cause the production of cytoplasmic, cylindrical (pinwheels and scrolls), and amorphous inclusions in host cells (Shukla et al., 1994).

Stability in Sap

For SrMV, the TIP is 58°C; the DEP is 1×10^{-2} to 2×10^{-4}; and LIV at 20°C is 24 hours (Tošić and Ford, 1974).

Serology

SrMV is moderately immunogenic. These antisera can be used in microprecipitation, SDS-agar gel immunodiffusion, enzyme-linked immunosorbent assay (ELISA), immunosorbent electron microscopy (ISEM), and electro-blot-immunosorbent (EBIA, Western Blots) tests (Shukla et al., 1994). SrMV is serologically related to ScMV, MDMV, and JgMV.

Strains

The H, I, and M strains, earlier described as strains of ScMV (Abbott, 1961; Tippett and Abbott, 1968; Koike and Gillaspie, 1976), are now designated as strains of SrMV (Shukla et al., 1994).

Strain H was isolated from sorghum (Giorda et al., 1986); thus, the virus was named SrMV (Shukla et al., 1989).

Geographic Distribution

SrMV is found mainly in sorghum (Giorda et al., 1986) and sugarcane crops (Abbott, 1961; Tippett and Abbott, 1968; Koike and Gillaspie, 1976) in the U.S. In addition to the U.S., this SrMV is recorded in India, Japan, and the Philippines (loc. cit. Shukla et al., 1994).

Host Range and Symptoms

Natural hosts of SrMV include sugarcane (Abbott, 1961; Tippett and Abbott, 1968; Koike and Gillaspie, 1976) and sorghum (Giorda et al., 1986). Symptom types occurring in sorghum plants depend greatly on the host variety and virus strain. Thus, the yellow-green short streaks most frequently appearing in sorghum plants quickly turn reddish-brown and become necrotic so that the leaves exhibit a reddish-brown necrosis. Some sorghum cvs. react solely with mosaic symptoms (Koike and Gillaspie, 1976; Giorda et al., 1986; Tošić et al., 1990).

The SrMV causes mild mottling in sugarcane, which delays growth and displays excessive tillering. Intensity of mottling usually decreases with aging of leaves. Strain I of this virus causes chlorosis and necrosis of young plants of C.P.31-294 sugarcane cv. In older leaves of the same sugarcane variety, chlorotic and necrotic streaks coalesce into broad, reddish stripes appearing frequently only on one leaf surface. Mottling is the only symptom that occurs in the C.P. 31-588 sugarcane cv. (Abbott, 1961; Tippett and Abbott, 1968; Koike and Gillaspie, 1976). Several sorghum and Sudan grass cvs. also susceptible to SrMV display various symptoms, whereas maize mostly reacts with mosaic symptoms. Johnsongrass and oat are not susceptible to SrMV (Tippett and Abbott, 1968; Koike and Gillaspie, 1976; Tošić et al., 1990).

Pathogenesis and Control

No other natural perennial host is known except sugarcane, in which SrMV infectivity is preserved overseason or between crops. SrMV is transmitted by plant aphids in nature, of which *Dactynotus ambrosiae* is a highly efficient vector (Koike and Gillaspie, 1976). As a potyvirus, SrMV is probably nonpersistent, but no data are available. SrMV transmission through seed has not been established.

Virus control measures should be directed toward selecting or developing resistant cvs. and destruction of the sources of infection (i.e., its perennial hosts and control of vectors).

Sorghum Stunt Mosaic Virus (SrSnMV)

Sorghum stunt mosaic virus was described as an apparently new disease causing serious economic losses in the Imperial Valley of California. The main symptoms in infected sorghum plants include chlorosis and streaking of leaves, as well as severe stunting of plants.

Bullet-shaped particles observed in infected plant cells are 63×95 nm in leaf dip preparations and 68×200 nm in cell viral aggregates (Mayhew, 1981). Symptomatological and morphological traits plus cytological changes suggest that SrSnMV is related to maize mosaic virus (a Rhabdovirus) (Mayhew and Flock, 1979; Mayhew, 1981). Its identity must be confirmed.

The leafhopper *Graminella sonora* is the vector that transmits SrSnMV onto maize, sorghum, and wheat plants. Acquisition of SrSnMV particles requires less than 1 day and the incubation period averages 11 days (Mayhew, 1981). Conditions favorable for vector development also influence the extent of disease spreading in nature. SrSnMV has not been transmitted mechanically with infected plant sap and other modes of transmission are also unknown.

Some Other Virus Diseases of Sorghum

Sorghum Yellow Banding Virus (SrYBaV)

Isometric viral particles, 25 nm in diameter, contain a single species of single-stranded RNA of 1500 kDa and a single capsid protein of about 29 kDa MW (Klassen and Falk, 1989). SrYBaV particles isolated from Sudan grass (*Sorghum sudanense*) plants infect only a few grasses (graminaceous plants).

Sorghum Chlorotic Spot Virus (SrCSptV)

According to their particle traits, SrCSptV belongs to the Potyvirus group. Virus hosts include some graminaceous and non-graminaceous plants. *Nicotiana benthamiana* is a suitable assay plant (Langenberg et al., 1989).

Sorghum: Natural Alternative Host for Some Viruses

Maize Dwarf Mosaic Virus (MDMV)

MDMV occurs widely in all sorghum production regions on all five continents (Toler, 1984). Natural virus hosts, *inter alia*, include all types of cultivated sorghum, Sudan grass, and Johnsongrass. Maize and sorghum are the most important natural hosts of MDMV (see the chapter on maize virus disease for detailed information).

FIGURE 27 Sorghum leaf mottling caused by maize dwarf mosaic virus strain A. (Courtesy R. Toler. Reproduced with permission from APS.)

Mosaic as a type of sorghum disease was first discovered in America in 1919 (Brandes, 1919; Brandes and Klaphaak, 1923). In Europe, it was first described in the sweet sorghum crop in Italy (Goidanich, 1939) and later also in other countries (Lovisolo and Aćimović, 1961; Šutić, 1983). MDMV is the most widely spread sorghum virus today. Mijavec (1989) studied symptom occurrence in 750 sorghum genotypes under field conditions and established that 719 (96.9%) displayed dwarf mosaic symptoms.

MDMV causes economically important damage in the production of all cultivated sorghum types. In grain sorghum production, MDMV can reduce grain yields from 15% (Toler, 1985) to 25% (Henzell et al., 1979), plus substantial reductions of the yields of silage. Dry matter and sugar content in particular can be significantly decreased in infected sweet sorghum plants (Zummo et al., 1970). Heads of infected broom corn plants are of poor quality, with yield losses up to 50% (Tošić and Mijavec, 1991).

MDMV causes systemic infection in sorghum plants. In some cases, depending on virus strain and sorghum genotype, the infection begins with the appearance of local reactions (spots) on inoculated leaves, later resulting in systemic infection of newly formed leaves (Tošić and Malak, 1973).

Systemic symptoms of MDMV infection on sorghum leaves differ with the large number of sorghum genotypes and the variability of virus isolates (Figure 27). Mijavec (1989) described four main symptom types on sorghum leaves: mosaic, red leaf, tan stripe, and red stripe. The sequence of disease symptoms on 750 sorghum genotypes in the field were recorded as follows: red stripe (33%), mosaic (27%), red leaf (25%), tan stripe (11%), and healthy—no symptoms (4%).

Some sorghum genotypes express infection by MDMV with panicle necrosis, small seed, and grain shriveling in infected heads, while others are stunted, partially or completely sterile, and die early in the season.

FIGURE 28   Symptoms on sorghum caused by maize streak virus. (Courtesy V. Damsteegt. Reproduced with permission from APS.)

Dwarf mosaic pathogenesis in sorghum is similar as in maize. The most frequent natural sources of infection are in perennial Johnsongrass plants. MDMV retains infectivity permanently in stolons and rhizomes underground, wherefrom each year it is newly transmitted to healthy sorghum plants and other natural hosts with the aid of vectors. Several active aphids (most frequently *Rhopalosiphum maidis*, *Myzus persicae*, *Schizaphis graminum*) act as vectors transmitting MDMV nonpersistently.

Preventive measures—such as the elimination of Johnsongrass, which is the most important source of infection plus chemical control of the aphid vector—are recommended to protect sorghum crops. Since these measures are insufficient due to the omnipresence of Johnsongrass and intensive vector activity, breeding cultivars and hybrids resistant both to MDMV and to vector feeding is a highly recommended method of sorghum protection.

Several sources of genetic resistance of sorghum to MDMV are known (Mijavec, 1989). A monogenic form of resistance derived from Johnsongrass (*Sorghum halepense*) was introduced into QL sorghum genotypes in Australia (Teakle and Pritchard, 1971; Teakle et al., 1972; Persley et al., 1972; Persley and Greber, 1982). Another multigenic (polygenic) form of resistance was discovered in the Australian line Q 7539 derived from a sorghum cultivar from Nigeria (Persley et al., 1972; Toler, 1985) and introduced into the CMS and QL of the A and the B lines. Field resistance was discovered in six grain sorghum lines in 1975 (Toler, 1985).

Among 750 sorghum genotypes tested under conditions of natural infection by MDMV, Mijavec (1989) found 11 resistant genotypes featuring Krish (monogenic) resistance determined to be dominant. Krish resistance was transmitted into several sorghum lines and hybrids, which should result in more successful sorghum production in the future.

Other maize virus diseases in sorghum include maize chlorotic dwarf virus (MCDV), maize chlorotic mottle virus (MCMtV), maize mosaic virus (MMV), maize streak virus (MSkV) (Figure 28), maize stripe virus (MSpV) (Figure 29), and maize vein enation virus (MVEnV).

Barley virus diseases in sorghum are barley yellow dwarf virus (ByYDV) (in some data for plant infection with artificial inoculation) and barley yellow striate virus (ByYStV).

Oat virus diseases in sorghum include oat mosaic virus (OtMV) and oat pseudo-rosette virus (OtPdRoV).

FIGURE 29   Symptoms on sorghum leaf caused by maize stripe virus. (Courtesy R. Greber. Reproduced with permission from APS.)

Sugarcane virus diseases in sorghum are sugarcane chlorotic streak virus (ScCSkV), sorghum mosaic virus (SrMV) formerly sugarcane mosaic virus (ScMV) (see Figure 11), and Fiji disease virus (FjDsV).

Viruses from other hosts in sorghum include brome mosaic virus (BrMV), cucumber mosaic virus (CMV), guinea mosaic virus (GuMV), Johnsongrass mosaic virus (JgMV), panicum mosaic virus (PcMV) (plant infection with artificial inoculations), peanut clump virus (PtCpV), and rice stripe virus (RSpV).

Details of these viruses are found in the chapters on their main hosts.

REFERENCES

Abbott, E.V. 1961. A new strain of sugarcane mosaic virus. (Abstr.) *Phytopathology* 51:621.
Afanasiev, M.M. 1956. Occurrence of barley stripe mosaic in Montana. *Pl. Dis. Reptr.* 40:142.
Agarkov, V.A. 1956. Virus diseases of wheat in the Vinnitsa district. *Plant Prot. Lening.* 31–34 (Abstr.: RAM 37:345, 1958).
Agarkov, V.A. 1969. Virusnaja tamnozelenaja karlikovost ovsa i rvzhi. *Zaschtschita rast.* 14(4):22–23.
All, J.N. et al. 1976. The changing status of corn virus diseases: potential value of a systemic insecticide. *Ga. Agric. Res.* 17:4–6, and 23.
All, J.N. et al. 1977. Influence of no-tillage cropping, carbofuran, and hybrid resistance of dynamics of maize chlorotic dwarf and maize dwarf mosaic diseases of corn. *J. Econ. Entom.* 70:221–225.
Anonymous. 1992. Northrup King and Sandoz Agro genetically engineer MDMV resistant corn. *Genetic Engineering News* Oct. 1, p. 28.

Anzola, S.D. et al. 1980. Evaluacion de hibridos de maiz dulce (*Zea mays* L.) tolerantes a la enfermedad mosaico enano del maiz (VMEM). *Agron. Tropical* 30 (1/6):17–28.
Appiano, A. and Conti, M. 1974. Some observations on barley yellow striate mosaic virus (BYSMV) ultrastructure. *J. Submicroscop. Cytol.* 6:103.
Atabekov, J.G. and Novikov, V.K. 1989. Barley stripe mosaic virus. *CMI/AAB Descriptions of Plant Viruses*, No. 344.
Atanasoff, D. 1965. Leaf fleck disease of maize and its possible relation in cytoplasmic inheritance. *Phytopathol. Z.* 52:89–95.
Atanasoff, D. 1966. Maize leaf fleck disease. *Phytopathol. Z.* 56:25–33.
Atanasov, D. and Dodov, D. 1961. Virusnie bolesti po žitnite (Virus diseases of the Gramineae). *Rast. Zast (Rast. Zasht.)* 9:13–19 (Abstr. RAM 40:671).
Bakker, W. 1974. *Characterization and Ecological Aspects of Rice Yellow Mottle Virus in Kenya*, Wageningen, Netherlands, Centre Agric. Publ. and Docum. Agric. Res. Reports 829, 152 pp.
Bakker, W. 1975. Rice yellow mottle virus. *CMI/AAB Descriptions of Plant Viruses*, No. 149.
Bancroft, J.B. et al. 1966. Some biological and physical properties of a midwestern isolate of maize dwarf mosaic virus. *Phytopathology* 56:474–478.
Banttari, E.E. and Zeyen, R.J. 1973. Oat blue dwarf virus. *CMI/AAB Descriptions of Plant Viruses*, No. 123.
Banttari, E.E. and Zeyen, R.J. 1970. Transmission of oat blue dwarf virus by the aster leafhopper following natural acquisition or inoculation. *Phytopathology* 60:399–402.
Barker, I. 1990. Barley yellow dwarf in Britain. In: Barnett, P.A., Ed. *1990—World Perspectives on Barley Yellow Dwarf*, 39–44, CIMMYT, Mexico, D.F., Mexico.
Baumgarten, G. von and Ford, R.E. 1981. Purification and partial characterization of maize dwarf mosaic virus strain A. *Phytopathology* 71:36–41.
Belli, G. et al. 1986. Purification and serological identification of an isolate of barley yellow dwarf virus causing rice giallume. *Rivista di Patologia Vegetale* 22:81–86.
Bellingham et al. 1957. Resistance to wheat streak mosaic virus in foreign and domestic wheats and various wheat crosses. (Abstr.) *Phytopathology* 47:516.
Bennett, C.W. 1959. Lychnis ringspot. *Phytopathology* 49:706–713.
Bennett, C.W. 1969. Seed transmission of plant viruses. *Adv. Virus Res.* 14:221–261.
Berger, P.H. et al. 1989. Properties and *in vitro* translation of maize dwarf mosaic virus RNA. *J. Gen. Virol.* 70:1845–1851.
Boccardo, G. and Milne, R.G. 1975. The maize rough dwarf virus. I. Protein composition and distribution of RNA in different viral fractions. *Virology* 68:79–85.
Boccardo, G. and Milne, R.G. 1980. Oat sterile dwarf virus. *CMI/AAB Descriptions of Plant Viruses*, No. 217.
Boccardo, G. and Milne, R.G. 1980a. Electrophoretic fractionation of the double-stranded RNA genome of rice ragged stunt virus. *Intervirology* 14:57–60.
Boccardo, G. and Milne, R.G. 1984. Plant Reovirus group. *CMI/AAB Descriptions of Plant Viruses*, No. 294.
Bock, K.R. 1974. Maize streak virus. *CMI/AAB Descriptions of Plant Viruses*, No. 133.
Bock, K.R. et al. 1974. Purification of maize streak virus and its relationship to viruses associated with streak disease of sugarcane and *Panicum. Ann. Appl. Biol.* 77:289–296.
Bradfute, O.E. et al. 1972. Tissue ultrastructure, sedimentation and leafhopper transmission of a virus associated with a maize dwarfing disease. (Abstr.) *J. Cell Biol.* 55:25a.
Bradfute, O.E. et al. 1972a. Isometric virus-like particles in maize with stunt symptoms. (Abstr.) *Phytopathology* 62:748.
Brakke, M.K. 1971. Soil-borne wheat mosaic virus. *CMI/AAB Descriptions of Plant Viruses*, No. 77.
Brakke, M.K. 1971a. Wheat streak mosaic virus. *CMI/AAB Description of Plant Viruses*, No. 48.
Brakke, M.K. 1979. Ultraviolet absorption spectra and difference spectra of barley stripe mosaic and tobacco mosaic viruses in buffer and sodium dodecyl sulfate. *Virology* 98:76–87.
Brandes, E.W. 1919. The mosaic disease of sugarcane and other grasses. *U.S. Dept. Agric. Tech. Bull.* 829.
Brandes, E.W. and Klaphaak, P.J. 1923. Cultivated and wild hosts of sugarcane or grass mosaic. *J. Agric. Res.* 24:257–262.
Brčak, L. 1979. Leafhopper and planthopper vectors of plant disease agents in central and southern Europe. In: Maramorosch, K. and Harris, K.F., Eds. *Leafhopper Vectors and Plant Disease Agents*, Academic Press, New York, p. 97–154.
Bridgmon, G.H. 1951. Gladiolus as a virus reservoir. *Phytopathology* 57:5.

Brown, J.K. et al. 1984. Irrigated corn as a source of barley yellow dwarf virus and vector in eastern Washington. *Phytopathology* 74:46–49.

Brysgalowa, W.A. 1945. Die Pseudoresettenkrankheit des Hafers und ihre Bekämpfung. Irkutsk (Russian).

Canova, A. 1964. Ricerche sulle malattie da virus delle Graminaceae. I. Mosaico del frumento trasmissibile attraverso il terreno. *Phytopath. Medit.* III:86–94.

Canova, A. 1966. Ricerche sulle malattie da virus delle Graminaceae. III. *Polymyxa graminis* Led. vettore del virus del mosaico del frumento. *Phytopathol. Medit.* V:53–58.

Carroll, T.W. 1980. Barley stripe mosaic virus: its economic importance and control in Montana. *Plant Disease* 64:136–140.

Conti, M. 1969. Investigations on a bullet-shaped virus of cereals isolated in Italy from planthoppers. *Phytopathol. Z.* 66:275–279.

Conti, M. 1972. Barley yellow striate mosaic virus isolated from plants in the field. *Phytopathol. Z.* 73:39–45.

Conti, M. and Plumb, R.T. 1977. Barley yellow striate mosaic virus in the salivary glands of its planthopper vector *Laodelphax striatellus* Fallen. *J. Gen. Virol.* 34:107–114.

Conti, M. 1980. Vector relationships and other characteristics of barley yellow striate mosaic virus (BYSMV). *Ann. Appl. Biol.* 95:83–92.

Conti, M. 1983. Maize viruses and virus diseases in Italy and other Mediterranean countries. In: Gordon, D.T. et al., Eds. *Proc. Int. Maize Dis. Colloq. and Workshop*, Aug. 2–6, 1982, The Ohio State Univ., Ohio Agr. Res. Dev. Center, Wooster, OH, p. 103–112.

Conti, M. 1988. Barley yellow striate mosaic virus. In: Smith, I.M. et al., Eds. *European Handbook Plant Diseases*, Blackwell Sci. Publ., London, p. 76.

Conti, M. et al. 1990. The "Yellow Plague" of cereals, barley yellow dwarf virus. In: Burnett, P.A., Ed. *World Perspectives on Barley Yellow Dwarf*, CIMMYT, Mexico, D.F. Mexico, p. 1–6.

Cooper, J.I. 1988. Brome mosaic virus (BMV). In: Smith, I.M. et al., Eds. *European Handbook Plant Diseases*, Blackwell Sci. Publ., London, p. 69.

Corbetta, G. 1967. La nuova malattia che colpisce il riso. *Il Risicoltore* 11(8):3.

Cornuet, P. 1987. *Eléments de Virologie Végétale*, INRA, Paris.

Damsteegt, V.D. 1981. Exotic virus and viruslike diseases of maize. In: Gordon, D.T. et al., Eds. Virus and viruslike diseases of maize in the United States. *South. Coop. Series Bull.* 247: June, p. 110–123.

D'Arcy, D.J. and Burnett, P.A. 1995. *Barley Yellow Dwarf: 40 Years of Progress*, APS Press, St. Paul, MN, 374 pp.

Derrick, K.S. 1975. Serological relationships among strains of sugarcane mosaic virus. *Proc. Am. Phytopathol. Soc.* 2:42.

De Zoeten, G.A. and Reddick, B.B. 1984. Maize white line mosaic virus. *CMI/AAB Descriptions of Plant Viruses*, No. 283.

Dowson, G.W. et al. 1986. Antifeedants: a new concept for control of BYDV in winter cereals. *Br. Crop Prot. Conf. Pests and Dis.: Proc.* 3:1001–1008.

Eslick, R.F. 1953. Yield reductions in Glacier barley associated with a virus infection. *Pl. Dis. Reptr.* 37:290–291.

Everett, T.R. and Lamey, H.A. 1969. Hoja Blanca. In: Maramorosch, K., Ed. *Viruses, Vectors and Vegetation*, New York, Interscience, p. 361–377.

Fenaroli, I. 1949. Il nanismo del mais. *Notiz. Malatt. Piante* 3:38–39.

Fenner, F. and Gibbs, A. 1983. Cryptograms—1982. *Intervirology* 19:121–128.

Ferro, D.N. et al. 1980. Effect of mineral oil and a systemic insecticide on field spread of aphid-borne maize dwarf mosaic virus in sweet corn. *J. Econ. Entom.* 73:730–735.

Finley, A.M. 1957. Wheat streak mosaic, a disease of sweet corn in Idaho. *Pl. Dis. Reptr.* 41:589–591.

Findley, W.R. et al. 1977. Breeding corn for resistance to virus in Ohio. In: Williams, L.E. et al., Eds. *Proc. Maize Virus Dis. Colloq. Workshop.* Ohio Agric. Res. Dev. Cent., Wooster, OH, p. 123–127.

Ford, R.E. et al. 1970. New hosts and serological identity of bromegrass mosaic virus from South Dakota. *Pl. Dis. Reptr.* 54:191–195.

Ford, R.E. and Tošić, M. 1972. New hosts of maize dwarf mosaic virus and sugarcane mosaic virus and a comparative host range study of viruses infecting corn. *Phytopathol. Z.* 75:315–348.

Ford, R.E. et al. 1989. Maize dwarf mosaic virus. *CMI/AAB Descriptions of Plant Viruses*, No. 341.

Fox, P.N. and Tola, J. 1990. Barley yellow dwarf in the Andean countries of South America. In: Burnett, P.A., Ed. *World Perspectives on Barley Yellow Dwarf.* CIMMYT, Mexico, D.F., Mexico, p. 25–28.

Francki, R.I.B. et al. 1981. Rhabdoviruses. In: Kurstak, E., Ed. *Hand. Pl. Virus Infec. Comp. Diag.* Elsevier/North-Holland Biomedical Press, p. 455–489.

Frederiksen, R.A. 1986. *Compendium of Sorghum Diseases.* APS Press, St. Paul, MN, 82 pp.

Fujii, S. 1975. Thesis, University of Tokyo.

Fujikawa et al. 1972. *Ann. Phytopathol. Soc. Japan* 38:213.

Futrell, M.C. and Scott, G.E. 1969. Effect of maize dwarf mosaic virus infection on invasion of corn plants by *Fusarium moniliforme. Pl. Dis. Reptr.* 53:600–602.

Galvez, G.E. 1967. Frecuencia de *Sogata oryzicola* and *S. cubana* en campos de arroz y *Echinochloa* en Colombia. *Agr. Trop., Bogota, Colombia* 23:384–389.

Galvez, G.E. 1971. Rice tungro virus. *CMI/AAB Descriptions of Plant Viruses,* No. 67.

Gámez, R. 1980. Maize rayado fino virus. *CMI/AAB Descriptions of Plant Viruses,* No. 220.

Gibbs, A.J. et al. 1963. The relationship between barley stripe mosaic and lychnis ringspot viruses. *Virology* 20:194–198.

Gildow, F.E. et al. 1983. Identification of double-stranded RNAs associated with barley yellow dwarf virus infections of oats. *Phytopathology* 73:1570–1572.

Gildow, F.E. 1990. Current status of barley yellow dwarf in United States: a regional situation report. In: Burnett, P.A., Ed. *1990—World Perspectives on Barley Yellow Dwarf,* CIMMYT, Mexico, D.F., Mexico, p. 11–20.

Gill, C.C. 1969. Annual variation in strains of barley yellow dwarf virus in Manitoba and the occurrence of green-bug-specific isolates. *Can. J. Bot.* 47:1277–1283.

Gill, C.C. 1976. Oat necrotic mottle virus. *CMI/AAB Descriptions of Plant Viruses,* No. 169.

Giorda, M. L. et al. 1986. Identification of sugarcane mosaic virus strain H isolate in commercial grain sorghum. *Plant Dis.* 70:624–628.

Gingery, R.E. et al. 1978. Maize chlorotic dwarf virus. *CMI/AAB Descriptions of Plant Viruses,* No. 194.

Gingery, R.E. 1981. Maize chlorotic dwarf virus. In: Kurstak, E., Ed. *Hand. Pl. Virus Infec. Comp. Diag.,* Elsevier/North Holland Biomedical Press, p. 19–21.

Gingery, R.E. 1985. Maize stripe virus. *CMI/AAB Descriptions of Plant Viruses,* No. 300.

Goidanich, G. 1939. Le piu importanti malattie del sorgo, con speciale riferimento a quelle del sorgo zuccherino. *Industr. Saccar. Ital.* 32:77–102 and 166–168.

Goodman, R.M. 1981. Geminiviruses. In: Kurstak, E., Ed. *Hand. Pl. Virus Infec. Comp. Diag.* Elsevier/North-Holland Biomedical Press, p. 879–910.

Gordon, D.T. and Nault, L.R. 1977. Involvement of maize chlorotic dwarf virus and other agents in stunting diseases of *Zea mays* in the United States. *Phytopathology* 67:27–30.

Gordon, D.T. et al. 1981. Introduction: history, geographical distribution, pathogen characteristics and economic importance. In: Gordon, D. T. et al., Eds. *Virus and Virus-like Dis. Maize United States, South. Coop. Ser. Bull. 247,* June, OARDC, OH, p. 1–12.

Gordon, D.T. et al. 1983. Introduction. In: Gordon, D. T. et al., Eds. *Proc. Int. Maize Virus Dis. Colloq. Workshop,* August 2–6, 1982. The Ohio State University, Ohio Agric. Res. Dev. Center, Wooster, p. v-x.

Gordon, D.T. et al. 1984. Maize chlorotic mottle virus. *CMI/AAB Descriptions of Plant Viruses,* No. 284.

Gorter, G.J.M.A. 1959. Breeding maize for resistance to streak. *Euphytica* 8:234–240.

Grancini, P. 1957. Un mosaico del mais e del sorgo in Italia. *Maydica* 2:83–104.

Greber, R.S. 1976. Rhabdovirus in Queensland grasses and cereals. *Aust. Pl. Pathol. Soc. Newsl.* 5:Suppl. Abstr. 228.

Greber, R.S. 1977. Cereal chlorotic mottle virus (CCMV) a rhabdovirus of Gramineae transmitted by the leafhopper *Nesoclutha pallida. Aust. Pl. Pathol. Soc. Newsl.* 6:17.

Greber, R.S. 1977a. A severe stunting virus disease of maize in Queensland. *Aust. Pl. Pathol. Soc. Newsl.* 6:18.

Greber, R.S. 1979. Cereal chlorotic mottle virus—a rhabdovirus of Gramineae in Australia transmitted by *Nesoclutha pallida* (Evans.). *Aust. J. Agr. Res.* 30(3):433–443.

Greber, R.S. 1981. Ecological aspects of cereal chlorotic mottle virus. *Aust. Pl. Pathol. Soc. Newsl.* 10(2):29–30.

Greber, R.S. 1982. Cereal chlorotic mottle virus. *CMI/AAB Descriptions of Plant Viruses,* No. 251.

Greber, R.S. 1982a. Maize sterile stunt—a delphacid transmitted rhabdovirus disease affecting some maize genotypes in Australia. *Aust. J. Agr. Res.* 33(1):13–23.

Gumpf, D.J. 1971. Purification and properties of soil-borne wheat mosaic virus. *Virology* 43:588–596.

Haber, S. 1990. Situation review of barley yellow dwarf in Canada. In: Burnett, P.A. Ed. *World Perspectives on Barley Yellow Dwarf,* p. 7–10.

Hamilton, R.I. 1965. An embryo test for detecting seed-borne barley stripe mosaic virus in barley. *Phytopathology* 55:798–799.
Hagborg, W.A.F. 1954. Dwarfing on wheat and barley by the barley stripe-mosaic (false stripe) virus. *Can. J. Bot.* 32:24–37.
Hariri, D. and Lapierre, H. 1988. Oat soil-borne stripe virus (OSBSV). In: Smith, I.M. et al., Eds. *European Hand. Pl. Dis.*, Blackwell Sci. Publ., London, p. 96.
Harpaz, I. and Applebaum, S.W. 1961. Accumulation of asparagine in maize plants infected by maize rough dwarf virus and its significance in plant virology. *Nature (London)* 192:780–781.
Harpaz, I. et al. 1965. Indagini comparative su *Javesella pellucida* (Fabricius) e *Laodelphax striatellus* (Fallen) quali vettori del virus del namismo ruvid"o del Mais (Maize rough dwarf virus). *Atti Accad. Sci. Torino* 99:885.
Harpaz, I. and Klein, M. 1969. Vector-induced modifications in a plant virus. *Ent. Ann. Appl.* 12:99–106.
Harpaz, I. 1972. *Maize Rough Dwarf Virus*, Israel University Press, Jerusalem.
Heagy, J. and Rochow, W.F. 1965. Thermal inactivation of barley yellow dwarf virus. *Phytopathology* 55:809–810.
Hebert, T.T. and Panizo, C.H. 1975. Oat mosaic. *CMI/AAB Descriptions of Plant Viruses*, No. 145.
Henry, M. 1987. Kinetics of barley yellow dwarf infections in maize. In: Burnett, P.A., Ed. *1990: World Perspectives on Barley Yellow Dwarf*, CIMMYT, Mexico, D.F., Mexico, p. 169–171.
Henzell, R.G. et al. 1979. The effect of sugarcane mosaic virus on the yield of eleven grain sorghum (*Sorghum bicolor*) cultivars. *Aust. J. Exp. Agric. and Anim. Husb.* 19(97):225–232.
Herold, F. 1972. Maize mosaic virus. *CMI/AAB Descriptions of Plant Viruses*, No. 94.
Hibi, T. et al. 1984. Double-stranded RNA of rice gall dwarf virus. *J. Gen. Virol.* 65:1585–1590.
Hibino, H. 1979. Rice ragged stunt, a new virus disease occurring in tropical Asia. *Rev. Pl. Prot. Res.* 12:98–110.
Hibino, H. et al. 1979. Reovirus-like particles associated with rice ragged stunt diseased rice and insect vector cells. *Ann. Phytopathol. Soc. Japan* 45:228–239.
Hibino, H. 1986. Rice grassy stunt virus. *CMI/AAB Descriptions of Plant Viruses*, No. 320.
Hill, J.H. et al. 1974. Seed transmission of maize dwarf mosaic and wheat streak mosaic viruses in maize and response of inbred lines. *Crop Sci.* 14:232–235.
Hill, A.S. 1980. Barley yellow mosaic in England. (Abstr.) *3rd Conf. Virus Dis. Gramineae Europe*, Rothamsted, May 28–30.
Hollings, M. and Brunt, A.A. 1981. Potyviruses. In: Kurkstak, E., Ed. *Hand. Pl. Virus Infec. Comp. Diag.*, Elsevier/North-Holland Biomedical Press, p. 731–807.
Hooper, G.R. and Wiese, M.V. 1972. Cytoplasmic inclusions in wheat affected by wheat spindle streak mosaic. *Virology* 47:664–672.
Hoppe, W. 1969. Badania nad wiroza psenici i ryta powodujaca paskowana mozaika na lisciach. *Ochrony Roslin* 44:101–108.
Hoppe, W. 1974. Researches on the virose of wheat caused by winter wheat mosaic virus. *Pr. Nauk. Inst. Ochr. Rosl. Poznan* 16:178–180.
Hsieh, S.P.Y., et al. 1970. Transmission of rice transitory yellowing virus by *Nephotettix impicticeps*. (Abstr.) *Phytopathology* 60:1534.
Huth, W. 1980. The occurrence of barley yellow mosaic virus in Germany. (Abstr.) *3rd Conf. Virus Dis. Gramineae Europe*, Rothamsted, May 28–30.
Huth, W. 1982. Evaluation of sources of resistance to barley yellow mosaic virus in winter barley. *Zeitschr. Pflanzenzüchtung* 89:154–164.
Huth, W. 1984. Die Gelbmosakvirose der Gerste in der Bundesrepublik Deutschland—Beobachtungen seit 1978. *Nachrichtenblatt des Deutschen Pflanzenschutzdienst* 36:49–55.
Huth, W. 1988. Barley yellow mosaic virus (BarYMV). In: Smith. I.M. et al., Eds. *European Hand. Pl. Dis.*, Blackwell Sci. Publ., London, p. 89–91.
Iida, T.T. et al. 1972. Rice dwarf virus. *CMI/AAB Descriptions of Plant Viruses*, No. 102.
Ikata, S. and Kawai, I. 1940. Studies on the wheat yellow mosaic. *Bull. Japan Min. Agr. Forest* 154:1–123.
Ikegami, M. and Francki, R.I.B. 1976. RNA-dependent RNA polymerase associated with subviral particles of Fiji disease virus. *Virology* 70:292–300.
Inouye, T. 1976. Wheat yellow leaf virus. *CMI/AAB Descriptions of Plant Viruses*, No. 157.
Inouye, T. et al. 1973. A new virus of wheat, barley and several other plants in Gramineae, wheat yellow leaf virus. *Nogaku Kenkyu* 55:1–15.

Inouye, T. and Fujii, S. 1977. Rice necrosis mosaic virus. *CMI/AAB Descriptions of Plant Viruses*, No. 172.
Inouye, T. and Omura, T. 1982. Transmission of rice gall dwarf virus by the green rice leafhopper. *Plant Dis.* 66:57–59.
Inouye, T. and Saito, Y. 1975. Barley yellow mosaic virus. *CMI/AAB Descriptions of Plant Viruses*, No. 143.
Ito, S. and Fukushi, T. 1944. Studies on northern cereal mosaic. *J. Sapporo Soc. Agric. For.* 36:62–89.
Ivanović, M. 1979. Uticaj virusnog mozaika kukuruza na osetljivost kukuruza prema gljivi Ustilago maydis D.C. Corda. *Zaštita bilja* 148:135–140.
Iwasaki, M. et al. 1980. Influence of rice grassy stunt disease on the yield components. *Proc. Assoc. Pl. Prot. Kyushu* 26:7–9.
Jackson, A.O. and Brakke, M.K. 1973. Multicomponent properties of barley stripe mosaic virus ribonucleic acid. *Virology* 55:483–494.
Jackson, A.O. et al. 1981. Rhabdovirus diseases of the Gramineae. In: Gordon, D. T. et al., Eds. *Virus and Virus-like Dis. Maize United States. South. Coop. Series Bull. 247*, OARDC, Wooster, Ohio, p. 51–76.
Jackson, A.O. and Lane, L.C. 1981. Hordeoviruses. In: Kurstak, E., Ed. *Hand. Pl. Virus Infec. Comp. Diag.*, Elsevier/North-Holland Biomedical Press, p. 565–625.
Janson, B.F. and Ellett, C.W. 1963. A new corn disease in Ohio. *Pl. Dis. Reptr.* 47:1107–1108.
Jarjees, M.M. and Uyemoto, J.K. 1984. Serological relatedness of strains of maize dwarf and sugarcane mosaic viruses as determined by microprecipitin and enzyme-linked immunosorbent assays. *Ann. App. Biol.* 104:497–501.
Jedlinski, H. 1976. Oat striate mosaic, a new virus disease in Illinois spread by the leafhopper, *Graminella nigrifrons* (Forbes). (Abstr.) *Proc. Am. Phytopathol. Soc.* 3:19.
Jensen, S.G. et al. 1986. Size variation among proteins induced by sugarcane mosaic viruses in plant tissue. *Phytopathology* 76:528–532.
Johnstone, G.R. et al. 1990. Epidemiology and control of Barley Yellow Dwarf in Australia and New Zealand. In: Burnett, P.A., Ed. *World Perspectives on Barley Yellow Dwarf*, CIMMYT, Mexico, D.F., Mexico, p. 228–239.
Jones, R.K. and Tolin, S.A. 1972. Factors affecting purification of maize dwarf mosaic virus from corn. *Phytopathology* 62:812–816.
Kerlan, C. et al. 1974. Observations sur l' apparition du virus de la mosaique nanisante du mais (maize dwarf mosaic) dans le nord de la France. *Ann. Phytopathol.* 6:455–470.
Khush, G.S. and Ling, K.C. 1974. Inheritance of resistance to grassy stunt virus and its vector in rice. *J. Hered.* 65:135–136.
Kimura, I. 1962. Further studies on the rice dwarf virus. *Ann. Phytopathol. Soc. Japan*, 27:197–203.
Kingsland, G.C. 1980. Effect of maize dwarf mosaic virus infection on yield and stalk strength of corn in the field in South Carolina. *Plant Dis.* 64:271–273.
Klaasen, V.A. and Falk, B.W. 1989. Characterization of a California isolate of sorghum yellow banding virus. *Phytopathology* 79:646–650.
Koike, H. and Gillaspie, A.G. Jr. 1976. Strain M, a new strain of sugarcane mosaic virus. *Pl. Dis. Reptr.* 60:50–54.
Knoke, J.K. et al. 1974. Distribution of maize dwarf mosaic and aphid vectors in Ohio. *Phytopathology* 64:639–645.
Kovačevski, I. et al. 1977. *Virusni i mikoplazmozni bolesti po kulturnite rastenija.* Zemizdat, Sofija.
Kurstak, E. 1981. Ed. *Handbook of Plant Virus Infections and Comparative Diagnosis.* Elsevier/North Holland Biomedical Press.
Lane, L.C. 1974. The components of barley stripe mosaic and related viruses. *Virology* 58:323–333.
Langenberg, W.G. and Schroeder, H.F. 1973. Endoplasmic reticulum-derived pinwheels in wheat infected with wheat spindle streak mosaic virus. *Virology* 55:218–223.
Langenberg, W.G. et al. 1989. Sorghum chlorotic spot virus binds to potyvirus cylindrical inclusions in tobacco leaf cells. *Ultrastr. Mol. Str. Res.* 102:47–52.
Lapierre, H. 1988. Cereal tillering disease virus (CTDV). In: Smith, I.M. et al., Eds. *European Hand. Pl. Dis.*, Blackwell Sci. Publ., London, p. 76.
Lapierre, H. and Maroquin, C. 1988. Barley yellow dwarf virus (BYDV). In: Smith, I.M. et al., Eds. *European Hand. Pl. Dis.*, Blackwell Sci. Publ., London, p. 20–21.
Lastra, R.J. 1977. Maize mosaic and other maize virus and virus-like diseases in Venezuela. In: Williams, L.E. et al., Eds. *Proc. Int. Maize Virus Dis. Coll. Workshop*, Ohio Agric. Res. Devel. Center, p. 30–39.

Lee, P.E. 1964. The extraction of wheat striate mosaic virus from diseased wheat plants. *Can. J. Bot.* 41:1617–1621.
Lee, P.E. and Bell, W. 1963. Some properties of wheat striate mosaic virus. *Can. J. Bot.* 41:767.
Lesemann, D. 1972. Electron microscopy of maize rough dwarf virus particles. *J. Gen. Virol.* 16:273–284.
Lindsten, K. 1959. A preliminary report on virus diseases of cereals in Sweden. *Phytopathol. Z.* 4:420–428.
Lindsten, K. 1961. Studies on virus diseases of cereals in Sweden I. On the etiology of a serious disease of oats (the "Bollnas disease"). *Ann. Royal Agr. Coll. Sweden* 27:137–197.
Lindsten, K. 1961a. Studies on virus diseases of cereals in Sweden. II. On virus diseases transmitted by the leafhopper *Calligypona pellucida* (F.). *Ann. Royal Agr. Coll. Sweden* 27:199–721.
Lindsten, K. 1973. Slökornsjuka" pa havre - en ny strasädevirus eller en svagare variant av dvargskottsjuka. *Växtskyddsnotiser* 37:55.
Lindsten, K. et al 1973. Cereal tillering disease in Sweden and some comparisons with oat sterile dwarf and maize rough dwarf. *Natl. Swed. Inst. Pl. Prot.* 15:151, p. 375–397.
Ling, K.C. et al. 1970. A mass screening method for testing resistance to grassy stunt disease of rice. *Pl. Dis. Reptr.* 54:565–569.
Lorenzoni, C. et al. 1987. Tolerance to barley yellow dwarf virus in maize. In: Burnett, P.A., Ed. *1990: World Perspectives on Barley Yellow Dwarf*, CIMMYT, Mexico, D.F., Mexico, pp. 401–403.
Lovisolo, O. 1957. Contributo sperimentale alla conoscenza ed alla determinazione del virus agenta dell arrossamento striato del Sorgo e di un mosaico del Mais. *Bell. Staz. Pat. Veg. Roma, Ser. 3*, 14 (1956), 2, p. 261–321 (Abstr. in RAM 37:232-233).
Lovisolo, O. and Aćimović, M. 1961. Sorghum red stripe disease in Yugoslavia. *FAO Plant Prot. Bull.* 9:99–102.
Lovisolo, O. 1971. Maize rough dwarf virus. *CMI/AAB Descriptions of Plant Viruses*, No. 72.
Luisoni, E. et al. 1973. Serological relationship between maize rough dwarf virus and rice black streaked dwarf virus. *Virology* 52:281–283.
Luisoni, E. et al. 1975. The maize rough dwarf virus II. Serological analysis. *Virology* 68:86–96.
MacFarlane, I. et al. 1968. A soilborne virus of winter oats. *Pl. Pathol.* 17:167–170.
MacFarlane, I. and Plumb, R.T. 1978. A new soil-borne virus of winter oats (oat golden stripe virus—OGSV). (Abstr.) *3rd Int. Cong. Pl. Pathol.*, München, August 16–18.
Mathre, D.E. 1997. *Compendium of Barley Diseases.* APS Press, St. Paul, MN, 90 pp.
Matthews, R.E.F. 1981. *Plant Virology*, 2nd edition. Academic Press, New York.
Matthews, R.E.F. 1982. Classification and nomenclature of viruses: Fourth Rept. Intl. Comm. Taxon. Viruses. *Intervirology* 15:1–119.
Mayhew, D.E. and Flock, R.A. 1979. A rhabdovirus isolated from sorghum in Imperial Valley, CA. (Abstr.) *Phytopathology* 69:917.
Mayhew, D.E. 1981. Sorghum stunt mosaic. *Plant Dis.* 65:84–86.
McDaniel, L.L. and Gordon, D.T. 1985. Identification of a new strain of maize dwarf mosaic virus. *Plant Dis.* 69:602–607.
McKinney, H.H. 1925. A mosaic disease of winter wheat and winter rye. *Bull. U.S. Dept. Agric.*, 1361, 1–10.
McKinney, H.H. 1953. Virus diseases of cereal crops. In: *Yearbook U.S. Dept. Agric.*, p. 250–360.
McKinney, H.H. 1953. New evidence on virus diseases in barley. *Pl. Dis. Reptr.* 37:292–295.
McKinney, H.H. and Greeley, L.W. 1965. Biological characteristics of barley stripe-mosaic virus strains and their evolution. *Tech. Bull. U.S. Dept. Agric.* 1324, 84 pp.
McLean, A.P.D. 1947. *Sci. Bull. S. Afr. Dept. Agr. Tech. For. Un. S. Afr.* 265, 39 pp.
McMillian, W.W. et al. 1962. Hoja blanca transmission studies on rice. *J. Econ. Entomol.* 55:796–797.
McNeal, F.H. et al. 1976. Barley stripe mosaic virus data from six infected spring wheat cultivars. *Pl. Dis. Reptr.* 60:730–733.
Mehrad, M. et al. 1979. RNA in potato leaf roll virus. *FEBS Lett.* 101:169–173.
Mijavec, A. 1989. Proučavanje otpornosti genotipova sirka *Sorghum bicolor* (L.) Moench prema virusu mozaične kržljavosti kukuruza. Doktorska disertacija, Poljoprivredni fakultet Beograd - Zemun, p. 1–230.
Mikel, M.A. et al. 1984. Seed transmission of maize dwarf virus in sweet corn. *Phytopathol. Z.* 110:185–191.
Mikel, M.A. et al. 1984a. Genetics of resistance of two dent corn inbreds to maize dwarf mosaic virus and transfer of resistance into sweet corn. *Phytopathology* 74:467–473.
Milne, R.G. et al. 1973. Partial purification, structure and infectivity of complete maize rough dwarf virus particles. *Virology* 53:130–141.
Milne, R.G. and Lovisolo, O. 1977. Maize rough dwarf and related viruses. *Adv. Virus Res.* 21:267–341.

Milne, R.G. and Luisoni, E. 1977. Serological relationships among maize rough dwarf-like viruses. *Virology* 80:12–20.

Milne, R.G. and Luisoni, E. 1977a. A serological investigation of some maize rough dwarf-like viruses. *Annales de Phytopathol.* 9:337–341.

Milne, R.G. 1980. Does rice ragged stunt virus lack the typical double shell of the reoviridae? *Intervirology* 14:331–336.

Milne, R.G. et al. 1982. Rice ragged stunt virus. *CMI/AAB Descriptions of Plant Viruses*, No. 248.

Milne, R.G. and Conti, M. 1986. Barley yellow striate mosaic virus. *CMI/AAB Descriptions of Plant Viruses*, No. 312.

Milne, R.G. et al. 1986. Serological relationships between the nucleocapsids of some planthopper-borne rhabdoviruses in cereals. *Intervirology* 25:83–97.

Moletti, M. and Osler, R. 1978. Determinazione della resistenza al "gialume" delle più importanti varietà di riso italiane mediante inoculazioni sperimentali con l'afide *Rhopalosiphum padi*. *Il Riso* 27:33–40.

Moletti, M. et al. 1979. Valutazione della resistenza al giallume mediante inoculazioni artificiali con l'afide *Rhopalosiphum padi* in linee di riso migliorate. *Il Riso* 28:53–61.

Moletti, M. et al. 1990. Some agronomic traits affected by rice giallume virus inoculated at two phenological stages in 11 Italian rice varieties. In: Burnett, P.A., Ed. *World Perspectives on Barley Yellow Dwarf*, CIMMYT, Mexico, D.F., Mexico, p. 464–467.

Moline, H.E. 1973. Mechanically transmissible viruses from corn and sorghum in South Dakota. *Pl. Dis. Reptr.* 57:373–374.

Montllor, C.B. and Gildow, F.E. 1986. Feeding responses of two grain aphids to barley yellow dwarf virus-infected oats. *Entomol. Exp. Appl.* 42:63–69.

Morales, F.J. and Niessen, A.I. 1985. Rice hoja blanca virus. *CMI/AAB Descriptions of Plant Viruses*, No. 299.

Morinaka, T. et al. 1982. Transmission of rice gall dwarf virus by cicadellid leafhoppers *Recilia dorsalis* and *Nephotettix nigropictus* in Thailand. *Plant Dis.* 66:703–704.

Mulligan, T.E. 1960. The transmission by mites, host-range and properties of ryegrass mosaic virus. *Ann. Appl. Biol.* 48:575–579.

Nault, L.R. et al. 1973. Semipersistent transmission of leafhopper-borne maize chlorotic dwarf virus. *J. Econ. Entomol.* 66:1271–1273.

Nault, L.R. and Bradfute, O.E. 1977. Reevaluation of leafhopper vectors of corn stunting pathogens. (Abstr.) *Proc. Am. Phytopathol. Soc.* 4:172.

Omura, T. et al. 1985. Location of structural proteins in particles of rice gall dwarf virus. *J. Gen. Virol.* 66:811–815.

Omura, T. and Inouye, H. 1985. Rice gall dwarf virus. *CMI/AAB Descriptions of Plant Viruses*, No. 296.

Osler, R. et al. 1977. Possible ways of controlling rice "giallume." *Ann. Phytopathol.* 9:347–351.

Osler, R. et al. 1980. *Leersia oryzoides*, a natural host and winter reservoir of the rice "giallume" strain of barley yellow dwarf virus. *Phytopathol. Z.* 97:242–251.

Osler, R. 1984. Caratterizzazione biologica di un ceppo del nanismo giallo dell'orzo (BYDV) agente causale del giallume del riso. *Rivista di Patologia Vegetale* 20:3–12.

Osler, R. et al. 1987. Transmission characteristics of a nonspecific isolate of barley yellow dwarf virus isolated from maize. In: Burnett, P.A., Ed. *World Perspectives on Barley Yellow Dwarf—Proc. Intl. Workshop*, July 6-11, Udine, Italy, p. 172–175.

Paliwal, Y.C. 1968. Changes in relative virus concentration in *Endria inimica* in relation to its ability to transmit wheat striate mosaic virus. *Phytopathology* 58:386–387.

Paliwal, Y.C. et al. 1968. Some properties and behaviour of maize mosaic virus in India. *Phytopathology* 58:1682–1684.

Palmer, L.T. et al. 1978. Rice yield losses due to brown planthopper and rice grassy stunt disease in Java and Bali. *Pl. Dis. Reptr.* 62:962–965.

Panarin, I.V. and Dubonosov. T.S. 1975. Iskhodnyi material pshenitsy, ustoichivyi k virusnym bolezyam. *Seleksiya i Semenovodstvo* 3:33–34 (Abstr. *1975 RPP* 54:5365).

Panarin, I.V. and Zabavina, E.S. 1977. Interrelation between Hungarian brome mosaic virus and the vector *Oulema melanopa* L. (Vzaimootnosheniya virusa mozaiki kostra bezostogo s perenoschikom Oulema malanopa L. *Sb. Nauch. Tr. Krasnodar. NII S.Kh.* (1977). No. 13, 156–157. From *Referativnyi Zhurnal, Biologiya* (1978) 1B 364.

Panić, M. and Tošić, M. 1969. Uticaj virozne crtičavosti na brojnost ektodezmi listova pšenice. *Zaštita bilja (Plant Protection) Beograd* 105:183–191.

Panić, M., Tošić, M. and Djurdjević, M. 1978. Uticaj prethodne infekcije virusom mozaika na stepen osetljivosti kukuruza prema gljivi *Helminthosporium turcicum*. (Abstr.) *Glasnik zaštite bilja* 10/11:315–316.

Panjan, M. 1960. Virusni mozaik kukuruza u Jugoslaviji. *Zaštita bilja (Plant Protection), Beograd* 62:3–8.

Panjan, M. 1966. About some manifestations of mosaic on corn in Yugoslavia. *Rev. Roum. Biol. Botanique* 11:159–162.

Persley, D.M. et al. 1972. Research notes: a new source of mosaic resistance in sorghum. *Austr. Pl. Pathol. Soc. Newsl.* 1:11–12.

Persley, D.M. and Greber, R.S. 1982. The reaction of sorghum genotypes to natural infection by sugarcane mosaic virus—Johnsongrass strain in Australia. *Sorghum Newslett.* 25:105.

Pešić, Z. et al. 1977. Influence of maize mosaic virus on total oil content and fatty acids ratio in corn grain. *Ann. Phytopathol.* 9:357–360.

Peters, D. 1981. Plant rhabdovirus group. *CMI/AAB Descriptions of Plant Viruses*, No. 244.

Pirone, T.P. et al. 1972. Virus-like particles associated with a leafhopper-transmitted disease of corn in Kentucky. *Plant Dis. Reptr.* 56:652–656.

Pitre, H.N. 1968. A preliminary study of corn stunt vector populations in relation to corn planting dates in Mississippi. Notes on disease incidence and severity. *J. Econ. Entomol.* 61:847–849.

Plumb, R.T. 1974. Problems and possibilities for control of BYDV in Britain. *Mikrobiologija* 11:41-47.

Pop, I. et al. 1966. Cercetari asupra rezistentei liniilor si hibrizilor de Porumb la mozaicul porumbului. *An. Inst. Cercet. Prot. Plant* 4:14–21.

Proeseler, G. 1986. Wheat soil-borne mosaic virus (WSBWV). In: Smith, I.M. et al., Eds. *European Hand. Pl. Dis.*, Blackwell Sci. Publ., Oxford, G.B., p. 98.

Proeseler, G. 1986a. Wheat streak mosaic virus (WSMV). In: Smith, I.M. et al., Eds. *European Hand. Pl. Dis.*, Blackwell Sci. Publ., Oxford, p. 99.

Pruša, V. 1958. Die sterile Verwerkung des Hafers in der Tschechlowakischen Republik. *Phytopathol. Z.* 33:99–107.

Pruša, V. et al. 1959. Oat sterile-dwarf virus disease. *Biol. Plantarum* 1:223–234.

Qualset, C.O. 1990. Genetics of host plant resistance to barley yellow dwarf virus. In: Burnett, P.A., Ed. *1990—World Perspectives on Barley Yellow Dwarf*, CIMMYT, Mexico, D.F., Mexico, p. 368–382.

Rao, S.A. 1968. Biology of *Polymyxa graminis* in relation to soil-borne wheat mosaic virus. *Phytopathology* 58:1516–1521.

Rao, S.A. and Brakke, K.M. 1969. Relation of soil-borne wheat mosaic virus and its fungal vector *Polymyxa graminis*. *Phytopathology* 59:581–587.

Razvyazkina, G.M. and Polyakova, G.P. 1967. Electron microscopy study of wheat mosaic virus transmitted by the cicada, *Psammotettix striatus* (in Russian). *Dokl. Akad. Nauk SSSR* 174:1435–1436.

Razvyazkina, G.M. and Polyakova, G.P. 1970. Electron microscopy study of winter wheat mosaic virus in vector *Psammotettix striatus* L. (in Russian). *Dokl. Akad. Nauk SSSR* 193:1170–1173.

Refatti, E. et al. 1987. Maize: a natural and experimental host of barley yellow dwarf virus in northern Italy. In: Burnett, P.A., Ed. *1990—World Perspectives on Barley Yellow Dwarf*, CIMMYT, Mexico, D.F., Mexico, p. 269–272.

Roane, C.W. et al. 1983. Inheritance of resistance to maize dwarf virus in maize inbred line Oh 7B. *Phytopathology* 73:845–850.

Rochow, W.F. 1969. Biological properties of four isolates of barley yellow dwarf virus. *Phytopathology* 59:1580–1589.

Rochow, W.F. 1970. Barley yellow dwarf virus. *CMI/AAB Descriptions of Plant Viruses*, No. 32.

Rochow, W.F. and Duffus, J.E. 1981. Luteoviruses and yellows disease. In: Kurstak, E., Ed. *Hand. Pl. Virus Infec. Comp. Diag.*, Elsevier/North-Holland Biomedical Press, p. 147–170.

Rose, D.J.W. 1978. Epidemiology of maize streak disease. *Annu. Rev. Entomol.* 23:259–282.

Rosenkranz, E. 1969. A new leafhopper-transmissible corn stunt disease in Ohio. *Phytopathology* 59:1344–1346.

Rosenkranz, E. 1981. Host range of maize dwarf mosaic virus. In: Gordon, D.T. et al., Eds. *Virus and Virus-Like Dis. Maize United States. South. Coop. Series Bull.*, Ohio Agric. Res. and Dev. Center, OH, pp. 152–160.

Rosenkranz, E. 1983. Susceptibility of representative native Mississippi grasses in six subfamilies to maize dwarf mosaic virus strains A and B and sugarcane mosaic virus strain B. *Phytopathology* 73:1314–1321.

Rosenkranz, E. and Scott, G.E. 1984. Determination of the number of genes for resistance to maize dwarf mosaic virus strain A in five corn inbred lines. *Phytopathology* 74:71–76.

Rosenkranz, E. 1987. New hosts and taxonomic analysis of the Mississippi native species tested for reaction to maize dwarf mosaic and sugarcane mosaic viruses. *Phytopathology* 77:598–607.

Ryden, K. 1990. Chemical control of barley yellow dwarf in spring oats. In: Burnett, P.A., Ed. *1990—World Perspectives on Barley Yellow Dwarf*, CIMMYT, Mexico, D.F., Mexico, p. 468–470.

Saito, Y. et al. 1962. Electron microscopy and purification of soil-borne wheat mosaic virus. *Ann. Phytopathol. Soc. Japan* 26:16–18.

Scalla, R. and Rochow, W.F. 1977. Protein component of two isolates of barley yellow dwarf virus. *Virology* 78:576–580.

Scheifele, G.L. and Wernham, C.C. 1969. Further evidence supporting the hypothesis that two genetic systems control disease reaction to maize dwarf mosaic virus strain A and strain B. *Pl. Dis. Reptr.* 53:150–151.

Scheller, H.V. and Shukle, R.H. 1986. Feeding behaviour and transmission of barley yellow dwarf virus by *Sitobion avenae* on oats. *Entomol. Exp. Appl.* 40:189–195.

Scott, H.A. 1961. Serological detection of barley stripe mosaic virus in single seeds and dehydrated leaf tissues. *Phytopathology* 51:200–201.

Sehgal, O.P. 1966. Host range, properties and partial purification of a Missouri isolate of maize dwarf mosaic virus. *Pl. Dis. Reptr.* 50:862–866.

Senboku, T. et al. 1978. *Ann. Phytopathol. Soc. Japan* 44:394.

Serjeant, P.E. 1967. The transmission of European wheat striate mosaic virus by *Javesella pellucida* (Fabr.) injected with extracts of plants and planthoppers. *Ann. Appl. Biol.* 59:39–48.

Shaskolskaya, N.D. 1962. Transovarial transmission of winter wheat mosaic virus by leafhopper *Psammotettix striatus* L. (in Russian). *Zool. Zh.* 41:717–720.

Shepherd, R.J. 1965. Properties of a mosaic virus of corn and Johnsongrass and its relation to the sugarcane mosaic virus. *Phytopathology* 55:1250–1256.

Shepherd, R.J and Holdeman, Q.I. 1965. Seed transmission of the Johnsongrass strain of the sugarcane mosaic virus in corn. *Pl. Dis. Reptr.* 49:468–469.

Shikata, E. 1972. Rice transitory yellowing virus. *CMI/AAB Descriptions of Plant Viruses*, No. 100.

Shikata, E. 1974. Rice black-streaked dwarf virus. *CMI/AAB Descriptions of Plant Viruses*, No. 135.

Shikata, E. 1981. Reoviruses. In: Kurstak, E., Ed. *Hand. Pl. Virus Infec. Comp. Diag.*, Elsevier/North-Holland Biomedical Press, p. 251–423.

Shukla, D.D. et al. 1989. A novel approach to the serology of potyviruses involving affinity-purified polyclonal antibodies directed towards virus-specific N termini of coat protein. *J. Gen. Virol.* 70:13–23.

Shukla, D.D. et al. 1989a. Taxonomy of potyviruses infecting maize, sorghum and sugarcane in Australia and the United States as determined by reactivities of polyclonal antibodies directed towards virus-specific N-termini of coat proteins. *Phytopathology* 79:223–229.

Shukla, D.D. et al. 1994. Sorghum mosaic virus. *CMI/AAB Descriptions of Plant Viruses*, (in press).

Shurtleff, M.C. 1986. *Compendium of corn diseases*, 2nd edition, APS Press, St. Paul, MN, 105 pp.

Signoret, P.A. et al. 1972. Particules de type viral chez Triticum durum Desf, presentant des symptomes de stries chlorotiques. *Ann. Phytopathol.* 4:45–53.

Signoret, P.A. 1974. Les Maladies à virus des Graminées dans le midi de la France. *Acta Biologica Yugoslavica* B 11:115–120.

Signoret, P.A. et al. 1976. Virus des stries chlorotiques du ble (wheat chlorotic streak virus). Identification, purification, transmission et èpidemiologie. *Poljopr. znan. smotra* 39:183–186.

Signoret, P.A. et al. 1977. Présence en France du wheat spindle streak mosaic virus. *Ann. Phytopathol.* 9:337–379.

Signoret, P.A. et al. 1977a. Donnés nouvelles sur la maladie des striés chlorotiques du blé (wheat chlorotic streak mosaic virus - WCSMV). *Ann. Phytopathol.* 9:381–385.

Signoret, P.A. 1983. Maize virus diseases in France. In: Gordon, D.T. et al., Eds. *Proc. Int. Maize Virus Dis. Colloq. Workshop*, Ohio State Univ. Ohio Agric. Res. Dev. Cent., Wooster, p. 133–116.

Signoret, P.A. 1988. Maize rough dwarf virus (MRDV). In: Smith, I.M., et al., Eds. *European Hand. Pl. Dis.*, Blackwell Sci. Publ., London, p. 76.

Signoret, P.A. 1988a. Wheat spindle streak mosaic virus (WSSMV). In: Smith, I.M. et al., Eds. *European Hand. Pl. Dis.*, Blackwell Sci. Publ., London, p. 98–99.

Sill, W.H. 1958. A comparison of some characteristics of soil-borne wheat mosaic virus in the Great Plains and elsewhere. *Pl. Dis. Reptr.* 42:912–924.

Sill, W.H., Jr. 1966. Maize dwarf mosaic virus in Kansas. *Pl. Dis. Reptr.* 50:11.

Sinha, R.C. 1968. Serological detection of wheat striate mosaic virus in extracts of wheat plants and vector leafhoppers. *Phytopathology* 58:452–455.

Sinha, R.C. and Behki, R.M. 1972. American wheat striate mosaic virus. *CMI/AAB Descriptions of Plant Viruses*, No. 99.

Sinha, R.C. and Chiykowski, L.N. 1967. Multiplication of wheat striate mosaic virus in its leafhopper vector *Endria inimica* Say. *Virology* 32:402–405.

Sinha, R.C. et al. 1976. Chemical composition and some properties of wheat striate mosaic virus. *Phytopathol. Z.* 87:314–323.

Slykhuis, J.T. 1953. Striate mosaic, a new disease of wheat in South Dakota. *Phytopathology* 43:537–540.

Slykhuis, J.T. 1958. A survey of virus diseases of grasses in northern Europe. *F.A.O. Pl. Prot. Bull.* 6:129–134.

Slykhuis, J.T. and Watson, M.A. 1958. Striate mosaic of cereals in Europe and its transmission by *Delphacodes pellucida* (Fab.). *Ann. Appl. Biol.* 46:542.

Slykhuis, J.T. 1962. Wheat striate mosaic, a virus disease to watch on the prairies. *Can. Plant Dis. Surv.* 42:135–142.

Slykhuis, T.J. 1962a. An international survey for virus disease of grasses. *F.A.O. Pl. Prot. Bull.* 10:1–16.

Slykhuis, J.T. 1963. Vectors and host relations of North American wheat striate mosaic virus. *Can. J. Bot.* 41:1171–1185.

Slykhuis, J.T. and Sherwood, P.L. 1964. Temperature in relation to the transmission and pathogenicity of wheat striate mosaic virus. *Can. J. Bot.* 42:1123–1133.

Slykhuis, J.T. and Polak, L. 1971. Factors affecting manual transmission, purification and particle lengths of wheat spindle streak mosaic virus. *Phytopathology* 61:569–574.

Slykhuis, J.T. 1972. Ryegrass mosaic virus. *CMI/AAB Descriptions of Plant Viruses*, No. 86.

Slykhuis, J.T. 1973. Agropyron mosaic virus. *CMI/AAB Descriptions of Plant Viruses*, No. 118.

Slykhuis, J.T. 1976. Wheat spindle streak mosaic virus. *CMI/AAB Descriptions of Plant Viruses*, No. 167.

Smith, M.K. 1972. *A Textbook of Plant Virus Diseases*, 3rd edition, Longman Group, Ltd.

Smith, R.P. 1980. Crop loss assessment and epidemiological studies on barley yellow dwarf virus in Victoria (Australia). (Abstr.) *3rd Conf. Virus Dis. Gramineae Europe*, Rothamsted, May 28–30, (Abstr.).

Spaar, D. and Schumann, K. 1977. Getreidearten und Gräser. In: Klinowski, M. und Mitarbeiter, Eds. *Pflanzliche Virologie*, Bd. 2, 3. Aufl. S., Akademie-Verlag, Berlin, p. 1–62.

Spaar, D. and Schumann, K. 1977a. Die Pseudorosettenkrankheit. In: Klinkowski, M. und Mitarbeiter, Eds. *Pflanzliche Virologie*, Bd. 3, 3. Aufl. 3, Akademie-Verlag, Berlin, p. 27–29.

Stoner, W.N. 1949. Virus Diseases of Corn with Special Reference to Insect Transmission of Leaf Fleck and Western Cucumber Mosaic. Ph.D. thesis, Univ. of California, Berkeley, 49 pp.

Stoner, W.N. 1952. Leaf fleck, an aphid borne persistent virus disease of maize. *Phytopathology* 42:683–689.

Stoner, W.N. et al. 1967. A virus infecting maize in South Dakota. *Pl. Dis. Reptr.* 51:705–709.

Stoner, W.N. 1977. Barley yellow dwarf virus infection in maize. *Phytopathology* 67:975–981.

Sukhov, K.S. and Vovk, A.M. 1938. Mosaic disease of oats (in Russian). *C.R. Acad. Sci. U.R.S.S.* 19:207–210.

Sukhov, K.S. and Vovk, A.M. 1938a. Mosaic of cultivated cereals and how it is communicated in nature (in Russian). *C.R. Acad. Sci. U.R.S.S.* 20:745–748.

Sukhov, K.S. and Sukhova, M.N. 1940. Interrelations between the virus of a new grain mosaic disease (Zakuklivanie) and its carrier, *Delphax striatella* Fallen. *C.R. Acad. Sci. U.R.S.S. N.S.* 26:479–482.

Sum, I. et al. 1979. A kukorica törpe mozaik virus hatasa 15 kukorica hibride. *Növenytermeles* 28:309–316.

Šutić, D. and Tošić, M. 1964. Prugasti mozaik ječma - nova viroza u Jugoslaviji. *Mikrobiologija* 1:61–67.

Šutić, D. and Tošić, M. 1964a. Virus crtičastog mozaika pšenice u našoj zemlji. *Zaštita bilja* Beograd 79:307–314.

Šutić, D. and Tošić, M. 1966. Recent investigations of wheat virus disease in Yugoslavia. *Savremena poljoprivreda* 11–12:385–394.

Šutić, D. and Tošić, M. 1966a. Transmission of mosaic virus by barley seeds. *Adv. Frontiers Pl. Sci., New Delhi* 17:205–208.

Šutić, D. and Tošić, M. 1966b. A significant occurrence of maize mosaic virus on Johnsongrass (*Sorghum halepense* Pers.) as a natural host plant. *Rev. Roum. Biol. Ser. Bot.* 11:219–224.

Šutić, D. et al. 1969. Značaj viroznog nozaike za proizvodnju kukuruza. *Savremena poljoprivreda* 5/6:449–462.

Šutić, D. 1983. *Viroze biljaka*, Nolit, Beograd.

Takahashi, R. et al. 1973. Studies on resistance to yellow mosaic disease in barley. I. Tests for varietal reactions and genetic analysis of resistance to the disease. *Berichte des Ohara Instituts für Landwirtschaftliche Biologie Okayama Universität* 16:1–17.

Tandon, J.P. et al. 1990. Status of barley yellow dwarf virus research in Asia, with special emphasis on India. In: Burnett, P.A., Ed. *World Perspectives on Barley Yellow Dwarf*, CIMMYT, Mexico, D.F., Mexico, p. 81–84.

Taylor, R.H. and Pares, R.D. 1968. The relationship between sugarcane mosaic virus and mosaic viruses of maize and Johnson grass in Australia. *Aust. J. Agric. Res.* 19:767–773.

Teakle, D.S. and Grylls, N.E. 1973. Four strains of sugarcane mosaic virus infecting cereals and other grasses in Australia. *Aust. J. Agric. Res.* 24:465–477.

Teakle, D.S. and Pritchard, A.J. 1971. Resistance of Krish sorghum to four strains of sugarcane mosaic in Queensland. *Pl. Dis. Reptr.* 55:596–598.

Teakle, D.S. et al. 1972. Transfer of sugarcane mosaic virus resistance from Krish sorghum to inbred lines. *Sorghum Newslett.* 16:2–3.

Teakle, D.S. et al. 1989. Sugarcane mosaic virus. *CMI/AAB Descriptions of Plant Viruses*, No. 342.

Teyssandier, E.D. et al. 1983. Maize virus diseases in Argentina. In: Gordon, D.T. et al., Eds. *Proc. Int. Maize Virus Dis. Colloq. Workshop*, 2–6 August 1982. The Ohio State Univ., Ohio Agric. Res. Dev. Center, Wooster, p. 93–99.

Timian, R.G. and Sisler, W.W. 1955. Prevalence, sources of resistance and inheritance of resistance to barley stripe mosaic (false stripe). *Pl. Dis. Reptr.* 39:550–552.

Timian, R.G. 1960. A virus of durum wheat in North Dakota transmitted by leafhoppers. *Pl. Dis. Reptr.* 44:771–773.

Timian, R.G. 1964. Dodder transmission of barley yellow dwarf virus. (Abstr.) *Phytopathology* 54:910.

Timian, R.G. 1974. The range of symbiosis of barley and barley stripe mosaic virus. *Phytopathology* 64:342–345.

Tippett, R.L. and Abbott, E.V. 1968. A new strain of sugarcane mosaic virus in Louisiana. *Pl. Dis. Reptr.* 52:449–451.

Toler, R.W. 1984. Status of international sorghum viruses and changes in resistance. *Proc. of the 39th Annu. Corn and Sorghum Int. Res. Conf.* A.S.T.A., Washington, No. 39:33–43.

Toler, R.W. 1985. Maize dwarf mosaic virus the most important virus disease of sorghum. *Plant Dis.* 69:1011–1015.

Toriyama, S. 1983. Rice stripe virus. *CMI/AAB Descriptions of Plant Viruses*, No. 269.

Toriyama, S. 1985. *Ann. Phytopathol. Soc. Japan* 51:358.

Tošić, M. 1962. Proućavanje viroza kukuruza u NRS. *Agron. Glas.* 12:418–421.

Tošić, M. 1962a. Proućavanje virozne prirode nekih patoloških pojava na pšenici u NRS. *Agron. Glas.* 5-6-7:384–386.

Tošić, M. 1965. Proućavanje viroznog mozaika kukuruza u Srbiji. *Zbornik radova Poljoprivrednog fakulteta Beograd*, XIII, No. 392, 19 pp.

Tošić, M., Mišović, M. 1967. Uticaj virusnog mozaika na porast i prinos nekih sorti i hibrida kukuruza. *Zaštita bilja* 93-95:173–181.

Tošić, M. and Šutić, D. 1968. Medjusobni odnos virusa mozaika kukuruza i nekih njegovih vektora. *Mikrobiologija (Belgr.)* 5:73–78.

Tošić, M. 1971. Virus diseases of wheat in Serbia. I. Isolation and determination of wheat streak mosaic virus and brome mosaic virus. *Phytopathol. Z.* 70:145–162.

Tošić, M. 1971a. Virus diseases of wheat in Serbia. II. Some changes in wheat plants infected with wheat streak mosaic virus (WSMV) and brome mosaic virus (BMV). *Phytopathol. Z.* 71:327–340.

Tošić, M. and Ford, R.E. 1972. Grasses differentiating sugarcane mosaic and maize dwarf mosaic viruses. *Phytopathology* 62:1466–1470.

Tošić, M. and Malak, J. 1973. Reakcija nekih sorti sirka prema izvesnim izolatima virusa mozaika kukuruza, mozaične kržljavosti i mozaika šećerne trske. *Zaštita bilja* 122:15–23.

Tošić, M. and Ford, R.E. 1974. Physical and serological properties of maize dwarf mosaic and sugarcane mosaic viruses. *Phytopathology* 64:312–317.

Tošić, M. and Šutić, D. 1974. Spreading of the mosaic virus on corn during vegetation. *Mikrobiologija (Belgr.)* 11:19–25.

Tošić, M. et al. 1977. Effect of previous mosaic virus infection on corn susceptibility to the stalk and ear rot caused by *Gibberella zeae*. *Ann. Phytopath.* 9:395–401.

Tošić, M. and Šutić, D. 1977. Investigations of maize mosaic virus transmission through corn seed. *Ann. Phytopathol.* 9:403–405.

Tošić, M. et al. 1978. Nasledjivanje otpornosti kukuruza prema viroznom mozaiku. *Savremena poljoprivreda XXVI*, 5-6:67–72.

Tošić, M. et al. 1979. Ocena osetljivosti nekih domaćih i evropskih hibrida kukuruza prema virusnom mozaiku. *Zaštita bilja* 148:125–133.

Tošić, M. 1988. Maize dwarf mosaic virus (MDMV). In: Smith, I.M. et al., Eds. *European Hand. Pl. Dis.*, Blackwell Sci. Publ., London, p. 41–42.

Tošić, M. et al. 1990. Differentiation of sugarcane, maize dwarf, Johnsongrass and sorghum mosaic viruses based on reactions of oat and some sorghum cultivars. *Plant Dis.* 74:549–552.

Tošić, M. et al. 1990a. Epidemijska pojava mozaične kržljavosti kukuruza u Jugoslaviji. *Zaštita bilja* 191:81–93.

Tošić, M. and Mijavec, A. 1991. Maize dwarf mosaic virus, an important pathogen on sorghum in Yugoslavia. *Acta Phytopathol. Entomol. Hung.* 26:147–151.

Trefzger-Stevens, J. and Lee, P.E. 1977. The structural properties of wheat striate mosaic virus, plant rhabdovirus. *Virology* 78:144–149.

Tu, J.C. et al. 1968. Comparisons of chloroplasts and photosynthetic rates of plants infected and not infected by maize dwarf mosaic virus. *Phytopathology* 58:285–288.

Tu, J.C. and Ford, R.E. 1971. Maize dwarf mosaic virus predisposes corn to root rot infection. *Phytopathology* 61:800–803.

Usugi, T. and Saito, Y. 1979. Relationship between wheat yellow mosaic virus and wheat spindle streak mosaic virus. *Ann. Phytopathol. Soc. Japan* 45:397–400.

Vacke, J. 1961. Wheat dwarf virus disease. *Biologia Plantarum* 3:228–233.

Vacke, J. 1966. Study of transovarial passage of the oat sterile-dwarf virus. *Biologia Plantarum* 8:127–130.

Webster, R.K. and Gunnell, P.S. 1992. *Compendium of Rice Diseases*, APS Press, St. Paul, MN, 62 pp.

Wechmar, B.M. von and Chauhan, R. 1984. Seedborne viruses of maize in South Africa. I. Sugarcane mosaic virus. *Maize Virus Dis. Newsl.* No. 1, 54–58.

Wellman, F.L. 1934. Infection of *Zea mays* and various other gramineae by the celery virus in Florida. *Phytopathology* 24:1035–1037.

Wernham, C.C. and MacKenzie, D.R. 1967. Field observation on resistance and susceptibility to a single isolate of maize dwarf mosaic virus. (Abstr.) *Phytopathology* 57:346.

Wetter, C. et al. 1969. Purification and serology of maize rough dwarf virus from plant and vector. *Phytopathol. Z.* 66:197–212.

White, D.G. 1999. *Compendium of Corn Diseases*, 3rd Ed., APS Press, St. Paul, MN.

Wiese, M.V. 1987. *Compendium of Wheat Diseases*, 2nd edition, APS Press, St. Paul, MN, 112 pp.

Williams, L.E. and Alexander, L.J. 1965. Maize dwarf mosaic virus a new corn disease. *Phytopathology* 55:802–804.

Williams, L.E. et al. 1967. A virus of corn and small grains in Ohio and its relation to wheat streak mosaic virus. *Pl. Dis. Reptr.* 51:207–211.

Williams, L.E. et al. 1968. Seed transmission studies on maize dwarf mosaic virus in corn. *Pl. Dis. Reptr.* 52:863–864.

Yossen, V. et al. 1983. Frequencia del virus del mosaico enanizante del maiz (MDMV) en la Republica Argentina. *Revista de Investigationes Agropecuarias* 18:225–230.

Zazhurilo, V.K. and Sitnikova, G.M. 1939. Mosaic of winter wheat. *C.R. Akad. Sci. U.R.S.S.* 25:798–801.

Zummo, N. et al. 1970. Reaction of sweet sorghum varieties to maize dwarf mosaic at two locations in Kentucky. *Pl. Dis. Reptr.* 54:223–224.

2 Virus Diseases of Forage Feed Plants

Many of the plants used both for animal feed and for human nutrition are infected by viruses not described here because they form integral parts of other chapters in this book. However, many plants used exclusively for animal feed and grown particularly for this purpose will be discussed in this chapter, as well as native or wild forage plants in natural prairies. In view of the economic importance of forages, virus diseases affecting these plants will be described. The exact cost is difficult to assess for various reasons. For example, perennial forage plants serve as reservoirs for viruses that spread to annual crops via insect vectors. These insects complete portions or all of their life cycles in the forage crop where they cause damage from feeding activity in addition to the reduction in tonnage yield or grazing capacity caused by the endemic virus.

Most plants used mainly for animal grazing and hay belong to grass (Gramineae) and legume (Papilionaceae) families.

VIRUS DISEASES OF GRAMINOIDS

Graminoid forage plants belong to the group of plants highly important for the nutrition of domestic and wild animals. Many of these plants are grown in plowed fields or represent cultivated and spontaneous inhabitants of meadow, grassland, pasture, hedgerow, and roadside plant populations.

Graminoid forage plants may consist of both annual and perennial plants that play particularly important roles in virus epidemiology in the Poaceae. As natural or alternative hosts, they represent a "green bridge" for virus transmission in the course of the same year, and therefrom onto winter cereals. Perennial grasses serve to maintain or preserve viruses during several years and provide the "green bridges" in their transmission from one year to the next (e.g., *Bromus inermis* for ByYDV transmission, *Agropyron repens* for WCSkV transmission, etc.).

Some graminoid plants are dangerous weeds in monocotyledonous and dicotyledonous plant crops that cause serious yield reductions. Weed plants are particularly damaging because they also represent natural and alternative hosts for several viruses and serve as reservoirs for the preservation and transmission of these viruses onto susceptible cultivated graminoid plants (e.g., *Agropyron repens* provides a reservoir for AgMV and WCSkV in wheat crops, as does *Leersia oryzoides* for ByYDV in rice crops, *Sorghum halepense* for MDMV in maize crops, etc.).

Forage graminoid plants serve as primary and alternate or secondary hosts of numerous viruses in nature. Some of these viruses cause economic damage through green matter and grain yield reductions; for example, cocksfoot streak virus (CkSkV) and cocksfoot mottle virus (CkMtV) on *Dactylis glomerata*, ryegrass mosaic virus (RgMV) on *Lolium* spp., brome mosaic virus (BrMV) on *Bromus inermis*, etc. (Šutić, 1983). Damages caused by these viruses regularly increase during the year in later harvests and grazings, and particularly in later years in the case of perennial plants.

Special attention should be dedicated to forage graminoid plant viruses in view of the economic importance of these plants and the importance of their role in the development cycle of several viruses, notably those occurring in species of the Gramineae.

VIRUS DISEASES OF BROMEGRASS (*BROMUS INERMIS*)

Brome plants are perennial, usually members of meadow and pasture plant populations that can also be grown in plowed fields. They are natural primary hosts of the brome mosaic virus and natural alternative hosts of viruses in several other grass species.

Brome Mosaic Virus (BrMV) [type]

BrMV, sometimes referred to as bromegrass mosaic virus, belongs to and is the "type virus" for the Bromovirus group. First isolated in the U.S. from *Bromus inermis*, after which it was named, it has been isolated also from *Poa pratensis* (the U.S.), *Triticum aestivum* and *Agropyron repens* (the former U.S.S.R. and former Yugoslavia), and *Avena sativa*, *Hordeum vulgare*, and *T. aestivum* (South Africa).

Several forage gramineous species listed as natural hosts include *Agropyron disertorum*, *A. repens*, *Avena fatua*, *B. inermis*, *Festuca pratensis*, *Hordeum murinum*, *Lolium multiflorum*, *Poa pratensis*, *Setaria* spp., as well as cereals (*Secale cereale*, *Triticum aestivum*) and *Zea mays* (Spaar and Schumann, 1977). Some researchers have mechanically inoculated several grass plants that became experimentally infected (*Agrostis gigantea*, *Alopecurus pratensis*, *Arrhenatherum elatius*, *Briza media*, *Bromus* spp., *Cynodon dactylon*, *Dactylis* spp., *Festuca* spp., *Phleum* spp., *Poa* spp., and others). Infections in some experimental hosts are latent (*Agropyron caninum*, *Phragmites australis*, *Poa compressa*, and others). Based on the large number of susceptible hosts, BrMV ranks as one of the most widely spread viruses of forage gramineous plants.

General symptoms on infected *Bromus inermis* leaves appear as light-green or clear-yellow mottling and striped spots. Its perennial nature serves as a permanent source of infection that allows virus transmission from year to year.

Dark-green stripes occurring first on the tips of infected Italian ryegrass leaves gradually become darker, leading to leaf tip necrosis. The leaves of infected *Hordeum murinum* wither quickly and turn paper-white. This virus may cause forage crop and wheat yield reductions of 50 to 100%, as well as decreases in germination of seed from infected plants (Spaar and Schumann, 1977).

A detailed description of this virus appears in the chapter section on brome mosaic virus on wheat.

Brome: Natural Alternative Host for Some Viruses

Barley Yellow Dwarf Virus (ByYDV)

Brome is one of numerous gramineous species that serves as an alternate natural host of ByYDV. As a perennial plant, it represents a perennial reservoir of this, otherwise most widespread virus of graminoid plants. The detailed description of ByYDV appears in the chapter containing virus diseases of barley.

Cocksfoot Mild Mosaic Virus (CkMdMV)

Brome is one of several graminoid plants that serves as a natural host of CkMdMV. Plant susceptibility depends primarily on virus strain. A detailed description of CkMdMV virus is included in the section on cocksfoot (*Dactylis glomerata*) virus diseases.

Other Brome Species, Hosts of Some Viruses

According to Spaar and Schumann (1977), natural virus hosts include *Bromus mollis* (European wheat striate mosaic virus (WEStMV), ByYDV), *B. rigidus* (ByYDV), and *B. secalinus* (wheat dwarf virus [WDV]). According to Rosenkranz (1981), some 15 *Bromus* species are susceptible to the maize dwarf mosaic virus (MDMV - strain A).

VIRUS DISEASES OF COCKSFOOTGRASS (*DACTYLIS GLOMERATA*)

Cocksfoot, a perennial plant regularly present in meadow plant populations, is found frequently in pastures, but also is grown in plowed fields. Three viruses, usually found associated in infected cocksfoot plants, include cocksfoot mild mosaic virus (CkMdMV), cocksfoot mottle virus (CkMtV), and cocksfoot streak virus (CkSkV).

Cocksfoot Mild Mosaic Virus (CkMdMV)

Properties of viral particles

The CkMdMV isometric particles, 28 nm in diameter, with a sedimentation coefficient of 105S, have a coat protein of 72 kDa and two RNA species of 500 and 1500 kDa (Paul and Huth, 1970).

Stability in Sap

The following physical properties were established in crude extracts obtained from *Setaria italica* leaves: a TIP of 80 to 85°C; a DEP of 5×10^{-5}; and infectivity after storage at –20°C of 5 years (Huth and Paul, 1972).

Serology

CkMdMV is strongly immunogenic. Based on serological and other traits, this virus is serologically related to the phleum mottle virus (PmMtV) (Huth and Paul, 1972).

Strains

CkMdMV strains designated based on susceptible host species and disease symptoms are brome stem leaf mottle, holcus transitory mottle, phleum mottle, and cocksfoot mosaic and necrosis (Torrance and Harrison, 1981).

Geographic Distribution

CkMdMV, first described in Germany (Huth and Paul, 1972), was later found widespread in other European countries (Huth, 1988).

Host Range

CkMdMV infects several gramineous species, depending on virus strains. *Dactylis glomerata*, *Festuca pratensis*, and *Lolium perenne* are natural hosts, and susceptible crop hosts include oats, barley, rye, plus several species in the genera *Agropyron*, *Andropogon*, *Avena*, *Bromus*, *Phleum*, *Poa*, and *Setaria* (Spaar and Schumann, 1977). *Digitalis glomerata* and *Setaria italica*, suitable propagation species, are used for diagnostic assays.

Symptoms

Natural hosts express a mild mosaic, for which the disease was named. This mild mosaic is diffused in the leaves, sometimes accompanied by chlorotic streaks. Symptom intensity decreases during periods of higher temperatures. Visible reduction of growth results from CkMdMV infection in *Lolium perenne* and *Festuca pratensis*.

Pathogenesis

CkMdMV retains infectivity in natural hosts, transmitted both by aphids and beetles (Huth, 1988). Vector activity seems low in view of the comparatively limited virus distribution. The virus is easily transmitted mechanically, but not through seed.

Control

Plant protection measures are insufficiently known. However, chemical control of vectors, if practicable, represents the main method for prevention of infection.

Cocksfoot Mottle Sobemovirus (CkMtV)

Properties of Viral Particles

The CkMtV isometric particles are 30 nm in diameter and consist of a 32-kDa coat protein comprising 75% of total particle weight and a single-stranded RNA of 1000 kDa, comprising 25% of total particle weight; its sedimentation coefficient is 118S (Catherall, 1970; Catherall et al., 1977).

Stability in Sap

The TIP for CkMtV is 60 to 65°C; the DEP is 1×10^{-3}; and the LIV at 20°C is 4 to 6 days. (Serjeant, 1967; Catherall, 1970).

Serology

CkMtV is weakly immunogenic. The ELISA test is more suitable for diagnosis of infection than symptomological host reactions (Lapierre and Hariri, 1988).

Strains

No data are available on virus strains.

Geographic Distribution and Economic Importance

The virus was first discovered in the U.K. (Serjeant, 1964; Catherall et al., 1977) and later in Germany, France, Japan, and New Zealand (Lapierre and Hariri, 1988). In the U.K., CkMtV sometimes causes serious damage in perennial cocksfoot crops (Lapierre and Hariri, 1988). Susceptible cultivars infected with CkMtV frequently die (Catherall, 1985).

Host Range

CkMtV infects a narrow range of hosts belonging to the Gramineae family. Serjeant (1967) lists *Dactylis glomerata* as the only natural virus host, whereas Spaar and Schumann (1977) list also *Triticum aestivum*. Other gramineous plants susceptible following artificial inoculation include oats, barley, rye, and other plants in the genera *Avena*, *Bromus*, *Dactylis*, *Hordeum*, *Lagurus*, and *Phalaris*.

Symptoms

Clearly visible symptoms of yellow streaked spots and mottling first appear on CkMtV-infected leaves of cocksfoot in the spring and early summer; then later as disease develops, leaves become whitish or necrotic and die prematurely. Erect young secondary stems are discerned in the mass of dead leaves. Infected plants flower poorly and produce seed of low viability, if any.

In early infections of wheat, the leaves display severe mottling, plants grow poorly, become chlorotic, and decay quickly. Similar symptoms, but less severe, occur on oats and barley. CkMtV causes symptoms quite similar to the cocksfoot streak virus (CkSkV), but otherwise differing markedly from it in particle shape, vectors, and ability to infect wheat, oats, and barley.

Pathogenesis

CkMtV infectivity is maintained in its natural hosts and transmitted therefrom by *Lema melanopa* and *L. lichenis*. Adult insects are more active vectors than larvae. By feeding on the infected plant, these insects acquire CkMtV in less than 5 minutes and transmit it to healthy plants in 5 minutes. Transmission is semipersistent and CkMtV remains infective in insect vectors for 2 weeks. Conditions for vector development and multiplication plus the production of susceptible plants on large acreages are most instrumental for spread of CkMtV in nature. Other modes of CkMtV transmission are unknown.

Control

Protection of plants against CkMtV has been ensured mainly with the control of vectors (*L. melanopa* and *L. lichenis*) as a key factor in disease epidemiology. Resistant cultivars to this virus have been described in the U.K. and in Japan (Lapierre and Hariri, 1988).

Cocksfoot Streak Potyvirus (CkSkV)

Properties of Viral Particles

The tubular, flexuous CkSkV particles are 752×13 nm, (Catherall, 1971), with a TIP of 55°C, DEP of 3×10^{-3}, and LIV at 20°C of 16 days (Ohmann-Kreutzberg, 1963). CkSkV has strong immunogenicity. Neither strains nor a closer relatedness to other viruses of grasses are known.

Geographic Distribution and Economic Importance

CkSkV has been reported in the U.S., England, France, Holland, Germany, Denmark, and Sweden (Slykhuis, 1958; Smith, 1972). Based on its mode of transmission and spread, CkSkV can be expected in many other countries.

CkSkV is an economically damaging plant pathogen. It infects a number of grass species and, depending on the host, can cause forage yield losses of 10 to 60%. Reductions in seed yield also occur in some fescues (*Festuca* spp.).

Host Range

Only plants in the Gramineae are susceptible to CkSkV. In addition to cocksfoot, other natural hosts are perennial ryegrass (*Lolium perenne*), Italian ryegrass (*L. multiflorum*), *Dactylis polygama*, and *Festuca gigantea*. Other experimental host species belong to the genera *Agropyron, Agrostis, Avena, Bromus, Cynosurus, Eragrostis, Festuca, Phalaris, Phleum, Poa, Sorghum*, and others (Spaar and Schumann, 1977).

Hosts that express tolerance to infection include *Avena strigosa, Festuca capillata, Hordeum murinum, Setaria viridis*, and others. *Setaria macrostachia* and *Avena sativa* display pronounced symptoms of longitudinal colored stripes and are therefore recommended as assay plants. CkSkV does not infect wheat, which differentiates it from the symptomatically similar CkMtV.

Symptoms

A characteristic symptom of disease caused by CkSkV on cocksfoot and other species in that genus are pronounced streaked stripes, for which the disease was named. Growth of infected plants is delayed. Infected plants form fewer secondary tillers, flower earlier, and produce fewer seeds. In other genera, CkSkV caused spots and streaks and a varying intensity of chlorosis. Tolerant hosts of CkSkV are symptomless. Some other viruses may cause similar symptoms in cocksfoot plants, which should be considered when virus isolations are made and the causal virus determined.

Pathogenesis

The aphid species that transmit CkSkV nonpersistently are *Myzus persicae, Macrosiphum avenae, M. euphorbiae, Metopolophium dirhodum*, and *Hyalopteroides humilis* (loc. cit. Catherall, 1971). Perennial ryegrass plant hosts play an important role in the long-term preservation of CkSkV in addition to annual ones. The wide distribution of this virus is readily explained because the aphid vectors are very active in local and long-distance movement. CkSkV is transmitted mechanically with infective plant sap, but it is not transmitted through seed.

Control

The most important plant protection measure is to control aphid vectors, particularly in cultivated cocksfoot fields. Reduction of infection sources can be achieved by spatial isolation of cocksfoot crops from other susceptible plants. The elimination of the sources of infection by rogueing diseased plants, particularly when they first appear, is, when possible, a most useful protection method. *Dactylis maritima* displays a certain degree of resistance to this virus (Spaar and Schumann, 1977).

Cocksfoot: Natural Alternative Host for Some Viruses

Cocksfoot is a natural alternative host of ByYDV. As a perennial plant, it represents a permanent source of inoculum of ByYDV for infecting susceptible crop plants in nature. The RPV type isolate occurs more frequently than the PAV isolate (Guy, 1990; Johnstone et al., 1990), possibly pointing to variations in plant susceptibility to individual isolates of this virus. A detailed description of the ByYDV can be found in the chapter on barley virus diseases.

Cocksfoot is also a natural host of RgMV that causes systemic mosaic symptoms on several gramineous plants. ELISA and ISEM tests are suitable for diagnostic differentiation of RgMV. A detailed description of this virus can be found in the chapter section on ryegrass mosaic virus.

After thorough testing, cocksfoot was found not susceptible to MDMV-A (Roane and Troutman, 1965).

VIRUS DISEASES OF COUCH GRASS (*AGROPYRON REPENS*)

Couch grass (*A. repens*), a widely grown crop in plowed fields, is a permanent resident in meadow and pasture plant populations. It is also known as a difficult weed to control in field crops. This omnipresence of *A. repens* favorably influences the natural spread of a virus, depending on the activity of the virus vectors. This primary natural host of agropyron mosaic virus is also known as the natural alternative host of some other viruses of the Gramineae.

Agropyron Mosaic Potyvirus (AgMV)

Long classified among ungrouped viruses, AgMV now belongs to subgroup 3 of the Potyviruses. The AgMV particle, 717×15 nm, is potyvirus-like and causes pinwheel inclusions to form in infected parenchyma host cells. Vectors of AgMV are eriophyid mites *Abacurus histrix*, thus distinguishing it from the aphid-transmitted type members of the Potyvirus group (Hollings and Brunt, 1981).

AgMV occurs in North America, Finland, and Germany (Polak, 1988). Natural hosts of AgMV are *Agropyron repens* and *Triticum aestivum*. Several gramineous plant species susceptible to it are *Agropyron* ssp., *Aegilops* spp., *Bromus* spp., *Festuca rubra*, *Hordeum* spp., *Lolium multiflorum*, *Poa pratense*, *Secale sereale*, *Setaria* spp., *Triticum* spp., and others (Spaar and Schumann, 1977).

The two main symptom types of streak mosaic that occur in AgMV-infected *A. repens* leaves are pale-green streaks that can be masked during plant maturation and yellow chlorotic streaks that become highly pronounced on older leaves. Symptoms of diffused streaks appear on infected wheat leaves. AgMV, which causes no significant alterations in plant development, is considered economically unimportant.

A detailed description of the virus features can be found in the chapter section on AgMV in wheat.

Couch Grass: Natural Alternative Host for Some Viruses

Couch grass has been described as a natural host of some economically damaging viruses infecting cereals. Spaar and Schumann (1977) list it as a natural host of BrMV, ByYDV, OtPdRoV, and CrTlDsV (Leclant and Signoret, 1974). Couch grass is not susceptible to MDMV-A (Roane and Troutman, 1965).

As a perennial natural host, couch grass poses a threat as a source of permanent viral infections that are preserved in the rhizomes, wherefrom they are transmitted from year to year. Detailed information on the properties of these viruses are found in their respective descriptions on main hosts.

VIRUS DISEASES OF FALSE OATGRASS (*ARRHENATHERUM ELATIUS*)

False oatgrass, which occurs spontaneously in meadow and pasture populations, can also be grown as a field crop. As a perennial crop, it is highly interesting in interrelations established among viruses of gramineous plants. False oatgrass is a natural host of the Arrhenatherum blue dwarf virus, and is also susceptible to some other plant viruses in the Gramineae.

Arrhenatherum Blue Dwarf Virus (AmBlDV)
(Member of the Reovirus subgroup 2)

AmBlDV particles are isometric, 70 nm in diameter (Shikata, 1981), and related to the maize rough dwarf (MRhDV) (Milne and Luisoni, 1977), oat sterile dwarf (OtSrDV), and cereal tillering disease (CrTlDsV) viruses (Spaar and Schumann, 1977).

AmBlDV has been observed in central Europe (Mühle and Kempiak, 1971; Milne et al., 1974).

Arrhenatherum elatius is the only natural host of AmBlDV, although *Lolium perenne* is also susceptible to AmBlDV. Infected *A. elatius* plants display pronounced dwarfing, as indicated by the name, which features a rosette-like appearance, blue-green color, poor flowering, and often white spikes. Viral particles located in the phloem cause characteristic hypertrophic swellings on plant parts. These pathological changes cause significant reductions in seed yield and green matter in infected plants.

Cicadellid leafhoppers *Javesella dubia* and *J. pellucida*, active vectors of AmBlDV, play the most important roles in disease pathogenesis. All developmental stages of insects actively transmit the virus. Leafhoppers acquire AmBlDV particles by feeding on infected plants for 1 hour; the latent period in cicadellid leafhoppers lasts 14 to 19 days, then they can transmit AmBlDV during a 4 hour feeding period (Mühle and Kempiak, 1971). Control of cicadellid leafhoppers, the only known vectors, represents the best method of preventing infection, thus spread.

False Oatgrass: Natural Alternative Host for Some Viruses

False oatgrass is an alternative natural host of OtSrDV (Spaar and Schumann, 1977), which suggests a possible relatedness with AmBlDV.

VIRUS DISEASES OF FESCUEGRASS (*FESTUCA* SPP.)

Fescues, wide spread perennials in meadows and pastures, can also be grown as field crops. They are the primary natural hosts of festuca necrosis virus (FeNV) and festuca leaf streak virus (FeLSkV), and they are also alternative natural hosts of several other Gramineae plant viruses.

Festuca Necrosis Closterovirus (FeNv)

FeNV particles are flexuous, 1725×18 nm, (Schmidt et al., 1963), and belong to the Closterovirus subgroup 1. FeNV resembles the wheat yellow leaf virus (WYLV), based on particle size and morphology, transmissibility by aphids, and the red leaf symptoms it causes on infected oats. However, since FeNV particles are of larger diameter and noninfective in barley, their relationship with WYLV remains unexplained (Lister and Bar-Joseph, 1981).

FeNV, first observed in central Europe (Schmidt et al., 1963), may be much more widely distributed because it is aphid-transmitted.

Natural virus hosts include *Festuca pratensis*, *F. arundinacea*, and *Lolium perenne*. *Avena sativa* is a suitable assay plant for FeNV because it is susceptible and dies 5 to 6 weeks after infection. Necrosis, reddening, and shoot- and plant-death from the roots upward characterize the symptoms of infected plants.

Vectors of FeNV are the aphids *Rhopalosiphum padi*, which transmit it from *F. pratensis* and *L. perenne* into oat plants (Schmidt et al., 1963). Vector control and elimination of viral overseason sources are useful plant protection measures.

Festuca Leaf Streak Rhabdovirus (FeLSkV) [possible]

FeLSkV particles are bacilliform measuring 61×330 nm in thin sections of plant cells and 61×286 nm *in vitro*. FeLSkV was discovered in Denmark in the cytoplasms of mesophyll and epidermal cells, but not in infected root tip and vascular bundle sheath cells (Lundsgaard and Albrechtsen, 1976; 1979).

The natural host of FeLSkV is *Festuca gigantea*, which exhibits symptoms of mosaic and mild leaf streaking. The host range of FeLSkV has not been established and its vectors are still unknown. Physical properties of FeLSkV have been insufficiently investigated to determine whether it represents a separate species or belongs to an already known rhabdovirus in the Gramineae.

Fescues: Natural Alternative Hosts for Some Viruses

Fescue, *Festuca pratensis*, has been described as an alternate natural host of ByYDV, BrMV, CkMdMV, and RgMV. *F. gigantea* is the natural host of the CkSkV and *F. rubra* of the ByYDV (Spaar and Schumann, 1977). Detailed information on these viruses can be found in their descriptions on main natural hosts.

VIRUS DISEASES OF FORAGE SORGHUM (*SORGHUM BICOLOR*)

Sorghum, an annual graminoid plant grown mostly as a field crop, also occurs spontaneously as a weed plant in some crops. It occurs worldwide, on nearly all continents.

Special types of sorghum (forage sorghum) are used exclusively for animal nutrition, both as green fodder and as silage. Some sorghum field crops can also be used for cattle and wild animal grazing. Some parts of sorghum plants can also be used as animal feed following processing (stalks, grain) for industrial uses. Sorghum is particularly important in animal nutrition because it provides high green matter and grain yields.

Viruses represent potentially limiting factors in sorghum production by significantly reducing green matter, grain yields, and quality. SrMV and MDMV have been described on sorghum as a natural host. These viruses, easily aphid transmitted in nature, are widely distributed and economically damaging. Detailed information about these viruses can be found in the chapter section on sorghum virus diseases included in the chapter on virus diseases of food grass plants.

VIRUS DISEASES OF FOXTAIL MILLET (*SETARIA* SPP.)

Annual foxtail plants frequently inhabit meadows and pastures as weeds in several cultivated crops. Foxtail is grown occasionally as a feed crop for animals. Foxtail mosaic virus has been described on these plants as its main hosts. However, foxtails are also listed as natural hosts of some other gramineous plant viruses.

Foxtail Mosaic Potexvirus (FxMV)

FxMV particles, slightly flexuous and filamentous, and 500 nm (Short, 1983), contain single-stranded viral RNA 2030 to 2240 kDa (Paulsen and Niblett, 1977), comprising 7% of the viral particle weight, plus a coat protein 21.2 kDa (Short, 1981). The TIP is 68 to 70°C; the DEP 1×10^{-6} to 1×10^{-7}; and LIV in plant sap at 24°C is 46 days and 105 days in buffered extract (Paulsen and Niblett, 1977). FxMV has strong immunogenic activity (Short, 1981).

Geographic Distribution
The virus has been observed only in the U.S.

Host Range and Symptoms
Setaria italica (foxtail millet) and *S. viridis* are main virus hosts with characteristic systemic mosaic symptoms. Some 56 gramineous and 41 dicotyledonous plant species belonging to 11 families have been experimentally infected with FxMV (Paulsen and Niblett, 1977).

Symptom reactions in FxMV-infected plants vary from symptomless to a type of systemic mosaic that occurs in wheat, oats, and barley. This virus causes local lesions in *Gomphrena globosa* and *Pisum sativum* cv. Meteor plants, and local lesions accompanied by systemic infection in *Chenopodium amaranticolor* (suitable assay plant), *Nicotiana clevelandii* and *Tetragonia expansa* (Short, 1983).

Pathogenesis
In nature, the infectivity of FxMV is preserved in its hosts and insufficient data are available on its distribution. Possible vectors are unknown. However, FxMV has been transmitted through seed

of *Briza maxima* (2%) and *Avena sativa* (1%) (Paulsen and Niblett, 1977). The mode of virus introduction from infected seed into its natural hosts is not known. Further studies are needed of viral epidemiology and interrelationships with other viruses.

Foxtails: Natural Alternative Hosts for Some Viruses

Setaria viridis has already been cited as a natural host of both oat pseudo-rosette (OtPdRoV) and wheat streak mosaic viruses (WSkMV), and *S. glauca* as a natural host of WSkMV (Spaar and Schumann, 1977). *Setaria* species, *S. glauca*, *S. italica*, *S. verticillata*, and *S. viridis* are considered important hosts of the MDMV. As reported by Rosenkranz (1981), 11 species in the *Setaria* genus are susceptible to MDMV-A.

Virus Diseases of Molinia (*Molinia coerulea*)

Molinia is a perennial plant spontaneously present in the flora of meadows, pastures, and grasslands. This naturally widespread plant may be susceptible to viruses that infect other species of Gramineae, but this has not been studied. Molinia is naturally susceptible to the molinia streak virus.

Molinia Streak Virus (MoSkV)

The MoSkV isometric particles are 28 nm in diameter and have a TIP of 85°C, a DEP of 4×10^{-4}, and a sedimentation coefficient of 112S. MoSkV is considered distantly related serologically to CkMdMV and PmMtV (Huth et al., 1974).

This virus, described first in Germany, suggests it may be distributed throughout central Europe (Huth, 1974; Huth et al., 1974).

Molinia coerulea is the only known natural host of MoSkV. Other gramineous plants (*Bromus* spp., *Setaria* spp., etc.) can be infected experimentally, and latent infections have been shown in plants of *Bromus racemosus*, *Hordeum vulgare*, *Setaria viridis*, *Zea mays*, and others (Huth et al., 1974).

General symptoms of disease in MoSkV-infected plants appear as chlorotic streaked spots plus a slight chlorosis of whole plants, delayed flowering, and occasional characteristic white spikes. The virus is transmitted with infected plant sap, but other modes of transmission are unknown.

Virus Diseases of Panic Grasses (*Panicum* spp.)

Panic grasses are annuals in the Gramineae family. Some are grown for human and animal nutrition, whereas others represent a spontaneous natural flora of weeds. Processed (prepared) true millet (*Panicum miliaceum*) grain is used for human nutrition and, in its unprocessed (crude) form, also as animal feed. Panic grasses have a high nutritive value for animals and are used both as green fodder and hay.

Several already described, economically damaging cereal, maize, and rice viruses occur in different species of *Panicum*. A separate Guinea grass mosaic virus has been described in Guinea grass (*Panicum maximum*) plants grown for fodder.

Guinea Grass Mosaic Potyvirus (GugMV)

GugMV particles are filamentous, 815×15 nm, and contain 6% RNA by weight (Thouvenel et al., 1978). GugMV has a TIP of 50°C, a DEP of 1×10^{-3}, and an LIV at 24°C in plant sap of several hours or at 4°C, 24 hours. GugMV is moderately immunogenic (Thouvenel et al., 1978).

GugMV is widely distributed in tropical Africa where Guinea grass is also widely grown.

Guinea grass, the natural host GugMV, exhibits symptoms of light green mosaic and stunting. GugMV infects only plants belonging to the Gramineae family. *Pancium maximum*, *Setaria italica*, and *Zea mays* are useful diagnostic plants.

GugMV is transmitted with infected plant sap. Other modes of transmission are unknown. GugMV has been classified as a Potyvirus, based on infective properties and serological relatedness.

Panicum spp.: Natural Alternative Host for Some Viruses

Several species of the *Panicum* may be alternative natural hosts of some cultivated gramineous plant viruses (Spaar and Schumann, 1977). Thus, *Pancium capillare* (witch grass) is a natural host of WSkMV. Also, MDMV, OtPdRoV, yellows (probably ByYDV), and WSkMV may infect *P. crusgalli* (barnyard grass) plants in nature. MDMV and OtPdRoV may occur in cultivated *P. miliaceum* (true millet) crops. *P. sanguinale* is an alternate natural host of MDMV and WSkMV. The *Panicum* spp. are very important as natural hosts of MDMV; that is, 24 species are susceptible to MDMV-A (Rosenkranz, 1981).

Virus Diseases of Ryegrass (*Lolium perenne* and *Lolium multiflorum*)

Perennial ryegrass (*Lolium perenne*) and italian ryegrasses, which may be perennial or annual, inhabit mostly meadows and also pastures. They are grown also as field crops.

Ryegrasses are natural hosts on which the following viruses have been described: ryegrass (*Lolium*) enation virus (RgEnV), ryegrass (*Lolium*) mottle virus (RgMtV), and ryegrass mosaic virus (RgMV). Both plants are also natural alternate hosts for cereal, maize, and several grass species viruses. Since susceptible to many viruses, ryegrasses play an important role, both as virus hosts and as reservoirs for the continuation of viral spread in nature and in crop productions.

Ryegrass Enation Virus (RgEnV)
(Originally named Lolium Enation virus–LoEnV)

RgEnV, a possible member of the Reovirus subgroup 2 (Shikata, 1981), infects *Lolium perenne* and *L. multiflorum*. Characteristic enations form on infected plant parts as the main disease symptom. This virus, described only in Germany, has a limited distribution and the modes of spread are unknown. Serological comparisons relate RgEnV to OtSrDV (Lesemann and Huth, 1975). RgEnV is little studied.

Ryegrass Mottle Virus (RgMtV)
(Originally named Lolium mottle virus–LoMtV)

RgMtV has isometric particles of 27-nm diameter, a TIP of 65°C, a DEP (in *Cynosurus cristatus* sap) of 1×10^{-3}, and an LIV of 10 days at 20°C or 6 months at –20°C. RgMtV is immunologically active and can be identified serologically (loc. cit. Spaar and Schumann, 1977).

RgMtV, which infects only plants in the Gramineae family is distributed only in western Europe. Natural virus hosts are *Lolium perenne* and *Cynosurus cristatus*. Other susceptible plants include *Avena sativa, Hordeum vulgare, Lagurus ovatus, Lolium multiflorum, Phleum pratense*, and *Secale cereale*.

Symptoms of disease appear first at the base of younger leaves as mild chlorotic mottling, which later encompasses the whole plant (Brook, 1972). The final symptom in mature plants is longitudinal necrosis on older leaves. RgMtV is easily transmitted with infective plant sap.

Ryegrass Mosaic Potyvirus (RgMV) [possible]

Properties of Viral Particles

RgMV particles have tubular, flexuous rods of 700 × 15 nm (Slykhuis, 1972), a TIP of 60°C, a DEP of 1×10^{-3}, and an LIV at 24°C of 24 hours (Mulligan, 1960). Biophysical properties differ somewhat in other reports, probably a consequence of the investigation of different strains. RgMV is rather weakly immunogenic. RgMV isolates in England differ in virulence, whereas all isolates from the U.S. and Canada constitute a group of strains considered mild (loc. cit. Slykhuis, 1972).

Geographic Distribution and Economic Importance

RgMV is widely distributed in all areas where ryegrass species are found; that is the U.S., Canada, England, France, Holland, Germany, Denmark, Sweden, and Finland. The activity of mite vectors contributes greatly to the wide distribution of RgMV. Mosaic of ryegrass species causes economic damage by reducing plant vigor and growth, resulting in losses of fodder yields up to 50%.

Host Range

In addition to Italian and perennial ryegrasses, other important natural hosts of RgMV are cocksfoot (*Dactylis glomerata*), fescue (*Festuca pratensis*), and oat (*Avena sativa*). Several experimental grass hosts include *Agrostis stemis, Alopecurus agrestis, Avena fatua, Bromus arvensis, B. inermis, B. sterilis, Oryza sativa, Poa pratensis,* and *P. trivialis*. Italian ryegrass cv. S22 is a suitable assay plant.

Symptoms

The RgMV causes chlorotic streaks and stripes between leaf veins differing in shape and size. Infected plants grow slowly and produce fewer secondary tillers. Symptoms of the RgMV-caused disease resemble those caused by other gramineous viruses, which makes it nearly impossible to rely on symptoms alone for practical diagnosis.

Pathogenesis

The virus retains infectivity in both annual and perennial hosts. The perennial *Lolium perenne* is particularly important for the preservation of infectivity and transmission of RgMV into the next cropping season. *Phytocoptes* mites (*Abacurus histrix*), vectors of RgMV, acquire the virus from infected plants in 2 hours, then lose their ability to transmit it within 24 hours of leaving the infected plant (Mulligan, 1960).

Development of mites from eggs to adults occurs in the same part of the plant, from which they migrate to other leaves and secondary tillers, developing into a large population on a plant by season end (Lewis and Heard, 1980). In some areas, such populations develop from May to October. Mites are capable of transmitting RgMV in all developmental stages. Adult mites are wind-borne from infected onto healthy plants, the primary means of RgMV spread (Gibson, 1980). RgMV is mechanically transmitted. No data exist on possible seed transmission from infected plants. Disease diagnosis can be performed by mechanical inoculation of *Lolium perenne, L. multiflorum, Dactylis glomerata* and *Avena sativa* plants, and/or with the aid of ELISA or ISEM serological tests.

Control

All measures that contribute to population reduction or elimination of mite vectors are useful for virus control. It is important to destroy older plants, which may remain in new crops, since they are possible hosts, both of RgMV and of the mite vectors. Initial healthy plants may be obtained both by meristem-tip culture and by selection of resistant plants (Kleinhempel, 1988).

Ryegrass Bacilliform Rhabdovirus (RyBcV)

Plumb and James (1975) observed bacilliform particles of a virus in ryegrass plants in England. RgBcV, one of several possible Rhabdoviruses, infects *Lolium multiflorum* plants and *L. multiflorum* × *L. perenne* hybrids, causing mild chlorotic symptoms on infected plants. The modes of natural virus transmission are unknown. The virus has not been transmitted experimentally with the aid of planthoppers (*Javesella pellucida*), beetles (*Oulema* spp.), and five aphid species. No descriptors of RgBcV have been published.

Ryegrass Cryptic Virus (RgCcV)

In Italian ryegrass (*Lolium multiflorum*), RgCcV was described as a cryptic virus for which little information is available. RgCcV particles are isometric and 29 nm in diameter (Luisoni and Milne,

1980). Symptoms of this disease, discovered in a less-known Italian ryegrass cultivar in England (Plumb, 1980), include chlorotic spots accompanied by necrosis, particularly on the oldest leaves. RgCcV is seed-borne in some hybrids up to 25 to 50%. Other modes of virus transmission are unknown. Its relatedness to other gramineous plant viruses has not been established.

Ryegrass: Natural Alternative Host for Some Viruses

Ryegrasses, *Lolium perenne* and *L. multiflorum*, are natural alternate hosts for some viruses of cereals (ByYDV, WEStMV, OtSrDV) and maize (MRgDV). *L. perenne* is a natural alternate host of CkSkV. *L. multiflorum* is known as a natural alternate host of BrMV and FeNV and others. Detailed information on these viruses are found in sections describing their main natural hosts.

VIRUS DISEASES OF SWEET VERNAL GRASS (*ANTHOXANTHUM ODORATUM*)

Anthoxanthum Mosaic Potyvirus (AxMV) [possible]

AxMV (Catherall, 1970a; Hollings and Brunt, 1981) has elongate particles of 750 × 13 nm and is easily seed transmitted. Other modes of transmission are unknown (Catherall, 1967; 1970a).

Anthoxanthum odoratum is the only known natural host of AxMV. Other species, *A. amarum*, *A. aristatum*, and *A. puelli*, plus *Avena* spp., *Hordeum vulgare*, and *Triticum aestivum*, are experimentally susceptible (Spaar and Schumann, 1977).

Annual *Anthoxanthum odoratum* plants inhabiting meadows, pastures, and forests, and known as a weed, might be expected to support infection of AxMV that infest these various habitats.

Symptoms of disease caused by AxMV occur as light-green striped spots at the base of infected leaves. Spots gradually merge and clearing occurs as disease develops. A streaked type of mottle or mosaic occurs on other plant parts. This virus causes no significant change in plant development, and thus is of little practical importance.

Sweet Vernal Grass: Natural Alternative Host for Some Viruses

A new virus with short, rod-shaped particles of 135 × 22 nm was discovered in sweet vernal grass in England. Its relatedness with other grass viruses has not been established (Catherall et al., 1980).

VIRUS DISEASES OF TIMOTHYGRASS (*PHLEUM PRATENSE*)

Timothygrass, frequent inhabitant of meadows and pastures, is grown as a field crop. It is a natural host for *Phleum* mottle virus and also an alternate natural host of other graminoid plant viruses. These perennial plants are particularly important as a potential reservoir for preserving and allowing further spread of *Phleum* mottle virus and possibly others in nature.

Timothy Mottle Virus (TtMtV) [Str. of CkMdMV]

Properties of Viral Particles

TtMtV has isometric particles of 30-nm in diameter, a TIP of 85°C, a DEP of 1×10^{-4}, and an LIV at 20°C of 30 to 35 days. TtMtV is serologically related closely to the CkMdMV, possibly strains of each other (Paul, 1974).

Geographic Distribution and Economic Importance

TtMtV and the related CkMdMV have been described only in Germany. No other data exist on its spread. TtMtV is most damaging by reduction of both the number of secondary tillers (30%) and green matter yields (40%) (Spaar and Schumann, 1977).

Host Range

Natural virus hosts are timothy *Phleum pratense* and the closely related specie *P. bertolonii*. Other hosts are oats (*Avena sativa*), barley (*Hordeum vulgare*), rye (*Secale cereale*), cocksfoot (*D. glomerata*), and others.

Symptoms

Symptoms of TtMtV infection appear as mild chlorotic spots on leaves. At high temperatures and reduced illumination, these symptoms become less apparent, even masked.

Pathogenesis

TtMtV is readily transmitted by mechanical means. Its natural vectors are leaf beetles *Lema* (*Oulema*) *melanopa* and *L. lichens*. Thus, control of beetles plays a key role in prevention of disease occurrence and virus spread.

Timothygrass: Natural Alternative Host for Some Viruses

Timothy may also be an alternate natural host of WEStMV, OtSrDV, and CrTlDsV (loc. cit. Spaar and Schumann, 1977), which has some importance for the cycle of development, preservation, and spread of these viruses in nature (Šutić, 1983). Detailed descriptions on the properties of these viruses are in their respective sections on primary natural hosts.

VIRUS DISEASES OF WEED GRASSES

Many grasses occur naturally in field crops competing as weed plants. Several genera include *Agropyron*, *Anthoxanthum*, *Panicum*, *Setaria*, *Sorghum*, and others. Numerous species among these genera are also cultivated crops. Whether weeds or crops, such plants provide effective reservoirs for preservation of viruses and sources of continuing infection in nature, especially perennial ones that play a particularly important role.

Weed grasses represent an integral part of the graminoid plant virus development cycle; thus, measures to control or eliminate them rank among the most important methods for prevention of virus infection in nature. Among these weeds, special attention should be devoted to *Agropyron repens*, *Leersia* (*Oryza*) *oryzoides*, and *Sorghum halepense*.

Agropyron repens (couch grass) is a generally widespread weed in several crops, especially cereals and maize. It is host to economically damaging viruses described in the chapter on couch grass (*A. repens*) virus diseases.

Leersia oryzoides is a widespread weed plant in rice crops, propagated both generatively by seed and vegetatively by stolons. It is the natural alternative host of the ByYDV (= rice yellows virus), which significantly damages rice. As a permanently established vegetatively propagated weed, it represents a key source of virus. Rogueing (elimination) is a most important measure for reduction of infection and potential damage in nature.

Sorghum halepense, commonly called Johnsongrass, is a perennial weed, and the most important alternative natural host as a reservoir that maintains MDMV infective in nature (Sehgal, 1966; Šutić and Tošić, 1966). Elimination of these plants is a most important control tactic to protect against virus infections and damages caused by them. A particular type of disease caused by Johnsongrass mosaic virus has been described recently in Johnsongrass plants. Some main properties of this new virus species will be briefly described.

VIRUS DISEASES OF JOHNSONGRASS (*SORGHUM HALEPENSE*)

Johnsongrass, genus *Sorghum* in the Gramineae family, is distributed worldwide and easily becomes a weed. Its vigorous rhizomes enable its perennial preservation during adverse conditions. Viruses are well preserved in infected Johnsongrass rhizomes and newly formed shoots therefrom are always infected. Therefore, infected Johnsongrass plants represent a perennial or permanent source of infection. Johnsongrass serves as the main natural host of JgMV and also as the well-known alternative host of MDMV, and other viruses described in other hosts, especially maize.

Johnsongrass Mosaic Potyvirus (JgMV)

Properties of Viral Particles

JgMV particles are flexuous filaments, 750 × 12 nm (Taylor and Pares, 1968; Teakle and Grylls, 1973), and contain a single species of single-stranded RNA of 3100 kDa with 9500 nucleotides, which account for 5% of the particle weight (Gough, 1987, loc. cit. Shukla and Teakle, 1989). The coat protein is a single polypeptide species of about 33.5 kDa, consisting of 303 amino acid residues. N and C termini of the polypeptide chain are surface located with 67 and 18 residues, respectively. The N terminus is acetylated (Shukla et al., 1987).

Stability in Sap

JgMV has a TIP of 60°C, a DEP of 1×10^{-4}, and LIV at 20°C of 2 days (Teakle and Grylls, 1973).

Serology

JgMV exhibits moderate immunogenic activity. Some monoclonal antibody clones have also been produced (Shukla and Teakle, 1989). Microprecipitin test, immunosorbent electron microscopy, gold labeling method, enzyme-linked immunosorbent assay (ELISA), electro-blot immuno assay (EBIA), and other serological methods are suitable for diagnosis of JgMV (Shukla and Teakle, 1989).

JgMV is closely related serologically to watermelon mosaic virus 2 (WmMV2) and distantly to BYMV, MDMV, SrMV, ScMV, and ClYVV (Shukla et al., 1988; 1989; 1989a).

Strains

JgMV was originally considered a strain of MDMV (Taylor and Pares, 1968; McDaniel and Gordon, 1985) or ScMV (Teakle and Grylls, 1973). It became evident after its identification as a distinct virus that it consists of several strains. The type strain is endemic in Australia on Johnsongrass, maize, and sorghum (Teakle and Grylls, 1973; Shukla et al., 1987). The Krish-infecting strain differs from the type strain by its ability to infect sorghum with the Krish resistance gene (Persley et al., 1987). The Texas strain isolated from maize in Texas has been assigned strain O of MDMV (McDaniel and Gordon, 1985). The type and the Texas strains infect oat, but not the Krish-infecting strain. These two strains were differentiated with the aid of monoclonal antibodies (Shukla and Teakle, 1989).

Geographic Distribution

JgMV has been reported in Australia and the U.S. (Shukla and Teakle, 1989).

Host Range and Symptoms

JgMV infects many species, most belonging to the Gramineae. Besides Johnsongrass (*Sorghum halepense*) and wild sorghum (*S. verticilliflorum*), which are perennial hosts and the primary virus reservoirs under field conditions; other species also established as natural JgMV hosts include *Brachiaria miliiformis, B. praetervisa, Cenchrus ciliaris, Dinebra retroflexa, Echinochloa colona, E. crus-galli, E. frumentacea, Eragrostis cilianensis, Panicum miliaceum, Paspalum orbiculare, Pennisetum typhoideum, Setaria anceps, S. italica*, and *S. verticillata* (Teakle and Grylls, 1973; Penrose, 1974; Persley and Greber, 1977). Many other grasses are susceptible to JgMV by mechanical inoculation, some of which do not express symptoms (Penrose, 1974).

Maize (McDaniel and Gordon, 1985), and commercial fodder and grain sorghums, *S. almum, S. bicolor*, and *S. sudanense*, (Teakle et al., 1970; Penrose, 1974; Persley et al., 1977) also represent natural JgMV hosts.

Symptoms caused by JgMV vary, mostly dependant on the host species. JgMV causes mosaic, red stripes, and ringspot symptoms in several *Sorghum* spp., whereas necrosis and stunting may occur also in cultivated sorghum (Teakle et al., 1970; Penrose, 1974a; Persley et al., 1977).

Differential sorghum cvs. following mechanical inoculation with JgMV in the greenhouse react with mild mosaic symptoms (Tošić et al., 1990). Symptoms in maize when infected with JgMV include mosaic, ringspot, and chlorosis (McDaniel and Gordon, 1985). JgMV also infects oat, causing mild mosaic accompanied by necrosis (Tošić et al., 1990).

Pathogenesis and Control

Perennial hosts provide the primary initial source of JgMV under field conditions. Both Johnsongrass and wild sorghum (*Sorghum verticilliflorum*), when infected, are permanent reservoirs for JgMV. Other hosts, including annuals, are also important for providing secondary inocula. JgMV is transmitted nonpersistently from infected to healthy plants by aphids, including *Aphis craccivora*, *A. gossypii*, *Myzus persicae*, and *Rhopalosiphum maidis* (Teakle and Grylls, 1973). Seed transmission has not been reported.

Control measures should be directed primarily toward the eradication of JgMV-infected perennial host plants.

VIRUS DISEASES OF FORAGE LEGUME PLANTS

Virus Diseases of Alfalfa (*Medicago sativa*)

Alfalfa, one of the oldest forage crops for green fodder in agricultural production on all continents, ranks first among legume plants. It is palatable feed for all domestic animals. Stems and leaves are used for processing several kinds of animal feed, including green fodder (by mowing or grazing), hay, silage, and meal (milled alfalfa). The many diseases that affect alfalfa quality and yield are summarized in the *Compendium of Alfalfa Diseases* (Stuteville and Erwin, 1990).

Alfalfa is the natural host of the alfalfa mosaic virus, and also the natural alternative host of several viruses usually infective and more damaging in other legume plants. Virus infections cause economic damage by reducing green matter, seed quality, and yields.

Viruses that reduce alfalfa yields vary, depending on the production year in which infection commenced, inasmuch as it is a perennial plant. When viral infection occurs in the first year, the greater opportunity to spread progressively mechanically or with the aid of vectors results in up to 50% infected plants in the fourth or fifth year (Šutić , 1983).

Alfalfa Mosaic Virus (AMV) [type]

Properties of Viral Particles

AMV, bacilliform in shape, features multicomponent particles of several sizes. The three longest components are infective, defined as B (58 × 18 nm), M (48 × 18 nm), and Tb (36 × 18 nm) (Dunez, 1988). Two or three noninfective components are labeled Ta, To, and Tz (loc. cit. Jaspars and Bos, 1980).

All sizes of AMV particles utilize the same protein (from the 425 strain), composed of 220 amino acids of 24.3 kDa (Jaspars and Bos, 1980). Total protein component molecular weights are 6900 (B), 5200 (M), 4300 (Tb), and 3800 (Ta) kDa, and sedimentation coefficients are 94S (B), 82S (M), 73S (Tb), 66S (Ta), and 60S (To), respectively (Jaspars and Bos, 1980).

AMV particles contain single-stranded RNA. Three positive-sense RNA molecules are encapsulated in B, M, and Tb particles as follows: RNA_1 (1100 kDa), RNA_2 (800 kDa), and RNA_3 (700 kDa), respectively. An isometric particle also contains RNA_4 of 300 kDa, consisting of two molecules as part of that 18-nm diameter isometric particle (Dunez, 1988).

Stability in Sap

The TIP for AMV is 50 to 70°C (usually between 60 and 65°C); the DEP is 1×10^{-3} to 1×10^{-4}; and LIV is 1 to 4 days (Jaspars and Bos, 1980).

Serology

AMV is weakly to moderately immunogenic; some differences in the intensity of this activity depend on the method of virus purification and the properties of individual isolates (strains) (Bancroft et al., 1960). Viral particles of varying length are not distinguishable serologically.

Strains

Numerous strains and variants of AMV have been described, most frequently based on differential symptoms and intensity of reactions in individual hosts (such as alfalfa yellow spot, chilli (*Capsicum*) mosaic, potato calico, potato tuber necrosis, and others) (Bos and Jaspars, 1971). Molecular and biological properties of some strains (AMV-S, AMV 425, AMV 15/64, and VRU) have been well studied (Jaspars and Bos, 1980). Symptomological differences among strains do not necessarily correlate directly either with amino acid composition or serological reactivity (Tremaine and Stace-Smith, 1969).

Geographic Distribution and Economic Importance

AMV, reported worldwide on all continents, essentially can be considered present in all areas of alfalfa production. Some authors suggest that, regardless of the widespread occurrence of AMV, it causes insignificant losses (Hull, 1969); yet others consider it quite damaging. Reactions in ten Yugoslav and foreign alfalfa cultivars show that AMV significantly reduced plant growth and yields, especially the second and subsequent years after sowing (Šutić et al., 1975). AMV infection increases susceptibility to drought and winter freezing (Bos and Jaspars, 1971). Henson and Diachun (1957) showed a 15 to 56% dry matter yield reduction in alfalfa caused by AMV. Frosheiser (1969) reported an 11% decrease in yield of alfalfa hay when 53% of plants were infected by AMV, and 17% in susceptible cultivars when 76% of plants were infected. Beczner (1968) reported 65% green matter yield reductions by AMV in infected alfalfa. In addition to alfalfa, AMV may cause losses in production of clover, pea, potato, tobacco, pepper, tomato, and celery (Van Regenmortel and Pinck, 1981).

Host Range

AMV, with one of the largest host ranges (i.e., more than 400 plant species belonging to 50 families (Dunez, 1988)), naturally infects many herbaceous and some woody plants. All are dicotyledonous, mostly 84 species in the Solanaceae family. About 165 species are natural hosts of AMV (Schmidt, 1977), which is particularly important for the unlimited opportunities for preservation and spread of AMV. Natural hosts other than alfalfa include *Apium graveolens, Beta saccharifera, B. vulgaris, Cannabis sativa, Capsicum annuum, C. frutescens, Daucus carota* ssp. *sativus, Dianthus caryophyllus, Dolichos lablab, Glycine max., Lactuca sativa, Linum usitatissimum, Lycoperisicon lycopersicum, Nicotiana tabacum, Phaseolus vulgaris, Pisum sativum, Ribes rubrum, Robinia pseudoacacia, Solanum melongena, S. tuberosum, Trifolium hybridum, T. incarnatum, T. pratense, T. repens, Vicia faba, Vigna sinensis*, and *Vitis vinifera.*

Natural hosts of AMV also include a large number of annual and perennial weeds such as *Atropa bella donna, Carthamus tinctorius, Chenopodium album, Lamium amplexicaule, Portulaca oleracea, Solanum dulcamara, Sonchus oleraceus, Stellaria media*, and *Trifolium repens*, all of which serve as important inoculum sources.

Suitable plants diagnostic for AMV include *Phaseolus vulgaris* (local and systemic reactions in different strains), *Vigna sinensis* (local spots for most strains; systemic infection for some strains), *Vicia faba* (black necrotic spots in most strains), *Pisum sativum* (necrotic local spots), *Chenopodium amaranticolor* and *C. quinoa* (local and systemic infection), and *Nicotiana tabacum* (local spots in some strains, systemic infection and possible enation growth).

Symptoms

Initial symptoms of AMV in young leaves at the initiation of growth usually are small, round, green-yellow spots. Chlorotic spots enlarge in the form of stripes and rings, or remain uneven.

FIGURE 30 Alfalfa mosaic. (Courtesy C. Hiruki. Reproduced with permission from APS.)

FIGURE 31 Symptoms on alfalfa caused by alfalfa mosaic virus. (Courtesy F. I. Frosheiser. Reproduced with permission from APS.)

Spots on older leaves remain light yellow and whitish. Leaves of plants infected early are crinkled and show retarded growth. Decreased plant vigor results in plant death up to 40% and shorter plants up to 30%, likely as a result of interaction with drought and/or winter frost. Symptoms of disease are striking until the first mowing; then later, particularly at higher temperatures, they become masked and apparently recover. Such symptoms vary within a broad range depending on virus strains, alfalfa cultivar susceptibility, and environmental factors. (See Figures 30 to 34.)

Pathogenesis

Infectivity of AMV is preserved in natural hosts such as *Medicago* spp., *Trifolium* spp., and *Stellaria media*. Potato tubers also serve as a reservoir from year to year.

AMV is transmitted nonpersistently by more than 15 aphid species, including *Acyrthosiphon pisum, A. solani, Aphis craccivora, A. fabae, A. medicaginis, Macrosiphum euphorbiae, Myzus*

FIGURE 32 Alfalfa mosaic virus-infected alfalfa.

ligustri, *M. persicae*, and *Phorodon cannabis*. Aphids can transmit AMV after having fed on the infected plants for only 1 second.

In nature, AMV is preserved and transmitted with the seed of alfalfa, pepper, and weeds (*Datura stramonium*, *Solanum nigrum*, *Chenopodium quinoa*, and possibly *Melilotus* spp.).

The extent of seed transmission (1 to 46%) depends on AMV strains and on alfalfa cultivars and the developmental stage of the plant at time of infection (Beczner and Manninger, 1975; Frosheiser, 1964). Babović (1976) showed 16 to 28% seed transmission of AMV in five alfalfa cultivars. Tošić and Pešić (1975) proved seed transmission in different stages of alfalfa seedling development with the highest percentage (50%) when infection by AMV occurred 10 to 14 days after germination, primary leaflet stage. AMV retains infectivity in alfalfa seed for at least 7 years (Šutić et al. 1975).

AMV is transmitted by several species of dodder (*Cuscuta campestris*, *C. epilinum*, *C. europaea*, *C. lupuliformis*, and *C. subinclusa*), which commonly inhabit alfalfa fields (Schmelzer, 1956). AMV is transmitted mechanically, but the extent of this mode of transmission depends on host plant species. Alfalfa is difficult to infect mechanically because it contains certain substances that can inhibit infection, and thus is probably not important for the spread of AMV in alfalfa fields during harvest activities. However, AMV is easily transmitted mechanically in pepper, where high plant densities and intensive handling occur during replant procedures in pepper nurseries.

Control

Early plant infections by AMV can be avoided by the use of healthy alfalfa seed, which should be produced in areas unfavorable for aphid development. New alfalfa fields for both seed and forage production should be established as far away as possible from established alfalfa and clover fields. One must also ensure adequate spatial isolation from pepper, tomato, potato, peas, celery, and other fields under naturally susceptible species.

FIGURE 33   Alfalfa mosaic virus-infected alfalfa.

FIGURE 34   Alfalfa mosaic virus-infected alfalfa.

Control of aphids in alfalfa fields is useful, particularly in the first and second years, to reduce initial infection and to ensure a long-term stand.

Alfalfa fields with high incidences of AMV infection should be ploughed under, then planted to produce species not susceptible to AMV (i.e., wheat, maize, or sunflower). Resistant or tolerant alfalfa cultivars should be used in production.

Alfalfa Latent Carlavirus (ALtV)

Properties of Viral Particles

ALtV particles are rod-shaped, slightly flexible, and 635 nm long (Veerisetty, 1979). They contain one species of protein (27 kDa) and single-stranded RNA (2450 kDa), with a sedimentation coefficient of 38S. The RNA composes 5.4% of the total viral particle weight (Veerisetty and Brakke, 1977). The TIP of ALtV is 65 to 75°C; the DEP is 1×10^{-3} to 1×10^{-4}; and the LIV at 25°C is 4 to 6 days. ALtV is moderately immunogenic. PeSkV has been reported to be serologically indistinguishable (Hampton, 1981). No information on strains is available.

Geographic Distribution

ALtV has been observed only in Nebraska and Wisconsin (U.S.).

Host Range and Symptoms

ALtV has a restricted host range and experimental infection succeeds mainly in legume plants. Development of disease symptoms depends primarily on temperature. ALtV-infected *Medicago sativa* and *Vicia villosa* plants display no symptoms at 20 and 26°C, and *Pisum sativum* cv. Lincoln plants develop symptoms only at 13°C (Veerisetty, 1979). Diagnostic plants include *Vicia faba* with dark red-brown lesions on inoculated leaves plus apical leaf necrosis, and *P. sativum* cv. Lincoln at 13°C with vein clearing, chlorosis of young leaves, and necrosis of old leaves (Veerisetty, 1979).

Pathogenesis

ALtV overwinters in alfalfa plants, representing a perennial source of inoculum for further infection. ALtV is transmitted mechanically to a few species, mostly in the Leguminosae family. *Acyrthosyphon pisum* aphids are vectors of ALtV, transmitting it nonpersistently. No information exists on seed transmission and insufficient data are available on the nature of spread of ALtV.

Lucerne (Alfalfa) Australian Latent Nepovirus (AAaLtV)

Properties of Viral Particles

AAaLtV is isometric, 25 nm in diameter, and has three sedimenting components in purified preparations: T (top), M (middle), and B (bottom). The protein of these components (T and M + B) consists of a 55-kDa polypeptide (Jones et al., 1979). The T particles contain no nucleic acid, whereas RNA_1 (2400 kDa) and RNA_2 (2100 kDa) have been isolated from B and M particles, respectively, with sedimentation coefficients of 32.5S (RNA_1) and 36S (RNA_2), respectively (Jones et al., 1979). AAaLtV is weakly immunogenic.

Geographic Distribution

This virus is reported only in Australia and New Zealand (Jones et al., 1979).

Host Range and Symptoms

Naturally or experimentally infected alfalfa plants are symptomless. It has been difficult to infect them mechanically with AAaLtV. A few plants in the Leguminosae (*Lupinus angustifolius*, *Medicago* spp., *Pisum sativum*, and *Trifolium* spp.), Chenopodiaceae, and Solanaceae family have been infected experimentally by mechanical inoculation with AAaLtV, and most of them also remained symptomless. *Chenopodium murale* with chlorotic or necrotic local rings followed occasionally by systemic necrotic spots, and *C. quinoa* with necrotic local lesions followed by systemic mottle or tip necrosis, are diagnostic plants (Jones and Forster, 1980).

Pathogenesis

The perennial alfalfa serves as a reservoir for AAaLtV, which is transmitted mechanically with infected plant sap and through seed, both of alfalfa and of *Chenopodium quinoa* (Jones and Forster, 1980). The vectors of AAaLtV have not been proven yet, which makes it difficult to explain its spread in alfalfa crops.

Lucerne (Alfalfa) Transient Streak Sobemovirus (ATtSkV)

Properties of Viral Particles

The ATtSkV isometric particles are 27 and 28 nm in diameter, with proteins of 29 and 32 kDa, respectively. Viral RNA, most probably single-stranded, is 1400 kDa with a sedimentation coefficient of 26S. The TIP of ATtSkV in *Nicotiana clevelandii* sap is 70°C; the DEP is 1×10^{-5}; and the LIV at 24°C is 4 weeks (Forster and Jones, 1980). ATtSkV is weakly immunogenic.

Geographic Distribution

ATtSkV has been described in alfalfa only in Australia and New Zealand. ATtSkV in field trials reduced dry matter yield by 18% (Blackstock, 1978).

Host Range and Symptoms

ATtSkV infects 19 plant species belonging to four families. Chlorotic stripes along lateral veins, accompanied in some cases by leaflet malformations, occur in ATtSkV-infected alfalfa leaves. Symptoms, sometimes transient, disappear gradually. Diagnostic plants include *Chenopodium quinoa* with chlorotic or necrotic local lesions and some plants develop scattered systemic chlorotic blotches; *Nicotiana clevelandii*, which reacts to New Zealand isolates with broad chlorotic mottle but is not susceptible to Australian ATtSkV isolates; and others (Forster and Jones, 1980).

Pathogenesis

The virus occurs commonly in alfalfa crops from which it likely spreads. It is transmitted mechanically, but not through seed of alfalfa, and its vectors are unknown (Forster and Jones, 1980). No reports exist on the mode of spread of ATtSkV in alfalfa crops.

Alfalfa: Natural Alternative Host for Some Viruses

Alfalfa is a possible host of some viruses with other primary hosts, mostly legume plants.

Cucumber Mosaic Virus (CMV)

The first data on significant CMV incidence in alfalfa were obtained in Yugoslavia (Babović, 1969; Šutić et al., 1975). Investigations during the last 10 years show CMV to be quite widespread in alfalfa crops. CMV occurred in 10 of 15 diseased alfalfa plant samples analyzed in 1972 and 1973, an incidence even greater than for AMV. Important for alfalfa production, these results allow a better understanding of CMV epidemiology, since alfalfa represents a perennial reservoir host for a long-term source of inoculum.

CMV causes light-green mosaic/mottle on alfalfa leaves. Leaflets are slightly malformed and plants have less vigor with retarded growth (Figure 34A). These symptoms are similar to those of other alfalfa viruses, especially AMV, and thus are not reliably diagnostic.

Research on modes of transmission of viruses in former Yugoslavia show that CMV is transmitted easily from one alfalfa plant to another with *Myzus persicae* and *Macrosiphon pisi* aphids. Experimental soil transmission was not achieved (Babović, 1969). Factors that influence the epidemiology of CMV in alfalfa are not fully determined.

FIGURE 34A Cucumber mosaic virus-infected alfalfa.

Legume Viruses

Alfalfa is a host of many legume viruses and most frequently BYMV and its pea common mosaic virus (PCmMV) strain, PeEnMV, bean ringspot virus (a strain of the TmBRV) (Schmidt, 1977), PeElBrV, peanut stunt virus (PtSnV), pea leafroll virus (PeLrV) (Figure 35), and PeSdBnMV (Dunez, 1988a,b; Schmidt, 1977). Their descriptions are found in the chapters on their primary hosts.

VIRUS DISEASES OF RED CLOVER (*TRIFOLIUM PRATENSE*)

Red clover as a forage legume plant ranks second immediately following alfalfa, based on production potential. A perennial plant used for 3 to 4 years per crop, it is grown usually as a field crop, but is an equally important component of meadow and pasture populations. Red clover produces high-quality fodder and thus is grown widely in America, Asia, and Europe.

Several viruses causing economically important losses in production have been described with red clover as their primary natural host. It is also the natural alternate host of some widespread viruses (e.g., AxMV, CMV, AMV, and others). In red clover crops, symptoms of disease appear particularly on young plants after mowing.

Red Clover Mottle Comovirus (RClMtV)

Properties of Viral Particles

RClMtV is isometric, 30 nm in diameter, and sediments in purified preparations as three components: the protein RNA-free coat (T) component and two components M and B, each with different RNA contents with sedimentation coefficients of 60S (T), 101S (M), and 127S (B) (loc. cit. Valenta and Marcinka, 1971). The single-stranded RNA accounts for 25% (M) and 36% (B) of the total particle weight. The noninfective M particles increase the infectivity of partially purified B particles (Valenta and Marcinka, 1971). The RClMtV TIP is 70 to 75°C; the DEP is 1×10^{-6}; and the LIV

FIGURE 35 Pea leafroll virus in alfalfa.

at 20°C is over two weeks. RClMtV is strongly immunogenic with serological resemblance to the cowpea mosaic virus group, and the broad bean stain virus in particular, pea green mottle virus, and cowpea mosaic virus (loc. cit. Valenta and Marcinka, 1971). Additionally, other isolates cause various symptoms on broad bean, pea, and red clover (Bos and Maat, 1965). Malak (1974) designated the former Yugoslavian isolate as Yugoslav in order to distinguish it from English, Dutch, and former Czechoslovakian isolates.

Geographic Distribution and Economic Importance

RClMtV is reported in England, Holland, Sweden, former Czechoslovakia, Germany, and former Yugoslavia. Its reported percentage incidences in regions of former Yugoslavia are 40% near Belgrade, 30% in the region of Kruševac, and 60% in Vinča, where it was described also as red clover mosaic based on characteristic disease symptoms (Šutić et al., 1975). RClMtV may cause significant delay of plant growth, causing yield losses that, depending on cultivar and the time of mowing, may exceed 40% (Schmidt, 1977).

Host Range

Natural hosts of RClMtV include red clover (*Trifolium pratense*) and peas (*Pisum sativum*), plus 50 other experimental legume hosts, except for *Chenopodium amaranticolor* and *Gomphrena globosa* (Schmidt, 1977).

Pisum sativum, which expresses mild mosaic followed by general chlorosis, and *Vicia faba*, which expresses local spots followed by systemic infection, are suitable diagnostic plants. RClMtV causes local spot without systemic infection on inoculated leaves of *Phaseolus vulgaris* cv. Topcrop, and others, *C. amaranticolor*, *C. quinoa*, and *G. globosa* (Valenta and Marcinka, 1971).

Symptoms

Initial symptoms of RClMtV usually appear as discolorations along leaf veins, accompanied by partial tissue death (Figure 36). Chlorotic rings, followed by diffused chlorosis spreading into contiguous tissues, are formed between leaf veins. The early occurrence of necrosis causes infected leaflets to twist and remain malformed.

FIGURE 36 Red clover mottle virus-infected red clover.

Pathogenesis

Infectivity of RClMtV is preserved in its annual and perennial hosts in nature. It is easily transmitted mechanically with infected plant sap, which may represent the primary mode of transmission in the nature. Its widespread distribution in nature suggests the existence of a natural vector, yet none have been discovered.

With sufficient knowledge of the primary modes of virus spread, its control can be ensured by the elimination of infection focal points.

Red Clover Necrotic Mosaic Bromovirus (RClNcMV)

Properties of Viral Particles

RClNcMV is isometric, 27 nm in diameter, and the protein capsid is a single polypeptide of 39.5 or 40.8 kDa, depending on the isolates. RClNcMV contains single-stranded RNA of 1550 kDa, accounting for 20% of the total virus particle weight (loc. cit. Hollings and Stone, 1977). The TIP of RClNcMV for most strains in *Nicotiana clevelandii* sap is 85°C (sometimes 90°C); the DEP is 5×10^{-4}; and LIV at 20°C is 10 weeks and at 2°C is 43 weeks. RClNcMV has strong immunogenic activity, which makes serological detection a useful tool. Different virus isolates of RClNcMV include the former Czechoslovakian and the Swedish isolates, serologically related and designated A and B serotypes (Hollings and Stone, 1977).

Geographic Distribution

RClNcMV is reported in Great Britain, former Czechoslovakia, Poland, and Sweden.

Host Range and Symptoms

Trifolium pratense and *Melilotus officinalis* are natural hosts; 47 plant species belonging to 15 families are susceptible to RClNcMV.

Symptoms on infected red clover vary by season. During winter, symptoms appear as severe leaf mottle, distortion, and local necrosis, accompanied by moderate or pronounced stunting of plants. In the summer, symptoms may remain masked. The characteristic symptom on *M. officinalis* is mosaic. RClNcMV causes local lesions on inoculated leaves, such as necrotic local lesions, not systemic on *Chenopodium quinoa*, whitish local lesions, not systemic on *Gomphrena globosa*, chlorotic or necrotic local lesions, not systemic on *N. glutinosa*, and necrotic brown local lesions, not systemic on *Ocimum basilicum*, among others (Hollings and Stone, 1977).

Pathogenesis

In nature, RClNcMV maintains infectivity in red clover as its perennial host. It is transmitted mechanically, but not through seed of red clover. Vectors of RClNcMV are unknown, but it has been transmitted experimentally in soil. No reliable information on its natural spread in red clover fields has been reported; therefore, recommendations of adequate measures for the prevention of infection and production losses are difficult to make.

Red Clover Vein Mosaic Carlavirus (RClVMV)

Properties of Viral Particles

RClVMV particles are tubular, 645×12 nm (Varma, 1970), and vary in length (500 to 740 nm) (Malak, 1974). The RClVMV capsid protein consists of some 2000 subunits, which accounts for 93.75% of total particle weight, and the single-stranded RNA accounts for 6.25% (Varma, 1970). The TIP of RClVMV is 60 to 65°C; the DEP is 1×10^{-3} (or 10^{-4} for some isolates in Serbia); and LIV at 24°C is 6 days. RClVMV is active immunogenically and easily identified serologically. RClVMV, a carlavirus, is distantly related serologically to cactus 2, carnation latent, chrysanthemum B, passiflora latent, potato M, and potato S viruses, and closely related to pea streak virus (Varma, 1970a). It is also related serologically to muskmelon vein necrosis virus (Freitag and Milne, 1970). Several strains of RClVMV have been described, based on symptoms. The P 42 strain, latent in red clover, causes pea and broadbean red leaf, and some isolates differ in virulence in *Chenopodium amaranticolor* and *C. quinoa*. Strains have been designated based on geographical origin (e.g., American, Canadian, and Yugoslav) (Malak, 1974).

Geographic Distribution

RClVMV is distributed worldwide in North America, South Africa, and Europe. In Serbia, it has been detected in 70% of plants (Malak, 1974; Šutić et al., 1975).

Host Range

RClVMV infects plants in Leguminosae, Amaranthaceae, Chenopodiaceae, and Solanaceae families, and natural hosts include *Trifolium pratense*, *T. hybridum*, *T. repens*, *Lathyrus odoratus*, *Medicago lupulina*, *Melilotus albus*, *M. officinalis*, and *Pisum sativum*. RClVMV, when experimentally inoculated, infected 50 species in the above genera (Schmidt, 1977).

Diagnostic plants are *Chenopodium amaranticolor* and *C. quinoa* (local chlorotic spots), *Gomphrena globosa* (violet local spots and systemic mosaic under specific conditions), *Pisum sativum* (systemic mosaic), *Vicia faba* (chlorotic mosaic of apical leaves, later symptomless), and *V. sativa* (dark necrotic local spots and sometimes systemic mosaic infection).

Symptoms

Symptoms of RClVMV on legume hosts include chlorosis along leaf veins, mosaic, streaks, and plant stunting. Lateral vein necrosis and vein yellowing occur on red clover leaves, for which this disease was named, but mosaic symptoms are not characteristic on red clover.

Pathogenesis

RClVMV, preserved in red clover and in other annual and perennial legume plants, is transmitted by the aphids *Acyrthosiphon pisum*, *Cavariella aegopodii*, *Myzus persicae*, *Therioaphis maculata*, and *T. anonidis*. Transmission occurs nonpersistently, wherein aphids acquire RClVMV by feeding on infected plants for at least 2 minutes, and then can inoculate healthy plants in 5 minutes. The virus is easily transmitted mechanically with infective plant sap, and thus probably spread mechanically during cutting of clover fields. RClVMV is not transmitted naturally through red clover seed in the field, as confirmed by studies in Yugoslavia; however, seed transmission of RClVMV has been demonstrated as experimentally possible (loc. cit. Varma, 1970).

Control

Red clover should be seeded in isolation away from old established clover fields, especially for seed production. Chemical control of aphids is recommended, particularly in young crops shortly following establishment of clover fields. Clover seed should be produced in areas unfavorable for aphid development and clover cultivars should be tested to select resistant and tolerant ones for production.

Red Clover Chlorotic Ringspot Virus (RClCRsV) n. sp.

Properties of Viral Particles

RClCRsV with filamentous particles, 779 nm long, has a TIP of 55 to 58°C, a DEP of 1×10^4, and an LIV at 24°C of 4 days. RClCRsV is not serologically related to BYMV and white clover mosaic viruses (WClMV). No strains of RClCRsV have been described.

Geographic Distribution and Economic Importance

RClCRsV was discovered in red clover near Čačak (Serbia) in 1974 with an infection incidence of 40% (Malak, 1974). No data on economic damage caused by RClCRsV have been reported; its distribution is comparatively limited.

Host Range

Red clover (*Trifolium pratense*) is the natural host of RClCRsV. Experimentally infected hosts include Leguminosae (*Pisum sativum* cv. *arvense* and *P. sativum* cv. *saccharatum*, *Melilotus officinalis*, and *Lens esculenta*), Chenopodiaceae (*Chenopodium amaranticolor*, *C. quinoa*, *C. murale*, *C. foetidum*, and *Spinacia oleracea* cv. Matador), and Solanaceae (*Nicotiana glutinosa*, *N. tabacum* cv. Samsun, *N. acuminata*, *N. clevelandii*, and *N. sylvestris*). Suitable assay plants are *Chenopodium* spp. (local spots) and *Nicotiana acuminata* (local spots and later systemic infection).

Symptoms

Chlorotic ringspots (Figure 37) appear abundantly on infected red clover leaves, making them clearly visible. Early symptom occurrence is accompanied by leaflet malformation and depression of plant growth.

RClCRsV causes systemic infection in plants belonging to the Leguminosae, characterized by general mosaic mottling, chlorotic ringspots, and more or less pronounced chlorosis (Figure 38). Infected *Nicotiana glutinosa* and *N. tabacum* cv. Samsun are tolerant to the virus.

Pathogenesis

Myzus persicae transmits RClCRsV nonpersistently, with no data on the role of other aphids. Transmission of RClCRsV by aphids into clover seedlings was done with experimental inoculations from infected red clover and *Melilotus officinalis* seedlings. Aphids appear as the main virus vectors in nature.

RClCRsV is mechanically transmitted with difficulty to red clover (about 15%), but more easily to other hosts. RClCRsV is transmitted through neither seed nor soil.

Relationships

RClCRsV isolates differ from possible related red clover viruses, based on infection, serological, and symptomological traits. The virus particle is much longer than BYMV, RClVMV, and ClYVV, and the primary symptoms differ. RClCRsV seems not related to the pea common mosaic virus BYMV-strain, which is not transmitted through red clover seed. The strains of RClCRsV are related

Virus Diseases of Forage Feed Plants

FIGURE 37 Red clover chlorotic ringspot virus-infected red clover.

FIGURE 38 Red clover chlorotic ringspot virus-infected sweet clover.

serologically to neither BYMV nor WClMV. Based on all comparative features tested, the investigated isolates have been identified as the red clover chlorotic ringspot virus n. sp., with the need for further investigations of the nature of this virus and its relationship to other viruses.

Control

Control of aphids in clover fields, particularly at the time of crop emergence, is highly recommended. Spatial isolation of the young crop from older established fields of clover is also a prerequisite for clover seed production.

FIGURE 39 Clover yellow mosaic virus-infected red clover.

Clover Yellow Mosaic Potexvirus (ClYMV)

ClYMV, a single-component virus, has elongate particles of 540 nm; no chemical composition data have been published (Bos, 1973). The TIP of ClYMV is 58 to 60°C; the DEP is 1×10^{-4}; and the LIV exceeds 6 months. ClYMV is moderately immunogenic (Pratt, 1961).

ClYMV is widespread in the western U.S. and Canada. It infects several legumes and some non-legumes. Its natural hosts are clovers with characteristic symptoms of plant stunt, leaf malformation, and reduced yields and winter hardiness. It has been isolated from alfalfa, sweet clover, *Chenopodium album*, and *Stellaria media* plants (Pratt, 1961), pea crops where it causes necrotic streaks (Ford and Baggett, 1965), and from apple trees with symptoms of leaf pucker disease (Welsh et al., 1973). Diagnostic species include *Antirrhinum majus* with local necrotic rings and systemic chlorotic mottling, *C. amaranticolor* with chlorotic local lesions and systemic chlorotic mottling, and *C. sativus* with transient chlorotic spots and systemic infection with some isolates (Bos, 1973).

ClYMV is transmitted mechanically and through seed of *Trifolium pratense* in low percentages (Hampton, 1963). No vectors have been reported; thus, insufficient information exists on its spread in nature.

Clover Yellow Vein Potyvirus (ClYVV)

ClYVV is filamentous, 760 nm, with a sedimentation coefficient of 159S, and composed of single-stranded RNA and one protein species 34.4 kDa (Hollings and Stone, 1974). The TIP of ClYVV in *Nicotiana clevelandii* sap is 55 to 60°C; the DEP is 1×10^{-4} to 1×10^{-5}; and the LIV at 18°C is 8 days and is 11 weeks at 0°C. ClYVV is strongly immunogenic, thus easily identified serologically and its strains differ slightly, depending on the host species (Hollings and Stone, 1974).

ClYVV, discovered in Britain and Canada, has been recorded in the U.S. (loc. cit. Hollings and Stone, 1974). ClYVV hosts are clover species—especially white clover, in which it causes vein yellowing and leaf mottle. ClYVV causes vein clearing and mosaic symptoms on *Coriandrum sativum* leaves. Virus hosts include plants belonging to some six families. *Chenopodium*

FIGURE 40 Bean yellow mosaic virus-infected red clover.

amaranticolor with local lesions and no systemic necrosis, and *Nicotiana clevelandii* with local lesions and systemic chlorotic mottle, among others, are suitable diagnostic plants (Hollings and Stone, 1974).

ClYVV is transmitted mechanically with infective plant sap and by *Acyrthosiphon pisum*, *Aulacorthum solani*, *Macrosiphum euphorbiae*, and *Myzus persicae*, which transmit nonpersistently (loc. cit. Hollings and Stone, 1974). Virus spread in nature depends on the presence of the sources of infection and aphid vector activity.

Red Clover: Natural Alternative Host for Some Viruses

Red clover is infected by damaging viruses that have as their primary hosts mostly legume plants described elsewhere. These viruses include AMV, BYMV, and its Pea strain PeCmMV, CMV, PeElBrV, PeEnMV, and WClMV (Schmidt, 1977; Šutić, 1983). Detailed descriptions of these viruses are found in chapters on their main hosts; thus, only certain features on red clover will be discussed here.

Alfalfa Mosaic Virus (AMV)

AMV occurs in nature, but seldom in red clover. Isolates of AMV from red clover are typical (Grbelja, 1974; Malak, 1974). Research shows that AMV is transmitted neither through infected red clover seed, nor through the soil containing infected plant residues. It is important to spatially isolate red clover from alfalfa fields because each may have a number of common virus pathogens.

Bean Yellow Mosaic Virus (BYMV)

BYMV, economically damaging in red clover, is widespread. In Serbia, for example, it was discovered in four of ten localities studied. It frequently occurs jointly with the related strain pea common mosaic virus (PeCmMV) with an incidence of 30 to 70%. BYMV infected red clover plants with delayed growth, 25 to 30% shorter stems, and 29% fewer flower heads (Babović, 1969a) (see Figure 40). Some investigations in Serbia showed that BYMV is transmitted at 0.4%

through red clover seed. Even at this low percentage, it becomes an important parameter in epidemiology because it is transmitted readily by aphids in red clover fields (Šutić , 1983). Interestingly, the PeCmMV strain is transmitted more often through red clover seed (0.1 to 5.3%); thus, red clover represents a permanent natural reservoir for viruses (Šutić , 1973).

Sowing virus-free seed, spatial isolation of production fields from susceptible plant hosts, and aphid control, particularly in newly seeded fields to avoid early infections, play an important role in the control of BYMV in clover fields. Earlier mowing may reduce green matter losses. In addition, more resistant red clover populations and cultivars should be used in production. Further research is needed.

Cucumber Mosaic Virus (CMV)

CMV infects red clover plants in nature and has been recovered readily from samples in parts of Serbia (Malak, 1974a; Šutić , 1983). The presence of CMV in red clover suggests the potential of red clover as a perennial reservoir source of infection. CMV is transmitted neither through infected red clover seed, nor in the soil containing infected plant residues. Precautionary measures, particularly spatial isolation from other susceptible crops and aphid control, may be recommended to prevent occurrence of infection in red clover fields.

Pea Early Browning (PeElBrV) and Pea Enation Mosaic Virus (PeEnMV)

PeElBrV and PeEnMV can be isolated from infected red clover often associated with other viruses. No data exist on damages caused by these viruses.

White Clover Mosaic Virus (WClMV)

WClMV occurs frequently in red clover, usually associated with other viruses (BYMV, CMV, and others). WClMV can be transmitted through seed of red clover up to 6% (Hampton, 1963), but this has not been confirmed by others (Šutić et al., 1975). The vicinity of red clover to other legume crops is important for potential spread of WClMV in nature.

Virus Diseases of Alsike Clover (*Trifolium hybridum*)

Alsike clover, which inhabits mostly meadows and pastures and is less widespread than red clover, is naturally infected most frequently with viruses that cause symptoms of leaf mosaic (Tanasijević and Šutić , 1958; Šutić , 1983), but also leaf vein mosaic.

Single or mixed viral infections are caused by AMV, the broad bean stain virus (BBSV), the PeCmMV strain of BYMV, and PeEnMV and WClMV (Schmidt, 1977; Šutić , 1983). Mosaic symptoms occur as mosaic and ringspot mottle patterns on leaves, plus depressed plant growth. Symptom type is dictated by the specific virus.

The alsike clover vein mosaic has been described based on specific symptoms in leaf veins, but no other information exists. CMV also has been isolated from naturally infected alsike clover plants (Babović , 1974). Alsike clover is quite important among perennial hosts to preserve and spread this virus. General instructions listed for protection of seed clover against CMV are also recommended for protection of alsike clover.

Virus Diseases of Crimson Clover (*Trifolium incarnatum*)

Crimson clover is an annual plant for use both as green fodder (for grazing and mowing) and as hay. It is a natural host of widely distributed legume viruses such as AMV, BYMV (and its PeCmMV strain), PeElBrV, and PeEnMV (Schmidt, 1977).

The crimson clover mosaic strain of BYMV is described, based on characteristic symptoms on infected plants. Symptoms in infected leaves are mosaic and chlorosis along the veins, and sometimes leaf malformations. Green matter yields are reduced by 40% in infected plants (Schmidt, 1977).

Crimson clover ringspot virus (CicRsV) described in northern Europe has isometric particles (Engsbro, 1972). Chlorotic spots occur first in infected leaves and gradually turn into brown or black ringspots. This virus is transmitted neither mechanically nor through the seed.

VIRUS DISEASES OF SUBTERRANEAN CLOVER (*TRIFOLIUM SUBTERRANEUM*)

Subterranean Clover Mottle Sobemovirus (SrcMtV)

SrcMtV occurs in epidemics on subterranean clover in Australia. Viral particles are isometric, 30 nm in diameter, and contain single-stranded RNA of 1500 kDa. Some isolates contain one or both of the two circular single-stranded RNA molecules (Francki et al., 1988). The viral coat protein consists of two polypeptide molecules of 29 (larger) and 26 (smaller) kDa, possibly a result of degradation of the larger protein.

Symptoms in infected plants include leaf mottle, distortion, and severe stunt. SrcMtV has been successfully isolated from *Trifolium globeratum* and transmitted mechanically into a few legume plants. *Pisum sativum* with necrotic local lesions and others are suitable diagnostic plants. Vectors of SrcMtV are unknown. Viral particles have been discovered in commercial clover seed (up to 10%) and isolated from 3% of seedlings grown from this seed (Francki et al., 1988).

VIRUS DISEASES OF WHITE CLOVER (*TRIFOLIUM REPENS*)

White clover, among the most widespread of legumes, is encountered mostly in plowed fields, but also frequently in meadows. Some cultivars are grown as field crops. Their high nutritive value for animal feed comes from both green fodder in grazing and dry hay. White clover is the primary host of the WClMV and WClCcV 2, and the alternate natural host of several viruses, playing an important role in their development and preservation cycles.

White Clover Mosaic Potexvirus (WClMV)

Properties of Viral Particles

The elongated WClMV particles are 480×13 nm and single component with a sedimentation coefficient of 119S (Varma et al., 1970). WClMV RNA, probably single-stranded, is 2400 kDa (Koenig, 1971), and accounts for 6% of the virus weight (Varma et al., 1970). Its coat protein subunit is 14 kDa (Miki and Knight, 1967). The TIP of WClMV is 55 to 60°C; the DEP is 1×10^{-5} to 1×10^{-6}; and LIV at 24°C is 10 to 99 days (Bercks, 1971). WClMV has strong immunogenic activity and, in liquid media, the antisera react with fluffy sedimentations. Based on host plant symptoms, some isolates have been identified as separate strains, such as the Canadian, with local lesions on *Gomphrena globosa*, and the Dutch with systemic infection of cowpea plants, isolates.

Geographical Distribution and Economic Importance

WClMV, widespread in diverse areas, is found in North America, New Zealand, and many European countries. In some regions, it occurs naturally in a large number of plants (Tanasijević and Šutić, 1958). WClMV stunts plant growth and causes damage only in the production of white clover cultivars.

Host Range

Some economically important plants infected in nature include *Trifolium repens*, *T. pratense*, *T. hybridum*, *Medicago sativa*, *M. lupulina*, *Phaseolus vulgaris*, *Pisum sativum*, and *Vicia faba*. Plants susceptible to artificial infection include *Chenopodium album*, *Citrullus vulgaris*, *Cucurbita pepo*, *Lycopersicum esculentum*, *Nicotiana sylvestris*, and *T. hybridum*. Suitable diagnostic plants include peas (wilting of inoculated leaves that leads to the name pea wilt virus), and cucumber (yellow green local lesions on inoculated cotyledons and systemic infection of permanent leaves) (Bercks, 1971).

Symptoms

Light- to yellow-green striped zones occur on infected white clover leaves. Chlorotic ringspots or irregular spots, later turning dark, frequently appear between lateral veins of infected leaves, which may be slightly distorted. Symptoms are more pronounced in the first half of the season and may be masked at high summer temperatures.

Pathogenesis

WClMV, preserved in several hosts including many of perennial habit, is transmitted mechanically in the sap of infected plants. Although some data about aphid transmission are contradictory, *Acyrthosiphon pisum* transmits WClMV nonpersistently. WClMV is transmitted through seed of neither white nor red clover (Šutić et al., 1975), although Hampton (1963) reported transmission through red clover seed of 6%.

Control

Due to the general distribution of WClMV, white clover should be grown with adequate isolation from other legume forage plants known as virus hosts. Infected plants should be destroyed early in production in order to eliminate them as sources of infection. For the same reason, naturally occurring white clover populations close to susceptible crops should be destroyed. Only white clover cultivars and populations resistant and tolerant to infection by WClMV should be grown.

White Clover Cryptic Cryptovirus 2 (WClCcV2)

WClCcV2 isometric particles, 38 nm in diameter, contain double-stranded RNA of 1490 or 1380 kDa, which account for 24% of the particle weight (Luisoni and Milne, 1988). No data exist on the composition of the viral protein. WClCcV2 is antigenic (loc. cit. Luisoni and Milne, 1988).

WClCcV2 has been discovered only in white clover plants which are symptomless with low virus concentrations. It has been discovered in white clover plants produced from the seed originating from Europe, Japan, New Zealand, and the U.S., and it is transmitted through white clover seed, being detected in 15 to 41% of the seedlings tested (Boccardo et al., 1985). Virus vectors are unknown. WClCcV2 is transmitted neither mechanically nor by grafts, which substantially complicates the experimental testing and completion of a host range.

White Clover: Natural Alternative Host for Some Viruses

White clover is the alternate natural host of some viruses very important in general crop production (Schmidt, 1977; Šutić, 1983). This group of viruses includes AMV, BYMV, pea leafroll virus (PeLRlV syn. BLRlV), CMV, and PeElBrV.

Cymbidium ringspot virus (CyRsV) was discovered in Cymbidium orchid and in white clover plants in nature (Hollings and Stone, 1977a). Vectors of CyRsV are unknown, but it is transmitted mechanically. CyRsV is exuded from infected plant roots into the surrounding soil from which other healthy plants can be infected. Chlorotic flecks, mottle, and slight stunt symptoms characterize CyRsV-infected white clover plants.

White clover ring mottle (WClRMtD), caused by *arabis mosaic virus* (ArMV) in nature, also occurs in white clover plants (Gibbs et al., 1966). Symptoms, which become clearly visible during periods of cool weather (mild chlorotic ringspots), remain mostly masked.

White clover enation disease (WClEnDs), believed caused by a bacilliform virus of 80 × 200 nm, has been described in white clover (Bos and Grancini, 1963; Rubio-Huertos and Bos, 1969). No information exists on the properties of this virus; it is transmitted by both grafting and *Cuscuta campestris*, but not through seed of white clover. Specific symptoms of irregular swellings appear along the veins on the lower surface of infected leaves, involving slower growth; resultant leaves become distorted and asymmetrical.

Peanut stunt virus (PtSnV) has been identified as being widely distributed in southern U.S. pasture legumes, especially in white clover. Infected plants are weakened and stunted in greenhouse and growth chamber studies. A recent well-designed field study measured losses of white clover yield in a complex of factors, including root knot nematode (*Meloidogyne incognita*), PtSnV infection, and water stress. McLaughlin and Windham (1996) found that the PtSnV infection component reduced cumulative herbage yield by 14% in the first year and by 28% in the second year.

Virus Diseases of Birdsfoot Trefoil (*Lotus corniculatus*)

Birdsfoot trefoil is a valuable perennial legume important as a forage for animal grazing. It is a widespread volunteer plant of diverse habitats, occurring regularly in all types of meadows and pastures.

Few investigations have been done with birdsfoot trefoil as a host of viruses. The first data on the occurrence of viruses in birdsfoot trefoil were from the U.S. for TRsV and TmRsV (Ostazeski, 1965; Ostazeski and Scott, 1966); in Europe, it was described as the host of TRsV (Bode and Klinkowski, 1968). CMV was isolated and identified from naturally infected birdsfoot trefoil in former Yugoslavia (Buturović, 1974). Symptoms of CMV in birdsfoot trefoil include mosaic, mottle, leaf distortion, and stunting of plants. Aphid (vector) control and spatial isolation from other crops susceptible to CMV are recommended to help reduce losses (Šutić , 1983).

Virus Diseases of Vetch (*Vicia* Spp.)

Vetch species, both annual and perennial, used for animal feed in broad geographic areas represent high-quality green fodder, hay, silage, or a prepared concentrate of grain. Vetches serve as alternative natural hosts of several viruses that occur in other legume plants in nature.

Although bearing specific names, many viruses that occur in vetch are caused by viruses primarily described already on other legume plants. *Narbonne vetch leafroll* is a virosis of *Vicia narbonensis* caused by PeLRIV in nature (Quantz and Völk, 1954). Severe chlorosis, erect leaves, and depression of plant growth are the main disease symptoms.

Vetch spotted crinkle disease, a virosis occurring in *Vicia narbonensis*, *V. sativa*, and *V. villosa* plants is caused by PeEnMV (Quantz, 1951). It first causes a mosaic, followed by leaf crinkle, which then results in the curled appearance of the whole plant. Both *V. villosa* and *V. vulgare* growing along roadways and in pastures in northwestern U.S. are naturally infected with PeEnMV, semipersistently transmitted by aphids (personal observations R.E. Ford, W.J. Zaumeyer, and R.O. Hampton). However, an intensive survey failed to confirm wild vetch in Idaho as a host of PeEnMV (personal communication R.O. Hampton). Vetch is not a natural host for BYMV or BCmMV, but BLRIV, and on rare occasions, RClVMV have been isolated (personal communication R.O. Hampton).

Vetch vein banding chlorosis disease, thus far described on naturally infected *V. villosa* plants, is caused by the broad bean stain virus (BBSV) (Tapio, 1970). Severe chlorosis and downward leaf twisting are main symptoms of disease. BBSV is transmitted through the infected seed of *V. villosa* (10%).

Vetch yellow mosaic disease, occurring in *V. sativa*, *V. villosa*, and other *Vicia* spp., is caused by BYMV (Quantz, 1961; Roland, 1969). Symptoms are dark- to yellow-green stripes along the veins on the lower leaf surface, accompanied by leaf and stalk necrosis, which gives plants a distorted appearance.

Detailed information on the properties of these viruses can be found in their descriptions on main hosts.

Other Forage Legumes: Hosts for Some Viruses

In conclusion, it should be stressed that several viruses cause serious economic damage in other forage legumes not yet discussed. Some deserve special mention as virus hosts.

Lupins, *Lupinus* spp.: Both annual and perennial, especially sweet lupins, are particularly useful as green fodder, hay, or silage. BYMV, known for its very broad host range, occurs naturally in *L. affinis*, *L. albus*, *L. angustifolius*, and *L. luteus*. Symptoms in infected plants vary, depending on the host species (Verhoyen and Meunier, 1988). A severe mottle accompanied by malformations occurs in infected white lupin (*L. albus*) leaves. Infected blue lupin, *L. angustifolius*, leaves first displaying bronze coloration and mild wilt, later show unilateral stem necrosis. On infected yellow lupin, *L. luteus*, initial symptoms include vein clearing and light mottle, then later fully developed leaves remain narrow, distorted, and usually erect. Auxiliary bud proliferations occur, resulting in a branchy plant appearance.

Sweetclover, *Melilotus* spp. Both annual and perennial, are important as green fodder and grazing.

Sweetclover necrotic mosaic virus (StcNcMV): *M. albus* and *M. officinalis* are natural hosts of StcNcMV, a member of the Dianthovirus group. Isometric StcNcMV particles, 34 nm in diameter, contain both single-stranded RNA (which accounts for 20% of the total particle weight) and a single protein of 38 kDa (Hiruki, 1986). Its TIP in *Phaseolus vulgaris* sap is 90°C; the DEP is 1×10^{-6} to 1×10^{-7}; and its LIV in dry leaves is 10 weeks at 22°C. StcNcMV has strong immunogenic activity (Hiruki, 1986).

It has been described as infecting both melilots and alfalfa in Alberta, Canada (Inouye and Hiruki, 1985). Symptoms in infected melilot leaves include mosaic, ringspots, and veinal necrosis. StcNcMV has been transmitted mechanically to 16 dicot species. Diagnostic hosts include *Chenopodium amaranticolor* with necrotic local lesions without systemic infection, *C. quinoa* with necrotic local lesions and no systemic infection, and *Nicotiana clevelandii* with necrotic local lesions and systemic chlorotic rings and mottle. StcNcMV is transmitted neither by plant-to-plant contact, nor through drainage water. Viral particles exuded from infected plant roots can infect neighboring healthy plants (Hiruki, 1986).

Sweetclover latent virus (StcLV) (also known as Melilotus latent virus): StcLV has been isolated and identified from naturally infected Melilot plants. StcLV particles, 80 × 300 to 350 nm, belong to the Rhabdovirus group (Kitajima et al., 1969). StcLV is widespread in North America, its vector is unknown, and it has been transmitted only by grafting.

M. albus and *M. officinalis* are usually infected with BYMV in nature (Schmidt, 1977; Verhoyen and Meunier, 1988). First symptoms on infected leaves are small light yellow spots that later coalesce to form light-green zones; then vein clearing and leaf streaking occurs frequently.

BLRlV has been observed in *Melilotus* spp. AMV (Schmidt, 1977) and TSkV (Fulton, 1988) has been found in *M. albus*. RClNcMV is infective in *M. officinalis*, in which it causes mosaic (Hollings and Stone, 1977).

Onobrychis spp.: Sainfoins are perennial plants used in animal nutrition for green fodder, grazing, or hay. BYMV and BLRlV, as widely spread legume plant viruses, have been discovered and identified in *O. sativa* (*O. viciaefolia*) plants (Schmidt, 1977).

REFERENCES

Babović, M. 1976. Stepen prenosivosti virusa mozaika lucerke njenim semenom. *Mikrobiologija* 13:83–88.

Babović, M. 1969. Viroze lucerke u Jugoslaviji. *Zaštita bilja* 102:335–410, Beograd.

Babović, M. 1969a. Uticaj virusa žutog mozaika pasulja na porast, bokorenje i formiranje cvasti u crvene deteline. *Prvi kongres mikrobiologa Jugoslavije* p. 710–713.

Babović, M. 1974. Pojava viroza hibridne deteline u Jugoslaviji. *Mikrobiologija* 11:159–164.

Bancroft, J.B. et al. 1960. The antigenic characteristic and the relationship among strains of alfalfa mosaic virus. *Phytopathology* 50:34–39.

Beczner, L. 1968. A lucerne mosaic virus elterjedése magyarországon es gazdasságy jeletosége. *Kül. a Kis. Közl. L. XI/A Növénytermesztés*, 1-3 sz., p. 51–65.

Beczner, L. and Manninger, S. 1975. A lucerne mosaic virus epidemiógiája levéltetüés megatoiteli vizgéilatok Kül. *A. Noven. Kutató Intézet Evkonyve*, Vol. XII, sz., p. 167–176.
Bercks, R. 1971. White clover mosaic virus. *CMI/AAB Descriptions of Plant Viruses*, No. 41.
Blackstock, J.Mck. 1978. Lucerne transient streak and lucerne latent, two new viruses of lucerne. *Aust. J. Agric. Res.* 29:291–304.
Boccardo, G. et al. 1985. Three seedborne cryptic viruses containing double-stranded RNA isolated from white clover. *Virology* 147:29–40.
Bode, O. and Klinkowski, M. 1968. Handelspflanzen. In: Klinkowski, M. et al., Eds., *Pflanzliche Virologie*, Vol. II/1, Akademie-Verlag, Berlin, p. 161–187.
Bos, L. and Grancini, P. 1963. Ein eigenaardige enatie - virus ziekte bij witte klaver. (Abstr.) *Neth. J. Plant Pathol.* 69:151.
Bos, L. and Maat, L.D. 1965. A distinct strain of the red clover mottle virus in the Netherlands. *Neth. J. Plant Pathol.* 71:8–13.
Bos, L. and Jaspars, E.M.J. 1971. Alfalfa mosaic virus. *CMI/AAB Descriptions of Plant Viruses*, No. 46.
Bos, L. 1973. Clover yellow mosaic virus. *CMI/AAB Descriptions of Plant Viruses*, No. 111.
Brook, A.J. 1972. Lolium mottle virus. *Plant Pathol.* 21:118–120.
Buturović, D. 1974. Neke karakteristike jednog virusnog izolata iz smiljkite. *Mikrobiologija* 11:173–180, Beograd.
Catherall, P.L. 1967. Anthoxanthum mosaic virus. *8th Rept. Welsh Pl. Breed.* 1966:116–117.
Catherall, P.L. 1970. Cocksfoot mottle virus. *CMI/AAB Descriptions of Plant Viruses*, No. 23.
Catherall, P.L. 1970a. Anthoxanthum mosaic virus. *Plant Pathol.* 19:125–127.
Catherall, P.L. 1971. Cocksfoot streak virus. *CMI/AAB Descriptions of Plant Viruses*, No. 59.
Catherall, P.L. et al. 1977. Host-ranges of cocksfoot mottle and cynosurus mottle viruses. *Ann. Appl. Biol.* 87:233–235.
Catherall, L.P. et al. 1980. A new virus disease of sweet vernal grass (*Anthoxanthum* L.). (Abstr.) *3rd Conf. Virus Dis. Gramineae Europe*, Rothamsted, 28–30 May.
Catherall, L.P. 1985. Resistance of grasses to two sobemoviruses, cocksfoot mottle and cynosurus mottle. *Mitteilungen aus der Biologishen Bundesanstalt für Land und Forstwirtschaft Berlin - Dahlen* 228:92–93.
Dunez, J. 1988. Alfalfa mosaic virus. *In*: Smith, I. M. et al., Eds. *European Hand. Pl. Dis.*, Blackwell Sci. Publ., London, p. 68–69.
Dunez, J. 1988a. Peanut stunt virus (PSV). *In*: Smith, I.M. et al., Eds. *European Hand. Pl. Dis.*, Blackwell Sci. Publ., London, p. 13.
Dunez, J. 1988b. Pea early browning virus (PEBV). *In*: Smith, I. M. et al., Eds. *European Hand. Pl. Dis.*, Blackwell Sci. Publ., London, p. 81.
Engsbro, B. 1972. Viroser hos landbrugsplanter. Plantesyngdomme i Denmark 1971 (Lyngby) Nr. 88, S. 26.
Ford, R.E. and Baggett, J.R. 1965. Reactions of plant introduction lines of *Pisum sativum* to alfalfa mosaic, clover yellow mosaic, and pea streak viruses, and to powdery mildew. *Pl. Dis. Reptr.* 49:787–789.
Foster, R.L.S. and Jones, A.T. 1980. Lucerne transient streak virus. *CMI/AAB Descriptions of Plant Viruses*, No. 224.
Francki, R.I.B. et al. 1988. Subterranean clover mottle virus. *CMI/AAB Descriptions of Plant Viruses*, No. 329.
Freitag, J.H. and Milne, K.S. 1970. Host range, aphid transmission and properties of muskmelon vein necrosis virus. *Phytopathology* 60:166–170.
Frosheiser, F.I. 1964. Alfalfa mosaic virus transmitted through alfalfa seed. (Abstr.) *Phytopathology* 54:839.
Frosheiser, F.I. 1969. Variable influence of alfalfa mosaic virus strains on growth and survival of alfalfa and on mechanical and aphid transmission. *Phytopathology* 59:857–862.
Fulton, R.W. 1988. Tobacco streak virus (TSV). *In*: Smith, I.M. et al., Eds. *European Hand. Pl. Dis.*, Blackwell Sci. Publ., London, p. 18–20.
Gibbs, A.J. et al. 1966. Viruses occurring in white clover (*Trifolium repens* L.) from permanent pastures in Britain. *Ann. Appl. Biol.* 58:231–240.
Gibson, W.R. 1980. Interactions between methods of cultivating ryegrass and the spread of ryegrass mosaic virus. (Abstr.) *3rd Conf. Virus Dis. Gramineae Europe*, Rothamsted, 28–30 May.
Grbelja, J. 1974. Alfalfa mosaic virus on clovers in Yugoslavia. *Acta. Bot. Croat.* 33:23–29.
Guy, P.L. 1990. Incidence of PAV-like and RPV-like viruses in four pasture species on two sites at Sandford, Tasmania. *In*: Burnett, P.A., Ed. *1990 World Perspectives on Barley Yellow Dwarf*, CIMMYT, Mexico, D.F., Mexico, p. 329–331.

Hampton, R.O. 1963. Seed transmission of white clover mosaic and clover yellow mosaic viruses in red clover. (Abstr.) *Phytopathology* 53:1139.

Hampton, R. O. 1981. Evidence suggesting identity between alfalfa latent and pea streak viruses. (Abstr.) *Phytopathology* 71:223.

Henson, L. and Diachun, S. 1957. Effect of a strain of alfalfa mosaic virus on the yield of clonally propagated Atlantic alfalfa. (Abstr.) *Phytopathology* 47:45.

Hiruki, C. 1986. Sweet clover necrotic mosaic virus. *CMI/AAB Descriptions of Plant Viruses*, No. 321.

Hollings, M. and Stone, O.M. 1974. Clover yellow vein virus. *CMI/AAB Descriptions of Plant Viruses*, No. 131.

Hollings, M. and Stone, O.M. 1977. Red clover necrotic mosaic virus. *CMI/AAB Descriptions of Plant Viruses*, No. 181.

Hollings, M. and Stone, O.M. 1977a. Cymbidium ringspot virus. *CMI/AAB Descriptions of Plant Viruses*, No. 178.

Hollings, M. and Brunt, A.A. 1981. Potyviruses. *In*: Kurstak, E., Ed. *Hand. Pl. Virus Infec. - Comp. Diag.*, Elsevier/North-Holland Biomedical Press, p. 731–807.

Hull, R. 1969. Alfalfa mosaic virus. *Adv. Virus. Res.* 15:365–433.

Huth, W. and Paul, L.H. 1972. Cocksfoot mild mosaic virus. *CMI/AAB Descriptions of Plant Viruses*, No. 107.

Huth, W. 1974. Molinia streak virus—a new virus on grasses. *Mikrobiologija* 11:195–196, Beograd.

Huth, W., et al. 1974. Molinia streak virus: a virus with isometric particles from *Molinia coerules*. *Intervirology* 2:345–351.

Huth, W. 1988. Cocksfoot mild mosaic virus (CMMV). *In*: Smith, I. M. et al., Eds. *European Hand. Pl. Dis.*, Blackwell Sci. Publ., Oxford, p. 70.

Inouye, T. and Hiruki, C. 1985. Sweet clover necrotic mosaic virus. *Ann. Phytopathol. Soc. Japan* 51:82.

Jaspars, E.M.J. and Bos, L. 1980. Alfalfa mosaic virus. *CMI/AAB Descriptions of Plant Viruses*, No. 229.

Johnstone, G.R., et al. 1990. Epidemiology and control of barley yellow dwarf viruses in Australia and New Zealand. *In*: Burnett, P.A., Ed. *World Perspectives on Barley Yellow Dwarf*, CIMMYT, Mexico, D.F., Mexico, p. 228–239.

Jones, A.T. et al. 1979. Purification and properties of Australian lucerne latent virus, a seed-borne virus having affinities with nepoviruses. *Ann. Appl. Biol.* 92:49.

Jones, A.T. and Forster, R.L.S. 1980. Lucerne Australian latent virus. *CMI/AAB Descriptions of Plant Viruses*, No. 225.

Kitajima, E.W. et al. 1969. Morphology and intracellular localization of a baclliform latent virus in sweet clover. *J. Ultrastruct. Res.* 29:141–150.

Kleinhempel, K. 1988. Ryegrass mosaic virus (RyMV). *In*: Smith, I.M. et al., Eds. *European Hand. Pl. Dis.*, Blackwell Sci. Publ., London, p. 48.

Koenig, Renate. 1971. Nucleic acids in the potato virus X group and in some other plant viruses: comparison of the molecular weights by electrophoresis in acrylamide-agarose composite gels. *J. Gen. Virol.* 8:111–114.

Lapierre, H. and Hariri, D. 1988. Cocksfoot mottle virus. *In*: Smith, I.M. et al., Eds. *European Hand. Pl. Dis.*, Blackwell Sci. Publ., London, p. 79–80.

Leclant, F. and Signoret, P. 1974. Premiers résultats concernant la transmission d'un rhabdovirus du blé le Delphacide laodelphax striatellus Fall. *In*: *Quatriemés, C.R. Journées de Phytiatrie - Phytopharmacie Circummediterranéennes*, Montpelier, p. 215–217.

Lesemann, D. and Huth, W. 1975. Nachweis von maize rough dwarf virus - ähnlichen Partikeln in Enationen von Loliumpflanzen aus Deutschland. *Phytopathol. Z.* 82:246–253.

Lewis, C.G. and Heard, J.A. 1980. The incidence of ryegrass crops of Eriophyid mites, vectors of ryegrass mosaic virus. (Abstr.) *3rd Conf. Virus Dis. Gramineae Europe*, Rothamsted, 28–30 May.

Lister, R.M. and Bar-Joseph, M. 1981. Closteroviruses. *In*: Kurstak, E., Ed. *Hand. Pl. Virus Infec. Comp. Diag.*, Elsevier/North-Holland Biomedical Press, p. 809–841.

Luisoni, E. and Milne, R.G. 1980. Some properties of ryegrass cryptic virus. (Abstr.) *3rd Conf. Virus Dis. Gramineae Europe*, Rothamsted, 28–30 May.

Luisoni, E. and Milne, R.G. 1988. White clover cryptic virus 2. *CMI/AAB Descriptions of Plant Viruses*, No. 332.

Lundsgaard, T. and Albrechtsen, S.E. 1976. Electron microscopy of rhabdovirus-like particles in *Festuca gigantea* with leaf mosaic. *Phytopathol. Z.* 87:12–76.

Lundsgaard, T. and Abrechtsen, S.E. 1979. Ultrastructure of *Festuca gigantea* with rhabdovirus-like particles. *Phytopathol. Z.* 94:112–118.

Malak, Y. 1974. Epidemiologija viroza crvene deteline u Srbiji. Doktorska diseracija, Poljoprivredni fakultet, Zemun.

Malak, Y. 1974a. Crvena detelina kao domaćin virusa mozaika krastavca. *Zaštita bilja* (Beograd) 25:219–226.

McDaniel, L.L. and Gordon, D.T. 1985. Identificaton of a new strain of maize dwarf mosaic virus. *Plant Dis.* 69:602–607.

McLaughlin, M.R. and Windham, G.L. 1996. Effects of peanut stunt virus, *Meloidogyne incognita*, and drought on growth and persistence of white clover. *Phytopathology* 86:1105–1111.

Miki, T. and Knight, C.A. 1967. Some chemical studies on a strain of white clover mosaic virus. *Virology* 31:55–63.

Milne, R.G. et al. 1974. Viral nature of Arrhenatherum blue dwarf virus (Blauverzwergung von *Arrhenatherum elatius*). *Phytopathol. Z.* 79:315–319.

Milne, R.G. and Luisoni, E. 1977. Serological relationships among maize rough dwarf-like virus. *Virology* 80:12–20.

Mulligan, T.E. 1960. The transmission by mites, host-range and properties of ryegrass mosaic virus. *Ann. Appl. Biol.* 48:575–579.

Mühle, E. and Kempiak, G. 1971. Zur Geschichte, Atiologie und Symptomatologie der Blauverzwergung des Glatthafers (*Arrhenatherum elatius* [L.] I. et C. Presl). *Phytopathol. Z.* 72:269–278.

Ohmann-Kreutzberg, G. 1963. Ein Beitrag zur Analyse der Gramineevirosen. III. Strichelvirus des Knaulgrasse. *Phytopathol. Z.* 47:113–122.

Ostaževski, S.A. 1965. The natural occurrence of tobacco ringspot virus in birdsfoot-trefoil (*Lotus corniculatus*). *Pl. Dis. Reptr.* 49:855–856.

Ostaževski, S.A. and Scott, H.A. 1966. Natural occurrence of tomato ringspot virus in birdsfoot-trefoil. *Phytopathology* 56:585–586.

Paul, L.H. and Huth, W. 1970. Untersuchungen über das cocksfoot mild mosaic virus. *Phytopathol. Z.* 69:1–8.

Paul, L.H. 1974. Properties of cocksfoot mild mosaic virus (CMMV) and its relationship to the other viruses. *Mikrobiologija (Beograd)* 11:193–194.

Paulsen, A.Q. and Niblett, C.L. 1977. Purification and properties of foxtail mosaic virus. *Phytopathology* 67:1346–1351.

Penrose, L.J. 1974. Distribution of a Johnsongrass strain of sugarcane mosaic virus in New South Wales and studies of the host range of the Johnsongrass and sugarcane strains. *Aust. J. Agric. Res.* 25:99–104.

Penrose, L.J. 1974a. Identification of the cause of red stripe disease of sorghum in New South Wales (Australia) and its relationship to mosaic virus in maize and sugarcane. *Pl. Dis. Reptr.* 58:832–836.

Persely, D.M. and Greber, R.S. 1977. Additional field hosts of sugarcane mosaic virus in Queensland. *APPS Newsletter* 6:54.

Persley, D.M. et al. 1977. The inheritance of the red leaf reaction of grain sorghum to sugarcane mosaic virus infection. *Aust. J. Agr. Res.* 28:853–858.

Persley, D.M. et al. 1987. Isolates of sugarcane mosaic virus—Johnsongrass strain infecting Krish resistant grain sorghum genotypes in Australia. *Sorghum Newsletter* 30:72–73.

Plumb, R.T. and James, M. 1975. A probable rhabdovirus infecting ryegrass (*Lolium* spp.). *Ann. Appl. Biol.* 80:181–184.

Plumb, T.R. 1980. Seed transmission and effects of ryegrass spherical virus (RGSV). (Abstr.) *3rd Conf. Virus Dis. Gramineae Europe*, Rothamsted, 28–30 May.

Polak, J. 1988. Ungrouped viruses and virus complexes. *In*: Smith, I.M. et al., Eds. *European Hand. Pl. Dis.*, Blackwell Sci. Publ., London, p. 89.

Pratt, M.J. 1961. Studies on clover yellow mosaic and white clover mosaic viruses. *Can. J. Bot.* 39:655–665.

Quantz, L. 1951. Eine Virose der Erbse und anderer Leguminosen. *Phytopathol. Z.* 17:472–477.

Quantz, L. and Völk, J. 1954. Die Blattrollkrankheit der Ackerbohne und Erbse, eine neue Viruskrankheit bei Leguminosen. *Nachrichtenbl. dtsch. PflSchDienst (Braunschweig)* 6:177–182.

Quantz, L. 1961. Investigations on virus diseases of leguminous forage crops in Germany. *Proc. 8th Int. Grassland Congr.* 1960, 8 A/4, S. 204–207.

Roane, C.W. and J.L. Troutman. 1965. The occurrence and transmission of maize dwarf mosaic in Virginia. *Pl. Dis. Reptr.* 49:665–667.

Roland, G. 1969. Etude d'une virose de la vesce (*Vicia sativa* L.). *Parasitica (Gembloux)* 26:48-51.

Rosenkranz, E. 1981. Host range of maize dwarf mosaic virus. *In*: Gordon, D.T. et al., Eds. *Virus and Virus-Like Dis. Maize United States.* Southern Coop. Series Bull. 247, Ohio Agric. Res. Dev. Center, Wooster, Ohio, p. 152–161.

Rubio-Huertos, M. and Bos, L. 1969. Morphology and intracellular localization of bacilliform virus particles associated with the clover enation disease. *Neth. J. Pl. Pathol.* 75:329–337.

Schmelzer, K. 1956. Beiträge zur Kenntnis der Übertragbarkeit von Viren durch Cuscuta-Arten. *Phytopathol. Z.* 28:1–56.

Schmidt, H.B. et al. 1963. Untersuchungen über eine virusbedingte Nekrose an Futtergräsern. *Phytopathol. Z.* 47:66–72.

Schmidt, H.E. 1977. Leguminosen. *In*: Klinkowski, M. et al., Eds., *Pflanzliche Virologie*, Vol. 3, 3rd edition, Akademie-Verlag, Berlin, p. 144–293.

Sehgal, P.O. 1966. Host range, properties and particle purification of a Missouri isolate of maize dwarf mosaic virus. *Pl. Dis. Reptr.* 50:862–866.

Serjeant, E.P. 1964. Cocksfoot mottle virus. *Plant Pathol.* 13:23–24.

Serjeant, E.P. 1967. Some properties of cocksfoot mottle virus. *Ann. Appl. Biol.* 59:31–38.

Shikata, E. 1981. Reoviruses. *In*: Kurstak, E., Ed. *Hand. Pl. Virus Infec. Comp. Diag.*, Elsevier/North-Holland Biomedical Press, p. 424–451.

Short, N.M. 1981. Rept. John Innes Inst. for 1980.

Short, N.M. 1983. Foxtail mosaic virus. *CMI/AAB Descriptions of Plant Viruses*, No. 264.

Shukla, D.D. et al. 1987. Coat protein of potyviruses. 3. Comparison of amino acid sequences of the coat proteins of four Australian strains of sugarcane mosaic virus. *Arch. Virol.* 96:59–74.

Shukla, D.D. et al. 1988. The N and C termini of the coat protein of potyviruses are surface located and the N terminus contains the major virus-specific epitopes. *J. Gen. Virol.* 68:1497–1508.

Shukla, D.D. et al. 1989. A novel approach to the serology of potyviruses involving affinity purified polyclonal antibodies directed towards virus-specific N termini of coat proteins. *J. Gen. Virol.* 70:13–23.

Shukla, D.D. et al. 1989a. Taxonomy of potyviruses infecting maize, sorghum and sugarcane in Australia and the United States as determined by reactivities of polyclonal antibodies directed towards virus-specific N-termini of coat proteins. *Phytopathology* 79:223–229.

Shukla, D.D. and Teakle, S.D. 1989. Johnsongrass mosaic virus. *CMI/AAB Descriptions of Plant Viruses*, No. 340.

Slykhuis, J.T. 1958. A survey of virus diseases of grasses in Nothern Europe. *FAO Plant Prot. Bull.*, VI, 9:129–134.

Slykhuis, J.T. 1972. Ryegrass mosaic virus. *CMI/AAB Descriptions of Plant Viruses*, No. 86.

Slykhuis, J.T. 1973. Agropyron mosaic virus. *CMI/AAB Descriptions of Plant Viruses*, No. 118.

Smith, K. 1972. *A Textbook of Plant Virus Diseases*, 3rd edition. Academic Press, New York, 684 pp.

Spaar, D. and Schumann, K. 1977. Getreidearten und Gräser. *In*: Klinkowski, M. et al., Eds. *Pflanzliche Virologie*, Vol. 3, third edition, Akademie-Verlag, Berlin, p. 1–62.

Stuteville, D.L. and Erwin, D.C. 1990. *Compendium of Alfalfa Diseases*, 2nd edition, APS Press, St. Paul, MN, 84 pp.

Šutić, D. and Tošić, M. 1966. A significant occurrence of maize mosaic virus on Johnsongrass (*Sorghum halepense* Pers.) as a natural host plant. *Rev. Roum. Biol. - Botanique* II, 1–3, 219–224, Bucarest.

Šutić, D., et al. 1975. Seed transmission of forage legume viruses, PL 480, Project E30-CR-102. Fin. Techn. Report.

Šutić, D. 1983. *Viroze biljaka*, Nolit, Beograd.

Tanasijević, N. and Šutić, D. 1958. Bolesti i štetočine lucerke i crvene deteline. Zadružna Knjiga, str. 92. Beograd.

Tapio, E. 1970. Virus disease of legumes in Finland and in the Scandinavian countries. *Ann. Agr. Finniae* 9:1–97.

Taylor, R.H. and Pares, R.D. 1968. The relationship between sugarcane mosaic virus and mosaic viruses of maize and Johnsongrass in Australia. *Aust. J. Agr. Res.* 19:767–773.

Teakle, D.S., et al. 1970. Inheritance of the necrotic and mosaic reactions in Sorghum infected with a "Johnsongrass" strain of sugarcane mosaic virus. *Aust. J. Agr. Res.* 21:549–556.

Teakle, D.S. and Grylls, N.E. 1973. Four strains of sugarcane mosaic virus infecting cereals and other grasses in Australia. *Aust. J. Agr. Res.* 24:465–477.

Thouvenel, J.C. et al. 1978. Guinea grass mosaic virus. *CMI/AAB Descriptions of Plant Viruses*, No. 190.

Torrance, L. and Harrison, B.D. 1981. Properties of Scottish isolates of cocksfoot mild mosaic virus and their comparison with others. *Ann. Appl. Biol.* 97:285–295.

Tošić, M. and Pešić, Z. 1975. Investigation of alfalfa mosaic virus transmission through alfalfa seed. *Phytopathol. Z.* 83:320–327.

Tošić, M., et al. 1990. Differentiation of sugarcane, maize dwarf, Johnsongrass and sorghum mosaic viruses based on reactions of oat and some sorghum cultivars. *Plant Dis.* 74:549–552.

Tremaine, J.H. and Stace-Smith, R. 1969. Amino acid analyses of two strains of alfalfa mosaic virus. *Phytopathology* 59:521–522.
Valenta, V. and Marcinka, K. 1971. Red clover mottle virus. *CMI/AAB Descriptions of Plant Viruses*, No. 74.
Van Regenmortel, M.H.V. and Pinck, L. 1981. Alfalfa mosaic virus. *In*: Kurstak, E., Ed. *Handb. Pl. Virus Infec. Comp. Diag.*, Elsevier/North-Holland Biomedical Press, p. 415–421.
Varma, A. 1970. Red clover vein mosaic virus. *CMI/AAB Descriptions of Plant Viruses*, No. 22.
Varma, A. et al. 1970. A comparative study of red clover vein mosaic virus and some other plant viruses. *J. Gen. Virol.* 8:21–32.
Veerisetty, V. 1979. Alfalfa latent virus. *CMI/AAB Descriptions of Plant Viruses*, No. 211.
Veerisetty, V. and Brakke, M.K. 1977. Differentiation of legume carlaviruses based on their biochemical properties. *Virology* 83:226–231.
Verhoyen, M. and Meunier, S. 1988. Bean yellow mosaic virus (BYMV). *In*: Smith, I.M., Ed. *European Hand. Pl. Dis.*, Blackwell Sci. Publ., London, p. 36–38.
Welsh, M.F. et al. 1973. Clover yellow mosaic virus from apple trees with leaf pucker disease. *Phytopathology* 63:50–57.

3 Vegetable Virus Diseases

Vegetables represent a widely diverse group of annual and perennial herbaceous plants in over ten families. Depending on the plant species, various parts of the plant are edible. They include roots, bulbs, stems, leaves, fruits, inflorescences, and unripe seeds for human nutrition. The nutritive composition of vegetables represents an irreplaceable complex of foodstuffs, providing the main caloric needs in carbohydrates, proteins and oils, the essential vitamins and minerals, as well as specific organic acids, ethereal oils, glucose, and others. A unique curative function is satisfied by such vegetable plants as garlic, horseradish, pepper, ginger, and others used either as fresh foodstuffs or as a raw material for medicinal preparations.

The American Phytopathological Society's APS Press has published six well-summarized and excellently illustrated *Compendia of Diseases* of pea (Hagedorn, 1989), bean (Hall, 1991), Cucurbits (Zitter et al., 1996), lettuce (David et al., 1997), tomato (Jones et al., 1991), and onion and garlic (Schwartz and Mohan, 1995). These are required reading for all persons and scientists interested in disease diagnosis, each of which contains a virus disease section. The treatment of peanut and potato sections with references to their Compendia are found under Industrial Crops in this Handbook.

Botanically different vegetables at one end of the spectrum are hosts of many viruses that have developed pathogenic specificity for related plants belonging only to one family; or at the other end of the spectrum, that are universally infective for taxonomically widely different species. Thus, for example, all solanaceous vegetables are susceptible to AMV, CMV, TMV, and TmSpoWV, which are generally widespread in numerous host families. On the other hand, a small group of specific viruses, including cabbage black ringspot, cauliflower mosaic, and radish mosaic viruses, infect only plants in Crucifereae. Likewise, all species in the Liliaceae are susceptible to onion yellow dwarf, shallot latent, and leek yellow stripe viruses. Viruses specific for the cucurbitaceous vegetable group are watermelon mosaic 2, squash mosaic, and zucchini yellow mosaic viruses. BYMV and BCmMV characteristically infect several legume vegetables. Although not an exhaustive monograph, his lifetime of work on legume viruses culminated after Bos' retirement in writing a historical reminiscence of the major scientific advances over 4 decades; this document was stimulated by his membership in the International Working Group on Legume Viruses (Bos, 1996). Lastly, carrot mosaic and celery mosaic viruses are specific for the parsley vegetables.

In conclusion, a large number of vegetable crops are infected by many viruses that potentially can limit production. Clearly, vegetable production relies on the use of economic practices to protect against virus infections.

VIRUS DISEASES OF SOLANACEOUS VEGETABLE PLANTS

Solanaceous vegetables (including tomato, pepper, and eggplant), which resemble each other in several traits, require similar cultural practices and production methods. They are susceptible to CMV, TMV, AMV, PVY, broad bean wilt virus (BBWV), and tomato spotted wilt virus (TmSpoWV), which cause economic losses. In addition, tomato black ring virus (TmBRV), tomato bushy stunt virus (TmBuSnV), tomato aspermy virus (TmAsV), and eggplant mottled dwarf virus (EMtDV) can also become economically important. An impressive two-volume monograph summarizes viruses infecting peppers and other solanaceous crops, and presents an exhaustive list of species susceptible to the viruses, in addition to some of the basic biological information presented herein (Edwardson and Christie, 1997).

Virus Diseases of Pepper (*Capsicum annuum*)

Pepper Mild Mottle Tobamovirus (PpMdMtV)

Properties of Viral Particles

The PpMdMtV rod-shaped particles closely resemble TMV in size, 312 × 18 nm. No published data exist on the viral nucleic acid. The coat protein has 158 amino acid residues (Wetter and Conti, 1988) and is a good immunogen.

Geographic Distribution

PpMdMtV infects pepper grown in fields, glasshouses, or in plastic tunnels. It has been reported from North America, Australia, Japan, Denmark, Iceland, England, France, Greece, Italy, the Netherlands, and Spain (Wetter and Conti, 1988).

Host Range and Disease Symptoms

According to Wetter and Conti (1988), this virus causes systemic infection both in sweet pepper cvs. Lamuyo and Yolo Wonder, and in hot peppers *Capsicum baccatum*, *C. chinense*, *C. frutescens*, *C. microcarpum*, and *C. pubescens*. A few isolates differ according to hypersensitive reactions in *Capsicum* spp., but display no other mutual differences (Wetter and Conti, 1988). Other solanaceous plants such as *Nicotiana clevelandii*, *N. debneyi*, *N. tabacum* cvs. Samsun, White Burley, and Xanthi - n.c., *N. sylvestris*, and *Datura stramonium*, are susceptible to PpMdMtV. Plants in the *Chenopodiaceae* (i.e., *C. amaranticolor* and *C. quinoa*) produce local lesions without systemic infection.

Symptoms of disease in pepper include mild leaf chlorosis and reduced growth of plants infected at early stages. Infected fruits are small, malformed, with necrotic depressed zones sometimes occurring on their surface. All pepper plants in a field may be infected, resulting in complete crop loss.

Diagnostic plants, among others, include *Datura stramonium* with necrotic local lesions and no systemic infection, and *C. amaranticolor* and *C. quinoa* with chlorotic local lesions and no systemic infection.

Pathogenesis and Control

The virus is preserved in infected pepper seed mostly in the outer seed coat, seldom in the endosperm (Wetter and Conti, 1988). They also report pepper seedlings to be infected with PpMdMtV during manual transplanting operations up to 41% incidence. PpMdMtV is also transmitted mechanically during cultivation. Virus transmission via insect vectors or in any other manner is unknown.

The key preventive measure to protect pepper from infection by PpMdMtV is to use only healthy virus-free pepper seed for production. All contact between hands or tools and infected plants should be avoided during cultivation. Demski (1981) recommends soaking of seed in 4.2% calcium hypochlorite for 15 minutes or in 10% trisodium phosphate for 30 minutes to eliminate PpMdMtV from the seed coat surface.

Pepper Mottle Potyvirus (PpMtV)

Properties of Viral Particles

PpMtV has flexuous, filamentous particles of 737 nm (Nelson et al., 1982). Lacking data about the viral nucleic acid, two coat protein electrophoretic components of 28 and 34 kDa have been measured (Hiebert and McDonald, 1973). Serologically, PpMtV is a good immunogen. Both chilli (*C. frutescens*) and bell (*C. annuum*) peppers are susceptible to the type PpMtV strain. Differences among virus strains (isolates) occur based on reactions in some hosts. One strain is characterized as not causing local lesions on pepper cv. Tabasco (Nelson et al., 1982).

Geographic Distribution

PpMtV occurs primarily in North America (Florida, New Mexico, Texas, Arizona, and California), but has been reported in Central America (El Salvador) (Nelson et al., 1982).

Host Range and Symptoms

Virus hosts include different *Capsicum* species, *C. annuum* and *C. frutescens*, plus other plants in the Solanaceae such as *Datura meteloides, Lycopersicum esculentum, Nicotiana tabacum, N. benthamiana, N. glutinosa*, and *N. glutinosa × N. clevelandii*. On PpMtV-inoculated *Chenopodium amaranticolor* leaves, local lesions develop (Nelson et al., 1982).

Mottle is the key symptom that occurs in *C. annuum* leaves. Some isolates cause pronounced fruit malformation. Symptom intensity depends on pepper cv. susceptibility; for example, *C. annuum* cv. Delray Bell has considerable tolerance.

Necrotic lesions followed by systemic necrosis and plant death occur in PpMtV-inoculated *C. frutescens* cv. Tabasco leaves. The green leaf Tabasco pepper type is resistant to PpMtV. Diagnostic plants include *C. annuum* with systemic mottling and tolerance in some cultivars, *C. frutescens* cv. Tabasco with necrotic lesions and systemic necrosis, and *Chenopodium amaranticolor* with chlorotic local lesions.

Pathogenesis and Control

During the growing season, PpMtV is preserved in infected pepper plants that represent natural sources of infection. Infected *Datura meteloides* plants that overwinter in Arizona represent important sources of infection.

Several aphid species transmit PpMtV nonpersistently in nature and *Myzus persicae* is probably the most efficient transmitter, both as nymphs and adults. PpMtV also is transmitted with infected plant sap, which enables spread of infective PpMtV during handling and cultivation. Other modes of transmission are unknown.

The main plant protection measure is to produce tolerant or more resistant pepper cvs. Chemical control of aphids may reduce the populations that spread PpMtV in the field. Healthy seedlings should be ensured for production, and special measures should be taken to avoid mechanical inoculation, which may occur during transplanting and cultivation.

Pepper Veinal Mottle Potyvirus (PpVMtV)

Properties of Viral Particles

PpVMtV particles are flexuous, 770×12 nm, but with special treatment slightly longer (Brunt and Kenten, 1972). Viral single-stranded RNA constitutes 6% of the particle weight. The coat protein is 32 to 33 kDa. The PpVMtV sedimentation coefficient is 155S (Brunt and Kenten, 1971); the TIP is 55 to 60°C; the DEP is 1×10^{-3} to 1×10^{-4}; and the LIV at 25°C is 7 to 8 days (Brunt and Kenten, 1972). PpVMtV is a good immunogen. No information on virus strains is available.

Geographic Distribution

The PpVMtV has been reported only in Ghana.

Host Range and Symptoms

Hosts of PpVMtV include 16 species, with 11 in the Solanaceae. Pepper, both *Capsicum annuum* and *C. frutescens*, are the primary natural hosts. Other solanaceous hosts include *Nicotiana clevelandii, N. megalosiphon, N. tabacum*, and *Petunia hybrida*.

Mottle and leaf blade distortion occurs in PpVMtV-infected *C. annuum* and *C. frutescens*, and likewise fruits are mottled and misshapen with significant yield reductions. Symptoms and severity depend on susceptibility of cvs. Leaf vein chlorosis, followed by interveinal chlorosis and mottle occur in infected *C. annuum* cv. Long Red (Brunt and Kenten, 1972).

FIGURE 41 Alfalfa mosaic virus in pepper.

Chlorotic local lesions followed later by systemic leaf mottle occur in PpVMtV-inoculated leaves of *Petunia hybrida* cv. Rosy Mon. *Chenopodium amaranticolor* and *C. quinoa* react to PpVMtV with local lesions, thus, are suitable assay plants (Brunt and Kenten, 1972).

Pathogenesis and Control

Vectors and mechanical virus transmission are key in the disease development cycle. The aphids *Myzus persicae* and *Aphis gossypii* transmit PpVMtV nonpersistently, with an acquisition period of 2 minutes (Brunt and Kenten, 1972). Mechanical transmission of contact between infected and healthy plants is especially effective in the spread of PpVMtV in both fields and glass houses.

The most important protective measures include chemical control of the vectors and applying any phytosanitary procedure necessary to avoid mechanical infection of seedlings during transplanting and of growing plants during production.

Pepper: Natural Alternative Host for Some Viruses

Alfalfa Mosaic Virus (AMV)

Pepper is one of the most important vegetable hosts of AMV, which causes substantial production losses, as reported in countries with large production (i.e., Yugoslavia, Bulgaria, and Hungary). AMV was described in former Yugoslavia in 1956, infecting pepper jointly with CMV in 36% of industrial pepper plants (Šutić, 1959). Another report of AMV widespread in pepper crops in several localities noted a 2 to 43% incidence of infected plants (Delević, 1961). AMV is economically important because it can reduce pepper yields by half, or even by 65% in severe cases.

Symptoms of infections that originate from infected seed are detected very early, already on cotyledonary leaflets. AMV-infected cotyledons have small (dotted) whitish spots evenly covering their surface (Figure 41). Such cotyledons are small and, based on these symptoms, the infected plantlets must be eliminated immediately from the nursery seedbed. Similar symptoms of white and yellowish irregularly shaped spots also occur on the first permanent leaf and continuing on newly developing leaves. With time, these spots enlarge and encompass the interveinal tissue. They usually, but not always, spread from the leaf base toward the tip.

Seed-borne infections by AMV cause growth retardation of plants, poor flowering and a reduced fruit set. Lines or narrow whitish and yellowish stripes appear along the still green fruits representing a diagnostic symptom. AMV-infected fruits are usually distorted in the apical third or half section. Symptoms are less pronounced when AMV infects later in the season.

Most importantly, AMV is transmitted with commercial pepper seed at 1-5% incidence (Šutić, 1959) and up to 69% (Delević, 1961).

Measures recommended to prevent AMV in pepper crops are 1) plant only healthy, virus-free seed and rogue all infected plantlets from the nursery immediately; 2) chemical control of aphid vectors is necessary; 3) fields on which alfalfa or clover were grown in the previous year should not be planted to pepper since residual forage plants serve as sources of infection; 4) spatial isolation between pepper and alfalfa or clover crops is recommended to avoid virus transmission, and 5) all pepper cvs. tested to date are susceptible to AMV, thus more effort is required for breeding of resistant or tolerant plants.

Cucumber Mosaic Virus (CMV)

The omnipresent CMV occurs wherever pepper is grown, especially in countries known for large pepper production, such as Bulgaria (Kovaćevski, 1940), Hungary (Szirmai, 1937/1940), and former Yugoslavia (Šutić, 1959). CMV is both widespread and present in nearly all pepper crops, and possibly the most damaging of all viruses in pepper production. Generally, CMV infects 20 to 30, or frequently 50%, of pepper plants; 100% infection has been reported in fields of highly susceptible pepper cvs.

CMV reduces quality significantly, and fruit yield by 60 to 100%, especially in the early plant infections that usually occur in nature.

Symptoms of CMV infection are conspicuous on all herbaceous plant parts, with severity dependant on susceptibility of pepper cvs., virulence of virus strain, plant age, and environmental conditions. Initial symptoms on leaves are a fine chlorotic mosaic, then later as disease develops, clearly visible irregular, circular yellow spots (Figures 42 and 43). Some cvs. become necrotic with brown coloration appearing along mosaic patterns and spots. Other changes occur in some infected plants such as narrower and elongate, willow-like leaf blades, or shorter and smaller ones especially near the plant apex. Occasionally, individual leaf blades may be larger than normal, with highly pronounced veins and midribs ordered in a zigzag position.

Flower malformation and pollen sterility cause very few, if any, fruits to form. Fruits that do form remain dwarfed, malformed, and frequently have necrotic spots on the surface. Some virus isolates cause abnormal (2 to 3X) elongation of pedicles compared to the fruits. Brown streaks are a sign of partial necrosis along stems and branchlets, particularly in susceptible cultivars.

Many branches have shortened internodes and compacted leaves in CMV-infected plants for which the disease is known as pepper proliferation and bushiness. In early infections, proliferation encompasses the entire plant; but when infections occur later in the season, only partial proliferation occurs, starting at the point of infection.

Under favorable conditions, CMV-infected plants continue to grow, develop, and remain vigorous, with only few dwarfed fruits by season's end, which confuses inexperienced pepper producers.

Several measures complementary of each other should be applied simultaneously for virus control. Only healthy seedlings should be procured for production. Seedlings and indoor production should be grown only in fresh soils with no infected plant residues. All infected plants should be rogued as soon as observed. Crop rotations should include crops not susceptible to CMV (wheat, barley, and others). It is important not to grow pepper the following year in fields where cucumber,

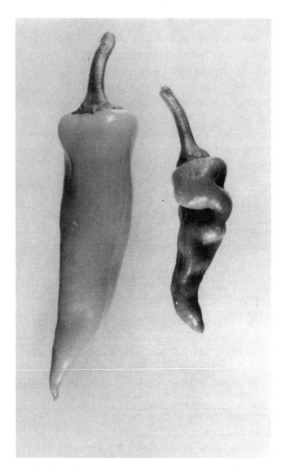

FIGURE 42 Cucumber mosaic virus in pepper; healthy (left).

melon, spinach, alfalfa, or red clover were grown previously, since residual plant parts may represent important sources of infection. Weed control is indispensable because many weed species are susceptible to CMV. Chemical control of aphid vectors is useful when pepper is grown on large acreages, isolated from other susceptible crops.

Potato Viruses

Potato virus X (PVX) and potato virus Y (PVY) occur frequently in pepper fields, but many investigations show that neither cause damage in pepper production. PVX occurs less often in pepper than PVY because PVY is aphid transmitted.

Depending on virus strains and pepper cvs., PVX in pepper is characterized by three main symptoms: necrotic, chlorotic, and mixed. The necrotic type occurs as primary and secondary necrotic spots on leaves and stems, and leaves fall off. Systemic infection may not occur. The chlorotic type features chlorotic spots, accompanied by systemic infection and general mosaic. In the mixed (chlorotic–necrotic) disease type, chlorotic spots first appear on the leaves, which later become necrotic, also causing plant defoliation.

PVY causes pepper mosaic, mottle, and crinkle of apical leaves with a dark vein banding in fully expanded leaves. No typical fruit symptoms appear on PVY-infected pepper.

Both PVX and PVY are transmitted easily with infective plant sap; thus, all measures to eliminate the mechanical spread must be used at all stages of production, starting from the nursery and transplanting stage and ending with crop cultivation in the field. Control of aphid vectors, especially where they multiply rapidly, is essential in crop protection against PVY. Pepper, tobacco,

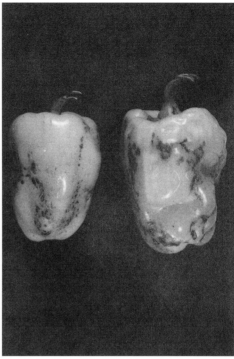

FIGURE 43 Cucumber mosaic virus in bell pepper; healthy (left).

tomato, potato, and other susceptible vegetable plants should be grown in isolation from each other in the same area.

Tobacco Viruses

A number of viruses originating from tobacco are of variable economic importance in pepper production. Least important are TLCuV—causing curling, yellowing, and puckering of pepper leaves—and TSkV, found occasionally in sweet pepper.

TMV causes severe damage to pepper in nearly all areas where grown. Recently, TMV has caused epiphytotics in several countries raising serious concern for pepper producers (Šutić, 1983).

For TMV, pepper cvs. are classified into three susceptibility groups: (1) the Šorok Šari group with systemic virus spread and mosaic as the main symptoms; (2) the Niška Šipka group with local necroses along the stems and shoot dessication; and (3) the Yolo Wonder group of pepper resistant to TMV.

Pepper reactions depend both on the strain of TMV and the cv. The disease caused by the virulent strain of TMV shows both acute and chronic symptoms. The acute stage causes abrupt growth reduction, and infected leaves turn yellow along the veins and frequently fall off (Figures 44 and 45). Plants die if necroses occur in the stems. In the chronic stage, plants that survive recover and grow to maturity, although of reduced growth and development. Fruit on such plants are smaller and of poor quality. The group of virulent strains also includes the strain, causing necrosis and death of the Niška Šipka pepper plants, particularly in indoor production (Šutić et al., 1978). The milder mottling strain of TMV causes a less damaging mosaic that spreads gradually in the leaves. No distinct disease development stages exist. The aucuba strain of this virus causes bright yellow mosaic spots on leaves.

Damages caused by TMV in pepper production are determined by factors such as cv. susceptibility, abundance of sources of infection, plant phenologic stage at the time of infection, and virulence of the strain in that production area. For example artificial infections in the Californian

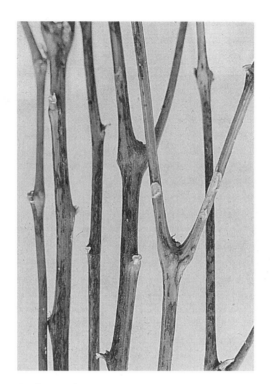

FIGURE 44 Tobacco mosaic virus strain N in hot pepper stems; symptoms never evident on leaves.

FIGURE 45 Tobacco mosaic virus strain N in hot pepper; healthy (top).

Wonder cv. ranged from 60 to 80% yield losses. Where 100% incidence of infection occurs, the tomato mosaic virus reduces average Šorok Šari pepper yields by some 78% (Jasnić, 1978). Aleksić and Marinković (1981) showed that under conditions of natural infection, reduced yields of the cv. Šorok Šari by 37 to 43%, which caused this high-yielding cv. to yield about the same as more resistant cvs. such as A1-12 and Zeleni Rotund. TMV has spread recently as epidemics in both pepper and tomato crops, a consequence of the ease of mechanical transmission, in addition to infected seed and plant residues in the soil. The sap of infected plants is very infective because

TMV occurs in high concentrations in plant cells. Generally, many TMV-infected plants originate from the infected seedlings. The number of infected plants is larger when transplanted than when direct sown.

Infections of seed by TMV may be endogenous or the infestation may be superficial. Superficial or exogenous external infestations on the seed coat, under experimental conditions, can result in up to 64% disease incidence in resultant seedlings, whereas endogenous internal infections within the seed coat (i.e., the endosperm and the embryo) can result in up to 15% diseased seedlings (Jasnić, 1978). Some authors have shown TMV infection rates in pepper seed in field trials of 20%, and in some cases even up to 100% (Demski, 1981).

Data on the rates of pepper seed infection or infestation differ markedly, a consequence of different test conditions. Practically, data on transmissibility assessed on the basis of the resultant number of infected seedlings from the seed samples are the most important ones. By this procedure, TMV was transmitted with the Niška Šipka pepper seed at 21% incidence (Tošić et al., 1980), and with the Šorok Šari pepper seed at about 1% incidence (Aleksić and Marinković, 1981).

TMV is easily preserved, as already mentioned, in infected plant residues, especially roots in the soil for a period of 4 months to 2 years, practically in infected plant roots for about 1 year.

Numerous host plants as sources of TMV infection play an important role in virus transmission, such as tobacco (*N. tabacum*), some floweres (*Begonia semperflorens*, *Petunia hybrida*, and others) and various weeds (*Chenopodium murale*, *Melandrium album*, *Plantago major*, and *P. lanceolata*).

The resistance of some pepper cvs. to TMV is based on hypersensitivity reactions. This resistance is controlled by the L gene (Holmes, 1937). This gene, for example, was introduced into the Yolo Wonder pepper, which was the first TMV-resistant cv. Several naturally bred pepper cvs. and populations contain this gene. Thus, one assumes that some widely used cvs., such as A1-12, Zeleni Rotund, Turišjara Ia, Beli Kalvil, Bela Kapija, Zlatna Medalja, and Kurtovska Kapija, contain this gene. The cv. Šorok Šari is highly susceptible to TMV, a clear indication that it does not contain the resistance gene (Aleksić and Marinković, 1981).

The occurrence and spread of TMV in pepper and tomato crops can be prevented in several ways. First, only healthy seed should be used for establishment of nurseries for production of seedlings. All seed should be disinfested. TMV particles are partially inactivated on the surface of tomato seed during natural fermentation of fruits from which the seed is obtained, by soaking the seed during extraction in 2% HCl solution for 24 hours, or by seed treatment at higher temperatures (85°C for a period of 24 hours). Pepper seed is disinfested by a 10-minute soak in a 2% NaOH solution. Following disinfestation, pepper and tomato seeds are rinsed and dried to avoid any phytotoxic action on embryos by the caustic chemical. Chemical disinfestation is efficient in prevention of superficial seed infestation, but it is only partially effective for prevention of infections of seedlings from internally borne TMV in pepper seed. Complete virus inactivation can be achieved by a 2-hour soak of pepper seed in a 10% trisodium phosphate (Na_3PO_4) solution or by heating at 70°C for 3 days. Such disinfestation techniques do not damage seed germination. Demski (1981) successfully disinfested pepper seed by soaking in the following solutions: 9% HCl (for 30 minutes in 100 ml, with a 1-hour rinse); 4.2% $Ca(OCl)_2$ (for 15 minutes, followed by a 15-minute rinse); 2.63% NaOCl (for 15 minutes, followed by a 15-minute rinse), and 10% Na_3PO_4 (for 30 minutes, followed by a 1-hour rinse).

Second, seed beds should be disinfested with steam at least for pricking seedling production. Infected seedlings should not be transplanted, but it is difficult to distinguish TMV-infected from healthy pepper and tomato seedlings. These differences become evident only in the well-developed pricked tomato seedlings. Prior to transplanting, such tomato plants should be incinerated.

During both seedling and plant production, indoors or outdoors, all implements, tools and clothes by which infective plant sap may be transmitted must disinfested. Workers should also disinfest their hands at regular intervals by washing with soap or detergent, especially following known contact with infected plants. Smoking should be forbidden during the work hours since TMV has been shown transmissible by contact with the tobacco used for smoking.

All previously present weeds should be destroyed since several natural hosts of TMV exist. Pepper and tomato should not follow susceptible crops of tobacco, potato and some flowers (petunia, begonia, etc.). Tolerant or resistant pepper and tomato cvs. or populations should be planted.

Viruses from Other Hosts in Pepper

Pepper is also the alternative natural host of other viruses and particularly those affecting vegetable plants, such as tomato bushy stunt virus (TmBuStV), tomato spotted wilt virus (TmSpoWV), and broad bean wilt virus (BBWV). BBWV is especially detrimental on bell pepper fruits, markedly decreasing their value. Initially, moist chlorotic spots appear, which later become necrotic. Concentric ringspots usually appear at points of initial infection (Giunchedi, 1972). Necrosis of these spots creates depressions in the fleshy fruits. Based on characteristic symptoms, this disease could be named bell pepper fruit ruggedness. Pepper golden mosaic geminivirus (PGMV) is especially serious in Mexico where bean golden mosaic (BGMV) and squash and tomato yellow leaf curl geminiviruses predominate (Figures 46 and 47).

VIRUS DISEASES OF TOMATO (*LYCOPERSICON LYCOPERSICUM*)

Tomato Aspermy Cucumovirus (TmAyV)

Properties of Viral Particles

The TmAyV isometric particles of 30 nm in diameter. (RNA probably single-stranded, but no data on protein composition) have a sedimentation coefficient of 98 to 100S (Hollings and Stone, 1971). The biological properties vary with isolates and virus hosts. The TIP of TmAyV in tobacco sap is 50 to 60°C; the DEP is 1×10^{-2} to 1×10^{-6}; and the LIV at 20°C is 2 to 6 days (Hollings and Stone, 1971). TmAyV has good immunogenicity, although some strains are poor antigens (Kaper and Waterworth, 1981). TmAyV is a serotype of the cucumovirus group (Devergne and Cardin, 1975), of which some isolates differ substantially depending on host range, virulence, symptom severity, aphid transmissibility, and serological relatedness.

Geographic Distribution and Economic Importance

TmAyV (synonym, Chrysanthemum aspermy virus), distributed worldwide in chrysanthemum plants, is the most important chrysanthemum virus disease in Europe. TmAyV is less widespread in tomato and celery (loc. cit. Kaper and Waterworth, 1981). In susceptible chrysanthemum cvs., TmAyV causes break, dwarf, and flower distortion, resulting in significant monetary losses in production. Infected tomato plants are dwarfed with seedless fruits.

Host Range

Chrysanthemum and tomato are the most well-known natural hosts of TmAsV, but other natural hosts include *Ageratum houstonianum*, *Apium graveolens*, *Canna indica*, *Capsicum annuum*, *Lilium longiflorum*, and *Zinnia elegans* (Schmelzer et al., 1977). TmAyV infects more than 100 experimental hosts of 24 dicot and monocot families. Suitable assay plants, *inter alia*, include *Chenopodium amaranticolor*, *C. quinoa*, and *Vigna sinensis*, which react with local spots (Hollings and Stone, 1971).

Symptoms

TmAyV causes pronounced bushiness in tomato. Apical meristems are frequently killed in TmAyV-infected plants, which causes secondary plant branching, and thus bushiness. Leaves and leaf petioles are twisted downward; dark-green protuberances appear on the leaves and enations grow on lower surfaces. The most important pathological consequence of TmAyV infection is that fruits remain small and seedless, a symptom after which this disease was named.

TmAyV, known as a chrysanthemum pathogen, occurs wherever *C. indicum* is grown. Symptoms in infected plants depend on virus strains and plant susceptibility. For example, petals may curl, develop irregularly, or join at their margins into tubular forms. Central florets in some cvs.

Vegetable Virus Diseases

FIGURE 46   Pepper golden mosaic-infected chili pepper (Mexico).

FIGURE 47   Pepper golden mosaic-infected pepper (Mexico).

remain underdeveloped and greenish. The disease is also known as chrysanthemum mosaic or chrysanthemum aspermy disease.

Pathogenesis

Chrysanthemum as a perennial plays an important role in virus preservation. In nature, *Aphis fabae*, *A. rumicis*, *Brachycaudus helichrysi*, *Macrosiphum euphorbiae*, *Myzus persicae*, *Phorodon humuli*, and other aphids easily transmit the virus nonpersistently (Schmelzer et al., 1977). Aphids acquire the virus in 15 seconds and can transmit it into healthy plants within 1 minute. Some strains cannot be transmitted by aphids (loc. cit. Hollings and Stone, 1971). Although not transmitted through tomato and chrysanthemum seed, it has been transmitted with some weed seeds (e.g., *Stellaria media*). Transmission with dodder (*Cuscuta subinclusa* and *C. europea*) is possible in low percentage (Schmelzer, 1975). Finally, the virus is easily transmitted with infective plant sap.

Control

Spatial isolation during growing of tomato and chrysanthemum and control of the primary aphid vectors are highly recommended to prevent TmAyV infection and spread. Heat treatment of chrysanthemum meristems will free plants of TmAyV (Monsion and Dunez, 1971). Kassanis (1954) obtained virus-free tomato and tobacco plants by growing infected plants at 36°C for 21 to 32 days.

Tomato Black Ring Nepovirus (TmBRV)

Properties of Viral Particles

TmBRV consists of three types of isometric particles (T, M, and B) of 30-nm diameter (Murant, 1970). T particles contain no RNA, whereas M and B particles contain different RNA quantities (RNA-1 and RNA-2). TmBRV (potato bouquet strain) contains RNA-1 (2500 to 2800 kDa) and RNA-2 (1500 to 1700 kDa). The TmBRV coat protein is 54 kDa. TmBRV has sedimentation coefficients of 55S (T), 97S (M), and 121S (B) (Murant, 1981). In addition to the two viral RNAs, some strains of TmBRV contain a satellite RNA of 500 kDa (Dunez, 1988).

RNA-1 and RNA-2 carry genetic information and, although both alone are infective, symptoms of infectivity are more pronounced when plants are inoculated with both RNAs together (Harrison et al., 1972). RNA-1 codes for the host range, virus seed transmissibility, and some symptoms, whereas RNA-2 codes information for other symptoms, serological specificity, and virus nematode transmissibility, and virulence is determined by both RNAs (Harrison et al., 1972, 1974; Harrison and Murant, 1977).

Stability of TmBRV in sap is characterized by a TIP of 60 to 65°C, a DEP of 1×10^{-3} to 1×10^{-4}, and a LIV at 24°C of 21 days. TmBRV is moderately immunogenic (Murant, 1970).

A number of strains of TmBRV have been described, based on symptoms and serological relationships. Type strains include tomato black ring strain (Smith, 1946), beet ringspot strain (Harrison, 1957), celery yellow vein strain (Hollings, 1965), lettuce ring strain (Smith and Short, 1959), potato bouquet strain (Harrison, 1958), and potato pseudo-aucuba strain (Bercks, 1962), all serologically related, and only cacao necrosis and Hungarian grapevine chrome mosaic strains belonging to the TmBRV subgroup differ substantially from them.

Geographic Distribution and Economic Importance

TmBRV is widespread in Europe, and the only strain outside of Europe is the cacao necrosis virus recorded in West Africa (Kenten, 1977).

Losses in tomato production are significant in young plant infections. Black necrotic rings evident on all plant organs coalesce on leaves and encompass apical meristems, and some plants die. TmBRV, which infects several herbaceous and woody plants, poses a potentially dangerous source of soil contamination because it is transmitted by soil nematodes.

Host Range

TmBRV readily infects many monocots and dicots (270 species), cultivated and natural (92 species) hosts (Schmelzer et al., 1977). Its host range also includes economically important woody plants: for example, *Prunus persicae*, *P. avium*, *Vitis vinifera*, *Rubus idaeus*, and *Robinia pseudoacacia*. A number of important TmBRV-susceptible vegetable species include *Lycopersicon lycopersicum*, *Allium cepa*, *A. porrum*, *Apium graveolens*, *Brassica oleracea*, *Lactuca sativa*, *Pastinaca sativa*, *Phaseolus vulgaris*, and *Solanum tuberosum*. TmBRV infects small fruits (strawberry, raspberry) and ornamental plants (narcissus, rose, and flower forsythia) in nature. Its weed hosts include *Arctium lappa*, *Capsella bursa-pastoris*, *Chenopodium album*, *Draba* sp., *Senecio vulgaris*, and *Stellaria media*. Suitable diagnostic plants are *C. amaranticolor* (local lesions) and *Petunia hybrida* (chlorotic local lesions sometimes with dark necrotic edges, systemic chlorosis, or vein necrosis).

Symptoms

Both acute and chronic symptoms characterize TmBRV in tomato. The acute stage in young infected plants develops quickly and severely, with black ringspots on the leaves, petioles, and stalks. These spots later enlarge, merge, and cause necrosis of whole plant parts and, when necrosis involves the apical meristem, plant death occurs quickly.

The chronic disease stage is characterized by symptom mitigation that disappears or remains visible only as dark-green zones on the leaves. Since no ringspots characterize this stage, it is difficult to diagnose the disease. Some infected plants are symptomless for both leaves and fruits.

TmBRV causes ringspots characteristic on lettuce, beans, sugar beet, etc. Yellow spots that may encompass entire leaves occur in infected celery leaves. Yellow mosaic and stunt characterize the symptoms of grapevine plants infected by the potato bouquet strain or by the Hungarian grapevine chrome mosaic strain.

Pathogenesis

TmBRV is preserved in the roots of infected herbaceous and woody plants. The nematodes *Longidorus elongatus* and *L. attenuatus* are vectors of TmBRV in the soil. Both nymphs and adults of the *L. elongatus* nematode transmit the virus. Viral particles are not passed from female nematodes to offspring. TmBRV is transmitted through seed of many plants, about 24 species in 15 families, and depending on individual hosts, the transmission rate ranges from 10 to 100% (Murant, 1970). TmBRV has been preserved in the seed of *Capsella bursa-pastoris* and *Stellaria media* for 6 years. Such seeds play an important role in virus preservation and survival in the soil. Many plants are infected, although seeds display no symptoms. The virus, easily transmitted by infected plant sap, is easily researched.

Control

It is difficult to destroy sources of infection in the soil because TmBRV is preserved in both vegetative plant parts and seed of several plants. Thus, crop rotations—including resistant crops—are highly recommended. Chemical control of nematodes is successfully applied on smaller acreages. Early destruction of weed reservoirs, particularly prior to flowering, ensures the prevention of the formation of TmBRV-infected seed. Virus-free planting material can be produced by heat treatment and/or meristem culture, which should be used in high-risk areas of production. More tolerant or resistant cvs. will help in quickly achieving adequate control of TmBRV.

Tomato Bushy Stunt Tombusvirus (TmBuSnV) [Type]

Properties of Viral Particles

TmBuSnV particles are isometric, 30 nm in diameter, with single-stranded RNA (1500 kDa) which accounts for 16 to 17% of the particle weight (Martelli et al., 1971). Four different polypeptides occur in the coat protein, the major one of 40 kDa and the minor one of 28 kDa. The TmBuSnV particles are 9600 kDa, with a sedimentation coefficient of 131 to 140S, and strain dependant (Martelli et al., 1971; Martelli, 1981). Again, depending on strains and hosts, the TIP is 80 to 90°C; the DEP is 1×10^{-2} to 1×10^{-6}; and the LIV at 20°C is 4 to 5 weeks. TmBuSnV is immunologically active, which produces high-titred antisera.

TmBuSnV comprises several strains. The tomato strain is the type strain and others isolated from various hosts include apple strain (apple), cherry strain (cherry, plum, grapevine), pepper strain (pepper), piggyback strain (piggyback), spinach strain (spinach), and tulip strain (tulip) (Martelli, 1981).

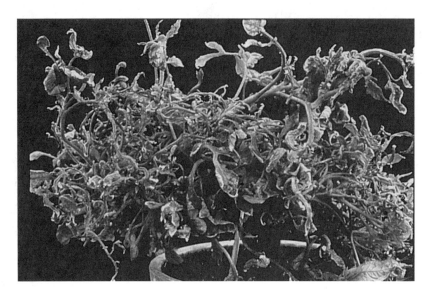

FIGURE 48   Tomato busy stunt on tomato. (Courtesy C.E. Fribourg. Reproduced with permission from APS.)

Geographic Distribution and Economic Importance

Other TmBuSnV strains have been described in Europe, Africa, and South and North America. TmBuSnV has no great influence in tomato production. The strains affecting woody plants, however, are important because they are permanently preserved and carried along with propagating material.

Host Range

The broad host range of TmBuSnV includes both herbaceous and woody plants. Natural herbaceous hosts include several vegetables (tomato, pepper, spinach, artichoke, melon), flowering plants (carnation, pelargonium, petunia, tulip), and weeds (*Chenopodium album*, *Plantago major*, *Stellaria media*). Natural hosts among woody plants include apple, cherry, plum, and grapevine; and experimentally infected hosts include some 126 species in several families (Schmelzer et al., 1977). Suitable diagnostic plants include *C. amaranticolor*, *Gomphrena globosa*, and *Nicotiana glutinosa* (local lesions and systemic infection with some strains).

Symptoms

The main disease symptom of TmBuSnV is a bushy appearance of infected tomatos, resulting from secondary shoots extensively formed following apical meristem necrosis. Yellow or violet-red spots and a line mosaic appear on the leaves, and older leaves become chlorotic, curl up, and fall off (Figure 48). Infection of young plants causes a decline and wilt due to basal stem necrosis. Fruits that remain seedless display visible bleaching plus ring-like spots.

TmBuSnV strains in other hosts display characteristic symptoms such as the artichoke mottled crinkle, pelargonium leaf curl, and petunia asteroid mosaic. Vein necrosis, leaf distortion, shoot stunting, and leaf rosette occur in infected sweet cherry. A type of "detrimental cancer" occurs in susceptible sweet cherry cvs. (Albrechtová et al., 1975).

Pathogenesis

TmBuSnV, although widespread and suggestive of biological vector transmission, has yet to reveal its natural vector. Preserved in the soil, TmBuSnV is probably soil borne (tomato, type strain). TmBuSnV is not seed transmitted in tomato, but is transmitted through seed of sweet cherry and apple (Schmelzer et al., 1977). Infected woody plants represent perennial sources of infection from which the virus is carried on through propagation. Although transmissible with infective plant sap, the role of mechanical transmission in nature is unknown.

Control

Young infected plants should be rogued quickly because they represent a continuing source of infection. Soil should be disinfested on smaller acreages and for indoor production. Crops should be rotated, including use of plants resistant to TmBuSnV. Only virus-free planting material should be used for orchard establishment. TmBuSnV-infected seed of sweet cherry and apple should be avoided for seedling production.

Tomato Golden Mosaic Geminivirus (TmGMV)

Properties of Viral Particles

TmGMV particles of 25×13 nm are geminate (twins), with single-stranded circular DNA composed of separate segments A and B of 840 kDa. The coat protein is a single polypeptide of 28.5 kDa (Buck and Coutts, 1985). The TIP of TmGMV is 40 to 45°C, and the LIV is 15 minutes (Goodman, 1981). TmGMV is moderately immunogenic.

Geographic Distribution

TmGMV is widespread in Brazil; a similar disease of tomato occurs in Venezuela, Africa, and Asia (Buck and Coutts, 1985).

Host Range

Virus hosts are solanaceous, including tomato, in which this disease is of little importance, *Datura stramonium*, and *Nicotiana* spp.

Symptoms and Pathogenesis

Golden mosaic, a characteristic symptom in tomato leaves, occurs in all susceptible hosts and described by yellow blight, stunt, and leaf/shoot deformation. A partial "recovery" phenomenon seems possible in older plants (Buck and Coutts, 1985). The whitefly, *Bemisia tabaci*, is the only known natural vector. Mechanical transmission of TmGMV depends greatly on infection sources and virus hosts.

Tomato Mosaic Tobamovirus (TmMV)

Properties of Viral Particles

TmMV particles are rigid rods of 300×18 nm that contain single-stranded RNA (2000 kDa) and a coat protein of a single polypeptide (21 kDa) (Hollings and Huttinga, 1976). The TIP of TmMV in tomato sap is 85 to 90°C; the DEP is 1×10^{-5} to 1×10^{-6}; and the LIV at 24°C is several months and in air-dried leaves beyond 24 years (Hollings and Huttings, 1976). TmMV particles are immunogenically active.

Strains

Different isolates resembling each other in amino acid composition, buoyant density, and serological reactivity have been described from TmMV-infected tomato. These isolates are defined as strains based on their symptoms in *Lycopersicon* spp. or in the isogenic line of *Graigella* tomato with the Tm-1, Tm-2, or Tm-2^2 resistance genes. Based on their ability to overcome the resistance coded by these genes, the strains have been classified into five Pelham groups: 0 (unable to overcome any resistance genes), 1 (gene Tm-1), 2 (gene Tm-2), 1.2 (genes Tm-1 and Tm-2), and 2^2 (gene Tm 2^2) (Hollings and Huttinga, 1976). Symptoms described in tomato plants include, *inter alia*, tomato aucuba mosaic strain (Pelham group 0, 1, or 2), Dalhemense strain (Pelham group 1), yellow ringspot strain (Pelham group 0), and others (Hollings and Huttinga 1976). Typically, strains of TmMV are related closely to TMV. However, these isolates differ, based on reactions of infected hosts, cross protection (in tomato and tobacco), serological relationships, and amino acid composition of the virus protein (Hollings and Huttinga, 1976).

FIGURE 49   Tomato mosaic on tomato. (Courtesy J.P. Jones. Reproduced with permission from APS.)

Geographic Distribution and Economic Importance

TmMV is widespread wherever tomato is grown. TmMV reduces plant vigor and yields by 3 to 23% (Broadbent, 1964). Also, under certain conditions, it causes necrosis and whole plant death.

Host Range

TmMV, in the wide-host-range category, infects many plants in the Solanaceae. It has infected experimentally plants in the Aizoaceae, Amaranthaceae, Chenopodiaceae, and Scrophulariaceae (Hollings and Huttinga, 1976).

Symptoms

Tomato is the natural host in which TmMV frequently occurs in epidemic proportions, both in glass houses and in fields. Symptoms in tomato differ, depending on environmental factors, virus strains, and susceptibility of cvs. (Figure 49).

Symptoms of TmMV in tomato leaves occur as pale- and dark-green mosaic, and young leaves are malformed, stunted, and sometimes fern-shaped. The most damaging symptom is necrosis in leaves, along the stems and petioles and on the fruits. Necrosis is much more severe in mixed infections in combination with other viruses, causing severe stunt and plant death.

Isolates of TmMV may cause similar symptoms in pepper and several *Nicotiana* species that are distinguishable from similar TMV strains. For example, *N. sylvestris* reacts to the TmMV strain with local lesions and to the TMV strain with systemic infections; most TmMV strains cause necrotic local lesions, but without systemic infection caused by the TMV strains in *N. tabacum* cv. White Burley or Paraguay and *N. rustica* plants.

Pathogenesis

Infested and infected seed and plants represent sources of inoculum in nature. Infected seed serves to transmit TmMV from year to year, ensuring a permanent virus cycle. Seed infections are primarily external in outer parts and less often endogenous in the endosperm. External infestations occur during seed extraction from the pulp of infected plants, when the seed is thoroughly coated with TmMV. The virus is then transmitted into fields with infected seedlings resulting from infested seed. See also TMV in the tobacco sections.

Infected plants in the field provide the sources of TmMV that can further spread mechanically into an epidemic by handling and cultural operations. This mode of transmission occurs readily, especially during transplanting of seedlings.

Infection has been documented through roots of tomato and pepper in contaminated soils (Broadbent, 1965). No transmission has been reported of TmMV by vectors, or in any other manner.

Control

Only clean and healthy seed should be used. Seed disinfestation by soaking for 20 minutes in 10% (W/V) Na_3PO_4 solution eliminates external seed infections. External infestations and internal infections can also be eliminated by heat treatment of dry seed for 24 hours at 82 to 85°C, which does not reduce seed germination significantly (Rana et al., 1988). The method of cross protection, in which plants previously infected with mild strains are protected from subsequent infections by a severe strain, has been used successfully for protection (Rast, 1972; 1975); but it is risky because it cannot provide absolute protection against virulent strains in nature (Marrou and Migliori, 1971). Successful disease control is ensured by the use of cvs. bred for resistance and tolerance (Pelham, 1972; Rast, 1975).

Tomato Ringspot Nepovirus (TmRsV)

Properties of Viral Particles

The TmRsV sediment contains three particles—T (top), M (middle), and B (bottom)—with sedimentation coefficients of 53S (T), 119S (M), and 127S (B) respectively. All particles are isometric, 28 nm in diameter, and contain two RNA species (bi-partite genome). The single-stranded RNA of one species is 2800 kDa and of the other 2400 kDa, which accounts for 41% (M) and 44% (B) of the total viral particle weight, and the single-component coat protein is 58 kDa (Stace-Smith, 1984). M and B particles carry the genetic code for infectivity. The TIP of TmRsV in crude sap at 58°C is 10 minutes; at 20°C is 2 days, at 4°C is 3 weeks, and at –20°C is several months. TmRsV is strongly immunogenic (Stace-Smith, 1984).

Strains

Several strains of TmRsV have been described, based on symptoms in different hosts, *inter alia*, including tobacco strain (the type strain occurring in tobacco seedlings in glasshouses), peach yellow bud mosaic strain (occurring in peach, almond and apricots in nature), and grape yellow vein strain (occurring in grapevine as its natural host) (Stace-Smith, 1984).

Geographic Distribution

TmRsV, spread widely in North America and locally in Europe, was reported in the former U.S.S.R. and Japan. It causes a severe disease of woody perennials, particularly frequent in North America.

Host Range and Symptoms

TmRsV infects several woody and herbaceous plants in nature and experimentally infected host species in over 35 dicot and monocot families. Natural hosts include economically important woody plants such as almond, apple, apricot, cherry, nectarine, peach, raspberry, red current, and grapevine. Herbaceous hosts include tomato, tobacco, and several ornamentals (gladiolus, narcissus, and pelargonium).

Individual hosts usually react to the virus with specific symptoms. Local necrotic ringspots, systemic mottle, and necrosis occur in infected tomato plantlets. Many diseases caused by TmRsV have been described, based on the diagnostic symptoms in other hosts; these include yellow bud mosaic and stem pitting in almond, peach, nectarine, and plum; rasp leaf in cherry; decline of apricot; union necrosis in apple, yellow vein in grapevine; crumbly fruits in raspberry; and chlorosis mosaic in red current (Williams and Holdeman, 1970). *Chenopodium amaranticolor, C. quinoa* (chlorotic necrotic local lesions and systemic apical necrosis), *Cucumis sativus* (local chlorotic spots, systemic mottle, and chlorosis), and others are suitable diagnostic plants (Stace-Smith, 1984).

Pathogenesis

Numerous woody and ornamental hosts, plus annual vegetatively propagated flowering plants, represent permanent sources of infection.

Virus vectors are the nematodes *X. americanum* and other *Xiphinema* spp., identified in several countries. They actively transmit TmRsV both as adults and larvae by acquiring the virus in 2 hours and inoculating in 7 hours. TmRsV is seed transmitted in several hosts: soybean, strawberry, raspberry, pelargonium, and dandelion (loc. cit. Stace-Smith, 1984).

TmRsV-infected plants in nature represent important sources for transmission and spread of TmRsV with infective plant sap.

Control

Only virus-free propagating material should be used for planting. Plant parts for vegetable propagation should be taken only from healthy plants. All seed used for production should pass a rigid phytosanitary control. Nematode-contaminated soils should be disinfested with nematicides. Early rogueing to eliminate infected plants from plantations and crops is important in soils contaminated with vector nematicides.

Tomato Spotted Wilt Virus (TmSpoWV) [Type and Sole Member]

Properties of Viral Particles

TmSpoWV isometric particles of 85-nm diameter contain a lipoprotein envelope and have a sedimentation coefficient of 520 to 530S (Joubert et al., 1974). TmSpoWV is composed of three genomic RNAs of 8200, 5400, and 3000 nucleotides. Matthews (1991) has identified four major polypeptides of 27 kDa for the coat protein inside the envelope; three proteins outside the envelope (52, 58, and 78 kDa); and up to three minor ones. RNA accounts for 5%, carbohydrate for 7%, and lipids for 20% of the particle weight (Best, 1968). The RNA alone is not infective.

Based on physical and chemical properties, TmSpoWV is unstable; its stability in sap for TIP is 40 to 46°C; its DEP is 2×10^{-2} to 1×10^{-3}; and its LIV at 24°C is 2 to 5 hours. TmSpoWV is weakly immunogenic and satisfactory titres of antisera require special preparations of antigens (Ie, 1970).

Strains

Isolates that differ in virulence and cause characteristic symptoms are classified as strains, including tip blight, necrotic, ringspot, mild, very mild plus other strains (Ie, 1970).

Geographic Distribution and Economic Importance

TmSpoWV is common in temperate and subtropical regions, having been recorded in Africa, Asia, Australia, North and South America, plus in warm climates and glasshouses in Europe. In former Yugoslavia, it is widespread in drier, warmer, and more sunny areas, primarily in Macedonia and southern Serbia, especially in glasshouses and plastic tunnels (Aleksić et al., 1990).

TmSpoWV prevents fruit setting (anthesis) or, if set, the fruits remain underdeveloped cause branchlet death and early plant decline. TmSpoWV causes severe damage in tomato and tobacco production.

Host Range

The very broad host range of TmSpoWV includes over 500 species (T. German, University of Wisconsin, personal communication) in at least 27 dicot and 7 monocot families (Ie, 1970). Natural hosts include perennial (oleander, papaya, and others) and annual herbaceous plants (Francki and Hatta, 1981; Schmelzer et al., 1977).

FIGURE 50 Tomato spotted wilt on tomato (Courtesy L.L. Black. Reproduced with permission from APS.)

Tomato, pepper, potato, and eggplant are the main hosts of TmSpoWV among vegetables; others susceptible include *Lactuca sativa*, *Phaseolus vulgaris*, *Pisum sativum*, *Vicia faba*, and *Vigna sinensis*. Tobacco (*Nicotiana tabacum*) is susceptible, which causes serious losses in production. Pineapple is the most frequently affected monocot. Natural virus hosts also include flower species, including *Begonia* sp., *Chrysanthemum* sp., *Dahlia* sp., *Pelargonium zonale*, *Petunia hybrida*, *Tropaeolum majus*, and *Zinnia elegans*. Weed plants as carriers of TmSpoWV in nature include *Datura stramonium*, *Plantago major*, and *Polygonum convolvulus*.

Suitable diagnostic plants include *Petunia hybrida* (local necrotic spots without systemic infection), *C. sativus* (local chlorotic spots with necrotic centres on cotyledons), *Lycopersicon lycopersicum* (inoculated leaves symptomless, systemic symptoms of bronze patching of leaves, and necrosis), and *Nicotiana tabacum* cv. Samsun NN (local necrotic spots accompanied by necrotic mottling and leaf distortion) (Francki and Hatta, 1981).

Symptoms

Vein swelling and necrotic ringspots in young leaves are initial symptoms of TmSpoWV infection in tomato (Figure 50). These leaves twist upward or downward and become brittle. The stems also twist, shorten, and the shoots provide for a bushy plant appearance. The typical bronzing on lower surfaces of leaves gradually spreads along the whole blade (tomato bronze leaf). Infected plants yield less. Pale red and sometimes white zones edged with concentric rings appear on infected fruits. Similar bronzing symptoms also characterize TmSpoWV-infected pepper and eggplant. However, being less susceptible than tomato to TmSpoWV, they display less pronounced symptoms.

Pathogenesis

TmSpoWV is preserved in many hosts that serve as natural sources of infection. Thrips species, including *Thrips tabaci*, *Frankliniella schultzei*, *F. occidentalis*, and *F. fusca*, are natural virus vectors (Ie, 1970). The insects acquire TmSpoWV only in their larval stages, and transmit it only in the adult stage. The acquisition by larvae require feeding on the infected plant at least 10 minutes. An incubation period in the insect, during which the virus replicates (T. German, personal communication), of 10 to 23 days is required before successful transmission, which then is sustained infective throughout the life of the thrips. TmSpoWV is not transmitted transovarially to the offspring (Ie, 1970). TmSpoWV infectivity is preserved in overwintered thrips, which can again transmit following activation in the spring. Thrips multiply excessively in years characterized by warm and dry weather when tomato bronze leaf and tobacco chlorosis and dwarfing occur as epidemics.

Control

Tomato and pepper seedlings should be produced in separate warm seed beds and isolated from tobacco seedlings to avoid initial infection potential. Likewise, they all should be grown in spatially isolated fields. Chemical control of thrips, effective to prevent their role as vectors, should begin in early spring, both in warm seed beds and in their vicinity. Chemical control in open fields should be adequately timed to prevent excessive insect multiplication. This type of virus vector control is more successful when performed in a larger area. Other susceptible plants, weeds in particular, in these crops and vicinity should be destroyed as potential sources of infection.

Other Virus Diseases of Tomato

Several virus species have been described in tomato, but the data are insufficient to determine their identity or relationships with other viruses. The following viruses provide a partial listing.

Peru Tomato Virus (TmPuV)

TmPuV, a potyvirus with flexuous elongated particles of 750 × 12 nm, has been discovered by its severe symptoms only in Peru, where it caused substantial reductions in both tomato and pepper yields in the irrigated coastal areas. Natural hosts include tomato (mosaic, epinasty, crinkle, vein banding), *Physalis peruviana*, *Solanum nigrum*, and *Nicandra physaloides*. The aphid *Myzus persicae* vector transmits TmPuV nonpersistently. TmPuV is also transmitted with infected plant sap (Fribourg, 1982).

Tomato Top Necrosis Virus (TmTpNV)

TmTpNV, a tentative nepovirus, has three isometric particles (T, M, B) with sedimentation coefficients of 52S (T), 102S (M), and 126S (B), respectively. Virus vectors are unknown (Bancroft, 1968).

Tomato Yellow Dwarf Virus (TmYDV)

TmYDV, with geminate particles, belongs to the geminivirus group. It develops in the phloem tissue of its natural host, tomato. It is reported in Japan (Osaki and Inouye, 1978). The whitefly *Bemisia tabaci* is the main virus vector. No other information has been published on TmYDV properties.

Tomato Yellow Leaf Curl Virus (TmYLCuV)

TmYLCuV, with geminate particles, was discovered in tomatos in Israel, Tunisia, and Senegal, and also occurs in vast areas in northwestern and eastern Africa, India, Mexico, Venezuela, and potentially in Europe. Tomato (*Lycopersicon lycopersicum*), *Datura stramonium* and *Malva nicaeensis* are natural virus hosts. Infected tomato leaves are small, malformed, curled upward, and severely chlorotic (Figures 51 and 52). Flowers prematurely dihisce and yield losses can reach 80% (Nitzani, 1975). A few experimental hosts are found, mostly among solanaceous plants. The whitefly *Bemisia tabaci* is the vector that transmits in a persistent and circulative manner. Instars acquire the virus and transmit it through the adult stage. Chemical control of vectors, cultural practices, and production of resistant cultivars are recommended to reduce damages caused by TmYLCuV (Cohen, 1982).

Tomato Yellow Net Virus (TmYNtV)

TmYNtV, a possible member of the luteovirus group, causes characteristic yellow net symptoms. The aphid *Myzus persicae* is the only known vector (Sylvester, 1954).

Tomato Yellow Top Virus (TmYTpV)

TmYTpV, also a possible member of the luteovirus group, exhibits chlorotic and yellow apical coloration, the characteristic disease symptom. The aphid *Macrosiphum euphorbiae* is the virus vector (Sutton, 1955).

FIGURE 51 Tomato yellow leaf curl-infected tomato (Mexico).

FIGURE 52 Tomato yellow leaf curl-infected tomato (Mexico).

Tomato: Natural Alternative Host for Some Viruses

Alfalfa Mosaic Virus (AMV)

Natural infections caused by AMV are less frequent in tomato than in pepper crops. Symptoms of varying intensities range from mild mosaic and leaf mottle to severe necrosis (Figure 53). The necrotic phase causes severe damage to production.

FIGURE 53 Alfalfa mosaic symptoms on tomato. (Courtesy D.G.A. Walkey. Reproduced with permission from APS.)

Cucumber Mosaic Virus (CMV)

CMV-infected tomato displays pronounced pathological changes. Mild mosaic and mottle first appears in leaves and becomes more evident with development of the disease in the formation of new leaves of abnormally narrowed/elongated thread-like formations (Figure 54). Shortened and compressed internodes alter the general plant appearance. Virulent strains cause necrosis along leaf veins, necrotic streaks along the stems, and death of shoot tips. Flower sterility results from mixed infections of CMV with TMV. The extent of damage to production is dictated by the number of infected plants.

Tobacco Mosaic Virus (TMV)

TMV is so widespread in tomato that it is difficult to find healthy plants at the end of the season. General virus spread characterizes indeterminant tomato cvs. resulting from activities of transplanting, pruning (cutting of stem apices), and successive fruit harvests required for these cultivars. This mode of virus spread characterizes indoor tomato production, where up to 60% of plants are infected with TmMV (TMV strain) (Juretić, 1978). A lower incidence of TMV-infected plants occurs in cvs. with simpler growing methods and less intensive associated cultural practices.

The severity of symptoms of TMV on tomato leaves depends on tomato cultivar susceptibility, virulence of virus strain, stage of plant development, and environmental conditions. Symptoms are more severe in glasshouse or plastic-protected crops than in those grown in the open. In the winter, with low light intensity, short days, and temperatures below 20°C, plants are mostly stunted and leaves distorted to fernleaf-shaped. Mosaic symptoms are most pronounced on leaves of young plants, which also become pointed. Necrotic strains and mixed infections with other viruses, such as potato mosaic virus and potato streak mosaic virus, cause necrosis and death of individual plant parts. Older plants tend to recover and develop further without greater changes. Some tomato cvs. are less susceptible, or tolerant, to TMV. Characteristically, symptoms become milder during the season so that infections have little impact on fruit yields. Tomato cvs. resistant or immune to TMV are available. Recently, TMV has reached epidemic proportions in pepper and tomato crops, basically a consequence of its easy transmission mechanically, through seed and via infective plant residues in the soil. Infective plant sap is very contagious due to the high concentrations of TMV in plant cells. Generally, many TMV-infected plants originate from infected seedlings and are much higher when transplanted than when direct seeded.

FIGURE 54 Cucumber mosaic virus causes shoestring of tomato leaves. (Courtesy S.K. Green. Reproduced with permission from APS.)

TMV is transmitted with tomato seed. Tomato seed infestations, mainly external (exogenous), occur during the extraction of seed from tomato pulp, wherein the TMV particles liberated from infected fruits remain adsorbed to the seed surface and remain ineffective for months. Tomato seed is infested as high as 94%.

The occurrence and spread of TMV in tomato is preventable by application of measures as recommended for protection of pepper against TMV.

Potato Viruses

Potato viruses X (PVX) and Y (PVY) occur frequently in tomato crops. PVX infection causes mild leaf mosaic and a slight growth reduction in tomato plants. Rare severe symptoms cause reduced yields. PVX in mixed infections with TMV is particularly damaging, characterized by both necrotic spots and stripes along stalks and stems and apical necrosis. Doubly infected plants are stunted, become rosette-like, and flower poorly, which is economically damaging.

PVX is transmitted mechanically with infective plant sap; thus, adequate preventive measures should be used in pepper and tomato production. Pepper and tomato should not be grown in the same field, and the separate production fields should be spatially isolated from other susceptible crops (i.e., potato and tobacco in particular).

PVY is widespread in indoor tomato crops, where damage often results. PVY-induced chlorosis along leaf veins and necrotic spots between leaf veins varying in size develop slowly, therefore causing no plant part malformations. No detriment to fruit set has been observed.

Mild strains of PVY cause initial symptoms of vein chlorosis and slight mosaic in leaves, which do not develop in leaves formed later in the season, making the disease less apparent. Virulent strains of PVY cause necrosis, particularly in apical leaves, and resultant plants develop poorly,

FIGURE 55 Curly top on tomato. (Courtesy A.S. Costa. Reproduced with permission from APS.)

fruit irregularly, and those fruits that do form have dark-brown sunken necrotic spots. PVY and PVX that occur frequently together cause pathological changes of a more pronounced and detrimental nature.

PVY is transmitted with infective plant sap and by plant aphids nonpersistently. Aphid transmission of PVY seems more important in tomato than in pepper.

Preventive measures recommended for PVX control are also effective for PVY; but, in addition, one must provide adequate control of vectors and other viruses that cause damage in tomato and pepper.

Viruses from Other Hosts

Various other viruses can infect tomato; for example, BBWV, BtCuTpV (Figure 55), TLCuV (induces a yellow pucker and curl of leaves), and TSkV. Detailed descriptions of these viruses can be found in the chapters on their main natural hosts.

VIRUS DISEASES OF EGGPLANT (*SOLANUM MELONGENA*)

Eggplant Mosaic Tymovirus (EMV)

Properties of Viral Particles

EMV particles are isometric, 30 nm in diameter, and sediment as two types: empty, without RNA (T), and those with RNA (B) with sedimentation coefficients of 53S and 111S, respectively (Gibbs and Harrison, 1973). The single-stranded RNA is 2500 or 2000 kDa (depending on the method of preparation); the coat protein is a single 20.5-kDa species (Gibbs and Harrison, 1973). The TIP of EMV is 90°C; the DEP is 1×10^{-7}; and LIV at 24°C is 7 days. EMV has strong immunogenicity (Gibbs and Harrison, 1973).

Virus strains differ in virulence, symptoms on host plants, and serological relatedness. Strains other than the types described include the Andean potato latent (PAnLtV) strain (from wild potatoes in tropical Andes) and the Trinidad strains (differing according to virulence).

Geographic Distribution

EMV reportedly is widespread in Trinidad (West Indies), Colombia (northwestern South America), and Peru (South America) (Gibbs and Harrison, 1973; Koenig and Lesemann, 1981).

Host Range

About 30 plant species, mostly in the Solanaceae and the Chenopodiaceae, are susceptible to EMV. Natural hosts with symptoms of systemic infection are eggplant, tomato, and potato (but symptomless for the PAnLtV strains). Other susceptible plants include (*Chenopodium amaranticolor*, *C. quinoa*, *Cucumis sativus*, *Datura stramonium*, *Gomphrena globosa*, *Nicotiana* spp., and *Petunia hybrida* (Koenig and Lesemann, 1981).

Symptoms and Pathogenesis

EMV causes mosaic and mottle on leaves of eggplant and pepper. Other plants that display systemic symptoms include *C. quinoa*, *D. stramonium*, *Nicotiana clevelandii*, *N. glutinosa*, and *P. hybrida*, and hosts with latent local infections are *C. sativus*, *Gomphrena globosa*, *N. tabacum*, and others.

The flea-beetle *Epitrix* sp., natural vectors, transmit EMV into nearby and some more distant areas. EMV is transmitted easily with infective plant sap, and thus important for local spread of EMV during cultural operations. The PAnLtV strain is seed transmitted in *N. clevelandii* and *P. hybrida*, which represent sources of natural infection. Chemical control of vectors and measures aimed at preventing mechanical transmission can reduce the damage caused by EMV.

Eggplant Mottled Dwarf Rhabdovirus (EMtDV)

Properties of Viral Particles

EMtDV particles are bacilliform and 220×60 nm. Chemical analysis suggests the presence of lipids in the outer envelope (Martelli and Russo, 1973). The TIP of EMtDV is 54°C; the DEP is 1×10^{-3} to 1×10^{-4}; and the LIV at 4°C is 30 to 44 hours.

Geographic Distribution and Economic Importance

EMtDV has been described in Spain and southern Italy (Peña Iglesias et al., 1972; Martelli and Russo, 1973). EMtDV causes severe stunt of eggplant, but since it affects only few plants (1 to 2%), it represents no great danger.

Host Range

Eggplant with many susceptible cvs. is the only natural host of EMtDV (Martelli and Cirulli, 1969). Experimental virus hosts belong to four dicot families. *Nicotiana tabacum* and *N. glutinosa* (local spots on inoculated leaves followed by systemic vein yellowing) are suitable assay plants.

Symptoms

Mosaic mottle and leaf roll occur in infected leaves. The growth of EMtDV-infected plants is stunted and they remain sterile.

Pathogenesis and Control

Natural vectors of EMtDV have not been identified. It is transmitted with infective plant sap, but other modes of transmission remain unknown. The use of healthy seed and the elimination of infection plants from production fields are recommended control measures.

Eggplant: Natural Alternative Host for Some Viruses

Eggplant is the natural host of viruses mainly originating from other, particularly, solanaceous vegetables (e.g., AMV, CMV, TMV, and TmSpoWV).

AMV causes yellow mottle characterized by pale and striking yellow-to-whitish spots over the leaf surface in infected eggplants. However, AMV causes no reductions in growth or yield of eggplants.

CMV, which causes a specific mosaic on eggplant, was described in Europe, and is called the European type to distinguish it from the mosaic on the American continent. Eggplant infections caused by CMV do little damage compared with pepper and tomato.

TMV, generally widespread and frequently recorded in eggplant crops, has no detrimental effects on plant development and yields.

TmSpoWV occurs in all eggplant crops. Many hosts serve as sources of infection from which thrips vectors transmit TmSpoWV, especially among a mixed production of susceptible plants of tomato, pepper, and others.

VIRUS DISEASES OF CRUCIFEROUS VEGETABLES

Brassica plants, the most numerous of this group, are common hosts of several viruses that cause damage to production (e.g., cabbage black ringspot virus, cauliflower mosaic virus, radish mosaic virus). Turnip, in the same genus, is particularly interesting as a primary host because several important viruses were described on it (turnip crinkle virus, cabbage black ringspot virus, turnip rosette mosaic virus, turnip yellow mosaic virus, and turnip yellows virus).

Virus Diseases of Cabbage (*Brassica oleracea* var. *Capitata*), Cauliflower (*B. oleracea* var. *Botrytis*), and Brussels Sprouts (*B. oleracea* var. *Gemmifera*)

Cabbage Black Ringspot Virus (Syn: Turnip Mosaic Potyvirus – TuMV)

Properties of Viral Particles

The TuMV filamentous particles are 750×12 nm, have a TIP of 62°C, a DEP of 1×10^{-3} to 1×10^{-4}, and a LIV at 20°C of 3 to 4 days. TuMV is highly immunogenic. Isolates of TuMV characterized by differences in virulence on hosts are classified as separate strains (loc. cit. Tomlinson, 1970).

Geographic Distribution and Economic Importance

TuMV, generally widespread, is observed most frequently in temperate zones of America, Europe, Asia, and Africa, and has been identified in cabbage in former Yugoslavia (Miličić et al., 1963; Tošić and Malak, 1974; Aleksić et al., 1980).

The TuMV detrimental effects differ on individual hosts, but seemingly are greatest in cabbage where stalk and leaf petiole death and defoliation occur. TuMV also reduces seed yields substantially.

Host Range

The TuMV that causes cabbage black ringspot is the most widespread of viruses in the Cruciferae. TuMV was isolated first from turnip (*Brassica rapa* var. *rapa*) plants, and thus named turnip mosaic virus; later it was established that it infects all important plants in this family, including cabbage (*B. oleracea* var. *capitata*), a most susceptible host displaying pronounced detrimental symptoms.

Natural hosts of TuMV among *Brassica* plants include cauliflower (*Brassica olearacea* var. *botrytis*), kale (*B. oleracea* var. *sabauda*), Brussels sprouts (*B. oleracea* var. *gemmifera*), cole wort (*B. oleracea* var. *acephala*), broccoli (*B. oleracea* var. *italica*), kohlrabi (*B. oleracea* var. *caulocapa*), rutabaga (*B. napus* var. *napobrassica*), rape (*B. napus* var. *napus*), and turnip (*B. rapa* var. *rapa*). Other natural criciferous hosts are radish (*Raphanus sativus*), horseradish (*Armoracia lapathifolia*), Chinese cabbage (*Brassica pekinensis*), and watercress (*Nasturtium officinale*).

TuMV also infects lettuce (*Lactuca sativa*), oilseed rape (*Brassica napus* var. *olifera*), poppy (*Papaver sominferum*), wall flower (*Cheiranthus cheiri*), and other plants. Several weeds represent sources of natural infection of TuMV (e.g., *Amaranthus caudatus*, *Capsella bursa-pastoris*, *Papaver rhoeas*, *Raphanus raphanistrum*, *Sinapis alba*, *S. arvensis*, and *Stellaria media*).

Suitable assay plants include *Nicotiana tabacum* cv. White Burley (chlorotic local spots with necrotic centers and chlorotic halos and no systemic infection), *Chenopodium amaranticolor* (chlorotic local spots with reddish or dark brown necrosis), and *C. quinoa* (chlorotic and necrotic spots, systemic lesions along the veins).

Symptoms

Symptoms of TuMV differ somewhat, depending on the host species, but common features are chlorotic spots between the veins as the primary symptom (these spots are sometimes bounded by small veins and have an angular appearance). On red cabbage, spots are green on the reddish leaf background or initially yellow with reddish and green edges. Bleaching followed by a dark green and striping occurs along the veins of kohlrabi leaves. Slight mottle and chlorosis of older leaves occur on kale. Circular chlorotic spots with green centers are later accompanied by diffuse coloration in cauliflower leaves.

The main secondary symptom of TuMV infection is the necrosis of tissues in the centers of spots with darker brown or black edges—thus the name of the disease. Necrosis within the spots also occurs on leaf petioles, causing them to gradually dessicate and fall off in severe infections. Concentric ringspots that occur around initial spots of infections, due to prolonged virus multiplication, are more frequently observed in red cabbage leaves. In leaves of Brussels sprouts, such spots merge together, and become elongated and surrounded by necrotic rings or ring parts to form arches. At this stage, spots elongate, parallel with small veins on cauliflower leaves. Turnip leaves display typical mosaic, as do horseradish leaves (horseradish mosaic virus) and watercress leaves (watercress mosaic virus). In mixed infections with CMV, mosaic plus curl occurs on radish leaves.

The vivid necroses characterize the symptoms in all plants considered highly reliable for disease diagnosis. Fungi of facultative pathogens (*Alternaria* sp., *Fusarium* sp., and *Botrytis* sp.), particularly with increased moisture, colonize necrotic plant parts, causing a wet rot and faster plant decline. TuMV is common in vegetatively propagated rutabaga (Tomlinson and Walkey, 1967) and in watercress crops in Europe.

Pathogenesis

The virus is preserved in several cultivated crops and weeds. Most crop plants are continuously grown indoors; this provides a permanent potential reservoir source of TuMV inoculum for succeeding crops.

Of over 40 aphid species that may contribute as vectors for general virus spread, *Myzus persicae* and *Brevicoryne brassicae* are the most common. Aphids acquire TuMV by feeding on infected plants for 1 minute, can transmit it within 1 minute to healthy plants, and retain the ability to transmit it, if not feeding actively, for 3 hours. (Tomlinson, 1970).

The virus is transmitted easily with infective plant sap, especially important for its spread in seedling production and indoor production. No evidence of seed transmission exists.

Control

Brassica seed stock plants should be produced only on noncontaminated soils and in regions where vegetables and flowers susceptible to TuMV are not grown. Individual plant species should be grown separately, in isolation, because mixed plant populations increase sources of infection. All weeds, both in the crops and in their immediate vicinity, should be destroyed. Sources of infection can be reduced further by destroying plant residues, particularly following cabbage harvest. The use of chemicals to control aphid vectors is a useful measure. Both aphid and weed control should also be done in neighboring fields. Tolerant or more resistant cabbage and other *Brassica* cultivars should be used in production, especially white cabbage (*Brassica oleracea* var. *capitata alba*) (Walkey and Neeley, 1980). TuMV can be eliminated successfully from infected rutabaga through meristem tip culture (Walkey, 1968). New plantations should be established from virus-free stocks, preferably checked in an indexing program.

Cauliflower Mosaic Caulimovirus (CfMV) [Type]

Properties of Viral Particles

CfMV is isometric, 50 nm in diameter, and 22800 kDa, with a sedimentation coefficient of 202S. The coat protein is likely a single polypeptide with high lysine and arginine (loc. cit. Shepherd and

Lawson, 1981). The CfMV genome is double-stranded DNA of about 5000 kDa, which is 17% of the viral weight (Shepherd, 1981). It is circular, with at least six ORFs. Viral DNA cloned in the *Escherichia coli* plasmid preserved its infectivity (Šutić and Sinclair, 1991). Its possible use as a gene vector has also been studied (Hohn et al., 1982; Howell, 1982).

The biophysical properties of CfMV are a TIP of 75 to 80°C, a DEP of 1×10^{-3}, and a LIV at 20°C of 5 to 7 days (Shepherd, 1981). CfMV is moderately immunogenic; it contains few variants differing in virulence in cauliflower and turnip plants (Shepherd, 1981).

Geographic Distribution and Economic Importance

CfMV is generally distributed in the U.S., New Zealand, British Isles, and several European countries, and widespread in temperate regions wherever various species of the cabbage are grown. The virus depresses cauliflower development, particularly in early infections when it also prevents formation of inflorescences. When CfMV infection occurs early, young plants wilt and die if water-stressed. Infection of seed stock plants causes low seed yields. Necrotic spots cause outer leaf fall on cabbage and red cabbage. In years favorable for disease development, the incidence of infection can exceed 70% and subsequent yields of certain cauliflower cvs. may be reduced 20 to 50%.

Host Range

CfMV ranks second after TuMV, based on the number of susceptible hosts among cruciferous vegetables. In addition to cauliflower (*Brassica oleracea* var. *botrytis*), CfMV infects kale (*B. oleracea* var. *sabauda*), Brussels sprouts (*B. oleracea* var. *gemmifera*), cole wort (*B. oleracea* var. *acephala*), broccoli (*B. oleracea* var. *italica*), kohlrabi (*B. oleracea* var. *gongyloides*), rutabaga (*B. rapa* var. *rapifera*), cabbage (*B. oleracea* var. *capitata*), turnip (*B. rapa* var. *rapa*), and others.

Vegetables naturally susceptible to CfMV in other genera include radish (*Raphanus sativus*) and horseradish (*Armoracea lapathifolia*). All host plants belong to the Cruciferae, except for *Nicotiana clevelandii* and *Datura stramonium* in the Solanaceae.

Symptoms

Disease begins with a mild vein chlorosis spread from the leaf base toward the outer edges. Pronounced vein yellowing and entire leaf chlorosis develops later and whitish spots first occurring on leaf blades gradually become necrotic and dark colored. A yellow hue appears in tissues surrounding veins, a symptom known as cauliflower leaf striping. These symptoms may disappear above 25°C, making it difficult to distinguish diseased from healthy plants.

CfMV causes damage in both cabbage and red cabbage where necrotic spots on outer leaves allow infection by facultative fungi causing leaf rot. CfMV causes radish (*Raphanus sativus*) leaf vein chlorosis—thus, the name radish vein mosaic. It also causes mosaic spots in broccoli (*Brassica oleracea* var. *italica*) plants. Cauliflower and tendergreen mustard, *Brassica campestris*, reacting with specific symptoms are suitable diagnostic plants.

Pathogenesis

CfMV, spread easily and naturally by aphids, is transmitted by some 26 aphid species, including *Aphis fabae, Brevicoryne brassicae, Macrosiphum euphorbiae*, and *Myzus persicae*, either nonpersistently or semipersistently.

Infected seed stock plants (cabbage, kale, cauliflower) serve as a reservoir to allow transmission of CfMV from year to year. CfMV is transmitted easily with infective plant sap, and is especially risky while handling seedling and in indoor production. CfMV has not been shown to be seed transmitted.

Control

The following preventive measures will reduce spread of CfMV: growing of annual and biennial susceptible plants separately to prevent chain infections; and control of aphids with insecticides, particularly in the production of seed crops. Chemical controls should also be applied simultaneously to the neighboring susceptible cultivated crops and cruciferous weed plants. All weeds susceptible to CfMV should be destroyed and all residues of previous crops eliminated. To ensure profitability, seed stock plants should be grown only where susceptible vegetables and flowers have not been grown. Virus-free plants can be obtained from inflorescence tissue through meristem-tip culture (Yot, 1988).

Broccoli Necrotic Yellows Rhabdovirus (BcNcYsV)

Properties of Viral Particles

BcNcYsV particles are bacilliform (675 × 75 nm) or bullet-shaped (266 × 66 nm), depending on preparation technique, with a sedimentation coefficient of 874 ± 41S. In addition to coat proteins, BcNcYsV has an outer envelope (Campbell and Lin, 1972). The BcNcYsV has a TIP in *Datura stramonium* sap of 50°C, a DEP of 1×10^{-3} to 1×10^{-4}, and a LIV at 23°C of 24 hours. BcNcYsV has poor immunogenicity, although serological analyses are useful for detection and diagnosis. No virus strains have been described (Campbell and Lin, 1972).

Geographic Distribution

BcNcYsV, first discovered in Great Britain, has been reported in Australia and North America.

Host Range and Symptoms

BcNcYsV, with a limited natural host range of cauliflower, headed broccoli, and Brussels sprouts, expresses first symptoms of mosaic and yellow leaf mottle. As disease develops, the yellowed areas of leaves develop tissue necrosis. *Datura stramonium* is the diagnostic plant; it displays characteristic symptoms on leaves of chlorotic and necrotic local lesions, followed by fine vein necrosis.

Pathogenesis and Control

BcNcYsV-infected cauliflower, headed broccoli and Brussels sprouts, where grown continuously in open fields and indoors, represent natural virus reservoirs. The vector of BcNcYsV is *Brevicoryne brassicae*, the aphid in whose organs (except the gut) viral particles have been observed (Garret and O'Loughlin, 1977). Not easily transmitted with infective plant sap, other modes of transmission are unknown. Losses to BcNcYsV can be reduced by the use of more resistant cvs. by controllng the vector, the ecology of which requires investigation.

Virus Diseases of Radish (*Raphanus sativus*)

Radish Mosaic Comovirus (RaMV)

Properties of Viral Particles

RaMV isometric particles are 30 nm in diameter. RaMV separates with centrifugation into top (T), middle (M), and bottom (B) components, with sedimentation coefficients of 57S (T), 97S (M), and 116S (B) (Campbell, 1973). The T component may be RNA-free, whereas M and B components contain RNA that has greatest infectivity jointly in a mixture. The M RNA comprises about 26% and the B RNA about 35% of the particle (Campbell, 1973).

RaMV TIP is 65 to 70°C; the DEP is 15×10^{-3}; the LIV at about 20°C is 2 to 3 weeks; and it has strong immunogenicity (Campbell, 1973). The Yugoslavian isolate of RaMV differs serologically and on some hosts from the typical American isolates, and it has been described as the European strain of the small radish mosaic virus (Štefanac, 1973).

Geographic Distribution and Economic Importance

RaMV has been reported in the U.S. (California), Japan, and Europe (Hungary, Austria, Germany, and former Yugoslavia). It was identified in turnip and other crucifers in former Yugoslavia (Mamula and Miličić, 1971; Štefanac, 1973). RaMV causes stunting of growth in seedlings and young plants, but infections later in the season cause no obvious damage. Inflorescence formation is rather poor in infected cauliflower plants. RaMV-infected turnip plants are smaller and have reduced seed yields.

Host Range

Most Cruciferae, including important vegetable crops, are susceptible to RaMV; these include radish (*Raphanus sativus*), cabbage (*Brassica oleracea* var. *capitata*), cauliflower (*Brassica oleracea* var. *botrytis*), Brussels sprouts (*B. oleracea* var. *gemmifera*), turnip (*B. rapa* var. *rapa*), and rutabaga (*Brassica napus* var. *napobrassica*). Other susceptible plants in this family include *Brassica chinensis*, *B. perviridis*, *Capsella burssa-pastoris*, *Sinapis alba*, *S. arvensis*, *Sisymbrium loeselii*, and *S. officinale*. Several experimental hosts of RaMV are in the Chenopodiaceae (*Chenopodium amaranticolor*, *C. foetidum*, *C. murale*, *C. quinoa*, and *Spinacia oleracea*), Cucurbitaceae (*Cucurbita sativus* cv. Delikates, and *C. maxima*), Solanaceae (*Nicotiana megalosiphon* and *N. tabacum* var. Samsun) and Tropaeolaceae (*Tropaeolum majus*). *C. amaranticolor* (local lesions on inoculated leaves) and *N. tabacum* cvs. Turkish and Samsun (chlorotic local lesions or necrotic rings) are virus diagnostic plants.

Symptoms

Primary symptoms of RaMV in radish leaves are small, circular or irregular chlorotic spots between and along the veins. These initial spots increase and merge to form secondary symptoms of enlarged spots and chlorotic zones between the veins. General chlorosis encompasses the leaves so that only small dark-green islands remain in them. Enations may develop on lower leaf surfaces (radish enation mosaic virus). Leaf shape is altered only slightly, no necrotic lesions form, and plant growth is not substantially disturbed.

Systemic infection of cabbage by RaMV is characterized by chlorotic and necrotic lesions on leaves, whereas in cauliflower it occurs as diffuse chlorotic spots. Necrotic spots encompassing the midribs and leaf petioles result in poor formation of inflorescences.

Pathogenesis

RaMV preservation in natural hosts is similar to other cabbage viruses. The *Epitrix hirtipennis*, *Phyllotreta cruciferae*, *P. undulata*, and *Diabrotica undecimpunctata* beetles are the primary virus vectors in nature (Campbell, 1973). RaMV is retained infectious in these beetles for 48 hours.

Transmitted mechanically with infective plant sap, this mode of transmission can be important for the natural spread of RaMV. It is not transmitted with seed from infected radish.

Control

All the measures recommended for the control of other viruses infective to the cabbage family are also recommended for RaMV, and particular emphasis should be placed on the chemical control of its specific beetle vectors.

Radish Yellow Edge Virus (RaYEgV)

Properties of Viral Particles

RaYEgV is isometric of 30-nm diameter, with a sedimentation coefficient of 118S. It contains double-stranded RNA, which readily distinguishes it from other viruses infecting cruciferous plants, and two polypeptides of a 63 kDa and 61 kDa, equally represented in the particle protein (Natsuaki, 1985). It has good immunogenicity. No information on strains is published.

Geographic Distribution

RaYEgV, common in Japan, has been detected in seed from Australia, China, England, Italy, and the U.S. (Natsuaki, 1985).

Host Range and Symptoms

Radish is the only known natural virus host; experimentally infected hosts are unknown. Most infected Japanese cultivars are symptomless, but the leaf edge yellowing in some cvs. gave impetus to the name. In infected plants, viral particles have been discovered only in the phloem tissue, sieve tubes, and companion cells.

Pathogenesis

RaYEgV is not transmitted with infective plant sap and its natural vectors are not yet known. It is transmitted in 80 to 100% incidence in seed obtained from naturally infected *Raphanus sativus* plants (Natsuaki, 1985), most likely the important method of spread.

VIRUS DISEASES OF TURNIP (*BRASSICA RAPA*)

Turnip Crinkle Tombusvirus (TuCrV)

Properties of Viral Particles

The TuCrV isometric particles of 28-nm in diameter are 8000 to 9000 kDa and have a sedimentation coefficient of 129S (Hollings and Stone, 1972). The RNA of TuCrV is probably single-stranded (1400 kDa) and accounts for some 17% of the particle weight. Particle protein components are 38, 28, and possibly also 80 kDa (Butler, 1970). The TIP of TuCrV in turnip or *Brassica pekinensis* sap is 90 to 95°C; the DEP is 1×10^{-6}; and LIV at 20°C is 6 weeks. TuCrV has strong immunogenicity (Hollings and Stone, 1972).

Geographic Distribution

TuCrV occurs in Scotland, England, and former Yugoslavia.

Host Range and Symptoms

Turnip, *Brassica rapa*, the main host of TuCrV among the natural *Brassica* hosts has been transmitted experimentally into many species in 20 dicot families (Hollings and Stone, 1972).

Mottle, the initial symptom on infected turnip leaves, later includes crinkle and distortion with plant growth. Leaf mottle is also common in other TuCrV-infected *Brassicas*. *Chenopodium amaranticolor* and *C. quinoa* (local lesions without systemic infection), *Datura stramonium* (local chlorotic spots without systemic infection), among others, are suitable diagnostic plants.

Pathogenesis

TuCrV-infected turnip and other *Brassicas* represent natural virus reservoirs. Both larvae and adults of flea beetles, in the *Phyllotreta* (nine species) and *Psylliodes* (two species) are involved in transmission and spread of TuCrV. Vectors acquire TuCrV in a few minutes and may remain infective for more than 1 day (Hollings and Stone, 1972). TuCrV transmitted with infective plant sap has a certain epidemiological importance. No data exist on other modes of its transmission.

Turnip Mosaic Virus (TuMV) (*See* Cabbage Black Ringspot Virus)

Turnip Rosette Mosaic (probable) Sobemovirus (TuRoMV)

The TuRoMV isometric particles of 28 nm diameter have a sedimentation coefficient of 112S and contain a 27.7-kDa coat protein (Hollings and Stone, 1973). The TIP of TuRoMV is 85 to 90°C (seldom 95°C); the DEP is 1×10^{-5} to 1×10^{-6}; and the LIV at 20°C is over 30 days. TuRoMV is strongly immunogenic (Hollings and Stone, 1973).

TuRoMV, found only in Scotland and Switzerland, infects primarily crucifers, but sometimes plants in the Compositae, Resedaceae, and Solanaceae families. It causes necrosis of petioles and veins, followed by leaf twisting and rosetting of turnip. Characteristic symptoms in *Brassica napus* are vein banding, rosette and stunt (Hollings and Stone, 1973). *B. pekinensis* with local necrotic and chlorotic spots and unusual systemic infection is the most important diagnostic plant.

Crysomelid beetles, especially *Phyllotreta nemorium*, are vectors of TuRoMV and the virus is transmitted with infective plant sap. No other mode of transmission is unknown.

Turnip Yellow Mosaic Tymovirus (TuYMV)

Properties of Viral Particles

TuYMV has isometric particles 28 nm in diameter. Three particle components are recovered from centrifugal preparations: T (top component, noninfective), B (bottom components of B_{1a} and B_{1b}, both infective), and noninfective particles containing subgenome RNA (Matthews, 1980). TuYMV particles weigh 3600 kDa (T) and 5400 kDa (B_1), with a sedimentation coefficient of 53 to 54S (T) and 116 to 117S (B_1). A single-stranded RNA (2000 kDa) accounts for 35% of the B_{1a} nucleoprotein component weight.

The TIP of TuYMV is 70 to 75°C; the DEP is 1×10^{-4} to 1×10^{-6}; the LIV at 24°C is 7 days (Matthews, 1980); and it is highly immunogenic. Isolates of TuYMV that differ serologically and are based on symptom reactions in assay plants are designated separate strains based on these characteristics.

TuYMV replicates to high concentrations in infected plants and has been isolated in large quantities from Chinese cabbage (0.5 to 2.0 mg/g fresh weight) (Matthews, 1980). Since easily obtained in pure crystalline form, the isometric TuYMV is a well-studied model, similar to the elongated virus model TMV.

Geographic Distribution and Economic Importance

TuYMV is known only in Europe. It was isolated and identified from *Brassica rapa* var. *silvestris* plants in former Yugoslavia (Mamula, 1968). The pale yellow leaves of TuYMV-infected plants have substantially reduced photosynthetic activity. No publication is available on the economic damage caused by TuYMV.

Host Range

Almost all natural hosts of TuYMV are in the Cruciferae. The most important vegetables include cabbage (*Brassica oleracea* var. *capitata*), cauliflower (*B. oleracea* var. *botrytis*), Brussels sprouts (*B. oleracea* var. *gemmifera*), rutabaga (*B. oleracea* var. *napobrassica*), kale (*B. oleracea - sabauda*), and turnip (*B. rapa*). Most experimental hosts of TuYMV are also in the Cruciferae (*Brassica campestris*, *B. oleracea* var. *italica*, and others). Chinese cabbage (*Brassica pekinensis*), the assay plant, reacts primarily with chlorotic local spots.

Symptoms

TuYMV in turnip first causes vein chlorosis in young leaves; then later, secondary symptoms of chlorotic, pale-yellow spots that later merge into yellow zones on larger portions of the leaf surface. Finally, the green chlorophyll disappears or remains only in traces. A similar process of symptom development occurs in other *Brassica* species. Also, the flower color in Chinese cabbage is altered and leaves have striped mottle, yellow zones, and occasional whitening of portions of leaves.

Pathogenesis

Similar to other *Brassica* viruses, TuYMV is preserved in its hosts. The most important vectors in the field are flea beetles in the genera *Phyllotreta* (*P. alta*, *P. cruciferae*, *P. nemorum*, *P. undulata*, etc.) and *Phylliodes* (*P. chrysocephala*, *P. cuprea*). The *Phoedon cochleariae* mustard beetles and their larvae are also active vectors. Also, a potential vector role is played by other animals (*Pieris brassicae*, the cabbage butterfly, *Mamestra brassicae*, the cabbage moth, and others).

TuYMV is easily transmitted with infective plant sap, which may have importance for its natural spread. No seed transmission has been reported.

Control

Preventive control measures recommended for other *Brassica* viruses apply for TuYMV. Chemical insecticides should be used to reduce the populations of known vectors of TuYMV and more resistant plant cvs. should be used.

Turnip Yellows Luteovirus (TuYsV)

The TuYsV, a luteovirus, was described in Belgium in 1950 (Vanderwalle, 1950), identified as turnip mild yellowing virus in England (Watson, 1963), and was reported in Germany (Burckhardt, 1960) and Denmark (Duffus and Russel, 1962).

TuYsV, similar to BtWsYV, is widespread in Europe. Serological studies show a close relationship between TuYsV isolates from turnip in England and Germany and BtWsYV isolates from America and England (Duffus and Russel, 1962). TuYsV antisera showed positive reactions with other luteoviruses; that is, malva yellows (MaYsV), barley yellow dwarf (ByYDV strains RPV, MAV, and PAV) and soybean dwarf (SyDV) (Rochow and Duffus, 1981), and legume yellows viruses (LgYsV).

The TuYsV narrow host range is restricted mostly to cruciferous plants. Turnip, its natural host, displays red coloration along leaf edges, followed by general chlorosis of leaves. Such leaves are rigid and brittle, and roots and whole plants are stunted.

Studies show that the English isolate infects 27 species in 9 families: Compositae (*Lactuca sativa, Senecio vulgaris*), Cruciferae (*Brassica oleracea capitata, B. rapa, Capsella bursa-pastoris*), Leguminosae (*Pisum sativum, Trifolium incarnatum*), Solanaceae (*N. clevelandii, Physalis floridana*), and others (Duffus and Russel, 1962).

In nature, TuYsV is preserved long term in its natural hosts to be transmitted persistently by *Myzus persicae*. It has not been transmitted mechanically and other modes of transmission are unknown.

CRUCIFEROUS VEGETABLES: NATURAL ALTERNATIVE HOST FOR SOME VIRUSES

Beet Western Yellows Virus (BtWsYsV)

Among numerous dicots, BtWsYsV also infects cruciferous vegetables: for example, broccoli (*Brassica oleracea italica*), Brussels sprouts (*B. oleracea gemmifera*), cauliflower (*B. oleracea botrytis*), turnip (*B. rapa*), and radish (*Raphanus sativus*) (Duffus, 1972). Symptoms of chlorosis and stunt in differing severities occur in BtWsYsV-infected plants.

BtWsYsV is permanently preserved over winter in *Capsella bursa-pastoris, Senecio vulgaris*, and possibly winter rape plants (Tomlinson, 1988), wherefrom in the spring aphids transmit it onto several other natural hosts.

Cucumber Mosaic Virus (CMV)

In addition to numerous natural hosts, CMV also naturally infects *Brassica* vegetables: for example, rape (*Brassica napus*), rutabaga (*B. napus napobrassica*), turnip (*B. rapa* var. *rapa*), and radish (*Ruphanus sativus*) (Schmelzer et al., 1977). Both single and mixed infections with other viruses that are more infective to *Brassica* can occur. The widespread nature of CMV is aided by many diverse annual and perennial hosts, plus its ready transmission with infective sap and by numerous aphid vectors.

VIRUS DISEASES OF LILY (*Allium* spp.) VEGETABLES

The genus *Allium* includes onion (*A. cepa*), shallot (*A. ascalonicum*), leek (*A. porrum*), and garlic (*A. sativum*), which are most widespread and important lily vegetables. All are natural hosts of the onion yellow dwarf virus and, except for *A. sativum*, of the shallot latent virus and leek yellow stripe virus. The garlic mosaic and garlic latent viruses infect only garlic. Based on biological and especially morphological traits, viral-infected *Liliaceous* vegetables react in quite a specific manner and with important consequences.

Onion Yellow Dwarf Potyvirus (OYDV)

Properties of Viral Particles

The OYDV elongated particles feature different reported lengths of 750 to 775 × 14 to 16 nm. The TIP of OYDV is 60 to 65°C (some reports of 75 to 80°C); the DEP 1×10^{-2} to 1×10^{-4}; and LIV is 2 to 3 days. (Bos, 1976; Tošić and Šutić, 1976). Antisera have low to medium titres.

Isolates of OYDV from different hosts have different levels of infectivity. Those from garlic and narcisus are less infective for onion than those from onion, and thus probably represent different virus strains. Garlic isolates of OYDV in France seem closely related serologically to the Netherlands onion strain (Delecolle and Lot, 1981).

Geographic Distribution and Economic Importance

OYDV, discovered in 1960 (Aleksić et al., 1980), has been described worldwide wherever onion, shallot, and leek are grown. It was observed in several localities in former Yugoslavia (Tošić and Šutić, 1976; Štefanac, 1977). OYDV especially damages onion seed crops since infected plants neither flower nor bear seeds. The damage level depends on the number of infected plants. Vegetable yields of OYDV-infected plants in annual crops may be reduced by 25%, whereas potential seed yield losses may be reduced 50 to 75%. Some areas of Yugoslavia recorded up to 40% incidence of OYDV-infected plants in the field.

Host Range

Natural hosts of OYDV include onion (*A. cepa*), shallot (*A. ascalonicum*), *A. moly*, *A. scorodoprasum*, and *A. vineale*. Some authors report garlic (Bos, 1976) and leek (Tošić and Šutić, 1976) as susceptible to OYDV, which has also been observed in ornamental species of *Allium* and *Narcissus* (*Narcissus pseudonarcissus*, *N. tazetta orientalis*, and *N. odorus regulosus*). *Chenopodium amaranticolor* and *C. quinoa* serve as assay plants for some isolates with local spots.

Symptoms

Symptoms of OYDV occur first on leaves, emerging directly from infected bulbs as numerous striking, short, yellow-streaked spots in the base of such leaves. Symptoms of primary infections become evident on leaves that form later in the season, while those formed earlier appear healthy (Figure 56). Based on the streaked mosaic mottle, this disease has also been referred to as onion mosaic.

As disease develops, OYDV-infected leaves become nearly completely chlorotic, partially flattened and crinkled, at which stage leaves abnormal in appearance wilt. Narrow striped spots also occur along infected flower stems, which twist and crinkle; this led to naming the disease onion yellow dwarf.

OYDV-infected plants from infected bulbs fail to form inflorescences or their inflorescences remain sterile, which ruins seed production.

Pathogenesis

The OYDV is preserved in the infected onion bulb or in the bulbs for onion seed plants. The percentage of OYDV in onion bulbs for seed plants is always greater than that for production of bulbs. The difference is that onions for bulb production are exposed to OYDV infection only one year, while onions for seed plants are exposed to OYDV for two years. Plants developed from such bulbs represent natural sources of infection during the vegetation period. Infected ornamental and wild *Allium* spp. plants are also natural sources of infection.

FIGURE 56 Advanced leaf twisting, flattening caused by onion yellow dwarf virus in onion. (Courtesy R.M. Davis. Reproduced with permission from APS.)

Some 60 aphid species transmit the virus nonpersistently in nature, and the most common are *Acrythosiphon pisum*, *Aphis craccivora*, *A. fabae*, *Brachycaudus cardui*, *Hyalopterus pruni*, *Myzus ascalonicus*, *M. cerasi*, *M. persicae*, *Rhopalosiphum maidis*, and *R. padi*.

OYDV easily transmits in infective sap, but this mode of transmission is important only for experimental purposes. OYDV transmission through infected seed has not been confirmed. The many aphid vectors play the most important role in epidemics.

Control

Production of healthy onion bulbs mandates the importance of prevention or reduction of infection by OYDV. During periodic inspections, all infected plants should be rogued and destroyed, particularly early in the season. To prevent sequential infections of bulb plants, seed bulbs for plant production and the onion seed plants should be given adequate spatial isolation. Shallot, which represents a regular source of natural infection, should not be grown near onion crops. Chemical control of aphids can be used to prevent infection in growing onion bulbs and seed bulbs. All discarded onion bulbs and commercial onion residues should be destroyed because they serve as potential virus reservoirs after onion harvest.

Shallot Latent Carlavirus (ShLtV)

Properties of Viral Particles

ShLtV is a slightly curved particle of 650 nm and has a sedimentation coefficient of 147S. It contains a coat protein of 23 kDa (Bos, 1982). The TIP of ShLtV in leek crude sap is 80°C; the DEP is 1×10^{-4} to 1×10^{-5}; and LIV at 24°C is 8 to 11 days (Bos, 1982). StLtV is moderately immunogenic. No strains have been designated.

Geographic Distribution

ShLtV, reported in Europe (Belgium, Denmark, England, France, and the Netherlands), can be considered distributed worldwide (Bos, 1982).

Host Range and Symptoms

Shallot (*A. ascalonicum*), onion (*A. cepa*), and leek (*A. porrum*) are natural virus hosts of ShLtV. ShLtV is latent in shallot and remains symptomless. Mild chlorotic streak occurs on infected leek leaves. Mixed infections with the leek yellow stripe virus cause severe chlorotic or white streaking of leaves and some mixed infected cvs. die (Paludan, 1980). *Chenopodium amaranticolor* and *C. quinoa* with local lesions, necrotic on older and chlorotic on younger leaves, are good diagnostic and assay plants for ShLtV.

Pathogenesis and Control

ShLtV is preserved in host plants that represent permanent sources of infection. *Myzus ascalonicus* and possibly *Aphis fabae* are vectors of StLtV, transmitting the virus nonpersistently (Bos, 1982). The virus can be transmitted mechanically into onion (*A. cepa*) and Welsh onion (*A. fistulosum*), which is symptomless. Infected bulbs and aphid vectors are the most important factors in the epidemiology of ShLtV; these factors should receive special attention in plant protection. Measures recommended for the control of other viruses infecting this group of plants are also effective for control of ShLtV.

Leek Yellow Stripe Potyvirus (LkYSpV)

Properties of Viral Particles

The LkYSpV particles are flexuous and 815 to 820 nm in length. It has a coat protein of 34 kDa (Bos, 1981). The TIP of LkYSpV is 50 to 60°C; the DEP is 1×10^{-2} to 1×10^{-3}; and the LIV is 3 to 4 days. LkYSpV is immunogenically active (loc. cit. Bos, 1981).

LkYSpV is—by morphological traits and host range—related to OYDV; but serologically, only distantly (Bos, 1981). Their marked similarities led researchers to consider them synonymous, but more study revealed the differences. LkYSpV does not infect Welsh onion (*A. fistulosum*) and only seldom infects shallot and onion. Likewise, OYDV seldom infects leek. Both also differ in local lesion reactions in *Chenopodium amaranticolor* and *C. quinoa*. Based on these differences, LkYSpV is classified as a distinct virus (Verhoyen, 1973; Verhoyen and Horvath, 1973; Bos et al., 1978).

Geographic Distribution and Economic Importance

LkYSpV, common wherever leek is grown, is considered one of the most important and damaging leek viruses in several European countries. In former Yugoslavia, it is observed regularly in several localities in high incidence. The stalks of LkYSpV-infected plants are thin, dwarfed, and lighter, with substantially reduced quality; as well, they are susceptible to frosts, and die either during the winter or following overwintering.

Host Range

Leek (*A. porrum*), shallot (*A. ascalonicum*), and onion (*A. cepa*) are natural virus hosts. Several leek cultivars of Dutch origin are susceptible to LkYSpV (e.g., Brabantse winter, Goliath, Luikse winter, Siegfried, and Winter reuzen) (Bos et al., 1978).

Experimental inoculations of LkYSpV cause local chlorotic and necrotic spots in *Chenopodium amaranticolor*, *C. album*, and *C. quinoa*, which makes them suitable for diagnosis and assay investigations.

Symptoms

Easily distinguishable chlorotic spots in LkYSpV-infected leaves first occur close to the leaf base and gradually spread toward the tip. Shorter or longer striped spots form mosaic intervals, alternating with the green; and during disease development, these yellow stripes spread and encompass entire leaves, which become completely chlorotic. Leaf blades become wrinkled and twisted, and their surface rough and somewhat more tender. Plants grow slowly, and since already stressed, die during the winter or in the spring.

Pathogenesis

Although LkYSpV infection occurs during the growing season, symptoms appear only in September and winter crops may be 100% infected. LkYSpV remains in infected plants during the winter, and then is transmitted by aphids to healthy plants in the spring. This cycle occurs frequently in vegetable-growing areas because leek is produced on small acreages. Aphids, *Aphis fabae* and *Myzus persicae* most commonly, transmit the virus nonpersistently. Transmission occurs with infective plant sap, but not through seed. Since leek is a perennial plant and aphids are an active group of vectors, their role seems obvious in this disease developing into epidemic proportions.

Control

The main preventive measure is to produce healthy seedlings and provide adequate isolation from overwintered onion crops to avoid sources of infection. Transplantation should not be close to any overwintered onion crops. All plants with symptoms should be rogued during summer, as well as all plant residues destroyed following harvest. Chemical control of vector aphids for seedling production is highly recommended, as is the protection of transplanted plants in the field. For leek crops destined for the green market, chemical protection is not recommended for reasons of possible residual chemicals.

Garlic Mosaic (possible) Potyvirus (GcMV)

Properties of Viral Particles

Elongated GcMV particles are 640 × 14 nm (Brčak, 1975). Yugoslavian isolates ranged from 400 to 1200 nm and averaged 750 nm (Šutić and Tošić, 1976). The TIP of GcMV is 70°C and the DEP is 1×10^{-4}. GcMV has not been shown to be transmissible by aphids, which differs from the OYDV and the LkYSpV.

Geographic Distribution and Economic Importance

Mosaic ranks among the most common garlic diseases. It occurs wherever garlic was grown in former Yugoslavia, thus making it impossible to select healthy plants necessary for experimental work (Šutić and Tošić, 1976). The GcMV virus causes general plant chlorosis, thereby reducing growth and vitality. Individual bulb parts (cloves) remain underdeveloped, and thus reduced bulb weight and yield losses of 30 to 40%. Such damage caused by GcMV disease makes it one of the most important causes of lower yields.

Host Range

Garlic (*A. sativum*) is the primary natural host of the GcMV. GcMV seldom infects onion, in which it causes mild symptoms, and leek can be experimentally infected. Other experimental hosts include *A. galanthum*, *A. hymenorrhizum*, *A. longicuspis*, *A. oschanini*, *A. proliferum*, and *A. rotundum*.

GcMV causes characteristic local spots in *Chenopodium amaranticolor*, *C. murale*, and *C. quinoa*, making them suitable as assay hosts and for diagnosis.

Symtoms

Symptoms of GcMV that appear on the first leaves arise from infected cloves as pale, chlorotic, narrow stripes and streaks; thus, the disease is also known as the streak mosaic. During disease development, these spots gradually spread from the leaf base toward the tip and merge together, causing general leaf chlorosis. Plants are stunted and form underdeveloped bulbs.

Pathogenesis

GcMV is preserved in infected bulbs which serve for garlic propagation. Due to vegetative propagation, and also to little past emphasis on selection of healthy bulbs, GcMV is now common in garlic production. It is also transmitted with infective plant sap, which may account for some of its natural spread. Natural vectors of GcMV are unknown, but they might be spider mites.

Control

The production and use of healthy seed bulbs is the primary—and possibly the only—good measure for control of infection. Thus, growers must ensure healthy parent plants for use exclusively in healthy seed bulb production. The use of meristem-tip culture and thermotherapy have proved successful in producing virus-free bulbs; thus, this method is recommended for production of healthy parent plants.

Lily Vegetables: Hosts for Some Other Viruses

Garlic Latent Virus (GcLtV) in Garlic

GcLtV, a probable member of the carlavirus group (Cadilhac et al., 1976), occurs commonly in apparently healthy plants of several cultivars. Possibly related serologically to the StLtV, GcLtV can be transmitted mechanically into onion and leek, causing symptomless sytemic infection (Delecolle and Lot, 1981).

Tomato Black Ring Virus (TmBRV) in Leek

TmBRV causes natural infections in leek (Bos, 1981). The *Longidorus atteneatus* and *L. elongatus* nematodes are its vectors and are an important factor in soil virus epidemiology.

Tomato Spotted Wilt Virus (TmSpoWV) in Onion

TmSpoWV is quite damaging in onion seed production. It infects the flower stem, causing pale-green and depressed yellow-encircled spots. The tissue within the spots becomes necrotic, as do greater portions of stems caused by merging of adjacent spots. The mechanical strength of stems diminishes and flower heads fall over. TmSpoWV is most damaging because infected onion plants bear no seeds (Šutić, 1983).

VIRUS DISEASES OF CUCURBITACEOUS VEGETABLE PLANTS

The most important vegetable cucurbits include *Cucumis sativus* (cucumber), *C. melo* (melon), *Citrullus lanatus* (watermelon), *Cucurbita maxima* (squash), and *C. pepo* (courgette, pumpkin, zucchini squash), all of which represent natural main or alternate hosts of CMV, WmMV2 and nearly all are natural hosts of cucumber green mottle virus (noninfective only for *C. pepo*), SqMV, ZYMV, and ZYFkV. In addition, cucumber is affected by cucumber leaf spot virus, cucumber necrosis virus, and melon necrotic spot virus. Cucurbits are alternate hosts of some viruses initially described in non-cucurbitaceous plants (e.g., ArMV on cucumber and courgette, and TNV on cucumber and melon).

Virus Diseases of Cucumber (*Cucumis sativus*)

Cucumber Leafspot (possible) Tombusvirus (CLsV)

Properties of Viral Particles

The CLsV isometric particles of 28-nm diameter have a sedimentation coefficient of 127 S (Weber et al., 1982). The type strain of CLsV contains a single-stranded RNA of 1650 kDa, accounting for 20% of the particle weight, and a single polypeptide coat protein of 44 kDa. CLsV has a TIP of 80 to 85°C, a DEP of 1×10^{-6} to 1×10^{-7}, and LIV at 22°C of 20 days (Weber et al., 1982).

CLsV has strong immunogenicity. Cucumber fruit streak virus is considered a separate strain of this virus (Weber, 1986).

Geographic Distribution

CLsV has been reported in Europe (Germany, Greece, and Great Britain) and Jordan (Weber, 1986).

Host Range and Symptoms

Cucumber is the only natural host known for CLsV. Irregularly shaped spots, at first pale-green to yellowish, which later have necrotic centers occur in infected cucumber leaves. Infected plants grow slowly. Experimentally infected hosts include 16 species in 5 dicot families (Weber et al., 1982). *Celosia argentea* with red-brown necrotic spots without systemic infection and *Chenopodium quinoa* with local necrotic spots without systemic infection are suitable for assay and diagnosis (Weber, 1986).

Pathogenesis

CLsV is preserved in cucumber plants continuously grown in fields and glasshouses. CLsV is naturally seed (1% incidence) and soil transmitted (loc. cit. Weber, 1986), and transmitted easily with infective plant sap, which is also important in its epidemiology. Animal virus vectors are unknown. Planting healthy seed and disinfestation of soil are the main preventive proteciton measures.

Cucumber Mosaic Cucumovirus (CMV)

Properties of Viral Particles

CMV consists of three types of isometric particles, each 28 nm in diameter and a sedimentation coefficient of 98S. The single-stranded RNA accounts for 18% of the particle weight. Four RNA species—1270 kDa (RNA-1), 1130 kDa (RNA-2), 820 kDa (RNA-3), and 350 kDa (RNA-4)—are encapsidated as RNA-1 and RNA-2 separately and RNA-3 and RNA-4 together in one particle (Francki et al., 1979). RNA-1, -2, and -3 are infective, whereas RNA-4 contains the gene for the coat protein. Some CMV isolates contain a small single-stranded RNA (10 kDa) known as a satellite; it depends completely on CMV for its replication. This satellite RNA is believed to be associated with special symptom expression. The satellite coat protein consists of a single polypeptide of 24.5 kDa (Kaper and Waterworth, 1981).

Stability in Sap

CMV, unstable in plant extract, has a TIP of 70°C, a DEP of 1×10^{-4}, and an LIV at 24°C of several hours to several days. These biophysical properties may differ somewhat, depending on the sap extracted from individual infected host species.

Serology

CMV immunogenicity is poor, yet many antisera are used successfully for diagnostic purposes. Serological differentiation among isolates is successful using gel diffusion and ELISA tests (Devergne and Cardin, 1974; Devergne et al., 1981).

Strains

Nearly 70 strains of CMV have been described, based on various differential criteria used; some designated strains are valid and others are not. Strains have been assigned two groups based on a combination of symptomological properties: *in vitro* thermosensitivity, and serological and electrophoretical behavior (Devergne and Cardin, 1975; Marrou et al., 1975). In most cases, strains were classified based on the host source of the strain: that is, C (spinach), G (melon), R (rape), T (tomato), and W (Wisconsin tobacco) strains; or by characteristic symptoms in infected plants: that is, celery southern mosaic, squash ringspot, W (white), yellow, and others (loc. cit. Kaper and Waterworth, 1981). Physical and chemical properties served to describe the S, Q, and

Y strains (Francki et al., 1979). Several strains were classified into two groups based on antigenic properties: ToRS and DTL (Devergne and Cardin, 1975). It is evident that CMV has numerous biological variants, isolates, strains, and serotypes, suggesting the full evolutionary development of this virus.

Geographic Distribution and Economic Importance

CMV is one of the most widespread viruses in the world, especially in temperate areas with conditions favorable for the development of aphids, which are the primary virus vectors. Numerous alternate annual and perennial hosts (pepper, tomato, tobacco, alfalfa, red clover, and others) naturally infected in high percentages also contribute to its general wide distribution (Delević, 1961; Mickovski, 1964; Šutić, 1959; Šutić et al., 1975).

CMV is particularly damaging to cucumber and melon production. Plants infected at early growth stages are stunted, sometimes resulting in a cucumber no longer than 40 cm, which results in substantial yield reductions. Infected plants are subject to root rot and rapid decline. CMV infections damage pickling cucumber production sown as a second crop or in the spring. Melon is highly susceptible to CMV, resulting in damages similar in magnitude to that in cucumber. CMV prevents flowering and fruit set in all Cucurbitaceae, thus reducing yields. CMV is among the most economically damaging pathogens in cucurbits, tomato, pepper, spinach, celery, tobacco, flowers (gladioli, lilies), and ornamentals.

Host Range

CMV has an almost unlimited host range. Nowath (1969) reported 470 species in 67 families as natural hosts to CMV. Schmelzer et al. (1977) reported that CMV infects over 700 species in 40 families, including dicots and monocots, ornamentals, weeds, and woody species. More than a half of these plants are natural virus hosts.

In addition to cucumber, watermelon, melon, and other cucurbit species, CMV also infects other vegetables; for example, *Apium graveolens*, *Daucus carota*, *Capsicum annuum*, *Lycopersicon lycopersicum*, *Lactuca sativa*, *Pisum sativum*, *Raphanus sativus*, *Spinacia oleracea*, and others. It causes a disease known as the downy mildew in spinach, thread-like leaf in tomato, and leaf mosaic in celery plants. Hosts of CMV include some well-known legume plants used for animal grazing, hay, and silage (e.g., *Medicago sativa*, *Trifolium pratense*, *Trifolium hybridum*), and many flower species (e.g., *Begonia semperflorens*, *Calendula officinalis*, *Cheiranthus cheiri*, *Gladiolus* sp., *Iris germanica*, *Lilium candidum*, *Petunia hybrida*, *Zinnia elegans*, and *Viola tricolor*). Weeds susceptible to CMV represent natural sources of infection, including *Amaranthus retroflexus*, *Arctium lappa*, *Capsella bursa-pastoris*, *Chelidonium majus*, *Convolvulus arvensis*, *Melandrium album*, *Polygonum persicaria*, *Rumex acetosa*, and *Sinapsis arvensis*.

CMV-infected bushy and woody, cultivated or volunteer plants represent other permanent sources of CMV, including *Ligustrum* sp., *Maclura pomifera*, *Musa* sp. (banana), *Sambucus nigra*, *Prunus* spp. (almond, cherry, peach, plum), *Ribes nigrum*, and *Rubus idaeus*. *Chenopodium amaranticolor*, *C. murale*, and *C. quinoa* all produce local lesions on CMV-inoculated leaves, which make them suitable as assay plants.

Symptoms

Symptoms of CMV infection occur on all plant parts, especially leaves and fruits. Initial symptoms on young leaves are small green-yellow spots, edged with small veins and initially only slightly transparent (Figure 57). Then later, a yellow mosaic forms with leaf blade distortion and stunting of plants that flower poorly and bear only few, if any, fruits. CMV plants branch abnormally at the soil surface, creating a bushy appearance. The green-yellow mosaic appearing first at the fruit base gradually spreads toward the tip. Wart-like protuberances occur late in the season on fruits, which then become distorted.

FIGURE 57 Cucumber mosaic virus in cucumber.

Pathogenesis

Numerous hosts, annual and perennial, herbaceous and woody, cultivated and volunteer, represent abundant natural sources for continued CMV infection. The presence of CMV in so many hosts is a major reason for its common spread.

Several aphids readily transmit CMV nonpersistently in nature. Among more than 60 aphid species vectors, the most efficient are *Myzus persicae*, *Aphis gossypii*, *A. craccivora*, and *A. fabae*. All instars of the vectors can acquire the virus within 5 to 10 seconds and remain capable of transmission for 2 hours. The large extant population of aphid vectors is one reason for the widespread nature of CMV.

CMV, easily transmitted with infective plant sap, is an important consideration during either transplanting or excessive handling during cultivation. Easy mechanical transmission represents a continuous risk for indoor vegetable production. The virus is naturally transmitted by about ten *Cuscuta* spp. CMV is not transmitted through cucumber seed, but it is seed transmitted in 20 species, including tomato (0.2%), pumpkins and beans (7%), blue lupins (6%), cowpeas (4 to 28%), *Stellaria* (21 to 40%), and *Erchinocystis* (9 to 95%) (loc. cit. Kaper and Waterworth, 1981). CMV preservation in seed contributes also to its spread from year to year.

Control

All CMV-infected plants in indoor cucumber production should be rogued immediately upon detection. Control of aphid vectors successfully reduces disease incidence in indoor production. The most susceptible plants must be grown in isolated plots. Chemical control of aphids in the field is risky, due to possible residual pesticides in the soil. Weeds in all cucurbit crops should be destroyed. One must sow very early in order for the crop to reach a more mature level of development before the first aphid flight. Cucumbers, *Datura*, and begonia bulbs can be made virus-free by subjecting CMV-infected plants to heat therapy (loc. cit. Kaper and Waterworth, 1981). The most cost-effective control of CMV is to plant resistant and tolerant varieties, which are available in cucumber, melon, and squash crops.

Cucumber Necrosis (possible) Necrovirus (CNV)

CNV isometric particles of 31-nm diameter are 8600 kDa and have a sedimentation coefficient of 133S. Viral RNA accounts for 16% of the particle weight (Dias and McKeen, 1972). The TIP of CNV in tobacco sap is 75 to 80°C; the DEP is 1×10^{-4} to 1×10^{-5}; and LIV at 24°C is 30 days

(Dias and McKeen, 1972). Serological tests enable reliable detection of CNV in the crude sap of infected plants.

CNV was discovered in Canada in glasshouse cucumbers, its only natural host. CNV causes necrotic spots, pronounced leaf malformations, and dwarfing of plants.

Experimental, mechanical inoculations were successful in many species in Amaranthaceae, Chenopodiaceae, Compositae, Leguminosae, and Solanaceae. Suitable assay plants include *Cucumis sativus*, *Gomphrena globosa*, *Chenopodium amaranticolor* (local necrotic lesions) (Dias and McKeen, 1972).

The soil-borne CNV is preserved in infected plant parts in the soil, wherefrom it is transmitted by zoospores of *Olpidium cucurbitacearum*, a chytrid fungus, which also transmits the related tobacco necrosis virus. The related fungus, *O. brassicae*, does not transmit CNV (Dias and McKeen, 1972).

Uncontaminated soil in glasshouses is the primary measure to protect against CNV.

Cucumber Green Mottle Mosaic Tobamovirus (CGnMtMV)

Properties of Viral Particles

CGnMtMV particles are straight rods, 300×10 nm, with a coat protein of 17.2 kDa and a single-stranded RNA that accounts for 6% of the particle weight (Hollings et al., 1975). The TIP of CGnMtMV in cucumber sap is 86 to 100°C (strain dependant); the DEP is 1×10^{-6} to 1×10^{-7}; and LIV at 24°C is several months, and several years at 0°C (loc. cit. Hollings et al., 1975). Many virus strains differ, based on serology and symptom reactions in *Chenopodium amaranticolor* and *Datura stramonium*. CGnMtMV has good immunogenicity.

Geographic Distribution and Economic Importance

CGnMtMV, widespread in Asia (India), Japan, and Europe, causes severe damage in indoor cucumber production. The quality of fruits is reduced; yield losses may be 15% and, in winter production, even 30%.

Host Range and Symptoms

In addition to cucumber, melon (*Cucumis melo*) and watermelon (*Citrullus vulgaris*) are natural hosts of CGnMtMV, which does not infect the *Cucurbita pepo*.

Initial symptoms of CGnMtMV on cucumber leaves include mild vein clearing and wrinkling, which develop into pale- to dark-green mosaic. Leaves are small and malformed, especially indoors. Some strains of CGnMtMV cause pale-yellow to white line, ring, or star-shaped spots. Infected fruits seldom have symptoms.

Pathogenesis and Control

The CGnMtMV virus is spread with cucumber, watermelon, and bottlegourd (*Lagenaria siceraria*), of which infective seed from these plants play an important role in virus preservation. Seed infestations are mostly superficial. CGnMtMV is transmitted easily with infective plant sap, during casual contact with other leaves, or by handling during cultivation. Infective plant parts in the soil may represent sources of infection. No insects or other animal organisms are known vectors.

Virus-free seed should be planted. Seed of unknown origin should be disinfested by heat treatment for 3 days at 70°C or by soaking in a 15% trisodium-phosphate (Na_3PO_4) solution for 3 days. (Schmelzer et al., 1977). It is also necessary to disinfest all tools used for tending the crop, either by heat treatment or with chemicals (formalin or strong detergents). Workers should disinfest their hands by washing at regular intervals with soap or detergent. Sowing for indoor production should be done so as to avoid all contact during growing. CGnMtMV-infected plants should be rogued immediately. Soil and fermented compost free of any virus infested plant residues should be used for production.

Cucumber: Natural Alternative Host for Some Viruses

Arabis Mosaic Virus (ArMV)

ArMV was described in cucumber in Europe (Van Dorst and Van Hoof, 1965). In seedlings, ArMV first causes yellowing, later leaf necrosis, growth reduction and death (Hollings, 1963). Lateral axillary dwarfed shoots form in ArMV-infected plants that fail to bear fruits.

Tobacco Necrosis Virus (TNV)

TNV causes cucumber necrosis in Europe. Numerous grey-brown spots occur in infected leaves; interveinal tissues become necrotic and drop out. Pale sunken spots surrounded by moist, dark-green edges occur on fruits. Roots also become necrotic. Symptoms are pronounced in low soil and air temperatures, whereas they disappear and plants appear to recover in sunny, warm weather. TNV is preserved in infected plant parts in the soil, wherefrom it is transmitted by *Olpidium brassicae*.

Only uncontaminated or disinfested soils should be used for cucumber production. Plants grafted onto *Cucurbita ficifolia* are less susceptible to this virus (Schmelzer et al., 1977). The symptoms in TNV-infected cucumbers are alleviated by maintaining constant temperature and moisture in glasshouses. The TNV necrosis should be distinguished from that in glasshouses in Canada caused by CNV (Dias and McKeen, 1972).

Tomato Black Ring Virus (TmBRV)

TmBRV occurs in cucumber field-grown in soils contaminated by the nematode vector (Schmelzer et al., 1977).

Cucumber is the potential natural host of other viruses, including BtCuTp, BtPdYs, PaRs, MnNSpo, WmM, ZYM, and ZYFk (Smith et al., 1988).

Hop Stunt Viroid (HSnVd)

HSnVd causes the disease cucumber pale fruit, observed occurring naturally in glasshouses in the Netherlands (Van Dorst and Peters, 1974). The disease is characterized by symptoms of pale green cucumber fruits that are small and pointed at both ends. Leaves formed following infection are small, bluish, and wrinkled, with downward turned tips. Short stem internodes cause a bushy plant appearance. HSnVd that occurs in indoor cucumber production is easily transmitted with infective plant sap by contact between the plants or during handling of plants and cultivation. No other mode of transmission is known. The main preventive measure is to eliminate all sources of mechanical transmission.

Tobacco Ringspot Virus (TRsV)

TRsV occasionally occurs naturally in cucumber causing dramatic symptoms, but seldom is damage to a crop serious.

VIRUS DISEASES OF MELON (*CUCUMIS MELO*), WATERMELON (*CITRULLUS VULGARIS*), SQUASH (*CUCURBITA MAXIMA*), AND COURGETTE (*CUCURBITA PEPO*)

Melon Necrotic Spot Virus (MnNcSpoV)

Properties of Viral Particles

The MnNcSpoV isometric particles of 30-nm diameter have a sedimentation coefficient of 134S. The coat protein consists of a single 46-kDa polypeptide, and the single-stranded RNA accounts for about 17.8% of the particle weight (Hibi and Furuki, 1985). The TIP of MnNcSpoV is 60°C; the DEP is 1×10^{-4} to 1×10^{-5}; and LIV at 24°C is 9 to 32 days and at 4°C, 131 days. MnNcSpoV is immunogenically active. Insufficient comparative data are available to designate strains of some isolates which differ slightly based on symptomology and host range (loc. cit. Hibi and Furuki, 1985).

FIGURE 58 Tobacco ringspot virus in cucumber foliage. (Courtesy R. Provvidenti. Reproduced with permission from APS.)

Geographic Distribution

MnNcSpoV is common in melon and cucumber crops worldwide and can substantially reduce yields (Bos et al., 1984).

Host Range and Symptoms

Melon and cucumber as natural hosts of MnNcSpoV express symptoms of disease as necrotic lesions and spots on leaves and stem necrosis on melon. Experimentally infected cucurbitaceous hosts that react differently to isolates of varying origin include *Citrullus lanatus*, *Cucurbita moschata*, *Lagenaria siceraria*, and others. Diagnostic plants include *Cucumis melo* with necrotic lesions in inoculated cotyledons followed by systemic infection, and *Citrullus lanatus* with dark local lesions on inoculated leaves, without systemic infection (Hibi and Furuki, 1985).

Pathogenesis and Control

MnNcSpoV is preserved in the soil and transmitted by zoospores of the chytrid fungus *Olpidium radicale* (*O. cucurbitacearum*). It is transmitted easily with infective plant sap, especially important in glasshouse production. The California isolate is seed borne (Gonzalez-Garsa et al., 1979) and also transmitted by cucumber beetles (*Diabrotica* species).

Preventive crop protection measures include growing of plants and production of healthy seed in uncontaminated soils, as well as elimination of mechanical virus transmission and spread.

Watermelon Mosaic Potyvirus 2 (WmMV2)

Properties of Viral Particles

The WmMV2 filamentous particles of 760 nm consist of a coat protein subunit of 34 kDa and RNA that sediments at 39S (Purcifull et al., 1984). The TIP of WmMV2 isolates is 60 to 65°C; the DEP is 1×10^{-2} (and 1×10^{-4} for some isolates); and the LIV at 18 to 24°C is 10 to 20 days (up to 60 days for some isolates) (Purcifull et al., 1984). Isolates differ based on host range, aphid transmissibility, and the degree of serological relatedness. WmMV2 has good antigenicity. WmMV2 was described separately to distinguish it from the earlier described watermelon mosaic virus 1, which is now designated as a strain of papaya ringspot virus (Purcifull and Hiebert, 1979).

Geographic Distribution and Economic Importance

WmMV2 is distributed worldwide and ranks as a serious problem in cucurbit production. WmMV2 causes economic damage in watermelon and other susceptible cucurbits, and is reflected mostly as reduced plant growth, yield losses, and a decrease in fruit quality.

Host Range

Most natural hosts of WmMV2 are in the gourd family. In addition to watermelon (*Citrullus lanatus*), they include cucumber (*Cucumis sativus*), melon (*C. melo*), courgette (*Cucurbita pepo*), pumpkin (*C. pepo*), and squash (*C. maxima*). Beans and peas are important natural hosts among legumes. WmMV2 in nature infects different leguminous, malvaceous, and chenopodiaceous weeds, crops, and ornamentals (Purcifull et al., 1984). More than 160 dicot species in 23 families are experimental virus hosts (Molnar and Schmelzer, 1964). Diagnostic plants include *Chenopodium amaranticolor* with chlorotic and necrotic local lesions, *Lavatera trimestris* with necrotic local lesions, and others.

Symptoms

The initial symptom of WmMV2 is mild chlorosis at watermelon leaves, followed later by mosaic and leaf distortion. Green mosaic along the veins and/or green bubble-like protuberances on chlorotic interveinal leaf parts occur as disease develops. WmMV2 infection early in the season causes poor development of young leaves, internode shortening, slow plant growth, and reduced fructification.

Pathogenesis

Among aphids that transmit WmMV2 nonpersistently in nature, 38 species are known vectors, including *Aphis citricola*, *A. craccivora*, *A. gossypii*, *Macrosiphum euphorbiae*, and *Myzus persicae* (Purcifull et al., 1984).

WmMV2 is transmitted with infective plant sap into susceptible plants of the gourd family, which may be of greater importance for its spread in indoor production. The presence and overpopulation of aphids in glasshouses and plastic tunnels may result in the spread of WmMV2 into the majority of plants. Virus transmission through seed has not been confirmed.

Control

Susceptible plants should be grown on isolated plots less favorable for aphid development. All weeds should be destroyed to eliminate an aphid habitat. Chemical control of aphids is more useful for indoor production, but special attention is needed for time of treatment and harvest to reduce or eliminate the presence of chemical residues in the produce. Significant resistance to WmMV2 has been observed in cucumber, beans, and peas (Schroeder and Provvidenti, 1971; Provvidenti, 1974).

Squash Mosaic Comovirus (SqMV)

Properties of Viral Particles

SqMV consists of three isometric particles (T, M, and B) of 30-nm diameter with sedimentation coefficients of 57S (T), 95S (M), and 118S (B) and molecular weights of 4500 (T), 6100 (M), and 6900 (B) kDa (Campbell, 1971). The presumably single-stranded RNA is 1600 kDa (M) and 2400 kDa (B), accounting for about 26.8% (M) and 34.8% (B) of the particle weight. The B component is infective, but other components require further investigation.

The TIP of SqMV is 70 to 80°C; the DEP is 1×10^{-4} to 1×10^{-6}; and LIV at 20°C is 4 weeks (Campbell, 1971). Strains are designated based on differences in host reaction, that is, muskmelon mosaic virus and cucurbit mosaic virus (Stace-Smith, 1981). Various isolates of SqMV belong to two serologically distant strains; strain 1 (but not strain 2) isolates experimentally infect cantaloupe melon (Nelson and Knuhtsen, 1973). SqMV is strongly immunogenic (loc. cit. Stace-Smith, 1981).

FIGURE 59 Various foliar symptoms of squash mosaic of melon. (Courtesy R. Provvidenti. Reproduced with permission from APS.)

Geographic Distribution and Economic Importance

The uncontrolled introduction of infected seed has contributed greatly to the wide distribution of SqMV in several countries.

SqMV causes economical damage resulting from severe systemic infections, noted by stunt and leaf and fruit malformations.

Host Range and Symptoms

SqMV occurs naturally in cucurbits, primarily in squash, cucumber, courgette, pumpkin, melon (Figure 59), and watermelon. Infected plants may remain symptomless or show different symptoms, such as mosaic, blister mottle, ring patterns, enations, and leaf deformations, and/in the case of early, severe infections the fruits are also malformed (Campbell, 1971). Experimental infections have been recorded in hosts in five families (Freitag, 1956).

Pathogenesis and Control

SqMV is permanently preserved and carried in infected seed into geographically distant areas. Seed transmission of SqMV varies; that is, the group 1 strains are transmitted in seed of melon, pumpkin, squash, and watermelon (Alvarez and Campbell, 1978), and those of the group 2 strains only in seed of pumpkin and squash (Nelson and Knuhtsen, 1973a).

Crysomelid beetles are natural vectors of SqMV in the field. Transmission with infective plant sap also occurs and is particularly important in glasshouse production.

Measures to prevent infection include the use of healthy, phytosanitarily controlled seed and chemical control of the vector population. All possible measures to prevent mechanical transmission of SqMV should be taken, especially in glasshouse production.

Zucchini Yellow Mosaic Potyvirus (ZYMV)

Properties of Viral Particles

The ZYMV filamentous particles of 750 nm consist of a 36-kDa coat protein and 2930-kDa single-stranded RNA (Lisa et al., 1981). The TIP of ZYMV is 55 to 60°C; the DEP is 1×10^{-4} to 1×10^{-5}; and LIV at 24°C is 3 to 5 days (Lisa and Lecoq, 1984). Several strains of ZYMV have been described, based on symptoms in individual hosts. ZYMV has good immunogenicity (Lisa and Lecoq, 1984).

Vegetable Virus Diseases

FIGURE 60 Zucchini yellow mosaic of summer squash. (Reproduced with permission from APS.)

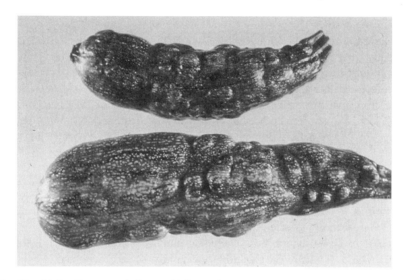

FIGURE 61 Zucchini yellow mosaic in zucchini squash. (Reproduced with permission from APS.)

Geographic Distribution

ZYMV, first described in the Mediterranean, is considered to be spread worldwide.

Host Range and Symptoms

Cucurbits are natural hosts of ZYMV, including cucumber, courgette, melon, watermelon, and zucchini squash. Symptoms in infected plants vary, depending on virus strain and plant cv. (Lecoq et al., 1981; Lesemann et al., 1983). Symptom types in infected leaves include vein clearing, yellowing, shoestring, and enations; infected fruits are malformed, with mixed pulp and distorted seeds; and severe systemic infections cause plant dwarfing (see Figures 60 to 62). Plants in 11 families have been experimentally infected (loc. cit. Lisa and Lecoq, 1984). Diagnostic plants include *Chenopodium amaranticolor* and *C. quinoa* with chlorotic local lesions, but no systemic infection; and *Gomphrena globosa* with local lesions, but no systemic infection.

FIGURE 62 Zucchini yellow mosaic in pumpkin; healthy (right). (Reproduced with permission from APS.)

Pathogenesis and Control

ZYMV is preserved in cucurbits continuously grown on the same plots, wherefrom aphids transmit it nonpersistently, that is, by *Aphis citricola*, *A. gossypii*, *Myzus persicae*, and *Macrosiphum euphorbiae* (Lisa and Lecoq, 1984). It is transmitted easily with infective plant sap, both through contact with infected plants and during cultural operations. No seed transmission is reported.

Preventive crop protection measures include chemical control of vector populations practical primarily for indoor production and procedures designed to prevent mechanical inoculations of plants. Many pathotypes of ZYMV make it difficult to incorporate tolerance or resistance into acceptable cvs.

Zucchini Yellow Fleck Virus (ZYFkV)

ZYFkV is less common in Spain and Greece (Martelli, 1988). In nature, it infects courgette plants, first causing various yellow spots on the leaves that later coalesce to produce a general yellowing and necrosis. Plants severely infected with ZYFkV early in the season grow slowly and yields are reduced substantially. Other natural hosts include cucumber, melon, and watermelon, and experimentally infected hosts belong to the Cucurbitaceae family. Marrow with necrotic lesions followed by systemic yellow flecks and *Lagenaria siceraria* with systemic vein clearing are suitable tests for assay (Martelli, 1988).

Aphid vectors transmit ZYFkV nonpersistently. ZYFkV can also be transmitted by contact with infective plant sap, probably important in intensive indoor production.

Squash Leaf Curl Geminivirus (SqLcV)

Squash is damaged seriously by SgLcV especially in Mexico where many other diseases of beans and tomatoes caused by geminiviruses result in severe yield losses.

Cucurbits: Natural Alternative Hosts for Some Viruses

Cucurbits are natural alternate hosts of some viruses whose primary hosts belong to the Cucurbitaceae family. Melon, squash, and watermelon represent alternate hosts of CMV and of CGnMtV, and courgette (*Cucurbita pepo*) is not susceptible. Squash is the alternate host of WmMV2. (Detailed information on the alternate cucurbit hosts can be found in the description of individual virus species in this group of plants.)

FIGURE 63 Squash yellow leaf curl virus (Mexico).

FIGURE 64 Squash leaf curl of summer squash. (Courtesy R. Provvidenti. Reproduced with permission from APS.)

Some cucurbits are also the alternate natural hosts of some viruses described primarily in noncucurbitaceous plants, that is, papaya ringspot virus (PaRsV) (Figures 65 and 66) on courgette, melon, and watermelon (Smith et al., 1988), ArMV and TmBRV on courgette, and TNV on melon (Schmelzer et al., 1977). Cucurbit plants can be infected with these viruses under such favorable conditions as ideal sources of infection, suitable vector species, and suitable environment. (Descriptions of these viruses can be found in the chapters on their main hosts.)

VIRUS DISEASES OF LEGUME VEGETABLES

Legume vegetables are widely cultivated, owing to their rich and specific oils and proteins composition, representing valuable and irreplaceable components of human nutrition. Cowpea is, *inter alia*, also a very important animal feed. More than 40 different viruses occur singly, or as mixed

FIGURE 65 Natural infection by papaya ringspot virus of squash (Australia).

FIGURE 66 Zucchini squash naturally infected by papaya ringspot virus (Australia).

infections, in this group of plants. They include bean (*Phaseolus vulgaris*), pea (*Pisum sativum*), broad bean (*Vicia faba*), and cowpea (*Vigna* spp.). These plants represent primary natural hosts for these viruses and alternate natural hosts for the remaining ones.

Importantly, BYMV, BLrV, AMV, and CMV infect all plant species belonging to this group. BCmMV, PeElBrV, and TmSpoWV infect the majority of these plants.

Most viruses cause important yield and quality reductions in legume vegetables; thus, all available measures must be used to obtain high yields.

VIRUS DISEASES OF BEAN (*PHASEOLUS VULGARIS*)

Bean Common Mosaic Potyvirus (BCmMV)

Properties of Viral Particles

BCmMV filamentous particles of 750 × 12 to 15 nm have a sedimentation coefficient of 154 to 158S, and consist of one coat protein polypeptide of 32 to 35 kDa. The single-stranded RNA of 3500 kDa accounts for 5% of the particle weight (Morales and Bos, 1988). The TIP of

FIGURE 67 Mosaic and leaf distortion in bean caused by bean common mosaic virus. (Courtesy J.R. Stavely. Reproduced with permission from APS.)

BCmMV is 60°C; the DEP is 1×10^{-3} to 1×10^{-4}; and LIV at 24°C is 1 to 4 days. (Aleksić, 1965; Morales and Bos, 1988).

Strains

BCmMV consists of several strains in Europe and the U.S. that vary depending on host susceptibility and symptoms. They are described on *Phaseolus* spp. based on resistance in two differential cultivars; that is, consisting of mosaic-susceptible cvs. possessing recessive resistance genes, and necrosis consisting of susceptible cvs. possessing the dominant necrosis gene; Drijfhout (1978) assigned 10 strains into 11 groups.

Geographic Distribution and Economic Importance

BCmMV, one of the most damaging bean viruses, occurs wherever French beans are grown. The extent of damage caused in any year depends on cv. susceptibility, earliness of plant infection, and the mode of plant infection (through seed, by vectors, or mechanically). Prior to development of resistant cvs., BCmMV caused yield losses of up to 15% (Nelson, 1932). In plants of cv. Red Mexican U.I.34 that were either moderately or severely infected, pod yields were reduced by 50 and 64%, respectively, and seed yields were reduced by 53 and 68%, respectively (Hampton, 1975). High losses in bean yields can be expected when crop densities favor the occurrence and spread of natural infections.

Host Range

Natural hosts of BCmMV include mostly *Phaseolus* species, thus predominantly *Phaseolus vulgaris*, and less often *Lupinus luteus* and wild legumes, *Rhynchosia minima*, and others. Experimentally infected hosts include 23 species (*Cicer aretinum*, *Glycine max*, *Trifolium incarnatum*, *Vicia faba*, and others). *Nicotiana clevelandii* and *N. benthamiana* are nonleguminous hosts.

Symptoms

Symptoms in bean may vary, depending on the virus strain and host genotype. Slight narrowing of systemically infected leaves characterizes vegetable cvs. Various types of mild to severe mosaic, vein banding with leaf malformation, and curling occur on the leaves of susceptible cvs. Their pods may also exhibit mosaic and malformation. (See Figure 67.)

In hypersensitive cvs., some strains of BCmMV, under certain conditions, cause systemic necrosis that spreads into various plant parts, leading to black root syndrome that results in plant death. Hypersensitive cvs. react with local vein necrosis at normal temperatures and with black root at high temperatures. This reaction, genetically dominant, represents the basis of field plant resistance (Verhoyen and Meunier, 1988). When investigating the susceptibility to BCmMV of 52 bean and string bean cvs., Aleksić (1965) observed three groups: (1) susceptible with mosaic plus leaf and fruit deformations; (2) hypersensitive phloem necrosis leads to death; and (3) hypersensitive localized necrosis of non-phloem tissues, (i.e., considered resistant).

Pathogenesis
BCmMV is naturally preserved in and transmitted by the seed from susceptible plants, but not in seed from hypersensitive plants. Some seed lots result in over 50% transmission (Šutić, 1983), up to 83% (Morales and Bos, 1988). Seed infection occurs only when BCmMV infects the plants before flower initiation, not afterward. This phenomenon seems linked with pollen transmission of BCmMV; that is, when the virus is introduced into the egg cell at the time of fertilization. BCmMV has been observed both in the ovule and in the cotyledonary leaves, but seldom in the seed envelope. The distribution of infected and noninfected seeds in any one pod is random. BCmMV is capable of maintaining infectivity in the seed for up to 30 years (loc. cit. Morales and Bos, 1988). In view of large percentages of infected seed, BCmMV is often found in many plants shortly following crop emergence. Winged aphids transmit BCmMV from infected into healthy seedlings, contributing to its significant natural spread.

Several aphid species such as *Acyrthosiphon pisum*, *Aphis fabae*, and *Myzus persicae* transmit BCmMV in nature. Other species reported as vectors include *Aphis gossypii*, *A. medicaginis*, and *Macrosiphum pisi* (Morales and Bos, 1988). Aphids transmit the virus nonpersistently. The dynamics of aphid flights characteristic for continental regions show virus transmission activity reaching a peak in May/June; then it decreases substantially during July and August.

BCmMV is transmitted easily with infected plant sap, enabling spreading by mechanical infection in susceptible cvs. especially when grown in dense stands.

Control
Only healthy seed should be sown since seed represents the most important natural source of infection. In hypersensitive cvs., practically no danger exists of transmission of BCmMV through seed. Plants for seed production should be grown in isolation from other susceptible legume crops and under conditions least favorable for aphid development. Each infected plant should be rogued from such crops immediately upon detection. Although chemical protection of the seed crop against aphid attacks has been recommended, it is not advisable.

In production of string bean and bean, susceptible cvs. should not be grown, but selection of more resistant (hypersensitive) ones is recommended, such as Top Crop, Harvester, and Panonija among string beans, and small grain bean, P-1, Biser, and Medijana among beans. Growing resistant cvs. may prevent BCmMV infection because bean (*P. vulgaris*) is considered its major host in nature.

Bean Yellow Mosaic Potyvirus (BYMV)

Properties of Viral Particles
BYMV has filamentous particles of 750×15 nm, a sedimentation coefficient of 140 to 144S (Huttinga, 1975), with a major coat protein subunit of 33 kDa and a minor one of 28 kDa (Moghal and Francki, 1976). The TIP of BYMV is 55 to 60°C; the DEP is 1×10^{-3} to 1×10^{-4}; and the LIV is 1 to 2 days (sometimes exceeds 7 days) (Aleksić, 1965; Bos, 1970). BYMV is relatively immunogenic. Although biophysical or serological properties show relatedness to BCmMV, BYMV can be distinguished by differential host range and lack of seed transmission from French bean (*P. vulgaris*).

Strains
BYMV comprises several strains. The bean top necrosis strain causes necrosis in several cvs. and mosaic in cvs. tolerant to normal strains. The pea mosaic strain, referred to as pea common mosaic

virus (PCmMV), causes pea mosaic and is serologically similar to BYMV. The cowpea strain causes systemic symptoms in *Vigna unguiculata* not characteristic of the common strains. The red clover necrosis strain causes red clover and pea necrosis and local and systemic necrosis in *Chenopodium amaranticolor* (Bos, 1970). Based on host range and symptoms in pea and some bean cultivars, the strains of BYMV were classified into three groups, which simplified breeding for resistance (Bos et al., 1974).

Geographic Distribution and Economic Importance

BYMV is distributed worldwide wherever legumes are grown. Many aphid species can account for most of the virus movement. BYMV can occur in high incidence in beans, although some localities record no more than 16% in a field (Aleksić et al., 1980). *Gladiolus* spp. harbor a high (nearly 100%) incidence of BYMV infection in Holland (Nagel et al., 1983), an eventual consequence of vegetative propagation.

BYMV causes economic damage in susceptible bean cvs. that react with apical bud necrosis, leading to plant death. Hamptom (1975) found reductions in pod yields of 33% and in seed yields of 41% in naturally infected plants of Red Mexican U.I. 34 cv. Severe infections caused yield losses up to 96% in *Vicia faba* beans (Frowd and Bernier, 1977).

Host Range

BYMV infects a large number of natural and experimental hosts, a trait substantially different from BCMV. Many hosts (77) are in the Leguminoseae and fewer in other families (Schmidt, 1977). The most important natural hosts among vegetable plants include bean, pea, and broad bean. Other spontaneous Leguminoseae hosts include species of the genera *Lathyrus*, *Lupinus*, *Medicago*, *Melilotus*, and *Vigna*, and such perennials as red clover (*Trifolium pratense*), alsike clover (*T. hybridum*), crimson clover (*T. incarnatum*), white clover (*T. repens*), and alfalfa (*Medicago sativa*); all such perennials can act as permanent sources of infection. BYMV also infects plants in the Iridaceae, including gladiolus, freesia, tritonia, and crocosmia (Derks and Van den Abeele, 1980). BYMV also infects globe artichoke in nature (Russo and Rana, 1978). Black locust (*Robinia pseudoacacia*), a natural perennial woody plant, can be important as a virus reservoir (Kovachevsky, 1968).

Chenopodium quinoa with chlorotic local spots, but no systemic infection, and *Nicotiana tabacum*, *Petunia hybrida*, and *Z. elegans* sometimes with local chlorotic spots, are suitable assay plants (Schmidt, 1977).

Symptoms

BYMV-infected bean and string bean display two main symptoms. The most frequent symptoms are mosaic or mosaic spots on leaves of most cvs. Mosaic spots and patterns are bright yellow, thus the name bean yellow mosaic (Figure 68a and b). The entire surface of infected leaves is characteristically wrinkled. Leaflets that emerge and develop after infection show no significant alterations in shape and size. Early infected plants grow slowly, often resulting in complete loss of grain or pod yields.

The second symptom type characterized by a pronounced apical bud necrosis occurs in fewer cvs. (small grain bean, P-4, Michigan pea). Apical necrosis causes plant death when early infections occur. When infections occur later, shoots that develop from plant buds display yellow mosaic and signs of heavy damage. Symptoms may vary due to differences in both string bean and bean susceptibility and virus strain virulence.

In pea, BYMV causes translucent vein chlorosis, usually accompanied by mild mosaic. Necrosis can occur in the stems and veins. Perfection-type peas are usually resistant (Bos, 1970).

Broad bean is highly susceptible and reacts to BYMV infection with translucent chlorosis of the veins and bright green or yellow mosaic spots on the leaves.

BYMV frequently infects gladiolus in nature, causing mottled spots and yellow striping in leaves and color-break in flowers.

FIGURE 68a Mosaic caused by bean yellow mosaic virus. (Courtesy R. Provvidenti. Reproduced with permission from APS.)

FIGURE 68b Bean yellow mosaic virus-infected bean.

Pathogenesis

BYMV is preserved and transmitted from year to year in its numerous perennial hosts of legumes and some flowers (freesia, gladiolus) and weeds. The yellow mosaic occurs in high percentages in beans grown near gladiolus (Van der Want, 1954; Aleksić, 1965). Plant breeders at Oregon State University interplanted BYMV-infected gladiolus every third row in bean breeding nurseries for virus reistance to assure maximum natural spread of BYMV via aphids into the rows of beans. In Bulgaria, for example, perennials black locust (*Robinia pseudoacacia*) and creeping thistle (*Cirsium arvense*) play an important role in preservation of BYMV because they are suitable hosts of the aphids *Doralis fabae* and *Aphis craccivora*, which serve as primary vectors of BYMV (Kovachevsky, 1968).

In nature, over 20 plant aphid species, including *Acyrthosiphon pisum*, *Macrosiphum euforbiae*, *Myzus persicae*, and *Aphis fabae*, transmit BYMV nonpersistently (Kennedy et al., 1962). BYMV is not transmitted through bean seed, but is seed transmitted in *Vicia faba*, *Melilotus alba* (3 to 5%) and *Lupinus luteus* (6%) (loc. cit. Hollings and Brunt, 1981), mostly in seed from plants infected very early in the season. Thus, some seed represents a natural source of infection. BYMV is transmitted easily with infective plant sap, so mechanical transmission may be important in virus spread among plants grown in dense populations.

Control

To ensure best control of BYMV, it is recommended to grow string beans and beans isolated from perennial virus hosts. It is mandatory to destroy weeds that host both BYMV virus and aphids.

Chemicals to control plant aphids, which play the most important role in the spread of BYMV, to be effective should be applied immediately upon appearance of aphids to prevent early infections. Unless intensive scouting or monitoring is done, this seldom works. Virus-free plants obtained by thermotherapy or meristem-tip culture are particularly important for vegetatively propagated crops.

Bean Leaf Roll Luteovirus (BLrV)

Properties of Viral Particles

The isometric BLrV particles are 27 nm in diameter, with a single polypeptide coat protein of 35 kDa and single-stranded RNA of 2400 kDa (Ashby, 1984). BLrV has good immunogenicity. The two reported isolates show a difference in aphid transmissibility.

Geographic Distribution

BLrV is reported in several regions of America, Europe, India, the Middle East, and North Africa (Ashby, 1984).

Host Range and Symptoms

BLrV naturally infects legumes, causing various symptoms (Ashby, 1984). French bean, chickpea, cowpea, and lentil plants react with chlorosis and stunt. Peas with stunt, chlorosis, and downward leaf roll, and broad bean with interveinal yellowing, upward leaf roll, and reduced pod set are economically important natural hosts.

Leaf roll, the characteristic diagnostic symptom in French bean, pea, and broad bean, results from the viral particles in the phloem in which sieve tube necrosis reduces the carbohydrate and sap movement from the leaves, thus the disease name pea leaf roll, bean leaf roll, etc.

Perennial alfalfa and white clover hosts, important in the disease cycle, are usually systemically infected with BLrV.

Pathogenesis and Control

Aphids transmit BLrV nonpersistently in nature. *Acyrthosiphon pisum*, the principal vector, is followed closely by *Myzus persicae* as the next most efficient. BLrV is transmitted neither with infective sap, nor through legume seeds.

Susceptible vegetable legumes should not be grown near alfalfa crops, which represent the primary natural sources of BLrV infection. Resistant cultivars should be used in pea production. Chemical control of vectors can be useful in crops isolated from severe focal points of infection.

Bean Southern Mosaic Sobemovirus (BSuMV) [Type]

Properties of Viral Particles

BSuMV is isometric, approximately 30 nm in diameter, with a sedimentation coefficient of 115S. The single polypeptide coat protein is 28 kDa. The single-stranded RNA is 1400 kDa, which accounts for 21% of the viral particle weight (Tremaine and Hamilton, 1983). The TIP of BSuMV is 90 to 95°C; the DEP is 1×10^{-5} to 1×10^{-8}; and LIV at 18 to 22°C is 20 to 165 days (Tremaine and Hamilton, 1983). The well-known strains include B (bean), G (Ghana), and M (Mexican severe), all serologically related. BSuMV is a good immunogen.

Geographic Distribution

BSuMV, generally widespread, occurs in the warm temperate and tropical regions of the Americas, Africa, Brazil, India, and France (Férault et al., 1969), which implies its possible presence in broader regions of Europe.

Host Range and Symptoms

The host range of BSuMV, with the exception of *Gomphrena globosa*, is restricted to the Leguminosae. Economically important natural hosts include bean (*Phaseolus vulgaris*), cowpea (*Vigna unguiculata*), green gram (*V. mungo*) and, to a lesser extent, soybean (*Glycine max.*). Symptoms are a general mosaic, mottle and puckering on leaves, veinal necrosis, defoliation, and dark green blotched spots on pods in infected bean.

Satisfactory assay plants include bean (cvs. Bountiful and Pinto) and cowpea (*Vigna unguiculata*), which form local lesions after mechanical inoculation with BSuMV (loc. cit. Smith et al., 1988).

Pathogenesis and Control

BSuMV is spread in the field by beetles (Chysomelidae). In North America, the B and C strains are transmitted by *Ceratoma trifurcata* and *Epilachna varivestis*; and in Nigeria, *Ootheca mutabilis* transmits the C strain. BSuMV is transmitted through bean seed at about 1.5% incidence and through cowpea seed at 5 to 40% incidence (Lamptey and Hamilton, 1974). SuBMV is transmitted easily through sap.

Healthy seed for planting and growing of hypersensitive cvs. Great Northern U.I. 123 and Corbett Refugee are recommended. Chemical control of virus vectors is of little importance to stop initial infections.

Other Virus Diseases of Bean

Bean Golden Mosaic Geminivirus (BGMV)

BGMV has particles of 19-nm diameter that occur mainly as pairs. The coat protein subunit is 31 kDa and the RNA is 750 kDa, which accounts for 29% of the particle weight (Goodman and Bird, 1978). BGMV is the first plant virus reported to have single-stranded DNA. The TIP of BGMV is 50 to 55°C; the DEP is 1×10^{-2} to 1×10^{-3}; and LIV is 48 to 72 hours. BGMV is weakly immunogenic (loc. cit. Goodman, 1981).

BGMV, discovered in Brazil (1965), was found later in tropical and subtropical Americas.

Hosts of BGMV are legumes in the genera *Phaseolus*, *Macroptilium*, *Vigna*, and *Calopogonium* (Goodman and Bird, 1978). Systemic infections of BGMV are characterized by initial symptoms of irregular spots, followed by vein chlorosis, and finally an overall golden mosaic in bean (*Phaseolus vulgaris*) (Figures 69 to 71). Growth of infected plants is slow and fewer pods form. The disease was named after the characteristic golden mosaic symptoms in French bean (*P.*

Vegetable Virus Diseases

FIGURE 69   Bean golden mosaic virus in bean (Mexico).

FIGURE 70   Bean golden mosaic virus in bean (Mexico).

vulgaris) and lima bean (*P. lunatus*) resulting from virus development in phloem sieve tubes and parenchymal cells.

The whitefly, *Bemisia tabaci*, both male and female, is the natural virus vector that can acquire BGMV and transmit it in less than 6 minutes (Goodman and Bird, 1978). Although transmitted experimentally with infective plant sap, other modes of transmission of BGMV are not known.

Bean Mild Mosaic Carmovirus (BMdMV)

The BMdMV isometric particle is 20 nm in diameter and has a sedimentation coefficient of 127S. The single-stranded 1700-kDa RNA accounts for 20% of the particle weight (Waterworth, 1981).

FIGURE 71 Bean golden mosaic virus in bean (Mexico).

The TIP of BMdMV is 84°C; the DEP is 1×10^{-8}; and the LIV at 20°C is 6 weeks. BMdMV is strongly immunogenic.

BMdMV is reported only in El Salvador and Columbia. It naturally infects various *Phaseolus* spp. and *Glycine max.*, causing mild mosaic or is occasionally symptomless. Several beetles serve as vectors, including *Epilachna varivestis, Diabrotica undecimpunctata howardi, D. balteata, Ceratoma ruficornis,* and *Cynandrobrotica variabilis* (Waterworth, 1981). BMdMV is transmitted with infective plant sap, but other modes of transmission are unknown.

Bean Pod Mottle Comovirus (BPoMtV)

BPoMtV has isometric particles of 30-nm diameter that, when purified, consist of an RNA-free protein shell (T) and two virions containing RNA (M and B), probably of 5000 kDa (T), 6500 kDa (M), and 7500 kDa (B) with sedimentation coefficients of 53S (T), 91S (M), and 112S (B). The single-stranded RNA components M and B account for 30 and 37% of the particle weight, respectively (Semancik, 1972), both necessary for infection. The TIP of BPoMtV is 70 to 75°C; the DEP is 1×10^{-4} to 1×10^{-5}; and the LIV at 18°C is 62 to 93 days; BPoMtV has strong immunogenicity (Zaumeyer and Thomas, 1948).

Common in the southern and eastern U.S., legumes are the only known hosts of BPoMtV, primarily *Phaseolus vulgaris* and *Glycine max*. BPoMtV causes severe systemic mottle and chlorosis in bean leaves, and pods are severely mottled, dark green, curled, twisted, and malformed (Zaumeyer and Thomas, 1948). Severe systemic mottle occurs on the leaves, pods, and seed coats of soybean.

The bean leaf beetles *Ceratoma trifurcata, Diabrotica balteata, D. undecimpunctata howardi,* and others transmit BPoMtV naturally (Semancik, 1972). It is also transmitted readily with infective plant sap.

Bean Rugose Mosaic Comovirus (BRgMV)

BRgMV has isometric particles 28 nm in diameter and sediments into three—top (T), middle (M), and bottom (B)—components. T is an RNA-free protein shell, and both M and B contain RNA. They have sedimentation coefficients of 59S (T), 97S (M), and 113S (B) and weights of 3600 (T), 5000 (M), and 5900 (B) kDa, respectively. The coat protein consists of polypeptides of 38 and 22 kDa. The single-stranded RNAs are 2300 and 1400 kDa (Gamez, 1982). The TIP of BRgMV is 65 to 70°C; the DEP is 1×10^{-4} to 1×10^{-5}; and LIV at 22°C is 2 to 4 days. It features strong

FIGURE 72 Systemic yellow spotting in bean caused by alfalfa mosaic virus. (Courtesy W. Kaiser. Reproduced with permission from APS.)

immunogenicity with antisera with a high titre (Gamez, 1982). Severe and mild strains are serologically identical or closely related.

BRgMV is common in Central America. Bean (*Phaseolus vulgaris*) is the natural host, but other legumes have been experimentally infected with BRgMV with symptoms varying, depending on cv. susceptibility. Severe mosaic, puckering, and malformations occur in infected leaves and also on pods. *Chenopodium amaranticolor* is diagnostic, reacting with local lesions upon mechanical inoculation.

The chrysomelid beetles *Ceratoma ruficornis*, *Diabrotica balteata*, and *D. adelpha* are natural vectors of BRgMV (Gamez, 1982). The virus is also transmitted by plant sap. BRgMV causes little damage because the most bean cvs. grown in Central America are resistant.

Bean Summer Death Geminivirus (BSmDhV)

BSmDhV, based on morphological and serological properties plus host range, is a strain of TYDV (Goodman, 1981). This virus was described in Australia in bean (*Phaseolus vulgaris*), and the leafhopper *Orosius argentatus* was shown to be the primary vector (Thomas and Bowyer, 1980).

Bean: Natural Alternative Host for Some Viruses

Alfalfa Mosaic Virus (AMV)

AMV causes lemon yellow spots of different shape and size on bean leaves, thus the disease name bean yellow fleck (Figure 72). The veins of infected leaves remain green, forming a greenish network on the yellowish background of the leaf blade. Leaf development is slowed only slightly with partial distortion of plants. AMV was described in bean in Europe (Beczner, 1973; Schmelzer et al., 1972).

FIGURE 73 Tobacco streak virus inoculated 'Bountiful' bean, healthy (top right).

FIGURE 74 Chlorosis and necrosis of veins and petioles of trifoliolate leaf caused by peanut mottle virus. (Courtesy C.W. Kuhn. Reproduced with permission from APS.)

Cucumber Mosaic Cucumovirus (CMV)

CMV, long-known in bean in Europe (Marrou et al., 1969), causes symptoms of leaf blade curling and chlorotic striping along the veins. Plants grow slowly, with severe stunting dependant on the plant developmental stage at the time of infection. CMV is transmitted through bean seed at up to 18% frequency; thus, procurement of virus-free seed for sowing is important.

Pea Green Mottle Virus (PeGnMtV) in Bean

Initial symptoms of PeGnMtV on bean leaves are translucent, poorly colored zones; later, pale colored ringspots appear. The leaves of some cvs. express chlorotic mosaic that gradually fades in later development stages.

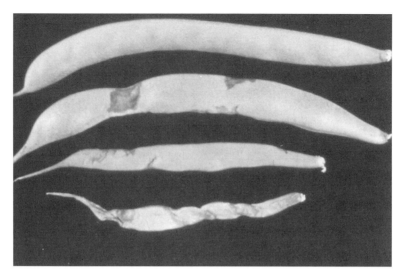

FIGURE 75 Reddish, sunken lesions and deformed pods caused by the bean red node strain of tobacco streak virus. (Courtesy W.J. Kaiser. Reproduced with permission from APS.)

FIGURE 76 Stunting of bean caused by peanut stunt virus. (Courtesy J.R. Stavely. Reproduced with permission from APS.)

Tomato Black Ring Nepovirus (TmBRV)
Yellow, brown, and sometimes ring and necrotic spots occur in young TmBRV-infected bean leaves. In some cvs., plants wilt and apices die. Infected plants may recover later in the season.

Tomato Necrosis Necrovirus (TmNV)
TmNV causes reddish-brown necrosis in bean leaf veins; thus, the disease is named bean vein necrosis. Severely infected leaves dry and fall off. Necrosis in leaf petioles and stalks frequently results in plant death. Violet or reddish spots appear on the pods.

FIGURE 77 Pea mosaic. (Courtesy R.O. Hampton. Reproduced with permission from APS.)

Bean (*Phaseolus vulgaris*) is also a natural host of other viruses (loc. cit. Smith et al., 1988), including artichoke yellow ringspot virus (AkYRsV), BtCuTpV, peanut stunt virus (PtSnV), peanut mottle virus (PtMtV), TSkV, TmSpoWV, and WmMV2.

Virus Diseases of Pea (*Pisum sativum*)

Pea (Common) Mosaic Virus (PeCmMV) – syn: *see* BYMV

Pea Early-Browning Tobravirus (PeElBrV)

Properties of Viral Particles

PeElBrV consists of rigid rods 21 to 22 nm wide and of two lengths: long (L) 215 nm and short (S) 105 nm, with sedimentation coefficients of 286S (L) and 210S (S) and a coat protein molecular weight of 24 kDa. The single-stranded RNA is 2500 kDa (L) and 1300 kDa (S) (loc. cit. Harrison and Robinson, 1981). Only long particles are infective, but both particles must be present for their replication in the hosts (Huttinga, 1969). The TIP of PeElBrV in *Nicotiana clevelandii* sap is 74 to 78°C; the DEP is 1×10^{-5}; and LIV at 20°C is 1 year (Harrison, 1973). PeElBrV is moderately immunogenic. The serologically differing strains are named after the countries in which they were isolated (Dutch strain, British strain, Italian No. 6 strain, etc.) (Harrison, 1973).

Geographic Distribution and Economic Importance

PeElBrV has been observed only in western Europe (Holland and England) (Bos and Van der Want, 1962; Gibbs and Harrison, 1964) and in Morocco (Harrison and Robinson, 1981). Necrosis occurs in PeElBrV-infected pea throughout plant parts, which causes the death of entire shoots. Economic damage occurs primarily when peas are grown on loamy sand soils that favor the development of nematode vectors.

Host Range

In addition to pea (*Pisum sativum*), other natural hosts of PeElBrV include alfalfa (*Medicago sativa*), French bean (*Phaseolus vulgaris*), and broad bean (*Vicia faba*). Nearly 20 species in 10 families have been infected experimentally by mechanical sap inoculation (Bos and Van der Want, 1962; Gibbs and Harrison, 1964).

Symptoms

Necrosis in the stipules and leaflets, necrotic streaks along the stem, and necrotic areas on pods occur in PeElBrV infected pea, with dead shoots representing the most important consequence of infection. Symptoms in other plants include leaf chlorosis and distortion in alfalfa, mosaic and malformation in French bean, and mild mosaic in broad bean. Diagnostic plants include *Chenopodium amaranticolor* with mostly necrotic lesions, and *Nicotiana clevelandii* with chlorotic and necrotic local lesions, plus tissue necrosis in systemically infected first leaves (Harrison, 1973).

Pathogenesis and Control

PeElBrV is permanently preserved in its perennial host, alfalfa, in nature. Trichodorid nematodes *Trichodorus teres*, *T. pachydermus* (in the Netherlands), *T. anemones*, *T. primitivus*, and *T. viruliferous* (in England) are the soil vectors of PeElBrV. Specific relationships exist between PeElBrV and its vectors; that is, *T. anemones* and *T. similis* do not transmit the Dutch isolates (Harrison, 1973). PeElBrV, also transmitted through pea seed, is the most important mode for its natural dissemination. Some Dutch isolates have been transmitted through peas cv. Rondo seed at 10% incidence, but the English isolate transmits through seed of other cultivars at only 1 to 2% frequency (Harrison and Robinson, 1981). Most isolates can be transmitted mechanically with plant sap.

Soils noncontaminated with the nematode vectors should be used for pea production. Chemical control of nematode vectors is not sufficiently reliable to recommend as a disinfestation method. Pea cvs. resistant to PeElBrV exist, but there is some risk of resistance-breaking isolates arising.

Pea Enation Mosaic Virus (PeEnMV) [Type]
(Pea Enation Mosaic Virus Group)

Properties of Viral Particles

PeEnMV isometric particles of 28-nm diameter consist of two sedimenting components: top (T) and bottom (B). They are 43800 kDa (T) and 55700 kDa (B), with sedimentation coefficients of 99S (T) and 112S (B), respectively (Peters, 1982). The single-stranded RNA is bipartite, with RNA-1 in the B component of 1700 kDa and RNA-2 in the T component of 1300 kDa. RNA-1 codes for the major polypeptide coat protein of 22 kDa (Martelli, 1988a). Aphid-transmissible isolates have an additional minor polypeptide of 2.8 kDa (Hull and Lane, 1973; Hull, 1981). A mixture of both components is necessary for PeEnMV infection.

The TIP of PeEnMV is 56 to 58°C (65°C according to some authors); the DEP is 1×10^{-3} (1×10^{-4} according to some sources); and LIV at 20°C is 4 days (Peters, 1982). PeEnMV has moderate immunogenicity. Virus isolates differ, based on symptoms, ease of mechanical transmission, vector transmissibility, and some properties of the purified virus (Peters, 1982).

Geographic Distribution and Economic Importance

PeEnMV, first discovered in New York, is considered to be economically important on pea, and occurs primarily in northern temperate regions. It has been observed in the warmer areas of Sicily and Iran (loc. cit. Peters, 1982) and in Europe (Schmidt, 1977). PeEnMV causes yield reduction via rolling and occasional necrosis of leaves and stems, stunting of plants, and deformation of pods. Yields of pea (*Pisum sativum*) and broad bean (*Vicia faba*) plants may be reduced to 50% (Hull, 1981).

Host Range

Forty-six legumes and six non-legumes are susceptible to PeEnMV (loc. cit. Hull, 1981). Garden and field pea, broad bean, and sweet pea (*Lathyrus odoratus*) are important natural hosts. The perennials alfalfa (*Medicago sativa*) and subterranean clover (*Trifolium subterraneum*) represent reservoir sources of infection. However, exhaustive research by R. Larsen (personal communi-

FIGURE 78 Distortion and enations on pea caused by pea enation mosaic virus. (Courtesy D.J. Hagedorn. Reproduced with permission from APS.)

cation, 1994) at Washington State University fails to confirm that alfalfa is naturally infected by PeEnMV. Assay plants reacting with local lesions include *Chenopodium album*, *C. amaranticolor*, and *C. quinoa*.

Symptoms

Pea is very susceptible to PeEnMV, resulting in severe symptoms and damage. Initial symptoms on leaves are mosaic and mottle; then leaves gradually curl, internodes are shortened, and plants appear bushy. The initial chlorotic spots on leaves become white or poorly translucent. In highly susceptible pea cvs., necrosis occurs in all leaf parts. Over growth, protuberances and enations form near veins on the lower leaf surface (Figures 78 and 79), and hyperplastic (proliferation) cell division are characteristic and diagnostic for PeEnMV. Infection prior to pod formation causes stunt and pod twisting, and seeds from such pods are chlorotic and smaller.

PeEnMV causes a mosaic in both broad bean and crimson clover plants, and leaf spots are pale, translucent, and occasionally with necrotic blotches and streaks. Plants are stunted.

Pathogenesis

PeEnMV is preserved primarily in its perennial natural hosts, either in open fields or in glasshouses. The aphid vector *Acyrthosiphon pisum* is the principal vector that transmits effectively in all instar stages, although nymphs more actively than adults. Nymphs acquire the virus in 15 minutes, whereas adults require 1 to 2 hours. Infective insects can transmit within 1 to 2 minutes. PeEnMV retains infectivity in the insect during molts up to 30 days, but it does not pass on to the offspring. PeEnMV is circulative and semipersistent (Nault et al., 1964).

FIGURE 79 Enations on pea leaf caused by pea enation mosaic virus. (Courtesy D.J. Hagedorn. Reproduced with permission from APS.)

PeEnMV is not transmitted through pea seed, but possibly through seed of broad bean (*Vicia faba*), *V. narbonensis*, and *V. sativa* (Schmidt, 1977). Although sap transmitted, this is of little importance in natural spread.

Control

Peas and broad bean should be grown neither near each other, nor near perennial reservoir hosts of PeEnMV. Special sanitation is required because broad bean seed can transmit PeEnMV. Chemical control of aphids, where feasible, for both crops should be applied as soon as the first colonies are observed to avoid early infections of young plants. Single dominant gene resistant cvs. (Schroeder and Barton, 1958) should be used in high-risk production areas.

Pea Seed-Borne Mosaic Potyvirus (PeSdBnMV)

Properties of Viral Particles

The PeSdBnMV filamentous particles are 770×12 nm and have a sedimentation coefficient of 154S. The coat protein subunit is 34 kDa (Huttinga, 1975). The presumably single-stranded RNA accounts for about 5.3% of the particle weight (Knesek et al., 1974). The TIP of PeSdBnMV is 55°C; the DEP is 1×10^{-3} to 1×10^{-4}; and the LIV is 24 to 96 hours. PeSdBnMV is moderately immunogenic (Hampton and Mink, 1975).

Geographic Distribution and Economic Importance

The worldwide distribution of PeSdBnMV undoubtedly has been through international exchanges of pea seed germplasm, some of which contain up to 90% incidence of PeSdBnMV (Knesek and Mink, 1975; Hampton and Mink, 1975). PeSdBnMV-infected plants are stunted, flowers are distorted, and pods twist irregularly, often bearing only one or two seeds.

Host Range

PeSdBnMV has a comparatively broad host range capable of infecting 47 species in 12 dicot families. Pea is the primary natural host of PeSdBnMV, and *Chenopodium amaranticolor* plus *C. quinoa*, which produce local lesions, are suitable assay plants.

Symptoms

A translucent clearing occurs on PeSdBnMV-infected pea leaf veins, leaflets roll downwards, shoots curl, internodes shorten, and plants are rosetted (Figure 80). Early infections reduce flower and fruit formation or eliminate their development. Broad bean has symptoms accompanied by a certain margin rolling and leaflet distortion.

Pathogenesis

The virus is preserved in and transmitted through infected seed. Plants infected at early stages of growth, prior to flowering, produce seed with 30 to 90% seed transmission (loc. cit. Hampton and Mink, 1975). About ten aphid species transmit the virus nonpersistently in nature. The most frequent aphid vectors include *Acyrthosiphon pisum*, *Macrosiphum euforbiae*, *Myzus persicae*, and *Rhopalosiphum padi* (Aapola and Mink, 1973). Alfalfa may be its natural perennial host, and thus a permanent source of infection.

Control

Highest priority to reduce risk of PeSdBnMV is to plant only healthy seed. The first infected plants in a crop must be rogued immediately to reduce potential secondary spread of the virus.

FIGURE 80 Pea seedling stunt and leafroll caused by pea seedborne mosaic virus. (Courtesy R.O. Hampton. Reproduced with permission from APS.)

Pea Streak Carlavirus (PeSkV)

Properties of Viral Particles

PeSkV filamentous particles are 620 × 12 nm with sedimentation coefficients of 160 and 137S, and its single-stranded RNA accounts for 5.4% of the particle weight (loc. cit. Bos, 1973). The TIP of PeSkV is 60°C (in rare cases 78 to 80°C); the DEP is 1×10^{-6} to 1×10^{-7}; and the LIV infectivity is 2 to 7 days (50 to 70 days for the Wisconsin strain) (Bos, 1973). PeSkV, serologically related to red clover mosaic virus, has strong immunogenicity. Virus isolates differ only slightly based on host range and disease symptoms.

Geographic Distribution

PeSkV occurs in pea in North America and in clovers in Canada and the U.S., and sweet clover in West Germany (Bos, 1973).

Host Range and Symptoms

Pea and clover are natural hosts of PeSkV and experimental hosts are mostly among the legumes, seldom in other families. PeSkV causes necrotic streaks on leaves and stripes on stems, leaf petioles, and leaves involving vein necrosis (Figures 81 to 84). Pods are spotted, malformed, and seedless in severe infections. The leaves and apices of infected plants wilt and die (Hagedorn and Walker, 1949). PeSkV-infected red clover is usually symptomless. *Chenopodium amaranticolor*, *Gomphrena globosa*, and others that react with necrotic local lesions are suitable assay plants (Bos, 1973)

Pathogenesis and Control

PSkV is preserved in annual and perennial plants. It is transmitted nonpersistently by the aphid vector *Acyrthosiphon pisum*. Individual isolates differ based on aphid transmissibility. Mechanical plant sap transmission is possible, but is of little natural importance in peas. Chemical control of aphids and the use of resistant pea cvs. help reduce infections in the field .

FIGURE 81   Pea streak on pea pods. (Courtesy A.S. Lawyer. Reproduced with permission from APS.)

FIGURE 82   Pea streak virus 15 days after inoculation into Perfected Wales pea.

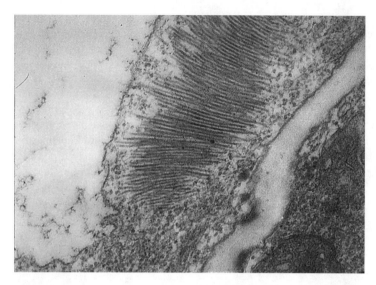

FIGURE 83 Electron micrograph of a pea streak virus "inclusion" of particles in Perfected Wales pea.

FIGURE 84 Pod necrosis and rupture in pea streak virus-infected vetch.

Other Virus Diseases of Pea

Pea Mild Mosaic Comovirus (PeMdMV)

PeMdMV causes vein clearing, later turning into vein etching and necrosis. Leaves roll downward and leaflets become necrotic and distorted. The chronic stage of infection exhibits a mild mosaic or remains nearly masked, but it causes reductions in seed yield (Stace-Smith, 1981).

Pea (Pisum) Rhabdovirus (PeRbV)

PeRbV in Brazil has particles of 240 × 45 nm and can be transmitted mechanically with infective plant sap. No vectors have been identified (Caner et al., 1976).

Pea Stunt Virus (PeStV)

(See *Pea Disease Compendium*, Hagedorn, Ed. 1984 and Figure 85.)

FIGURE 85 Terminal rosette of pea stunt. (Courtesy R.O. Hampton. Reproduced with permission from APS.)

Pea: Natural Alternative Host for Some Viruses

Alfalfa Mosaic Virus (AMV)

AMV infects most pea cultivars, causing mild mosaic, translucent veins, and a diffuse green-yellow of tender parts. Spots may also involve necrosis of tissue. Infected plants grow slowly, and malformed pods produce fewer ovules; therefore, yield losses reach 50%. AMV is observed widespread in Europe (Buturović and Pešić, 1976).

Beet Mosaic Virus (BtMV)

Dark, necrotic stripes and diffuse clearing and wilting of leaflets occur on the stems, leaf petioles, and veins of BtMV-infected pea (Quantz, 1958).

Bean Leafroll Virus (BLrV)

The initial symptom of BLrV infection of pea is a chlorotic clearing between lateral veins at the leaf margins, then the diffuse green coloration spreads along veins and the leaf blades roll upward. Leaf yellowing begins from the leaf margins, gradually spreading and encompassing the entire plant. In susceptible cvs., necroses affect the phloem tissue along the stems, also encompassing plant apices. BLrV is common in peas and has also been named pea leafroll virus (Quantz and Volk, 1954). Pea yield losses of 67 to 85% have been recorded.

Broad Bean Mild Mosaic Virus (BBMdMV)

BBMdMV-infected peas have small curled leaves, shortened internodes, and they grow slowly. Seed from these plants is discolored and malformed.

Broad Bean Stain Comovirus (BBSV)

Leaflets of BBSV-infected peas are small, crinkled, and mottled, and these plants are severely stunted. BBSV has been reported in Northern Europe (Tapio, 1970).

Broad Bean True Mosaic Virus (BBTMV)

Symptoms including vein clearing, translucent chlorosis, diffuse spots and leaf distortion occur in BBTMV-infected peas (Quantz, 1953). These symptoms are later accompanied by necrotic stem streak and a bushy growth habit.

Broad Bean Wilt Virus (BBWV)

BBWV causes translucent chlorosis and line vein mosaic in infected pea leaves. Leaf blades roll upward and die. Shortened internodes in BBWV-infected plants create a dwarfed appearance. During disease development, necrosis also affects apices of stems, which in early infections kills the plants. Necrosis of pods prevents normal seed formation and causes losses in yield up to 90% (Schmidt, 1977).

Cucumber Mosaic Cucumovirus (CMV)

CMV causes chlorotic mosaic and necrotic spots on pea leaves. Leaves of plants infected early display various degrees of malformation. CMV has been reported in pea crops in some European countries (Bos, 1969; Quantz, 1957; Pešić, 1977).

Bean Yellow Mosaic Potyvirus (Strain: Pea Common Mosaic Virus–PeCmMV)

PeCmMV occurs frequently in peas and also infects other legumes, including red clover and especially broad bean.

PeCmMV causes yellowing along the veins with a pronounced yellowish network on the leaves in susceptible pea cvs. A mosaic mottle of alternate chlorotic/green patterns occurs in the interveinal leaf tissues. Symptom severity depends on virus strains and the susceptibility of pea cvs., with variable impacts on yield (i.e., reductions in grain yield may be 5 to 50%). PeCmMV is not transmitted through pea seed, but may be transmitted in red clover seed at 0.1 to 5.3% incidence (Šutić et al., 1975).

Red Clover Mottle Comovirus (RClMtV)

Peas infected with RClMtV are symptomless, but diffuse chlorotic mottle occurs in the differential host *Pisum jomardi* (Mahmood et al., 1972).

Red Clover Vein Mosaic Carlavirus (RClVMV) in Pea

Peas infected with RClVMV have a pronounced rosette-like shape wherein leaves are compacted due to internode shortening, especially in apices. Plants appear mildly chlorotic green and whitish toward the apices (Figure 86). Violet-streaked spots occur in the stems, leaf petioles, and veins. If apices exhibit necrosis, plants die. RClVMV-infected pods form no seeds, causing losses even greater than 50% in pea yields.

Tobacco Necrosis Necrovirus (TNV)

TNV was isolated from pea (*Pisum sativum*) roots (Mowat, 1962).

Tobacco Black Ring Nepovirus (TBRV)

TBRV causes yellowing and dark brown ringspots in infected pea leaves. Severe symptoms include vein, leaf, and stem necrosis, which stop growth, and plants dessicate (Schmidt, 1977).

Tomato Spotted Wilt Tospovirus (TmSpoWV)

TmSpoWV causes several symptoms in infected peas, including purple to brown leaf petiole and stem coloration, leaflet spotting, vein necrosis, and concentric ringspots on pods (Pelet, 1959).

Pea is also an alternate natural host of other viruses, including peanut stunt virus (PtSnV), TSkV, and WmMV2 (loc. cit. Smith et al., 1988).

Virus Diseases of Broad Bean (*Vicia faba*)

Broad Bean Mottle Bromovirus (BBMtV)

Properties of Viral Particles

BBMtV has isometric particles 26 nm (Gibbs, 1972) or 29.4 nm in diameter (White and Fischbach, 1973). Like other viruses in this group, BBMtV consists of three different particles, each with a sedimentation coefficient of 85S (Matthews, 1981). BBMtV particles contain four single-stranded RNA molecules of 1100 (RNA_1), 1030 (RNA_2), 900 (RNA_3), and 360 (RNA_4) kDa, respectively

FIGURE 86 Vein chlorosis in pea leaves caused by red clover vein mosaic virus. (Courtesy R.O. Hampton. Reproduced with permission from APS.)

(Hull, 1972). RNA_4 is monocistronic for the coat protein, and the other RNA (1, 2, and 3) genomes are required for infection. RNA accounts for about 22% of the particle weight (Gibbs, 1972). The TIP of BBMtV is above 95°C; the DEP is greater than 1×10^{-3}; and LIV at 15°C is greater than 20 days (Bawden et al., 1951). BBMtV is moderately immunogenic and serological differences among isolates exist.

Geographic Distribution

BBMtV is restricted to areas where broad bean is grown in England, Portugal, and Sudan (loc. cit. Lane, 1981).

Host Range and Symptoms

BBMtV infects broad bean in nature, and its experimental hosts are mostly legumes (e.g., French bean and pea) and less often *Chenopodium amaranticolor*, *Nicotiana clevelandii*, and recently *Cyamopsis psoraloides* (loc. cit. Lane, 1981).

Characteristic symptoms of BBMtV in the primary broad bean host (Gibbs, 1972) include symptomless in the first inoculated leaves, then vein clearing, accompanied by the more distinct symptom of mottley mosaic on the apical leaves. All broad bean cvs. tested are susceptible. *C. amaranticolor* with chlorotic local lesions, *N. clevelandii* with mild systemic mosaic, and others are suitable diagnostic plants.

Pathogenesis

BBMtV is transmitted easily with infective plant sap, but probably of little importance in the field. Vectors of BBMtV include *Acalymma trivittata* (striped cucumber beetle), *Colaspis flavida* (grape colaspis), *Diabrotica undecimpunctata* (spotted cucumber beetle), and *Sitona lineata* (pea leaf weevil) (loc. cit. Lane, 1981). Transmission through French bean and broad bean seed has not been confirmed (Bawden et al., 1951). Information on virus spread under natural conditions is lacking.

Broad Bean True Mosaic Comovirus (BBTMV)

Properties of Viral Particles

The isometric BBTMV particles of 25-nm diameter sediment into two components, with sedimentation coefficients of 98 and 119S and weights of 6100 and 7500 kDa, respectively. The probably single-stranded RNAs of 1700 kDa (98S) and 2800 kDa (119S) account for 26% (98S) and 35% (119S) of the particle weights, respectively (Gibbs and Paul, 1970). The TIP of BBTMV is 65 to 75°C; the DEP is 1×10^{-4}; and LIV at 24°C is 6 to 7 days (Quantz, 1953). BBTMV is moderately immunogenic. Isolates differ based on symptom severities.

Geographic Distribution and Economic Importance

BBTMV, of limited distribution, occurs in Africa and Europe only sporadically. BBTMV causes economic damage to broad bean. Early infections may cause losses of grain of 30 to 76% (Schmidt, 1977).

Host Range

The host range is comparatively restricted, all in the Leguminosae family, among which broad bean and pea are of primary importance.

Symptoms

Variable mosaic patterns appear in BBTMV-infected broad bean leaves; that is, small leaves gradually curl becoming variably malformed and internodes are severely shortened with substantial growth reduction. Symptom severity depends on cv. susceptibility and virulence of virus strains.

BBTMV-infected peas display severe chlorotic mosaic and occasional necrosis of individual plant parts. Leaves distort and plants remain dwarfed.

Pathogenesis and Control

BBTMV was transmitted through the seed obtained from individually infected broad bean plants in 4 to 26% incidence and through seed originating from the field at 2% incidence (Schmidt, 1977). Infected seed represents the most important source of infection by BBTMV in nature. Natural vectors are unknown, but some authors suggest the importance of *Apion vorax* and *Sitona lineatus* weevils as vectors.

The production and use of healthy broad bean seed is the primary way to prevent infection occurrence. Timely control of the vectors of BBTMV will reduce spread of the virus.

Broad Bean Wilt (possible) Comovirus (BBWV)

Properties of Viral Particles

BBWV, an isometric particle of 25-nm diameter, consists of three components: empty protein shells without RNA (T) and two with RNA labeled (M and B). Their sedimentation coefficients are 63S (T), 100S (M), and 126S (B). The probably single-stranded RNA accounts for 0, 22, and 33% of each particle weight, respectively (Taylor and Stubbs, 1972). The TIP of BBWV is 58°C (or 66 to 70°C); the DEP is 1×10^{-4} to 1×10^{-5} (or 1×10^{-6}); and LIV at 21°C is 2 to 3 days (or 4 to 9 days) (Stubbs, 1947; Schmelzer, 1960). BBWV is strongly immunogenic (Taylor et al., 1968).

BBWV consists of several strains described from different hosts (e.g., petunia ringspot, nasturtium ringspot, pea streak, and others) (loc. cit. Taylor and Stubbs, 1972).

Geographic Distribution and Economic Importance
BBWV has been reported in the U.S., Japan, southeastern Australia, Africa, and Europe. Leaves and flowers fall off, infected broad bean plants wilt and die, all of which cause reductions in yield directly correlated with the number of infected plants.

Host Range
BBWV infects many plants; for example, the garden nasturtium (*Tropaeolum majus*) isolate has experimentally infected 63 species in 19 dicot families (Schmelzer, 1960).

Broad bean and pea are the most important natural hosts among legumes of BBWV. BBWV can occur in red clover and other legumes. Other vegetables naturally susceptible are pepper (*Capsicum annuum*, *C. frutescens*), carrot (*Daucus carota*), lettuce (*Lactuca sativa*), tomato (*Lycopersicon lycopersicum*), and spinach (*Spinacia oleracea*). Also, some common flowers are natural hosts of BBWV, including foxglove (*Digitalis* sp.), gladiolus (*Gladiolus* sp.), narcissus (*Narcissus* sp.), petunia (*Petunia hybrida*), and nasturtium (*Tropaeolum majus*). Diagnostic plants include *Chenopodium* spp. with chlorotic lesions, systemic mottle, and top necrosis; and *Vicia faba* with systemic vein clearing and necrosis of terminal leaves (Taylor and Stubbs, 1972).

Symptoms
Scattered chlorotic spots and vein clearing are initial symptoms on BBWV-infected leaves. As disease develops, necrosis affects apical leaves. BBWV markedly stunts infected plants. The general wilting symptom is accompanied by defoliation and flower loss.

In petunia and other solanaceous plants, BBWV causes ringspots—thus the disease name petunia ringspot disease, and similarly also the nasturtium ringspot disease.

Pathogenesis
BBWV is maintained in numerous perennial hosts, which serve as reservoirs for transmission from year to year. It is transmitted in nature nonpersistently by over 20 aphids, and *Myzus persicae*, *Aphis craccivora*, and *Macrosiphum euphorbiae* are most frequent. BBWV is transmitted with infective plant sap, but not through seed of broad bean or nasturtium.

Control
To prevent infection by BBWV, broad beans must be grown isolated from other, particularly perennial, hosts. Control of aphid vectors simultaneously protects the broad bean crop from other aphid-transmitted viruses. Sowing of broad bean in autumn helps avoid aphid activity. Inbred lines are used as resistant parents in breeding programs (Schmidt, 1988).

Other Virus Diseases of Broad Bean

Broad Bean Necrosis Tobamovirus (BBNV)
BBNV has rod-shaped particles of 150 and 250×25 nm. The TIP of BBNV in pea sap is 55 to 60°C; the DEP is 1×10^{-3} to 1×10^{-4}; and LIV at 20°C is 8 days (loc. cit. Inouye and Nakasone, 1980). BBNV has moderate immunogenicity.

BBNV was discovered in Japan. Broad bean is its only natural host, but it can experimentally infect ten species in four dicot families (Inouye and Nakasone, 1980). Narrow necrotic local lesions plus systemic infection of necrotic spots and rings, veinal necrosis, and streaks on stems and petioles occur in infected broad bean leaves (Inouye and Nakasone, 1980). Although soilborne, the vector of BBNV is unknown. It is also transmitted mechanically with infective sap.

Broad Bean Stain Comovirus (BBSV)
BBSV isometric particles are 25 nm in diameter. Centrifugation separates three particles of equal size with sedimentation coefficients of 60, 100, and 127S. The single-stranded RNA, present only in the two heavier particles, probably accounts for about 25% and 30% of the particle weight in

the 100S and 127S particles, respectively (Gibbs and Smith, 1970). The TIP of BBSV is 60 to 65°C; the DEP is 1×10^{-5}; and LIV at 4°C is 31 days. It is moderately immunogenic.

BBSV is common in Europe, North Africa, and Australia. It infects broad bean (*Vicia faba*), and has been recovered also from vetch (Stace-Smith, 1981). Other legume hosts include *Phaseolus vulgaris* (French bean), *Pisum sativum* (pea), and *Trifolium incarnatum* (crimson clover). Chlorotic mottle and mosaic of leaves and frequently discolored seeds are the main symptoms on broad bean. *P. vulgaris* cv. Prince with local infection and Tendergreen with chlorotic local lesions and systemic mosaic, among others, are suitable diagnostic plants (Gibbs and Smith, 1970).

Concerning pathogenesis and control, apparently, BBSV-infected seed is the main source of natural spread of the virus; for example, in broad bean seed, up to 10% incidence. BBSV is easily transmitted mechanically, but no vector has been established. Thus, the use of virus-free seed is the primary preventive measure.

Broad Bean Mild Mosaic Virus (BBMdMV)

BBMdMV particles are 25 nm in diameter. BBMdMV-infected broad bean exhibit symptoms of mild mosaic to partial necrosis (Devergne and Cousin, 1966). Infected peas have shortened internodes, and thus plant stunting. Discolored and malformed seeds characterize both broad bean and pea. Inoculated *Chenopodium quinoa* have local lesions and mosaic. BBMdMV is transmitted both through seed and mechanical sap inoculations.

Broad Bean: Natural Alternative Host for Some Viruses

Alfalfa Mosaic Virus (AMV)

Grey-green to grey-brown mottle occurs on AMV-infected broad bean leaves, and necrotic spots and zones become evident on the stems. Both plant growth and pod formation are poor.

Bean Leafroll Luteovirus (BLrV) (Syn: Pea Leafroll Virus)

Yellowing gradually spreading into interveinal tissues begins at infected leaf margins. Leaflets become thickened, brittle, bent slightly upward, and erect on the petioles. Necrosis occurs in the phloem sieve tubes. Early infections alter the appearance of plants. The number of pods is reduced and, in early infections, seed yield losses can amount to 93% (Schmidt, 1977).

Bean Yellow Mosaic Potyvirus (BYMV) (Strain, Pea Common Mosaic Virus)

A typical symptom of BYMV infection in broad bean leaves is vein clearing accompanied by mottle and mosaic. Diffuse chlorosis and dark green zones occur later along the veins. Young leaves curl and bend slightly downward. Early infections of BYMV reduce seed yields by 50% (Schmidt, 1977).

Cucumber Mosaic Cucumovirus (CMV)

Numerous red-brown elongated spots merge gradually into necrotic zones, spreading into leaf veins and branchlets and petioles of CMV-infected broad bean plants; thus, the disease is named the broad bean reddish ringspot. Necrosis of vascular bundles causes leaves to wither and fall off.

Pea Enation Mosaic Virus (PeEnMV)

PeEnMV-infected broad beans exhibit characteristic leaf blade malformation and changes in growth. Enation protuberances appear on and near veins on the lower leaf surface, which is diagnostic symptom.

Pea Seed-borne Mosaic Potyvirus (PeSdbMV)

PeSdbMV causes rolling and yellowing of broad bean leaves. The growing points of stems become necrotic, causing stunt or cessation in growth.

Artichoke yellow ringspot virus (AkYRsV) and PeElBrV also naturally infect broad bean plants.

Virus Diseases of Cowpea (*Vigna* spp.)

Cowpea Aphid-Borne Mosaic Potyvirus (CpApBnMV)

Properties of Viral Particles

CpApBnMV particles are filamentous of 750 nm, have a sedimentation coefficient of 150S (Bock and Conti, 1974), and the coat protein subunits are 34 to 35 kDa. The TIP of CpApBnMV is 57 to 60°C; the DEP is 1×10^{-3} to 1×10^{-4}; and LIV at 20°C is 1 to 3 days. CpApBnMV is moderately to strongly immunogenic. Closely related to several legume potyviruses, it is now considered a strain of Blackeye Cowpea Mosaic potyvirus (Edwardson and Christie, 1991). Virus strains are classified based on disease symptoms, host, and regions of origin: that is, European (type) strain, African (neo-type) strain, African mild strain, and African vein banding strain (Bock and Conti, 1974).

Geographic Distribution

CpApBnMV occurs generally where cowpea is grown in Africa, Asia, and Europe. Viruses related to CpApBnMV occur in the U.S.

Host Range and Symptoms

Natural hosts of CpApBnMV include *Vigna sinensis*, *Vigna unguiculata*, and *Gladiolus* sp. (Schmidt, 1977). This virus experimentally infects both legumes and non-legumes. Yellowish vein clearing is the first symptom occurring in infected cowpea plants, later followed by mosaic of varying severity that depends on cv. susceptibility and virulence of strain. Finally, dark-green vein banding, interveinal necrosis, and leaf malformations with stunting of the plant occur. Flowering and pod formation are reduced in highly susceptible cvs. Diagnostic plants include *Chenopodium amaranticolor* with local chlorotic or necrotic lesions, and *Ocimum basilicum* with local necrotic lesions (Bock and Conti, 1974).

Pathogenesis and Control

CpApBnMV is transmitted from year to year through cowpea seed from infected plants. The extent of transmission, dependant on susceptibility of cvs. and virus strains, is from 0 to 3% to a high of 20% (Bock and Conti, 1974). Several aphid species, among them *Aphis craccivora*, *A. fabae*, and *A. gossypii* in particular, serve as vectors to transmit CpApBnMV nonpersistently. Mechanical transmission with infective plant sap is possible.

In view of pathogenesis of CpApBnMV, the important crop protection measures include the use of virus-free seeds, chemical control of vectors (of limited importance for nonpersistent viruses), and earlier planting to encourage adequate plant growth before aphids move into the crop—that is, avoid early infections.

Cowpea Chlorotic Mottle Bromovirus (CpCMtV)

Properties of Viral Particles

The CpCMtV isometric particles are 25 nm in diameter, and as a member of the bromoviruses, its purified preparations consist of three particles of equal size and the same sedimentation coefficient, 85S (Matthews, 1981). Each particle contains four RNA genomes, respectively, of 1100 (RNA_1), 1000 (RNA_2), 700 (RNA_3), and the 300 (RNA_4) kDa coat protein gene. The coat protein subunits are 19.4 kDa (Matthews, 1981). Its multicomponent structure makes it easier to study genetics and molecular biology.

Although different investigations of biological properties of CpCMtV vary, generally the TIP is 67 to 76°C, the DEP is 1×10^{-4} to 1×10^{-5}, and LIV is 1 to 2 days or even up to 44 days (loc. cit. Lane, 1981). CpCMtV is moderately immunogenic. CpCMtV seems the most variable of the bromoviruses. Others, in addition to type strain, based on host are the soybean strain and the Desmodium strain (Bancroft, 1971). Serologically distinct strains have also been described (Lane, 1981).

Geographic Distribution and Economic Importance

CpCMtV, of limited distribution, is reported in the southern U.S. and Central America. Although the bromoviruses generally do not cause economically damaging diseases, the soybean strain of CpCMtV reduced soybean yields by 20 to 30% (Harris and Kuhn, 1971).

Host Range and Symptoms

CpCMtV has been isolated from natural legume hosts, including bean (*Phaseolus vulgaris*), cowpea (*Vigna unguiculata*), soybean (*Glycine max.*), and *Desmodium laevigatum*. Hosts experimentally infected by CpCMtV are mostly legumes, plus a few in families of Chenopodiaceae, Cucurbitaceae, and Solanaceae (Bancroft, 1971).

Mosaic and leaf mottle are the main symptoms in CpCMtV-infected plants; occasionally, local lesions and vein necrosis occur, which are also dependant on host susceptibility and virus strains. Diagnostic plants are *Chenopodium album* and *C. hybridum* with local lesions (Bancroft, 1971).

Pathogenesis and Control

Beetles that transmit CpCMtV naturally include *Ceratoma ruficornis* (beetle), *C. trifurcata* (bean leaf beetle), *Diabrotica balteata* (banded cucumber beetle), and *D. undecimpunctata* (spotted cucumber beetle) (loc. cit. Lane, 1981). The ease of mechanical transmission has enabled numerous experimental investigations of CpCMtV.

Chemical control of vectors and any other measures to reduce vector population are useful in plant protection. Resistant cowpea (Sowell et al., 1965) and soybean cvs. (Harris and Kuhn, 1971) are reported. Unfortunately, some CpCMtV isolates have overcome this host genetic resistance in cowpea.

Cowpea Mosaic Comovirus (CpMV)

Properties of Viral Particles

CpMV isometric particles of 24-nm diameter consist of three equal-size particles: T (RNA-free), and M and B containing 25 and 36%, respectively, RNA. These particles have sedimentation coefficients of 58S (T), 95S (M), and 115S (B) and molecular weights of 3800 (T), 5150 (M), and 5870 (B) kDa, respectively (Van Kammen and De Jager, 1978). The coat protein consists of two polyproteins (37 kDa and 22 kDa) (Wu and Bruening, 1971; Geelen et al., 1972). The single-stranded RNA contained in each of the M and B particles has a sedimentation coefficient of 26S (M) and 34S (B) and weighs 1370 kDa (M) and 2000 kDa (B), respectively (Reijnders et al., 1974). CpMV is infective only when both M and B particles are present. Biophysical properties of CpMV in sap vary, depending on origin and assay plants. The TIP is 55 to 65°C; the DEP is $1 \times 10^{-4.7}$ to $1 \times 10^{-6.7}$; and LIV is 4 to 10 days (Van Kammen and De Jager, 1978). Agrawal (1964) initially described the yellow and severe strains. However, the severe strains have also been named cowpea severe mosaic virus. Serological tests do distinguish these two.

Geographic Distribution and Economic Importance

CpMV has been reported in the U.S., Cuba, the Netherlands, Nigeria, Kenya, and Surinam (Van Kammen and De Jager, 1978). CpMV causes disease by slowing or stopping growth and reducing the number of flowers in *Vigna* spp. These changes cause yield reductions as high as 95% (Van Kammen and De Jager, 1978).

Host Range and Symptoms

CpMV has comparatively few hosts, mostly legumes, and some non-legumes. *Vigna* spp. is the primary natural host of CpMV, with symptoms that vary depending on cv. susceptibility (Van Kammen and De Jager, 1978). The initial symptom on inoculated leaves is a faint, diffuse mosaic, followed by a systemic mosaic on leaflets that develops during the season. Malformed leaves have a yellow mosaic and flower numbers are reduced. Cowpea cvs. show widely different susceptibilites to CpMV, that is, immunity, tolerance, or hypersensitivity. *Chenopodium amaranticolor* with local yellow and necrotic lesions, and *Phaseolus vulgaris* cv. Pinto with local lesions, are suitable diagnostic hosts (Van Kammen and De Jager, 1978).

Pathogenesis and Control

Perennial legume and infected seeds represent continuous sources of infection. Pigeon pea (*Cajanus cajan*) may be a perennial host in coastal East Africa. Seed transmission of CpMV at 1 to 5% incidence is reported in Nigeria (Van Kammen and De Jager, 1978). Several chrysomelid beetle species in the different geographic areas are important natural vectors. Reportedly, two thrips and two grasshopper species also can transmit CpMV (Whitney and Gilmer, 1974), which is mechanically transmitted easily with infective plant sap.

Useful plant protection measures include the elimination of perennial hosts that serve as permanent sources of infection and procurement of healthy, virus-free seed for production. Cowpea lines immune to CpMV should be grown in high-risk areas (Beier et al., 1977).

Cowpea Severe Mosaic Comovirus (CpSeMV)

Properties of Viral Particles

The CpSeMV isometric particles are 25 nm in diameter. On centrifugation, they sediment into three bands of morphologically/serologically identical particles. The top (T) one contains no RNA and the middle (M) and bottom (B) ones contain RNA (Shepherd, 1964). Their sedimentation coefficients are 58S (T), 95S (M), and 115S (B) (De Jager, 1979). The two RNA species of different molecular weights, by analogy, occur separately in the M and B components. Infectivity depends on a mixture of both M and B. Although somewhat variable, the biophysical traits of CpSeMV (Agrawal, 1964) are a TIP of 65 to 70°C, a DEP of 1×10^{-4} to 1×10^{-5}, and a LIV at 23 to 24°C of 1 to 5 days. CpSeMV is strongly immunogenic, producing antisera of high titre. Isolates originating from various geographic areas differ enough based on infectivity, symptomology, and antigenicity to represent separate virus strains.

Host Range and Symptoms

Natural hosts of CpSeMV are legumes and include cowpeas (*Vigna unguiculata sesquipedalis*) and soybean (*Glycine max.*). Symptoms in CpSeMV-infected cowpea vary with cv. and include local lesions of inoculated leaves that become necrotic and systemic symptoms on leaflets of vein clearing and chlorotic mottle, and finally light- and dark-green mosaic areas followed by severe leaf distortion.

Natural hosts of CpSeMV include *Phaseolus lathyroides* and some legume weeds such as hoary tick clovers, *Desmodium canescens*, perennial hosts that represent potential reservoirs of permanent sources of infection. Experimentally infected hosts comprise a number of legume and non-legume species (loc. cit. Stace-Smith, 1981). *Chenopodium amaranticolor* is a suitable diagnostic plant, reacting only with necrotic local lesions on inoculated leaves and no systemic infection, a trait that differentiates CpSeMV from CpMV.

Pathogenesis and Control

Perennial legume hosts and infected seeds are natural sources of infection. Perennial hosts represent primary inocula, wherefrom the virus is transmitted in the field by leaf-feeding beetle vectors mainly in the Crysomelideae family. *Ceratoma ruficornis* and *C. trifurcata* are important among the ten species (De Jager, 1979). The results of studies of seed transmission of CpSeMV vary, but confirm that it occurs at up to 10% incidence (loc. cit. De Jager, 1979), and thus represents a ready source of natural infection. The virus is transmitted easily with infective plant sap, which may contribute to some spread of CpSeMV by direct contact among plants or by contact during cultural operations.

Useful plant protection measures include rogueing perennial hosts, especially weeds, as primary sources of infection, plus spatial isolation of the cultivated crop from these hosts, the use of only healthy, virus-free seed in production, and chemical control or cultural operations aimed at reduction or prevention of vector development.

Other Cowpea Virus Diseases

Blackeye Cowpea Mosaic Potyvirus (BeCpMV)

BeCpMV has filamentous particles of 750 nm and a sedimentation coefficient of 157 to 159S. The coat protein subunit is 34 to 35 kDa, and the single-stranded RNA is 2900 kDa (Purcifull and Gonzalves, 1985). The TIP of BeCpMV is 60 to 65°C; the DEP is 1×10^{-3} to 1×10^{-5}; and LIV is 1 to 2 days. The virus isolates that differ based on symptomology and host range probably represent strains. BeCpMV occurs wherever cowpeas are grown.

Cowpea (*Vigna unguiculata* subs. *sesquipedalis*), *Crotolaria spectabilis*, and *Desmodium* sp. are natural hosts of BeCpMV. Mosaic is the main symptom in cowpea leaves, but some cvs. are symptomless. Experimental hosts include 36 species in 7 dicot families (Purcifull and Gonzalves, 1985). *Chenopodium amaranticolor* and *C. quinoa* with local lesions are suitable diagnostic hosts.

The seed transmission of BeCpMV up to 30.9% represents the primary permanent source of infection. Many aphids (e.g., *Aphis craccivora* and *Myzus persicae*) transmit BeCpMV nonpersistently and naturally in the field. BeCpMV also is transmitted mechanically with infective plant sap (Purcifull and Gonzalves, 1985). The use of healthy, virus-free seed is the main prerequisite to eliminate sources of BeCpMV infection.

Cowpea Mild Mottle Carlavirus (CpMdMtV)

CpMdMtV has filamentous particles of 650×13 nm, consisting of a coat protein polypeptide of 32 kDa and an RNA that accounts for 5% of the particle weight (Brunt and Kenten, 1974). CpMdMtV is immunologically active with antisera of high titre. CpMdMtV is limited to Eastern Ghana (Brunt and Kenton, 1974) and India in groundnut (Iizuka et al. 1984).

The natural hosts of CpMdMtV are cowpea (*V. unguiculata*), and soybeans. Cowpea exhibits symptoms of infection of mild leaf mottle, and sometimes systemic chlorosis and necrosis (Brunt and Kenten, 1974). Experimental hosts, mostly legumes, also include some non-legumes. Necrotic local lesions without systemic infection occur in inoculated leaves of *Beta vulgaris* cv. Detroit.

CpMdMtV is seed transmitted by 2 to 90% in *Vigna unguiculata*, *Glycine max.*, and *Phaseolus vulgaris* (Brunt and Kenten, 1974); thus, virus-free seed is mandatory for sowing new crops. It is mechanically transmitted quite easily, and whitefly vectors have been reported (Muniyappa and Reddy 1983; Naidu et al., 1997).

Cowpea Mottle Virus (CpMtV)

The CpMtV isometric particles of 30-nm diameter constitute a single component of 122S with coat protein subunits of 44.5 kDa and a single-stranded RNA of 1400 kDa that accounts for about 20% of the particle (Bozarth and Shoyinka, 1979). CpMtV has been observed only in Nigeria.

Natural hosts of CpMtV include *Vigna unguiculata* with yellow mosaic, leaf distortion, and witches broom symptoms; and Bambarra groundnut, *Voandezia subterranea*, with green mottle, stunt, and shortened petioles. Many legumes and some non-legumes have been infected experimentally. Chlorotic local lesions occur in inoculated leaves of *Chenopodium amaranticolor* and *C. quinoa* plants.

CpMtV is transmitted through cowpea (up to 10%), Bambarra groundnut, and French bean seed. Vectors of CpMtV include *Ootheca mutabilis*, the galerucid beetle, and *Paraluperodes lineata* (Bozarth and Shoyinka, 1979). It is transmissible with infective plant sap. Planting of healthy, virus-free seed is the most important plant protection measure.

Other Virus Diseases of *Vigna* spp.

Blackgram Mottle Virus (BgMtV) in Blackgram (Vigna mungo)

The BgMtV has isometric particles 28 nm in diameter and a sedimentation coefficient of 122S (Scott and Hoy, 1981). Its protein coat is a single species of 38.2 kDa and its single-stranded RNA accounts for 20% of the particle weight. BgMtV has been found only in India and Thailand (Scott and Hoy, 1981).

Blackgram is the natural host of BgMtV, which is seed transmitted at 8% incidence. Only legumes are experimental hosts. Local lesions without systemic infection result in inoculated *Cyamopsis tetragonoloba* leaves.

The bean leaf beetle *Ceratoma trifurcata* and the Mexican bean beetle *Epilachna varivestis* are vectors of BgMtV in nature (Scott and Hoy, 1981). The virus is transmitted with infective plant sap.

Mung Bean Yellow Mosaic Geminivirus (MbYMV) in Mung Bean (Vigna radiata)

MbYMV has paired-isometric particles of 30×18 nm. Its coat protein is a 27.5-kDa polypeptide, and its DNA is circular, single-stranded, and 800 kDa, which accounts for 20% of the particle weight (Honda and Ikegami, 1986). MbYMV is common in Asia. Its natural hosts include mung bean with yellow specks on the lamina and thin, upward curled pods; and blackgram with yellow mottle and necrotic mottle, which can cause severe losses. MbYMV is mechanically transmitted and has infected legume hosts. It has been transmitted by whiteflies also into Gramineae and Compositae.

In *Bemisia tabaci*, which transmit nonpersistently, the adult females are 3X more efficient transmitters than males. The whitefly nymphs acquire the virus from diseased leaves (Honda and Ikegami, 1986). MbYMV is not seed transmitted.

Cowpea: Natural Alternative Host for Some Viruses

Bean viruses in cowpea include BCmMV, BLRV (syn.: pea leaf roll virus), BYMV, and BSmMV.
 Tobacco viruses in cowpea include TRsV and TSkV.
 Other viruses in cowpea include AMV, CMV, and TmSpoWV (loc. cit. Smith, et al., 1988; Klinkovski et al., 1977).

VIRUS DISEASES OF UMBELLIFEROUS VEGETABLES

The primary umbelliferous vegetables include carrot (*Daucus carota*), celery (*Apium graveolens*), parsnip (*Pastinaca sativa*), and parsley (*Petroselinum crispum*). These biennial root crops, common worldwide, are important in human nutrition based on their high nutritive and dietetic values. Their specific composition of vitamins and minerals, especially in parsnip, have curative effects. Parsnip and parsley are valuable as aromatic spices.

Numerous viruses occur in umbelliferous vegetables as primary or alternate hosts. Some of them, such as celery mosaic virus and CMV, infect all species; whereas others, such as carrot mottle virus, AMV, TmBRV, etc., infect most of them.

Virus damage is highly detrimental in umbelliferous vegetables because they cause both yield and quality reductions, particularly unfavorable for the curative components of these plants.

Virus Diseases of Carrot (*Daucus carota*)

Carrot Mottle Virus (CoMtV)

Properties of Viral Particles

CoMtV is isometric, of 52 nm diameter, with a sedimentation coefficient of 270S. It has a single-stranded RNA (Murant, 1974). The TIP of CoMtV is 67.5 to 70°C; the DEP is 1×10^{-3}; and LIV at 24°C is 9 to 24 hours. (Schmelzer et al., 1977). CoMtV is immunogenically active, but more research is required (loc. cit. Murant, 1974).

Geographic Distribution and Economic Importance

CoMtV is widespread in Australia, New Zealand, Japan, the U.S., and throughout Europe (Murant, 1974).

CoMtV in carrots causes substantial reduced growth and significant yield losses in some years.

Host Range

The host range of CoMtV is restricted to the Amaranthaceae, Chenopodiaceae, Solanaceae, and Umbellifereae families.

Some natural hosts that belong exclusively to the Umbellifereae include some economically most important ones of carrot, parsnip (*Pastinaca sativa*), parsley (*Petroselinum crispum*), and dill (*Anethum graveolens*), and hemlock (*Conium maculatum*) as a common weed plant quite important for both preservation and spread of CoMtV (Schmelzer et al., 1977). Diagnostic plants of CoMtV are French bean cv. Prince with dark-brown local lesions; and *Chenopodium quinoa* with chlorotic spots and neither infected systemically.

Symptoms

CoMtV-infected leaves turn yellow-green, petioles become erect (epinasty), and small leaves curl veil-like. CoMtV occurs regularly in mixed infections with carrot red leaf virus. Such mixed infections in carrot cause variable leaf mottle and plant dwarfing, thus known as the carrot mottley dwarf disease. Such mixed infections also cause similar symptoms in parsley.

Pathogenesis

CoMtV is preserved in biennial Umbellifereae, from which it is transmitted year to year. Vectors ensure further transmission of the virus from infected plants.

The carrot-willow aphid *Cavariella aegopodii*, the vector of CoMtV, is believed to transmit the virus only when its helper, the carrot red leaf virus, is present (Matthews, 1981). The aphid vector in Japan is *Brachycolus (Semiaphis) heraclei* (loc. cit. Murant, 1974). CoMtV is transmitted with infective plant sap, but not through seed from infected plants.

Control

All residual plants representing sources of infection should be destroyed to prevent infection. Spring crops must be isolated spatially from winter crops in which the virus is preserved. Weeds, especially in the Umbellifereae (hemlock and others), which serve both as a CoMtV reservoir and to harbor aphid vectors, should be destroyed immediately. Tolerant carrot cvs. should be grown.

Carrot Red Leaf (possible) Luteovirus (CoReLV)

Properties of Viral Particles

CoReLV has isometric particles of 25-nm diameter and a sedimentation coefficient of 104S. The coat protein is 25 kDa and the single-stranded RNA is 1800 kDa, which accounts for 28% of the particle weight (Waterhouse and Murant, 1982). CoReLV is immunogenically active.

Geographic Distribution

In complex with the CoMtV, CoReLV has been found in Australia, Canada, Europe, Japan, and New Zealand (Waterhouse and Murant, 1982).

Host Range and Symptoms

Natural hosts of CoReLV include cultivated and wild carrot, parsley (*Petroselinum crispum*), some cow parsley (*Anthriscus sylvestris*) isolates, and hodweed (*Heracleum sphondylium*).

CoReLV has been experimentally transmitted into some umbelliferous plants (Waterhouse and Murant, 1982).

CoReLV infection causes reddening and yellowing of carrot leaves, then stunt of plants. When naturally infected with the CoMtV/CoReLV complex, plants display the mottley dwarf symptoms. CoReLV seems confined to the host phloem tissue, certainly correlated with leaf coloration and stunt.

Pathogenesis and Control

The carrot-willow aphids *Cavariella aegopodii* are vectors in the field that transmit CoReLV in a persistent and circulative manner. The minimum period for acquisition is 30 minutes and for inoculation is 2 minutes. Aphids continue to transmit CoReLV after a molt (Waterhouse and Murant, 1982). The CoReLV is a helper for aphid transmission of CoMtV and its multiplication depends on this complex (Elnagar and Murant, 1978). CoReLV is transmitted neither mechanically nor through seed. All measures recommended for control of CoMtV are useful for CoReLV.

Carrot Thin Leaf Potyvirus (CoTnLV)

Properties of Viral Particles

The CoTnLV filamentous particles of 736×11 nm contain 4% RNA (Howell and Mink, 1980). The TIP of CoTnLV is 50 to 55°C; the DEP is 1×10^{-5}; and the LIV at 22°C is 2 days. CoTnLV is weakly immunogenic.

Geographic Distribution and Economic Importance

CoTnLV has very limited distribution and is observed only in semi-desert irrigated zones of northwestern U.S., where it can cause yield losses in carrot up to 25% (Howell and Mink, 1980).

Host Range and Symptoms

Cultivated and wild carrot and the weed hemlock (*Conium maculatum*) are natural hosts of CoTnLV that can experimentally infect some species in the Chenopodiaceae, Solanaceae, and Umbellifereae (Howell and Mink, 1980). Twisting and narrowing of leaves characterize the symptoms in CoTnLV-infected carrot plants, thus the disease name. *Chenopodium amaranticolor* and *C. quinoa* with local lesions are diagnostic plants.

Pathogenesis

CoTnLV-infected carrot roots remaining in the field after harvest represent an important source of infection, and aphid vectors transmit it from these plants into the newly established crops. Both *Myzus persicae* and *Cavariella aegopodii* transmit CoTnLV nonpersistently (Howell and Mink, 1980). It is transmitted mechanically with infective plant sap, but not through seed from infected carrots.

Other Carrot Virus Diseases

Carrot Latent Rhabdovirus (CoLtV)

The CoLtV, a rhabdovirus, is widespread in Japan. *Semiaphis heraclei* is the natural vector, and CoLtV is not transmitted with infective plant sap (Ohki et al., 1978).

Carrot Mosaic (possible) Potyvirus (CoMV)

CoMV, in Europe and the U.K. (Chod, 1988), causes spots with unclearly defined margins and slight malformations in infected leaves of carrot, its main host. *C. gigantea* with yellow spotting and systemic infection, and *Chenopodium quinoa* with yellow local lesions, are suitable diagnostic plants (Chod, 1965). Both *Myzus persicae* and *Cavariella aegopodii* aphids are natural field vectors that transmit CoMV nonpersistently.

Carrot Yellow Leaf Closterovirus (CoYLV)

CoYLV has flexuous particles of 1600×12 nm (Yamashita et al., 1976). Carrot, the primary natural host, reacts to CoYLV infection with yellow-red discoloration. Other hosts are found mostly in the Umbellifereae.

Carrot: Natural Alternative Host for Some Viruses

Other viruses, described as non-Umbellifereae, such as AMV, AkMV, CMV, and BBMtV, frequently infect carrots in nature.

Virus Diseases of Celery (*Apium graveolens*)

Celery Mosaic Potyvirus (CeMV)

Properties of Viral Particles

The CeMV filamentous particles are 760×12 to 15 nm (Schmelzer et al., 1977). The TIP of CeMV is 55 to 60°C; the DEP is 1×10^{-2} to 1×10^{-3}; and LIV at 24°C is 6 days (Shepherd and Grogan, 1971). Some strains of CeMV have strong immunogenicity.

Geographic Distribution and Economic Importance

CeMV has been described in some U.S. states and in some European countries (Britain, France, and Germany). Based on symptoms, the authors believe this virus occurs in former Yugoslavia (Aleksić et al., 1990). Early infections by CeMV cause economic reductions of celery yield up to 50% (Schmelzer et al., 1977).

Host Range

A limited range of hosts in the Umbellifereae are susceptible to CeMV. The economically important hosts include celery, carrot (*Daucus carota*), parsnip (*Pastinaca sativa*), parsley (*Petroselinum crispum*), dill (*Anethum graveolens*), and caraway (*Carum carvi*). Hemlock (*Conium maculatum*) weed is the important natural weed host. *Chenopodium amaranticolor* and *C. quinoa*, which react with chlorotic and partial necrotic local lesions, are diagnostic.

Symptoms

CeMV causes mosaic in infected celery leaves; the leaves are malformed and plants grow slowly. Chlorotic spots, particularly evident in young leaves, occur in CeMV-infected carrot. Parsnip resists the common strain, but is susceptible to the crinkle leaf strain.

Pathogenesis

CeMV is preserved in its biennial winter hosts of Umbellifereae from which it is transmitted into next year's crop. Infected Umbelliferae weeds also represent important sources of natural infection.

Natural vectors of CeMV include 26 aphid species, including *Acyrthosiphon pisum*, *Aphis apigraveolens*, *A. fabae*, *Cavariella aegopodii*, *C. pestinacae*, and *Myzus persicae* (Wolf and Schmelzer, 1972). Aphid transmission of CeMV is nonpersistent. CeMV is transmitted mechanically, but not through seed from infected plants.

Control

A new celery crop should be adequately isolated from other susceptible biennial plants. All umbelliferous weeds in the crop and near vicinity should be destroyed. Control of aphids helps reduce the risks of transmission of CeMV in the crop.

Other Virus Diseases of Celery

Celery Latent Virus (CeLtV)

CeLtV causes no visible symptoms of disease in celery, thus the name latent celery virus. CeLtV particles are filamentous and 885 nm long (Bos et al., 1978a). The celery cvs. investigated are tolerant to the virus. Tolerant hosts also include *Nicotiana megalosiphon*, *Pisum sativum*, *Spinacea oleracea*, and *Trifolium incarnatum*. *Chenopodium amaranticolor* and *C. quinoa* are suitable assay plants. Some pea cvs. react to infection by CeLtV with systemic infection. The prince's feather (*Amaranthus caudatus*) weed plant is also a virus host.

CeLtV is transmitted through seed of celery up to 34% incidence, of *C. quinoa* up to 67%, and prince's feather. No aphids investigated to date have been shown to be a vector. The use of virus-free celery seed is of paramount importance in production of celery.

Celery Ringspot Virus (CeRsV)

CeRsV was described first in western Europe in celery cv. Dulce as its only natural host (Hollings, 1964). Characteristic symptoms in celery leaves are pale green-yellow ringspots and various ring-like patterns.

Celery Yellow Net Virus (CeYNtV)

CeYNtV has been reported in western Europe in celery cv. Dulce, its only natural host, plus some Umbellifereae can be infected experimentally. Some properties of isolates of CeYNtV have been studied (Hollings, 1964), but current information is sketchy. Characteristic symptoms of CeYNtV in celery leaves are pale yellow spots, line patterns along the midribs, some veins retain a diffuse color, and leaves are malformed. Infected plants grow slowly.

Celery Yellow Spot (Possible) Luteovirus (CeYSptV)

CeYSptV is reported in England and the U.S. Celery with circular whitish spots is a natural host of CeYSptV, which also infects parsnip with mild mottle. The aphid *Hyadaphis foeniculi* transmits CeYSptV nonpersistently (Smith, 1972).

Celery: Natural Alternative Host for Some Viruses

A number of common and economically damaging viruses occur naturally in the celery crops, including AMV, ArMV, CMV, TmAsV, TmBRV, and strawberry latent ringspot virus (Smith et al., 1988; Schmelzer et al., 1977; Šutić, 1983).

Virus Diseases of Parsnip (*Pastinaca sativa*)

Parsnip Mosaic Potyvirus (PnpMV)

The PnpMV filamentous particles of 730×14 nm have a sedimentation coefficient of 149S (Murant, 1972). The TIP of PnpMV is 55 to 58°C; the DEP is 1×10^{-3} to 1×10^{-4}; and the LIV at 18°C is 7 to 10 days (Murant et al., 1970). Some suggest that PnpMV may be a celery mosaic virus strain (Schmelzer et al., 1977). Reported in England, PnpMV probably occurs in other countries.

Parsnip is the natural host of PnpMV, which causes mild mosaic symptoms early, but then disappears later in the season, after which plants remain symptomless. PnpMV infects other Umbellifereae and some in the Amaranthaceae, Chenopodiaceae, and Scrophulariaceae (Murant, 1972). *Chenopodium amaranticolor* and *C. quinoa* are diagnostic, reacting with local lesions.

PnpMV is transmitted with infective plant sap and by the aphids *Cavariella aegopodii*, *C. theobaldi*, and *Myzus persicae* nonpersistently (Murant et al., 1970).

Parsnip Yellow Fleck Virus (PnpYFkV)

PnpYFkV isometric particles of 30-nm diameter with a sedimentation coefficient of 148S contain a single-stranded RNA of 3700 kDa (Murant, 1974a). The TIP of PnpYFkV is 57 to 65°C; the DEP is 1×10^{-3} to 1×10^{-4}; and the LIV is 4 to 7 days. PnpYFkV has moderate immunogenicity (Murant, 1974a).

PnpYFkV, reported in England (Murant and Gold, 1968) and Germany (Wolf and Schmelzer, 1972), probably is more widespread.

Parsnip and chervil (*Anthriscus*) are natural hosts of PnpYFkV. Most experimental hosts are in the Umbellifereae, plus only a few in other families. General chlorosis along larger and smaller veins occurs first in PnpYFkV-infected parsnip leaves, resulting in a net-like appearance of the entire leaf. Yellow spots and flecks, leading to partial or general chlorosis of leaves, appear between the veins. Later, the chronic stage symptoms are hardly noticeable on the newly emerged leaves. Early infected plants become rosette-like in appearance.

PnpYFkV is transmitted mechanically with infective plant sap. The aphid *Cavariella aegopodii* transmits PnpYFkV semipersistently in a circulative manner. Aphid transmission occurs only when it also transmits simultaneously the anthriscus yellows helper virus (Elnagar and Murant, 1976). The helper virus is 22 nm in diameter and causes yellows on chervil plants.

Parsnip: Natural Alternative Host for Some Viruses

Several viruses, including CMV, PnpMV (probably a CoMtV strain), and TmBRV, can occur in parsnip as their alternate host.

VIRUS DISEASES OF PARSLEY (*PETROSELINUM CRISPUM*)

Parsley is naturally infected by 10 viruses. Among these, Parsley latent virus (PyLtV) and Parsley virus 5 (PyV5) have been described in parsley as their main (primary) host. Data available are sparse. PyLtV, a rhabdovirus, described in England, is transmitted both mechanically and by the aphid *Cavariella aegopodii* (loc. cit. Francki et al., 1981). PyV5 is a possible member of the potexvirus group (loc. cit. Purcifull and Edwardson, 1981).

Most other viruses infect parsley as their alternate natural host causing specific disease symptoms. These viruses originate from related infected umbellifereae such as CoMtV (parsley symptomless infection) and CeMV (parsley mosaic disease). Other viruses include AMV (parsley yellow mottle), BtClTpV, BBWV (parsley ring mosaic), chicory yellow mottle virus (parsley foliar alterations), CMV (irregular yellow mottling) and TmBRV (parsley yellow vein disease).

VIRUS DISEASES OF COMPOSITEAE VEGETABLES

VIRUS DISEASES OF ARTICHOKE (*CYNARA SCOLYMUS*)

Artichoke Italian Latent Nepovirus (AkILtV)

Properties of Viral Particles

The AkILtV isometric particles are 30 nm in diameter and consist of three different components: top (T), which is RNA-free with a sedimentation coefficient of 55S, and two middle (M) and bottom (B), which contain RNA with sedimentation coefficients of 96S (M) and 121S (B), respectively (Martelli et al., 1977). RNA (M) accounts for 34% and RNA (B) for 41% of their respective particle weights. Only the mixture of M and B particles is infective. The coat protein composition is a single polypeptide of 54 kDa (Jankulova et al., 1976). The TIP of AkILtV in *Chenopodium quinoa* sap is 55 to 60°C; the DEP is 1×10^{-2} to 1×10^{-5}; and the LIV at 4°C is 3 to 14 days. AkILtV has good immunogenicity.

Geographic Distribution

AkILtV has been reported only in Bulgaria, Greece, and Italy.

Host Range and Symptoms

Natural hosts of AkILtV include cultivated species of chicory, gladiolus, grapevine, lettuce, and pelargonium, and several symptomless weeds of *Crespis neglecta*, *Lactuca virosa*, *Lamium amplexicaule*, and *Sonchus* spp. Sixty-three species can be experimentally infected with AkILtV (Martelli et al., 1977).

In some artichoke cvs., AkILtV causes general leaf yellowing and stunted plants, but most cvs. are symptomless. Symptoms in other hosts vary; for example, *Cichorium intybus* with chlorotic mottle and bright yellow leaf spot; *Pelargonium zonale* with leaf malformation and stunt; and grapevine with fanleaf-like symptoms; *Gomphrena globosa* with necrotic, whitish local lesions; and *Nicotiana tabacum* with chlorotic, necrotic ringspots and line patterns are diagnostic plants (Martelli, 1988).

Pathogenesis and Control

AkILtV is preserved in the soil and transmitted with the nematodes *Longidorus apulus* (serological variant of the virus in Italy) and *L. fasciatus* (serological variant of the virus in Greece) (Martelli, 1988b). It is transmitted easily with infective plant sap, but no other mode of transmission is known. Planting the crop in nematode-free soils and the use of cultural operations aimed at avoiding mechanical virus transmission are the most useful measures for prevention of infection.

Artichoke Vein Banding (Tentative) Nepovirus (AkVBaV)

Properties of Viral Particles

The AkVBaV isometric particles are 30 nm in diameter and consist of three particles: one RNA-free (T) and two containing RNA (M and B) with sedimentation coefficients of 56S (T), 92S (M), and 124S (B); the single-stranded RNA accounts for 24% (M) and 37% (B) of each particle weight, respectively (Gallitelli et al., 1984). Only the mixture of M and B particles together is infective. Protein preparations consist of three polypeptides of 22, 24, and 27 kDa (Gallitelli et al., 1978), and one additional polypeptide has also been marked.

Geographic Distribution

AkVBaV was identified in artichoke cvs. growing in southern and central Italy that had originated in Turkey (Gallitelli et al., 1984).

Host Range and Symptoms

Artichoke is the only natural host of AkVBaV, which causes chlorotic discoloration in some cvs. It has been transmitted experimentally into 20 species. Satisfactory local lesion hosts are *Chenopodium amaranticolor* and *C. quinoa* (Gallitelli et al., 1984).

Pathogenesis

AkVBaV has been insufficiently studied, but is transmitted with infective plant sap; no other mode of transmission is known.

Artichoke Yellow Ringspot Nepovirus (AkYRsV)

Properties of Viral Particles

The AkYRsV is isometric, 30 nm in diameter, and consists of three particles: one RNA-free and two containing single-stranded RNA (RNA-1 and RNA-2) that are infective only when mixed together. The coat protein is a single polypeptide of 53 kDa (Rana and Martelli, 1983). The TIP of AkYRsV is 60°C; the DEP is 1×10^{-3} to 1×10^{-4}; and LIV at 22 to 24°C is 48 to 72 hours. AkYRsV is weakly immunogenic.

Geographic Distribution

AkYRsV, with a very restricted distribution, occurs only in northeastern Peloponnesus and Sicily (Rana and Martelli, 1983).

Host Range and Symptoms

Artichoke and cardoon (*C. cardunculus*) are the main natural hosts of AkYRsV, which causes bright yellow blotches, ringspots, and line patterns on leaves. Cultivated natural hosts include tobacco with yellow necrotic rings and oakleaf patterns, and French bean and broad bean with diffuse yellowing and leaf malformation. AkYRsV naturally infects several weeds and shrubs, causing yellow discoloration of leaves (Rana and Martelli, 1983). The *Gomphrena globosa* and *Chenopodium quinoa* experimental hosts that react with local lesions and systemic infection can serve as assay hosts.

Pathogenesis

AkYRsV is naturally transmitted at high percentages through both pollen and seed of several infected species, and through vegetative propagation stocks (Martelli, 1988c). It is transmitted easily with infective plant sap. No information is known about vectors.

Other Virus Diseases of Artichoke

Artichoke Curly Dwarf Virus (AkCuDV)

AkCuDV, a possible potexvirus, has flexuous rod-like particles of 582×15 nm (Morton, 1961). The TIP of AkCuDV is 55 to 60°C; the DEP is 1×10^{-3} to 1×10^{-4}; and the LIV is 2 to 3 days in *Zinnea* and artichoke sap. AkCuDV is common in California.

Artichoke is the natural host for AkCuDV, and experimental hosts involve only composites (*Centaurea cyanus*, *Helianthus annuus*, *Tagetes* sp., and others). Common symptoms in artichoke are curled leaves, necrotic pockets in shoots, veinal necrosis in young leaves, shortened internodes, and general stunting of plants. Infected *Zinnia* sp. have local lesions at 13 to 18°C) (Morton, 1961). AkCuDV is transmitted mechanically with infective plant sap.

Artichoke Latent Nepovirus (AkLtV)

AkLtV occurs in Mediterranean countries, California, and Brazil. Artichoke is the natural host of AkLtV; some infected cvs. remain symptomless, but others express mild mottle and stunt. AkLtV, aphid transmitted nonpersistently, is transferred through propagating material but not through seed from infected plants (Rana et al., 1982). Healthy plants can be obtained by meristem-tip culture (Harbaoui et al., 1982).

Artichoke Mosaic Disease (AkMD)

Mosaic in artichoke was described first in Sicily (Gigante, 1951). In addition to mosaic, bright yellow spots and blotches of irregular shape occur on the leaves. The causal agent of the disease has been mechanically transmitted but not yet defined ethiologically (Martelli, 1988d).

Artichoke Mottled Crinkle Tombusvirus (AkMtCrV)

AkMtCrV, observed in Greece, Italy, Morocco, and Tunisia, is a tombusvirus. Artichoke is its natural host that may remain symptomless or display symptoms dependant on cv. and environmental conditions. Crinkle, chlorotic blotches, leaf malformation, and smaller and fewer flower heads are the most frequent symptoms of AkMtCrV (Martelli, 1965; 1981). AkMtCrV has a broad host range. *Ocimum basilicum* with dark brown local lesions, and *Datura stramonium* with chlorotic and necrotic local lesions, are suitable assay hosts. AkMtCrV is transmitted mechanically and through propagating material. Potential virus vectors in soil have not been discovered.

Artichoke: Natural Alternative Host for Some Viruses

Artichoke is naturally susceptible to the viruses in noncomposite plants, including BYMV, TmBuSnV, and TSkV.

VIRUS DISEASES OF ENDIVE (*CHICORIUM ENDIVIA*)

Endive is the alternative natural host of several viruses, some with a restricted host range, including bidens mottle virus (BdMtV) (Purcifull et al., 1976) and LBiVV (Schmelzer et al., 1977), plus others common in many species, including AMV, BtWsYsV, CMV, and LMV, the properties of which are noted in their descriptions on the main hosts.

VIRUS DISEASES OF LETTUCE (*LACTUCA SATIVA*)

Lettuce Mosaic Potyvirus (LMV)

Properties of Viral Particles

LMV filamentous particles are 750×13 nm, a single component with a sedimentation coefficient of 143S. The single-stranded RNA is 3000 kDa and the coat protein is a single polypeptide of 33kDa (Lot, 1988). The TIP of LMV is 55 to 60°C; the DEP is 1×10^{-1} to 1×10^{-2}; and the LIV is 1 to 2 days (Tomlinson, 1970). LMV is moderately immunogenic. Serological analyses (ELISA tests) can be used to test greater numbers of plant samples, also including the seed (Jafarpour et al., 1979; Falk and Purcifull, 1983). Variants strains of LMV that differ in aggressiveness in cvs. were identified based on symptoms in lettuce, peas, and weeds (McLean and Kinsey, 1962).

Geographic Distribution and Economic Importance

LMV is distributed worldwide wherever the lettuce crop is grown. LMV is economically damaging in lettuce production. If infection occurs early, plants remain dwarfed with heads half the size of healthy ones, resulting in yield losses up to 50%. Seed yields of lettuce are also reduced in severely infected acreages. Overall damage has been assessed, based on the number of infected plants in the field.

Host Range

LMV with a broad host range naturally infects 21 species, mostly Compositae, and can be transmitted mechanically into 121 species (Horvath, 1980). Economically important natural hosts include lettuce, spinach (*Spinacia oleracea*), and pea (*Pisum sativum*). In nature, LMV infects some common weeds, including *Chenopodium album*, *C. murale*, *Lamium amplexicaule*, *Senecio vulgaris*, *Sonchus asper*, and *Stellaria media*. *Chenopodium amaranticolor* and *C. quinoa* with local lesions on inoculated leaves, and systemic infection in younger leaves, are suitable assay plants.

Symptoms

Vein clearing is the primary symptom of LMV infection, followed by secondary symptoms of pale- and dark-green mosaic. Striped mosaic occurs along and around the veins. Scattered tissue necrosis also occurs in susceptible cvs. and infected plants remain smaller than healthy ones. Early LMV infections prevent head formation, whereas heart leaves crinkle and compact to form a rosette.

Pathogenesis

LMV is transmitted through seed of lettuce (3 to 10%), grinsel (*Senecio vulgaris*) (Phatak, 1974), and wild lettuce *Lactuca serriola* (loc. cit. Tomlinson, 1970), and through pollen and ovules of LMV-infected plants. Biennial hosts play an important role in preserving and transmitting LMV season to season. In warmer areas, LMV is continuously preserved in lettuce that is grown without rotation of other crops on the same acreages, especially for indoor production.

LMV is transmitted nonpersistently by 15 aphid species in nature; *Myzus persicae*, which colonizes lettuce plants readily throughout the growing season, is a primary vector in France. *Macrosiphum euphorbiae*, frequently observed in lettuce crops, is an active vector, and *Aphis gossypii* is an important summer vector in southern regions. The root aphid *Pemphigus bursarius* is also an active vector of LMV (Messiaen and Lafon, 1965). LMV is transmitted easily with infective plant sap and represents a high risk in seedling production and densely populated lettuce crops.

Control

The use of healthy, virus-free seed is mandatory to prevent LMV infection. For a lettuce seed crop, aphids should be closely monitored and systematically sprayed with insecticides as soon as the first focal points of infection and/or aphid colonies appear. Lettuce crops must be inespected on a regular schedule, and all infected plants rogued immediately. Any plants remaining in the field after harvest should be destroyed. Susceptible weeds in lettuce crops, *Senecio vulgaris* in particular, and in their vicinity should be rogued.

Lettuce Necrotic Yellows Rhabdovirus (LNcYsV)

Properties of Viral Particles

LNcYsV, a bacilliform particle of 227 × 66 nm, has a sedimentation coefficient of 940S, and the RNA genome has an outer envelope covered with uniformly arranged projections (Francki and Randles, 1970). The TIP of LNcYsV is 52°C; the DEP is 1×10^{-2}; and the LIV is 8 to 24 hours. LNcYsV has a certain immunogenic activity. LNcYsV isolates differ, based on symptom severity in *Nicotiana glutinosa*.

Geographic Distribution

LNcYsV, comparatively restricted, is reported only in Australia and New Zealand.

Host Range and Symptoms

LNcYsV has a narrow host range. Lettuce is its only economically important natural host. Sowthistle (*Sonchus oleraceus*), dune thistle (*Embergeria megalocarpa*), and the weed *Reinchardia tingitana* represent natural sources of infection as virus hosts of LNcYsV (Matthews, 1981). The virus can be transmitted mechanically into noncomposites, valuable as diagnostic plants: that is, *Datura stramonium* with interveinal chlorosis and *Nicotiana glutinosa* with downward curl and faint mosaic of leaves and plant stunt.

Pathogenesis

Infected weeds and lettuce grown continuously on the same acreages in fields and indoors represent primary sources of LNcYsV infection in nature. The aphids *Hyperomyzus lactucae* and *H. carduellinus* are vectors that transmit LNcYsV nonpersistently (loc. cit. Francki and Randles, 1970). LNcYsV, a rhabdovirus, replicates in its vector *H. lactucae* (Matthews, 1981). It is transmitted mechanically with infected plant sap. Other modes of transmission are unknown.

Other Virus Diseases of Lettuce

Lettuce Big Vein Virus (LBiVV)

The viral nature of the LBiVV needs more research. It may be an elongated particle (350 × 18 nm) containing double-stranded RNA (Dodds and Mirkov, 1983). The disease has been observed in Europe and North America.

LBiVV occurs frequently in lettuce crops with characteristic symptoms of swollen leaf veins much larger than normal, especially prominent on the midrib (Figure 87a). The midrib is substantially shortened, causing crumple and distortion of the leaf blade. Chlorotic stripes accompany and appear near the veins. Interveinal tissues remain green. Leaf blades are swollen, brittle, and break

FIGURE 87a Lettuce big vein in romaine lettuce (right) and healthy (left). (Courtesy F.F. Laemmlen. Reproduced with permission from APS.)

FIGURE 87b Lettuce infectious yellow virus in tolerant (green foliage) and susceptible (yellow foliage) cultivars of crisp-head lettuce. (Courtesy F.F. Laemmlen. Reproduced with permission from APS.)

easily. Lettuce heads form slowly. Cool weather during most active plant growth favors the occurrence/development of this disease. Endive (*Cichorium endivia*) and sowthistle (*Sonchus oleraceus*) are also important hosts of LBiVV.

LBiVV retains infectivity in the soil in which it is transmitted from plant to plant through the infective activity of *Olpidium brassicae* zoospores (Lin et al., 1970). Control of *O. brassicae* by soil disinfestation is the key preventive measure known.

Lettuce Necrosis Virus (LNV)

Lettuce leaf necrosis, caused by LNV whose properties have been little studied, was discovered and described in England (Kassanis, 1944, 1947), although it is known to occur in other parts of Europe. Early infections of LNV cause leaf necrosis and death; thus, LNV has been placed among the economically damaging viruses. Primary symptoms are bronze necrosis along leaf veins, forming a net-like pattern on some leaf parts, plus ringspots of irregular shape and size in interveinal tissues. Secondary symptoms typically include leaf blackening, curl, and decline, plus substantial reductions in plant growth. Necrosis also occurs in floral parts of seed plants, causing reduced seed set.

LNV also infects dandelion (*Taraxacum officinale*). It is seldom transmitted mechanically. In nature, the aphids *Aulacorthum solani*, *Myzus ascalonicus*, *M. ornatus*, and *Nasanovia ribis-nigri* transmit LNV nonpersistently. Control of dandelion in and around the lettuce crop and aphids represent the two main measures required to protect against an epidemic of LNV.

Lettuce Infectious Yellows Closterovirus (LInYsV)

A problem in the Southwest U.S. near desert lands, LInYsV transmitted by *Bemisia tabaci*.

Lettuce: Natural Alternative Host for Some Viruses

Alfalfa Mosaic Virus (AMV)

Vein clearing and mottle of varying severity are the most frequent symptoms in AMV-infected lettuce leaves.

Arabis Mosaic Nepovirus (ArMV)

ArMV-infected plants are chlorotic and dwarfed; thus, the disease is known as lettuce chlorotic dwarf. Infected plants fail to form heads. ArMV is seed transmitted in lettuce at 60 to 100% incidence, and is transmitted mechanically and by soil nematodes. The use of healthy, virus-free seed and planting in nematode-free soils, preferably disinfested for insurance of nematode control, are the two most important measures to ensure healthy plant production.

Beet Western Yellows Luteovirus (BtWsYsV)

BtWsYsV causes slight or intense interveinal yellowing, followed by dark-brown marginal leaf necrosis (Figure 88). This desease causes severe economic damage to lettuce crops.

FIGURE 88 Beet western yellows virus in crisp-head lettuce. (Courtesy J.E. Duffus. Reproduced with permission from APS.)

FIGURE 89 Tobacco rattle virus in romaine lettuce. (Courtesy D.E. Mayhew. Reproduced with permission from APS.)

Cucumber Mosaic Cucumovirus (CMV)

Lettuce is one of many natural hosts of CMV. Chlorosis along the veins, followed by mosaic and necrotic changes, occur in CMV-infected lettuce. Plants infected early remain dwarfed and fail to head. Aphid vectors spread CMV continuously throughout the entire vegetation period.

Tobacco Necrosis Necrovirus (TNV)

TNV is preserved in the soil, wherefrom it is transmitted by the zoospores of *Olpidium brassicae*. Soils with generally high moisture levels allow *O. brassicae* to transmit TNV from plant to plant, thus contributing to the continuation of virus spread. Root and leaf vein necrosis are the main symptoms in infected plants.

Tobacco Rattle Tobravirus (TRtV)

TRtV-infected lettuce leaves curl, malform, and compact, and infected plants grow slowly and develop irregularly (Figure 89). Nematode vectors are of primary importance in virus epidemiology; thus, one must select noncontaminated s oils or they should be disinfested before planting the lettuce crop.

Tomato Spotted Wilt Tospovirus (TmSpoWV)

TmSpoWV infection usually occurs on and affects one side of lettuce plants, then it gradually spreads into the heart leaves. General chlorosis, followed by small brown spots and tissue necrosis in older leaves, characterize symptoms in infected lettuce leaves (Figure 90). Plants infected early in the season grow slowly, wilt, and then die prematurely. The risks of virus spread are particularly high when lettuce seedlings are produced continuously in the same seed beds or near tomato, pepper, or tobacco nurseries, all of which are susceptible to TmSpoWV.

FIGURE 90 Tomato spotted wilt virus in crisp-head lettuce. (Courtesy J.J. Cho. Reproduced with permission from APS.)

FIGURE 91 Vascular discoloration caused by beet yellow stunt virus. (Courtesy J.E. Duffus. Reproduced with permission from APS.)

Other Viruses

Several other viruses can affect the lettuce crop in nature, including AkILtV, BtPdYsV, BtYSnV (Figure 91), BiMtV, BBWV, SlYVV (Figure 92a), TSkV (Figure 92b), TRsV, TmBRV, and TuMV. (Detailed information about these viruses are in their descriptions on their primary natural hosts.)

VIRUS DISEASES OF CHENOPODACEAE VEGETABLES

VIRUS DISEASES OF SPINACH (*SPINACIA OLERACEA*)

Spinach is also susceptible to several viruses, including the spinach latent virus described here as its main host. Other viruses originate from other hosts, mostly sugar beet, which belongs with spinach in the Chenopodiaceae.

Vegetable Virus Diseases

FIGURE 92a Sowthistle yellow vein virus in crisp-head lettuce. (Courtesy J.E. Duffus. Reproduced with permission from APS.)

FIGURE 92b Tobacco streak virus in crisphead lettuce. (Courtesy J.J. Cho. Reproduced with permission from APS.)

Spinach Latent (Possible) Ilarvirus (SLtV)

Properties of Viral Particles

The SLtV particles are quasi-isometric, 27 nm in diameter, and sediment into three components with sedimentation coefficients of 87, 98, and 103S (Bos, 1984). SLtV contains five single-stranded RNA species of 1300 (RNA-1), 1180 (RNA-2), 910 (RNA-3), 350 (RNA-4), and 270 (RNA-5) kDa, respectively, and a single polypeptide coat protein of 28 kDa. The TIP of SLtV is 60 to 65°C; the DEP is 1×10^{-3} to 1×10^{-4}; and LIV at 24°C is 6 to 8 days (Bos et al., 1980; Štefanac and Wrischer, 1983). SLtV is weakly immunogenic.

Geographic Distribution

SLtV is restricted in distribution to Holland (Bos et al., 1980) and former Yugoslavia (Štefanac and Wrischer, 1983).

Host Range and Symptoms

Spinach is the only natural known host of SLtV and remains either symptomless or displays transient disease symptoms. SLtV has been transmitted experimentally into 24 dicot species with systemic infection, but without symptoms (Bos et al., 1980; Štefanac and Wrischer, 1983). Species with local reactions and without systemic infection, including *Chenopodium amaranticolor* and *Phaseolus vulgaris* cv. Bataaf, are suitable assay hosts for experimental studies.

Pathogenesis

Infected spinach seed is a primary source of infection because it transmits SLtV in high percentages. Seed from commercial spinach samples contained over 50% SLtV infection and seed from some other infected plants had 30 to 90% incidence (Bos et al., 1980). SLtV is transmitted easily with infective plant sap, and no other mode of transmission is known. The use of healthy, virus-free seed is of primary importance to avoid SLtV infections.

Spinach: Natural Alternative Host for Some Viruses

Beet Mosaic Virus (BtMV)

BtMV causes mosaic symptoms in spinach leaves; thus, the disease was named spinach mosaic. Inner leaves in BtMV-infected spring spinach curl upward, become distorted, and small golden yellow spots are first noticed, later becoming mosaic. Infected plants are stunted. Winter spinach, although infected with BtMV at a lower incidence, carries the BtMV into the next summer season. Because BtMV causes economic damage in spinach, new production fields must have adequate isolation from a sugar beet crop. Control of aphid vectors is also a useful preventive measure.

Beet Yellows Virus (BtYsV)

Spinach is highly susceptible to BtYsV, a disease easily recognized by the large, irregular, yellow spots and asymmetric, distorted leaf blades. Both sugar beet and spinach are important sources of BtYsV, which is transmitted actively and spread by aphids. Winter spinach carries BtYsV through the winter into the next summer season. Spinach should be grown far away from a sugar beet crop to prevent infection by BtYsV. Control of aphids is essential to reduce the level of field infections by BtYsV.

Other Beet Viruses

BClTpV, BtMdYV, BtNcYVV, and BtWsYsV all infect spinach.

Cucumber Mosaic Cucumovirus (CMV)

CMV, very common in spinach production, causes yellow spots on infected leaves; thus the name spinach yellow spot. The first symptoms, easily noticed on young leaves, curl and distort leaf margins. CMV-infected plants grow slowly. Necrosis that occurs on leaves during disease development frequently causes their death. CMV-infected winter spinach plants decline during the winter. Several aphid species transmit CMV nonpersistently.

Spinach production acreages must be spatially isolated from other susceptible plants—vegetables in particular—a useful measure for infection prevention in production. Control of aphids with proper insecticide applications contributes to reducing transmission of CMV in nature. Spinach cultivars with resistance to CMV should be planted.

Lettuce Mosaic Potyvirus (LMV)

Spinach is an important host of LMV. Symptoms in infected leaves are bright yellow circular spots, followed by chlorotic mottle. Older LMV-infected leaves wilt and the entire plant grows slowly.

Adequte spatial isolation is mandatory between spinach, lettuce, and other cultivated plants and weeds susceptible to this virus.

Other viruses include BBWV, TRsV, and TmBuSnV.

REFERENCES

Aapola, A.A. and Mink, G.I. 1973. Potential aphid vectors of pea seed-mosaic virus in Washington. *Pl. Dis. Reptr.* 57:552.

Agrawal, H.O. 1964. Identification of cowpea mosaic virus isolates. *Meded. Landbourwhogesch. Wageningen* 64 (5):

Albrechtová, Z. et al. 1975. Nachweis des Tomatenzwergbusch - Virus (Tomato bushy stunt virus) in Susskirschen, die mit virösem Zweigkrebs befallen waren. *Phytopathol. Z.* 82:25–34.

Aleksić, Z. 1965. Proučavanje viroza pasulja sa naročitim osvrtom na potrebe selekcije na otpornost. Doktorska disertacija, Poljoprivredni fakultet, Zemun.

Aleksić, Z. et al. 1980. *Bolesti povrća i njihovo suzbijanje.* Nolit, Beograd.

Aleksić, Z. and Marinković, N. 1981. Virus mozaika duvana (tobacco mosaic virus) na nekim sortama paprike u uslovima prirodne zaraze. *Zaštita bilja* 32 (1), br. 155:45–54.

Aleksić, Z. et al. 1990. *Bolesti povrća i njihovo suzbijanje.* Nolit, Beograd.

Alavarez, M. and Campbell, R.N. 1978. Transmission and distribution of squash mosaic virus in seeds of cantaloupe. *Phytopathology* 68:257–263.

Ashby, J.W. 1984. Bean leaf roll virus. *CMI/AAB Descriptions of Plant Viruses*, No. 286.

Bancroft, J.B. 1968. Tomato top necrosis virus. *Phytopathology* 58:1360–1363.

Bancroft, J.B. 1971. Cowpea chlorotic mottle virus. *CMI/AAB Descriptions of Plant Viruses*, No. 49.

Bawden, F.C. et al. 1951. Some properties of broad bean mottle virus. *Ann. Appl. Biol.* 38:774–784.

Beczner, L. 1973. A lucerne mosaic virus gazda növenyköre. *Növenyvédelem Korszerüsitése* 7:103–122.

Beier, M. et al. 1977. Survey of susceptibility to cowpea mosaic virus among protoplasts and intact plants from *Vigna sinensis* lines. *Phytopathology* 67:917–921.

Bercks, R. 1959. Serologische Untersuchungen über das Phaseolus - Virus 1. *Phytopathol. Z.* 35:105.

Bercks, R. 1962. Serologische Überkreuzreaktionen zwischen Isolaten des Tomaten Schwarzringflecken-virus. *Phytopathol. Z.* 46:97.

Best, R.J. 1968. Tomato spotted wilt virus. *Adv. Virus Res.* 13:65–146.

Bock, K.R. and Conti, M. 1974. Cowpea aphid-borne mosaic virus. *CMI/AAB Descriptions of Plant Viruses*, No. 134.

Bos, L. and Van der Want, J.P.H. 1962. Early browning of pea, a disease caused by a soil- and seed-borne virus. *T. Plantenziekt* 68:368–390.

Bos, L. 1969. Enige otwikkelingen bij het onderzoek over virus-ziekten van peulfruchten. *Jubileummiuitgave 30 jahre P.S.C.S.* 139–149.

Bos, L. 1970. Bean yellow mosaic virus. *CMI/AAB Descriptions of Plant Viruses*, No. 40.

Bos, L. 1973. Pea streak virus. *CMI/AAB Descriptions of Plant Viruses*, No. 112.

Bos, L. et al. 1974. The identification of bean mosaic, pea yellow mosaic and pea necrosis strains of bean yellow mosaic virus. *Neth. J. Pl. Pathol.* 80:173–191.

Bos, L. 1976. Onion yellow dwarf virus. *CMI/AAB Descriptions of Plant Viruses*, No. 158.

Bos, L. 1981. Leek yellow stripe virus. *CMI/AAB Descriptions of Plant Viruses*, No. 240.

Bos, L. et al. 1978. Leek yellow stripe virus and its relationships to onion yellow dwarf virus; characterization, ecology and possible control. *Neth. J. Pl. Pathol.* 84:185–204.

Bos, L. et al. 1978a. Further characterization of celery latent virus. *Neth. J. Pl. Pathol.* 84:61–79.

Bos, L. et al. 1980. Spinach latent virus, a new ilarvirus seed-borne in *Spinacea oleracea*. *Neth. J. Pl. Pathol.* 86:79.

Bos, L. 1982. Shallot latent virus. *CMI/AAB Descriptions of Plant Viruses*, No. 250.

Bos, L. et al. 1984. Further characterization of melon necrotic spot virus causing severe disease in glasshouse cucumbers in the Netherlands and its control. *Neth. J. Pl. Pathol.* 90:55–69.

Bos, L. 1984. Spinach latent virus. *CMI/AAB Descriptions of Plant Viruses*, No. 281.

Bos, L. 1996. Research on viruses of legume crops and the International Working Group on Legume Viruses; historical facts and personal reminiscenses. *Intl. Work. Group Legume Vir.*, c/o ICARDA, P.O. Box 5466, Aleppo, Syria.

Bozarth, R.F. and Shoyinka, S.A. 1979. Cowpea mottle virus. *CMI/AAB Descriptions of Plant Viruses*, No. 212.

Brćak, J. 1975. Garlic mosaic virus particles and virus infections of some wild *Allium* species. *Sbornik UVTI - Ochrana rostlin* 11:(XLVIII)237–242.

Broadbent, I. 1964. The epidemiology of tomato mosaic. VII. The effect of TMV on tomato fruit yield and quality under glass. *Ann. Appl. Biol.* 54:209–224.

Broadbent, I. 1965. The epidemiology of tomato mosaic. XI. Seed-transmission of TMV. *Ann. Appl. Biol.* 56:177–205.

Brunt, A.A. and Kenten, R.H. 1971. Pepper veinal mottle virus, a new member of the potato virus Y group from peppers (*Capsicum annuum* L. and *C. frutescens* L.) in Ghana. *Ann. Appl. Biol.* 69:235–243.

Brunt, A.A. and Kenten, R.H. 1972. Pepper veinal mottle virus. *CMI/AAB Descriptions of Plant Viruses*, No. 104.

Brunt, A.A. and Kenten, R.H. 1974. Cowpea mild mottle virus. *CMI/AAB Descriptions of Plant Viruses*, No. 140.

Buck, K.W. and Coutts, R.H.A. 1985. Tomato golden mosaic virus. *CMI/AAB Descriptions of Plant Viruses*, No. 303.

Burckhardt, F. 1960. Untersuchungen über eine viröse Vergilbung der Stoppel Rübe. *Mitt. Biol. Bundesanst. Berlin - Dahlem* 99:84–96.

Butler, J.P.G. 1970. Structures of turnip crinkle and tomato bushy stunt viruses. *J. Mol. Biol.* 52:589–593.

Buturović, D. and Pešić, Z. 1976. Prilog proučavanju viroza graška u *Jugoslaviji*. *III Kongres mikrobiologa Jugoslavije*, 6-9 Oktobar, Beograd.

Cadilhac, B. et al. 1976. Mise en évidence en microscope électronique de deux virus differents infectant l'ail et l'échalotte. *Ann. Phytopathol.* 8:65–72.

Campbell, R.N. 1971. Squash mosaic virus. *CMI/AAB Descriptions of Plant Viruses*, No. 43.

Campbell, R.N. and Lin, M.T. 1972. Broccoli necrotic yellows virus. *CMI/AAB Descriptions of Plant Viruses*, No. 85.

Campbell, R.N. 1973. Radish mosaic virus. *CMI/AAB Descriptions of Plant Viruses*, No. 121.

Caner, J. et al. 1976. Characteristics of a rhabdovirus isolated from pea plants (*Pisum sativum* L.). Characteristicas de un rhabdovirus isolado de plantas de ervilha (*Pisum sativum* L.). *Summa Phytopathol.* 2:264–269.

Chod, J. 1965. Studies of some ways in which carrot mosaic virus can be transmitted. *Biologia Plantarum* 6:463–468.

Chod, J. 1988. Carrot mosaic virus. In: Smith, I.M. et al., Eds. *European Handb. Pl. Dis.*, Blackwell Sci Publ., London, p. 38.

Cohen, S. 1982. Control of whitefly vectors of viruses by color mulches. In: Harris, K.F. and Maramorosch, K., Eds. *Pathogens, Vectors and Plant Diseases, Approaches to Control*, Academic Press, New York. p. 45–46.

Davis, R.M. et al. 1997. *Compendium of Lettuce Diseases*. APS Press, St. Paul, MN, 79 pp.

De Jager, C.P. 1979. Cowpea severe mosaic virus. *CMI/AAB Descriptions of Plant Viruses*, No. 209.

Delécolle, B. and Lot, H. 1981. Viroses de l'ail. I. Mise en évidence et essais de caractérisation par immuno-elec-tromicroscopie d'un complexe de trois virus chez différentes populations d'ail atteintes de mosaique. *Agronomie* 1: 763–770.

Delević, B. 1961. Viroze peprike u Srbiji. Doktorska disertacija, Poljoprivredni fakultet, Zemun.

Demski, J.W. 1981. Tobacco mosaic virus is seedborne in Pimiento peppers. *Plant Dis.* 65:723–724.

Dercks, A.F.L.M. and Van den Abeele, J.L. 1980. Bean yellow mosaic virus in some iridaceous plants. *Acta Hort.* 110:31–38.

Devergne, J.C. and Cousin, R. 1966. Le virus de la mosaïque de la féve et les symtpômes d'ornementation sur le grains. *Ann. Epiphyties* 17 (No Hors-Série, Etudes de Virologie):147–161.

Devergne, J.C. and Cardin, L. 1975. Relation sérologique entre cucumovirus (CMV, TAV, PSV). *Ann. Phytopathol.* 7:255–276.

Dias, H.F. and McKeen, C.D. 1972. Cucumber necrosis virus. *CMI/AAB Descriptions of Plant Viruses*, No. 82.

Dodds, J.A. and Mirkov, T.E. 1983. Association of double-stranded RNA with lettuce big vein disease. In: *4th Int. Cong. of Plant Pathology*, Abstract 113, Melbourne.

Drijfhout, E. 1978. Genetic Interaction Between *Phaseolus vulgaris* and Bean Common Mosaic Virus, Doctoral thesis, Centre Agric. Publ. Docum, 98 pp. Also in *Agric. Res. Rep. Wageningen* 872.
Duffus, E.J. and Russel, G.E. 1962. Serological relationship between beet western yellows and turnip yellows viruses. *Phytopathology* 62:1274–1277.
Duffus, J.G. 1972. Beet western yellows virus. *CMI/AAB Descriptions of Plant Viruses*, No. 89.
Dunez, J. 1988. Tomato black ring virus (TBRV). In: Smith, I.M. et al., Eds. *European Hand. Pl. Dis.*, Blackwell Sci. Publ., London, p. 31.
Edwardson, J.R. and Christie, R.G. 1991. *CRC Handbook of Viruses Infecting Legumes*, CRC Press LLC, Boca Raton, FL, 504 pp.
Edwardson, J.R. and Christie, R.G. 1997. Viruses Infecting Peppers and Other Solanaceous Crops. Univ. Florida Agr. Expt. Stn., Gainesville, FL. Monograph 18-I and II, 766 pp.
Elnagar, S. and Murant, A.F. 1976. Relations of the semi-persistent viruses parsnip yellow fleck and anthriscus yellows with their vector, *Cavariella aegopodii*. *Ann. Appl Biol.* 84:169–181.
Elnagar, S. and Murant, A.F. 1978. Aphid-injection experiments with carrot mottle virus and its helper virus, carrot red leaf. *Ann. Appl. Biol.* 89:245–250.
Falk, B.W. and Purcifull, D.E. 1983. Development and application of an enzyme-linked immunosorbent assay (ELISA) to index lettuce seeds for lettuce mosaic virus in Florida. *Plant Dis.* 67:413–416.
Férault, A.C. et al. 1969. Identifiecation dans la région Parisienne d'une marbure du haricot comparable au bean southern mosaic virus (Zaumeyer and Harter). Deus Congr. 1 Union Phytopath. medit. Avignon-Antibes, *Ann. Phytopathol.* 1:619–626.
Francki, R.I.B. and Randles, J.W. 1970. Lettuce necrotic yellow virus. *CMI/AAB Descriptions of Plant Viruses*, No. 26.
Francki, R.I.B. et al. 1979. Cucumber mosaic virus. *CMI/AAB Descriptions of Plant Viruses*, No. 213.
Francki, R.I.B. and Hatta, T. 1981. Tomato spotted wilt virus. In: Kurstak, E., Ed. *Hand. Pl. Virus Infec. and Comp. Diag.*, Elsevier/North-Holland and Biomedical Press, p. 491–512.
Francki, R.I.B. et al. 1981. Rhabdoviruses. In: Kurstak, E., Ed. *Hand. Pl. Virus Infec. Comp. Diag.*, Elsevier/North-Holland Biomedical Press, p. 455–489.
Freitag, J.H. 1956. Beetle transmission, host range, and prospectives of squash mosaic virus. *Phytopathology* 46:73–81.
Fribourg, C.E. 1982. Peru tomato virus. *CMI/AAB Descriptions of Plant Viruses*, No. 255.
Frowd, J.A. and Bernier, C.C. 1977. Virus diseases of faba beans in Manitoba and their effects on plant growth and yield. *Can. J. Pl. Sci.* 57:845–852.
Gallitelli, D. et al. 1978. Il virus della scolorazine perinervale del carciofo. *Phytopathol. Mediterr.* 17:1–7.
Gallitelli, D. et al. 1984. Aritchoke vein banding virus. *CMI/AAB Descriptions of Plant Viruses*, No. 285.
Gamez, R. 1982. Bean rugose mosaic virus. *CMI/AAB Descriptions of Plant Viruses*, No. 246.
Garrett, R.G. and O'Loughlin, G.T. 1977. Broccoli necrotic yellows virus in cauliflower and in the aphid *Brevicoryne brassicae* L. *Virology* 76:653–663.
Geelen, J.L.M.C. et al. 1972. Structure of capsid of cowpea mosaic virus. The chemical subunit: molecular weight and number of subunits per particle. *Virology* 49:205–213.
Gibbs, A.J. and Harrison, B.D. 1964. A form of pea early-browning virus found in Britain. *Ann. Appl. Biol.* 54:1–11.
Gibbs, A.J. and Paul, H.L. 1970. Echtes Ackerobohnemosaik-virus. *CMI/AAB Descriptions of Plant Viruses*, No. 20.
Gibbs, A.J. and Smith, H.G. 1970. Broad bean stain virus. *CMI/AAB Descriptions of Plant Viruses*, No. 29.
Gibbs, A.J. 1972. Broad bean mottle virus. *CMI/AAB Descriptions of Plant Viruses*, No. 101.
Gibbs, A.J. and Harrison, B.D. 1973. Eggplant mosaic virus. *CMI/AAB Descriptions of Plant Viruses*, No. 124.
Gigante, R. 1949. Il mosaico del carciofo. *Bolletino della Stazione di Patologia Vegetale di Roma* 7:177–181.
Giunchedi, L. 1972. Ricerche virologiche sulla buttatura del peperone. *Inf. Fitopatologico* 4:3–8.
Gonzalez-Garza, R. et al. 1979. Identification, seed transmission and host range pathogenicity of a California isolate of melon necrotic spot virus. *Phytopathology* 69:340–345.
Goodman, R.M. and Bird, J. 1978. Bean golden mosaic virus. *CMI/AAB Descriptions of Plant Viruses*, No. 192.
Goodman, R.M. 1981. Geminiviruses. In: Kurstak, E., Ed. *Handb. Pl. Virus Infec. Comp. Diag.*, Elsevier/North-Holland Biomedical Press, p. 879–910.
Hagedorn, D.J. 1984. *Compendium of Pea Diseases*, APS Press, St. Paul, MN, 57 pp.
Hagedorn, D.J. and Walker, J.C. 1949. Wisconsin pea streak. *Phytopathology* 39:837–847.
Hall, R. 1991. *Compendium of Bean Diseases*, APS Press, St. Paul, MN, 73 pp.

Hampton, R.O. 1975. The nature of bean yield reduction by bean yellow and bean common mosaic viruses. *Phytopathology* 65:1342–1346.

Hampton, R.O. and Mink, G.I. 1975. Pea seed-borne mosaic virus. *CMI/AAM Descriptions of Plant Viruses*, No. 146.

Harbaoui, Y. et al. 1982. Assainissement viral de l'artichaut (*Cynara scolymus* L.) par la culture *in vitro* d'apex méristématiques. *Phytopath. Mediter.* 21:15–19.

Harris, H.B. and Kuhn, C.W. 1971. Influence of cowpea chlorotic mottle virus (soybean strain) on agronomic performance of soybeans. *Crop Sci.* 11:71–73.

Harrison, B.D. 1957. Studies of the host range, properties, and mode of transmission of beet ringspot virus. *Ann. Appl. Biol.* 45:462–472.

Harrison, B.D. 1958. Relationship between beet ringspot, potato bouquet and tomato black ring viruses. *J. Gen. Microbiol.* 18:450–460.

Harrison, B.D. et al. 1972. Two properties of raspberry ringspot virus determined by its smaller RNA. *J. Gen. Virol.* 17:137–141.

Harrison, B.D. 1973. Pea early-browning virus. *CMI/AAB Descriptions of Plant Viruses*, No. 120.

Harrison, B.D. et al. 1974. Distribution of determinants for symptom production, host range and nematode transmissibility between the two RNA components of raspberry ringspot virus. *J. Gen. Virol.* 22:233–247.

Harrison, B.D. and Murant, A.F. 1977. Nematode transmissibility of pseudo-recombinant isolates of tomato black ring virus. *Ann. Appl. Biol.* 86:209–212.

Harrison, B.D. and Robinson, D.J. 1981. Tobraviruses. In: Kurstak, E., Ed. *Handb. Pl. Virus Infec. Comp. Diag.*, Elsevier/North Holland Biomedical Press, p. 515–540.

Hibi, T. and Furuki, I. 1985. Melon necrotic spot virus. *CMI/AAB Descriptions of Plant Viruses*, No. 302.

Hiebert, E. and McDonald, J.G. 1973. Characterization of some proteins associated with viruses in the potato Y group. *Virology* 56:349–361.

Hohn, T. et al. 1982. Cauliflower mosaic virus on its way to becoming a useful plant vector. *Curr. Topics Microbiol. Immunol.* 96:193–236.

Hollings, M. 1963. Cucumber stunt mottle, a disease caused by a strain of arabis mosaic virus. *J. Hort. Sci.* 38:138–149.

Hollings, M. 1964. Some properties of five viruses of celery (*Apium graveolens* L.) in Britain. *J. Hort. Sci.* 39:130–141.

Hollings, M. 1965. Some properties of celery yellow vein, a virus serologically related to tomato black ring virus. *Ann. Appl. Biol.* 55:459–470.

Hollings, M. and Stone, O.M. 1971. Tomato aspermy virus. *CMI/AAB Descriptions of Plant Viruses*, No. 79.

Hollings, M. and Stone, O.M. 1972. Turnip crinkle virus. *CMI/AAB Descriptions of Plant Viruses*, No. 109.

Hollings, M. and Stone, O.M. 1973. Turnip rosette virus. *CMI/AAB Descriptions of Plant Viruses*, No. 125.

Hollings, M. et al. 1975. Cucumber green mottle mosaic virus. *CMI/AAB Descriptions of Plant Viruses*, No. 154.

Hollings, M. and Huttinga, H. 1976. Tomato mosaic virus. *CMI/AAB Descriptions of Plant Viruses*, No. 156.

Hollings, M. and Brunt, A.A. 1981. Potyviruses. In: Kurstak, E., Ed. *Handb. Pl. Virus Infec. Comp. Diag.*, Elsevier/North-Holland Biomedical Press, p. 731–807.

Holmes, F.O. 1937. Inheritance of resistance to tobacco-mosaic disease in the pepper. *Phytopathology* 27:637–642.

Honda, Y. and Ikegami, M. 1986. Mung bean yellow mosaic virus. *CMI/AAB Descriptions of Plant Viruses*, No. 323.

Horvath, J. 1980. Viruses of lettuce. II. Host ranges of lettuce mosaic virus and cucumber mosaic virus. *Acta Agron. Scient. Hungaricae* 29:333–352.

Howell, W.E. and Mink, G.I. 1980. Carrot thin leaf virus. *CMI/AAB Descriptions of Plant Viruses*, No. 218.

Howell, S.M. 1982. Plant molecular vehicles: potential vectors for introducing foreign DNA into plants. *Annu. Rev. Pl. Phys.* 33:609–650.

Hull, R. 1972. The multicomponent nature of broad bean mottle virus and its nucleic acid. *J. Gen. Virol.* 17:111–117.

Hull, R. and Lane, L.C. 1973. The unusual nature of the components of pea enation mosaic virus. *Virology* 55:1–13.

Hull, R. 1981. Pea enation mosaic virus. In: Kurstak, E., Ed. *Handb. Pl. Virus Infec. Comp. Diag.*, Elsevier/North-Holland Biomedical Press, p. 239–256.

Huttinga, H. 1969. Interaction between components of pea early-browning virus. *Neth. J. Pl. Pathol.* 75:338–342.

Huttinga, H. 1975. Properties of viruses of the potyvirus group. 3. A comparison of buoyant density, S value, particle morphology and molecular weight of the coat protein subunit of 10 viruses and virus isolates. *Neth. J. Pl. Pathol.* 81:58–63.

Ie, T.S. 1970. Tomato spotted wilt virus. *CMI/AAB Descriptions of Plant Viruses*, No. 39.

Iizuka, N. et al. 1984. Natural occurrence of a strain of cowpea mild mottle virus on groundnut (*Arachis hypogaea*) in India, *Phytopathol. Z.* 109, 245.

Inouye, T. and Nakasone, W. 1980. Broad bean necrosis virus. *CMI/AAB Descriptions of Plant Viruses*, No. 223.

Jafarpour, B. et al. 1979. Serologic detection of bean common mosaic and lettuce mosaic viruses in seed. *Phytopathology* 69:1125–1129.

Jankulova, M. et al. 1976. *Abstr. Proc. 6th Conf. Int. Counc. Grapevine Viruses*, Cordova, 24.

Jasnić, S. 1978. Proucavenje epidemiologije virusa mozaika paradajza na paprici. Doktorska disertacija, Poljoprivredni fakultet, Novi Sad.

Jones, J.B. et al. 1991. *Compendium of Tomato Diseases*, APS Press, St. Paul, MN. 73 p.

Joubert, J.J. et al. 1974. Purification and properties of tomato spotted wilt virus. *Virology* 57:11–19.

Juretić, N. 1978. Tomato mosaic virus infection of glasshouse tomato crops in Yugoslavia. *Acta Bot. Croat.* 29:17–26.

Kaper, J.M and Waterworth, H.E. 1981. Cucumoviruses. In: Kurstak, E., Ed. *Handb. Pl. Virus Infec. Comp. Diag.*, Elsevier/North-Holland Biomedical Press, p. 257–332.

Kassanis, B. 1944. A virus attacking lettuce and dandelion. *Nature* 154:16.

Kassanis, B. 1947. Studies on dandelion yellow mosaic and other virus diseases of lettuce. *Ann. Appl. Biol.* 34:412–421.

Kassanis, B. 1954. Heat-therapy of virus-infected plants. *Ann. Appl. Biol.* 41:470–474.

Kennedy, J.S. et al. 1962. A Conspectus of Aphids as Vectors of Plant Viruses. Commonw. Inst. Entom., London.

Kenten, R.H. 1977. Cacao necrosis virus. *CMI/AAB Descriptions of Plant Viruses*, No. 173.

Klinkovski, M. et al. 1977. *Pflanzliche Virologie*, Vols. 2 and 3, Akademie-Verlag, Berlin.

Knesek, J. and Mink, G.I. 1970. Incidence of seed-borne virus in peas grown in Washington and Idaho. *Pl. Dis. Reptr.* 54:497.

Knesek, J. et al. 1974. Purification and properties of pea seed-borne mosaic virus. *Phytopathology* 64:1076–1081.

Koening, Renate and Lesemann, D.E. 1981. Tymoviruses. In: Kurstak, E., Ed. *Handb. Pl. Virus Infec. Comp. Diag.*, Elsevier/North-Holland Biomedical Press, p. 33–60.

Kovachevski, I. 1940. Die Reisigkrankheit der Paprikapflanze (*Capsicum annuum* L.). *Z. Pflanzenkrankh. u. Pflanzenschutz. Bd.* 50:298–308.

Kovachevski, I. 1968. Das Bohnengelbmosaik-Virus in Bulgarien. *Phytopathol. Z.* 61:41–48.

Lamptey, P.N.L. and Hamilton, R.I. 1974. A new cowpea strain of Southern bean mosaic virus from Ghana. *Phytopathology* 64:1100–1104.

Lane, L.C. 1981. Bromoviruses. In: Kurstak, E., Ed. *Handb. Pl. Virus Infec. Comp. Diag.*, Elsevier/North-Holland Biomedical Press, p. 333–376.

Lecoq, H. et al. 1981. Identification et caractuant la maladie du rabougrissement jaune du melon. *Agronomie* 1:827–834.

Lesemann, D.E. et al. 1983. Natural infection of cucumbers by zucchini yellow mosaic virus in Lebanon. *Phytopathol. Z.* 108:304–313.

Lin, M.T. et al. 1970. Lettuce big vein virus transmission by single-sporangium isolates of *Olpidium brassicae*. *Phytopathology* 60:1630–1634.

Lisa, V. et al. 1981. Characterization of a potyvirus that causes zucchini yellow mosaic. *Phytopathology* 71:668–672.

Lisa, V. and Lecoq, H. 1984. Zucchini yellow mosaic virus. *CMI/AAB Descriptions of Plant Viruses*, No. 282.

Lot, H. 1988. Lettuce mosaic virus (LMV). In: Smith I.M. et al., Eds. *European Handb. Pl. Dis.*, Blackwell. Sci. Publ., London, p. 40–41.

Mahmood, K. et al. 1972. Pea symptomless virus, a newly recognized strain of red clover mottle virus. *Neth. J. Pl. Pathol.* 78:203–211.

Mamula, Dj. 1968. Virus _utog mozaika postrne repe (Turnip yellow mosaic virus) u Jugoslaviji. *Acta Bot. Croat.* 26–27:85–100.

Mamula, Dj. and Mili_i_, D. 1971. Prilog poznavanju rasprostranjenosti virusa krsta_ica u ju_noj Europi. *Acta Bot. Croat.* 30:41–52.

Marrou, J. et at. 1969. Mise en évidence d'une nouvelle maladie à virus du haricot dans le sud-est de la France. Deus. Congr. l'Union Phytopath. Medit., Avignon-Antibes, *Ann. Phytopathol.* 1:400–413.

Marrou, J. and Migliori. 1971. Essai de protection des cultivars de tomate contre virus de la mosaïque du tabac: mise en évidence d'une specificite étroite de la prémunition entre souches de ce virus. *Ann. Phytopathol.* 3:447–459.

Marrou, J. et al. 1975. Caracterisation par la symptomatologie de quatorze souches du virus de la mosaïque du cocombre et de deux autres cucumovirus: tentative de la classification. *Medelingen van de Faculteit Landbourwetenschapen Rijksumversiteit Gent* 40:107–121.

Martelli, G.P. 1965. L'avviccinamento maculato del carciofo. *Phytopathol. Mediter.* 4:58–60.

Martelli, G.P. and Cirulli, M. 1969. Mottled dwarf of eggplant (*Solanum melongena*) a virus disease. *Ann. Phytopathol. No. hors série*, 393–397.

Martelli, G.P. et al. 1971. Tomato bushy stunt virus. *CMI/AAB Descriptions of Plant Viruses*, No. 69.

Martelli, G.P. and Russo, M. 1973. Eggplant mottled dwarf virus. *CMI/AAB Descriptions of Plant Viruses*, No. 115.

Martelli, G.P. et al. 1977. Artichoke Italian latent virus. *CMI/AAB Descriptions of Plant Viruses*, No. 176.

Martelli, G.P. 1981. Tombusviruses. In: Kurstak, E., Ed. *Handb. Pl. Virus Infec. Comp. Diag.*, Elsevier/North-Holland Biomedical Press, p. 61–90.

Martelli, G.P. 1988. Zucchini yellow fleck virus. In: Smith, I.M. et al., Eds. *European Handb. Pl. Dis.*, Blackwell Sci. Publ., London, p. 51.

Martelli, G.P. 1988a. Pea enation mosaic virus (PEMV). In: Smith, I.M. et al., Eds. *European Handb. Pl. Dis.*, Blackwell Sci. Publ., London, p. 75.

Martelli, G.P. 1988b. Artichoke Italian latent virus (AILV) In: Smith, I.M. et al., Eds. *European Handb. Pl. Dis.*, Blackwell Sci. Publ., London, p. 25–26.

Martelli, G.P. 1988c. Artichoke yellow ringspot virus. In: Smith, I.M. et al., Eds. *European Handb. Pl. Dis.*, Blackwell Sci. Publ., London, p. 26.

Martelli, G.P. 1988d. Artichoke mosaic. In: Smith, I.M. et al., Eds. *European Handb. Pl. Dis.*, Blackwell Sci. Publ., London, p. 103.

Matthews, R.E.F. 1980. Turnip yellow mosaic virus. *CMI/AAB Descriptions of Plant Viruses*, No. 230.

Matthews, R.E.F. 1991. *Plant Virology*, 3rd edition, Academic Press, New York, 835 p.

McKinney, H.H. 1968. Further study of the latent strain of the tobacco mosaic virus. *Pl. Dis. Reptr.* 52:919–922.

McLean, D.L. and Kinsey, M.G. 1962. Three variants of lettuce mosaic virus and methods utilized for differentiation. *Phytopathology* 62:403–406.

Messiaen, M.C. and Lafon, R. 1965. *Les Maladies des Plantes Maraicheres*, Vol. II, INRA, 272–276.

Mickovski, J. 1964. Prilog poznavanju viroza duvana u SR Makedoniji. Doktorska disertacija, Poljoprivredini fakultet, Zemun.

Miličić, D. et al. 1963. Rasparostranjenost nekih vrsta virusa krucifera u Jugoslaviji. *Agron. Glasnik*, 1–2:91–100.

Moghal, S.M. and Francki, R.I. B. 1976. Towards a system for the identification and classification of potyviruses. I. Serology and amino acid composition of six distinct viruses. *Virology* 73:350–352.

Molnár, A. and Schmelzer, K. 1964. Beiträge zur Kenntnis des Wassermelonenmosaik-virus. *Phytopathol. Z.* 51:361–384.

Monsion, M. and Dunez, J. 1971. The status of French studies on virus diseases of chrysanthemums. État des recherches poursuivies en France sur les maladies à virus du chrystantheme. *Rev. Zool. Agric.* 71:95–103.

Morales F. J. and Bos, L. 1988. Bean common mosaic virus. *CMI/AAB Descriptions of Plant Viruses*, No. 337.

Morton, D.J. 1961. Host range and properties of the globe artichoke curly dwarf virus. *Phytopathology* 51:731–734.

Mowat, W.P. 1962. Other soil-borne viruses. *Annu. Rep. Scot. Hort., Res. Inst.* 9:69.

Muniyappa, V. and Reddy, D.V.R. 1983. Tranmission of cowpea mild mottle virus by *Bemisia tabaci* in a nonpersistent manner. *Plant Dis.* 67:391.

Murant, A.F. and Goold, R.A. 1968. Purification, prospectives and transmission of parsnip yellow fleck, a semi-persistent, aphid-borne virus. *Ann. Appl. Biol.* 62:123–137.

Murant, A.F. 1970. Tomato black ring virus. *CMI/AAB Descriptions of Plant Viruses*, No. 38.

Murant, A.F. et al. 1970. Parsnip mosaic virus, a new member of the potato virus Y group. *Ann. Appl. Biol.* 65:127–135.

Murant, A.F. 1972. Parsnip mosaic virus. *CMI/AAB Descriptions of Plant Viruses*, No. 91.

Murant, A.F. 1974. Carrot mottle virus. *CMI/AAB Descriptions of Plant Viruses*, No. 137.
Murant, A.F. 1974a. Parsnip yellow fleck virus. *CMI/AAB Descriptions of Plant Viruses*, No. 129.
Murant, A.F. 1981. Nepoviruses. In: Kurstak, E., Ed. *Handb. Pl. Virus Infec. Comp. Diag.*, Elsevier/North-Holland Biomedical Press, p. 197–238.
Nagel, J. et al. 1983. Strains of bean yellow mosaic virus compared to clover yellow vein virus in relation to gladiolus production in Florida. *Phytopathology* 73:449–454.
Natsuaki, T. 1985. Radish yellow edge virus. *CMI/AAB Descriptions of Plant Viruses*, No. 298.
Nault, L.R. et al. 1964. Biological relationship between pea enation mosaic virus and its vector, the pea aphid. *Phytopathology* 54:1269–1272.
Nelson, R.M. 1932. Investigations in the Mosaic Disease of Bean (*Phaseolus vulgaris* L.). *Mich. Agric. Exp. Sta. Techn. Bull.*, No. 118.
Nelson, M.R. and Knuhtsen, H.K. 1973. Squash mosaic virus variability: Review and serological comparisons of six biotypes. *Phytopathology* 63:920–926.
Nelson, M.R. and Knuhtsen, H.K. 1973a. Squash mosaic virus variability: epidemiological consequences of differences in seed transmission frequency between strains. *Phytopathology* 63:918–920.
Nelson, M.R. et al. 1982. Pepper mottle virus. *CMI/AAB Descriptions of Plant Viruses*, No. 253.
Nitzany, F. 1975. Tomato yellow leaf curl virus. *Phytopathol. Mediter.* 14:127–129.
Ohki, S.T. et al. 1976. Beet yellows virus and beet western yellows virus isolated from spinach and fodder beet plants affected by yellowing diseases. *Ann. Phytopathol. Soc. Japan* 43:46.
Ohki, S.T. et al. 1978. Carrot latent virus, a new rhabdovirus of carrot. *Ann. Phytopathol. Soc. Japan* 44:202.
Osaki, T. and Inouye, T. 1978. Resemblance in morphology and intranuclear appearance of viruses isolated from yellow dwarf diseased tomato and leaf curl diseased tobacco. *Ann. Phytopathol. Soc. Japan* 44:167–178.
Paludan, N. 1980. Virus attack on leek: survey, diagnosis, tolerance of varieties and winter hardiness. *Tidsskrift Plantearl* 84:371–385.
Pelham, J. 1972. Strain-genotype interaction of tobacco mosaic virus in tomato. *Ann. Appl. Biol.* 72:219–228.
Pelet, F. 1959. Les maladies à virus des légumineuses. I. Virus s'attaquant aux légumineuses et à d'autres plantes. *Rev. Romande Agric. Vitic. Arboric.* 15:91–94.
Pena-Iglesias, A. et al. 1972. A virus of the bacilliform type in-artichoke (*Cynara scolymus* L.) Un virus de tipo baciliforme en *Annales del I.N.I.A. (Spain), Serie: Prot. Veg.*, No. 2:123–127.
Pešić, Z. 1977. Identifikacija nekih virusa graška. *Zaštita bilja* 140:155–159.
Peters, D., 1982. Pea enation mosaic virus. *CMI/AAB Descriptions of Plant Viruses*, No. 257.
Phatak, H.C. 1974. Seed-borne plant viruses - identification and diagnosis in seed health testing. *Seed Sci. and Tech.* 3:155.
Provvidenti, R. 1974. Inheritance of resistance to watermelon mosaic virus 2 in *Phaseolus vulgaris*. *Phytopathology* 64:1448–1450.
Purcifull, D.E. and Hiebert, E. 1979. Serological distinction of watermelon mosaic virus isolates. *Phytopathology* 69:112–116.
Purcifull, D.E. et al. 1976. Bidens mottle virus. *CMI/AAB Descriptions of Plant Viruses*, No. 161.
Purcifull, D.E. and Edwardson, J.R. 1981. Potexviruses. In: Kurstak, E., Ed. *Handb. Pl. Virus Infec. Comp. Diag.*, Elsevier/North-Holland Biomedical Press, p. 627–693.
Purcifull, D.E. et al. 1984. Watermelon mosaic virus 2. *CMI/AAB Descriptions of Plant Viruses*, No. 293.
Purcifull, D.E. and Gonzalves, D. 1985. Blackeye cowpea mosaic virus. *CMI/AAB Descriptions of Plant Viruses*, No. 305.
Quantz, L. 1953. Untersuchungen über samen Übertragbares Mosaikvirus der Ackerbohne (*Vicia faba*). *Phytopathol. Z.* 20:421–448.
Quantz, L. and Volk, J. 1954. Die Blattrollkrankheit der Ackerbohne und Erbse, eine neue Viruskrankheit bei Leguminosen. *Nachrichtenbl. dt. Pflanzenschutzd. (Braunscheweig)* 6:177–182.
Quantz, L. 1957. Über das Auftreten des Gurkenmosaikvirus auf Erbsen. *Angew. Bot.* 31:166–173.
Quantz, L. 1958. Ein Beitrag zur Kenntnis der Erbsenvirosen in Deutschland. *Narchrichtenbl. dt. Pflanzenschutzd. (Braunschweig)* 10:65–70.
Rana, G.L. et al. 1982. Artichoke latent virus: characterization, ultrastructure and geographical distribution. *Ann. Appl. Biol.* 101:279–289.
Rana, G.L. and Martelli, G.P. 1983. Artichoke yellow ringspot virus. *CMI/AAB Descriptions of Plant Viruses*, No. 271.

Rana, N.H. et al. 1988. Prilog poznavanju zaraženosti semena paradajza virusom mozaika i mogućnosti dezinfekcije. *Zaštita bilja* 394. N.186:393–399, Beograd.

Rast, A.T.B., 1972: M II-16, an artificial symptomless mutant of tobacco mosaic virus for seedling inoculation of tomato crops. *Neth. J. Pl. Pathol.* 78:110–112.

Rast, A.T.B. 1975. Variability of tobacco mosaic virus in relation to control of tomato mosaic in glasshouse tomato crops by resistance breeding and cross protection. *Agric. Res. Rep. (Neth.)* 834:1–76

Reijnders, L. et al. 1974. Molecular weights of plant viral RNAs determined by gel electrophoresis under denaturing conditions. *Virology* 60:515–521.

Rochow, W.F. and Duffus, J.E. 1981. Luteoviruses and yellows diseases. In: Kurstak, E., Ed. *Handb. Pl. Virus Infec. Comp. Diag.*, Elsevier/North-Holland Biomedical Press, p. 147–170.

Russo, M. and Rana, G.L. 1978. Occurrence of two legume viruses in artichoke. *Phytopathol. Mediter.* 17:212–216.

Schmelzer, K. 1960. Untersuchungen über Ringmosaikvirus der Kapucinerkresse. *L. Pflz. Krankh.* 67:193–210.

Schmelzer, K. et al. 1972. Gelbfleckung eine fur Europa neue, vom Lucernemosaikvirus verursachte Krankheit der Gartenbohne (*Phaseolus vulgaris* L.). *Nachrichtenbl. Pflanschutzd. DDR, n.F.* 26:45–47.

Schmelzer, K. 1975. Versuche zur Übertragung des Tomatenaspermivirus mit Cuscuta-Arten. *Phytopathol. Z.* 30:449–452.

Schmelzer, K. et al. 1977. Gemüsepflanzen. In: Klinkowski, M. et al., Eds. *Pflanzliche Virologie*, Vol. 3, 3rd edition, Akademie-Verlag, Berlin, p. 1–138.

Schmidt, H.E. 1977. Leguminosen. In: Klinkowski, M. et al., Eds. *Pflenzliche Virologie*, Vol. 3, 3rd edition, Akademie-Verlag, Berlin, p. 144–293.

Schmidt, H.E. 1988. Broad bean wilt virus (BBWV). In: Smith, I.M. et al., Eds. *European Handb. Pl. Dis.*, Blackwell, Sci. Publ., London, p. 93.

Schroeder, W.T. and Barton, D.W. 1958. The nature and inheritance of resistance to the pea enation mosaic virus in garden pea, *Pisum sativum* L. *Phytopathology* 48:628–632.

Schroeder, W.T. and Provvidenti, R. 1971. A common gene for resistance to bean yellow mosaic virus and watermelon mosaic virus 2 in *Pisum sativum*. *Phytopathology* 61:846–848.

Schwartz, H.F. and Mohan, S.K. 1995. *Compendium of Onion and Garlic Diseases*, APS Press, St. Paul, MN, 54 pp.

Scott, H.A. and Hoy, J.W. 1981. Blackgram mottle virus. *CMI/AAB Descriptions of Plant Viruses*, No. 237.

Semancik, J.S. 1972. Bean pod mottle virus. CMI/AAB Descriptions of Plant Viruses, No. 108.

Shepard, J.F. and Grogan, R.G. 1971. Celery mosaic virus. *CMI/AAB Descriptions of Plant Viruses*, No. 50.

Shepherd, R.J. 1964. Properties of a mosaic virus of cowpea and its relationship to the bean pod mottle virus. *Phytopathology*, 54:466–473.

Shepherd, R.J. 1981. Cauliflower mosaic virus. *CMI/AAB Descriptions of Plant Viruses*, No. 243.

Sheperd, R.J. and Lawson, R. 1981. Caulimoviruses. In: Kurstak, E., Ed. *Hand. Pl. Virus Infec. and Comp. Diag.*, Elsevier/North-Holland and Biomedical Press, p. 848–878.

Smith, I.M. et al. 1988. *European Handb. Pl. Dis.*, Blackwell Sci. Publ., London.

Smith, K.M. 1946. Tomato black ring, a new virus disease. *Parasitology* 37:126–130.

Smith, K.M. and Short, E.M. 1959. Lettuce ringspot, a soil-borne virus disease. *Pl. Pathol.* 8:52.

Smith, K.M. 1972. Celery yellow spot virus. In: *Textbook of Plant Virus Diseases*, Longman, London, p. 159–160.

Sowell, G., et al. 1965. Resistance of southern pea, *Vigna sinensis*, to cowpea chlorotic mottle virus. *Proc. Am. Soc. Hort. Sci.* 86:487–490.

Sutton, W.S. 1955. Yellow top disease of tomatoes. *Agric. Gaz. N.S.W.* 66:655–658.

Stace-Smith, R. 1981. Comoviruses. In: Kurstak, E., Ed. *Handb. Plant Virus Infec. Comp. Diag.*, Elsevier/North-Holland Biomedical Press, p. 171–195.

Stace-Smith, R. 1984. Tomato ringspot virus. *CMI/AAB Descriptions of Plant Viruses*, No. 290.

Stubbs, L.L. 1947. A destructive vascular wilt disease of broad bean (*Vicia faba* L.) in Victoria. *J. Dept. Agr. Victoria* 46:323–332.

Sylvester, E.S. 1954. Yellow-net virus disease of tomato. *Phytopathology* 44:219–220.

Szirmai, J. 1937/40. A fűszerpaprika leromlását megindidó újhitüsegegenek nevezett virusbetegségröl. *Növényegészégüyi Evkönyv* 1:109–133.

Štefanac, Z. 1973. Svojstva evropskog soja virusa mozaika rotkve (radish mosaic virus) Habilitacijski rad, Prirodoslovno matematički fakultet Sveučilišta, Zagreb.

Štefanac, Z. 1977. Virus žute kržljavosti crnog luka u Jugoslaviji. *Acta Bot. Croat.* 36:39–45.

Štefanac, Z. and Wrischer, M. 1983. Spinach latent virus: some properties and comparison of two isolates. *Acta Bot Croat.* 42:1.
Šutić, D. 1959. Die Rolle des Paprikasamens bei der Virüsubertragung. *Phytopathol. Z.* 36:84–93.
Šutić, D. et al. 1975. Seed Transmission of Forage Legume Viruses. PL 480, Project E 30-CR-102, Fin. Tech. Report.
Šutić, D. and Tošić, M. 1976. Virozna infekcija belog luka u našoj zemlji. *III Kongres Mikrobiologa Jugoslavije*, 184-185, 5-9 Oktobar, Bled.
Šutić, D. et al. 1978. Virus mozaika duvana prouzrokovač nekroze paprike. *Zaštita bilja*, XXIX, 146:309–315.
Šutić, D. 1983. *Viroze biljaka*, Nolit, Beograd.
Šutić, D. and Sinclair, J.B. 1991. *Anatomy and Physiology of Diseased Plants*, CRC Press, Boca Raton, FL.
Tapio, E. 1970. Virus diseaes of legumes in Finland and in the Scandinavian countries. *Ann. Agr. Fennial* 9:1–97.
Taylor, R.H. et al. 1968. Purification and properties of braod bean wilt virus. *Aust. J. Biol. Sci.* 21:929–935.
Taylor, R.H. and Stubbs, L.L. 1972. Broad bean wilt virus. *CMI/AAB Descriptions of Plant Viruses*, No. 81.
Thomas, J.E. and Bowyer, J.W. 1980. Properties of tobacco yellow dwarf and bean summer death viruses. *Phytopathology* 70:214–217.
Tomlinson, J.A. and Walkey, D.G.A. 1967. The isolation and identification of rhubarb viruses in Britain. *Ann. App. Biol.* 59:415–427.
Tomlinson, J.A. 1970. Turnip mosaic virus. *CMI/AAB Descriptions of Plant Viruses*, No. 8.
Tomlinson, J.A. 1970. Lettuce mosaic virus. *CMI/AAB Descriptions of Plant Viruses*, No. 9.
Tomlinson, J.A. 1988. Beet western yellows virus (BWYV). In: Smith, I.M. et al., Eds. *European Handb. Pl. Dis.*, Blackwell Sci. Publ., London, p. 21–23.
Tošić, M. and Malak, J. 1974. Proučavanje jednog viroznog oboljenja kupusa u Srbiji. *Zaštita bilja* 130:203–212.
Tošić, M. and Šutić, D. 1976. Prouzrokovač žute kržljavosti crnog luka. *III Kongres Mikobiologa Jugoslavije*, 5-6 Oktobra, Bled, p. 190–191.
Tošić, M. et al. 1980. Transmission of tobacco mosaic virus through pepper (*Capsicum annuum*) seed. *Phytopathol. Z.* 97:10–13.
Tremaine, J.H. and Hamilton, R.I. 1983. Southern bean mosaic virus. *CMI/AAB Descriptions of Plant Viruses*, No. 274.
Vanderwalle, R. 1950. La jaunisses des navets. *Parasitica* 6:111–112.
Van der Want, 1954: *Onderzoekingen over virusziekten van der boon (Phaseolus vulgaris L.)* H. Veenman and Zonen, Wageningen, p. 84.
Van Dorst, H.J.M. and Van Hoof, H.A. 1965. Arabis-mosaikvirus buj komkommer in Nederland. *Neth. J. Pl. Pathol.* 71:176–179.
Van Dorst, H.J.M. and Peters, D. 1974. Some biological observations on pale fruit, a viroid incited disease of cucumber. *Neth. J. Pl. Pathol.* 80:85–96.
Van Kammen, A. and De Jager, C.P. 1978. Cowpea mosaic virus. *CMI/AAB Descriptions of Plant Viruses*, No. 197.
Verhoyen, M. 1973. La "struire chlorotique du Poireau." I. Identification de l'agent causal. *Parasitica* 29:16–28.
Verhoyen, M. and Meunier, S. 1988. Bean common mosaic virus (BCMV). In: Smith, I.M. et al., Eds. *European Handb. Pl. Dis.*, Blackwell Sci. Publ., London, p. 26–35.
Walkey, D.G.A. 1968. The production of virus-free rhubarb by apical tip-culture. *J. Hort. Sci.* 43:283–287.
Walkey, D.G.A. and Neeley, H.A. 1980. Resistance of white cabbage to necrosis caused by turnip and cauliflower mosaic viruses and pepper spot. *Ann. Appl Biol.* 95:703–713.
Waterhouse, P.M. and Murant, A.F. 1982. Carrot red leaf virus. *CMI/AAB Descriptions of Plant Viruses*, No. 249.
Waterworth, H. 1981. Bean mild mosaic virus. *CMI/AAB Descriptions of Plant Viruses*, No. 231.
Watson, M.A. 1963. *Annual Report. Rothamsted Exp. Sta. Rep. 1962*.
Weber, I. et al. 1982. Charakterisierung des Gurkenblattfleckenvirus (cucumber leaf spot virus) eines bisher nicht bekannten Virus an Gewächshausgurken (*Cucumis sativus*). *Archiv. für Phytopathol. Pflanzenschutz* 18:137–154.
Weber, I. 1986. Cucumber leaf spot virus. *CMI/AAB Descriptions of Plant Viruses*, No. 319.
Wetter, C. and Conti, M. 1988. Pepper mild mottle virus. *CMI/AAB Descriptions of Plant Viruses*, No. 330.
White, R.A. and Fischbach, R.A. 1973. An X-ray scattering investigation of broad bean mottle virus in solutions of various electron densities. *J. Mol. Biol.* 75:549–558

Whitney, W.K. and Gilmer, R.M. 1974. Insect vectors of cowpea mosaic virus in Nigeria. *Ann. Appl. Biol.* 77:17–21.

Williams, H.E. and Holdeman, G.L. 1970. American Currant mosaic. In: Frazier, N.W., Ed. *Virus Dis. Small Fruits and Grapevine*. University of California, Berkeley.

Wolf, P. and Schmelzer, K. 1972. Untersuchungen an Viruskrankheiten der Umbelliferen. *Zbl. Bakt. II Abt.*, 127:665–672.

Wu, G.J. and Bruening, G. 1971. Two proteins from cowpea mosaic virus. *Virology* 46:596–612.

Yamashita, S. et al. 1976. *Ann. Phytopathol. Soc. Japan* 43:46.

Yot, P. 1988. Cauliflower mosaic virus (CaMV). In: Smith, I.M. et al., Eds. *European Handb. Pl. Dis.*, Blackwell Sci. Publ., London, p. 71.

Zaumeyer, W.J. and Thomas, H.R. 1948. Pod mottle, a virus disease of beans. *J. Agr. Res.* 77:81–96.

Zaumeyer, W.J. and Goth, R.W. 1964. A new severe symptom-inducing strain of common bean mosaic virus. *Phytopathology* 54:1378–1385.

Zitter, T.A. et al. 1996. *Compendium of Cucurbit Diseases*, APS Press, St. Paul, MN, 87 pp.

4 Viral Diseases of Industrial Plants

Industrial plants are cultivated and generally processed into value-added food items or byproducts, some are used as materials in other industries, and the majority are important for international trade. For this reason, we separated industrial plants into a special group. Among them are hop (*Humulus lupulus*), potato (*Solanum tuberosum*), tobacco (*Nicotiana tabacum*), sugar beet (*Beta vulgaris*), sugarcane (*Saccharum officinarum*), and soybean (*Glycine max.*). Peanut (*Arachis hypogaea*), potato, and soybean are also used directly as vegetable plants for human nutrition. Several compendia provide outstanding summaries of all diseases of beet (Whitney and Duffus, 1986), peanut (Nokalis-Burelle, 1997), potato (Hooker, 1981), soybean (Sinclair and Backman, 1989), and tobacco (Shew and Lucas, 1991).

Among these industrial plants, potato and tobacco are taxonomicallly related, belonging to the family of Solanaceae. Many viruses of tobacco (TMV, TNV, TRtV, and TSkV) are infectious to potato and vice versa. Some viruses described on potato (e.g., PVX and PVY) are infectious to tobacco and other related plants.

Many of the industrial plants, which markedly differ taxonomically, serve primarily as alternative natural hosts of some very widespread viruses like CMV (potato, tobacco, sugar beet, and sunflower) and TmSpoWV (potato, tobacco, sunflower, and soybean). The hop plant, both herbaceous and perennial, is a host and a reservoir source of ApMV and PsNcRsV, which are widespread and economically important viruses for various fruit trees.

Industrial plants are natural hosts of numerous viruses, which, based on the severity of disease caused, are limiting factors to production: for example, PLrV and PVY in potato, TMV and TmSpoWV in tobacco, BtNcYVV and BtYsV in sugar beet, and ScMV in sugarcane production. Therefore, the basic prerequisite for profitable and successful production of industrial plants is to provide sustainable management practices to protect these plants against virus infections. Recently, numerous studies of transgenic resistance in various "genetically recalcitrant" hosts have used pathogen-derived genes, such as the coat protein gene, with success (Grumet, 1994). These technologies provide new hope for control of many virus diseases.

VIRUS DISEASES OF PEANUT (*Arachis hypogaea*)

This peanut disease section was prepared by international virologist D.V.R. Reddy, Ph.D., ICRISAT, Hyderabad, India (Address: Principal Virologist, Patancheru, AP 502324, India). He was one of the authors of the *Compendium of Peanut Diseases, Part I. Biotic Diseases: Diseases Caused by Viruses* (Demski and Reddy, 1997), in addition to numerous other review articles. Their list of viruses which infect peanut crops in nature is informative (Table 3).

DISEASES OF MAJOR ECONOMIC IMPORTANCE

Peanut Bud Necrosis (PtBdNV) and Tomato Spotted Wilt (TmSpoWV) Tospoviruses

Properties of Viral Particles

Virus particles are spherical, 80 to 100 nm in diameter, and surrounded by a glycoprotein-rich lipid membrane. They encapsulate three species of RNA: small (S), medium (M), and large (L). The complete nucleotide sequence of all the three RNA species of the two major tospoviruses, peanut

TABLE 3
Viruses That Naturally Infect Peanut[a]

Name	Taxonomic Group	Family	Distribution
Cowpea chlorotic mottle virus	Bromovirus	Bromoviridae	United States
Cowpea mild mottle virus	Carlavirus	NA[b]	China, India, Indonesia, Ivory Coast, Nigeria, Thailand, Philippines, Papua New Guinea, Sudan
Groundnut crinkle virus	Carlavirus	NA	Ivory Coast
Peanut chlorotic streak virus	Caulimovirus	Pararetroviridae	India
Peanut chlorotic streak virus (veinbanding isolate)	Caulimovirus	Pararetroviridae	India
Cucumber mosaic virus	Cucumovirus	Bromoviridae	China
Peanut stunt virus	Cucumovirus	Bromoviridae	Sudan, Japan, Spain, United States
Peanut clump virus	Furovirus	NA	Niger, Burkina Faso, Ivory Coast, Senegal
Indian peanut clump virus	Furovirus	NA	India, Pakistain
Groundnut yellow mosaic virus (bean golden yellow mosaic virus)	Geminivirus	Geminiviridae	India
Tobacco streak virus	Ilarvirus	NA	Brazil
Groundnut rosette assistor virus	Luteovirus	NA	All of Africa south of the Sahara
Sunflower yellow blotch virus	Luteovirus	NA	Malawi, Kenya, Zambia, Tanzania
Groundnut veinal chlorosis virus	Rhabdovirus	Rhabdoviridae	India, Indonesia
Groundnut chlorotic spotting virus	Potexvirus	NA	Ivory Coast
Bean yellow mosaic virus	Potyvirus	Potyviridae	United States
Groundnut eyespot virus	Potyvirus	Potyviridae	Ivory Coast
Passion fruit woodiness virus	Potyvirus	Potyviridae	Australia
Peanut green mosaic virus	Potyvirus	Potyviridae	India
Peanut mottle virus	Potyvirus	Potyviridae	Worldwide
Peanut stripe virus	Potyvirus	Potyviridae	Brazil, China, India, Indonesia, Japan, Malaysia, Philippines, Myanmar, Thailand, Taiwan, Vietnam, United States
Peanut bud necrosis virus	Tospovirus	Bunyaviridae	India, Nepal, Sri Lanka, China, Taiwan, Indonesia, Thailand
Peanut yellow spot virus	Tospovirus	Bunyaviridae	India, Thailand
Tomato spotted wilt virus	Tospovirus	Bunyaviridae	North America, South America, South Africa, Nigeria
Groundnut yellow mottle virus	Tymovirus	NA	Nigeria
Groundnut rosette virus	Umbravirus	NA	All of Africa south of the Sahara

[a] Listed alphabetically by taxonomic group.
[b] Not yet assigned.

From *Peanut Disease Compendium*, 2nd Ed., APS Press, St. Paul, MN, 1997, P. 58. With permission

bud necrosis virus (PtBdNV) and tomato spotted wilt virus (TmSpoWV), infecting peanut has been published. S-, M-, and L-RNAs of PtBdNV contain 3057, 4801, and 8911 nucleotides, respectively. TmSpoWV contains 2916 (SRNA), 4821 (MRNA), and 8897 (LRNA) nucleotides (Peters and Goldbach, 1995).

The L-RNA is of negative polarity, while M- and S-RNAs are ambisense in character. They are bound by nucleocapsid protein molecules, and code for six proteins that include two glycoproteins G1 and G2 by M-RNA, RNA polymerase (L-RNA), nucleocapsid protein (S-RNA), and two nonstructural proteins, NSs and NSm (S-RNA and M-RNA). Purified virus particles are not infective. Both TmSpoWV and PtBdNV have a low TIP (45°C for 10 minutes) and a short LIV (less than 5 hours at room temperature). Nucleotide sequence of the S- and M-RNA, the deduced aminoacid sequence of the nucleocapsid protein gene, host range and species of the thrips vector are used to differentiate PtBdNV from TmSpoWV.

Geographic Distribution and Economic Importance

TmSpoWV is widely distributed in the Americas, Australia, Africa, Europe, and Japan. PtBdNV appears to be restricted to Asia (with the exception of western Asia). PtBdNV in India alone has been estimated to cause nearly $89 million in crop losses. In 1994, crop losses due to TmSpoWV in Georgia alone were estimated at $26 million.

Host Range and Symptoms

Both TmSpoWV and PtBdNV have an extremely wide host range. Symptoms on peanut vary. Initial symptoms appear as chlorotic spots or a mild mottle. Subsequently, necrotic and chlorotic rings and streaks develop. Necrosis of terminal buds is usually observed when ambient temperatures are above 30°C, and can lead to death of the plant due to spread of necrosis to other parts of the plant (see Figures 93 and 94). Secondary symptoms include stunting and proliferation of axillary shoots, with leaflets reduced in size, which show puckering, mosaic, mottling, and chlorosis. Early infected plants are stunted, and seeds from them are small and shriveled, with mottled seed coats (Figure 95). Late infected plants can produce seed of normal size, yet seed coats are often mottled and cracked.

Diagnostic hosts are cowpea on which both TmSpoWV and PtBdNV produce concentric necrotic or chlorotic lesions, and necrotic lesions on *Petunia hybrida*.

Pathogenesis and Control

TmSpoWV is transmitted mainly by two thrips species, *Frankliniella occidentalis* and *F. fusca*. PtBdNV is transmitted by *Thrips palmi*. For mechanical transmission, it is essential to follow various precautions (Reddy, 1991).

Cultural practices—such as date of planting, maintenance of optimum plant population (approximately 2– to 300,000 per hectare), vigorous crop growth to develop a closed canopy, and intercropping with fast-growing cereal crops—can greatly reduce the incidence of PtBdNV and TmSpoWV. Good sources of field resistance to PtBfNV and TmSpoWV have been identified. Genes governing resistance have been transferred to well-adapted and widely grown peanut cultivars (Reddy, 1998).

FIGURE 93 Late infection of peanut by tomato spotted wilt virus (India). (Courtesy D.V.R. Reddy.)

FIGURE 94 Chlorosis, leaf distortion, and line pattern caused by tomato spotted wilt virus. (Courtesy D.H. Smith. Reproduced with permission from APS.)

FIGURE 95 Peanut seeds from healthy plants (left) and from plants infected with tomato spotted wilt virus (right). (Courtesy D.H. Smith. Reproduced with permission from APS.)

Peanut Clump Pecluvirus (PtCpV)

Properties of Viral Particles

The genus *Peclus* contains peanut clump virus (PtCpV), which causes peanut clump disease in western Africa and Indian peanut clump virus (PtIdCpV), which causes clump disease in the Indian subcontinent (Figure 96). These viruses contain two rod-shaped particles, length approximately 245 nm and 190 nm, and diameter approximately 21 nm. RNA from one of each isolate of PtIdCpV and PtCpV has been fully sequenced. RNA-1 contains approximately 6000 nucleotides (5841 nucleotides for PtIdCpV-H) and RNA-2 contains approximately 4000 nucleotides (4290 nucleotides for PtIdCpV-L).

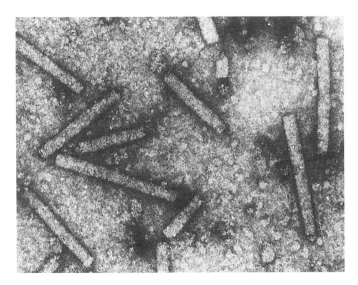

FIGURE 96   Peanut Indian clump virus particles from negative stain electron micrography (India). (Courtesy D.V.R. Reddy.)

RNA 1 contains three open reading frames which encode for methyl transferase domain, polymerase domain and a protein of ca 15 kDa. RNA 2 contains five open reading frames one of which is the coat protein gene (24 kDa polypeptide). The three ORF's form a triple gene block and encode three polypeptides, 51 kDa, 14 kDa and 17 kDa. They are assumed to play a role in the virus movement (Reddy et al. 1998).

Geographic Distribution and Economic Importance

In southern Asia, peanut clump disease occurs in India and Pakistan. In western Africa, the disease has been reported from Burkina Faso, Niger Republic, Mali, Cote d'Ivoire, Gambia, and Senegal. Losses due to clump disease on a global scale were estimated at $38 million.

Host Range and Symptoms

Peanut plants affected by clump disease are conspicuous in the field because they are stunted and dark green. They often occur in patches in the field and the disease recurs more or less in the same position in successive peanut crops. Early symptoms occur on young quadrifoliates as mottling, mosaic, and chlorotic rings. The same leaflets after maturity turn dark green, with or without faint mottling. Early infected plants are severely stunted and seldom produce well-developed pods.

PtIdCpV and PtCpV have extremely wide host ranges. They readily infect many monocotyledonous and dicotyledonous plants, including wheat, barley, maize, sorghum, pearl millet, finger millet, peanut, and pigeonpea. PtIdCpV isolates can be distinguished by symptoms produced in *Canavalia ensiformis*, *Nicotiana clevelandii*, and *N. benthamiana*. PtIdCpV isolates were shown to produce different symptoms in *Chenopodium amaranticolor*.

Pathogenesis and Control

Peanut clump disease is soil borne. PtIdCpV and PtCpV are transmitted by the fungus *Polymyxa* sp. Dicotyledonous hosts do not permit fungal multiplication, and are unlikely to contribute to perpetuation of virus inoculum. Monocotyledonous hosts were found to be preferred hosts for multiplication of *Polymyxa* species. PtIdCpV is transmitted through seed of peanut and several monocotyledonous hosts. Seeds of millets, corn, wheat, and rhizomatous grasses such as *Cynodon dactylon* are likely to contribute to clump disease establishment in soils containing *Polymyxa* sp. All PtIdCpV and PtCpV isolates can be transmitted by mechanical sap inoculation. However, the virus—although present in peanut extracts—was difficult to transmit mechanically.

FIGURE 97 Severe stunt and dark green foliage characterize peanut infected by peanut Indian clump virus (note scattered larger uninfected plants) (India). (Courtesy D.V.R. Reddy.)

FIGURE 98 Effect of treatment before sowing (right) with Nemagon (DBCP) on peanut Indian clump virus; untreated (left) (India). (Courtesy D.V.R. Reddy.)

Application of soil biocides and soil solarization has been shown to reduce clump disease incidence. However, they are not economical to adopt and the biocides that proved effective are known to be hazardous. Although more than 9000 peanut genotypes have been tested, none was found to resist PtIdCpV. Through an understanding of the epidemiology of the disease, cultural practices are shown to be effective in reducing the disease incidence. These include early planting, trap cropping with pearl millet, and avoiding crop rotation with highly susceptible cereal hosts such as maize or wheat (Reddy et al., 1998).

Groundnut Rosette Disease (GtRoDs)

Properties of Viral Particles

Groundnut rosette disease is caused by a complex of two viruses and a satellite RNA. Of the two, one is a mechanically transmitted umbravirus, coat-protein deficient, and referred to as groundnut rosette virus (GtRoV). GtRoV-RNA contains 4019 nucleotides and contains four large open reading frames (Taliansky et al., 1996). GtRoV depends on the groundnut rosette assistor luteovirus (GtRoAsV) for aphid transmission. The satellite RNA is approximately 900 nucleotides in length and largely responsible for the symptoms of rosette disease. Additionally, the satellite plays an essential role in the GtRoAsV-dependant aphid transmission of GtRoV. Therefore, in symptomatic field-infected peanut plants, all three components—GtRoV, GtRoAsV, and satellite RNA—will be present.

Geographic Distribution and Economic Importance

The disease is known to occur only in African countries, south of the Sahara. Reports on the occurrence of rosette in Argentina, India, Indonesia, and Philippines are considered inaccurate. The 1975 epidemic in Nigeria caused crop losses exceeding $250 million. An epidemic in 1996 caused over $4 million worth of crop losses in Southern Africa. Rosette is the most important peanut disease in Africa.

Host Range and Symptoms

At least three types of rosette (chlorotic, mosaic, and green) are recognized on the basis of symptoms. Chlorotic rosette appears on young leaflets as mild chlorosis interspersed with a few green islands. Leaflets produced subsequently are yellow with green veins. Early infected plants show progressively smaller, chlorotic, curled, and distorted leaflets. When plants are infected at later stages of growth, symptoms may be restricted to a few branches or to the apical portion of the plants.

Early infected plants are stunted and seldom produce pods. Young leaflets affected by green rosette show mild chlorotic mottling and isolated flecks. Older leaflets are reduced in size (not distorted), show outward rolling of leaf edges, and symptoms are largely masked. Plants infected early are severely stunted and appear darker than healthy plants. It is important to mention that external symptoms of green rosette resemble PtCpDs.

With the exception of conspicuous mosaic on leaflets, symptoms of mosaic rosette resemble those of chlorotic rosette.

Currently, peanut is the only host known to be infected by both GtRoV and GtRoAsV. Some other species are infected with only one virus or the other. Diagnostic hosts have been identified (Reddy, 1991).

Pathogenesis and Control

The viruses associated with all three types of rosette disease are transmitted by *Aphis craccivora* in a persistent or circulative manner. Currently, peanut is considered the main source of inoculum from which the initial spread of the rosette disease occurs.

Rosette disease can be controlled effectively by adopting such cultural practices as early planting, planting at high density, and maintenance of plant population, as in the case of PtBdNV. Excellent sources of resistance to groundnut rosette were located in peanut land races in Africa. Many long-duration and medium-duration peanut cultivars have been bred. All currently known rosette-resistant cultivars are susceptible to GtRoAsV but are resistant to GtRoV and its satellite. Early-maturing, rosette-resistant peanut genotypes have recently been identified and are currently being used to produce early-maturing rosette resistant cultivars (Reddy, 1998). Since coat protein gene of GtRoAsV has been sequenced, prospects for producing GtRoAsV-resistant peanut cultivars are good, using nonconventional approaches.

Peanut Mottle Potyvirus (PtMtV)

Properties of Viral Particles

Particles are flexuous rods (750 × 12 mm) that contain a single-stranded RNA (3×10^6 kDa). RNA is infectious. The molecular weight of coat protein is 32 to 36 kDa. The coat protein gene was shown to contain 1247 nucleotides (Teycheney and Dietzgen, 1994). The TIP is 55 to 64°C and LIV is 1 to 2 days at room temperature. High-quality polyclonal antibodies that can distinguish PtMtV from another widely distributed peanut stripe potyvirus are available.

Geographic Distribution and Economic Importance

PtMtV is currently known to occur in all the peanut-growing countries. Because of wide distribution and its potential to cause significant crop losses, PtMtV is considered of global economic importance.

Host Range and Symptoms

Symptoms first appear on young leaflets as mild mottle or a mosaic of irregular, dark-green islands (Figure 99). Older leaflets often do not show the typical mosaic symptoms but they can be seen by transmitted light. In some genotypes, characteristic interveinal depression with upward rolling of leaf edges appears (Figure 100). Plants are slightly stunted. However PtMtV has been shown to reduce yields up to 40%.

Diagnostic host is *Phaseolus vulgaris* Topcrop, which produces reddish-brown local lesions.

FIGURE 99 Peanut infected by peanut mottle virus (India). (Courtesy D.V.R. Reddy.)

Pathogenesis and Control

PtMtV is readily transmitted by mechanical sap inoculations and in a nonpersistent manner by several aphid species (Reddy, 1991). The virus is also seed transmitted, up to 8% in some genotypes. In the majority of peanut genotypes, it is transmitted to peanut seed at approximately 1%. PtMtV is also seed transmitted in mungbean and cowpea, but not in soyabeans. Infected seed appears to be the primary source of inoculum. Therefore, planting of virus-free seed is important. Few genotypes are tolerant to PtMtV and many genotypes that do not transmit the virus to seed have been identified. They are yet to be fully exploited in generating cultivars tolerant to PtMtV and, at the same time, do not transmit the virus to seed. Resistance to PtMtV has been located in wild *Arachis* species (Reddy, 1998).

FIGURE 100   Inward curling of leaf edges and interveinal depression of peanut leaves caused by peanut mottle virus. (Courtesy D.V.R. Reddy. Reproduced with permission from APS.)

Peanut Stripe Potyvirus (PtSpV)

Properties of Viral Particles

PtSpV is a potyvirus; the particles resemble PtMtV. The genome contains 10,059 nucleotides. As with other potyviruses, it contains a single open reading frame from which eight proteins are processed (Gunasinghe et al., 1994). PtSpV is currently considered a strain of BCmMV, based on high homology of their coat protein peptide profile (McKern et al., 1992).

Geographic Distribution and Economic Importance

Although PtSpV was first reported in the U.S. in 1983, it was known to be endemic in China. Widely distributed in the peanut-growing countries in southeast Asia, it also occurs in India and Senegal. The virus poses a serious threat to peanut production in southern and southeastern Asia. Crop losses due to PtSpV in China have been estimated to exceed $100 million.

Host Range and Symptoms

Isolates of PtSpV that produce distinct symptoms in peanuts and other hosts have been identified. The name "stripe" was given to the PtSpV isolate that produced discontinuous, dark-green stripes along the lateral veins of young leaflets. The most widely distributed isolate, referred to as the blotch isolate, causes irregular green blotches on young leaflets that persist as the leaflets age. The isolate that produces a mild mottle symptom is widely distributed in China. PtSpV can cause stunting and yield reductions up to 50%.

Natural hosts are *Centrosema pubescens, C. macrocarpum, Calopogonium coeruleum, Crotalaria striata, Desmodium siliquasum,* and *Pueraria phaseoloides.* PtSpV does not produce local lesions on *Phaseolus vulgaris* Topcrop, and therefore can be distinguished from PtMtV.

Pathogenesis and Control

It is readily transmitted by mechanical sap inoculations. PtSpV is transmitted nonpersistently by many aphids. It is seed-transmitted in peanut to a higher frequency than PtMtV, often exceeding 1%. Seed appears to be the primary source of inoculum. More than 10,000 peanut genotypes were evaluated for resistance to PtSpV in Indonesia and none were found resistant. Planting with virus-free seed has been shown to reduce PtSpV incidence. Since PtSpV genome has been fully sequenced, the potential exists for utilizing its genes in developing transgenic peanuts.

Diseases of Minor Economic Importance

Peanut Stunt Cucumovirus (PtSnV)

Properties of Viral Particles

The particles are 25 to 30 nm in diameter and encapsidated within are four single-strand RNAs (ca. 3300, 3000, and 2200 1000 nucleotides). The fourth RNA is subgenomic and codes for the coat protein gene. From the U.S., two serologically distinct isolates occur: PtSnV-E from the eastern region and PtSnV-W from the western region. Three serotypes have been reported from China.

Geographic Distribution and Economic Importance

PtSnV is reported from the U.S., China, and Sudan. It can cause crop losses up to 75%. However, its incidence is relatively low.

Host Range and Symptoms

In the U.S., PtSnV causes severe stunting of the entire plant or of one or more branches. Isolates occurring in China do not cause severe stunting. Characteristic symptoms are shortening of the petioles, reduction in leaflet size, chlorosis, and malformation. To some extent, symptoms resemble those of early infected plants by tospoviruses. Very few pods are produced by early infected plants and the seeds are not marketable.

PtSnV causes epinasty, with systemic mosaic and malformation in cowpea (*cv.* Blackeye). Systemic symptoms produced by PtSnV in peanut, *Phaseolus* beans, and cowpea can be used to distinguish it from CMV.

Pathogenesis and Control

PtSnV is transmitted by *Aphis craccivora*, *A. spiraecola*, and *Myzus persicae*. The virus is seed transmitted up to 0.2%. Some of the alternate hosts, such as white clover, provide the primary source of inoculum. Therefore, peanuts should not be planted in fields located near leguminuous hosts. The disease is of minor importance and no efforts have been made to search for sources of resistance (Demski and Reddy, 1997).

Cucumber Mosaic Cucumovirus (CMV)

Properties of Viral Particles

Similar to those described for PtSnV, two strains of CMV which naturally infect peanuts in China have been reported. A strain of minor importance, CMV-CS, is serologically related to PtSnV.

Geographical Distribution and Economic Importance

The virus is reported only from China. The disease caused by CMV in China is referred to as peanut yellow mosaic and is currently recognized as economically important in the northern regions of China. The virus reportedly causes losses up to 40%.

Host Range and Symptoms

Initial symptoms are chlorotic spots and upward rolling of young leaflets. Leaflets produced subsequently show a yellowing of the lamina with green stripes along the lateral veins. Leaflets are sometimes deformed and plants are slightly stunted. The severe mottling and yellow symptoms observed on young leaflets are not apparent on older plants.

CMV has a very wide host range.

Pathogenesis and Control

CMV is transmitted by sap inoculation and a number of aphids. Seed transmission occurs in peanut up to 4%. Seed from infected plants should not be planted. Cultural measures such as mulching with transparent plastic sheets and removal of diseased seedlings at early stages of crop growth can reduce disease incidence. Peanut cultivars resistant to CMV have yet to be developed. (Demski and Reddy, 1997).

Cowpea Mild Mottle Carlavirus (CpMdMtV)

Properties of Viral Particles

The physicochemical properties of CpMdMtV resemble carlaviruses. Particles are rod shaped, approximately 650 nm in length, and 15 nm in diameter, consisting of a coat protein of 31 to 33 kDa and a single-stranded RNA of 2500 kDa. Carlavirus-specific primers have been used to demonstrate that CpMdMtV belongs to carlavirus. The genome contains an 11.7-k open reading frame at the 3' terminus, containing a "zinc-finger" motif, unique to carlaviruses (Badge et al., 1996). The TIP is 75 to 80°C. (See Figure 101.)

FIGURE 101 Dwarfing and mottle of peanut caused by cowpea mild mottle virus (India). (Courtesy of D.V.R. Reddy.)

Geographic Distribution and Economic Importance

Although CpMdMtV is widely distributed in Asia and Africa, its incidence seldom exceeds 5%. Currently, it is considered a minor disease of peanuts.

Host Range and Symptoms

Initial symptoms appear on young leaflets as vein clearing, followed by downward rolling of the leaflet edges and vein banding. Subsequently, necrosis of petioles and leaflets occurs (Figure 102). Plants are often stunted and are conspicuous because of the rolled edges and vein banding of the leaflets (Reddy, 1991).

CpMdMtV produces necrotic local lesions on *Beta vulgaris* and *Chenopodium amaranticolor*.

Pathogenesis and Control

CpMdMtV is transmitted very efficiently by the whitefly *Bemisia tabaci* in a nonpersistent manner. Any preferred host for the vector can result in high disease incidence. Cowpea and soybeans, preferred hosts for *B. tabaci*, are susceptible to CpMdMtV and show relatively high virus incidence. Therefore, sowing of peanuts near cowpea and soybean fields or intercropping peanut with such highly susceptible hosts should be avoided, especially in seasons when the whitefly populations are likely to be high.

Peanut genotypes resistant to CpMdMtV have not yet been identified.

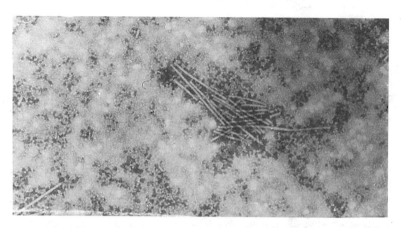

FIGURE 102 Electron micrograph of negatively stained cowpea mild mottle virus particles (India). (Courtesy of D.V.R. Reddy.)

Peanut Chlorotic Streak Caulimovirus (PtCSkV)

Properties of Viral Particles

PtCSkV particles are icosahedrical, approximately 50 nm in diameter. It has a double-stranded DNA of 8174 base pairs. The genome encodes eight major open reading frames. The proteins involved in viral movement—a major capsid, reverse transcriptase, and a transactivating function—have been identified.

Geographic Distribution and Economic Importance

PtCSkV is widely distributed in India. However its incidence is very low except for one isolate, which occurred at a high incidence in one location in South India.

Host Range and Symptoms

In peanut, the virus causes reduction in the size of leafalets, chlorotic streaks, and stunting.
Host range is restricted largely to Leguminosae and Solanaceae.

Pathogenesis and Control

PtCSkV is readily transmitted by mechanical sap inoculations. Attempts to transmit PtCSkV by various aphids have failed. No efforts have been made to devise control measures for PtCSkV (Reddy and Richins, 1998).

OTHER VIRUS DISEASES

Among the 15 other viruses recorded as having been isolated from naturally infected peanut are cowpea chlorotic mottle, groundnut crinkle, bean yellow mosaic, passion fruit woodiness, tobacco streak, and sunflower yellow blotch viruses, plus other peanut/groundnut viruses (Figure 103). These viruses can be important because they occur in peanut at least occasionally in nature. Therefore, a potential exists—given the right environmental conditions and disturbance of the agroecosystem—for them to become economically important and even, although remote at present, to become epidemic (see Table 3).

VIRUS DISEASES OF POTATO (*Solanum tuberosum*)

Potato is a generally widespread plant of average importance for both human and animal nutrition and for industrial processing. An important raw material for the chemical industry, it has become increasingly important for the food industry—thus the main reason for including potato in this section of the book.

FIGURE 103 Yellow spots on peanut leaves caused by peanut yellow spot virus. (Courtesy D.V.R. Reddy. Reproduced with permission from APS.)

Potato is seemingly the first herbaceous food plant on which symptoms of an infectious virus disease were positively determined: that is, the potato degeneration which was observed in Germany and England in the 18th century when it seriously endangered potato production. Based on information available then, it was impossible to diagnose the causal agent that resulted in several different explanations. The first plausible explanation of the disease agent resulted from a study of potato leaf roll as a highly damaging potato disease (Quanjer, 1913). The data from these investigations contributed to the determination of the viral nature of potato degeneration two centuries after the first observation. Discovery of the first potato virus disease marked the beginning of research of viruses as the causal agents of these diseases. Since then, worldwide potato research has identified 33 virus and phytoplasma species (Martyn, 1968; 1971); and in Europe 22 pathogen species, among which only about 10 cause serious economic damages are known.

All potato viruses are preserved in and transmitted through infected tubers. These viruses differ according to other modes of transmission by which they can be grouped:

1. Viruses transmitted by aphid vectors (potato leafroll virus, PLrV)
2. Viruses transmitted mechanically and by aphid vectors (potato viruses A and S, PVA and PVS)
3. Viruses transmitted mechanically and by aphid vectors in the presence of "helper" viruses such as PVY and PVA (potato aucuba mosaic virus, PAcMV)
4. Viruses transmitted mechanically and by non-aphid vectors (potato yellow dwarf virus, PYDV; potato spindle tuber viroid, PSlTbVd, which is also transmitted through potato seed and pollen)
5. Viruses transmitted mechanically and in soil by fungal vectors (potato mop-top virus, potato virus X, PVX).

Potato Leafroll Luteovirus (PLrV)

Properties of Viral Particles

PLrV particles are isometric, 24 nm in diameter, and develop in and are restricted to phloem tissues (Kojima et al., 1969). They consist of ssRNA to which a small genome-linked protein of 7kDa (Mayo et al., 1982) and a single coat protein of 26 kDa (Rowhani and Stace-Smith, 1979) are bonded. RNA comprises 30% and the protein 70% of the total viral particle weight. The virus is strongly immunogenic (Harrison, 1984).

The TIP of PLrV in *Physalis floridana* sap is 70 to 80°C; the DEP is 1×10^{-4}; and the LIV at 2°C is 5 to 10 days (Murayama and Kojima, 1965). Some isolates differ in virulence and transmissibility by vectors. Strains described in former Yugoslavia are mostly mild and occasionally moderate in virulence (Kus, 1964).

Geographic Distribution and Economic Importance

PLrV is widespread and probably occurs wherever potato is grown. Virus transmission occurs with movement of infected tubers and by flight of aphid vectors, which contributes to its general distribution.

Reductions in potato yield caused by PLrV depend on the presence of the sources of infection, but even more on cv. susceptibility; occasionally, incidence of disease is 60 to 70% (Šutić, 1983). Average yield losses caused by PLrV estimated at 10% represent an annual loss of about 20 million tonnes of tubers worldwide (Kojima and Lapierre, 1988).

Host Range

PLrV has a comparatively narrow host range. Potato (*Solanum tuberosum*) is its only known natural host (Spaar and Hamann, 1977). The majority of experimentally infected hosts are in the family Solanaceae, and include *Atropa belladonna*, *Capsicum annuum*, *Datura stramonium*, *Lycopersicon lycopersicum*, and *Nicotiana* spp. Susceptible plants in other families include *Amaranthus caudatus*, *Celosia argentea*, and *Gomphrena globosa*. Among solanaceous plants, *Physalis floridana* is a suitable assay plant; it reacts as chlorosis between leaf veins and varying growth delay, depending on strains and temperatures.

Symptoms

Both primary and secondary (Figures 104 and 105) symptoms have been described on infected potato plants. Primary symptoms occur on plants grown from healthy tubers and infected during the vegetation period. Secondary symptoms occur on plants originating from infected tubers.

Primary symptoms, most pronounced on apical leaves, include erect, lighter green, and the leaf base turns blue-violet or reddish, depending on the cv. Leaflets twist upward markedly, which led to the disease name potato leafroll. Symptoms are more pronounced in earlier infections. Secondary symptom intensity depends on both virulence of virus strains and susceptibility of potato cvs. In some cvs. infected with mild strains, upper leaves are erect and chlorotic; whereas lower leaves roll upward with necrotic patches appearing along leaf margins later with age. Infected plants exhibit delayed growth. Virulent strains cause discoloration and severe leaflet roll, plus pronounced stunt in susceptible cvs.

Phloem necrosis occurs in stems and tubers of infected plants (thus, the earlier description of potato phloem necrosis), and the abundant callus that forms in the phloem of these organs is diagnostic. Phloem necrosis hinders the normal movement of carbohydrate assimilates, and their accumulation in leaves causes leaf roll. Disorders in the phloem functions cause the formation of filamentous germ threading from infected tubers, which also characterizes the disease. Net-like tissue necrosis occurs in tubers of some cultivars. Infected plants produce some small tubers, worthless either for consumption or for planting.

Pathogenesis

PLrV-infected tubers represent the main source of viral inoculum, which is preserved and transmitted through a series of generations. Vectors of PLrV in nature are aphids, including *Myzus persicae*, *M. ascalonicus*, *M. ornatus*, *Macrosiphum euphorbiae*, *Brachycaudus helichrysi*, and *Phorodon humuli*. *M. persicae* probably plays the most important role in vector activity.

Aphids that transmit PLrV persistently require 48 to 72 hours to transmit it (Spaar and Hamann, 1977). The aphids acquire PLrV in 12 to 14 hours, which then takes 12 to 24 hours to move through

FIGURE 104   Secondary symptoms on potato plants from tuber-borne potato leafroll virus infection. (Courtesy E.R. French. Reproduced with permission from APS.)

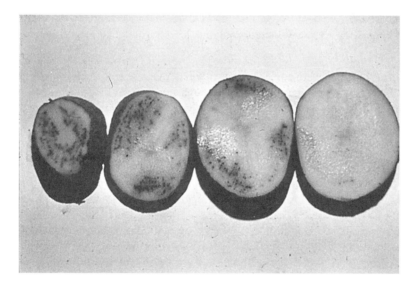

FIGURE 105   Potato leafroll virus-infected tubers show net necrosis. (Courtesy Clemson University.)

the intestinal canal to the salivary glands and into the mouth apparatus; and finally, they remain capable of transmitting PLrV, infecting healthy plants for 12 to 14 hours. No information exists on the multiplication of PLrV in aphids.

The virus is transmitted neither mechanically nor through potato seed from infected plants. Dodder (*Cuscuta subinclusa*) transmits the virus in experimental tests.

Control

A general control measure in commercial production is the use of healthy tubers as planting material, which are produced under special conditions. Lacking the availability of healthy tubers, producers should substitute their planting material every second or third year with seed tubers originating from mountainous regions where aphid activity is minimal. Control of aphids with the use of highly active insecticides is recommended. The most important factor in prevention of primary infection is complete protection of the potato crop prior to the first aphid flight, which usually occurs from winter hosts to the new crop. In order to prevent secondary infections intra- and intercrop, one must apply insecticides shortly before the second flight of aphids, usually in early July in continental regions. This protection against the second aphid flight is useful only if the incidence of primary infection is low.

In case of high incidence of primary infections, one must destroy infected above ground plant parts either mechanically or with defoliants at the beginning of the summer aphid flight which eliminates inoculum sources of further infections and allows earlier harvest. Reduction of yields due to earlier harvest can be avoided partially by planting sprouted tubers which enables up to 2 weeks earlier emergence. In such cases early and medium-early potato cultivars may provide high yields of both seed and commercial potatoes. Finally, only cultivars more resistant to PLrV should be planted.

Potato Potyvirus A (PVA)

Properties of Viral Particles

PVA particles are elongate, filamentous, 730 × 15 nm, and contain an ssRNA, plus a single polypeptide coat protein of 33kDa. PVA has a sedimentation coefficient of 150S (Weidemann, 1988); a TIP of 44 to 52°C; a DEP of 1/10 to 1/40; and an LIV at 18°C of 12 to 18 hours. It has moderate immunogenicity (Bartels, 1971).

Based on virulence on potato and *Nicandra physalodes*, strains of PVA have been classified as very mild, mild, moderately severe, and severe (Calvert, 1960; loc. cit. Bartels, 1971). Virulence of isolates in Yugoslavia vary, and therefore may belong to more than one group of strains.

Geographic Distribution and Economic Importance

PVA occurs in all regions where potato is grown, which ranks it widespread among potato viruses. Some potato cvs. (Allefrüheste) are tolerant, although infected with PVA. Virulent strains of PVA causing severe mosaic may reduce potato yields 40% (Dedić, 1975). Yield losses vary depending on the susceptibility of cvs. and the incidence of infected plants in production.

Host Range

Potato is the only known natural host of PVA. Most experimental hosts belong to the family Solanaceae, including *Datura metel*, *Lycopersicon lycopersicum*, *Nicotiana clevelandii*, *Petunia hybrida*, and *Physalis floridana* plus numerous Leguminoseae (e.g., *Melilotus italicus* and *Trigonella foenum-graecum*) (Edwardson, 1974; Spaar and Hamann, 1977). Suitable assay plants include *Solanum demissum* and the A6 hybrid (*S. demissum* × Aquila), which react with star-like, dark local spots on inoculated leaves.

Symptoms

Symptoms of infection vary, depending on virulence of the strain and potato cv. susceptibility. In most cvs., PVA causes a mosaic prominent along the veins, but not in interveinal tissue. Veins remain shallow, causing disorders in leaf development and primarily leaflet wrinkling, especially in Mercur and Oneida cultivars. In some cultivars, instead of the typical mosaic, chlorosis is evident—particularly on apical leaves. Partial necrosis as streaks occurs along leaf veins on the lower surface of older leaves. Leaflets remain smaller, especially in cultivar Cvetnik. A synergistic reaction occurs when PVA + X are inoculated simultaneously (Figure 106).

Some cultivars react to PVA infection with severe internode shortening (i.e., dwarfing). Dark necrotic spots, a hypersensitive reaction, occur in the tubers during storage of cvs. Bintje and Saskia.

Pathogenesis

Tubers serve as the only source of viral inoculum in preserving and transmitting PVA into the next season. Seven species of aphids, including *Myzus persicae*, *Aphis frangulae*, *A. nasturtii*, and *Macrosiphum euphorbiae*, transmit PVA in nature nonpersistently. *M. persicae* is the primary virus vector. Aphids require only 20 seconds to inoculate healthy plants; they remain infective for about 20 minutes.

FIGURE 106 Potato viruses X and A interaction in potato; healthy (left).

PVA is transmitted mechanically with sap from infected plants, which allows its natural spread by contact, both between the plants and with cultivation implements. No information is available on seed transmission.

Control

Prevention measures recommended for PVY control are equally important for PVA because these two viruses are similarly preserved and transmitted in nature. Virus-free plants have been obtained with the use of thermotherapy (Thomson, 1956) and through meristem-tip culture (Morel and Martin, 1955).

Potato Carlavirus M (PVM)

Properties of Viral Particles

PVM is elongate, filamentous, and 650×12 nm. The TIP is 65 to 71°C; the DEP is 1×10^{-2} to 10^{-3}; and the LIV at 20°C is several days. PVM has good immunogenicity (Wetter, 1972).

Strains of PVM differ, based either on symptoms on infected plants or on virulence of strains. Symptoms include mosaic, leaf roll, interveinal leaf mosaic, abnormal leaf crinkle (paracrinkle isolate), plus others also described according to disease symptoms. Strains of PVM have been classified based, on virulence ranging from mild to severe, with various distributions in individual countries.

Geographic Distribution and Economic Importance

PVM, which occurs worldwide, has a relatively low incidence in western Europe and North America, whereas its incidence in eastern Europe is quite high (Gabriel, 1960; Sinnema, 1978). PVM may cause reduced tuber yields of 20 to 50%. In view of the lower incidence in the potato crop, it may be considered less important than other potato viruses.

Host Range

The host range of PVM is narrow because potato is the only known natural host and nearly all experimental hosts are in the Solanaceae (e.g., *Datura stramonium*, *Lycopersicon lycopersicum*,

Solanum demissum, S. melongena, and others). Few experimental hosts outside this family are *Chenopodium quinoa, Vigna sinensis*, and others. *Datura metel* and *Gomphrena globosa* are suitable for assay purposes because they react to infection by producing local spots (Spaar and Hamann, 1977).

Symptoms

Symptom appearance and severity depends on the virulence of potato strains and on cultivar susceptibility. Mild chlorosis on portions of apical leaves are discerned with difficulty. The stem apex is compacted slightly, due to internode shortening, which occurs on cvs. Saskia, Bintje, and others. Mosaic symptoms, accompanied by leaflet malformation and shortened petioles that overlap and exhibit spoon-like twisting, occur on cvs. Cvetnik and Voran. The third symptom type includes leaf vein necrosis accompanied by many axillary shoots on infected plants. Various size fissures appear on infected tubers and the eyes protrude, especially on cultivar Mercur.

Pathogenesis

PVM is preserved in and transmitted by potato tubers. Aphids transmit PVM nonpersistently in nature with *Myzus persicae* the most active vector; others (e.g., *Aphis frangulae, A. nasturtii, Aulacorthum solani*, and *Macrosiphum euphorbiae*) are less active. The abnormal leaf crinkle strain, not aphid transmitted, is transmitted mechanically via infective plant sap. Other modes of transmission remain unknown (Šutić, 1983).

Control

General preventive measures recommended for PVY are equally important for control of PVM because they both have comparable modes of preservation and transmission. Virus-free plants can be obtained by thermotherapy and meristem-tip culture (Stace-Smith and Mellor, 1968).

Potato Carlavirus S (PVS)

Properties of Viral Particles

PVS is elongate at 650 × 12 nm; has a TIP of PVS 55 to 60°C; a DEP of 1×10^{-2} to 1×10^{-3}; and an LIV at 20°C of 3 to 4 days (Wetter, 1971). PVS has strong immunogenicity. Although certain differences among individual isolates exist, PVS strains are not yet sufficiently defined.

Geographic Distribution and Economic Importance

PVS is widespread and occurs in all potato cvs., generally wherever potato grows. Severe infections, some 100%, occur in many potato cvs. The average infection rate in Italy in 25 cvs. was 41% incidence (Lovisolo and Bennetti, 1960); and in southern Germany in 28 cvs., it was 41 to 61% (Hunnius, 1976). In former Yugoslavia, 46% of the Prof. Wohltmann cv. was infected (Kus, 1964) and up to 80% of the Bintje, Urgenta, Saskia, and other cvs. were infected (Buturović, 1970). Cvs. Allerfrüheste gelbe, Ella, Mercur, Oneida, Weckaragis, and others were 100% infected.

Although the most widely spread among all potato viruses, PVS is considered least damaging in potato production. Pathological changes are minor and thus, the consequences of infection are less severe. Some data show that PVS reduces tuber yields by 10 to 20% (Wetter, 1971) and yield losses in former Yugoslavia ranged from 6 to 15% (Kus, 1964).

Host Range

Potato and melilots (*Melilotus albus*) are the only reported natural hosts of PVS (Spaar and Hamann, 1977). The small group of susceptible experimental hosts belonging mostly to the Solanaceae and Chenopodiaceae (e.g., *Chenopodium album, C. amaranticolor, C. hybridum, C. murale, Datura metel, Solanum demissum, S. villosum*, and others. *C. amaranticolor, C. quinoa*, and *C. album* are suitable as biological assay plants because they react to infection with local lesions.

Symptoms

The mild symptoms caused by PVS on all potato cvs. can be grouped into three classes.

The first include cvs. that react with a mild mosaic between leaf veins, which later disappears and thus the infection is masked. Infected plants of cvs. Saskia, Cvetnik, and Prof. Wohltmann, are a slightly lighter green without other visible changes in appearance.

The second class of cvs. includes those symptoms of class one plus a leaflet wrinkle and mild crinkle. Leaf growth is reduced, thus, stipples remain less covered with leaves in cvs. Bintje and Voran.

The third class includes those tolerant cvs., Mercur and Oneida, which when infected are symptomless.

Pathogenesis

As for all potato viruses, infected tubers represent the main source of reinfection. In nature, PVS is transmitted mechanically with infective plant sap such as with implements and machines used for potato cultivation. Aphids transmit PVS readily and nonpersistently, the primary reason for its extensive and intensive spread in nature (Turska, 1980). *Aphis nasturtii* is the most efficient among other active vectors (such as *Myzus persicae*, *Aphis fabae*, and *Rhopalosiphum padi*) (Gabriel and Kowalska-Noordam, 1988).

Control

The best virus control is planting healthy tubers. Tuber production and sanitary control is done as for all other potato viruses. Because of the ease of the mechanical transmission of PVS in commercial potato production, adequate spatial isolation between individual plots, especially in production of planting materials, is essential. Tolerant potato cvs. should be used in severely contaminated areas. Virus-free plants can be obtained from infected potato cultivars with the use of thermotherapy and meristem-tip culture methods (Stace-Smith and Mellor, 1968).

Potato Potyvirus Y (PVY) [Type]

Properties of Viral Particles

PVY filamentous particles are 730×11 nm, have a sedimentation coefficient of 145S (Huttinga, 1975), and contain single-stranded RNA that accounts for 6% of the particle weight, plus a coat protein unit of 34kDa (De Bokx and Huttinga, 1981). The TIP of PVY in tobacco sap is 50 to 62°C; the DEP is 1×10^{-2} to 1×10^{-6}; and the LIV at 18 to 22°C is 7 to 50 days. PVY has strong immunogenicity (De Bokx and Huttinga, 1981).

Strains

Several strains of PVY have been classified into three groups based on characteristic symptoms on potato, tobacco, and *Physalis floridana*: the common strain (PVY^O), the stipple streak strain (PVY^C), and the tobacco veinal necrotic strain (PVY^N). Serological differences among individual strains are not consistent among the group members.

Geographic Distribution and Economic Importance

PVY ranks among the generally widespread potato pathogens. De Bokx and Huttinga (1981) pointed out the irregular presence of individual strain groups in different parts of the world. PVY is one of the most damaging viruses in the world, based on its distribution, rapid rate of spread by aphids, and severity of reactions on infected plants. Tuber yields from PVY-infected plants are often reduced 50 to 90% (Spaar and Hamann, 1977); in Yugoslavia, tuber yield losses were 33 to 77% (Šutić 1983).

Host Range

In addition to potato, other natural hosts of PVY include tobacco, tomato, pepper, petunia (*Petunia hybrida*), and henbane (*Hyosciamus niger*). Experimental hosts among numerous susceptible plants include *Chenopodium amaranticolor*, *C. quinoa*, *Convolvulus arvensis*, *Datura metel*, *Hyoscyamus albus*, *Lycium chinense*, *Nicotiana acuminata*, *N. clevelandii*, *N. glutinosa*, *N. rustica*, *Portulaca oleracea*, *Solanum melongena*, *S. nigrum*, and *S. villosum* (Spaar and Hamann, 1977).

FIGURE 107 Potato virus Y^o-infected Ackersegen potato. (Courtesy Clemson University.)

C. amaranticolor, *C. quinoa*, *N. rustica*, and *S. demissum* are useful as assay plants because they react by displaying local spots (Delgado-Sanchez and Grogan, 1970).

Symptoms

PVY causes various symptoms, depending primarily on the specificity of virus strains and susceptibility of potato cvs. (Kus, 1964; De Bokx and Huttinga, 1981).

The group of common strains (Y^o) rarely cause mottling or yellowing of leaflets (Figure 107). More often, they cause severe symptoms including crinkle, rugosity, necrosis, leaf-dropping, and premature death of infected potato plants. Necrosis starts usually as spots or rings on the leaflets, followed by collapse and dropping of, or hanging on the stem, of leaves ("leaf drop streak"). Secondary symptoms consist of dwarfing of plants, wrinkling and brittleness of leaves, and rarely necrosis.

Y^o strains cause systemic necrosis in *Physalis floridana* and mottling in tobacco.

The group of "stipple streak" strains (Y^C) cause hypersensitive or sensitive reaction in potato, depending on cv. susceptibility. The most common symptoms on hypersensitive cvs. are necrotic lesions and streaks on leaves, petioles, and stems. Necrosis can also occur in tubers. Crinkling and mottling are characteristic symptoms on susceptible potato cvs. Secondary symptoms are somewhat milder.

Y^C group of strains cause symptoms similar to those caused by Y^o strains on *P. floridana* and tobacco.

The group of necrotic strains (Y^N), named after its reaction in tobacco, cause usually less severe symptoms on potato. On inoculated leaves of potato, these strains very rarely cause necrotic spots or rings, but more often mild to very mild mottle (Figures 108 and 109). Secondary symptoms are somewhat more obvious.

These strains cause veinal and leaf necrosis in tobacco and mottling in *P. floridana*.

Pathogenesis

Potato tubers, the vegetative reproduction organ, represent the primary sources of infection. In nature, PVY is transmitted nonpersistently by plant aphids, and *Myzus persicae* and *Aphis nasturtii* are most well known. Other vector aphid species include *M. ornatus*, *Aphis frangulae*, *A. frangulae gosypii*, and *Macrosiphum euphorbiae*. Following a 2- to 7-hour starvation period, aphids acquire PVY in 0.5 to 5 minutes and remain infective for 1 minute to 24 hours. (Spaar and Hamann, 1977).

Aphid transmission of PVY and other potyviruses is mediated by a "helper component" present in infected plant sap. This helper component is a virus-coded protein (Thornbury and Pirone, 1983).

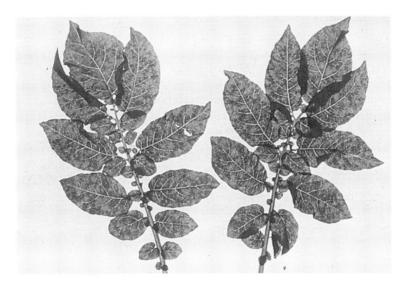

FIGURE 108 Potato virus Y^N-infected King Edward potato showing mottle and rugosity. (Courtesy Clemson University.)

FIGURE 109 Potato virus Y^N in potato.

The virus is easily transmitted mechanically, and thus can be spread naturally by contact with infected plants or with machinery during crop cultivation. No information is available on its transmission through seed or in any other manner. Multiple infection of PVX and PVY react synergistically to cause severe symptoms.

Control

The ease of transmission of PVY causes difficulty in designing control measures. Clonal selection of potato is a primary recommendation for control, based on the isolation of symptomless tubers, which are subject to sanitary control. Such tubers are primarily used as planting material for reproduction and production of commercial tubers.

Control of aphids with insecticides is useful, that is, the same methods as recommended for PLrV.

Early survey of potato fields and the destruction of all infected plants prior to the spring aphid flight markedly reduces that potential source of infection. It is especially recommended for production of potato seed in high (mountainous) regions where environmental conditions reduce aphid development and vector activity.

In situations where severe primary infections occur, earlier tuber harvest is recommended in order to eliminate sources of infection for the first summer aphid flight. To aid earlier harvest, planting of sprouted tubers enables a more rapid-start for plant development at the beginning of the vegetation period, and when combined with use of early and medium-early maturity cvs., potential damage is greatly alleviated. PVY can be eliminated successfully from infected plants by thermotherapy (Thomson, 1956).

Some potato cvs. are less susceptible to individual virus strains under field conditions and should therefore be used for production in areas where these strains predominate.

Potato Aucuba Mosaic [Possible] Potexvirus (PAcMV)

Properties of Viral Particles

PAcMV particles are elongate, 580 × 11 nm, and the viral particle sedimentation coefficient is 130S (Kassanis and Govier, 1972). Its RNA consists of 5% of the particle weight and coat protein units are 26 kDa. The virus has strong immunogenicity. Potato virus F (PVF), the tuber blotch strain of PAcMV, causes mild symptoms on leaves and a typical tuber necrosis (loc. cit. Kassanis and Govier, 1972).

Geographic Distribution and Economic Importance

PAcMV occurs on few plants, but it is present in all areas where potato is grown. The PAcMV virus causes necrosis and death of individual tuber parts. Damage is most severe in storage at higher temperatures. Since this disease occurs sporadically, no great economic loss occurs.

Host Range

Potato is the only known natural host of PAcMV. Experimental hosts are mostly in the Solanaceae (e.g., *Datura metel, D. tatula, Hyoscyamus niger, Nicotiana clevelandii, Solanum melongena*, and others). Only a few susceptible species exist outside this family; they include *Amaranthus retroflexus, P. vulgare, P. sativum, Trifolium incarnatum*, etc. (Spaar and Hamann, 1977). Pepper with necrotic local lesions and epinasty symptoms, and *N. glutinosa* with green mottle and dark vein banding, are suitable assay plants.

Symptoms

Symptoms on PAcMV-infected plants of some cultivars vary and depend mostly on the strains of the virus. The most common symptoms in some cultivars include light-yellow spots on lower and middle leaves, after which this disease was named (Figure 110). The potato pseudo-aucuba disease has similar symptoms on potato leaves, but is caused by a strain of the tomato black ring virus. Leaflet malformation and tip necrosis occurring in some cvs. cause stunt. The occurrence of bronze-colored necrotic spots, especially on lower leaves, is a unique symptom. PAcMV causes tuber necrosis typified by dark brownish spots and other net-like shapes, leading to tuber malformation and extinction.

Pathogenesis

PAcMV remains in infected potato tubers and is transmitted by aphids nonpersistently. *Myzus persicae* and *Aphis nasturtii*, among several other species, are the most well-known vectors. Aphid transmission of PAcMV to potatoes is aided by the presence of other "helper" viruses in infected plant sap: for example, PVY and PVA. Transmission succeeds only if the vector acquires both PAcMV and the "helper" simultaneously from an infected plant or from different infected plants if it first acquires the "helper" virus, then PAcMV.

FIGURE 110 Potato aucuba mosaic virus-infected Sieglinde potato. (Courtesy Clemson University.)

PAcMV is also transmitted experimentally with infected plant sap, whereas other modes of transmission are still unknown.

Control

The key preventive measure is to plant healthy tubers. Infected plants easily diagnosed in the field by characteristic yellow spots should be rogued immediately. Tubers with necrotic spots should be discarded, and should not be used for planting. General preventive measures effective for the control of other potato viruses are recommended to protect against PAcMV.

Potato Yellow Dwarf Rhabdovirus (PYDV)

Properties of Viral Particles

PYDV is bacilliform or bullet-shaped and 380×75 nm. It is a single-particle type with a sedimentation coefficient of 810 to 950S. PYDV has a single-stranded RNA, at least three proteins, and 20% lipids (Black, 1970). The TIP of PYDV in *Nicotiana rustica* sap is 50°C; the DEP is 1×10^{-3}; and the LIV at 23 to 27°C is 2.5 to 12 hours (Black, 1970). PYDV has moderate immunogenicity and is easily detected serologically. Virus isolates differ, based on virulence and transmissibility by vectors.

Geographic Distribution

PYDV, discovered in the U.S. in 1917, has spread naturally into neighboring Canada. Its appearance in different localities depends on the planting of infected tubers.

Host Range and Symptoms

Potato, the natural host on which PYDV was first described, has other solanaceous relatives that have been infected experimentally, plus plant species in the families Compositae, Cruciferae, Labiatae, Leguminosae, Polygonaceae, and Scrophulariaceae (Younkin, 1942; Hansing, 1942).

Under favorable conditions, PYDV causes serious losses in tuber yield (Walker and Larson, 1939).

Characteristic symptoms in infected plants appear as leaf chlorosis and pit necrosis along the stem, which results in severe stunt. Fewer tubers are formed with lower germination potential. The intensity of symptoms increases in warm weather (Slack, 1979). *D. stramonium* with interveinal chlorosis and *Nicotiana rustica* with chlorotic local lesions, mottle, and leaf chlorosis are suitable diagnostic plants (Black, 1940; 1970).

Pathogenesis

PYDV is preserved in infected tubers, which represents an important source of natural infection. Younkin (1943) suggests that *Chrysanthemum leucanthemum* var. *pinnatifidum* represents the main source of PYDV for infection of the potato crop. Several species of agallian leafhoppers *Aceratagallia sanguinolenta, Aceratagallia spp, Agallia constricta,* and *A. quadripunctata* transmit PYDV (Black, 1970). The transmissibility of PYDV has been classified as either PYDV-Sm (sanguinolenta yellow dwarf virus transmitted by *A. sanguinolenta,* but not by *A. constricta*) or PYDV-Cm (constricta yellow dwarf virus transmitted by *A. constricta,* but not by *A. sanguinolenta*). The virus multiplies in nymphs with a minimal incubation period in the vectors of 6 days (Black, 1970); then adult male and female insects transmit the virus. PYDV is experimentally transmitted mechanically into *Nicotiana glutinosa* and *N. rustica,* but no other mode of transmission is yet known.

Control

Only healthy tubers should be used for planting, and potato crops should be grown as far away as possible from potential sources of infection. Vector control may reduce infection intensity, but one must remember that clover represents a suitable host for vectors (Watkins, 1941). Some potato cultivars show differences in susceptibility to PYDV under field conditions.

Potato Spindle Tuber Viroid (PSlTbVd)

Properties of Viral Particles

PSlTbVd consists only of naked, single-stranded RNA, without coat protein. It was first described as a plant viroid by Diener and Raymer (1967).

PSlTbVd RNA contains 359 nucleotides, covalently closed in circular molecules of 120 kDa (Sänger, 1982; Sänger et al., 1976). Nucleotide seqeunce comparisons show that PSlTbVd is grouped with citrus exocortis and chysanthemum stunt viroids. Antisera of the viroid alone have not been obtained (Diener and Raymer, 1971).

Stability in Sap

Stability in leaf sap depends on the method of extraction from infected potato. The TIP of PSlTbVd is 75 to 80°C (or higher), and the DEP is 1×10^{-2} to 1×10^{-3} (Singh and Bagnall, 1968).

Strains

In addition to the type strain, two other strains are the unmottled curly dwarf strain with symptoms in tomato identical to type strain plus severe cracking of potato tubers, and the mild strain with less severe stunting and necrosis of tomato plants based on differential symptoms in infected potato and tomato (loc. cit. Diener and Raymer, 1971).

Geographic Distribution

PSlTbVd has been described in North America and Canada, observed in eastern Europe (Poland, former U.S.S.R., and Turkey), but it has not been recorded in western Europe (Salazar, 1988).

Host Range and Symptoms

Potato is the only natural viroid host known. PSlTbVd may infect tomato (*L. lycopersicum*), which causes characteristic symptoms on some cvs. PSlTbVd is also infective to many species, mostly Solanaceae, which usually remain symptomless (O'Brien et al., 1963).

Characteristic symptoms on potato are smaller leaflets with fluted margins, sharper angles between stems and petioles, and elongate tubers with both more eyes and more prominent tuber surfaces. Susceptible cvs. form fewer tubers, which significantly reduces the yields of PSlTbVd-infected plants up to 40% (Figure 111).

Diagnostic symptoms on PSlTbVd-infected tomato plants cv. Rutgers are epinasty, rugosity, and lateral twisting of leaflets, plus a progressive necrosis in the vascular tissue. Tubers are narrow and plants visibly stunted (Raymer and O'Brien, 1962).

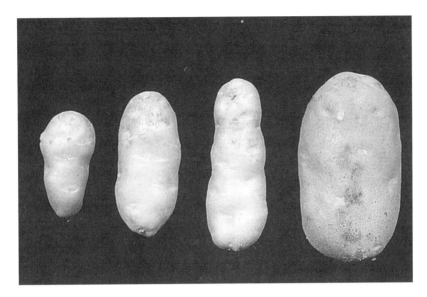

FIGURE 111 Potato spindle tuber viroid-infected Russet Burbank potato; healthy (right). (Courtesy Clemson University.)

Pathogenesis and Control

PSlTbVd is preserved in infected tubers, which represent permanent sources of infection. It is transmitted mechanically during tuber-cutting for planting and by natural contact between healthy and PSlTbVd-infected plants. Vectors of viroids, several species of chewing insects (Salazar, 1988), are minimally important in nature. PSlTbVd is transmitted through both seed and pollen of potato and other plants, a unique situation among potato viruses (Diener and Raymer, 1971).

Selection and production of healthy tubers and planting of uncut tubers represent the main measures for protection of the potato crop against introduction of this pathogen. Viroid-free plants are obtained by low-temperature treatments and meristem culture (Lizárraga et al., 1980).

Potato Mop-Top Tobamovirus (PMpTpV)

Properties of Viral Particles

PMpTpV particles are rigid tubular structures of two sizes: 18 to 20 nm × either 100 to 150 or 250 to 300 nm. Viral RNA is probably single-stranded, and the coat protein consists of one polypeptide of 18.5 to 20 kDa (Harrison, 1974). Infectivity is associated with the 300-nm particles. The TIP of PMpTpV in tobacco sap is 75 to 80°C; the DEP is 1×10^{-2} to 1×10^{-4}; and the LIV at 20°C is 1 day, although possibly maintained partially up to 10 weeks (Harrison, 1974). PMpTpV has moderate immunogenicity. Virus isolates differ in virulence.

Geographic Distribution

PMpTpV seems widespread wherever potato is grown, but it is noted specifically in the Andean region of South America, in North America, and in Europe.

Host Range and Symptoms

Potato is the natural host of PMpTpV, which has been transmitted experimentally into 26 species of the Solanaceae and Chenopodiaceae and to *Tetragonia expansa*. Both *Chenopodium amaranticolor* (which produces only necrotic ringspot local lesions) and *N. debneyi* (which produces local necrotic spots and ringspots plus systemically infected leaves that produce chlorotic or necrotic line patterns) are suitable diagnostic plants (Harrison, 1974).

Symptoms in potato differ, depending on virus strains and cv. susceptibility. Primary symptoms include shortened internodes toward plant apices, providing a bushy appearance of the plant after which the disease was named (Figures 112 and 113). Accompanying symptoms are leaf wrinkle and/or roll, plus differently shaped mosaics. Necrosis spreads irregularly and affects other tissues, but first appears on the tuber surface.

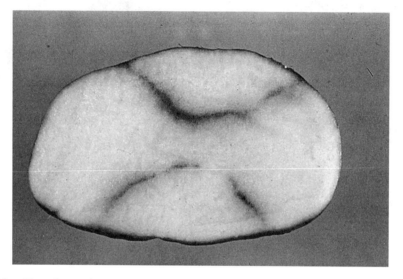

FIGURE 112 Necrotic arcs in Arran Pilot potato infected with potato mop-top virus. (Courtesy Clemson University.)

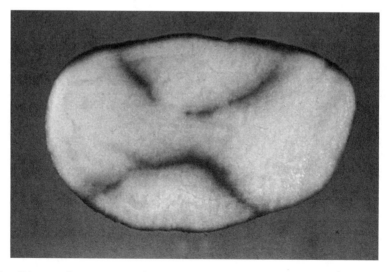

FIGURE 113 Primary tuber symptoms of potato mop-top virus infection. (Courtesy R.A.C. Jones. Reproduced with permission from APS.)

Pathogenesis and Control

Infectivity of PMpTpV is maintained in potato tubers, from which it is transmitted by the zoospores of the fungus *Spongospora subterranea* to potato roots in the soil. The resting spores of *S. subterranea* carry infectious virus up to 2 years (Harrison, 1974).

Planting of virus-free tubers and of tubers noninfected with *Spongospora subterranea* is the main measure to control infection. Soils not contaminated with *S. subterranea* should be used for potato production. Treatment with Zn salts and reduction of soil pH with sulfur enables good control of the soil fungus vector (Cooper et al., 1976).

Potato Potexvirus X (PVX) [Type]

Properties of Viral Particles

PVX particles are flexuous rods 515 × 13 nm, with a sedimentation coefficient of 117S (Bercks, 1970). PVX RNA is single-stranded (2100 kDa) and accounts for 6% of the particle weight. The coat protein capsid consists of 1400 protein subunits, each 27k Da (Koenig and Lesemann, 1989). The TIP of PVX is 68 to 76°C; the DEP is 1×10^{-5} to 1×10^{-6}; and LIV at 20°C is several weeks. PVX is strongly immunogenic (Bercks, 1970).

Strains

Several strains of PVX have differing host ranges, symptoms, serological reactions, tryptophan and tyrosin contents in the coat protein, isoelectric points, and pH stability (Leiser and Richter, 1980). Individual isolates differ also in TIP and tolerance of infected plants to heat treatment.

Serological reactions of strains of PVX have caused classification into the genuine group and the B group (Rozendaal, 1954). PVX strains are related, but do not display complete cross protection reactions.

Four strains differ in their ability to assimilate the Nx and Nb genes that code hypersensitive reactions of individual cvs. (Cockerham, 1955). Alternatively, four different potato cvs. display hypersensitive reactions characterized by infected plant apical necrosis; thus, they are useful for differentiation of strains.

Geographic Distribution and Economic Importance

PVX is found worldwide wherever potato is grown. PVX causes a mild disease, and thus is less damaging, and it reduces yields from 10 to 20%, in rare cases up to 50% depending on virus strains/cv. interaction. High infection rates have been recorded in susceptible cvs. Prof. Wohltmann (25%), Mercur (51%, Oneida (85%), and Weckaragis (84%) (Kus, 1960). Mixed infections with PVA and PVY are particularly damaging in production.

Host Range

PVX naturally infects many plants, primarily in the Solanaceae [e.g., potato, (*S. tuberosum*), tobacco (*Nicotiana tabacum*), tomato (*Lyperscion lycopersicum*), pepper (*C. annuum*), *C. frutescens*, *Solanum demissum*, *S. laciniatum*, and *S. sarachoides*—plus those in other families [e.g., *Amaranthus retroflexus*, *Chenopodium album*, *Euphorbia helioscopia*, *Trifolium pratense*, and *Vitis vinifera*] (loc. cit. Purcifull and Edwardson, 1981). PVX has a broad range of hosts that can be experimentally infected. *Gomphrena globosa*, which reacts to infection with local lesions, is a suitable assay plant.

Symptoms

Symptoms of infection vary, depending on PVX strain and cv. susceptibility. Differences in symptom appearance range from masked to severe mosaic, in addition to leaflet malformation and apical necrosis.

The mild mosaic is the general symptom clearly visible on young plants at the beginning of infection (Figure 114). As the season progresses, with increasing temperatures, the mosaic becomes

FIGURE 114   Potato virus X in potato.

less apparent and more difficult to discern. In some cultivars, mosaic becomes more severe with increased temperatures, leading to leaflet malformation. Infection of cvs. with strains that activate hypersensitive genes causes plant tip necrosis, which occurs only rarely. Cloudy weather and temperatures of 16 and 20°C favor symptom development.

Pathogenesis

Infectivity of PVX is preserved in and transmitted by infected tubers. PVX easily transmits mechanically or by contact between infected and healthy plants. Mechanical spread of PVX occurs aided by machines used for cultivation, workers' clothing, and animals carrying traces of infected plant sap on their skins. Where plant populations are dense, PVX can also be transmitted by contact between infected and healthy plant roots.

Nienhaus and Stille (1965) showed that PVX can be soil transmitted with the aid of the zoosphores of the fungus *Synchytrium endobioticum*. Following development on the roots of infected plants, they continue to spread the infection in the soil. However, this was not corroborated in later investigations (Lange, 1978). Other data show the fungus *Spongospora subterranea* as a possible soil vector of PVX. PVX is not transmitted through infected potato true seed, but it can be transmitted with the aid of dodder (*Cuscuta campestris*).

Control

Only virus-free tubers should be used for planting. Any tuber infected, or suspected infected, with *Synchytrim endobioticum* or *Spongospora subterranea* should be discarded. Soils used for production should be free of residual tuber parts and resting spores of these vector fungi. A longer crop rotation is a useful measure in this respect. Chemical sterilization with formaldehyde, pyrolidine, or slaked lime of propagating and cultivating tools is highly recommended (loc. cit. Purciful and Edwardson, 1981).

Potato cvs. which are hypersensitive are rated as field resistant, thus recommended for production. Meristem or auxiliary bud culture combined with heat treatment effectively eliminates PVX from several potato cultivars (Stace-Smith and Mellor, 1968).

Potato: Natural Alternative Host for Some Viruses

Ten viruses—the natural hosts of which are species other than potato, mostly among solanaceous plants—have been recorded on potato as their natural host (loc. cit. Smith et al., 1988; Spaar and Hamann, 1977; Šutić, 1983). The most important viruses in this group are:

- Tobacco viruses: TMV, TNV, TRtV (Figures 115 to 117), and TSkV;
- Tomato viruses: TmBRV and TmSpoWV; and
- Viruses of other hosts: AMV (Figure 118), BtCuTpV, carnation latent virus and CMV

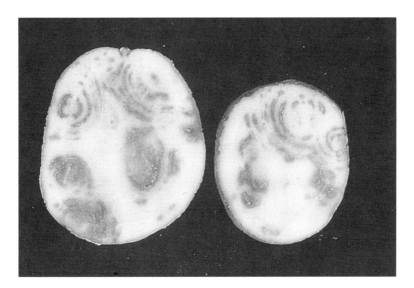

FIGURE 115 Brown arcs sometimes as corky tissue or cracks in tobacco rattle virus-infected potato. (Courtesy of Clemson University.)

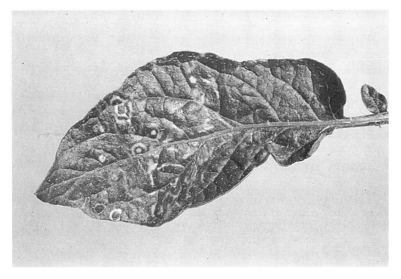

FIGURE 116 Yellow ring (spots) in Rector potato caused by tobacco rattle virus. (Courtesy Clemson University.)

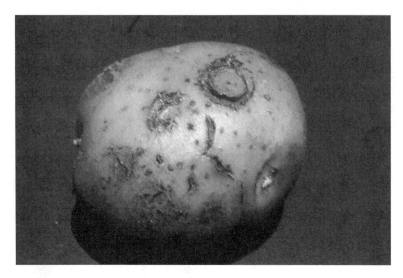

FIGURE 117 Symptoms on tuber of tobacco rattle virus infection. (Courtesy D.P. Weingartner. Reproduced with permission from APS.)

FIGURE 118 Alfalfa mosaic virus in potato; healthy (right).

All these viruses can cause damage in potato with AMV and TmSpoWV as potentially most damaging. The vegetative nature of reproduction via tubers makes potato, a particularly important host for the long-term preservation of these viruses. Therefore, virus incidence, spread, and damage caused by natural infections can be reduced significantly by planning an adequate crop rotation ratio between potato and non-hosts.

Clonal Selection of Healthy Plants

The main feature of all potato viruses is their preservation in and transmission by tubers. When infected tubers are planted, viruses are transmitted concomitantly generation to generation as long as such tubers are used for production. Infections accumulated in tubers year by year are reflected in a progressive decrease in yields of tubers. Yield losses occur in the second year of production; and by the third and fourth years, depending on cv. susceptibility, they become significant, and highly significant respectively, at which time production becomes unprofitable. Viruses represent the central issue in potato production; this can be systematically solved by clonal selection of healthy plants. Clonal selection is usually performed as follows.

1st year: Initial material selection. Methods: visual selection and sanitary control (biotests on diagnostic plants; plus ELISA tests, which aid in the serological control of a large number of plants). Infected cvs. may be freed from the virus(es) with the use of meristem-tip culture and/or combined with thermotherapy.

2nd year: Initiation of clones starting with known healthy plants. Serological assays of all plants. Eliminate infected plants.

3rd year: Multiplication of clones starting with healthy plants. Serological assays of most plants. Eliminate infected plants.

4th year: Multiplication of healthy clones and serological assays of a predetermined number of plants. Eliminate infected plants.

5th year: Multiplication of clones in the field grown under isolated conditions. Serological assay of a number of plants evenly distributed in a given space. Eliminate both infected and suspected infected plants. Selected plants represent the super elite.

6th year: Multiply super-elite clones in isolated fields. Serological assays of a predetermined number of plants. Eliminate infected or suspected infected plants. Continue to multiply elite plants, which represent the basic potato seed for continued reproduction; that is, the selection of virus-free seed potato is an ongoing process, an irreplaceable procedure for modern potato production.

VIRUS DISEASES OF TOBACCO (*Nicotiana tabacum*)

The tobacco plant is especially interesting for the study of plant virology. It was of great importance in the historical development of the scientific discipline of plant virology because it is both the natural and experimental host of numerous viruses. Most significant, however, is the host of tobacco mosaic virus (TMV). Tobacco holds the foremost position in the development of virology in general and of plant virology in particular. The disease caused by TMV, one of the first discovered virus diseases, was described in the mid-19th century (loc. cit. Šutić, 1983). Shortly thereafter, scientists investigated its transmissibility and infectivity (Mayer, 1886; Ivanovski, 1892), which stimulated the initiation of many experimental studies in plant virology. By the end of the 19th century, Beijerinck (1898) had already separated and suggested the difference of the tobacco mosaic agent from among fungi and bacteria already known as plant pathogens and named it *Contagium vivum fluidum* (infective live fluid). This pivotal discovery marked the beginning of the development of plant virology as a separate scientific discipline. The Royal Society of Edinburgh Scotland held a special Centenary Symposium in 1998 to celebrate this discovery of TMV and to review the impact it has had on the science of virology.

Tobacco and tobacco viruses, especially TMV, have long served as valuable biological subjects for teaching and the investigations of which enabled discoveries of revolutionary importance for the development of plant virology. Now, in modern plant virology, tobacco again pioneered by being the first plant into which certain virus genes have been transferred via genetic engineering. The virus gene that determines the code for viral capsid protein synthesis was introduced into tobacco cells with the aid of *Agrobacterium tumefaciens* Ti-plasmids. Such plants, with the introduced "foreign TMV coat protein genetic material" now reproducing that gene in them, are labeled transgenic plants. Proteins in transgenic plants, whose synthesis is coded by foreign, introduced virus genes, increase the resistance of these plants to the virus that was the donor of that gene.

Such transgenic tobacco plants partially or fully resist infection by AMV, CMV, PVX, and TMV (Šutić and Sinclair, 1991).

Tobacco is the natural host of nearly two dozen viruses. Some viruses are economically important for the production of tobacco, both as an industrial crop and as a crop used for smoking. Half of these viruses were first discovered in tobacco (e.g., TMV, TNV, TRtV, TRsV). Other viruses of significant importance in tobacco production were first described in other plants, their primary hosts (e.g., AMV, CMV, PVX, PVY, and TmSpoWV).

Characteristically, viruses described on tobacco differ in morphological traits, host range, mode of preservation, and transmission in nature. Based on this broad knowledge, tobacco viruses can be classified into the following characteristic groups:

1. Transmitted mechanically with TMV-infected plant sap
2. Transmitted mechanically and by aphid vectors nonpersistently (tobacco etch virus also transmitted by dodder, tobacco vein mottling virus)
3. Transmitted by aphid vectors persistently (tobacco necrotic dwarf virus)
4. Transmitted persistently by cicadellid leafhoppers (tobacco yellow dwarf virus)
5. Transmitted persistently by the whitefly *Bemisia tabaci* (tobacco leaf curl virus)
6. Transmitted mechanically, by thrips and soil nematodes, via the seed of some plants, and also both tobacco streak and tobacco rattle viruses transmitted by dodder and also tobacco ringspot virus transmitted by some insect species
7. Transmitted mechanically or by the soil fungus *Olpidium brassicae* (tobacco necrosis virus, tobacco stunt virus)

TOBACCO MOSAIC TOBAMOVIRUS (TMV) [TYPE]

Properties of Viral Particles

The TMV rod-shaped viral particles (300 × 18 nm) contain single-stranded RNA composed of 6300 nucleotides. The capsid consists of 2130 identical protein subunits, each 17.5 kDA (Van Regenmortel, 1981). The protein:RNA ratio is 95%:5% (Kado and Agrawal, 1972) and the viral particle sediments at 194S.

Stability in Sap

TMV, one of the most stable plant viruses, reaches a high concentration in cells of tobacco and other solanaceous plants. The TIP of TMV is above 90°C; the DEP is 1×10^{-6} (Zaitlin and Israel, 1975); and the LIV is several months to 5 years at 24°C with preservatives. The lengthy preservation of infectivity of TMV in fermented tobacco differs from other viruses; thus, smokers easily transmit it by contact with susceptible plants.

Serology

TMV exhibits strong immunogenic activity and its high titre antiserum allows various types of serological investigations, including detection and diagnosis. Some virus strains differ markedly based on serological analyses (tobacco U1 and U2 strains).

Strains

Many TMV strains have been described based on biological virulence and symptomological, serological, and chemical traits. Yellow instead of green mosaic spots characterize the first TMV strain described in 1926 (McKinney, 1929). Strains are classified as severe or mild, depending on virulence. In addition, yellow, golden yellow (aucuba), white, dark green, and light green strains have been based on symptoms on tobacco leaves. Isolates of TMV have been categorized based on cytological and histological changes in infected plants as the ringspot, variegated, necrotic, and enation mosaic strains.

Comparative investigations of isolates have been described according to their host-source such as tobacco mosaic virus - strain vulgare, tomato mosaic virus strains, plantago strains, legume strains, cucurbit strains (cucumber green mottle mosaic virus), orchid strains, etc. (Van Regenmortel, 1981).

Geographic Distribution and Economic Importance

TMV, among the most widely spread of all viruses, occurs wherever tobacco is grown. It is usually the most common virus in countries where tobacco, tomato, and pepper are produced in fields and greenhouses. TMV causes substantial reductions in tobacco yield and quality. Data on yield losses show that the level of damage depends on cv. susceptibility, virus strains, and time of infection.

On average, TMV reduces tobacco yields by 30 to 35% and the market value is reduced by 50%. Investigations in former Yugoslavia showed a reduction of yield in tobacco infected 20 days after transplanting, dependant on cv., by 31 to 62% concomitant with a market value decrease of 47 to 81% (Mickovski, 1965).

Quality reduction involves a decrease of soluble carbohydrates, starch, and polyphenol content. Such reductions are reflected in lower contents of nicotine, nitrogen, and protein nitrogen, and an increase in protein content. Higher protein content greatly influences tobacco quality, which results in poor smoking quality, including an acrid taste and throat irritant, plus an off flavor. The lower ratio of carbohydrate to protein content (Schmuck number) is the best indicator of how TMV infection reduces tobacco quality.

Host Range

TMV has a broad host range: approximately 550 species of flowering plants, 60 natural, and all others experimental (Schmelzer and Schmidt, 1977).

Natural hosts important for commercial production include several tobacco species (*Nicotiana* spp.), plus pepper (*Capsicum annuum*), tomato (*Lycopersicon lycopersicum*), potato (*Solanum tuberosum*). Interestingly, TMV somehow infects woody plants such as wild apple (*Malus sylvestris*), cherry (*Prunus avium*), plum (*P. domestica*), and grapevine (*Vitis vinifera*). It infects several flower species in nature (*Petunia hybrida*, *Gerbera* spp., and others). *Chenopodium murale*, *Melandrium album*, *Plantago lanceolata*, *Digitalis lanata*, and *Erigeron canadensis* are natural virus hosts among weeds. Some strains of TMV infect several species of fungi (e.g., *Erysiphe graminis*, *E. polygoni*, *Phyllactinia corylea*, *Sphaerotheca lanestris*, *Phragmidium* spp., *Puccinia iridis*, *P. pelargonii-zonalis*, *Uromyces phaseoli*, *U. fabae*, and others) (Yarwood and Hecht-Poinar, 1973). The occurrence of TMV in fungi probably represents an exception as one of its specific traits.

Holmes discovered local reactions to TMV in datura (*Datura stramonium*) and in *Nicotiana glutinosa*, which are used now as suitable assay plants and as an established method of quantification in virological investigations (Holmes, 1929; 1932).

Symptoms

TMV infections in primary tobacco leaves cause mild clearing of leaf veins. Clearly visible mottle and mosaic occur in younger leaves later in the season. Further symptom development depends on tobacco cv. and virus strain (Figures 119 and 120). For example, golden-yellow mosaic or light-green spots are seen on *Nicotiana sylvestris* cv. Otlja and tissue necrosis occurs within these spots in Jaka and Jebel cvs.

Various changes of shape occur in younger leaves; for example, wrinkling, crinkling, twisting of margins, and narrowing of leaflets. Leaflets and whole plants infected by TMV early in the season show marked reductions in growth, resulting ultimately in substantial yield reductions.

Pathogenesis

TMV is preserved in nature in herbaceous and woody plants. Vegetables (tomato, pepper) grown entirely indoors are particularly important because TMV is easily transferred from crop to crop, either by mechanical transmission or in the roots/soil organic matter from infected plants; that is, infectivity is maintained for about 2 years in soil not exposed to freezing and drying. TMV is well

FIGURE 119   Tobacco mosaic virus in Turkish tobacco.

FIGURE 120   Tobacco mosaic virus in tobacco.

preserved in plant residues in the soil and in composts of tomato plants. Thus, contaminated plant residues represent long-lasting sources of infection, particularly for vegetable production in glass houses (Šutić, 1980).

Fermented tobacco in which the virus remains infective for months to years represents an important source of infection. Smokers who contact TMV-infected/contaminated tobacco, particularly on fingers when rolling cigarettes, easily transmit it to susceptible plants while working the fields. This mechanical transmission of TMV is unique among plant viruses. Thus, the labor-intensive nature of tobacco culture (weeding, steckling uprooting, transplanting, inflorescence cutting, periodical harvests, etc.) creates the environment for the ready transmission of TMV from existing sources onto healthy plants in the field.

Insect vectors are not known with certainty, but possibly some animals (e.g., grasshoppers, butterfly caterpillars, snails) transmit TMV through mechanical contact with plants. Natural virus carriers include several dodder species, such as *Cuscuta campestris*, *C. subinclusa*, and *C. japonica*.

TMV is not transmitted through tobacco seed. It is transmitted superficially on the surface, plus internally in seeds both of tomato and pepper. TMV infectivity on the surface of tomato seeds, depending on its strain, may last up to several months; while when the virus is endogenous, it lasts several years (Broadbent, 1961; 1965). Thus, seed of these vegetables represents an important source of infection for the natural preservation and transmission of the virus. TMV is transmitted through the seed of natural hosts of stone fruit and the grapevine.

Control

Because of the marked stability of TMV, it is nearly impossible to prevent infection of plants in nature. Some preventive measures can reduce substantially the number of infected plants, and thus enable an important reduction of field losses. Recommended measures of prevention are classified in a number of categories (Šutić, 1983).

Plant only clean seed, free of plant residues, for seedling production.

Maintain seedlings healthy up to the time of transplanting: (1) use only noncontaminated soils for seedling production in flower beds; (2) disinfest all soils of unknown origin prior to sowing; (3) produce only tobacco seedlings in flower beds; (4) do not use the same flower beds for tomato or pepper seedlings; (5) disinfest all components of warm flower beds, frames, and other parts prior to sowing; (6) forbid smoking by workers tending the seedlings; (7) all workers should disinfest their hands at reguler intervals (washing with soap, detergent, or skim milk); and (8) remove all infected or suspected infected plants from the nurseries.

Take all measures necessary to prevent infection by TMV prior to transplanting the seedlings in the field, such as: (1) transplant into soils free of tobacco, tomato and pepper plant residues; (2) select fields far from acreages under solanaceous vegetable crops; and (3) prohibit smoking during work; (4) ask workers periodically to disinfest hands.

Practice cultural operations that substantially reduce plant infection. (1) Grow tobacco in a three-year crop rotation with maize and wheat, avoiding tomato and pepper crops. (2) Contact with plants during crop cultivation and soil tillage should be avoided. (3) Leaves from healthy plants should be harvested first, then those from TMV-infected plants later. (4) All known weed hosts of TMV should be destroyed in tobacco and in immediately contiguous fields. (5) Infection by mechanical inoculation can be markedly reduced by spraying plants with skim milk or buttermilk, but such an application on a large scale represents added expense; in view of the potential results, it may be envisaged only for first sprayings on limited acreages.

Plant resistant or tolerant tobacco cvs. available on the market.

TOBACCO ETCH POTYVIRUS (TEhV)

Properties of Viral Particles

The TEhV particles are flexuous filaments, 12 to 13 × 730 nm, with a sedimentation coefficient of 154S and contain single-stranded RNA of 3200 kDa, which accounts for 5% of the particle weight. The capsid protein consists of 30 to 32 kDa subunits (Purciful and Hiebert, 1982). The TIP is 55°C; the DEP is 1×10^{-4}; and the LIV in sap at 20°C is 5 to 10 days (McKinney et al., 1965). TEhV is strongly immunogenic (Hollings and Brunt, 1981). TEhV is related serologically to potato virus A, potato virus Y^O, potato virus Y^N, and henbane mosaic virus (Bartels, 1964). Both "severe etch" and "mild etch" strains exist based on virulence, and some virus variants are transmitted by aphid vectors, while others are not.

Geographic Distribution

TEhV, widespread in North (U.S. and Canada) and South America (Mexico, Puerto Rico, and Venezuela) (Purciful and Hiebert, 1982), is also encountered in southeastern Europe, but with no definitive record of its occurrence in western and eastern Europe (Schmelzer and Schmidt, 1977).

Host Range

The natural host range of TEhV is broad in addition to a large number of experimentally susceptible hosts. Natural hosts of TEhV include *Nicotiana tabacum*, *Capsicum annuum*, *Linaria canadensis*, *Lycopersicon lycopersicum*, *Physalis angulata*, *P. ciliosa*, *Solanum acuelatissimum*, and *S. nigrum* (Schmelzer and Schmidt, 1977), plus several weed species: *Datura stramonium* (the first known natural host of this virus), *Cassia obtusifolia*, *Cirsium vulgare*, *Chenopodium album*, *Linaria canadensis*, *Physalis* spp., and *Solanum* spp. (loc. cit. Purciful and Hiebert, 1982).

Diagnostic plants are *Nicotiana tabacum* with chlorotic spots or necrotic rings and arcs in inoculated leaves and vein clearing and necrotic etching in systemic infection; *Datura stramonium* with systemic infection of leaf mottle, distortion, and vein banding; *Cassia obtusifolia* with necrotic local lesions and systemic infection of chlorosis, necrosis, and stunting of infected leaves, and others.

Symptoms

Disease symptoms of TEhV infection differ, primarily depending on the plant cv., virus strain, and the growth stage of the plant at time of infection. Generally, symptoms are most evident at flowering. Initial symptoms of TEhV in tobacco leaves usually include vein clearing, later becoming chlorotic mosaic and necrotic etching (Figure 121). Necrosis accompanied by yellowing may occur along the veins of older leaves. Plants infected with virulent strains early in the season grow slowly.

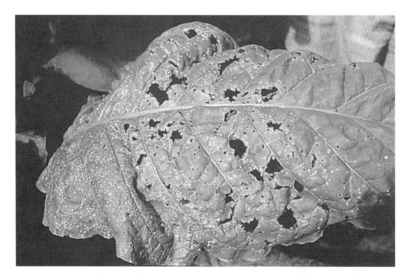

FIGURE 121 Etching on flue-cured tobacco caused by the tobacco etch virus. (Courtesy H.D. Shaw. Reproduced with permission from APS.)

Pathogenesis

TEhV retains infectivity in both natural cultivated hosts (tobacco, pepper, tomato) and susceptible weeds. Transmission of TEhV with infected plant sap plays an important role in the spread of the virus because several cultural field operations used in solanaceous crops allow contact of machinery with the plants. Important vectors transmitting the virus nonpersistently are several aphid species, including *Myzus persicae*, *Macrosiphum euphorbiae*, and *Aphis fabae*. Some TEhV strains not normally transmitted by aphids are transmitted with the aid of "helper" factors from potato virus

Y (Govier and Kassanis, 1974; Pirone, 1979). Although transmission of TEhV by dodder, *Cuscuta californica* and *C. lupuliformis*, can occur, it probably has little importance in natural spread. No seed transmission of TEhV has been reported.

Control

Spatial isolation of susceptible plants effectively reduces infection. Weed control is also recommended, as is chemical control of aphid vectors. Hands, clothing, implements, and tools used for cultivation should be disinfested at regular intervals during work to reduce the probability of mechanical inoculation.

TOBACCO VEIN MOTTLING POTYVIRUS (TVMgV)

Properties of Viral Particles

TVMgV particles are flexuous, 13×765 nm (Sun et al., 1974)., and consist of single-stranded RNA of 3200 kDa, which probably accounts for 5% of the viral mass, and the capsid protein of 29 kDa (Pirone and Shaw, 1988). The TIP of TVMgV is 60 to 70°C; the DEP is 1×10^{-2} to 1×10^{-3}; and the LIV at 22 to 25°C is 48 hours (Sun et al., 1974). TVMgV is a good immunogen. Virus isolates differ in symptomology based on differences in tobacco cvs.

Geographic Distribution, Host Range, and Symptoms

The virus discovered in southeastern U.S. where the Burley tobacco is grown has limited distribution. TVMgV, with a narrow host range among solanaceous plants, is economically important in Burley tobacco (Pirone et al., 1973; Sun et al., 1974). Characteristic symptoms on leaves include discontinuous spots beside the veins and necrosis in the veins and leaves (Figure 122). Most cvs. of tomato, also a natural host of TVMgV, are symptomless.

FIGURE 122 Severe tobacco vein mottling virus on susceptible Burley cultivars (left); resistant (right). (Courtesy P.B. Shoemaker. Reproduced with permission from APS.)

Pathogenesis

TVMgV, which retains infectivity in tobacco and tomato, is transmitted easily with infective plant sap, the principle mode of spread in the field. In nature, TVMgV can also be transmitted naturally and nonpersistently by different aphids, including *Myzus persicae*, *Macrosiphum euphorbiae*, *Aphis craccivora*, *A. gosypii*, *Rhopalosiphum maidis*, and *R. padi* (Pirone and Shaw, 1988). Preventive measures employed successfully for other tobacco viruses are recommended for protection against TVMgV.

TOBACCO NECROTIC DWARF LUTEOVIRUS (TNcDV)

Properties of Viral Particles

TNcDV particles are isometric, 25 nm in diameter, and are composed of single-stranded RNA of 2000 kDa and a capsid consisting of 25.7-kDa coat protein subunits (Kubo, 1981). The TIP of TNcDV is 80°C; the LIV at 4°C is over 6 months. TNcDV has strong immunogenic activity with antisera titre of 1/4096 (Kubo, 1981).

Geographic Distribution, Host Range, and Symptoms

TNcDV exists only in Japan in tobacco, spinach and *Capsella bursa-pastoris* as natural hosts, but experimentally can infect 20 species in 5 families (Kubo, 1981). TNcDV causes stunt, early yellowing, and death of lower and middle leaves in infected plants. Phloem-limited, it causes phloem necrosis, depending on cv. susceptibility and mild vein yellowing in infected spinach.

Pathogenesis

Natural hosts represent permanent sources of infection in nature; that is, it overwinters in winter spinach, thus providing infective virus for the new spring crop. *Myzus persicae*, the main vector, transmits TNcDV nonpersistently. Other modes of transmission are unknown.

TOBACCO YELLOW DWARF GEMINIVIRUS (TYDV)

Properties of Viral Particles

TYDV particles are geminate, 20 × 35 nm, with a sedimentation coefficient of 76S (Goodman, 1981) and the nucleic acid, although still unknown, is probably single-stranded DNA. The coat protein subunit is 27.5 kDa. The TIP of TYDV, determined by leafhopper feeding on infective tobacco preparations, is 50 to 60°C; the DEP is 1/50 to 1/500; and the LIV at 4°C is 26 days. The virus is moderately immunogenic (Thomas and Bowyer, 1984). TYDV is considered serologically related to the bean summer death virus strain (Goodman, 1981). Some isolates differ, based on symptom intensity on *Datura stramonium* and *Nicotiana tabacum* leaves.

Geographic Distribution, Host Range, and Symptoms

TYDV is widespread only in Australia. *Nicotiana tabacum* and *P. vulgaris* are economically important natural hosts of TYDV, which has been transmitted experimentally into 30 species in 7 dicotyledonous families (loc. cit. Thomas and Bowyer, 1984). The virus develops in phloem cells, causing severe tobacco stunt and necrosis in susceptible bean cvs.

Pathogenesis

Leafhoppers, *Orosius argentatus*, transmit TYDV persistently in nature. Other modes of transmission are unknown.

TOBACCO LEAF CURL GEMINIVIRUS (TLCuV)

Properties of Viral Particles

TLCuV features geminate particles, 15 to 20 × 25 to 30 nm, which probably have circular single-stranded DNA molecules. No published information exists about the coat protein (Osaki and Inouye, 1981). TLCuV is moderately immunogenic. Individual virus isolates differ, based on host susceptibility and symptom intensity on infected tobacco leaves.

Geographic Distribution, Host Range, and Symptoms

TLCuV is widespread both tropically, subtropically, and in certain temperate areas in Japan, Europe, and the U.S. (Osaki and Inouye, 1981). Natural hosts of TLCuV belong to five families, and include *Ageratum conyzoides*, *Carica papaya*, *Eupatorium odoratum*, *Euphorbia hirta*, *Lonicera japonica*, *Sida rhombifolia*, *Solanum nigrum*, *Vernonia citrerea*, and *Withania som-*

nifera. TLCuV has been transmitted to the Caprifoliaceae, Compositae and Solanaceae by vectors (loc. cit. Osaki and Inouye, 1981).

TLCuV-infected tobacco plants are stunted, with both twisted stems and small, curled/rolled leaves. Hyperplastic swellings or enations occur along the veins on lower leaf surfaces. Symptoms in infected plants differ in appearance and severity, depending on plant susceptibility and virus strains (Figure 123). The virus, which is phloem restricted, causes the characteristic symptoms mentioned plus yellowing of infected plant parts.

FIGURE 123 Stunting and leaf distortion of tobacco leaves caused by early infection by tobacco leaf curl virus. (Courtesy G.B. Lucas. Reproduced with permission from APS.)

Pathogenesis

TLCuV in Japan is preserved in *Lonicera japonica*, which represents a potential perennial virus reservoir. Whitefly *Bemisia tabaci*, the primary vector, transmits TLCuV persistently (Bird and Maramorosch, 1978). The virus is transmitted neither mechanically with infective sap nor through seed from infected plants.

TOBACCO RATTLE TOBRAVIRUS (TRtV)

Properties of Viral Particles

TRtV consists of two sizes of interdependent particles—long (180 to 200 × 22 nm) and short (50 to 110 × 22 nm)—with sedimentation coefficients of 300S and 155 to 243S, respectively (Harrison, 1970). Single-stranded RNA accounts for 5% of the viral particle; RNA_1 (long particles) has a molecular weight of 2400 kDa and RNA_2 (short particles) 600 to 1400kDa (Harrison, 1970; Harrison and Robinson, 1978; 1981). The sedimentation coefficients of RNA_1 and RNA_2 are 26S and 18S, respectively (Semancik and Kajiyama, 1967). The protein capsid consists of coat protein subunits of 24kDa. RNA_1 extracted from long particles is infective because it contains genetic information for replication, but not for production of the coat protein. RNA_2 contains the genetic code for the synthesis of both the protein and the long and short particles, which enables complete synthesis of both when simultaneously present in the inoculum.

Stability in Sap

The TIP of TRtV in *Nicotiana clevelandii* sap is 80 to 85°C; the DEP is 1×10^{-5} to 1×10^{-6}; and the LIV at 20°C is 6 weeks (Harrison, 1973).

Serology

TRtV is weakly immunogenic, which results in antisera of low titre. Antigenic properties of isolates of TRtV place them in three serotypes (I, II, III). Serotype III, which comprises Brazilian isolates, is serologically distant from isolates in serotypes I and II (Harrison and Robinson, 1981).

Strains

TRtV strains are based both on viral properties and symptoms in some hosts. Strain M contains both long and short particles with pronounced antigenic variability, whereas strain NM contains only RNA_1 unprotected by coat protein. Other strains based on origin and hosts include the California strain (isolated from *Capsicum frutescens* in California), Oregon strain (isolated from potato in Oregon), and PRN strain (isolated from potato plants in Scotland) (Harrison, 1970).

Geographic Distribution and Economic Importance

TRtV, probably present wherever tobacco is grown, occurs in North and South America, Europe, and Japan. TRtV causes severe damage in tobacco, including leaflet distortion, often resulting in perforations of leaves and stem and pit necrosis. Infected plants are small and greatly reduced in yield and quality.

Host Range

The many hosts of TRtV—836 species, of which 130 represent natural and experimental hosts—rank it as a major virus group (Schmelzer and Schmidt, 1977). In addition to tobacco, other economic crops susceptible to TRtV include *Beta vulgaris*, *Brassica oleracea*, *Helianthus annus*, *Solanum tuberosum*, *Medicago* sp., *Secale cereale*, *Daucus carota*, *Lactuca sativa*, and *Cucurbita pepo*. The same authors list natural hosts of TRtV to include many ornamental species, such as *Calendula officinalis*, *Petunia hybrida*, *Tropaeolum majus*, *Viola tricolor*, *Pelargonium zonale*, *Salvia* sp., *Gladiolus* sp., *Hyacinthus* sp., and *Narcisus* sp. Weed hosts include *Capsella bursa-pastoris*, *Convolvulus arvensis*, *Polygonum aviculare*, *Rumex acetosella*, *Stellaria media*, *Senecio vulgaris*, *Sonchus arvensis*, *Cirsium arvense*, *Chenopodium* spp., *Lamium purpureum*, and *Agropyron repens*.

Chenopodium amaranticolor and *Phaseolus vulgaris* with local spots and without systemic infection are suitable assay plants. *Nicotiana tabacum* and *Cucumis sativus*, which support systemic infection, are suitable for transmission of the virus with the aid of vectors.

Symptoms

Primary symptoms of TRtV infection on leaves include large grey spots that later turn into dark necrotic zones. Infected plants become distorted and characteristically necrotic tissues fall, resulting in visible perforations on the leaves. Systemic infection results in necrotic spots along leaf veins and petioles, and the leaves wilt and gradually die. Blackish striped spots with necrotic tissues occur along the stem in early infections which then spread toward the center of the stem and plants die.

TRtV causes mosaic and mottle on stems, corky ringspots on potato tubers, and yellow spots on sugarbeet leaves. Similar ringspots also occur in lettuce, hyacinth, narcissus, and tulip. Due to the varying symptoms on many different plants, TRtV has been described often under different names.

Pathogenesis

TRtV retains infectivity in numerous hosts in nature. Biennial and perennial hosts, which provide reservoirs as sources of infection, hold and transmit TRtV season to season.

Vectors of TRtV are plant parasitic nematodes belonging to the *Trichodorus* and *Paratrichodorus* genera. Eleven *Trichodorus* species are reportedly active vectors and *T. pachydermis* and *T. primitivus* are the most important ones in Europe. Both larvae and adults actively transmit TRtV. For example, *T. allius* can acquire TRtV in 1 hour, requires 1 hour to inoculate the plant, and it remains infective in the soil for 20 weeks. Information is available on the specific relationship between individual nematode species and virus strains.

TRtV is not transmitted through tobacco seed. TRtV has been transmitted through seed in 5 of 15 infected weeds (Cooper and Harrison, 1973). Seed transmissibility of TRtV has been reported at 1 to 6% incidence in the weeds *Capsella bursa-pastoris* and *Myosotis arvensis* (loc. cit. Harrison, 1970). Dodder (*Cuscuta campestris*, *C. europaea*, *C. subinclusa*, etc.) can transmit TRtV, which contributes to its spread and preservation in many host species. TRtV is transmitted mechanically with infective plant sap, which allows its spread in tobacco and other crops during cultivation.

Control

General recommended measures of prevention already mentioned for TMV are equally useful to control TRtV. Also recommended are (1) careful and timely control of weeds as an important source of infection; (2) soil treatment with nematicides, when possible, to destroy nematode vectors; and (3) introduction of resistant plants into the crop rotation, which contributes to soil sanitation.

Tobacco Ringspot Nepovirus (TRsV)

Properties of Viral Particles

TRsV is isometric, 28 nm in diameter, and consists of three components, top (T), middle (M), and bottom (B), based on their sedimentation rates in density gradient columns. Their sedimentation coefficients are 53S (T), 91S (M), and 126S (B) (Murant, 1981). The T component consists of empty protein shells, whereas the M and B components contain different amounts of single-stranded RNA (Stace-Smith, 1985). The B component particles are of two types: RNA-1 or RNA-2 molecules of 2730 kDa and 1340 kDa, respectively (Harrison et al., 1972). Inoculations with a mixture of RNA-1 and RNA-2 are more infective than each alone. Some isolates of TRsV contain satellites of RNA that produce additional pairs of particles. The protein subunits of the type strain are 57 kDa (loc. cit. Murant, 1981).

Stability in Sap

The TIP of TRsV in infective sap of petunia, tobacco, and French bean is 60 to 65°C; the DEP is 1×10^{-3} to 1×10^{-4}; and the LIV at 24°C is 1 to 2 weeks (Stace-Smith, 1985).

Serology

TRsV has good immunogenicity and some virus strains differ serologically (Stace-Smith, 1985).

Strains

Many TRsV isolates differ based on symptoms. For example, strains of tobacco green ringspot and tobacco yellow ringspot occur in tobacco, anemone necrosis in anemone plants, etc. Some natural variants differ based on their antigenic traits.

Geographic Distribution

TRsV has been discovered in many different host species in Africa (Nigeria), Asia (Japan, India, Iran), Australia, North America, Canada, and Europe (Great Britain, Germany, The Netherlands, former Yugoslavia) (loc. cit. Stace-Smith, 1985). The virus in some countries possibly originated from imported infected seed and/or planting materials.

Economic Importance

TRsV can cause serious damage in various plant production systems, depending on infection potential and plant susceptibility. Tobacco plants infected by TRsV early in the season are severely stunted, resulting in loss of yield and quality. In North America, it can cause 100% loss of a soybean crop plus serious losses in peach (yellow bud mosaic and stem pitting), cherry (rasp leaf), and other fruit-tree crops (Murant, 1981).

Host Range

The infectivity of TRsV occurs widely both naturally and experimentally in at least 38 genera and 17 families (Price, 1940). Some 90 natural host species are reported (Schmelzer and Schmidt, 1977). TRsV occurs in many cultivated and ornamental plants. Naturally susceptible species include *Nicotiana tabacum, Helianthus annuus, Humulus lupulus, Mentha spicata, Solanum melongena, S. tuberosum, Lactuca sativa, Phaseolus vulgaris, Daucus carota, Cucumis sativus, Cucurbita pepo, Citrullus vulgaris, Glycine max., Spinacia oleracea, Trifolium pratense,* and *T. repens*. Naturally susceptible ornamental species include *Anemone* sp., *Gladiolus* sp., *Iris* sp., *Petunia violacea, Tulipa* sp., *Pelargonium* sp., and *Zinnia elegans*. Weeds that might represent natural hosts include *Amaranthus retroflexus, Chenopodium album, C.* sp., *Erigeron canadensis, Portulaca oleracea, Rumex acetosella,* and *Taraxacum officinale*. Woody plants that are natural hosts of TRsV include almond, blackberry, blueberry, cherry, dogwood, elderberry, grapevine, nectarine, *Fraxinus americana,* peach, and rose. *Chenopodium amaranticolor* and *C. quinoa*, which react with local necrotic spots to infection by TRsV, are good assay plants, and *C. sativus* is suitable for demonstrating nematode transmission of TRsV (loc. cit. Stace-Smith, 1985).

Symptoms

Initial symptoms of infection by TRsV on tobacco leaves are individual necrotic ringspots, which gradually become larger, merge, and form wavy necrotic zones, particularly along the midrib and other larger leaf veins (Figure 124). Isolated ringspots characterize interveinal tissues. Infected plants grow slowly and are of poor quality. Symptoms may disappear with time. Plants recover, that is, new growth is symptomless, and become resistant to repeated infections.

FIGURE 124 Tobacco oakleaf caused by tobacco ringspot virus. (Courtesy G.V. Gooding. Reproduced with permission from APS.)

Different types of symptoms caused by TRsV include those observed for American spearmint, blackberry, cherry, apple, grapevine, pelargonium, ash, dogwood, anemone, gladiolus, and elderberry (loc. cit. Stace-Smith, 1985). TRsV may cause union necrosis and apple decline at root/scion junctures (Lana et al., 1983).

Pathogenesis

TRsV retains infectivity in many herbaceous and woody hosts, and perennial ones are particularly important in the epidemiological cycle.

The *Xiphinema americanum* nematode serves as a vector by feeding first on TRsV-infected roots, after which they become infective within 24 hours. Other vectors include nymphs of *Thrips tabaci* that transmit TRsV to soybean, the spider mite *Tetranychus*, the grasshopper *Melanoplus*, the tobacco flea-beetle *Epitrix hirtipennis*, and possibly aphids (loc. cit. Stace-Smith, 1985).

TRsV is effectively seed transmitted in soybean, petunia, *N. glutinosa*, *Gomphrena globosa*, and *Taraxacum officinale*, and inefficiently in tobacco, cucumber, and lettuce. The seed transmission rate of TRsV in 13 species varies broadly from 3% in *Cucurbita melo* to 100% in *Glycine max*. (loc. cit. Stace-Smith, 1970; 1985). TRsV is transmitted with infective plant sap. Its worldwide distribution may be due to the exchange of infected ornamental plants that are frequent virus hosts.

Control

Control is most difficult due to the large host range and the wide diversity of vectors. In addition to those recommended for the control of other tobacco viruses, special measures should be taken to eliminate sources of TRsV, especially infected perennial plants. Concomitantly, only nematode-free soils should be used for tobacco production; smaller acreages should be disinfested with nematicides.

In high-risk situations, suitable pesticides should be used to control above-ground vectors. Rogueing or killing susceptible weeds is useful. Only healthy seed and virus-free seedlings should be used to avoid early infections that cause the most damage.

TOBACCO NECROSIS VIRUS (TNV) [TYPE]

Properties of Viral Particles

TNV is isometric, 28 nm in diameter, has sedimentation coefficients of 142 and 133S, and the single-stranded RNA which accounts for 18 to 21% of the weight of the viral particle is 1300 to 1600 kDa with a sedimentation coefficient of 24 to 27S. The molecular weight of the TNV particle is 6300 to 7600 kDa and the protein coat subunit 30 to 33 kDa (Uyemoto, 1981). The TIP of TNV is 85 to 95°C; the DEP is 1×10^{-6}; and the LIV at 20°C is several days. Viral particles are good immunogens.

TNV is usually accompanied by a satellite virus (SV) that is isometric, 17 nm in diameter, and has a sedimentation coefficient of 50S. SV is composed of an RNA of some 400 kDa and coat protein subunits of 22 kDa (Uyemoto, 1988). The satellite TNV, an active immunogen, is serologically related to TNV.

TNV replicates independently in cells of host plants. However, the SV particles replicate only in the presence of TNV; that is, this satellite virus RNA is dependant on the TNV ribonucleic acid. SV is capable of synthesizing only its coat protein.

Strains

Several TNV and SV strains differentiate based on disease symptoms, host types, and their ability to activate SV multiplication (Kassanis and Phillips, 1970). Two serotype groups of TNV include serotype A containing A, B, C, and S strains, and serotype D containing D, E, AC 43, AC 36 pear, grapevine, and citrus strains (Babos and Kassanis, 1963). Some strains in the D serotype, for example, activate only the replication of SV_1 and SV_2 strains, whereas other serotypes activate the SVe strain. All strains in the A serotype activate SV_1 and SV_2, but not SVe (Kassanis and Phillips, 1970).

Geographic Distribution and Economic Importance

TNV generally appears wherever tobacco is cultivated, owing to the large number of susceptible hosts. TNV frequently occurs in nonsterilized soils in glasshouses and in irrigated fields suitable for the development of the fungal vector. TNV causes early death of infected leaves in susceptible plants, most frequently in plant production under irrigation and indoors. In addition to tobacco, serious damage occurs in ornamental and vegetable plants.

Host Range

TNV, with a comparatively broad host range, has been transmitted experimentally into 88 species in 37 dicot and monocot families (Price, 1940). Among 25 species of natural hosts, several are important cultivated herbaceous (*Nicotiana tabacum, Cucumis sativus, C. melo, Fragaria vesca, Lactuca sativa, Phaseolus vulgaris, Pisum sativum,* and *Solanum tuberosum*) and woody (*Citrus* sp., *Malus sylvestris, Prunus domestica,* and *Pyrus domestica*) plants. *Chenopodium amaranticolor* is a suitable assay plant, which reacts to infection with local spots, and *Lactuca sativa* and *N. tabacum* are suitable for the transmission of the virus by soil vectors.

Symptoms

In nature, TNV usually affects first the roots of susceptible plants, many of which are tolerant without obvious disease symptoms. Then, TNV moves systemically in infected tobacco plants, causing symptoms initially on the lowest leaves, including necrotic spots on the leaf midrib followed by mosaic and mottling plus grey oak-leaf mosaic along larger leaf veins. Young plants die quickly if infected with TNV early in the season. Leaves of several plants directly inoculated display symptoms of local necrotic reactions. TNV causes dotted streaks on bean leaves (bean stripe disease), tulip necrosis, and one strain causes cucumber necrosis in Europe.

Pathogenesis

TNV retains infectivity in roots of infected plants, some perennial. Zoospores of *Olpidium brassicae* transmits TNV (Teakle, 1962). The fungus also represents a serious pathogen on many plants. The zoospores infect roots immediately following root colonization and also quickly acquire TNV. TNV is not preserved in the resting spores of the fungus. The extent of TNV transmission depends on the strains, the fungal vector activity, and the susceptibility of the host (Kassanis and MacFarlane, 1965). Soil water movement contributes to zoospore and viral spread; for example, TNV has been isolated from river water (Tomlinson et al., 1983).

TNV is also transmitted mechanically, but other modes of transmission are unknown.

Control

Crop rotation, which also includes plants resistant to TNV, is recommended. Irrigation of fields in which *Olpidium brassicae* and TNV are known to occur creates high risk for virus spread, and thus, one should grow resistant plants. Soil sanitation prior to sowing and adequate soil aeration and moisture regulation are necessary to create unfavorable conditions for the development of the fungal vector. Soil sterilization of indoor production areas is beneficial.

TOBACCO STUNT VIRUS (TSnV)

Properties of Viral Particles

The rod-shaped TSnV particles are 300 to 340 × 18 nm, and consist of a double-stranded RNA and a single species 48-kDa protein (Kuwata and Kubo, 1986). The TIP of TSnV in tobacco sap is 40 to 50°C and the DEP is 1×10^{-3} to 1×10^{-4}. TSnV has moderate immunogenicity.

Geographic Distribution, Host Range, and Symptoms

TSnV is reported only in Japan. Tobacco is its natural host and it has been transmitted experimentally by *Olpidium brassicae* into 35 species in 15 families, and mechanically into 41 species in 9 dicot families (loc. cit. Kiwata and Kubo, 1986). Primary symptoms on tobacco include vein clearing, followed by vein necrosis, ring-like patterns, and mottle, plus stunt later in the season. The type of symptoms and their intensity depend on cv. susceptibility.

Pathogenesis and Control

TSnV is preserved 1 to 10 years (Kuwata and Kubo, 1986) in resting spores of *Olpidium brassicae*, which represents permanent natural soil sources of infection. It is transmitted into healthy plants by the zoospores that contain the TSnV particles. In view of this cycle, crop rotations including nonhost species, noncontaminated soils, and the creation of conditions unsuitable for the soil-inhabiting *O. brassicae* vector all represent important measures to protect against TSnV.

TOBACCO STREAK ILARVIRUS (TSkV)

Properties of the Virus Group

The isometric TSkV consists of three sizes—27 (T), 30 (M), and 35 (B) nm in diameter (Lister et al., 1972)—each with sedimentation coefficients of 90, 98 and 113S, respectively (Lister and Bancroft, 1970), and molecular weights of 4720, 5920, and 7450 kDa, respectively. They contain 142 (T), 179 (M), and 225 (B) protein subunits, respectively (Ghabrial and Lister, 1974), and a single-stranded RNA of 850, 1100 and 1350 kDa, respectively, each in an individual particle. Additionally, a subgenomic RNA of 400 kDa has been discovered (Jones and Mayo, 1975).

The TIP is 53 to 64°C; the DEP 1/30 to 1/15625; and LIV in diluted tobacco sap is 36 hours. TSkV is moderately immunogenic (Fulton, 1985). Several virus strains of TSkV differ both serologically and symptomatically.

All four RNA types are required for viral infection. A mixture of the three largest RNAs is not infective. The addition of the smallest 400-kDa RNA, or a small quantity of coat protein belonging to TSkV or to AMV, stimulates infectivity (Fulton, 1981).

Geographic Distribution

TSkV is widespread in Europe, North and South America, Japan, Australia, and New Zealand. Usually found incidentally in individual hosts, it seldom causes epidemics (Fulton, 1985).

Host Range and Symptoms

TSkV has a broad host range in both natural and experimental hosts, including monocots and dicots and herbaceous and woody plants (Fulton, 1971; 1981). Necrotic ring symptoms occur in inoculated leaves of *Nicotiana tabacum* var. *Xanthi-nc* and other plants (see Figure 125). Necrotic lines form rapidly along small veins of newly emerging leaves, which may spread further and cause plant death. Leaves that develop later are symptomless. Variable symptoms occur on dahlia (mottle or symptomless), cotton, red clover, *Melilotus alba* (mottle), tomato (yellow ringspots, malformation), asparagus (stunt), rose (vein yellowing), bean (red node), pea, soybean (mimicked bud blight caused by TmRsV; first report in U.S. Fagbenle and Ford, 1970), potato (systemic necrosis), and strawberry (necrotic streak). TSkV has been isolated from grapevine, papaya, black raspberry, red raspberry, blackberry, groundnut, pepper, globe artichoke, alfalfa, sunflower, and numerous ornamentals (loc. cit. Fulton, 1985).

FIGURE 125 Tobacco streak virus in Samsun NN tobacco; healthy (right).

Pathogenesis

TSkV permanently infects perennial hosts. *Thrips tabaci* or *Frankliniella* sp. are reported vectors (Costa and Lima Neto, 1976; Kaiser et al., 1982). Transmission of TSkV in seed occurs in *Datura stramonium, Chenopodium quinoa, Melilotus alba, Glycine max, Gomphrena globosa, N. clevelandii, Vigna unquiculata*, black raspberry, and *Nicandra physaloides*. Seed transmission may range from 90% (*G. max.*) to less than 1% (*V. unguiculata*) (loc. cit. Fulton, 1985). TSkV is transmitted easily with infective plant sap and also via *Cuscuta campestris* (Fulton, 1948; loc cit. Fulton, 1985).

Measures important to protect against TSkV include use of seed from healthy plants and virus-free seedlings for woody plant production, plus control of vectors in fields where infection is detected.

OTHER VIRUS DISEASES OF TOBACCO

Tobacco vein distortion luteovirus (Smith, 1946) and tobacco yellow net luteovirus (Abeygunawardena et al., 1967) are frequently cited as causing tobacco diseases. Both have *Myzus persicae* as their vector. Adams and Hull (1972) have described the tobacco yellow vein disease caused by a mixed infection with tobacco yellow vein assisted virus and tobacco yellow vein assistor virus, transmitted by *M. persicae* and also a possible member of the luteovirus group.

TOBACCO: NATURAL ALTERNATIVE HOST FOR SOME VIRUSES

Alfalfa Mosaic Virus (AMV)

AMV is a well-known, widespread virus with a broad host range. In addition to legumes, important cultivated solanaceous plants include pepper, tomato, potato, eggplant, and others. Tobacco represents a frequent natural host of AMV in many countries.

Many strains of AMV with various symptoms are observed in tobacco plants (Figure 126). Intensively yellow or whitish spots are quite characteristic; other symptoms occur, such as mosaic, mottle, occasional dark-green bands along leaf veins, and finally, leaf necrosis and/or abaxial enations along veins.

FIGURE 126 Alfalfa mosaic virus in tobacco. (Courtesy G.V. Gooding. Reproduced with permission from APS.)

AMV is transmitted mechanically with plant sap and naturally by aphids. Control of aphids and weeds is essential to prevent infection. Tobacco should not be grown near either alfalfa fields or other natural hosts such as pepper, tomato, and eggplant.

Beet Curly Top Geminivirus (BtCuTpV)

The primary host of BtCuTpV is sugar beet, and other natural hosts include tobacco plus several herbaceous, ornamental, and woody plants. Vectors of BtCuTpV, which is transmitted nonpersistently, are some leafhopper species. Since tobacco is grown worldwide, it represents an important source of natural BtCuTpV infection.

Cucumber Mosaic Cucumovirus (CMV)

CMV ranks second in the world immediately following TMV as both widespread wherever tobacco is grown and one of the most economically damaging tobacco viruses. Symptoms vary, depending on virus strains and tobacco cvs., and the most common symptom is mild to severe mosaic. Spots of different kinds such as chlorotic blotches, chlorotic etched surfaces, oak-leaf patterns of mosaic etc. occur on leaves.

CMV frequently occurs in mixed infections with other viruses where symptoms become more pronounced and the resultant damages increase substantially. General preventive measures recommended for the control of TMV apply for control of CMV. In addition, the control of aphids as natural carriers is especially recommended.

Potato Potexvirus X (PVX)

PVX occurs regularly in tobacco and is usually widespread in areas where potato and tobacco are grown in close proximity. It is widespread in potato and other solanaceous crops. PVX is transmitted easily mechanically into tobacco with sap from infected plants by contact with plants during cultivation and tillage. Because PVX is relatively stable, it maintains infectivity in the soil for months or years. Symptoms in infected plants depend on viral strains and host plant cultivars. Chlorotic and occasional necrotic spots and rings that differ in size appear most frequently over the entire surface of infected leaves. PVX is observed mostly in mixed infections and is most damaging in combination with potato virus Y, causing dessication and death of entire leaves. General preventive measures for the control of other tobacco viruses are useful here. Special attention should be made to spatially isolate potato and tobacco crops. Replanting fields on which tobacco was grown the previous year should be avoided.

Potato Potyvirus Y (PVY)

PVY causing veinal and leaf necrosis and a streak mosaic ranks among the widest spread tobacco viruses, present in most countries where tobacco is grown. PVY is transmitted readily with sap from infected plants and the major natural vector is aphids. Both common and necrotic strains have been described in tobacco (Figure 127). The common strain causes symptoms of vein clearing and mild mosaic, whereas the necrotic strain causes leaf vein necrosis. PVY is most severe in mixed infections, especially with CMV and with PVX. Spatial isolation of tobacco and potato fields is recommended to prevent infection. Control of aphids enables the reduction of disease severity. Resistant or tolerant cultivars should be planted.

Tomato Ringspot Nepovirus (TmRsV)

TmRsV occurs in tobacco and many other herbaceous and woody plants. In many respects, it resembles TRsV, although they are different viruses (Stace-Smith, 1970). TmRsV is transmitted occasionally through tobacco seed (Kahn, 1956). Fulton (1967) showed that both TRsV and TmRsV can be transmitted by the same single nematode, *Xiphinema index*.

Tomato Spotted Wilt Tospovirus (TmSpoWV)

TmSpoWV is observed worldwide in tobacco. It has caused occasional epidemics in several countries, infecting up to 90% of plants in a crop and seriously endangering tobacco production

FIGURE 127 Initial symptom of potato virus Y^N on tobacco leaf. (Reproduced with permission from APS.)

(Mickovski, 1969). Periodic epidemics are caused by abnormally large increases in populations of thrips, the natural vector of TmSpoWV.

Necrotic mottle resembling fern leaves and concentric rings around spots of infection occur on young tobacco leaves of the Otlja and Jaka cvs. Apical necrosis in the field prevents apical bud development, severely stunting plants (Figure 128). Necrosis occurs also on the flowers and seed balls. Typical cell growth stops, except around the midrib where leaf tissue continues to grow and causes leaf blade irregularity that results in a wavy leaf surface and loss of leaf blade flexibility. Such leaves turn black following drying and lose their processing value (see Figure 129).

TmSpoWV can be transmitted with infective plant sap, but its natural vectors, thrips, should be controlled. Since many solanaceous plants are susceptible to TmSpoWV, tobacco fields should be spatially isolated from them in all stages of growth. General prevention measures as for TMV control are also useful.

VIRUS DISEASES OF SUGAR BEET (*Beta vulgaris*)

One century ago, the appearance of virus-diseased sugar beet plants that were heavily damaged attracted the attention of producers and agricultural experts. Thus, the first data on the yellows disease were recorded in France, although then believed caused by phytopathogenic bacteria (Gaillot, 1895; Prillieux and Delacroix, 1898). One now presumes that that sugar beet mosaic was then considered a type of the virus yellows disease as identified currently.

As the science of virology developed, sugar beet viruses became an increasingly important subject of study; thus, in the last half century, sugar beet mosaic and sugar beet yellows viruses rank among the most intensely investigated of plant viruses.

Sugar beet viruses clearly divide into two main groups. The first group includes the earliest described viruses known to cause economic damage of sugar beet: beet curly top virus (Ball, 1909) known in northwestern North America, from Canada to Mexico, and in the eastern Mediterranean basin; beet mosaic virus (Lind, 1915) and the beet yellows virus (Roland, 1936), which are severely damaging and widespread wherever sugar beet is grown.

The second group of viruses was described on sugar beet after 1960: beet mild yellowing virus (Russell, 1962), beet western yellows virus (Duffus, 1961), and beet necrotic yellow vein virus (Tamada and Baba, 1973), among others.

Viral Diseases of Industrial Plants

FIGURE 128 Typical symptom on bud of plant infected with tomato spotted wilt virus. (Courtesy G.V. Gooding. Reproduced with permission from APS.)

FIGURE 129 Tomato spotted wilt virus infected tobacco

Beet Curly Top (possible) Geminivirus (BtCuTpV)

Properties of Viral Particles

BtCuTpV is isometric, 20 nm in diameter, and occurs singly or geminate. The sedimentation coefficient initially reported for single particles was 82S. A later report, which include both single and geminate particles, showed them to have sedimentation coefficients of 55S and 86S, respectively (loc. cit. Thomas and Mink, 1979). The virus likely contains single-stranded DNA with no information on its protein. Both, single- and double-component viral particles are infective.

The TIP of BtCuTpV in sugar beet root sap is 80°C; the DEP 1×10^{-3}; and the LIV at 24°C is 8 days (Thomas and Mink, 1979). BtCuTpV is serologically strongly immunogenic and, since phloem restricted, it occurs in low concentrations in the plant. Several genetically stable strains exist, based on virulence, symptom reactions, and host range.

FIGURE 130 Beet curly top. (Courtesy J.E. Duffus. Reproduced with permission from APS.)

Geographic Distribution and Economic Importance

BtCuTpV is widespread in arid and semiarid western parts of North and South America, Iran, India (Polak, 1988), Turkey, and in the Mediterranean countries from which it seemingly originated (Matthews, 1981). In western U.S., BtCuTpV greatly reduces sugar yield, and thus remains economically important in sugar beet production (Magyarosy and Duffus, 1977).

Host Range and Symptoms

BtCuTpV has a broad host range. In addition to sugar beet, natural hosts include spinach, tomato, potato, cucumber, tobacco, bean, flax, clover, parsley, and both ornamental and woody plants (Polak, 1988). The virus has been experimentally transmitted via leafhoppers into over 300 species in 44 dicot families (Bennett, 1971), which include Chenopodiaceae, Compositae, Cruciferae, Leguminosae, and Solonaceae.

Typical symptoms of BtCuTpV infection in young sugar beet leaves appear as vein clearing; during growth, leaves twist upward and inward with general dwarfing and yellowing. Wart-like formations develop on the veins on the lower leaf surface. Exudates emanate from the phloem on larger leaf veins and petioles. BtCuTpV is limited only to the phloem, tissues and necrosis of phloem, as dark concentric circles, is evident in cross sections of roots.

Diagnostic plants for BtCuTpV include sugar beet (vein clearing in young leaves followed by leaf rolling), cucumber (seedlings killed and new growth on older plants dwarfed), French bean cv. Bountiful (epinastic bend at the base of the first trifoliolate leaf), among others (Thomas and Mink, 1979).

Pathogenesis and Control

BtCuTpV is maintained naturally in perennial hosts. Leafhopper vectors that transmit BtCuTpV nonpersistently are *Circulifer tenellus* (U.S.), *C. opacipennis* (Iran), and *Agallia albidulla* (India). BtCuTpV, not shown to replicate in these vectors, simply passes through as circulative (Bennett,

1971). Cicadellid leafhoppers are most important for spread of BtCuTpV, so its natural and local distribution depends primarily on the presence of these leafhoppers. Since BtCuTpV is phloem limited, it can be transmitted mechanically only experimentally by a special procedure. Bennett (1944) transmitted it via three *Cuscuta* species, but they have no importance for natural spread. Measures enabling the reduction of losses due to this virus include chemical control of vectors, early sowing of sugar beet seed, and the use of resistant sugar beet cultivars.

BEET MOSAIC POTYVIRUS (BTMV)

Properties of Viral Particles

BtMV particles measure 730×13 nm (Jafri, 1972) and physical properties include a TIP of 55 to 60°C, a DEP of 4×10^{-3}, and LIV at 20°C of 24 to 48 hours. BtMV is immunogenic and serologically related to BYMV and PVY (Bercks, 1960; Grüntzig and Fuchs, 1979).

Strains of BtMV differing in virulence on sugar beet and described in the U.S. (Shepherd et al., 1966) are unknown in Europe (Grüntzig and Fuchs, 1979).

Geographic Distribution and Economic Importance

BtMV, first reported in 1898 in garden beets in northern France and near Paris, was soon noted in other European countries (Denmark, Germany, Sweden) and in the U.S. This disease occurs regularly now in all areas where sugar beet is grown, especially in temperate zones. BtMV, although less damaging than beet yellows, reduces industrial sugar beet yields up to 30%, and beet seed crops infected in the first year result in 50% lower seed yields (Schmelzer and Hartleb, 1977). Research in former Yugoslavia showed that BtMV reduced yields of roots by 1.5 to 10.5%, green matter by 12%, and sugar by 8 to 19%, and sugar content by 0.65 to 1.06% (Nikolić et al., 1958). Sugar beet mosaic often occurs with beet yellows, which together cause severe damage both to industrial and seed yields.

Host Range

The moderate host range for BtMV includes all cultivated sugar beet cvs., *Beta vulgaris* subsp. *maritima*, *Amaranthus retroflexus*, *Campanula ranunculoides*, *Chenopodium album*, *Delphinum consolida*, *Melilotus officinalis*, *Papaver somniferum*, *Spinacia oleracea*, *Stellaria media*, *Trifolium incarnatum*, and *T. repens* (Schmelzer and Hartleb, 1977) and 100 species in 26 dicot families have been infected experimentally (Grüntzig, 1988). *A. retroflexus*, *A. caudatus*, *Chenopodium quinoa*, and *Gomphrena globosa* are suitable assay plants that react to infection by producing local spots.

Symptoms

Initial symptoms of BtMV infection are numerous small yellow spots and blotches on one or several central leaves. A light mosaic and mottle occurs on young leaves as disease develops. Leaflets with initial symptoms are stunted, with curling and rolling of leaf margins and leaf tip necrosis. In severe cases, diseased leaves roll into a tubular shape. In contrast to beet yellows, no phloem necrosis or starch accumulation occurs in leaves of BtMV-infected plants.

Numerous small yellow spots gradually merge together and form chlorotic zones on young leaves of infected spinach, older leaves turn chlorotic and necrotic, and the plant usually remains stunted.

Pathogenesis

BtMV is maintained in natural hosts, including biennial seed crops and winter spinach for year-to-year spread.

Natural vectors of BtMV include over 24 aphid species, with *Myzus persicae* and *Aphis fabae* the most important. Aphids transmit BtMV nonpersistently; while feeding on infected plants, they acquire the virus in 6 to 10 seconds and can transmit it to healthy plants in like time. Aphids can transmit BtMV in all developmental stages, although winged forms are the most active. The virus does not persevere through aphid moulting, and it is not passed on to the offspring (loc. cit. Russel, 1971). BtMV is transmitted easily with infective plant sap. It neither remains infective in sugar beet seed nor does it have any other modes of transmission.

Control

With the close resemblance of these two diseases, all measures recommended for beet yellows control are also recommended for BtMV. One must provide ample spatial isolation between industrial and sugar beet seed crops, chemical control of vectors, and elimination of weeds known to harbor BtMV.

BEET NECROTIC YELLOW VEIN TOBAMOVIRUS (BTNCYVV)

Properties of Viral Particles

The BtNcYVV 20 nm-wide, rod-shaped, multicomponent particles are characterized by groups based on length. Isolates from Japan are placed in three groups of 65 to 105, 270, and 390 nm long (Tamada and Baba, 1973). Isolates from France are placed in four groups of 85, 100, 265, and 390 nm long. Isolate from Yugoslavia are placed in five groups of 44 to 125, 133 to 259, 276 to 400 (the largest group), 418 to 569, and 586 to 769 nm long (Ivanović, 1979; Ivanović et al., 1979).

BtNcYVV particles have a single coat protein of 21 kDa and four species of single-stranded RNA of 2300, 1800, 700, and 600 kDa, respectively (Putz, 1988). Lemaire et al. (1988) showed a decrease in RNA_3 and RNA_4 during mechanical inoculations of *Chenopodium quinoa* leaves, or they are lost, suggesting that these two nucleic acids may be required only during natural infections (i.e., specific fungal vector involvement?).

Stability in Sap

Stability of BtNcYVV in sap for Japanese isolates are: TIP is 65 to 70°C; DEP is 1×10^{-4}; and LIV at 20°C is 5 days (Tamada and Baba, 1973). For Yugoslavian isolates TIP is 70 to 75°C; DEP is 1×10^{-5}; and LIV is 6 days.

Serology

BtNcYVV has moderate immunogenicity. The antiserum titre of the Yugoslavian isolate is 1/512, which has reacted positively with different isolates obtained from numerous sugar beet samples. The plate immunoenzyme (ELISA) and the paper immunoenzyme (Dot-blot) methods have been used successfully for diagnostic identification and for mass testing of plants (Krstić, 1990).

Strains

Four groups of isolates of BtNcYVV have been assigned, based on reactions in infected *Tetragonia expansa* plants: isolates causing (1) concentric rings; (2) yellow spots; (3) pale chlorotic spots or rings; and (4) necrotic spots or rings (Tamada, 1975). European isolates may differ in similar manners.

Geographic Distribution and Economic Importance

The BtNcYVV disease was first recorded in 1954 and 1957 in Italy (Dona Dalle Rose, 1954; Piolanti et al., 1957). Canova (1959) named it sugar beet rhyzomania, based on symptoms. The causal virus was soon described in sugar beets in Japan, France, Italy, Germany, former Yugoslavia, and in many European countries as well as in China, the U.S., and the U.S.S.R. Whereever it occurs, it causes a serious sugar beet disease (Tošić, 1982; Asher 1993).

BtNcYVV, a very widespread virus, has caused damage so great that fields in contaminated areas have been abandoned. BtNcYVV reduces yields of sugar beet more than 50% and sugar content of beets 3 to 4% (Šutić and Milovanović, 1978; Milovanović, 1989). BtNcYVV causes deterioration of beets and reduces their quality by causing increases in sodium and nitrate nitrogen and reductions in potassium, and amino and amido nitrogen contents (Giunchedi et al., 1987; Milovanović, 1989).

Host Range

Both sugar and fodder beets are susceptible to BtNcYVV, which also infects spinach and species of the Chenopodiaceae. Seventeen other plant species have been inoculated artificially with BtNcYVV (Putz, 1977), of which *Amaranthus hybridus, Beta macrocarpa, Beta vulgaris* subsp. *cicla, B. vulgaris* subsp. *esculenta, Chenopodium album, C. amaranticolor, C. foetidum, C. hybridum, C. quinoa, Nicotiana clevelandii, N. tabacum* (cvs. Havana and White Burley), *Stellaria media*, and others have been confirmed (Tošić et al., 1985; Šutić, 1983).

Only spinach (*Spinacia oleracea*) and *Beta macrocarpa* among the experimental hosts react with a systemic infection, and the systemic infection in *Tetragonia expansa* is caused only by the necrotic spot or ring strain. All other experimental hosts react only with chlorotic spots (lesions) on the inoculated leaves. Suitable assay plants for BtNcYVV are *T. expansa* (three strains cause local lesions and one strain causes both local lesions and systemic infection), *Chenopodium amaranticolor* (yellow chlorotic local lesions), and *Beta vulgaris* seedlings (for virus transmission with soil vectors).

Symptoms

Symptoms in BtNcYVV-infected sugar beets include yellow, roundish patches in the fields, occurring in smaller depressions in which the water remains following heavy rains. During development of disease in the summer, these yellow patches merge and enable easy detection of infected spots. Symptoms of vein clearing, yellowing, and necrosis occur on leaves infected early in the season, with ultimate severity dependant on cv. susceptibility, leaf development stage, and climatic factors (temperature, light, etc.) (Figure 131). As symptoms develop, infected leaves become chlorotic and clearly visible in the fields. Infections later in the season result in only slight color changes, and are quite difficult to detect. Other characteristic symptoms are elongation of leaf petioles and narrowing of leaf blades. These narrower leaf blades are erect in the center of the plant, and thus easily observed. Leaf firmness is similar to that of healthy plants, although leaf wilting occurs under conditions of severe infection.

FIGURE 131 Beet necrotic yellow vein virus in sugar beet.

Morphological symptoms are highly pronounced in roots as malformations that are classified either as hairy, dwarfed roots, or narrowed roots. Hairy roots result from the overdevelopment of numerous hairs that form a dense hair-like network, for which this symptom was named rhizomania. The numerous white root hairs become dark brown and, following death of root tips, new secondary root hairs branch from the base and create a dense, hairy network (Figures 132 to 135). Small

FIGURE 132 Rhizomania, caused by beet necrotic yellow vein virus, on sugar beet roots. (Courtesy J.E. Duffus. Reproduced with permission from APS.)

FIGURE 133 Beet necrotic yellow vein virus causes rhizomania in sugar beet.

Viral Diseases of Industrial Plants

FIGURE 134 Beet necrotic yellow vein virus causes rhizomania in sugar beet.

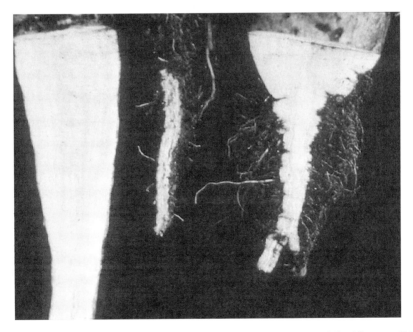

FIGURE 135a Root constriction and necrosis caused by rhizomania; healthy (left). (Courtesy E.D. Whitney. Reproduced with permission from APS.)

tumors clearly visible along the margins on the longitudinal root sections occur at the base of root hairs as a result of proliferative cell division (Figure 135a). These tumors are diagnostic of the disease caused by BtNcYVV. The abnormally large number of root hairs led to the naming of this disease rhizomania and/or sugar beet proliferation.

The dark coloration in vascular bundles visible in cross and longitudinal sections of diseased roots results from necrosis of infected cells. Hair death causes disorders in translocation of nutrients in sugar beet plants; thus, infected underdeveloped roots remain dwarfed (smaller than healthy ones). Roots from early infected plants weigh as much as 300 g less than healthy ones, and these pronouncedly narrowed roots are pointed, wedge-, and spindle-shaped.

FIGURE 135b Rhizomania of sugar beet caused by beet necrotic yellow vein virus.

FIGURE 135c Rhizomania of sugarbeet caused by beet necrotic yellow vein virus.

Pathogenesis

The virus transmission of BtNcYVV occurs in the soil by the fungus *Polymyxa betae* as the vector. *P. betae*, a Plasmodiophoraceae, is widespread on sugar beet roots (Keskin et al. 1962) and possibly also on other related plants. BtNcYVV penetrates the talus (plasmid) of the fungus during its development in epidermal cells or in infected root hairs, wherefrom it further reaches the zoospores of the fungus during their formation. Liberated zoospores swim through water films on soil particles and upon penetrating healthy root hairs they simultaneously inoculate the cells with BtNcYVV. BtNcYVV infectivity is preserved in resting spores of *P. betae*. The resting spores gather in bunches in root hairs and they are liberated following root-hair decomposition, which represents the long-term sources of infection from which BtNcYVV is carried into the next crop, or remain inactive for more than 1 year until another sugar beet crop is planted. The durable nature of the resting spores of *P. betae* allows the occurrence of this disease in the soil even several years after the sugar beet had been grown in those fields.

The dynamics and intensity of *Polymyxa betae* development are influenced by a complex of several soil factors (Šutić and Milovanović, 1980).

Soil moisture is the indispensible ecological factor for the growth of *Polymyxa betae*. It plays a decisive role in the occurrence and spread of infection and epidemics of BtNcYVV occur in years of abundant rainfall. Some investigations have shown that soil temperatures above 10°C immediately following sowing, as well as average soil temperatures of 17°C during the month of May, are optimal for fungus development, and thus virus spread. Soil acidity (pH) is a limiting factor for fungus development. The optimal soil pH for fungal growth is neutral. Soil pH values of 5.6 to 7.3 are recorded in areas of severe infections. Thick soil layers impervious to water due to the use of heavy machines for soil tillage, beet harvest, and transport contribute to fungal survival, and thus to infection preservation and spread of BtNcYVV.

High-risk focal points occur in areas where sugar beet has been grown for many years, especially in monoculture or with insufficient crop rotation that enables the preservation of infected root residues in the soil and in fields close to the beet processing plants. Sugar processing factories contribute to the spread of BtNcYVV if they return the processing residues to the growers, who may spread the organic matter back to their fields to help maintain soil tilth.

Control

No practice or economically justifiable measures currently enable substantial reductions of damage due to this disease. Several measures that jointly contribute to reduction of virus occurrence and spread are recommended.

Sugar beet plants should not be grown on moderately or severely contaminated soils. Fields with water drainage problems should not be selected for sugar beet production. Irrigation should be avoided in areas of known disease incidence.

Deep tillage (subsoiling) should be done as a normal cultural practice. Fields with high underground water levels should be drained to eliminate excess water and to reduce the dangers of infection. Plots with permanent drainage are more suitable for sugar beet production; but remember the possibility of water contamination with BtNcYVV. A minimum 4-year crop rotation is recommended only for less contaminated fields. Soil disinfestation chemicals may substantially reduce damage caused by BtNcYVV, with best results from methyl bromide, dasomet, or metham (Panić et al., 1978). However, their cost-effectiveness is questionable for large production acreages. During both harvest and transport, healthy, symptomless beets should be kept separated from those that are yellowed; those from contaminated fields should also be kept separate. Producers should not use beet, soil, or water residues from the processing plant during soil preparation, irrigation, and sugar beet cultivation. Sugar beet cvs. of better processing characteristics and with a greater level of tolerance to BtNcYVV have been introduced into production in recent years. In some countries, several sugar beet cvs. tolerant to rhizomania have been selected. In France, for example, the tolerance to this disease has been shown in cvs. Gabriella, Golf, Résis, Riposte, Rizor, and

Rexane (Merdinoglu et al., 1993). The sugar yield of these cvs. was between 12 to 13 tonnes per hectare, while the yield of the susceptible cv. Carat was 6 tonnes per hectare. Among sugar beet cvs. tested to rhizomania under field conditions in Serbia Ritmo group, Rima, Rizor, Hill mono 4086, Gabriella, KW Dora, Alvita, and others were shown as tolerant (Milovanović, 1994). Breeding research for tolerant cvs. is a continuous process due to the ready biological changeability of both the BtNcYVV and its fungal vector *Polymyxa betae*.

Finally, these measures may help protect the sugar beet crop: sow early; deep tillage; any tillage aimed at ensuring better soil aeration; use tolerant cvs.; and fertilization for best possible crop growth. Via genetic engineering, sugar beet transgenic plants were created that were tolerant to rhizomania disease (Le Buanec, 1988). Practical application of the transformed plants will depend on their regeneration ability and agronomic value in production.

BEET WESTERN YELLOWS LUTEOVIRUS (BtWsYsV)

Properties of Viral Particles

BtWsYsV is isometric, 26 nm in diameter (Duffus, 1972), and has a single-strand RNA species of 1900 kDa (two RNA species have been suggested in the particles of other isolates; loc. cit. Rochow and Duffus, 1981). The TIP of BtWsYsV is 65°C; the DEP is 1/8; and the LIV at 24°C is 16 days (Duffus, 1972). It has strong immunogenicity. The concentration of BtWsYsV in plants is low due to its phloem-limited nature.

Strains

BtWsYsV isolates differ, based on host range and symptom severity. Most American isolates infect both sugar beet and lettuce, whereas European isolates generally do not infect lettuce, or those infective for lettuce do not infect sugar beet, with only one exception recorded (Tomlinson, 1988). The common feature of all isolates is that they infect *Capsella bursa-pastoris*, *Claytonia perfoliata*, and *Senecio vulgaris*. Differences noted above are not detected serologically between American and European isolates (Duffus and Russel, 1975). BtWsYsV isolates are serologically related with the isolates of other luteoviruses (malva yellows virus, turnip yellows virus, and beet mild yellowing virus), all originally considered separate viruses (Rochow and Dufus, 1981).

Geographic Distribution and Economic Importance

BtWsYsV, thus far recorded in North America, Europe and Asia, is probably spread worldwide. It belongs to the group of viruses known as agents of sugar beet yellows important losses in sugar beet production. Heathcote (1978) found reductions in root yields of BtWsYsV-infected plants in England from 1970 to 1975 of 1 to 18%. Significant yield and quality losses due to BtWsYsV have been recorded in Europe (Hungary, The Netherlands, Germany, Sweden, the former U.S.S.R.), and in Turkey (Tomlinson, 1988).

Host Range

BtWsYsV infects a broad host range of 150 species in 23 dicot families (Rochow and Duffus, 1981). Economically important hosts include sugar beet, cabbage, cauliflower, flax, lettuce, pepper, radish, spinach, sunflower, tomato, and watermelon. Oat is susceptible in the monocot group (Duffus and Rochow, 1978). Important weed hosts include *Capsella bursa-pastoris*, *Senecio vulgaris*, *Stellaria media*, *Sinapis arvensis*, *Plantago lanceolata*, *P. major*, *Veronica* spp., *Cardamine hirsuta*, *Gallium aparine*, and *Portulaca oleracea* (Tomlinson, 1988).

Diagnostic plants are *Capsella bursa-pastoris* with severe chlorosis and leaf curl, *Senecio vulgaris* with purple coloration of the leaf margins, and *Claytonia perfoliata* with pink coloration of leaf edges (Duffus, 1972).

Symptoms

BtWsYsV infects many plants, but causes the most severe symptoms on sugar beet and lettuce.

Sugar beet plants infected with BtWsYsV expresses initially partial, and later general, leaf yellowing. This leaf yellowing is the now-known characteristic result of viruses limited to the phloem tissue, which leads to degeneration and chloroplast collapse. Yellows symptoms first appear as mild chlorotic spots between the veins, and then all interveinal tissues become markedly yellow (Figure 136). Symptoms develop acropetally from lower toward apical leaves. Symptoms seldom occur on the youngest leaves (in contrast to the beet yellows virus, which causes vein clearing on young leaves). Old leaves become thick, brittle, and completely yellow, except the green zone near the veins.

FIGURE 136 Resistant (left) and susceptible (right) sugar beet cultivars inoculated with beet western yellows virus. (Courtesy J.E. Duffus. Reproduced with permission from APS.)

Lettuce leaves express different degrees of yellowing in the field. Initial symptoms of mild interveinal yellowing become more pronounced and finally result in marginal leaf necrosis.

Pathogenesis

BtWsYsV maintains infectivity in numerous hosts in nature and bridges crops from year to year in infected overwintered weeds such as *Capsella bursa-pastoris* and *Senecio vulgaris*, and in winter rape (Tomlinson, 1988).

Eight aphid vectors transmit BtWsYsV persistently, and *Myzus persicae* is most important. BtWsYsV remains infective in vectors for 50 days, but it is not transmitted maternally to the offspring (Duffus, 1972). It is not transmitted mechanically in plant sap and seed transmission has been observed at only very low levels (Fritzsche et al., 1983).

Control

Several measures applied simultaneously may reduce losses due to this virus. (1) Eliminate all weed plants, especially overwintering ones. (2) Chemical control of vectors with an appropriate aphicide. (3) Apply biological control measures aimed at preventing development of aphid populations. (4) Grow more tolerant or resistant sugar beet cultivars. Likewise, for lettuce, more resistant cvs. should be grown (e.g., crisp and cos cvs. are more tolerant than butterhead cvs.) (Tomlinson and Ward, 1982).

Beet Yellows Closterovirus (BtYsV)

Properties of Viral Particles

BtYsV is a flexuous filament, 1250 × 10 nm, composed of a single-stranded RNA (4150 kDa), which accounts for 5% of total particle weight (Russel, 1970) and a single polypeptide coat protein of 23 kDa. The sedimentation coefficient of BtYsV is 130S; the TIP is 50 to 55°C; the DEP is 1×10^{-4}; and LIV at 24°C is 1 day. Although immunogenic, antisera have low titres.

BtYsV strains cause symptoms in sugar beet that vary from mild yellows to leaf necrosis, but resemble each other serologically. Strain differences are subtle except for the two classified mild and necrotic, which differ markedly (Smith, 1972).

Geographic Distribution and Economic Importance

BtYsV is widespread wherever sugar beet is grown in Europe, Asia, North and South America, Africa, and Australia. Severe epidemics occur in western and central Europe where sugar beet holds a central place in agricultural production.

Damage caused by BtYsV is primarily due to substantial reductions of root and green matter yield, sugar content, and total sugar quantity. BtYsV can reduce root yields by 29%, green matter yields by 21%, sugar content by 1.4%, and sugar yields by 35% (Šutić et al., 1959). In addition, BtYsV causes reductions of seed yield amounting to 47%; seed size, which results in an increase in number of seeds per gram by 19%; and finally, mean 1000-seed-weight by 16% (Šutić et al., 1959a). The extent of damage depends primarily on the incidence of infected plants per field per year.

Host Range

A large number of plants are susceptible to BtYsV: 130 species, of which 23 are natural hosts (Schmelzer and Hartleb, 1977). Cultivated plants among natural hosts include beet (*Beta vulgaris*), poppy (*Papaver somniferum*), and spinach (*Spinacia oleracea*). Most other natural hosts are Chenopodiaceae; for example, *Atriplex patula, Chenopodium album, C. bonushenricus, C. ficifolium, C. glaucum, C. murale,* and *C. polyspermum*, and others are *Amaranthus retroflexus, Papaver rhoeas, Plantago lanceolata, P. major, Polygonum convolvulus, Stellaria media, Tetragonia tetragonoides,* and *Thlaspi arvense*.

Suitable assay plants for BtYsV are *C. capitatum* with vein clearing, leaves twisted downward, reddening, and death of older leaves; *C. foliosum* with vein clearing, crinkling and necrosis of intermediate leaves; *Montia (claytonia) perfoliata* with red to dark brown necrosis on leaves and leaf petioles, among others.

Symptoms

Different from other virus diseases, beet yellows symptoms appear first on older beet leaves. Initial yellows symptoms occur at apical and external portions of the leaf blade and then spread toward the leaf base (Figures 137 and 138). Chlorosis of certain leaf parts differs, depending on the cv., and ranges from pale or greenish-yellow to orange and even red. Chlorotic surfaces have a waxy appearance, and leaves of severely affected plants become all yellow, dry out, and rustle when touched. Brittle leaf blades split into small fragments and crackle when bent. Leaves infected early in the season become necrotic. Sugar content decreases in phloem cells and starch accumulates in leaves, which results in leaf blade thickening and brittleness. Leaf yellowing caused by magnesium deficiency distinguishes BtYsV, from the BtWsYsV which has broad green zones along leaf veins.

Symptoms similar to these in sugar beets appear in BtYsV-infected spinach, easily recognized by the yellow color on leaf tips and margins. Central leaves have vein clearing and crinkling and, when necrotic, the whole plant dies.

Pathogenesis

BtYsV occurs in natural hosts, some biennials, such as sugar beet seed plants both transplants and direct sown, infected stecklings, and winter seedlings, which enable its transmission from year to year. BtYsV is transmitted similarly in fodder beet, garden beet, mangold production, and winter spinach crops.

Aphids actively vector BtYsV semipersistently and *Myzus persicae* is most important among 35 species reported. All developmental stages of aphid transmit the virus, with adults being most active. BtYsV neither maintains infectivity through the molt nor is passed on to the offspring (Russel, 1970).

BtYsV can be transmitted mechanically with infected plant sap only under experimental conditions of keeping the plants in the dark and inducing leaf injury by rubbing carborundum or celite during inoculation. Reports on seed transmission of BtYsV has not been confirmed. *Cuscuta gronovii* can transmit BtYsV.

FIGURE 137 Yellowing of beet leaf caused by beet yellows virus. (Courtesy R.T. Lewellen. Reproduced with permission from APS.)

FIGURE 138 Symptoms of beet yellows virus in sugar beet plants.

Control

Seed crops that can harbor BtYsV from year to year should be grown in adequate isolation (i.e., over 1 km from the commercial sugar beet crop). Seed crops should also be isolated from susceptible crops of fodder beet, garden beet, mangold, and spinach. Seed production is recommended in areas with less favorable conditions for aphid infestation to reduce their role as vectors. Only healthy stecklings should be transplanted.

Chemical control of aphids, both in the commercial and seed sugar beet crop, is obligatory and of vital importance when their flight activity is initiated, and insecticide spray should be repeated two or three times, especially for seed crops. Chemical aphid control is necessary when seedlings emerge and for the stored beet stecklings. Volunteer beets and weeds plants that serve as virus reservoirs and hosts of aphid should be rogued.

Losses caused by BtYsV can be reduced both by earlier spring sowing of commercial beet and later winter sowing; this ensures fewer infected seedlings, and thus fewer infected seed plants in the next year. Incidence of infection is lower in dense plant populations.

Šutić and Spasić (1963) reduced BtYsV damages by foliar and root top dressing. Sugar yields obtained after plants were artificially infected with BtYsV were higher by 6.19 mc/ha after application of foliar top dressing of fertilizer than without top dressing, and root top dressing improved yields up to 7.17 mc/ha.

Tolerant sugar beet cultivars should be used; for example, in eastern England, the use of cv. Maris Vanguard cut losses by 50%. Two American sugar beet cultivars, USH_9A and USH_9B, have moderate resistance to BtYsV (loc. cit. Smith, 1972).

According to recent results, the application of some new systemic insecticides provide control of vector activity for long periods of time and thus beet yellows infection is controlled (IIRB, 1994).

SOME OTHER VIRUS DISEASES OF BEET

Beet Leaf Curl Rhabdovirus (BtLCuV)

BtLCuV, with bacilliform particles of 225 × 80 nm and unknown chemical composition (Proeseler, 1983), has a TIP of 54 to 58°C; a DEP of 1×10^{-4} to 1×10^{-5}; and an LIV at 25°C of 24 hours. There is no information about antisera preparation or about virus strains (Proeseler, 1983). It is widespread in Europe (Germany, Poland, and sporadically in the former Czechoslovakia and the former U.S.S.R.) and of little economic importance.

The BtLCuV restricted host range includes the Chenopodiaceae and Azioaceae. *Beta vulgaris* (sugar and fodder beet) and *Spinacea oleracea* are the main natural hosts. A typical symptom on sugar beet plants, for which this disease was named, is vein clearing in young leaves that later curl inward, causing the bushy, stunted appearance of infected plants.

The bug *Piesma quadratum* nonpersistently transmits BtLCuV, which multiplies in its vector and then can transmit it for life. BtLCuV is not transmitted with infected plant sap and no other mode of transmission is known.

Beet Mild Yellowing Luteovirus (BtMdYgV)

BtMdYgV, with isometric particles 28 nm in diameter, is serologically related to BtWsYsV, malva yellows virus (MaYsV), and TuYsV (Rochow and Duffus, 1981), which are, it seems, strains of the same virus. BtMdYgV, widespread in northern, western, and central Europe, reduces yields of sugar by over 20% when infection occurs early.

Beta vulgaris and *Spinacea oleracea* are economically important natural hosts. Some weeds (e.g., *Capsella bursa-pastoris*, *Lamium amplexicaule*, *Plantago* spp., *Senecio vulgare*, and *Stellaria media*) as natural hosts are important in the disease cycle (Schmelzer and Hartleb, 1977). Characteristic orange-yellow symptoms occur on intermediate sugar beet leaves and leaf blades become brittle, but not necrotic. BtMdYgV has been transmitted experimentally onto a few plants, and those suitable for assay include *Montia perfoliata* with yellow and red coloration of the intermediate leaf margins, and *Sinapis alba* with yellow and red coloration of older leaves.

The aphids *Myzus persicae*, *Rhopalosiphoninus tilipaellus*, and *Aulacorthum curcumflexum* transmit BtMdYgV persistently and no other mode of transmission is known (Schmelzer and Hartleb, 1977).

Beet Pseudo Yellows Virus (BtPdYsV)

Few facts of BtPdYsV are known, but thread-like particles in ultra-thin sections suggest that it may belong to the Closterovirus group. BtPdYsV has been reported in North and South America, western and southern Europe, and Australia (Polak and Lot, 1988).

Hosts of BtPdYsV include sugar beet, greenhouse cucumbers, lettuce, and weeds in the Chenopodiaceae, Compositae, Umbelliferae, and Solanaceae families. *Capsella bursa-pastoris*, lettuce, and *Taraxacum officinale* are susceptible assay plants. Symptoms of interveinal yellowing, swelling, and brittleness occur in infected sugar beet leaves, and plants remain dwarfed. The whitefly *Trialeurodes vaporariorum* transmits BtPdYsV in nature (Duffus, 1965). No other mode of transmission is known.

Beet Yellow Net Virus (BtYNtV)

BtYNtV particles are straight rods, 300 × 32 nm in size (Schmelzer and Hartleb, 1977).

Clearing and yellowing of veins, which create the net symptom, plus chlorotic spots result from BtYNtV infection of sugar beet leaves (Figure 139). The only natural host of BtYNtV is sugar beet. *Myzus persicae* and *Aphis fabae* transmit the virus persistently (Watson, 1962). Polak (1988) believes the yellow net may be an initial symptom caused by severe or necrotic strains of BtYsV.

FIGURE 139 Beet yellow net. (Courtesy J.E. Duffus. Reproduced with permission from APS.)

Beet Yellow Stunt Closterovirus (BtYSnV)

BtYSnV is filamentous, 1250 × 12 nm (Duffus, 1972a), and occurs in North America and possibly in Europe. Economically important hosts of BtYSnV, in addition to sugarbeet, include red beet, mangold, spinach, and 11 other species in 5 families (Duffus, 1973). Symptoms on sugar beet include leaf mottle and yellowing, severe twisting and epinasty of intermediate leaves, petiole shortening, and plant malformation. Assay plants to differentiate BtYSnV from BtYsV include *Chenopodium capitatum*, *Claytonia perfoliata*, *Lactuca sativa*, *Senecio vulgaris*, and

Sonchus oleraceus (Lister and Bar-Joseph, 1981). Aphids transmit BtYSnV semipersistently (Duffus, 1972a).

BEET: NATURAL ALTERNATIVE HOST FOR SOME VIRUSES

Cucumber Mosaic Cucumovirus (CMV) in Beet

CMV, found in western, southern, and southeastern Europe (Avgelis and Vovlas, 1972; Kovachevsky, 1961), probably occurs in other beet growing areas. Infected leaves are small and twisted downward, and vein clearing is typical with yellow-green, then bronze coloration later in the season. CMV-infected external leaves wither and die. Root hair discoloration is mild and lateral hairs occur early. CMV-infected seed crops have reduced seed yields.

Tobacco Rattle Tobravirus (TRtV) in Beet

TRtV found in sugar beet in western Europe (Gibbs and Harrison, 1964) caused red-yellow ringspots on leaves. TRtV is transmitted mechanically and by the nematode *Trichodorus pachydermis*.

Tomato Black Ring Nepovirus (TmBRV) in Beet

TmBRV found in sugar beet in western Europe (Gibbs and Harrison, 1964) caused either irregular ringspots or infected leaves remain symptomless. TBRV is transmitted in sugar beet seed and pollen (Schmelzer and Hartleb, 1977).

Detailed descriptions of these viruses can be found in the chapters on their primary natural hosts.

VIRUS DISEASES OF SUGARCANE (*Saccharum officinarum*)

Sugarcane, a perennial plant in the Gramineae family that accounts for 68% of the world sugar production, also provides raw material for the production of rum, molasses, and denatured alcohol. Its leaf and stalk residues are used as animal feed. This tropical plant is grown widely in Asia (India) and South America (Cuba, Puerto Rico, and Brazil). This vegetatively propagated perennial plant serves inadvertently to transmit and spread any virus that infects the crop. Reviews in *Current Trends in Sugarcane Pathology. III. Virus and Phytoplasma Diseases* (Rao et al., Eds. 1999) cover the latest information and report the newest virus diseases.

SUGARCANE MOSAIC VIRUS (ScMV) (MEMBER OF THE POTYVIRUS GROUP)

Properties of Viral Particles

ScMV particles are flexuous, filamentous, 750 × 13 nm (Teakle and Grills, 1973; Teakle et al., 1989), and contain a 3000-kDa, single-stranded RNA constituting 5% of the viral particle, plus a single polypeptide, 35-kDa coat protein (Jilka and Clark, 1989). The sedimentation coefficient of the ScMV-MB strain is 176 ± 5S (Tošić and Ford, 1974).

Stability in Sap

The TIP of ScMV is 50 to 55°C; the DEP 1×10^{-2} to 1×10^{-3}; and the LIV at 20°C is 1 day (Teakle and Grills, 1973).

Serology

ScMV is moderately immunogenic with antisera titer usually about 1/125. Serological diagnosis has been successful with immunosorbent electron microscopy (ISEM), enzyme-linked immunosorbent assay (ELISA), and electro-blot immunoassay (EBIA) (Shukla et al., 1983; 1989).

Strains

ScMV was for several decades regarded as a complex of several virus strains, often described as they were found in and isolated from different gramineous crops. Now, with ample chemical, physical, and serological data, these strains have been reclassified as four separate viruses: ScMV, maize dwarf mosaic, sorghum mosaic, and Johnsongrass mosaic viruses (Shukla et al., 1989). The following groups of virus strains are based on host susceptibility and serological relatedness: (1) *sugarcane mosaic virus* strains (Brandes, 1919) causing mosaic in sugarcane plants, but seldom infecting Johnsongrass (*Sorghum halepense*); (2) *Queensland blue couch grass* and *Sabi grass* strains (Teakle and Grills, 1973) causing mosaic in *Digitaria didactyla* and *Urochloa mosambicensis* plants, but not infecting sugarcane; (3) *Maize dwarf mosaic virus* - strain B (MacKenzie et al., 1966) causing maize mosaic, but not infective in Johnsongrass or sugarcane C.P. 31-294 and C.P. 31-588 cvs. (Snazelle et al., 1971); and (4) *Abaca mosaic virus* (Eloja and Tinsley, 1963) causing mosaic in abaca (*Musa textilis*) and maize plants, but serologically related, and thus a sugarcane mosaic virus strain.

Geographic Distribution

ScMV was first observed on Java in 1890 (Abbott, 1953). ScMV, identified as the causal agent of this disease, occurs wherever sugarcane is grown worldwide (Teakle and Grills, 1973). Sugarcane mosaic is economically the most damaging disease of the sugarcane crop.

Host Range and Symptoms

Virus hosts include primarily numerous Gramineae plants, except for the abaca mosaic virus that infects the monocot *Musa textilis* in the Musaceae. In addition to sugarcane, natural virus hosts include maize, sorghum, and several wild grasses (Teakle et al., 1989). Cultivated cereals of wheat, barley, rye, and rice seldom are infected naturally (Teakle et al., 1989). Suitable assay plants include *Saccharum* spp. with mosaic and variable striping; *Zea mays* with mosaic in seedlings infected with various strains; and *Sorghum bicolor* with mosaic or necrosis on Atlas, Rio, and other cvs.

The main symptom on sugarcane is leaf mosaic, occasionally forming yellow striping or necrosis, depending on both cv. and virus strain. Also depending on both cv. and prevalent strains, ScMV markedly affects plant development and yields.

Pathogenesis and Control

ScMV is preserved in numerous perennial gramineous species, especially sugarcane, and has been distributed broadly with population expansion into remote areas. Aphid species such as *Dactynotus ambrosiae*, *Hysteroneura setariae*, *Rhopalosiphum maidis*, and *Toxoptera graminis* are active natural vectors transmitting ScMV nonpersistently (Teakle and Grills, 1973). ScMV is transmitted in infective plant sap. No other modes of transmission have been reported.

Important measures to reduce damage caused by ScMV include the control of weed host plants in and near sugarcane fields and the control of aphids (vectors) with the use of appropriate aphicides. Spatial isolation of sugarcane from susceptible hosts and any cultural operations used to reduce aphid populations are recommended. ScMV can be eliminated from mosaic-infected ratoons by thermotherapy (Benda, 1970) and by meristem-tip culture (Hendre et al., 1975). An important preventive measure is to use virus-free seed beds to establish sugarcane crops. The most efficient crop protection measure is to produce cultivars more resistant or tolerant to ScMV. Australia has already cloned the coat protein gene of ScMV and inserted it into sugarcane. The first field test one of us (Ford) observed in Queensland in 1994 and now genetically engineered transgenic resistance with the coat protein gene from ScMV is successful, as reported recently (Joyce et al., 1997a; 1997b; 1998; Mirkov et al., 1997; Smith and Harding, 1999).

Fiji Disease Reovirus (FjDsV)

Properties of Viral Particles

FjDsV polyhedral particles are 70 nm in diameter, with double-stranded RNA (Hutchinson and Francki, 1973). RNA consists of 10 individual segments ranging from 1100 to 2900 kDa, for a total molecular weight of 20,000 kDa (Matthews, 1981). No information is available on the biophysical and serological traits of this virus. Susceptibility of sugarcane to FjDsV varies from region to region, probably a consequence of different virus strains or leafhopper vector species.

Geographic Distribution

The Fiji disease was discovered in sugarcane in 1910 on the Fiji islands where it caused crop losses until resistant cultivars were introduced (Abbott, 1953). Following Fiji islands, FjDsV was also discovered in Australia, Madagascar, New Britain, New Guinea, New Hebrides, and Sampa (Hutchinson and Francki, 1973).

Host Range and Symptoms

FjDsV infects hosts only in the Gramineae. Sugarcane is the natural host of FjDsV, which has been transmitted into *Saccharum* spp., *Zea mays*, and *Sorghum* spp. by leafhopper vectors (Hutchinson et al., 1972).

Elongated galls that develop along the veins on the underside of sugarcane leaves are the diagnostic symptom of Fiji disease (Figures 140 and 141). These galls, which result from cell proliferation in vascular bundles, contain two types of abnormal cells; lignified gall xylem and nonlignified gall phloem cells (Hatta and Francki, 1976). FjDsV-infected sugarcane plants are stunted and dwarfed and shoots are bushy. Sugarcane cultivars Fiji 10 and H 47-991, which develop galls within 18 days after inoculation with leafhopper vectors, are suitable assay plants.

Pathogenesis and Control

FjDsV, permanently preserved in perennial sugarcane, is spread into other geographic areas with the transport of infected planting material (shoots). FjDsV is transmitted in nature persistently by Delphacid planthoppers *Pekinsiella saccharicida*, *P. vitiensis*, and *P. vastatrix*. Infected leafhoppers are

FIGURE 140 Galls on midribs of sugarcane caused by sugarcane Fiji disease virus. (Qld., Australia).

FIGURE 141 Galls on midribs of sugarcane caused by sugarcane Fiji disease virus. (Qld., Australia).

life-long vectors, but it is unknown if FjDsV can be transmitted transovarially (Hutchinson and Francki, 1973).

Selecting healthy planting material to establish a sugarcane crop and leafhopper control whenever possible is recommended, but the most important prevention measure is to develop sugarcane cultivars tolerant or resistant to FjDsV.

SOME OTHER VIRUS DISEASES OF SUGARCANE

Several virus or virus-like diseases have been described in sugarcane, such as serch, streak, and chlorotic streak disease (Abbott, 1953; Corbett and Sisler, 1964) and some new and more challenging, such as the sugarcane (bacilliform) badnavirus (Braithwaite, 1998).

Sugarcane Serch Disease (ScShDs)

ScShDs is found in the Philipines, Formosa, India, and Java. It results in severe plant stunting, development of lateral buds as leafy shoots, and abnormal formation of aerial roots on nodes. Vascular bundles are reddened by colored gum. The vector is unknown. This ScShDs caused serious damage until resistant cvs. were introduced.

Sugarcane Streak Virus (ScSkV)

ScSkV has been reported and is restricted to South Africa, Egypt, India, and Mauritius. Characteristic symptoms on infected plants are narrow, elongated, white streaks on leaves. Infected plants exhibit reduced growth and yields. *Cicadulina mbila* is the virus vector. Damage in sugarcane production is reduced with the use of resistant cultivars.

Sugarcane Chlorotic Streak Disease (ScCSkDs)

ScCSkDs, first observed on Java, has been recorded in Hawaii, Australia, Mauritius, Puerto Rico, Louisiana, and British Guyana. Characteristic symptoms of infection are long, narrow, longitudinal chlorotic streaks on leaves and necrosis along the spots that eventually affect the vascular bundles. Germination (in some cases), growth, and ratooning are reduced in ScCSkDs-infected plants.

ScCSkDs is soil borne, but the vector species and mode of transmission are unknown (Harrison, 1964). Disease spread depends on agents associated with roots and is water dependent (loc. cit. Harrison, 1964).

ScCSkDs can be eliminated by immmersing the shoots into water at 52°C for 20 minutes (Martin and Conant, 1939). However, the most effective protection measure is to produce tolerant and resistant plants.

Dwarf disease and ratoon stunting, earlier considered as virus diseases, belong to the disease group caused by the xylem-limited bacterium *Clavibacter xyli* subsp. *xyli* (Agrios, 1988).

SUGARCANE: NATURAL ALTERNATIVE HOST FOR SOME VIRUSES

Maize Dwarf Mosaic Potyvirus (MDMV)

MDMV was recovered from four commercial sugarcane cultivars grown in Louisiana (U.S.) (Gillaspie, 1967). MDMV is similar to and long was considered a strain of ScMV (Pirone, 1972), but is now classified as a separate virus (Ford et al., 1989; Shukla et al., 1989) so it may now be found more often on sugarcane. MDMV causes only mild mosaic on sugarcane compared to that caused by ScMV (Gillaspie, 1967).

Peanut Clump Furovirus (PtCpV)

PtCpV was found on sugarcane in Senegal, Burkina Faso, and Ivory Coast (Baudin and Chatenet, 1988). The symptoms on sugarcane caused by PtCpV include chlorotic mottle, developing into red or white streaks. Slight stunt may also occur. Easily transmissible mechanically, virus transmission is mainly through cuttings.

Sorghum Mosaic Potyvirus (SrMV)

SrMV was isolated first from sugarcane in the U.S. and identified as ScMV-H (Abbott, 1961), I (SCMV-I) (Tippett and Abbott, 1968), and M (ScMV-M) (Koike and Gillaspie, 1976). ScMV-H was also found on sugarcane in Japan, Philippines, and India; and ScMV-I in Japan (Gillaspie and Mock, 1979.; Kondaiah and Nayudu, 1984). More recently, ScMV-H was isolated from sorghum (Giorda et al., 1986). These three strains—H, I, and M—after differentiation from ScMV, were reclassified and named sorghum mosaic virus (SrMV) (Shukla et al., 1989).

Symptoms of SrMV in sugarcane differ, depending on sugarcane cvs. and virus strain. Strains H and M usually cause mild mosaic. These two strains can be differentiated based on their reaction in sorghum "Rio" (Koike and Gillaspie, 1976). Strain I usually causes severe mosaic and, in some cvs. like C.P. 31-294, also severe stunting. On secondary shoots, the mosaic is more mild. Some sugarcane clones recover after expressing mosaic early.

Sugarcane Striate Mosaic Disease (ScStMDs)

ScStMDs was first noticed in 1956 and reported in Australia (Hughes, 1961; Hughes et al., 1968; Currie, 1971), then in India (Nayudu et al., 1971).

Symptoms of ScStMDs appear on young, expanding leaves that exhibit short, fine, barely discernible striations, lighter than the normal leaf color. Yellowing of leaf tips occurs, in addition to severe stunting of diseased plants. Although yet unproven, it was suggested that two viruses are involved, one which causes striate mosaic and the other stunt (Hughes, 1961).

The control of striate mosaic on sugarcane can be achieved by resistant cvs. (Currie, 1971).

Sugarcane Mild Mosaic Closterovirus (ScMdMV)

All clones infected were also co-infected with ScBcV. ScMdMV causes mild or no symptoms. Symptoms may be due to the double-infection only. ScMdMV can be detected best by ISEM (Braithwaite 1998).

Sugarcane Bacilliform Badnavirus (ScBcV)

ScBcV was found in some sugarcane clones grown in Morocco and Hawaii (Rodriguez et al., 1985), plus in some genotypes of sugarcane in the USDA germplasm collections at Canal Point and Miami and in some samples of sugarcane from Florida and Texas (Comstock and Lockhart, 1990). Data are accumulating to determine the relationship among different isolates from sugarcane and similar bacilliform viruses from other plants such as banana and its streak virus (BaSkV) (Lockhart 1986). ScBcV is symptomless on Noble sugarcane although occasional white, chlorotic flecks or streaks are observed. BaSkV is integrated into the genome of host *Musa* sp. (LaFluer et al. 1996), thus limiting the efficacy of PCR. The most reliable method for detection is the use of ISEM (Autrey et al. 1990).

Sugarcane Yellow Leaf Luteovirus (ScYLV)

ScYLV first observed 10 years ago in Hawaii and Brazil, is now known to be widespread in sugarcane growing regions (Irey et al. 1997). It causes leaf yellowing in sugarcane, typical of symptoms in cereals caused by other luteoviruses. It is still uncertain whether all the symptoms are caused by ScYLV because a phytoplasma co-infection occurs, especially in South Africa. ScYLV is not related to other luteoviruses, including ByYDV, although the coat protein region has low homology with the PAV serotype. Most severe losses in yield of recoverable sugar are reported in Brazil. Heat treatments have not controlled the disease.

Two new virus-like diseases of sugarcane have been described in the Ramu Valley, Papua New Guinea. The *Ramu stunt disease (ScRamuSnDs)* killed 'Ragnar' plants within 1 year (Waller et al. 1987), all of which were replaced with the resistant 'Cadmus' variety. Hot water treatments do not eradicate RamuSnDs, and infected canes do not recover (Magarey 1996). The *Ramu streak disease (ScRamuSkDs)* symptoms are similar, but not identical to ScCSkV, because it rarely causes leaf necrosis (Magarey et al. 1996). Little else is known about ScRamuSkDs.

VIRUS DISEASES OF SUNFLOWER (*Helianthus annuus*)

Sunflower is an annual plant in the Compositae family. The sunflower seed contains a natural raw oil that ranks high in the quality of plant oils used in human nutrition. Sunflower oil is also used for margarine and as an industrial raw material for paint, varnish, stearine in the production of candles, etc. Sunflower also serves as animal feed, either as silage or green fodder. Sunflower, grown on nearly all continents, is most widely produced in Europe and South America.

Several viruses and virus-like diseases, some highly damaging, have been described in sunflower. Among seed-transmitted pathogens, the most important and widespread mosaic is that caused by the sunflower mosaic virus. Other mosaic diseases also occur in sunflower plants. Leaf symptoms quite similar to mosaic can result from toxins upon feeding of aphids in the sunflower crop (Šutić, 1960).

Sunflower Mosaic Virus (SfMV)

SfMV, the most widespread sunflower virus disease, occurs in Africa—Kenya, Mauritius, and Uganda (Anon., 1956; Wiltshire, 1955; Anon., 1971); Asia—India (Battu and Phatak, 1965); South America—Argentina, Brazil, and Uruguay (Muntanola Maria, 1948); North America—Florida, Maryland, and Texas (Weber, 1932; Arnott and Smith, 1967; Gill, 1965); and Europe—the former Soviet Union (Šutić, 1983). Some SfMV isolates are highly transmittable in sunflower seed, up to 43%; therefore, seed from infected plants is the main method of worldwide distribution of sunflower mosaic. Since the origins of mosaic diseases differ based on hosts, symptoms, vectors, and other modes of transmission, mosaic is considered a complex disease, the epidemiology of which requires further investigation.

Kolte (1985) summarized the descriptions of sunflower mosaic as follows:

SfMV in Argentina and Kenya

SfMV is transmitted easily, both by infective plant sap and by infected seed. *Thrips tabaci*, *Myzus persicae*, and *Trialeurodes vaporiorum* are natural vectors. The TIP is 98 to 99°C and its DEP is 1×10^{-5}. Widespread and highly damaging, the main disease symptoms include mottle, elongation of leaves, vein banding, and twisting of petioles. At 18 to 24°C, the disease develops slowly; whereas, at 26 to 28°C, symptoms appear quickly and plants die. This mosaic most likely was imported into Argentina with infected sunflower seed from the former Soviet Union.

SfMV in Austin (Texas)

The rod-shaped viral particles of 480×13 nm observed in infected sunflower cells are believed to belong to the Y group, which also infects wild sunflower. Initial symptoms of infection are mild mosaic patterns, followed by leaf and stalk necrosis, and then plant death. Fewer distorted leaves form on those parts of the stalks with necrosis. Distorted leaf blades form from axillary buds. The vector(s) is unknown, but the rapid spread of this disease in nature suggests aphid involvement.

SfMV in India

A different origin of this virus is suggested because it infects only sunflower and not *Nicotiana tabacum*, *N. glutinosa*, or *Capsicum annuum*. The initial symptom forms a mosaic pattern on leaves, followed by ringspots and chlorotic lesions that eventually coalesce.

This virus is further characterized by a TIP of 46 to 48°C; a DEP of 5×10^{-3} to 5×10^{-4}; and an LIV at 25°C of 2 to 3 days. The virus is transmitted with infective plant sap, by grafting, and through seed from infected plants. The *Bemisia tabaci* vector may transmit the virus persistently because it retains infectivity for 8 days. This virus causes systemic infections in sunflower and *Amaranthus* sp. Symptoms of disease in sunflower are mosaic patterns composed of circular spots; leaf deformations; and reduced plant rooting, yields, seed vitality, and pollen fertility. Chlorotic local lesions occur in *Chenopodium amaranticolor*.

In some countries—the U.S. (Florida), Africa, Mauritius, Uganda, and Uruguay—other types of sunflower mosaic exist and for which their etiology and other biological features require separate investigations.

OTHER VIRUS DISEASES OF SUNFLOWER

The following mosaics have been described, based on characteristic symptom type.

Rugose Mosaic in Kenya (Singh, 1979)

Leaf crinkling and rugosity characterize the symptoms of this mosaic. Leaf margins may roll downward, whereas upper leaves exhibit severe mottle and reduced size. Severely infected plants are stunted and produce few or no seed. This virus is transmitted mechanically and possibly through seed from infected plants.

Yellowing Mosaic in India (Gupta, 1981)

Young, infected leaves are malformed, and infected plants are dwarfed and rarely flower. Chlorotic local lesions occur on infected *Chenopodium amaranticolor* leaves.

Yellow Mosaic in Belgium (Roland, 1960)

Vein clearing and yellow mosaic-like symptoms occur on infected sunflower leaves. The virus also infects chrysanthemum plants.

Sunflower: Natural Alternative Host for Some Viruses

Beet Western Yellows Luteovirus (BtWsYsV)

Tomlinson (1988) reported strains of BtWsYsV that infect many cultivated plants, sunflower among them, causing symptoms of varying intensity. A sunflower yellowing disease in the U.K. is caused by a persistent aphid-transmitted virus of the BtWsYsV group (Russel et al., 1975). Pale-yellow zones appearing between the veins on infected leaves are followed by necrotic spots on leaf tips and margins.

Cucumber Mosaic Cucumovirus (CMV)

CMV infects sunflower in nature causing mosaic symptoms (Smith, 1972). Orellana and Quacquarelli (1968) described a sunflower mosaic caused by the CMV-SF strain with symptoms of mosaic and chlorotic rings on young leaves, increasing in severity on older leaves. Pale brown spots occur along the stalks and leaf petioles. Growth of plants is stunted. CMV-SF is mechanically transmitted and causes systemic infection in *Vigna sinensis*, but does not infect *Phaseolus* spp.. Perennial sunflower *Helianthus decapetalus L.V. multiformis* is the alternate host of CMV.

Tobacco Rattle Tobravirus (TRtV)

TRtV identified in sunflower in Central Europe (Schmelzer, 1970) causes discoloration and mild necrosis on infected leaves. TRtV is transmitted both mechanically and by *Trichodorus* nematodes.

Tomato Spotted Wilt Tospovirus (TmSpoWV)

TmSpoWV causes symptoms on all parts of the sunflower plant (Wallace, 1947). Yellow-to-pale green interveinal mottle occurs on the dwarfed, malformed infected leaves. Ringspots occur on stalks. Grooves and fissures form on upper parts of the stalks and on leaf petioles and the vascular tissues turn black. Internodes shorten and flower heads do not form or remain small and bent over.

Turnip Mosaic Potyvirus (TuMV)

TuMV occurred in sunflower in Bulgaria causing yellowing of lower and intermediate leaves beginning at the margins, gradually spreading into the interveinal tissues. This disease seems not to adversely affect plant growth and yields.

Tobacco Ringspot Nepovirus (TRsV) and Tobacco Streak Ilarvirus (TSkV)

These also occur in sunflower plants as their natural hosts (Schmelzer and Schmidt, 1977).

VIRUS DISEASES OF SOYBEAN (*Glycine max.*)

Soybean, an annual plant, belongs to the Leguminosae family. Soybean seed provides both protein and oil for human nutrition as milk, butter, cheeses, and meat substitutes. Its proteins provide raw materials for the processing industry in production of artificial rubber, plastics, paint, and varnish. Soybean is important as green fodder or in processed forms for animal nutrition. The symbiotic soybean *Rhizobium* sp. nodulation process enriches the soil by "fixing" nitrogen, which greatly enhances soil fertility.

Soybean is susceptible to infection by several viruses that substantially reduce both yield and product quality. Soybean mosaic virus, one of the most economically damaging viruses, is transmitted through seed, sometimes at a high percentage, and is generally widespread wherever this crop is grown. It is also the alternate natural host of several other viruses, including BPdMtV, BYMV, TRsV, and TSkV, which are also naturally widespread and occasionally represent a threat to production. Therefore, profitable production of soybean requires a complex, long-term program of protection against virus infections.

Soybean Mosaic Potyvirus (SyMV)

Properties of Viral Particles

SyMV flexuous rods, 750 × 15 nm, vary from 300 to 900 nm, and those of 656 nm are most infective. The single-stranded RNA of 3250 kDa accounts for 5.3% of the particle weight. The SyMV TIP is 55 to 70°C; LIV in infective sap is 2 to 5 days and at 25 to 33°C in dried leaves is 7 days. It is quite immunogenic. Several strains are based on their pathogenicity and virulence in soybean cvs., *Lespedeza stipulacea* and Japanese soybean cvs. (loc. cit. Bos, 1972). Many strains have been indexed by local lesions resulting from mechanical inoculations on detached leaves of *Phaseolus vulgaris* cv. Top Crop.

Geographic Distribution and Economic Importance

SyMV can be found wherever soybean is grown. Its ease of distribution via movement of seed often results in a significant number of infected plants primarily dependent on the susceptibility of individual cvs. SyMV ranks as economically damaging because it can reduce yields naturally by 50%, and in experimentally inoculated plots by 93% (Demski and Kuhn, 1989). Seed from infected plants is of lower quality for processing. SyMV-infected plants of some *Glycine max.* cultivars are predisposed as more susceptible to infection by *Phomopsis sojae*.

SyMV reduces soil fertility because it decreases the nitrogen fixing activity of nodules of *Rhizobium* sp., thus reducing nitrogen accumulation in the soil. Reduction of nodulation can be 81%, depending on the time of infection by SyMV (Tu et al., 1970); such nodules on soybean roots are fewer, smaller, and lighter.

Host Range

SyMV infects a limited host range with only soybean and its wild relatives serving as natural hosts. Most experimentally infected hosts belong to the Leguminosae. Some have systemic symptoms (*Cassia occidentalis, Dolichos folcatus, Glycine gracilis, Lespedeza stipulaceae, Lupinus albus, Phaseolus lathyroides*, and *P. vulgaris*); others are symptomless (*Lotus tetragonolobus, Phaseolus speciosus* and some *P. vulgaris* cvs.); and some have only local lesions (*Dolichus biflorus, D. lablab*, and some *P. vulgaris* cvs).

SyMV experimentally may infect other families; for example, *Amaranthus* sp., *Chenopodium album, C. quinoa, Physalis longifolia, P. virginiana, Setaria* sp., and *Solanum carolinense* (Demski and Kuhn, 1989).

Symptoms

Symptoms of infection by SyMV vary, depending on the cv., time of infection, strain of virus, and environmental factors. Most soybean cvs. react to SyMV infection with systemic vein clearing, primarily in youngest leaves. As the disease develops, leaves gradually curl, leaf margins twist downward, and leaf blades are smaller and somewhat brittle (see Figure 142). Infected plants become dwarfed, progressive necroses occur in some cvs., and they form fewer pods, sometimes deformed and seedless. Symptoms of disease are more pronounced at 18 to 20°C than at 27 to 30°C (loc. cit. Bos, 1972). Hilum bleeding causes discoloration on the surface of seed from infected plants, which provide an approximate estimate of the potential for seed transmission of SyMV (Figure 143).

Necrotic local lesions occur in experimentally infected bean plants (some cultivars) and local lesions also characerize infection by SyMV of inoculated *Chenopodium quinoa* and *C. album* leaves; thus, both are suitable assay plants.

Pathogenesis

SyMV infectivity is preserved naturally in and transmitted by soybean seed (Gardner and Kendrick, 1931); the degree of seed infection and transmission depends on cultivar susceptibility and the time

Viral Diseases of Industrial Plants

FIGURE 142a Soybean mosaic virus on cotyledonary leaves on a soybean seedling grown from an infected seed (above) and a trifoliolate leaf of an inoculated plant (below). (Courtesy G.R. Bowers. Reproduced with permission from APS.)

FIGURE 142b

of infection before flowering (i.e., the earlier, the higher the rate of seed infection). Seed harvested from plants that became infected after flowering occurred do not transmit SyMV. Viral particles are found equally in the seed coat and in the embryo, and more in the green than in the mature seed (loc. cit. Bos, 1972). The virus may remain infective in the seed for 2 years.

Aphids transmit the virus nonpersistently in nature; 31 species reported as vectors include *Acyrthosiphon pisum*, *Aphis craccivora*, *A. fabae*, *Myzus persicae*, *Rhopalosiphum maidis*, and *R. padi*. SyMV is mechanically transmitted easily with infective plant sap, although it probably plays a small role in disease epidemiology.

Control

Only virus-free seed should be used in the production of susceptible varieties. Soybeans intended for sale of seed for planting new crops should be harvested only from healthy plants or, with a certain amount of risk, only from plants infected following flowering. Small, mosaic-affected seeds should not be used for sowing. Soybean seed crops should be isolated from other production fields and any SyMV-infected plants should be rogued immediately upon detection. Chemical control of aphid vectors is recommended.

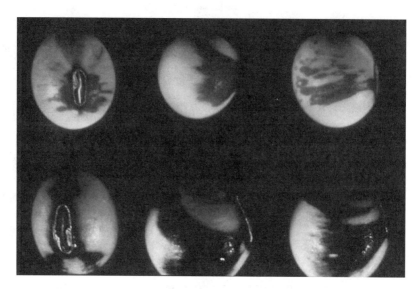

FIGURE 143 Common soybean mosaic seed discoloration. (Courtesy J.H. Johnson. Reproduced with permission from APS.)

Resistant or tolerant soybean cvs. primarily should be used in production. Among 110 soybean cvs. tested in Japan, 27 were classified resistant (e.g., Minnesota, Manchu, Holland 11, and F.P.J. 79610.)

SOYBEAN DWARF LUTEOVIRUS (SyDV)

SyDV particles are isometric, 25 nm in diameter, with a protein subunit of 22 kDa (Tamada and Kojima, 1977), a TIP of 45 to 50°C, a DEP of 0.5, and an LIV at 15°C for 20 days, or at 4°C for 4 months. SyDV is strongly immunogenic. Its dwarfing and yellowing strains are distinguishable based on characteristic symptoms in various hosts. The dwarfing strain causes stunt, downward leaf curl, and internode shortening (Figure 144). The yellowing strain turns leaves yellow between the veins, which sometimes become brittle.

SyDV, widespread in Japan, causes significant soybean crop losses (Tamada et al., 1969). The SyDV host range is in the Leguminosae and, in addition to soybean, *Phaseolus vulgaris* and *Pisum sativum* also represent economically significant hosts. Red clover (*Trifolium pratense*) and white clover (*T. repens*) are natural symptomless hosts. When 50% of soybeans are infected, 40% yield reductions occur (Tamada and Kojima, 1977).

SyDV is permanently preserved in perennial hosts such as *Trifolium pratense* and *T. repens*. *Aulacorthum solani* transmits SyDV persistently. SyDV is transmitted neither mechanically nor through soybean seed (Tamada et al., 1969; Hewings, 1989).

SOYBEAN CHLOROTIC MOTTLE CAULIMOVIRUS (SyCMtV)

SyCMtV is isometric, 50 nm in diameter, and consists of a 5450-kDa double-stranded DNA; no information exists for a coat protein (Hibi and Kameya-Iwaki, 1988). The TIP of SyCMtV is 85 to 90°C; the DEP is 1×10^{-3} to 1×10^{-4}; and the LIV at 20°C is 1 to 3 days (Iwaki et al., 1984). SyCMtV is moderately immunogenic and no strains have been identified.

SyCMtV, currently restricted to Japan, has soybean as its only natural host, although it has been transmitted mechanically into other legumes (e.g., *Dolichos lablab*, *P. vulgaris*, and *Vigna unguiculata*) (Iwaki et al., 1984). Symptoms vary, depending on the soybean cv., and include vein clearing and chlorosis on young leaves, which develop into mottle and plant stunting. SyCMtV is

Viral Diseases of Industrial Plants

FIGURE 144 Soybean dwarf virus: plant and leaf dwarfing (above); leaf symptoms (below). (Courtesy A.D. Hewings. Reproduced with permission from APS.)

transmissible in infective plant sap, natural vectors are unknown, and it is not transmitted through soybean seed from SyCMtV-infected plants (Iwaki et al., 1984).

SOYBEAN: NATURAL ALTERNATIVE HOST FOR SOME VIRUSES

The soybean plant is naturally susceptible to many viruses; 111 viruses or strains can infect soybean in nature or under experimental conditions (Demski et al., 1989). Soybean is a natural host for 33 potentially important viruses. Most viruses pathogenic for soybean have been described for other plants in the Leguminosae. For example, AMV causes soybean yellow spot; BPoMtV (Figure 145), PtMtV (Figure 148), and BYMV cause soybean yellow mosaic (Figure 146); ClYVV, CpCMtV, CMV, PeMV, PeSkV, BSuMV, and WClMV syn: pea wilt virus PeWV and Wisconsin PeSkV. Another important group of viruses that infect soybean have natural hosts in the Solanaceae TLCuV, TNcV, TRsV (Figure 147) TSkV and TmSpoWV. Finally, viruses, naturally infective for soybean, that originate from other families are BtWsYV (Chenopodiaceae), CMV (Cucurbitaceae), and AnMV = Abutilion mosaic (Malvaceae).

Mixed infections of several viruses that occur in soybean are more damaging than individual infections. For example, BPoMtV individually causes soybean yield losses of 10 to 15%, but losses of over 60% occur in mixed infections of BPoMtV and SyMV. Age of plant at time of infection greatly influences the amount of damage; usually earlier is more serious. For example, CpCMtV

FIGURE 145 Bean pod mottle virus in 'Williams' soybean. (Courtesy G.H. Hartman, USDA/IL.)

FIGURE 146 Yellow mosaic on soybean in India. (Courtesy J.B. Sinclair. Reproduced with permission from APS.)

does not markedly damage soybean, but early infection causes yield losses of 20 to 30%. Viruses AMV, BYMV, and CMV, with many natural hosts, especially perennial ones and transmitted by aphids often lead to epidemics of soybean viral diseases. However, the most damaging are viruses transmitted through soybean seed; for example, TSkV, which causes Brazilian bud blight and which Fagbenle and Ford (1970) first reported naturally infecting soybean in the U.S., may be seed transmitted at 30%, and TRsV, which causes bud blight, can reduce soybean yields by 25 to 100% under certain conditions.

Soybean is exposed naturally to many virus infections; therefore, several protective measures should be employed to ensure profits. Recommended are adequate cultural practices for normal plant development, reduction of vector populations, awareness of proximity to other susceptible plants, and the practice of wise crop rotation. Chemicals are used to control aerial vectors. Soybean production on uncontaminated soils enables one to avoid virus infections transmitted by nematode

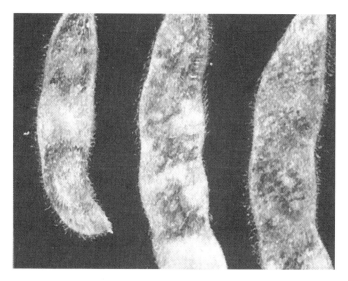

FIGURE 147 Soybean pods from a tobacco ringspot virus-infected plant. (Courtesy M.C. Shurtleff. Reproduced with permission from APS.)

Figure 148 Peanut mottle. (Courtesy G.W. Kuhn). Reproduced by permission from APS.

vectors. It is essential to select virus-free seed for planting and to plant cvs. more tolerant or resistant to infection by soybean viruses.

Information about viruses affecting soybean can be found in the section "Virus Disease" in *Compendium of Soybean Diseases* (J.B. Sinclair and P.A. Backman, 1989), in addition to the viral characteristics and their primary hosts described here.

VIRUS DISEASES OF HOP (*Humulus lupulus*)

Hop, a perennial in the Cannabinaceae family, is grown for the floral cones used extensively in the production of beer and in the pharmaceutical industry. Although grown worldwide, Europe and the former Soviet Union are the largest producers.

Hop is the primary or alternate host of several viruses, some particularly damaging to production. Hop is propagated vegetatively from shoots; thus, viruses are transported with infected planting material. This perennial crop is continuously exposed to airborne (hop mosaic virus, American hop latent virus, hop latent virus) and soilborne (arabis mosaic virus, tobacco necrosis virus) infections. Mechanical transmission during certain cultural operations may play an important role in spreading some viruses (hop stunt viroid). Hop gardens are expensive plantations; thus, their establishment mandates virus-free planting stock.

Hop Mosaic Carlavirus (HMV)

Properties of Viral Particles

The HMV particles are slightly flexuous, 650 × 14 nm, and are composed of a single-stranded RNA of 3000 kDa, which accounts for 5 to 7% of the particle weight, and a coat protein of 34 kDa (Adams and Barbara, 1980). HMV is an active immunogen. Both severe and mild isolates have been described (Legg, 1959).

Geographic Distribution and Economic Importance

HMV is widespread, having been described in Europe, Tasmania, and North America (loc. cit. Barbara and Adams, 1981). The perennial nature of hop suggests that viral infections may occur naturally in a high percentage of plants (e.g., 35% of plants in Bulgaria) or possibly entire crops if plantations were established with infected planting stock. This disease in the U.S. causes losses in hop yields up to 40% (Hoerner, 1949). Susceptible cv. yields are reduced 70% (Schmelzer and Schmidt, 1977).

Host Range and Symptoms

Hop, the main natural host of HMV, has been discovered in Europe in *Urtica urens* (Adams and Barbara, 1980), and in wild plants *Chenopodium album*, *Plantago major*, *Polygonum aviculare*, and *Stellaria media* (Eppler, 1980). Experimental infections are reported in 11 species of 5 families (Barbara and Adams, 1981). *Nicotiana clevelandii* is suitable for HMV preservation and transmission.

Disease symptom severity varies with cv. susceptibility. Tolerant hop cvs. are available. Initial symptoms are leaf vein clearing and pale-green to yellow spots near leaf veins; then leaf margins twist downward and leaves become yellow-green, brittle, and distorted; and internodes are shortened and plants become bushy in appearance. Poor shoot formation results in the absence of or twisted and poorly closed floral cones (Nuber, 1962). The roots of infected plants gradually decline, with a resultant early death.

Pathogenesis and Control

HMV is permanently preserved in infected perennial hop plants and transported naturally in these planting stocks (shoots) obtained from the infected parent. *Macrosiphum euphorbiae*, *Myzus persicae*, and *Phorodon humuli* transmit HMV nonpersistently (Adams and Barbara, 1980); HMV is transmitted mechanically with infective plant sap, but not through hop seed.

The main protective measure against this virus is to use virus-free planting stock (shoots) for establishment of plantations. Infected plants in newly established plantations should be rogued immediately, and further viral spread can be prevented with chemical control of vectors. Tolerant cvs. substantially reduce HMV damage.

Hop Latent Carlavirus (HLtV)

Properties of Viral Particles

HLtV filamentous particles are 675 × 14 nm, and contain single-stranded RNA of 2900 kDa, which accounts for 5 to 7% of the particle weight, and a coat protein of 33 kDa (Adams and Barbara,

1982). The TIP is 70 to 75°C; the DEP is 1×10^{-2} to 1×10^{-3}; and the LIV is 25 days (Barbara and Adams, 1983). HLtV has moderate immunogenicity. Some isolates differ, based on host range.

Geographic Distribution

HLtV is widespread in hop gardens in Europe, the U.S., and Australia (Probasco and Scotland, 1978; Adams and Barbara, 1982).

Host Range and Symptoms

Hop is the only natural host of HLtV, and 11 species in 4 families have been infected experimentally. Infected plants are symptomless, and thus represent a dangerous primary source of infection. Suitable assay plants include *P. vulgaris* cv. Kinghorn for necrotic local lesions, *Chenopodium murale* for chlorotic local lesions, and *Humulus lupulus* for aphid transmission.

Pathogenesis and Control

HLtV is both permanently preserved in its perennial hop host and transmitted into newly established hop gardens with infected shoots. Under certain conditions, it can be transmitted by *Phorodon humuli* aphids nonpersistently in the field and with infective plant sap (Barbara and Adams, 1983). All measures recommended for control of HMV are also recommended for HLtV.

HOP AMERICAN LATENT CARLAVIRUS (HALtV)

Properties of Viral Particles

HALtV filamentous particles are 680×15 nm and contain single-stranded RNA of 3000 kDa, which accounts for 5 to 7% of the particle weight, and a coat protein of 32.7 kDa (Barbara and Adams, 1983a). Although HMV and HLtV also belong to the Carlavirus group, HALtV differs serologically from them. HALtV has a wider host range and causes different symptoms on *Datura stramonium* and *Chenopodium quinoa* plants.

Geographic Distribution

HALtV occurs in many U.S. hop gardens, with the potential danger of introduction into England and Germany (Barbara and Adams, 1983a).

Host Range and Symptoms

Hop is the only natural host HALtV, and 17 species in 7 families have been experimentally infected (Adams and Barbara, 1982). Some British cvs. do not display symptoms of the HALtV disease. Symptoms in American seedlings are ring and line patterns in leaves. *Chenopodium quinoa* expresses chlorotic spots and systemic vein banding, and *Datura stramonium* expresses local lesion reactions suitable for diagnosis.

Pathogenesis and Control

HALtV is permanently preserved in infected hop plants and retained in infected planting stock. *Phorodon humuli* transmits HALtV nonpersistently. HALtV is also transmitted with infected plant sap (Barbara and Adams, 1983a). Recommended protection measures are similar to those recommended for HMV.

HOP STUNT VIROID (HSNVD)

Properties of the Viroid

HSnVd contains covalently closed circular, single-stranded RNA composed of 297 nucleotides (Sano and Shikata, 1988). In infective crude cucumber sap, the TIP of HSnVd is 84°C and the LIV at 4°C is 3 days (Sasaki and Shikata, 1978). The grapevine viroid (297 nucleotides) and the cucumber pale fruit viroid (303 nucleotides) are similar to HSnVd, based on host ranges and disease symptoms; thus, they are regarded as the grapevine and the cucumber strains of HSnVd (Sano and Shikata, 1988).

Geographic Distribution

HSnVd is reportedly widespread only in Japan. The cucumber strain, reported occurring naturally in glass houses in Holland, causes the cucumber pale fruit disease (Van Dorst and Peters, 1974). The grapevine strain is present in many grapevine cultivars occurs worldwide.

Host Range and Symptoms

H. lupulus and *H. japonicus* are natural hosts of HSnVd. Its strains have also been transmitted mechanically into the Cucurbitaceae, Moraceae, and Solanaceae (Sasaki and Shikata, 1977).

Symptoms of yellowing and curling occur on upper infected hop leaves. Shortened internodes along the main and lateral vines dwarf the plants. Cones from HSnVd-infected plants remain small and 50% fewer are formed compared with healthy plants (i.e., at least a 50% yield reduction) (Sano and Shikata, 1988). Diagnostic plants include white gourd *Benincasa hispida* (leaf curl, mosaic, top necrosis, plant stunt) and *Cucumis sativus* cv. Suyo (cotyledon stunt, vein clearing, leaf curl).

Pathogenesis

HSnVd is preserved in infected hop plants and transmitted by propagative shoots when planted in new gardens. All strains of HSnVd are transmittable mechanically by routine cultural operations of hop growing. The HSnVd cucumber strain is transmitted onto cucumber plants grown in glass houses in Holland during plant pruning, whereas the grapevine strain is transmitted with infected vegetative planting stock. Neither natural vectors nor seed transmission (hop, tomato, cucumber) of HSnVd is known. The cucumber strain of HSnVd is transmitted by dodder, *Cuscuta subinclusa* (loc. cit. Sano and Shikata, 1988).

Infected planting stock is the main factor in transmission and spread of HSnVd; thus, it is essential to use virus-free plant materials to establish new hop gardens. Close attention is required to prevent the mechanical transmission of HSnVd with infected plant sap during hop garden cultivation.

OTHER VIRUS DISEASES OF HOP

Hop Chlorotic Disesase Virus (HCDsV)

Viral hop chlorosis described in western Europe is probably also present in other European countries (Průša, 1961; Schmidt, 1965). The characteristics of this virus disease have been insufficiently studied.

Hop is the only natural virus host and hop seedlings are suitable assay plants. Pale or green-yellow spots first appear on primary leaves of infected hop plants, followed by chlorotic stripes and interveinal necrosis. Toothed leaf margins may become pronounced, but sometimes undiscernible in chlorotic leaves. Chlorotic leaf blades are wrinkled, with tops bent downward, resembling parrot beaks. Dwarfed plants and poorly developed inflorescences result in substantial reductions in yield.

The virus is transmitted both through seed from infected plants and mechanically, and thus assumed transmissable by plant-to-plant contact during cultivation. Virus vectors are unknown. All prevention measures recommended for the control of other hop viruses are recommended here.

Hop Mosaic Chlorosis Virus (HMCsV)

HMCsV particles are 680×18 to 20 nm. The TIP is $43°C$; DEP is 2.5×10^{-2}; and LIV at $24°C$ is 15 hours (Schmelzer and Schmidt, 1977).

Described in eastern Europe, hop is its natural host and *Urtica urens* its experimental host. Hop seedlings and *Chenopodium album* with pale spots on inoculated leaves are suitable assay plants. Chlorotic clearing and lesions appear on the interveinal tissue of infected hop leaves. Plants are stunted and the cones of infected plants are smaller.

HMCsV-infected hop plants are a perennial source of infection. In nature, this virus is transmitted by the aphid *Phorodon humuli* and mechanically by infected plant sap.

Hop Yellow Net Virus (HYNtV)

Hop yellow net disease has been described in western and central Europe (Legg and Ormerod, 1961; Schmidt, 1968). The viral nature of this disease has been confirmed; the physico/chemical properties of HYNtV remain unreported.

Hop is the only known host. Veins in the third row turn yellow, whereas those of the first and second rows, as a rule, maintain their usual green color. Leaf blades appear normal, with no visible changes in plant growth.

The virus has been graft transmitted only. Healthy planting stock for establishment of hop gardens and rogueing of infected plants are the main measures for disease prevention.

HOP: NATURAL ALTERNATIVE HOST FOR SOME OTHER VIRUSES

Arabis Mosaic Nepovirus (ArMV)

ArMV is economically damaging on hop in Europe. Infected plants grow more slowly. Internodes are shorter with poorly developed vines, causing plants to appear bushy. Pale-green to yellow spots irregular in shape appear on infected leaves. Thin, parchment-like, transparent leaf blades become necrotic, split, and break, thus known as the hop split leaf blotch (Bock, 1966; Keyworth, 1951). This disease can be economically damaging, reducing cone yields by 50%.

Mixed infections of ArMV and prunus necrotic ringspot virus (PuNcRsV) and potentially other viruses cause the hop nettle head disease (Legg, 1964; Thresh et al., 1972), which may result in economic damage. The key symptoms are smaller leaves, pale to yellow-green, partially very narrowed with toothed margins resembling nettle leaves. As the disease develops, leaf blades wrinkle and curl and some veins become highly pronounced and/or enations form along the midrib, and poorly closed cones remain underdeveloped. This disease may cause hop yield losses of 75%.

ArMV is transmitted from plant to plant mechanically by grafting, and/or by *Xiphinema diversicaudatum* nematodes in the soil. Other virus species may also be transmitted by these nematodes. All measures recommended to control other hop viruses apply to this disease. Nematode control in contaminated soils and rogueing weeds in hop gardens are indispensable.

Apple Mosaic Ilarvirus (ApMV)

ApMV (syn.: rose mosaic virus) occurs naturally in hop. The hop A and hop C strains are frequently symtomless, but certain temperature variations cause plants to display chlorotic and necrotic line mosaic patterns (Barbara, 1988). ApMV is serologically related to PuNcRsV, and similar isolates in hop and rose are considered serotypes of ApMV (Barbara, 1988).

Prunus Necrotic Ringspot Ilarvirus (PuNcRsV)

PuNcRsV, which infects many species of *Prunus* and causes a rose mosaic, is widespread in hops. PuNcRsV causes the ring and band pattern mosaic and the necrotic crinkle mosaic disease of hops (Schmelzer and Schmidt, 1977).

The ring disease symptoms are yellow-green rings and ringspots irregularly distributed on the leaf. Necrotic strains cause death of tissues and plant parts. In mid-summer, symptoms on less severe infected plants disappear. Some hop cvs. are tolerant to PuNcRsV, which in severe cases causes yield losses of 30%.

Leaves showing symptoms of the necrotic crinkle disease are small, twisted downward, and distorted. The tissue surrounding leaf veins first becomes yellow, then dark brown and necrotic. Tissues surrounding the necroses continue to grow and cause the characteristic crinkle symptom. Necrotic spots also appear along the vines, whereas cones remain underdeveloped and distorted. These changes somtimes cause yield losses of 50%.

General preventive measures recommended for other hop viruses are also applied to this virus.

Tobacco Necrosis Necrovirus (TNV)
Hop is one of several natural herbaceous and woody hosts of TNV (Uyemoto, 1988). TNV is preserved in and transmitted by zoospores of *Olpidium brassicae* in the soil (Teakle, 1962). The movement of the soil water is particularly important for the epidemiology of this virus.

REFERENCES

Abbott, E.V. 1953. Sugarcane and its diseases. In: *Plant Diseases: The Yearbook of Agriculture*, U.S. Dept. Agric., Washington, D.C., p. 526–535.

Abbott, E. V. 1961. A new strain of sugarcane mosaic virus. (Abstr.) *Phytopathology* 51:642.

Abeygunawardena, D.V.W. et al. 1967. Yellow net virus disease of flue-cured tobacco. *Trop. Agric.* 123:37–50.

Adams, A.N. and Hull, R. 1972. Tobacco yellow vein dependent on assistor viruses for its transmission by aphids. *Ann. Appl. Biol.* 71:135–140.

Adams, A.N. and Barbara, D.J. 1980. Host range, purification and some properties of hop mosaic virus. *Ann. Appl. Biol.* 96:201–208.

Adams, A.N. and Barbara, D.J. 1982. Host range, purification and some properties of two carlaviruses from hop (*Humulus lupulus*): hop latent and American hop latent. *Ann. Appl. Biol.* 101:483–494.

Agrios, G.N. 1988. *Plant Pathology*, 3rd edition, Academic Press, Inc., New York, 803 pp.

Anon. 1956. *Dept. of Agric., Kenya, Ann. Rept.* 1954 Vol. I, 1955, [Abstr. in *Rev. Appl. Mycol.* 35:276].

Anon. 1971. *Plant Pathology*, Report 1967 Dept. Agr., Mauritius.

Arnott, H.J. and Smith, K.M. 1967. Electron microscopy of virus infected leaves. *J. Ultrastructure Res.* 19:173.

Asher, M.J.C. 1993. Rhizomania. In: Cooke, D.A. and Scott, R.K., Eds. *The Sugarbeet Crop: Science into Practice*, Chapman Hall, London.

Autrey, L.J.C. et al. 1990. Occurrence of sugarcane bacilliform virus in Mauritius. *Proc. S. Afr. Sugar Tech. Assn.* 64:34–39.

Avgelis, A. and Vovlas, C. 1972. Le virosi delle piante ortensi in Puglia. IX. La bieticoltura fogliare gialla della bietola. *Phytopathol. Mediter.* 11:122–123.

Babos, P. and Kassanis, B. 1963. Serological relationship and some properties of tobacco necrosis virus strains. *J. Gen. Microbiol.* 32:135–144.

Badge, J. et al. 1996. A carla-specific PCR primer and partial nucleotide sequence provides further evidence for the recognition of cowpea mild mottle virus as a whitefly-transmitted carlavirus. *Eur. J. Pl. Pathol.* 102:305–310.

Ball, E.E. 1909. Leafhoppers of the sugar beet and their relation to the "curly-top" condition. U.S. Dept. *Agr. Ent. Bull.* 66:33–52.

Barbara, D.J. and Adams, A.N. 1981. Hop mosaic virus. *CMI/AAB Descriptions of Plant Viruses*, No. 241.

Barbara, D.J. and Adams, A.N. 1983. Hop latent virus. *CMI/AAB Descriptions of Plant Viruses*, No. 261.

Barbara, D.J. and Adams, A.N. 1983a. American hop latent virus. *CMI/AAB Descriptions of Plant Viruses*, No. 262.

Barbara, D.J. 1988. Apple mosaic virus (ApMV). In: Smith, I.M. et al., Eds. *European Hand. Pl. Dis.*, Blackwell Sci. Publ., London, p. 13–14.

Bartels, R. 1964. Untersuchungen über serologische Beziehungen zwischen Viren der "tobacco-etch-Virus-Gruppe". *Phytopathol. Z.* 49:257–265.

Bartels, R. 1971. Potato virus A. *CMI/AAB Descriptions of Plant Viruses*, No. 54.

Battu, A.N. and Phatak, H.C. 1965. Observations on a mosaic disease of sunflower. *Ind. Phytopathol.* 18:317.

Baudin, P. and Chatenet, M. 1988. Detection serologique du PCV, isolat canne a sucre, agent de la marbrure rouge des feuilles. *Agron. Tropicale* 43:228–235. (Abstr. in *Rev. Pl. Pathol.* (1990)69:4254)

Beijerinck, M.W. 1898. Over een contagium vivum fluidum als oorzaak van de vlekziekte der tabaksbladen. *Verhandel. Koninkl. Akad. Wetenschap., Afdel. Wis-Natuurk.* 7:229–235.

Benda, G.T.A. 1970. Heat-cure of sugarcane infected with sugarcane mosaic virus. (Abstr.) *Phytopathology* 60:1284.

Bennett, C.W. 1944. Latent virus of dodder and its effect on sugar beet plants. *Phytopathology* 34:77–91.

Bennett, C.W. 1971. The curly top disease of sugar beet and other plants. *Phytopathol. Monogr. No. 7.*, Am. Phytopathol. Soc., St. Paul, MN.

Bercks, R. 1960. Serological relationships between beet mosaic virus, potato virus Y and bean yellow mosaic virus. *Virology* 12:311–313.
Bercks, R. 1970. Potato virus X. *CMI/AAB Descriptions of Plant Viruses*, No. 4.
Bird, J. and Maramorosch, K. 1978. Viruses and virus diseases associated with whiteflies. *Adv. Virus Res.* 22:55–110.
Black, L.M. 1940. Strains of potato yellow-dwarf virus. *Am. J. Bot.* 27:386–392.
Black, L.M. 1970. Potato yellow dwarf virus. *CMI/AAB Descriptions of Plant Viruses*, No. 35.
Bock, K.R. 1966. Arabis mosaic virus and *Prunus* necrotic ringspot virus in hop (*Humulus lupulus* L.). *Ann. Appl. Biol.* 57:131–140.
Bos, L. 1972. Soybean mosaic virus. *CMI/AAB Descriptions of Plant Viruses*, No. 93.
Brandes, E.W. 1919. The mosaic disease of sugarcane and other grasses. *Tech. Bull. U.S. Dept. Agric.* 829, 26 pp.
Broadbent, L. 1961. The epidemiology of tomato mosaic: A review of the literature. *Rep. Glasshouse Crops Res. Inst.* 1960 (1961), 96–116. (Abstr. in *RAM* 41/1962/:271).
Broadbent, L. 1965. The epidemiology of tomato mosaic. XI. Seed transmission of TMV. *Ann. Appl. Biol.* 56:177–205.
Buturović, D. 1970. Virusi i virusne bolesti krompira u Bosni i Hercegovini. Zbornik radova, Institut za poljoprivredna istraživanja, 7:97–104, Sarajevo.
Canova, A. 1959. Appunti di patologia della barbabietola. *Inf. Fitopat.* 9:390–396.
Cockerham, G. 1955. Strains of potato virus X. *Proc. Second Conf. Pot. Virus Dis.*, Wageningen, p. 89–90.
Comstock, J.C. and Lockhart, B.E. 1990. Widespread occurrence of sugarcane bacilliform virus in U.S. sugarcane germ plasm collection. *Plant Dis.* 74:530.
Cooper, J.I. and Harrison, B.D. 1973. The role of weed hosts and the distribution and activity of vector nematodes in the ecology of tobacco rattle virus. *Ann. Appl. Biol.* 73:53–66.
Cooper, J.I., Jones, R.A.C., and Harrison, B.D. 1976. Field and glasshouse experiments on the control of potato mop-top virus. *Ann. Appl. Biol.* 83:215–230.
Corbett, M.K. and Sisler, H.D. 1964. *Plant Virology*. University of Florida Press. 527 pp.
Costa, A.S. and Lima Neto, V. da C. 1976. Transmissao de virus da necrose branca do fumo per*Frankriniella* sp. *Fitopatologia* 11:1–35.
Currie, J.A. 1971. A short history of striate mosaic disease. *Cane Grow. Q. Bull.* 35(1):12–13.
De Bokx, J.A. and Huttinga, H. 1981. Potato virus Y. *CMI/AAB Descriptions of Plant Viruses*, No. 242.
Dedić, P. 1975. Effect of virus A on yield in some varieties of potato. *Ochrana Rostlin* 11:127–133.
Delgado-Sanchez, S. and Grogan, R.G. 1970. Potato virus Y. *CMI/AAB Descriptions of Plant Viruses*, No. 37.
Demski, J.W. and Kuhn, G.W. 1989. Soybean mosaic virus. In: Sinclair, J.B. and Backman, P.A., Eds. *Compendium of Soybean Diseases*, 3rd edition, APS Press, St. Paul, MN, p. 55–57.
Demski, J.W. and Reddy, D.V.R. 1997. Biotic Diseases: diseases caused by Viruses. In (Nokalis-Burelle et al. 1997-which see) pp. 53–59.
Demski, J.W. et al. 1989. Virus Diseases. In: Sinclair, J.B. and Backman, P.A., Eds. *Compendium of Soybean Diseases*, 3rd edition, APS Press, St. Paul, MN, p. 50–51.
Diener, T.O. and Raymer, W.B. 1967. Potato spindle tuber virus. A virus with properties of a free nucleic acid. *Science* 158:378–381.
Diener, T.O. and Raymer, W.B. 1971. Potato spindle tuber "virus." *CMI/AAB Descriptions of Plant Viruses*, No. 66.
Dona Dalle Rose, A. 1954. A campagna bieticola conclusa gravi sintomi di "stanchezza" dei bietolae. *Agricoltura delle Venezie* 11:609–619.
Duffus, J.E. 1961. Economic significance of beet western yellows (radish yellows) on sugar beet. *Phytopathology* 51:605–607.
Duffus, J.E. 1965. Beet pseudo-yellows virus, transmitted by the greenhoue whitefly (*Trialeurodes vaporiorum*). *Phytopathology* 55:450–453.
Duffus, J.E. 1972. Beet western yellows virus. *CMI/AAB Descriptions of Plant Viruses*, No. 89.
Duffus, J.E. 1972a. Beet yellows stunt, a potentially destructive virus disease of sugar beet and lettuce. *Phytopathology* 62:161–165.
Duffus, J.E. 1973. The yellowing virus diseases of beet. *Adv. Virus Res.* 18:347–386.
Duffus, J.E. and Russell, G.E. 1975. Serological relationship between beet western yellows and beet mild yellowing viruses. *Phytopathology* 65:811–815.

Duffus, J.E. and Rochow, W.F. 1978. Neutralization of beet western yellows virus by antisera against barley yellow dwarf virus. *Phytopathology* 68:45–49.

Edwardson, J.R. 1974. Host range of viruses in the PYV-group. *Florida Agr. Exp. Stns. Monograph Series*, No. 5, 94.

Eloja, A.L. and Tinsley, T.W. 1963. Abaca mosaic virus and its relationship to sugarcane mosaic. *Ann. Appl. Biol.* 51:253–258.

Eppler, A. 1980. Dissertation zum Doktor der Naturwissenschaften der Fakultfät für Biologie der Eberhard —Karls Universität Tübingen.

Fagbenle, H.H. and Ford, R.E. 1970. Tobacco streak virus isolated from soybeans, *Glycine max. Phytopathology* 60: 814–820.

Frankel, M.J. et al. 1991. Sequence diversity in the surface-exposed amino-terminal regions of coat proteins of sugarcane mosaic virus strains correlated with natural host specificity. (Abstr.) *Phytopathology* 81:1155.

Ford, R.E. et al. 1989. Maize dwarf mosaic virus. *CMI/AAB Descriptions of Plant Viruses*, No. 341.

Fritzsche, R. et al. 1983. Nachweis der Samenübertragbarkeit der milden Rübenvergilbungs—Virus. *Phytopathol. Z.* 106:360–364.

Fulton, J.P. 1967. Dual transmission of tobacco ringspot virus and tomato ringspot virus by *Xiphinema americanum. Phytopathology* 57:535.

Fulton, R.W. 1948. Hosts of the tobacco streak virus. *Phytopathology* 38:421–428.

Fulton, R.W. 1971. Tobacco streak virus. *CMI/AAB Descriptions of Plant Viruses*, No. 44.

Fulton, R.W. 1981. Ilarviruses. In: Kurstak, E., Ed. *Handb. Pl. Virus Infec. Comp. Diag.*, Elsevier/North-Holland Biomedical Press, p. 377–413.

Fulton, R.W. 1985. Tobacco streak virus. *CMI/AAB Descriptions of Plant Viruses*, No. 307.

Gabriel, W. 1960. Poln. 1:41–51. Postepy Nauk.

Gabriel, W. and Kowalska-Noordam, A. 1988. Potato virus S (PVS). In: Smith, I. M. et al., Eds. *European Hand. Pl. Dis.*, Blackwell Sci. Publ., London, p. 4.

Gaillot. 1895. *Bull. Stn. Agronomique de l'Aisne.*

Gardner, M.W. and Kendrick, J.B. 1921. Soybean mosaic. *J. Agric. Res.* 22:111–114.

Ghabrial, S.A. and Lister, R.M. 1974. Chemical and physiochemical properties of two strains of tobacco streak virus. *Virology* 57:1–10.

Gibbs, A.J. and Harrison, B.D. 1964. Viruses in sugar beet. *Rep. Rothamsted Exp. Sta. for 1963*, p. 106.

Gill, C.C. 1965. Increased multiplication of viruses in rusted bean and sunflower tissues. *Phytopathology* 55:141.

Gillaspie, A.G. 1967. Maize dwarf mosaic virus recovered from commercial varieties of sugarcane. *Pl. Dis. Reptr.* 51:761–763.

Gillaspie, A.G., Jr. and Mock, R.G. 1979. Recent survey of sugarcane mosaic virus strains from Colombia, Egypt and Japan. *Sugarcane Pathol. Newsl.* 22:21–23.

Giorda, L.M. et al. 1986. Identification of sugarcane mosaic virus strain H isolate in commercial grain sorghum. *Plant Dis.* 70:624–628.

Giunchedi, L. et al. 1987. Correlation between tolerance and beet necrotic vein virus in sugar beet genotypes. *Phytopathol. Mediter.* 26:23–28.

Goodman, R.M. 1981. Geminiviruses. In: Kurstak, E., Ed. *Hand. Pl. Virus Infec. Comp. Diag.*, Elsevier/North-Holland Biomedical Press, p. 879–910.

Govier, D.A. and B. Kassanis. 1974. A virus-induced component of plant sap seeded with aphids acquire potato virus Y from purified preparations. *Virology* 61:420–426.

Grumet, R. 1994. Development of virus resistant plants via genetic engineering. *Plant Breed. Rev.* 12:47–79.

Grüntzig, M. and Fuchs, E. 1979. Untersuchungen zur serologischen Verwandschaft von beet mosaic virus, potato virus Y, bean yellow mosaic virus and plum pox virus. *Archiv für Phytopath. und Pflanzenschutz* 15:153–159.

Grüntzig, M. 1988. Sugar beet mosaic virus (SuBMV). In: Smith, I.M. et al., Eds. *European Hand. Pl. Dis.*, Blackwell Sci. Publ., p. 48–49.

Gupta, K.C. 1981. Studies on the identity of sunflower mosaic virus. *Abstr. 3rd Int. Symp. on Plant Pathol.*, New Delhi, December 14–18, p. 117.

Hansing, E.D. 1942. A study of the control of the yellow dwarf disease of potatoes. (Abstr.) *Phytopathology* 32:7.

Harrison, B.D. 1964. The transmission of plant viruses in soil. In: Corbett, M.L., and Sisler, H.D., Eds. *Plant Virology*, University of Florida Press, p. 118–147.

Harrison, B.D. 1970. Tobacco rattle virus. *CMI/AAB Descriptions of Plant Viruses*, No. 12.

Harrison, B.D. et al. 1972. Evidence for two functional RNA species in raspberry ringspot virus. *J. Gen. Virol.* 16:339–348.
Harrison, B.D. 1973. Pea early-browning virus. *CMI/AAB Descriptions of Plant Viruses*, No. 120.
Harrison, B.D. 1974. Potato mop-top virus. *CMI/AAB Descriptions of Plant Viruses*, No. 138.
Harrison, B.D. and Robinson, D.J. 1978. The tobraviruses. *Adv. Virus Res.* 23:25–77.
Harrison, B.D. and Robinson, D.J. 1981. Tobravirus. In: Kurstak, E., Ed. *Hand. Pl. Virus Infec. Comp. Diag.*, Elsevier/North-Holland Biomedical Press, p. 515–540.
Harrison, B.D. 1984. Potato leaf roll virus. *CMI/AAB Descriptions of Plant Viruses*, No. 291.
Hatta, T. and Francki, R.I.B. 1976. Anatomy of virus-induced galls on leaves of sugarcane infected with Fiji disease virus and the cellular distribution of virus particles. *Physiol. Pl. Pathol.* 9:321–330.
Heathcote, G.D. 1978. Review of losses caused by virus yellows in English sugar beet crops and the cost of partial control with insecticides. *Plant Pathol.* 27:12–17.
Hendre, R.R. et al. 1975. Growth of mosaic virus-free sugarcane plants from apical meristems. *Ind. Phytopathol.* 28:175–178.
Hewings, A.D. 1989. Soybean dwarf virus. In: Sinclair, J.B. and Beckman, P.A., Eds. *Compendium of Soybean Diseases*, 3rd edition, APS Press, St. Paul, MN, p. 54–55.
Hibi, T. and Kameya-Iwaki, M. 1988. Soybean chlorotic mottle virus. *CMI/AAB Descriptions of Plant Viruses*, No. 331.
Hoerner, R.G. 1949. Hop diseases in the United States. *Brew. Dig.* 24:45–51.
Hollings, M. and Brunt, A.A. 1981. Potyviruses. In: Kurstak, E., Ed. *Hand. Pl. Virus Infec. Comp. Diag.*, Elsevier/North-Holland Biomedical Press, pp. 731–807.
Holmes, F.O. 1929. Local lesions in tobacco mosaic. *Bot. Gaz.* 87:39–55.
Holmes, F.O. 1932. Symptoms of tobacco mosaic diesease. *Contr. Boyce Thomp. Inst.* 4:323–357.
Hooker, W.J. 1981. *Compendium of Potato Diseases*, APS Press, St. Paul, MN. 125 pp.
Hughes, C.G. 1961. Striate mosaic virus: a new disease of sugarcane. *Nature* 190(4773):366–367.
Hughes, C.G. et al. 1968. Division of Pathology. Rep. Bur. Sug. Exp. Stns Qd 68:65–80. (RAM (1969)48:908b).
Hunnius, W. 1976. Zum Problem der Interferenzen zwischen verschiedenen Virusarten der Kartoffel. *Bayerisches Landwirtsch. Jahrbuch* 53:525–564.
Hutchinson, P.B. et al. 1972. Corn, sorghum and Fiji disease. *Sugarcane Pathol. Newsl.* 9:12.
Hutchinson, P.B. and Francki, R.I.B. 1973. Sugarcane Fiji disease virus. *CMI/AAB Descriptions of Plant Viruses*, No. 119.
Huttinga, H. 1975. Properties of viruses of the potyvirus group. 3. A comparison of buoyant density, S value, particle morphology, and molecular weight of the coat protein subunit of 10 viruses and virus isolates. *Neth. J. Pl. Pathol.* 81:58–63.
IIRB. 1994. *Pest and Diseases Group Meeting*, 5–7 September. Broom's Barn Exp. Sta., Higham, Bury St. Edmunds, Suffolk, U.K.: Informations.
Ivanović, D. 1979. Investigation of Beet Necrotic Yellow Vein Virus. M. Sc. thesis, Fac. of Agric. Univ. of Belgrade, Beograd, Zemun, p. 1–47.
Ivanović, D. et al. 1979. Contribution to the Study of beet necrotic yellow vein virus Morphology. *XII Congresso della Societa Italiana di Microscopia Electronica*, Ancona, Italy, Sept. 20–22, Paper No. 104.
Iwaki, M. et al. 1984. Soybean chlorotic mottle, a new caulimovirus on soybean. *Plant Dis.* 68:1009–1011.
Iwanowski, D. 1892. Über die Mosaikkrankheit der Tabakspflanze. *Bull. Acad. Imp. Sci. St. Petersburg (Izv. Imp. Akad. Nauk SSSR)*, N.S. III, 35:65–70.
Jafri, A.M. 1972. Localization of virus and ultrastructural changes in sugar beet leaf cells associated with infection by a necrotic strain of beet mosaic virus. *Phytopathology* 62:1103.
Jilka, J. 1990. Cloning and Characterization of the 3'-terminal Regions of RNA from Select Strains of the Maize Dwarf Mosaic Virus and Sugarcane Mosaic Virus. Ph.D. dissertation, Univ. Ill., Urbana, 218 p.
Jones, A.T. and Mayo, M.A. 1975. Further properties of black raspberry latent virus, and evidence for its relationship to tobacco streak virus. *Ann. Appl. Biol.* 79:297–306.
Joyce, P.A. et al. 1997a. Mosaic resistant transgenic sugarcane. *Proc. ISSCT 5th Pathol.* and *2nd Mol. Biol. Workshop*.
Joyce, P.A. et al. 1997b. Engineering for resistance to SCMV in sugarcane. Proc. Int. Symp. Biotech. Tropical and Subtropical Species. *Acta Hort.* (in press).
Joyce, P.A. et al. 1998. Transgenic sugarcane resistant to sugarcane mosaic virus. *Proc. ASSCT* 20:204–210.
Kado, I.C. and Agrawal, O.H. 1972. *Principles and Techniques in Plant Virology*, Van Nostrand Reinhold Company.

Kahn, R.P. 1956. Seed transmission of the tomato ringspot virus in the Lincoln variety of soybeans. (Abstr.) *Phytopathology* 46:295.

Kaiser, W.J., Wyatt, S.D. and Pesho, G.R. 1982. Natural hosts and vectors of tobacco streak virus in eastern Washington. *Phytopathology* 72:1508–1512.

Kassanis, B. and MacFarlane, I. 1965. Interaction of virus strain, fungus isolate and host species in the transmission of tobacco necrosis virus. *Virology* 26:603–612.

Kassanis, R. and Phillips, M.P. 1970. Serological relationships of tobacco necrosis virus and their ability to activate strains of satellite virus. *J. Gen. Virol.* 9:119–126.

Kassanis, B. and Govier, D.A. 1972. Potato aucuba mosaic virus. *CMI/AAB Descriptions of Plant Viruses*, No. 98.

Keskin, B., Gaertner, A. and Fuchs, W.H. 1962. Über eine die Wurzel von *Beta vulgaris* befallende Plasmodiophoraceae. *Berichte der Deutsche Botanischen Gesellschaft* 75:275–279.

Keyworth, W.A. 1951. Split leaf blotch disease of the hop (*Humulus lupulus* L.). *J. Hort. Sci.* 26:163–168.

Koenig, R. and Lesemann, D.-E. 1989. Potato virus X. *CMI/AAB Descriptions of Plant Viruses*, No. 354.

Koike, H. and Gillaspie, A.G., Jr. 1976. Strain M, a new strain of sugarcane mosaic virus. *Pl. Dis. Reptr.* 60:50–54.

Kojima, M. et al. 1969. Purification and electron microscopy of potato leaf roll virus. *Virology* 39:162–171.

Kojima, M. and Lapierre, H. 1988. Potato leaf roll virus (PLRV). In: Smith, I.M. et al., Eds. *European Hand. Pl. Dis.*, Blackwell Sci. Publ., London, p. 23–24.

Kolte, S.J. 1985. *Diseases of Annual Edible Oilseed Crops. Vol. III: Sunflower, Safflower and Nigerseed Diseases*, CRC Press LLC, Boca Raton, FL.

Kondaiah, E. and Nayudu, M.V. 1984. A key to the identification of sugarcane mosaic virus (SCMV) strains. *Sugar Cane* 6:3–8.

Kovachevsky, J. 1961. Gurkenmosaic (*Marmor cucumeris*) der Zuckerrübe (Bulg.) Rastit. *Zaštita, Sofija* 9, Nr. 3, 11–16.

Krstić, B. 1990. Pogodnost nekih metoda za serološko dokazivanje virusa nekrotičnog žutila nerava repe (Suitability of Some Serological Methods for Beet Necrotic Yellow Vein Virus Detection.) M.Sc. thesis, Faculty of Agriculture, Beograd, 1–64.

Kubo, S. 1981. Tobacco necrotic dwarf virus. *CMI/AAB Descriptions of Plant Viruses*, No. 234.

Kus, M. 1960. Dosadašnji rezultati serološkog testiranja krompira na X-virus na selekcijskoj stanici Poljane. *Zaštita bilja* 62:73–81.

Kus, M. 1964. Problem viroza kod selekcije krompira u Sloveniji i Gorskom Kotaru, Doktorska disertacija, Poljoprivredni fakultet, Zemun.

Kuwata, S. and Kubo, S. 1986. Tobacco stunt virus. *CMI/AAB Descriptions of Plant Viruses*, No. 313.

LaFluer, D.A. et al. 1996. Portions of the banana streak badnavirus genome are integrated in the genome of its host *Musa* sp. (Abstr.) *Phytopathology* 86:S100.

Lana, A.F. et al. 1983. Association of tobacco ringspot virus with a union incompatibility of apple. *Phytopathol. Z.* 106:141–148.

Lange, L. 1978. *Synchytrium endobioticum* and potato virus X. *Phytopathol. Z.* 92:132–142.

Le Buanec, B. 1988. Example de la betterave dans le domaine de la protection des plantes. *C.R. Acad. Agric. Fr.* 74, 7:41–48.

Legg, J.T. 1959. A mild strain of hop mosaic virus. *Annu. Rept. East Malling Res. Sta.* 46:116–117.

Legg, J.T. and Ormerod, P.J. 1961. Yellow net—a virus disease of hop. *Annu. Rev. East Malling Res. Sta. 1960*, 47:105.

Legg, J.T. 1964. Virus causing nettle head symptoms. *Annu. Rep. East Malling Res. Sta.* 51:174–176.

Leiser, R.M. and Richter, J. 1980. Differenzierung von Stämmen des Kartoffel - X virus. *Tagungsbericht der Akademie der Landwirtschaftswissenschaften der DDR* 184:107–113.

Lemaire, O. et al. 1988. Effect of beet necrotic yellow vein virus RNA composition on transmission by *Polymyxa betae*. *Virology* 162:232–235.

Lind, J. 1915. Runkelroernes Mosaiksye. *Tijdsskrift for Planteavl* 22:444–457.

Lister, R.M. and Bancroft, J.B. 1970. Alteration of tobacco streak virus component ratios as influenced by host and extraction procedure. *Phytopathology* 60:689–694.

Lister, R.M. et al. 1972. Evidence that particle size heterogeneity is the cause of centrifugal heterogeneity in tobacco streak virus. *Virology* 49:290–299.

Lister, R.M. and Bar-Joseph, M. 1981. Closteroviruses. In: Kurstak, E., Ed. *Handb. Pl. Virus Infec. Comp. Diag.*, Elsevier/North-Holland Biomedical Press, p. 809–844.

Lizárraga, R.E. et al. 1980. Elimination of potato spindle tuber viroid by low temperature and meristem culture. *Phytopathology* 70:754–755.

Lockhart, B.E.L. 1986. Purification and serology of a bacilliform virus associated with banana streak disease. *Phytopathology* 76:995–999.

Lovisolo, O. and Benetti, M.P. 1960. Sulla presenza in Italia dei virus S ed M della Patata (On the presence in Italy of Potato viruses S and M). *Boll. Staz. Patol. Veget. Roma 7*, Ser. 3, 211–220.

MacKenzie, D.R. et al. 1966. Differences in maize dwarf mosaic virus isolates of the northeastern United States. *Pl. Dis. Reptr.* 50:814–818.

Magarey, R.C. 1996. Downy mildew, Ramu stunt and Ramu orange leaf. Pp. 113–115. In: B.J. Croft et al., Eds. *Sugarcane Germplasm Consesrvation and Exchange*, ACIAR Proc. No. 67, Canberra, 134 pp.

Magarey, R.C. et al. 1996. New sugarcane diseases observed in commercial cane at Gusap, PNG. *Proc. XXII ISSCT Congress* 2:472–476.

Martin, F. P. and Conant, R. K. 1939. *Hawaiian Planters' Record* 43, 277.

Martyn, E.B. (Ed.) 1968. Plant virus names. An annotated list of names and synonyms of plant viruses and diseases. *Phytopathol. Pap.* 9, p. 30.

Martyn, E.B., Ed. 1971. Plant virus names. Supplement No. 1. Additions and corrections to Phytopathological Paper No. 9, 1968 and newly recorded plant virus names. *Phytopat. Pap.* 9, Suppl. 1, p. 41.

Magyarosy, A.C. and Duffus, J.E. 1977. The occurrence of highly virulent strains of the beet curly top virus in California. *Pl. Dis. Reptr.* 63:248–251.

Matthews, R.E.F. 1981. *Plant Virology*, 2nd edition, Academic Press, New York.

Mayer, A. 1886. Über die Mosaikkrankheit des Tabaks. Landwirtschaft. *Versuchstationen* 32:451–467.

Mayo, M.A. et al. 1982. Evidence that potato leaf roll virus RNA is positive-stranded, is linked to a small protein and does not contain polyadenylate. *J. Gen. Virol.* 59:163–167.

McKern, N.M. et al. 1992. Coat protein properties suggest that azuki bean mosaic virus, blackeye cowpea mosaic virus, peanut stripe virus and three isolates from soybean are strains of the same potyvirus. *Intervirology* 33:121–134.

McKinney, H.H. 1929. Mosaic diseases in the Canary Islands, West Africa and Gibraltar. *J. Agr. Res.* 39:557–578.

McKinney, H.H. et al. 1965. Longevity of some plant viruses stored in chemically dehydrated tissues. *Phytopathology* 55:1043.

Merdinoglu, D. et al. 1993. Amélioration génétique de la betterave pour la résistance à la rhizomanie. *C.R. Acad. Agric. Fr.* 79:6:85–98.

Mickovski, J. 1965. Prilog poznavaja viroza duvana u SR Makedoniji: Doktorska disertacija, Poljoprivredni fakultet, Zemun.

Mickovski, J. 1969. Tomato spotted wilt virus na duvanu u Jugoslaviji (Lycopersicum virus 3 - Smith). *Zaštita bilja* 105:203–214.

Milovanović, M. 1989. O uticaju virusa rizomanije na proizvodne i anatomsko-histološke osobine korena nekih sorata šećerne repe. Doktorska disertacija. Poljoprivredni fakultet, Zemun (Beograd).

Milovanović, M. 1994. Personal communication, unpublished data.

Mirkov, T.E. et al. 1997. Transgenic virus resistant sugarcane. *Proc. ISSCT 5th Pathol.* and *2nd Mol. Biol. Workshop*.

Morel, G. and Martin, C. 1955. Guérison de pommes de terre atteintes de maladies à virus (Cure of potatoes infected by virus diseases). *C.R. Acad. Agric. Fr.* 41:472–475.

Muntanola, M. 1948. Description de una nueva enfermedad del girasol. *Rev. Invest. Agric., Buenos Aires* 2:205.

Murant, A.F. 1981. Nepoviruses. In: Kurstak, E., Ed. *Hand. Pl. Virus Infec. Comp. Diag.*, Elsevier/North-Holland Biomedical Press, p. 197–238.

Murayama, D. and Kojima, M. 1965. Studies on the properties of potato leaf roll virus by the aphid-infection method. *Ann. Phytopathol. Soc. Japan* 30:209.

Nayudu, M.V. et al. 1971. Striate mosaic of sugarcane in India. *Sugarcane Pathol. Newsl.* 7:24. (Abstr. in *Rev. Pl. Pathol.* (1972)51:2810).

Nienhaus, F. and Stille, B. 1965. Übertragung des Kartoffel-X-Virus durch Zoosporen von *Synchytrium endobioticum*. *Phytopathol. Z.* 54:335–337.

Nikolić, V. et al. 1958. Ispitivanja štetnosti mozaika na industrijskoj šećernoj repi. *Zaštita bilja* 46:69–73.

Nokalis-Burelle, T. et al. 1997. *Compendium of Peanut Diseases*, 2nd edition, APS Press, St. Paul, MN, 94 pp.

Nuber, K. 1962. Die Mosaikkrankheit der Hopfens. *Zeits. für Pflanzenkrankheiten (Pflanzenpathologie) und Pflanzenschutz* 69:598–602.

Orellana, R.G. and Quacquarelli, A. 1968. A sunflower mosaic caused by a strain of cucumber mosaic virus. *Phytopathology* 58:1439–1440.

O'Brien, M.J. et al. 1963. Symptomless hosts of the potato spindle tuber virus (PSTV). (Abstr.)*Phytopathology* 53:884.

Osaki, T. and Inouye, I. 1981. Tobacco leaf curl virus. *CMI/AAB Descriptions of Plant Viruses*, No. 232.

Panić, M. et al. 1978. Uticaj nekih sredstava za dezinfekciju zemljišta na pojavu i štetnost infektivne kržljavosti korena šećerne repe. *Agrohemija (Beograd)* 11–12:447–453.

Peters, D. and Goldbach, R. 1995. The biology of tospoviruses. In: R.P. Singh et al., Eds. *Pathogenesis and Host Specificity in Plant Diseases. Vol. III. Viruses and Viroids.* Pergamon Press. p. 199–210.

Piolanti, G. et al. 1957. Osservazioni sul fenomeno dei bassi titolidelle bietole in alcune provincie venete. *Il Giornale del Bieticoltore* IV:12.

Pirone, T.P. 1972. Sugarcane mosaic virus. *CMI/AAB Descriptions of Plant Viruses*, No. 88.

Pirone, T.P. et al. 1973. Tobacco vein mottling virus on burley tobacco in Kentucky.*Pl. Dis. Reptr.* 57:841–844.

Pirone, T.P. 1979. Comparative aphid transmission properties of isolates of tobacco etch virus. (Abstr.) *Phytopathology* 69:531.

Pirone, T.P. and Shaw, J.G. 1988. Tobacco vein mottling virus.*CMI/AAB Descriptions of Plant Viruses*, No. 325.

Polak, J. 1988. Beet curly top virus (BCTV). In: Smith, I.M. et al., Eds.*European Hand. Pl. Dis.*, Blackwell Sci. Publ., London, p. 73–74.

Polak, J. and Lot, H. 1988. Beet pseudoyellows virus. In: Smith, I.M. et al., Eds.*European Hand. Pl. Dis.*, Blackwell Sci. Publ., London, p. 8.

Price, W.C. 1940. Comparative host ranges of six plant viruses.*Am. J. Bot.* 27:530–541.

Prillieux and Delacroix. 1898. Jaunisse, maladie bacteriénne de la betterave.*Bull. Ass. Chim.* 235, and *C.R. Acad. Sci. Paris* 127:338–339.

Probasco, E.G. and Scotland, C.B. 1978. Host range, general properties, purification, and electron microscopy of hop latent virus. *Phytopathology* 68:277–281.

Proeseler, G. 1966. Beziehungen zwischen der Rübenblattwanze *Piesma quadratum* Fieb. und dem Rübenkräuselvirus II. Injektionsversuche. (Relationship between the leaf *Piesma quadratum* Fieb. and the leaf crinkle virus.) *Phytopathol. Z.* 56:213–237.

Proeseler, G. 1983. Beet leaf curl virus. *CMI/AAB Descriptions of Plant Viruses*, No. 268.

Pruša, V. 1961. Virova chloroza ehmele. *Chmelažstri* 34:78.

Purcifull, E. and Edwardson, J.R. 1981. Potexviruses. In: Kurstak, E., Ed.*Hand. Pl. Virus Infec. Comp. Diag.*, Elsevier/North-Holland Biomedical Press, p. 627–693.

Purcifull, D.E. and Hiebert, E. 1982. Tobacco etch virus. *CMI/AAB Descriptions of Plant Viruses*, No. 258.

Putz, C. 1977. Composition and structure of beet necrotic yellow vein virus.*J. Gen. Virol.* 35:397–401.

Quanjer, H.M. 1913. Meded R. Hoog, Land-tuin, en Boschbouwschool, 6:41–80.

Rao, G.P. et al. 1999. *Current Trends in Sugarcane Pathology. III. Viruses and Phytoplasma Diseases*, Oxford Press, India (in press).

Raymer, W. B. and O'Brien, M. J., 1962. Transmission of potato spindle tuber virus to tomato.*Am. Potato J.* 39:401–408.

Reddy, D.V.R. 1991. Groundnut viruses and virus diseases: distribution, identification and control.*Rev. Pl. Pathol.* 70:665–677.

Reddy, D.V.R. 1998. Control measures for the economically important peanut viruses. In: A. Hadidi et al., Eds. *Plant Virus Disease Control*, APS Press, St. Paul, MN. p. 541–546.

Reddy, D.V.R. et al. 1998. Pecluviruses. *Encyclopedia of Virology.* 2nd edition. Academic Press, New York, (in press).

Reddy, D.V.R. and Richins, R.D. 1998. Plant pararetroviruses: Legume caulimoviruses. *Encyclopedia of Virology*, 2nd edition: Academic Press, New York, (in press).

Richardson, D.E. and French, W.M. 1978. Evidence for the possible transmission of PVX by *Spongospora subterranea*, in *7th Triennial Conf. Eur. Assn. Potato Res.*, Warsaw, p. 254–255.

Rochow, W.E. and J.E. Duffus. 1981. Luteovirus and yellows diseases. In: Kurstak, E., Ed.*Hand. Pl. Virus Infec. Comp. Diag.*, Elsevier/North-Holland Biomedical Press, p. 147–170.

Rodriguez, L.E. et al. 1985. Report of a new sugarcane virus. Reporte de un nuevo virus de la cana de azucar. *Ciencias de la Agricultura* No. 23:130. (Abstr. in *Rev. Pl. Pathol.* (1989) 68:259.)

Roland, G. 1936. Recherches sur la jaunisse de la betterave et quelques observations sur la mosaïque de cette plante. *Publ. Inst. Belge Amél. Betterave* 2:35–60.

Roland, G. 1960. A contribution to the study of the viruses of chrysanthemum. (Abstr.)*Rev. Appl. Mycol.* 39:584.

Rowhani, A. and Stace-Smith, R. 1979. Purification and characterization of potato leaf roll virus. *Virology* 98:45–54.

Rozendaal, A. 1954. De Betekenis van verschillende virusgroepenvoor de teelt van pootgoed. *Landbouwvoorlichting* 11:299–308.

Russell, G.E. 1962. Sugar beet mild yellowing virus: a persistant aphid-transmitted virus. *Nature* 195:4847, 1231.

Russell, G.E. 1970. Beet yellows virus. *CMI/AAB Descriptions of Plant Viruses*, No. 13.

Russell, G.E. 1971. Beet mosaic virus. *CMI/AAB Descriptions of Plant Viruses*, No. 53.

Russell, G.E. et al. 1975. An aphid transmitted yellowing virus disease of sunflower.*Plant Pathol. (U.K.)* 24:58.

Salazar, L.F. 1988. Potato spindle tuber viroid. In: Smith, I.M. et al., Eds. *European Hand. Pl. Dis.*, Blackwell Sci. Publ., London, p. 57.

Sano, T. and Shikata, E. 1988. Hop stunt viroid. *CMI/AAB Descriptions of Plant Viruses*, No. 326.

Sasaki, M. and Shikata, E. 1977. Studies on the host range of hop stunt disease in Japan. *Proc. Japan Acad.* 53B:103–108.

Sasaki, M. and Shikata, E. 1978. Studies on hop stunt disease. II. Properties of the casual agent, a viroid.*Ann. Phytopath. Soc. Japan* 44:570–577.

Sänger, H.L. et al. 1976. Viroids are single-stranded covalently closed circular RNA molecules existing as highly base-paired rod-like structures. *Proc. U.S. Natl. Acad. Sci.* 73:3852–3856.

Sänger, H.L. 1982. Biology, structure, functions and possible origin of viroids. In: Parthier, B. and Boulter, D., Eds. *Nucleic Acids and Proteins in Plants*, Vol. 2, Springer-Verlag, Berlin, p. 368–454.

Schmelzer, K. 1970. Die Ursache des Kräuselmosaiks an Sonnenblumen. *Nachrichtenbl. dt. Pflanzenschutzd. (Berlin)* N.F. 24:1–5.

Schmelzer, K. and Schmidt, H.E. 1977. Spezialkulturen. In: Klinkowski, M. et al., Eds. *Pflanzliche Virologie*, Vol. 3, 3rd edition, Akademie-Verlag, Berlin, p. 294–364.

Schmelzer, K. and Hartleb, H. 1977. Beta-und Brassica-Rüben. In: Klinkowski, M. et al., Eds. *Pflanzliche Virologie*, Vol. 3, 3rd edition, Akademie-Verlag, Berlin, p. 114–143.

Schmidt, H.E. 1965. Untersuchungen über Virosen des Hopfens (*Humulus lupulus*) 1. Mitteilung, Symtome und Propfübertragung. *Phytopathol. Z.* 53:216–248.

Schmidt, H.E. 1968. Analyse von Hopfenviren und Aufbau virus-getester Hopfenklone. *Forschungasbschlussbericht*, 43S.

Semancik, J.S. and Kajiyama, M.R. 1967. Properties and relationships among RNA species from tobacco rattle virus. *Virology* 33:523–532.

Shepherd, R.J. et al. 1966. A severe necrotic strain of the beet mosaic virus. *J. Am. Soc. Sugar Beet Tech.* 14:97–105.

Shew, H.D. and Lucas, G.B. 1991. *Compendium of Tobacco Diseases*, APS Press, St. Paul, MN, 68 pp.

Shukla, D.D. et al. 1983. Characteristics of the electro-blot radioimmunoassay (EBRIA) in relation to the identification of plant viruses. *Acta Phytopathol. Acad. Sci. Hung.* 18:79–84.

Shukla, D.D. et al. 1989. Taxonomy of potyviruses infecting maize, sorghum and sugarcane in Australia and the United States as determined by reactivities of polyclonal antibodies directed towards virus-specific N-termini of coat proteins. *Phytopathology* 79:223–229.

Sinclair, J.B. and Backman, P.A. 1989. *Compendium of Soybean Diseases*, 3rd edition, APS Press, St. Paul, MN. 106 pp.

Singh, J.P. 1979. Mechanical and seed transmission of the causal agent of sunflower rugose mosaic in Kenya. *Sunflower Newsl.* 3:13.

Singh, R.P. and Bagnall, R.H. 1968. Infectious nucleic acid from host tissues infected with the potato spindle tuber virus. *Phytopathology* 58:696–699.

Sinnema, J.A. 1978. Summary of papers *Virology Section, European Assn. Potato Res., Congress*, Warsaw, 1977.

Slack, S.A. 1979. Potato yellow dwarf virus. In: Hooker, W.J., Ed. *Compendium of Potato Diseases*, p. 82–84.

Smith, G.R. and Harding, R.M. 1999. Genetic engineering for virus resistance in sugarcane. In Rao, G.P. et al. *Current Trends in Sugarcane Pathology III. Virus and Phytoplasma Diseases.* Oxford Press (in press).

Smith, I.M. et al., Eds. 1988. *European Handb. Pl. Dis.*, Blackwell Sci. Publ., London, p. 583.

Smith, K.M. 1946. The transmission of a plant virus complex by aphids. *Parasitology* 37:131–134.
Smith, K. 1972. A Textbook of Plant Virus Diseases, 3rd edition, Academic Press, New York, 684 p.
Snazelle, T.E. et al. 1971. Purification and serology of maize dwarf mosaic and sugarcane mosaic viruses. *Phytopathology* 61:1059–1063.
Spar, D. and Hamann, U. 1977. Kartofel. In: Klinkowski, M. et al., Ed. *Pflanzliche Virologie*, Vol. 3, 3rd edition, Akademie-Verlag, Berlin, p. 63–113.
Stace-Smith, R. and Mellor, F.C. 1968. Eradication of potato viruses X and S by thermotherapy and axillary bud culture. *Phytopathology* 58:199–203.
Stace-Smith, R. 1970. Tomato ringspot virus. *CMI/AAB Descriptions of Plant Viruses*, No. 18.
Stace-Smith, R. 1985. Tobacco ringspot virus. *CMI/AAB Descriptions of Plant Viruses*, No. 309.
Sun, M.K.C. et al. 1974. Properties of tobacco vein-mottling virus, a new pathogen of tobacco. *Phytopathology* 64:1133–1136.
Šutić, D. et al. 1959. Uticaj virusa žutice na prinos i sadržaj šećerne repe. *Zaštita bilja* 55:15–22.
Šutić, D. et al. 1959a. Über den Einfluss des Gelbsuchtvirus auf den Samenertrag und die Samengüte der Beta - Rüben. *Z. Pflkrankh. u. Pflanschutz* 66:681–684.
Šutić, D. 1960. Occurrence of a new sunflower disorder in Yugoslavia. *FAO Plant Protect. Bull.* 8:129–131.
Šutić, D. and Spasić, P. 1963. Einfluss der Düngung auf die Verringerung der schädlichen Wirkung der Gelbsucht auf den Zuckerrübenertrag. *Z. Pflkrankh. U. Pflschutz* 70:339–342.
Šutić, D. and Milovanović, M. 1978. Pojava i značaj kržljavosti korena šećerne repe. *Agrohemija (Beograd)* 9–10:363–368.
Šutić, D. 1980. *Biljni virusi*. Nolit, Beograd.
Šutić, D. and Milovanović, M. 1980. Some Facts Affecting Epidemiology of Sugar Beet Rhizomania-like Diseases, Vortrag Heraklea/Kreta.
Šutić, D. 1983. *Viroze biljaka*. Nolit, Beograd.
Šutić, D. and Sinclair, J.B. 1991. *Anatomy and Physiology of Diseased Plants*, CRC Press, LLC, Boca Raton, FL.
Taliansky et al. 1996. Complete nucleotide sequence and organization of the RNA genome of groundnut rosette umbravirus. *J. Gen. Virol.* 77:2335–2345.
Tamada, T. 1975. Beet necrotic yellow vein virus. *CMI/AAB Descriptions of Plant Viruses*, No. 144.
Tamada, T. and Baba, T. 1973. Beet necrotic yellow vein virus from Rizomania-affected sugar beet in Japan. *Ann. Phytopathol. Soc. Japan* 30:325–331.
Tamada, T. et al. 1969. Soybean dwarf, a new virus disease. *Ann. Phytopathol. Soc. Japan* 35:282–285.
Tamada, T. and Kojima, M. 1977. Soybean dwarf virus. *CMI/AAB Descriptions of Plant Viruses*, No. 179.
Teakle, D.S. 1962. Transmission of tobacco necrosis by a fungus *Olpidium brassicae*. *Virology* 18:224–231.
Teakle, D.S. and Grills, N.E. 1973. Four strains of sugar-cane mosaic virus infecting cereals and other grasses in Australia. *Aust. J. Agr. Res.* 24:465–477.
Teakle, D.S. et al. 1989. Sugarcane mosaic virus. *CMI/AAB Descriptions of Plant Viruses*, No. 342 (88 revised).
Teycheney, P.Y. and Dietzgen, R.G. 1994. Cloning and sequence analysis of the coat protein genes of an Australian strain of peanut mottle and an indonesian "blotch" strain of peanut stripe potyviruses. *Virus Res.* 31:235–244.
Thomas, P.E. and Bowyer, J.W. 1984. Tobacco yellow dwarf virus. *CMI/AAB Descriptions of Plant Viruses*, No. 278.
Thomas, P.E. and Mink, G.I. 1979. Beet curly top virus. *CMI/AAB Descriptions of Plant Viruses*, No. 210.
Thomson, A. 1956. Heat treatments and tissue culture as a means of freeing potatoes from virus. *Nature* 177:709.
Thornbury, D.W. and Pirone, T.P. 1983. Helper component of two potyviruses are serologically distinct. *Virology* 125:487–490.
Thresh, J.M. et al. 1972. The spread and control of nettlehead and related diseases of hop. *Annu. Rep. East Malling Res. Sta.* 59:155–162.
Tippett, R.L. and Abbott, E.V. 1968. A new strain of sugarcane mosaic virus in Louisiana. *Pl. Dis. Reptr.* 52:449–451.
Tomlinson, J.A. and Ward, C.M. 1982. Selection for immunity in swede (*Brassica napus*) to infections by turnip mosaic virus. *Ann. Appl. Biol.* 101:43–50.
Tomlinson, J.A. et al. 1983. Chenopodium necrosis: a distinctive strain of tobacco necrosis virus isolated from water. *Ann. Appl. Biol.* 102:135–147.

Tomlinson, J.A. 1988. Beet western yellows virus. In: Smith, I.M. et al., Eds. *European Hand. Pl. Dis.*, Blackwell Sci. Publ., London, p. 21–23.

Tošić, M. 1982. Prilog poznavanju virusne prirode infektivne kržljavosti korena šećerne repe (Contribution to the study of virus etiology of sugar beet infectious root dwarfing (rhizomania). *Zaštita bilja* 162:465–470. Beograd.

Tošić, M. and Ford, R.E. 1974. Physical and serological properties of maize dwarf mosaic and sugarcane mosaic viruses. *Phytopathology* 64:312–317.

Tošić, M. et al. 1985. Investigations of sugar beet rhizomania in Yugoslavia. *48th Winter Congress of IIPB*, Feb. 13–14, p. 432–445, Brussels.

Tu, J.C. et al. 1970. Effects of soybean mosaic virus and/or bean pod mottle virus infection on soybean nodulation. *Phytopathology* 60:518–523.

Turska, E. 1980. Ausbreitung des S-virus durch Blattläuse. *Kartoffelbau* 11:384–385.

Uyemoto, J.K. 1981. Tobacco necrosis and satellite viruses. In: Kurstak, E., Ed. *Hand. Pl. Virus Infec. Comp. Diag.*, Elsevier/North Holland Biomedical Press, p. 123–146.

Uyemoto, J.K. 1988. Tobacco necrosis virus. In: Smith, I.M. et al., Eds. *European Hand. Pl. Dis.*, Blackwell Sci. Publ., London, p. 80–81.

Van Dorst, H.J.M. and Peters, D. 1974. Some biological observations on pale fruit, a viroid incited disease of cucumber. *Neth. J. Pl. Pathol.* 80:85–96.

Van Regemmortel, M.H.V. 1981. Tobamoviruses. In: Kurstak, E., Ed. *Hand. Pl. Virus Infec. Comp. Diag.*, Elsevier/North-Holland Biomedical Press, p. 541–564.

Walker, J.C. and Larson, R.H. 1939. Yellow dwarf of potato in Wisconsin. *J. Agric. Res.* 59:259–280.

Wallace, G.B. 1947. Krommek disease. *East Afr. Agric. J.* 13(2):103.

Watkins, T.C. 1941. Clover leafhopper (*Aceratagalia sanguinolenta* Prov.). *Bull. Cornell Univer. Agric. Exp. Stn.* 758, 24 pp.

Watson, M. 1962. Yellow-net virus of sugar beet. I. Transmission and some properties. *Ann. Appl. Biol.* 50:451–460.

Weber, G.F. 1932. Mosaic, a virus disease of many plants. *Florida Growers* 40(12):10.

Weidemann, H.L. 1988. Potato virus A (PVA). In: Smith, I.M. et al., Eds. *European Handb. Pl. Dis.*, Blackwell Sci. Publ., London, p. 46.

Wetter, C. 1971. Potato virus S. *CMI/AAB Descriptions of Plant Viruses*, No. 60.

Wetter, C. 1972. Potato virus M. *CMI/AAB Descriptions of Plant Viruses*, No. 87.

Whitney, E.D. and Duffus, J.E. 1986. *Compendium of Beet Diseases and Insects.* APS Press, St. Paul, MN, 76 pp.

Wiltshire, S.P. 1955. Plant diseases in British Colonial Dependencies, *FAO Plant Prot. Bull.* 3, 140.

Yarwood, C.E. and Hecht-Poinar, E. 1973. Viruses from rusts and mildews. *Phytopathology* 63:1111–1115.

Younkin, S.G. 1942. Weed suscepts of the potato yellow dwarf virus. *Am. Potato J.* 19:6.

Younkin, S.G. 1943. *Fm. Res.* 9:6.

Zaitlin, M. and Israel, W.H. 1975. Tobacco mosaic virus (type strain). *CMI/AAB Descriptions of Plant Viruses*, No. 151.

5 Virus Diseases of Fruit Trees

Many fruit tree species distributed worldwide contribute to production of fruit for human consumption. Although adapted to different ecological conditions, they are mostly grown in temperate and subtropical climate regions; for example, pome fruits (apple, pear), stone fruits (apricot, cherry, peach, plum), shell fruits (walnut, chestnut, hazelnut), small fruits (strawberry, raspberry, gooseberry, blueberry), etc. The fruit tree species grown predominantly in tropical areas are known as tropical plants (avocado, banana, citrus, papaya).

Fruits represent natural reservoirs rich in various complexes of elements of structural, physiological, and biochemical importance for the humans. They contain caloric compounds (carbohydrates, lipids), structural and reactive elements (proteins, minerals), stimulators of biological processes (vitamins, enzymes), curative components (aromatic compounds, organic acids, glycosides), and others. Owing to their complex and heterogeneous composition, fruits represent a valuable raw material for the food and soft drink industries, irreplaceable in modern human nutrition.

Viruses have undoubtedly infected fruit trees and caused diseases centuries before the virus was described and before it was proven as a causal agent. The first records are of sour cherry yellows symptoms, which were observed as early as 1768, long before the discovery of viruses (*cf.* Gilmer, 1961). The viral nature of apple mosaic, which was transmitted by grafting in 1825 (*cf.* Bradfort and Joely, 1933), was established by Blodgett in 1923. Although plum pox symptoms were first observed by Bulgarian fruit growers in 1917/1918, the first scientific report on the plum pox virus (PlPV) and the disease it causes appeared 15 years later (Atanasoff, 1932).

Important discoveries relating to fruit tree virus disease identification began after 1930, stimulated by the development and improvement of research methodology used in plant virology. Only five stone fruit virus diseases were known in North America in 1930; 20 years later, this number increased to 48, and in 1976, to 95 virus diseases as reported and described in the agriculture handbook *Virus Diseases and Infectious Disorders of Stone Fruits in North America*, No. 437, 1976. In Britain in 1944, only apple mosaic virus and pear stony pit diseases were known. Then 20 years later, research showed that many orchards were infected with one or more viruses, some seriously affecting fruit yields (Cutting and Montgomery, 1973). Thirty virus diseases of stone and pome fruits have been described in former Yugoslavia since 1936 when the first fruit tree disease, plum pox, was recorded (Josifović, 1937). Comparable records of the dynamic discoveries of fruit virus disease exist in many other countries. More than 160 virus species (Šutić, 1983) occur in stone and pome fruits, and about 20 affect citruses (Agrios, 1988).

When considering the nomenclature of viruses, it is noted that the viral agents of some fruit tree virus diseases have been isolated from infected plants and identified with certainty, such as apple mosaic virus (ApMV), Prunus necrotic ringspot virus (PsNcRsV), citrus tristeza virus (CsTzV), and raspberry ringspot virus (RbRsV). Other viruses that cause some diseases have not yet been fully characterized as established in fact; thus, researchers use the intuitive process of observing similarities and correlations with other well-known virus-caused fruit tree diseases to assume temporarily while awaiting proof that the causal agents are viruses. Thus, this group of diseases is defined only with the common host name and characteristic symptoms (apple green crinkle disease, pear stony pit disease, cherry rusty mottle disease, raspberry curly dwarf disease, etc.) The etiological nature of these diseases will be identified and clarified with further research.

Finally, virus diseases have become the most dangerous of all diseases in modern fruit production, an industry that has developed enormously in the past few decades. Their control requires intensive, specialized, complex techniques. The viruses are distributed inadvertantly by man on a massive scale through vegetative propagation by rooting cuttings, budding, and grafting, in which infected mother plants perpetuate the viruses. Once infected with a virus, these perennial woody plants in production cannot be cured. Perennial, virus-infected fruit trees are subject to permanent damage of fruit quality and yields, a general decline of tree growth, and even death.

The Influence of Viral Infections on Fruit Tree Development and Yields

Viruses that infect fruit trees, as with other plant viruses, cause various structural, physiological, and biochemical changes evident as symptoms in internal (anatomical) and external (morphological) plant structures. The most easily observed changes pertain to plant growth, development, and yields; thus, these parameters are usually used to demonstrate the detrimental effects of individual viruses. For example, Németh (1972) demonstrated the detrimental effect of the ringspot viruses PsNcRsV, ChNcRsV, and ChRsV [the latter two are strains of PsDV] on the development of 2- to 3-year-old sour cherry trees. PsNcRsV, dependant on the year, reduced tree height by 2 to 8%, the length of shoots by 2 to 22%, and the trunk diameter by 5 to 27%. These same reductions caused by ChNcRsV were 11 to 36%, 19 to 43%, and 13 to 28%, respectively. Mixed infections of these viruses were more detrimental—27 to 47%, 44 to 72%, and 38.9%, respectively. The growth of peach plants cv. Lottie infected with PsNcRsV in North America was reduced 12% (Pine, 1964). The growth of peach plants cv. Springtime and Robin infected with PsNcRsV in France were reduced 24 to 28% and 24 to 33%, respectively, when compared with healthy ones (Saunier, 1970).

Detrimental effects of virus infections on growth and yields also occur in apple trees. ApMV can reduce plant growth by 50%, trunk diameter by 20%, and fruit yield by 30% in susceptible cvs. (Posnette and Cropley, 1959; Meijneke et al., 1963). Latent virus infections of apple trees influence their development, depending on how many latent viruses are involved in the infection, the virus strain, the types of rootstocks, and plant age at infection. Growth reductions of apple trees infected with latent viruses do not occur for 2 to 3 years following infection, but the infection usually becomes evident by the fourth year (Posnette and Cropley, 1965; Campbell and Bould, 1970). The trunk girth of Golden Delicious cv. trees grafted to M 26 rootstocks and infected with latent viruses was reduced by 16% (Meijneke et al., 1975), but this reduction did not occur when this cv. was grafted onto seedling rootstocks (Hickey and Shear, 1975).

Detrimental effects of viruses on fruit yields are most important because yield is one of the key parameters in fruit production. Damages by virus infections to fruit yield and quality ranges from drastic to nearly negligible. This implies the need to deliver adequate and precise information about virus diseases to fruit growers.

PlPV, the most detrimental stone fruit virus on plum, causes important morphological, anatomical, and chemical changes in fruits of susceptible cvs. While testing 21 plum cvs. in Yugoslavia, Jordović and Janda (1963) showed that size of the majority of PlPV-infected fruits was reduced by 0.3 to 33%, ranging from 8 to 33% in the most susceptible cvs. to 0.3 to 4% in the most tolerant cvs. Virus infections decreased the total sugar content over all cvs. by 0.1 to 2.0%, ranging from 1.0 to 2.1% in the most susceptible cvs. to 0.4 to 0.8% in the most tolerant cvs. The changes caused by virus infection in the cytological and histo-anatomic fruit structure resulted in a gummy, slimy texture, tasteless fruits, and unsuitable (sour) fruits for processing as marmelade or brandy due to increased acid and reduced total sugar content.

In addition to these quality changes, plum pox virus may cause reductions in fruit yield, depending on the cv. Research in Yugoslavia from 1954 to 1956 showed the plum pox virus to cause yield reductions in the Požegača cv. of 10 to 27%, plus early fruit fall of up to 41% (Jordović and Nikšić, 1957). PlPV caused early fruit fall of 80 to 100% in Bulgaria. In a PlPV-infected region in former Czechoslovakia from 1949 to 1953, the reduced fruit yield per tree was 83%. In Poland, average tree yields were reduced 22%, and the average 100-fruits weight was 12% lower (loc. cit. Bull. OEPP, 1974; Blattny and Heger, 1965; Zawadzka, 1965).

Ringspot viruses are particularly damaging in stone fruit production. PsNcRsV, the strecklenberg disease strain, may reduce sour cherry yields by 91 to 98% (Kegler et al., 1972). Such high losses characterize the acute stage of the disease, while losses of 36 to 43% occur in the chronic stage (Kunze, 1969). Posnette et al. (1968) established that five cherry viruses caused at least 35% yield losses in cherry plantations in England. PsNcRsV and PsDV were shown to reduce fruit set by 37% in four peach cvs., and some strains prevented fruit set (Topchiiska, 1983). Way and Gilmer (1963) observed 10 to 75% less fruit set with pollen from trees with the sour cherry yellows disease.

Virus infections also detrimentally affect the yields of pomaceous fruits. Susceptible apple cvs. infected with the ApMV severe strain produced 30 to 40% smaller fruits (Posnette and Cropley, 1956). Meijneke et al. (1975), who investigated detrimental effects of latent viruses, found a 12% reduced yield in a 4-year-old Golden Delicious cv. grafted onto M 26 rootstock, and when grafted onto M 9 rootstock, a 30% reduced yield (Van Oosten, 1976). Virus-free peach cvs. Conference and Doyenne du Comice, grafted onto quince C rootstock provided 40 to 45% higher yields than in virus-infected trees (Post et al., 1979).

It is important to note that one virus can differ symptomologically and cause varying yield reductions in different fruit tree species, depending on individual cvs. For example, PlPV causes severe symptoms in fruits of Požegača cv. and no symptoms in Anna Spät, Washington, Ruth Gerstetter and other cvs. (Šutić, 1977). Symptoms of plum pox disappear or are alleviated after some time in PlPV-infected Požegača cv. grafted onto the resistant sloe (*Prunus spinosa* var. *dasyphylla*) clone, which may have some practical importance (Šutić, 1965; Macovei et al., 1970).

In addition to apple, the chlorotic leaf spot virus (ApCLsV) infects other stone fruit species. It is latent in the plum cv. Požegača. ApCLsV strains may cause symptoms in fruits of some plum cvs. similar to those of plum pox, thus the name plum pseudopox (Cropley, 1968; Marénaud, 1971). Strains of ApCLsV cause bark split and severe stem cracking and splitting in some plum cvs.; for example, French prune d'Ente P 707 is the most susceptible (Marénaud, 1971; Dunez et al., 1972). ApCLsV destroys ornamental wild apple but cultivated apple, cvs. are tolerant. ApCLsV causes ring pattern mosaic in infected pear leaves, resulting in growth and yield reductions.

ApMV is generally widespread, causing symptoms in stone fruit and rose cultivars. Some strains cause yellow spots in certain apple cvs. (e.g., Golden Delicious, Lord Lambourne and others), whereas many other cvs. (e.g., Cox's Orange) remain symptomless. However, reductions in yield of 20% or more occur for all infected cvs., regardless of symptoms (Cutting and Montgomery, 1973).

Finally, with these examples of the destructive nature and complex influences of viruses on fruit tree growth and development, no fruit tree virus infection should be underestimated.

Fruit tree viruses are transmitted variously in nature, with vegetative propagation, infected seed and pollen, and animal vectors being most important.

Transmission by Vegetative Propagation

The different plant parts used for vegetative propagation include buds (budding), scions (grafting), and rootstocks (sprouts and cuttings). If they originate from virus-infected mother plants, the viruses are readily transmitted and permeate all offspring, spreading from scion varieties into rootstocks and vice versa. As the only mode of fruit tree propagation, vegetative propagation represents the highest risk mode for virus transmission. The viruses that cause obvious symptoms can be avoided during mother plant selection, which is unfortunately impossible with latent viruses which cause no symptoms. Therefore, the danger exists of uncontrolled latent virus spread and accumulation in several generations that may become more detrimental than other viruses.

The international exchange and sales of vegetative planting material have inadvertently contributed greatly to the worldwide distribution of these viruses, thus the reason that numerous fruit viruses are encountered in American, European, and Asian countries.

PlPV, first observed in Bulgaria (1917/1918) and which remained for two decades restricted to Balkan countries (Bulgaria, Romania, Yugoslavia), is now widespread throughout Europe and in Turkey, Cyprus, and Syria, most likely imported with infected planting material. The propagation of plum cv. Požegača from suckers, which has been practiced for several decades and which develop up to 100 per tree, greatly influenced the intensive spread of PlPV in former Yugoslavia when suckers were taken from infected trees. Some 20% of the total 84 million plum trees are currently infected with PlPV (Šutić, 1983).

Transmission Through Infected Seed and Pollen

Several stone fruit viruses are transmitted mostly through infected seed and pollen; for example, prunus necrotic ringspot virus (PsNcRsV), an ilarvirus; prune dwarf virus (PsDV) (syn.: cherry chlorotic ringspot virus and sour cherry yellows virus), possible member of the ilarvirus group; and cherry leaf roll virus (ChLRlV), a nepovirus. These are the most important viruses affecting stone fruits. Animal vectors seemingly play a less important role in the epidemiology of these viruses. The biological importance of mite vectors and nematodes in PsNcRsV development is not clear (Barbara, 1988). No vectors have been reported for PsDV, and ChLRlV transmission by nematode vectors is probably relatively unimportant in orchards (Cooper, 1988). The frequently massive transmission of these viruses through infected seed and pollen probably represents their most reliable preservation in nature.

The percentage of virus transmission through infected fruit tree seed varies with several factors: host species, modes of plant infection (naturally or experimentally), plant age at which viral infection occurs, disease development stage (acute or chronic), etc. PsNcRsV is transmitted in different percentages by cherry (55%), sour cherry (30%), mahaleb cherry (10 to 70%), peach (3.6%), and myrobalan (0 to 4%) (Schmelzer 1980). Seed transmission of PsDV has been established for cherry (6 to 25%), mahaleb cherry (2 to 41%), plum (1.4%), and myrobalan (0 to 2%). Mink and Aichele (1984) reported on noncertified seed lots infected to levels of 10% in *Prunus persicae*, 1% in *Prunus avium*, and 1 to 53% in *Prunus cerasifera*; and Cation (1949; 1952) showed PsDV in 46% of seedlings obtained from mahaleb tree seed and 56% from Montmorency sour cherry. Gilmer and Kamalsky (1962) reported 91% infection in the progeny of Mazzard and mahaleb trees infected by sour cherry yellows (PsDV), and 9% infection in the progeny of trees infected with PsNcRsV. Healthy Montmorency trees, pollinated by the pollen from infected SChYV and PsNcRsV viruses produced 4% and 77% of infected seed, respectively (George and Davidson, 1966), and they reported fewer infected seed from trees in the acute (38%) than in the chronic (78 to 93%) stage of disease. Jones (1985) summarized several authors' results, showing that ChLRlV is transmitted through seed at 1 to 35% of numerous hosts, including, interalia, *Prunus* spp., *Betula* spp., American elm, and walnut. Some ChLRlV strains were also transmitted through seed of herbaceous hosts at 1 to 100%.

Seed infection occurs through pollination with ChLRlV-infected pollen or from infected mother plants. Transmission of ChLRlV by infected pollen was established with the aid of ELISA and agar-gel diffusion for: *Prunus avium* (PsNcRsV 5.7% and PsDV 6.2%); *Prunus persicae* (PsNcRsV 6.6% and PsDV 5.1%); *P. cerasifera* (PsNcRsV 11.3% and PsDV 6.1%) (Poggi-Pollini et al., 1983), and in *Corylus avellana* (ApMV 2%) (Cameron and Thomson, 1986). ChLRlV is transmitted through walnut pollen, and the number of plants infected each year can increase at a 20 to 30% annual rate (Mircetich et al., 1980; 1982).

The data available on transmission of PlPV through seed and pollen are contradictory. Jordović (1963) and Kegler and Schade (1971) could not demonstrate PlPV transmission through plum seed. On the contrary, others reported transmission of PlPV through plum, peach, and apricot seed (Coman and Cociu, 1976; Németh and Kölber, 1983), and through pollen of apricot and peach (Szirmai, 1961; Savulescu and Macovei, 1965) and in Agen plum-trees (Macovei, 1970). This controversy should stimulate new experimentations of practical importance for fruit growing.

Pollen infection represents a high risk in fruit production because this ultimately results in permanent annual increases in numbers of infected trees; thus, nearly all trees in older orchards will become infected. The probability of seed and tree infection through the pollen continuously endangers the phytosanitary condition of orchards, regardless of their establishment with virus-free planting material. Infected seedlings, resulting from myrobalan, peach, apricot, and mahaleb seed, and used as generative rootstocks, thus introduce infections—often in large percentages—into new plantations, a potentially costly mistake for growers.

Transmission by Animal Vectors

Animal vectors of fruit tree viruses are classified phyla *Arthorpoda* (*Arachnoidea* and *Insecta*) and *Nemathelminthes* (*Nematoda*).

Among the *Arachnoidea*, only mites (*Acarina*) are proven vectors of a few stone fruit viruses, for example, peach mosaic disease-vector *Phytoptus* (*Eriophyes*) *insidiosus* (Wilson et al., 1955), cherry leaf mottle disease-vector *Phytoptus* (*Eriophyes*) *inaequalis* (loc. cit. Németh, 1986), and one isolate of prunus latent virus-vector *Aculus* (*Vasates*) *fackeui* (Proeseler, 1958; 1968). In southern Texas and northern Mexico, *P. insidiosus* has contributed to the dangerous spread of the causal agent of peach mosaic disease (PhMDs) transmitted from wild *Prunus americana* and *P. munsoniana*, which serve as natural hosts of both the peach mosaic disease and the mite vector (Pine, 1976). The *Eriophyes phloecoptes* mite occurs in massive populations on plum cv. Požegača trees heavily infected with PlPV in former Yugoslavia, leading one to believe this mite might transmit it, but experimental results do not confirm this belief. Therefore, it is evident that *E. phloeocoptes* is not a vector of PlPV, only that plum is their common host (Šutić, 1975).

Aphids, both abundant and active, represent the most important vectors of fruit-tree viruses. Some representatives of leafhoppers and psyllas are well-known vectors of damaging fruit-tree phytoplasmal diseases. Transmission of PlPV by *Empoasca flavescens* leafhoppers (Jordović, 1963) has not been confirmed.

Aphids in stone fruits are known only as vectors of PlPV and plum ochre mosaic disease. Interestingly, no aphid has been proven as a vector of pome fruit viruses. Vectors of PlPV include *Aphis craccivora*, *A. fabae*, *A. spiraecola* (Leclant, 1973), *Brachycaudus helichrysi* (Christoff, 1947), *B. cardui* (Kunze and Krczal, 1971; Šutić et al., 1972), *B. persicae* (Telvad et al., 1970; Németh, 1986), *Myzus persicae* (Kassanis and Šutić, 1965), *M. varians* (Leclant, 1973), and *Phorodon humuli* (Vaclav, 1960). *M. persicae* is especially important in transmitting PlPV to plum, peach, and less important in apricot. *M. persicae* has also been identified as the vector of plum ochre mosaic virus (Blattny, 1961) and peach latent mosaic (Desvignes, 1981), whose causal agent was recently discovered as a viroid (Flores et al., 1991).

Aphids are the most widespread vector of small fruit viruses, including strawberry crinkle, strawberry mild yellow edge, strawberry vein banding, black raspberry necrosis, raspberry leaf mottle, and raspberry vein chlorosis viruses (see Chapter 6). Aphids also efficiently transmit tropical fruit viruses, including citrus tristeza, banana bunchy top, and papaya ringspot viruses. (More details about these vector aphids can be found in the descriptions of individual virus species.) Owing to their general distribution, abundance, mobility, and rapidity of transmission, aphids undoubtedly represent the most active fruit-tree virus vectors.

Nematodes in the genera *Xiphinema*, *Longidorus*, and *Criconemoides* represent an important group of vectors of fruit tree viruses. All viruses transmitted by nematodes, except for carnation ringspot virus, are members of the nepovirus group. Such viruses described primarily from fruit trees include cherry rasp leaf (American) virus (vector *X. americanum*), peach rosette mosaic virus (vectors *X. americanum*, *Longidorus diadecturus*, *Criconemoides xenoplax*), raspberry ringspot virus - cherry strain (vectors *L. macrosoma*, *X. diversicaudatum*, *X. brevicolle*), and strawberry latent ringspot virus (vectors *X. diversicaudatum*, *X. coxi*).

Fruit trees represent natural alternate hosts of several nepoviruses described in herbaceous plants; for example, ArMV (in peach, raspberry, strawberry), TmBRV strains (in peach, almond, raspberry, strawberry), TmRsV (in apple, apricot, cherry, peach, plum), and SbLtRsV (in peach). Isolates of carnation ringspot virus (dianthovirus group), transmitted also by nematodes, have been identified in apple, pear, plum, and cherry. (More detailed information about the nematodes cited in this chapter can be found in the descriptions of individual fruit tree viruses.)

Finally, nematodes represent dangerous vectors because, on large acreages, it is nearly impossible to control them since often it is not cost-effective to fumigate for soil-borne organisms. Also, nematode-borne viruses generally have a broad range of herbaceous and woody host plants, and thus a high soil contamination potential, which more than doubles the effort needed to effect control. Therefore, soils contaminated with nematode-borne viruses should simply be eliminated from consideration for the production of susceptible plants. Only virus-resistant and nonhost plants should be grown on nematode-infested soils.

We recommend obtaining the *Compendium of* both *Stone Fruit Diseases* (Ogawa et al. 1995) and of *Apple and Pear Diseases* (Jones and Aldwinkle 1990) for a comprehensive treatment of all diseases.

CONTROL OF FRUIT TREE VIRUSES

Virus control is one of the main prerequisites for the improvement of fruit production yields and fruit quality. Prevention is the most important measure in virus control since no methods are known for the treatment of infected trees under regular production conditions. Integrated preventive measures are successful in the control of pome fruit virus diseases and also for stone and other fruit tree viruses that are not transmitted by vectors or through pollen. Curative measures, including thermotherapy, meristem culture, and micrografting (shoot-tip grafting), are of irreplaceable importance for obtaining virus-free plants to be used for the production of healthy initial planting material.

Preventive measures that can be used in fruit growing include production and use of healthy planting material, prevention of virus spread in nature, use of resistant and tolerant fruit tree cultivars, and biological virus control.

Healthy Planting Material

Phytosanitary aspects concerning planting material can be divided into two groups: virus-free and virus-tested. The virus-free material, by definition, cannot contain any virus detectable by the existing virological methods. Such planting material implies the highest quality and is produced for the most important fruit tree cultivars and rootstocks.

The virus-tested material does not contain the most damaging viruses; for example, PlPV-free plum, peach, and apricot scions and rootstocks; ApMV- and ApStmGgV-free scions and rootstocks; pear stony pit disease-free, and pear blister canker viroid-free pear scions and rootstocks, etc. Such planting material is considered the best available for fruit production.

The production of virus-free and virus-tested planting material must pass certain rigorous tests. First, a visual selection of plants fulfills all the phytosanitary and pomological requirements. Whenever possible, the selected materials are subjected to thermotherapy to destroy any potential virus in the plants. In some cases, healthy initial planting material can be obtained by apical meristem isolation and culture, and micrografting techniques.

The virus-free nature of potential planting materials should be further tested prior to increasing the germplasm. Both serological analyses and indicator plants are the primary procedures for these tests. Serological analyses are used specifically to diagnose virus-infected plants. Antisera are usually prepared for all commonly known viruses that have a high probability of being isolated from woody hosts. The serological analyses required depend on the goal of the investigation. Besides the traditional agar gel-diffusion test, other methods used for quick routine diagnosis include latex, ISEM (immunosorbent electron microscopy), and ELISA (enzyme-linked immunosorbent) techniques. The ELISA test is widely used because it is highly sensitive, which enables the positive detection of small quantities of virus and also the ease of simultaneously testing many plant samples. It has been useful for diagnosis of several pome fruit (ApCLsV, ApMV, and ApStmGgV), stone fruit (ChLRlV, PlPV, PsDV, and PsNcRsV) and small fruit viruses (RbBuDV, RbRsV, and

SbLtRsV), and also to diagnose the nepoviruses, TBRV and TmRsV, which are important in fruit trees, plus other herbaceous and woody plant viruses.

Indicator plants are recommended by the ISHS Working Group on Fruit-Tree Virus Diseases for diagnosis of viruses that cause disease because of their unique, quick, characteristic reactions to individual viruses. Attempts are made to transmit potential viruses from the material selected for the establishment of future mother plants by grafting, plant sap, and with vectors onto indicator plants. If present in the tested mother stock material, indicator plants reveal the virus by their reactions, and then those virus-positive materials are excluded from further use and should be destroyed. Woody plants may be suitable virus indicators in field trials and in glasshouses, whereas herbaceous plants are used only for glasshouse testing. Herbaceous plants are used whenever possible because glasshouse trials are more easily performed.

The indicator plant method is used for virus testing wherever virus-free planting material is being produced. Because of the many viruses with the potential to infect fruit trees and numerous suitable indicator plants, this testing proceedure is large-scale and quite complex, frequently avoided by the fruit growers who prefer traditional production methods. The initial mother stock planting material, in good virus-free condition as confirmed by the tests, is then increased under conditions of adequate protection against subsequent virus infections. Based on its quality, these plants present the super elite, "the mother plant nucleus," from which are established elite clones and mother plants for nursery production.

A standard indicator plant list has been established and registered by the International Committee for Cooperation in Fruit Tree Virus Disease Research; this will allow the production of planting materials and indicator plants on an international scale. The usual indicator plants are noted in the descriptions of individual virus species.

Prevention of Virus Spread

Measures contributing to the prevention of the spread of viruses in orchards play a very important role in fruit tree protection. One measure is to establish new plantations spatially isolated from established orchards that usually are infected with several viruses. Such isolation is quite important in order to avoid viruses transmitted either by aphids or by pollen originating from infected trees. Jordović (1968) showed that the aphid-transmitted PlPV within 10 years had infected 100% of healthy plum trees 100 m or less from an infective source, but into only 1.5% at 500 to 800 m from the source. Agrios (1988) observed that to avoid infection of cherry by ChLRlV, PsDV, and PsNcRsV, new plantings of susceptible fruit trees should be established at least 200 m from older orchards. In order to lower the risk of virus transmission from older orchards, even more new nurseries should be established in noninfected regions. Spatial isolation of fruit trees is difficult to do in areas densely populated with orchards; thus, special regulations exist in individual countries.

Acreages earmarked for nursery establishment and new plantations and their immediate surroundings should be cleared of all plants susceptible to economically damaging viruses, such as *Prunus* species (*P. cerasifera*, *P. triloba*, *P. insititia*, *P. spinosa*, etc.), known as hosts of PlPV and PsNcRsV; wild growing *P. americana* and *P. munsoniana*, known as natural hosts of the peach mosaic disease and its vector *Eriophyes insidiosus*, and others. Then, all nurseries and new plantations should be monitored regularly for any possible disease spread and the infected trees should be destroyed immediately. Destruction of infected trees is justified only in orchards infected with viruses that spread slowly, whose vectors are unknown.

Chemical control of vector-transmitted viruses in fruit trees is difficult. However, aphicides are recommended to control aphids, known as active virus vectors of PlPV. Mites that transmit the causal agent of PhMDs can be controlled with the aid of miticides, etc. Nematicides used to control nematode-borne viruses can be effective during the process of soil preparation for establishment of plantations and to prevent spread of infection in orchards. Chemical applications to protect against this complex virus-host system cannot ensure lasting positive effects and are thus of secondary importance only.

When the virus PsNcRsV or PsDV in different species is transmitted by pollen, timely elimination of flowers from mother plants is strongly recommended in order to avoid their infection during pollination. Likewise, one should also eliminate the flowers from the stone fruit stocks developed from casually transmitted flower buds during grafting.

Resistant and Tolerant Fruit Tree Cultivars

Different mechanisms of tolerance and resistance to virus infection exist in individual fruit tree species and cvs. Many cvs. are immune to virus infection, while some resistant cvs. overcome the infection often reacting in a hypersensitive manner; whereas, tolerant cultivars, although infected, remain symptomless or with only mild symptoms. In susceptible cvs., the viruses move systemically throughout infected plants, as evidenced by distinct symptoms. Each of these reaction types represents the result of a specific host/virus strain relationship. Any change among the factors in this system results in changes of reactions within the plants expressed in their resistance or susceptibility.

Virus-immune or virus-resistant cvs. represent the rare ideal in fruit growing. Šutić and Ranković (1981; 1981a) discovered two plum cultivars, Scoldus No. 1 and Jelta Boutil Covidna, immune to PPV, that could not be infected either artifically or by natural inoculations. They found the apricot Hybrid Banaesa 33/13 also immune to PlPV. Kegler and Grüntzig (1991) reported the plum hybrid "K4" (Kirke × Persikovaja) to be resistant to PlPV, evident in its hypersensitive reaction. Hybrids with hypersensitivity were both not systemically infected and symptomless. Hybrids, which are tolerant in contrast to immune, are more frequently encountered among stone and pome fruit species. Damage caused by virus infections in tolerant cvs. are generally economically negligible. Tolerant cvs. do represent important sources of inoculum from which viruses can be transmitted by animal vectors (PlPV virus and PhMDs), infected pollen (RbBuDV, PsDV, PsNcRsV), and infected vegetative rootstocks (ApCLsV and ApMV). Thus, the control of the pollen-borne RbBuDV in raspberry, for example, can be achieved only by breeding RbBuDV-resistant cvs. (Jones et al., 1988).

Kegler (1977) reported many cvs. tolerant to some economically damaging viruses among both pome and stone fruit species. Thus, apple cvs. Klarapfel and McIntosh are tolerant to ApMV; Jonathan and Granny Smith to apple rough skin disease, and Red Delicious, Staymann Winesap, and Rome Beauty to apple Leaf Pucker (russet ring) disease. Pear cultivars Beurré Hardy, Conference, and Fondante d'Automne are tolerant to pear blister canker viroid, and vs. Conference and Madame Verté are tolerant to pear stony pit disease. Peach cvs. Halehaven and Richhaven are tolerant to the stem pitting disease caused by TmRsV.

Particularly important in fruit growing are symptomless tolerant cvs., or those with only mild symptoms on fruits, which can be used or marketed normally. This group also comprises a number of plum cvs. tolerant to some strains of PlPV, for example, Stanley (Jordović and Janda, 1963a; Šutić, 1977; Ranković and Paunović, 1988), Washington, Anna Spät, Ruth Gerstetter, California, and others (Šutić, 1977), Mirabelca Jaune de Plovdiv, Hybride Cristoff No. 1, Edra jelta afiska, as well as Andress and Everts peach cultivars (Ranković, 1983; Ranković and Paunović, 1988; Šutić and Ranković, 1981; 1981a; 1983). (More details about the tolerance of stone fruit species to PlPV can be found in the *Bulletin OEPP*, Vol. 4, No. 1, 1974.)

Immune and tolerant fruit tree cvs. are interesting, both for production and as genetic material in selection (breeding) for crossing and creating new hybrids of economic importance. However, one must be cautious of the risk that infected tolerant plants represent permanent sources of infection from which the viruses are transmitted with the aid of vectors, where vectors exist, and vegetative planting material, which is characteristic for all fruit tree viruses.

Biological Control

Cross protection, vegetative protection, and the creation of genetic resistance are included as biological control measures of plant viruses.

Vegetative Protection

This type of protection is based primarily on the influence of resistant rootstocks on the increase of general plant resistance to virus infections. The concept of vegetative protection was first introduced by Šutić (1965; 1975b; 1976), based on the disappearance symptoms of the plum pox disease under the influence of resistant mahaleb (*Prunus mahaleb*) rootstocks and one sloe (*P. spinosa*) clone. These results motivated Šutić to suggest a broad investigation of resistant and tolerant sloe clones as rootstocks for plum production. While analyzing these and other similar results, Boxus (1969) stressed the possibility of new prospectives in fruit tree virus control resulting from this discovery and that these first observations may help revolutionize the traditional methods of virus control.

The principle of vegetative protection against virus infections has been confirmed (Macovei et al., 1971) and accepted by others (*Bull. OEPP*, 1974; Nemeth, 1986). Although only preliminary data are available on how to apply practical vegetative protection against virus infections, one expects its use will increase in both research and practical applications in the future because it is based on established and proven influences of rootstocks on tree production.

Cross Protection

Cross protection relies on a previous infection of the host plant with a mild virus strain that prevents symptom development or delays symptom occurrence in the case of later infections with a more virulent strain of that virus.

Posnette and Cropley (1956) succeeded in cross protecting among strains of ApMV. Lord Lambourne trees were inoculated with a mild strain that prevented symptom development following later infections with intermediate and virulent strains. Through a previous infection of the P 707 *Prunus domestica* cv. with a wild strain of the ApCLsV, causal agent of plum bark split disease, Marénaud et al. (1976) showed adequate protection against later infections with the virulent strain of the same virus. The Pera sweet orange cv. previously infected with a mild strain of tristeza virus was similarly protected against later infections with more virulent strains (Müller and Costa, 1968).

Šutić (1975a) investigated cross protection against the necrotic, intermediate, and yellow strains of PlPV, based on local lesions occurring on *Chenopodium foetidum* leaves with subsequent double inoculations. The protection afforded in these studies was partial, and the level of protection in individual strains differed. Kerlan et al. (1980) observed protection of the mild strain of PlPV against its virulent strain when subsequently inoculated into GF 305 peach seedlings. Although fully confirmed by many experimental results, primarily due to its complexity, cross protection has not yet been broadly applied as a principle in fruit production.

Creating Genetic Resistance

New research horizons have revealed that genetic resistance can be created by introducing the viral coat protein gene, and potentially other viral genes, into the genome of the host plant. This process is done by genetic engineering techniques to obtain genetically transformed plants, called transgenic plants. The viral coat protein gene in a transgenic plant affects both plant resistance to viral infection and viral spread within the plant. In fact, the capsid protein that is synthetized in the plant along with other normal plant proteins now protects transgenic plants from infections by the virus that was the donor of that coat protein gene. Since protection against infection in transgenic plants is obtained by the mediation of genetic transformation, which differs from the traditional breeding methods, Šutić and Sinclair (1991) refer to this type of plant protection as genetic cross protection.

Using these and similar genetic engineering techniques, scientists have *inter alia* created numerous transgenic plants resistant to virus infections, including tobacco resistant to TMV (Abel et al., 1986) and to CMV (Cuozzo et al., 1988); alfalfa, tobacco, and tomato resistant to AMV (Turner et al., 1987; Loesch-Fries et al., 1987; Van Dun et al., 1987); potato resistant to TmSpoWV

(T. German, U. WI., personal communication); sugarcane reistant to ScMV (Smith and Harding, 1999; Mirkov et al., 1997); and several others.

PlPV is one of the first fruit tree viruses whose coat protein gene has been introduced into transgenic tobacco *Nicotiana xanthi* and *N. benthamiana* via the gene transfer technique (Ravelonandro et al., 1991). Their initial results show that the progeny of regenerated plants exhibit variation of high to poor resistance. Transgenic tobacco *N. clevelandii* and *N. benthamiana* plants with the introduced coat protein gene of PlPV react to mechanical inoculation quite differently, ranging from complete absence to severe symptoms (da Câmara Machado et al., 1991).

The creation of transgenic fruit trees is most difficult, and the major obstacle lies in how to regenerate the genetically transformed plantlets. Laimer da Camara Machado et al. (1991) were first to develop a system to transfer foreign genes into apricot plants. A successful transformation and regeneration of *Prunus armeniaca* was achieved with the use of *Agrobacterium tumefaciens* plasmids, carrying the coat protein gene of PlPV, the first fruit tree virus now with the coat protein-mediated resistance against one of the most dangerous pathogens of stone fruit trees.

Exciting results have now begun paying benefits to papaya growers (*see* PaRsV section).

Research in creating a new approach to genetic resistance will overcome the main obstacles of transformation and regeneration of transformed plants. This approach to the control of fruit tree viruses will add an exciting dimension to the 21st century, with strong potential to revolutionize fruit growing.

VIRUS DISEASES OF POME FRUITS

Virus Diseases of Apple (*Malus sylvestris*)

Thirty virus and virus-like diseases known under different names have been described in apple trees, the majority of which occur in apple as their main host; also included are apple scar skin and dapple apple caused by viroids. Some of the disease names are synonyms normally used in disease nomenclature. Thus, for example, the apple stem grooving virus is considered synonymous with apple E 36 virus, apple latent virus type 2, apple dark green epinasty virus, Virginia Crab stem grooving virus, and apple brown line virus (Polak, 1988). The apple latent virus (ApLtV) type 1 is also a synonym of apple chlorotic leaf spot virus (ApCLsV) (Delbos and Dunez, 1988), which is believed the causal agent of apple leaf pucker disease (Kegler et al., 1979) and, at low temperatures in the early summer, also causes apple russet ring disease on cv. Golden Delicious or apple blotch disease on cv. Stayman (Smith et al., 1988). Apple Spy epinasty and decline and apple stem pitting diseases are probably caused by the same agent, which also causes pear stony pit (pear vein yellows disease) (Kegler and Schimanski, 1988).

Posnette (1963) grouped all apple and pear virus diseases according to characteristic leaf, fruit, and stem symptoms. Such a scheme is undoubtedly suitable, both for diagnostic purposes and for didactic motives, and will be used in further elaborations.

Virus Diseases of Apple With Characteristic Leaf Symptoms

Apple Mosaic Ilarvirus (ApMV)

Properties of Viral Particles

ApMV has isometric particles of about 26-nm diameter (Fulton, 1972); that is 25 and 29 nm in diameter, with sedimentation coefficients of 88S and 117S, respectively (De Sequeira, 1967). The 88S particles are not infective. Morphologically, ApMV is similar to PsNcRsV; thus, ApMV probably contains 16% RNA. The particles of ApMV are quite unstable in extracted plant sap; but in a buffer, infectivity is maintained for a few hours. The TIP is 54°C and the DEP is 2×10^{-3}. ApMV has moderate immunogenicity (Fulton, 1972). ApMV and PsNcRsV are distantly related serologically, and positive heterologous reactions occur with the use of high-titre antisera. Some authors consider ApMV a serotype of the PsNcRsV (Kegler, 1977). ApMV comprises a number of strains based on virulence (severely, moderately, or mildly virulent) and on reactions in herbaceous plants (Posnette and Cropley, 1956; Kristensen and Thomsen, 1962; Thomsen, 1975). Several symptoms described in differnt hosts include apple with infectious variegation, rose with mosaic and infectious chlorosis, plum with line pattern and European plum line pattern, and hop, which is a symptomless host.

Geographic Distribution and Economic Importance

ApMV occurs wherever apple and other hosts are grown, and is most frequently observed in susceptible cvs. such as Jonathan and Golden Delicious (Šutić, 1983). ApMV causes economic damage in susceptible apple cultivars, reducing growth by 50%, trunk diameter by 20%, and yield by 30% (Posnette and Cropley, 1959; Meijneke et al., 1963). Reduced vigor and flowering occur in infected rose.

Host Range

Virus hosts include over 65 species in 19 families (Kristensen and Thomsen, 1962; Fulton, 1965; Thomsen, 1975). Natural hosts in the Rosaceae are *Prunus amygdalus*, *P. domestica*, *P. triloba*, *P. virginiana*, and *Rosa* sp., plus other plants such as birch (*Betula alleghaniensis*, *B. papyrifera*), hazel-nut (*Coryllus avellana*), and hop (*Humulus lupulus*).

Plants that serve as diagnostic indicators of ApMV include *Cucumis sativus* with primary chlorotic spots on cotyledons, systemic infection, and severe stunt; *Torenia fournieri* with light yellow mosaic; *Vinca rosea* with systemic infection, chlorotic ringspots, and line patterns on leaves; *Cyamopsis tetragonoloba* most isolates with necrotic spots; *Phaseolus vulgaris* with dark local spots; and *Malus pumila* cv. Lord Lambourne with prominent leaf mosaic.

Symptoms

Mosaic symptoms occur on ApMV-infected apple leaves as light yellow and sharply bordered spots, line patterns, and chlorotic veins (Figure 149). Susceptible cvs. express symptoms on nearly all leaves, but symptoms in tolerant or resistant cvs. are limited to individual leaves as small yellow spots, which during disease development gradually turn to dark brown. The final result is premature leaf fall in susceptible cvs.

FIGURE 149 Apple mosaic virus-infected apple.

Symptoms of disease depend on cv. and rootstock susceptibility, plus virulence of strains. Virulent strains cause very severe symptoms in the Allington, Pippin, and Lord Lambourne apple cvs.; severe symptoms in the Beauty of Bath, Miller's Seedling, October Pippin, Worcester Pearmain, Bramley's Seeding, and Cox's Orange Pippin cvs.; severe, but localized symptoms in Newton Wonder; mosaic symptoms, but without necroses, in Laxton's Fortune; and finally, mild symptoms in the Lane's Prince Albert apple cv. (Posnette and Cropley, 1956).

Also, virulent strains of ApMV cause very severe symptoms in clones from rootstocks M 15, M 9, MM 104, and MM 105; severe symptoms in MM 102, MM 7, M 1, M 4, MM 109, MM 106, and other rootstocks; symptoms without necrosis in MM 3 and MM 2 rootstocks; and finally, localized symptoms in the M 16 rootstock.

Pathogenesis and Control

In nature, ApMV is transmitted most frequently with infected rootstocks or buds and scions taken from infected trees during grafting. Some virus spread in fruit tree plantations occurs through natural root grafts between infected and healthy trees. Other modes of natural spread are unknown. The virus is transmitted mechanically into herbaceous hosts, most easily from apple petals.

Healthy propagating material production and its use for the establishment of new apple plantations represents a successful virus control measure. All infected plants should be eliminated from nurseries and young dense plantations immediately upon detection to avoid transmission of ApMV by root grafts. Healthy material for reproduction can be obtained through thermotherapy treatment by exposing infected plants to "dry" heat at 37°C during 3 to 4 weeks.

Apple Chlorotic Leafspot Closterovirus (ApCLsV)

Properties of Viral Particles

ApCLsV consists of filamentous particles of 600×12 nm (Lister, 1970), or 720 to 740×12 nm (Lister and Bar-Joseph, 1981; German et al., 1991). The sedimentation coefficient of ApCLsV is 96S, with no accessory virus particles known. The single polypeptide coat protein is 23.5 kDa (Delbos and Dunez, 1988). The single-stranded RNA of 7555 nucleotides is 2500 kDa (German et al., 1991), or 2200 to 2400 kDa (Lister and Bar-Joseph, 1981).

The TIP of ApCLsV is 52 to 55°C; the DEP is 1×10^{-4}; and LIV is 1 day (Lister, 1970). ApCLsV is moderately antigenic (Fuchs, 1980).

ApCLsV consists of many strains with differences based on symptoms in various hosts, which makes biological indexing relatively complex and difficult. For example, some strains infect all hosts, but some from apple are transmitted into stone fruits only with difficulty. Researchers have attempted to establish natural groups of strains, based both on symptoms in individual hosts and serological traits (Marénaud et al., 1976). Two serotypes of ApCLsV have been described (Chairez and Lister, 1973). Among 9 strains isolated from 13 fruit tree species, Detienne et al. (1980) observed few antigenic variations.

Geographic Distribution and Economic Importance

ApCLsV, distributed worldwide, can infect all pome and stone fruit species in both nurseries and orchard, as well as in woody ornamental and wild plants in the Rosaceae family. ApCLsV occurs in many apple cvs. and clonal rootstocks, but because it causes no leaf and fruit symptoms, it is quite difficult to monitor any damages. However, regardless of its latency, decreased plant vigor and reduced cv. and rootstock compatibility are indisputable. No other information exists on the influence of other viruses on vigor and yields of apple, which would be helpful to know. ApCLsV causes some economically damaging diseases in stone fruits of peach dark green sunken mottle, plum pseudo-pox, and plum bark split. It may also cause incompatibilities in grafts of some *Prunus* spp. combinations, which are especially harmful in nurseries.

Host Range

Most natural hosts of ApCLsV are in the Rosaceae family. Natural pome fruit hosts include *Cydonia oblonga*, *Malus pumila*, *Mespilus germanica*, and *Pyrus communis*, and natural stone fruit hosts include *Prunus armeniaca*, *P. avium*, *P. cerasus*, *P. cerasifera*, *P. domestica*, *P. persica*, and *P. salicina*. Others of importance as sources of infection include *Amelanchier canadensis*, *Chaenomeles* sp., *Crateagus monogyna*, *C. oxyacantha*, *Prunus spinosa*, and *P. tomentosa*. Other hosts include 15 species from 8 dicot families (loc. cit. Lister, 1970).

Herbaceous hosts that have been infected experimentally with ApCLsV include several families: Amaranthaceae (*Celosia argentea*, *C. plumosa*, *Gomphrena globosa*); Chenopodiaceae (*Beta vulgaris flavescens*, *Chenopodium album*, *C. amaranticolor*, *C. botrys*, *C. capitatum*, *C. murale*, *C. quinoa*, and *Tetragonia expansa*); Cucurbitaceae (*Cucumis sativus*); and Papilonaceae (*Phaseolus vulgaris* and *Pisum sativum*) (loc. cit. Németh, 1986). Šutić (1981) reported the possible infection of *Stellaria media* with several isolates.

Indicator plants are *Chenopodium amaranticolor* and *C. quinoa* with local lesions and systemic spots; *P. vulgaris* cv. Top Crop with the brown-violet local lesions; *Prunus persica* (cv. Elberta or GF 305), plus *P. tomentosa* seedlings with dark green sunken mottle and line pattern; and *Cydonia vulgaris* C 7/1 with chlorotic leaf spots, deformation, and epinasty. Cuttings of *Pyronia veitchii* are sensitive to all strains of ApCLsV (Desvignes et al., 1991).

Symptoms

ApCLsV, latent in infected apple, can be detected by specific symptoms in several indicators: *Malus platycarpa*, Russian apple clone R 12740-7 A, and apple cv. Spy 227. The initial diffuse chlorotic mosaic on *M. platycarpa* leaves gradually become small roundish spots. The leaves with altered shapes are smaller and fall off before season-end. Leaf crinkle and chlorotic spots varying in size occur on the Russian apple clone R 12740-7 A, with young leaf blades unilaterally distorted. Apple cv. Spy 227 does not display stunting and resembles the *M. platycarpa* indicator. ApCLsV causes a ring pattern mosaic in leaves of naturally infected pear.

Pathogenesis

ApCLsV is permanently preserved in infected fruit trees, and ornamental and wild woody hosts. Infected apple rootstocks play the key role in virus dissemination. Investigations in England confirm its spread in apple rootstock clones. Among those tested, M 2, M 4, M 7, M 9, and M 15 are completely infected; others (M 1, M12, and M 16) are only partially infected, and infection has not been detected in M 26. All MM 109 plus some MM 106 and MM 111 rootstocks are infected. Infection does not occur in MM 104 rootstock (Luckwill and Campbell, 1963). Such high incidence of ApCLsV-infected rootstock has undoubtedly contributed much to its being widespread among apple cvs.

Field-spread ApCLsV has been observed in orchards, but the mode(s) remains unknown. Suggested transmission by an Eudosylaimoid nematode (Fritzsche and Kegler, 1968) or by *Myzus persicae* (Šutić, 1981) require confirmation.

Delbos and Dunez (1988) reported possible transmission of ApCLsV through apricot seed. It is mechanically transmitted from woody into herbaceous plants with plant sap from buds, young leaves, and fruits. This is the method used to test for fruit tree infection, both in orchards and in various research experiments.

Control

Only virus-free propagating material should be used to establish orchards. Such material can be obtained through heat therapy by exposing young plants or plant parts to temperatures of 37 to 38°C (Campbell and Best, 1964; Welsh and Nyland, 1965). ApCLsV can be eliminated from desired initial mother stocks by the meristem culture method. Good results have also been obtained with shoot-tip grafting (Navarro et al., 1982).

Apple Tulare Mosaic Ilarvirus (ApTrMV)

Properties of Viral Particles

The ApTrMV isometric particles are 33 nm in diameter (Fulton, 1971) and consist of two types, with sedimentation coefficients of 85 and 91S. Only 91S particles are infective (Mink et al., 1963). RNA accounts for 12% of the mixed particle weight, and their coat protein subunits are 19 kDa (Barnett and Fulton, 1969). ApTrMV particles are unstable in extracted plant sap, but when stabilized in buffer, the TIP is 60 to 62°C (Fulton, 1971). ApTrMV is moderately immunogenic (Fulton, 1967).

Geographic Distribution

ApTrMV was discovered first in apple in Tulare County, California, thus its name (Yarwood, 1955). After the orginal infected tree no longer existed, it was believed that ApTrMV existed only as an experimental virus (Fulton, 1971). However, later, ApTrMV was isolated from hazelnut trees displaying mosaic (Marenaud and Germain, 1975; Cardin and Marénaud, 1975) with very limited distribution; it poses no threat to fruit production.

Host Range and Symptoms

Apple *Malus pumila* and hazelnut *Corylus avellana* are natural virus hosts. Symptoms on apple resemble those caused by ApMV, thus the earlier belief that these were identical diseases, but comparative investigations have proven substantive differences (Gilmer, 1958; Fulton, 1971).

ApTrMV was transmitted experimentally into other woody Rosaceae hosts, for example, *Chaenomeles japonica*, *Crataegus chlorosarca*, *Cydonia oblonga*, *Pyrus communis*, and *Sorbus aucuparia* (Gilmer, 1958). Symptom-bearing infected herbaceous hosts include *Nicotiana tabacum* with local angular and roundish ringspots plus a systemic necrotic acute stage, followed by recovery to a chronic stage; *Phaseolus vulgaris* with necrotic leaf lesions; and *Tithonia speciosa* with chlorotic mosaic on leaves. *Chenopodium foetidum*, *Gomphrena globosa*, *Helianthus annuus*, *Vicia faba*, and others are symptomless herbaceous hosts.

Pathogenesis

This virus is transmitted experimentally into woody plants by grafting and into herbaceous plants mechanically with infective plant sap and by the dodders *Cuscuta subinclusa* and *C. campestris* (Yarwood, 1955; Gilmer, 1958; Fulton, 1971). Other modes of virus transmission are unknown.

Diseases of Apple Assumed Caused by Viruses

Apple Leaf Pucker Disease (ApLPkDs)

ApLPkDs has been described in various apple cvs. in several countries. Although the causal agent has not been identified, it is most certainly a virus. Apple leaf pucker is caused either by different strains or by several viruses (Welsh and Keane, 1963). Clover yellow mosaic virus (ClYMV) has been isolated from infected trees but its etiology has not been elucidated. Some suggest that it may be caused by ApCLsV (Kegler et al., 1979).

Symptoms on leaves are vivid yellow or green spots, vein chlorosis, ringspots, and line patterns, particularly pronounced in cool weather. Variously shaped depressions, russet, and ring patterns on the fruit skin are easily recognized as violet or most frequently reddish-brown. Fruit symptoms are also pronounced at lower temperatures during the summer.

Leaf and fruit symtpoms vary and are characteristic for individual apple cvs., thus known under different names; for example, apple russet ring disease (cv. Golden Delicious), apple Stayman blotch disease (cv. Stayman), and apple McIntosh leaf pucker (cv. McIntosh)—all synonyms for the same disease. The indicators *Malus pumila* cv. Golden Delicious (Dunez et al., 1982) and cv. Spartan (Welsh and Keane, 1963; Cropley, 1968a) aid in disease identification.

The causal agent was transmitted with infected buds and vegetative rootstocks during grafting. No other mode of transmission is known. The use of pathogen-free planting material is the only effective measure.

DISEASES OF APPLE WITH CHARACTERISTIC FRUIT SYMPTOMS ASSUMED CAUSED BY VIRUSES

Apple Green Crinke Disease (ApGnCrDs)

Apple green crinkle fruit disease is among the most frequent apple fruit diseases. It has been described in North and South America, Australia, Asia (Isreal, Turkey, Japan), New Zealand, and Europe.

The causal agent of ApGnCrDs has not been transmitted into herbaceous plants, thus its properties are unknown, although believed to be a virus. Some relatedness is suspected, based on similar symptoms of apple green crinkle with other apple diseases caused by viruses. Apple *Malus sylvestris* is the only known host into which the causal agent of ApGnCrDs can be successfully transmitted experimentally. Natural infections have been observed in 30 apple cvs., including Alexander, Baldwin, Belle de Boscoop, Cox's Orange, Golden Delicious, Granny Smith, and Gravenstein (Kristensen, 1963). Granny Smith, Gravenstein, and Guldborg are the most susceptible.

Symptoms evident only on fruits occur first a few weeks after flowering as depressions on young fruits about 2 cm in diameter and later wart-like crinkles occur. Uneven growths and swellings form irregular crinkles on the surface as the infected fruits mature. Sometimes, crinkle tissues split and the vascular bundles under the skin become irregular and greenish (Figure 150). Fruits are severely distorted and stunted. Symptoms may occur on only some or on most fruits; this dictates the extent of damage. The cv. Golden Delicious is the only disease inidicator (Dunez et al., 1982).

The causal agent is transmitted with infected planting material during grafting. Natural spread occurs in the field and, although observed, the mode is unknonwn. The use of healthy planting material is the main measure for control. Elimination of infected trees from young orchards is also recommended to reduce the risks of potential natural disease spread.

FIGURE 150 Green crinkle disease on Golden Delicious apples. (Courtesy A.J. Hansen. Reproduced with permission from APS.)

Apple Ringspot Disease (ApRsDs)

ApRsDs is presumably caused by a virus, although no transmission mechanically into herbaceous hosts has been accomplished with plant sap. Not particularly widespread in any one location, it is widely distributed in Australia, New Zealand, South Africa, and Europe (former Czechoslovakia, Germany, Poland, Spain).

Symptoms occur only on fruits of apple *Malus sylvestris*, which is the natural and the only host of ApRsDs. The initial symptoms appear on the skin of nearly all fruits of an infected tree, usually as pale-brown or olive-green spots on a somewhat roughened skin. At near-complete growth, brown rings or semi-rings and line patterns appear on their surface and continue to develop also following harvest. The skin around initial brown spots or between brown concentric rings is usually pale.

Fruit spots result from superficial tissue necrosis immediately beneath the skin and of the adjacent superficial fleshy tissue layers. Reddish zones of anthocyanin accumulate in those superficial layers just below the skin (Canova, 1963).

Abbondanza and Granny Smith cvs. are quite susceptible to ApRsDs, and symptoms also occur on Cox's Orange Pippin, Golden Delicious, and Sturmer Pippin fruits. Golden Delicious is a sutiable indicator plant (Dunez et al., 1982).

The causal agent of ApRsDs is transmitted only with infected vegetative parts used for grafting in seedling production. Healthy planting material ensures prevention of infection.

Apple Rough Skin Disease (ApRgSknDs)

ApRgSknDs, a frequent apple disease, has been discovered in India, North America, and several European countries. The causal agent is unknown.

Symptoms of ApRgSknDs occur only on the fruits of apple *Malus sylvestris*, the only natural host, appearing soon after initial development of fruits and easily observed during maturation at harvest (Katwijk and Meijneke, 1963). The characteristic brown-corky patches on the fruit skin vary in size and shape (i.e., roundish, elongated, striped). The superficial corky patch-affected tissue splits, creating a rough fruit surface, and thus the disease name. Based on symptoms, this disease resembles the star crack disease. Severely affected fruits grow and develop slowly. Symptoms usually appear first on fruits born on one branch, then gradually spread through the entire crown, more so in wet, cloudy seasons. Under dry, warm, and sunny conditons, these symptoms abate.

Belle de Boscoop, Cox's Orange Pippin, Golden Delicious, Reinette de Canada, Martini, Ontario, and Weisser Winter-glocken cvs. are highly susceptible to the virus, but not Jonathan and Granny Smith, which remain symptomless. Indicators are cvs. Belle de Boscoop (rough skin) and Golden Delicious (star crack) double budded on the M 9 rootstock (Dunez et al., 1982).

The causal agent of ApRgSknDs is transmitted with infected vegetative parts used for grafting. The natural mode of spread is unknown. The use of healthy planting material is the primary measure for preventing infection.

Apple Star Crack Disease (ApStrCkDs)

ApStrCkDs of apple tree, whose causal agent(s) is unidentified, causes similar symptoms in Australia, Canada, Mexico, New Zealand, the U.S., the former U.S.S.R., and 15 European countries. Occurring only sporadically, it does not spread in orchards. It causes no serious economic damage. Some damage may occur only in the most susceptible cvs.

Characteristic fruit symptoms occur as cracked spots (scabs) with star-like margins. Smaller or larger cracks of irregular shape later form within these spots. Scabs appear at random on the fruit surface, but more frequently near the sepals. Diseased fruits are of irregular shape.

The assumed causal agent of ApStrCkDs comprises strains differing in virulence. Virulent strains cause symptoms on leaves, shoots, and fruits, whereas mild ones cause only fruit symptoms. Symptom severity is also influenced by cv. susceptibility; for example, Cox's Orange Pippin is the most diseased cv. in England, in which changes of the canker-type around the buds on infected annual shoots are followed by frequent shoot-tip death in winter. Buds break and foliation and flower development are delayed for 7 to 21 days, depending on strain and temperatures (Cropley, 1963). Young leaves become chlorotic and cone-shaped. Symptoms of disease occur on the leaves, bark, and fruits of this cultivar and trees grow slowly. Symptoms on fruit have been observed in cvs. Bernack Beauty, Charles Ross, Early Victoria, Laxton's Fortune, and Monarch. Granny Smith, Lord Lambourne, Sunset, and Worcester Pearmain cvs. appear immune. Similar symptoms, described in apple in many countries, cannot be confirmed until the causal agent is defined. Cox's Orange Pippin or Golden Delicious are suitable indicator plants. The causal agent of the ApStrCkDs is graft transmitted with infected vegetative parts. Other modes of transmission are unknown, although a certain natural slow spread seems to occur. The use of healthy planting material is the key to disease control.

Apple Scar Skin Viroid (ApScSknVd) and Apple Dapple Viroid (ApDlVd)

Apple scar skin and apple dapple were earlier considered as two distinct diseases presumably caused by viruses. However, more recent studies show them as homologous variants of the disease caused by a viroid (Koganezawa et al., 1982; Hadidi et al., 1990; Hurtt et al., 1991).

Based on characteristic symptoms, both diseases were described in Canada, North America, Australia, Japan, and Europe. Apple was the only known natural host until it was recently shown that ApScSknVd and ApDlVd in nature cause latent infections in several pear cvs. (Hurtt et al., 1991). Both diseases were recognized in nature only by fruit symptoms.

Apple scar skin symptoms appear on fruits 6 weeks after fruit set as pale spots close to the fruit stem (Millikan, 1963), later spreading radially or irregularly on the fruit surface that may have the corky texture, accompanied by an absence of natural fruit pigmentation (Millikan, 1963) (see Figure 151). ApScSknVd-infected fruits remain small, and never reach full maturity, have an unpleasant taste, poor quality, and in some cvs. (e.g., Golden Delicious, Jonathan, and Turley), are also unsuitable for juice production.

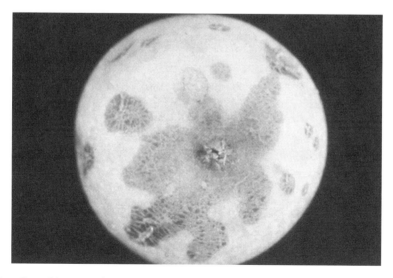

FIGURE 151 Scar skin on 'Ralls Janet' inoculated with the apple scar skin viroid. (Courtesy H. Koganezawa. Reproduced with permission from APS.)

The first symptoms of the ApDlVd type occur on apple fruits toward mid-July as small, pale, round spots (McGrun et al., 1963). During disease development, these spots gradually spread, merge, and cause discoloration of larger superficial fruit parts, a symptom observed on cvs. Golden Delicious, Cortland, McIntosh (Figure 152), Starking Delicious, Virginia Crab, and others.

FIGURE 152 Dapple apple symptoms around the calyx of a McIntosh apple. (Courtesy A.J. Hansen. Reproduced with permission from APS.)

Stark's Earliest and Sugar Crab apple cvs. are suitable indicators for rapid biological detection of these viroids (Howell and Mink, 1991). The ApScSknVd also has been detected by using a reverse transcription-polymerase chain reaction (RT-PCR) assay (Hadidi, 1991). The RT-PCR assay enables the transcription and amplification of the viroid RNA in nucleic acid extracts obtained from infected tissues with a high specificity and fidelity. A conserved 330 nocleotide sequence has been determined in the structure of this viroid (Jang et al., 1991).

In nature, ApScSknVd and ApDlVd are transmitted with infected vegetative plant parts, common for this group of pathogens. Hadidi and Parish (1991) proved that apple seed is also infected with these viroids. They observed the presence of the viroids in anthers, petals, receptacles, seed coats, cotyledons, and embryos, and also in leaves, bark, and roots, showing that infected apple trees probably present sources of infection for transmission and spreading of these viroids in orchards. The use of viroid-free planting material is the main method for elimination of infection in apple production.

VIRUS DISEASE OF APPLE WITH CHARACTERISTIC STEM SYMPTOMS

Apple Stem Grooving Capillovirus (ApStmGgV)

Properties of Viral Particles

The ApStmGgV is filamentous of 600 to 700 × 12 nm (Yoshikawa and Takahashi, 1988) that contains a single coat protein of 27 kDa. The sedimentation coefficient of ApStmGgV is 112S; theTIP is 60 to 63°C; the DEP is 1×10^{-4}; and the LIV at 20°C is 2 days (Lister, 1970a). It is moderately immunogenic. ELISA and ISEM methods are used for mass scale virus testing (Kalashjan et al., 1988).

The virus strains of ApStmGgV are based on varying symptoms in indicator plants: C-431 strain or ApLtV2 (*Chenopodium quinoa*, necrotic lesions and mild systemic infection; Virginia Crab, stem grooving graft union necrosis) and dark green epinasty virus strain (*C. quinoa*, necrotic lesions and systemic chlorotic ring mottle; Virginia Crab, most isolates no stem grooving) (loc. cit. Lister, 1970a).

Geographic Distribution and Economic Importance

ApStmGgV has been discovered in North America, Australia, Asia, South America, New Zealand, and over 20 European countries, probably occurring wherever apple is grown. Most apple cvs. are tolerant to ApStmGgV, thus it causes minimal damge to plant growth, development, and fructification. Infected Virginia Crab *Malus pumila* apple trees, which serve as virus indicators, grow slowly and break easily at the rootstock and scion union.

Host Range

ApStmGgV has been transmitted experimentally into 20 species in 9 dicot families. Natural hosts of ApStmGgV include apple (*Malus sylvestris*), pear (*Pyrus communis*), and quince (*Cydonia oblonga*). Experimentally infected woody hosts include *Aronia* sp., *Cotoneaster* sp., *Sorbus aucuparia*, and *Pyronia veitchii* (Van der Meer, 1976); and herbaceous hosts include *Amaranthus retroflexus, Antirrhinum majus, C. amaranticolor, C. quinoa, Cucumis sativus, Cucurbita maxima, Gomphrena globosa, Nicotiana glutinosa, N. rustica*, and *P. vulgaris*. Symptomless hosts are *Datura stramonium, Nicotiana clevelandii, N. tabacum*, and *Petunia hybrida* (Lister et al., 1965; Waterworth and Gilmer, 1969).

Symptoms

ApStmGgV-infected commercial rootstocks and cvs. display no symptoms, but ApStmGgV causes symptoms only on indicator plants; for example, *Malus sylvestris* cv. Virginia Crab is the most susceptible indicator, with severe characteristic reactions. Specific symptoms include sunken areas and longitudinal grooves, visible when the bark is peeled off the trunk (Lister, 1970a; Pleše et al., 1975). Elongated grooves and sunken areas are covered with marginal furrows. The E-36 strain causes graft union necrosis and deformations (brownline virus disease syndrome), at which point breakage occurs easily especially serious during the second year. By contrast, C-431 strain causes only the stem grooving symptoms. Virus strains from pear trees do not cause stem grooving symptoms. Most strains also cause leaf chlorosis. Symptoms on herbaceous indicator plants vary, depending on the virus strain; for example, *Chenopodium quinoa*, with primary necrotic lesions followed by systemic infection, chlorotic ring patterns, epinasty, etc.; *Nicotiana glutinosa*, with yellow mosaic, line patterns, and occasional necrosis; *Phaseolus vulgaris*, with chlorotic lesions or purple-brown spots and rings; and most cvs. with systemic veinal necrosis (Lister and Bar-Joseph, 1981).

Pathogenesis and Control

The ApStmGgV is graft transmitted with infected vegetative parts. Infected stocks represent the greatest source of virus spread. Thus, one must use virus-free stocks to prevent further spread of infection. Natural disease spread has been observed, but the mode(s) of transmission is unknown. ApStmGgV is seed transmitted in both apple *Malus platycarpa* and *Chenopodium quinoa* (Van der Meer 1976). ApStmGgV is transmitted mechanically from apple and pear trees (most easily from flower petals) into susceptible herbaceous plants. Mother stocks can be cured by incubating infected plantlets to 37°C for weeks (Campbell, 1968).

Apple Stem Pitting Disease (ApStmPgDs)

Although the causal agent of ApStmPgDs is not yet identified, it is a presumed virus (Kegler et al., 1979). Many studies suggest that the same agent also causes Spy epinasty and decline, quince sooty ringspot, pear stony pit, and pear vein yellows. This disease, widespread in several regions, is reported in North America, Australia, South America, Asia, New Zealand, and 20 European countries.

Apple (*Malus pumila*) including *M. sieboldii* and *M. sieboldii* var. *arborescens*, pear (*Pyrus communis*), and quince (*Cydonia oblonga*) are natural disease hosts. ApStmPgDs was also reported in woody hosts *Crateagus monogyna* and *Sorbus discolor* (Sweet and Campbell, 1976; Sweet et al., 1978; Sweet, 1980). The causal agent of ApStmPgDs was transmitted experimentally into *Aronia floribunda, Malus sylvestris*, and *Pyronia veitchii* plants (Nemeth, 1986).

ApStmPgDs-diseased apple cvs. and rootstocks display no symptoms. Symptoms on susceptible indicators are apple Spy 227 with stem pitting and decline, and *Pyronia veitchii* with chlorotic spots, epinasty of leaves, plus elongated grooves on the xylem cylinder. The indicator apple Virginia Crab has stem and fruit symptoms of pits beneath the trunk bark and, in severely infected trees, the bark becomes abnormally thick with narrow longitudinal fissures. Fruits from infected trees are small with grooves from calyx to the stems. Susceptible cvs. Golden Delicious and Virginia Crab remain underdeveloped and fruits mature prematurely (Posnette and Cropley, 1963).

The apple Spy 227 is a specific ApStmPgDs symptom indicator, that is, grafted seedlings express characteristic leaf and bark symptoms. Leaf blades curl downward, resulting in severe epinasty, followed by a general decline. Bark tissues become necrotic, and trees grow slowly and are susceptible to early death. The disease was described as apple Spy epinasty and decline.

Similar to other diseases, ApStmPgDs is graft transmitted with infected vegetative parts. Thus, the use of healthy stocks represents the only logical preventive measure. Natural disease spread occurs but the mode of spread is not known. The causal agent can be eliminated by incubating young plantlets at 37 to 38°C for 4 to 5 weeks (Kegler and Schimanski, 1988), the exact length of time depending on the type of planting material.

Apple Platycarpa Scaly Bark Disease (ApPcSyBkDs)
(Syn: Apple Platycarpa Dwarf Disease – ApPcDDs)

Symptomologically, ApPcSyBkDs and ApPcDDs were described initially in England as two separate diseases (Luckwill and Campbell, 1963a; 1963b), but it has been established that these symptoms are of the same disease (Reffati and Osler, 1975). The causal agent is not known.

ApPcSyBkDs has been described in the U.S., Canada, Australia, Israel, Japan, New Zealand, South Africa, and 20 European countries. Apple (*Malus pumila*) is the natural host of the causal agent, which has been transmitted experimentally into 10 species in Malus. ApPcSyBkDs is latent in most commercial apple cvs. and clonal rootstocks (Luckwill and Campbell, 1959).

Symptoms of ApPcSyBkDs occur in apple *Malus platycarpa* trees as a rough, scaby bark on young trees, frequently accompanied by swellings. Secondary fungal infections occur in bark fissures, which complicates disease diagnosis. A dwarfing symptom has been described in *M. platycarpa*, with tree size reduced by 70% (Luckwill and Campbell, 1963a). Stunt is severe if the trees are simultaneosly infected also with other diseases. ApPcDDs is lethal to *Malus floribunda*, *M. sargenti*, and Robert Crab, severely stunts *M. glaucescens*, and slightly stunts *M. lancifolia*, *M. atrosanguinea*, *M. prattii*, *M. prunifolia Rinki*, *M. pumila Niedzwetsyana*, and *M. lemoinei*. The Cowichan Crab and Hopa Crab cvs. are tolerant ApPcDDs (Campbell, 1962).

The causal agent of ApPcSyBkDs is transmitted by grafting infected buds and rootstocks from mother stocks. The use of healthy seedlings is the only known measure to prevent infection. The disease agent can be eliminated from the shoot tips by incubating planlets at 37°C for 2 to 3 weeks.

Apple: Natural Alternative Host for Some Viruses

Cherry Rasp Leaf Virus (American) (ChRpLV)

ChRpLV, discovered and described in apple trees in the U.S. (Cheney et al., 1967; Parish, 1977), causes flat fruits and reduces their market value. ChRpLV is transmitted with infected vegetative parts during grafting, but also by nematodes present in the orchards.

Carnation Ringspot Virus (CaRsV)

CaRsV occurs wherever carnation is grown. Apple (*Malus pumila*), pear (*Pyrus communis*), plum (*Prunus domestica*), and sour cherry (*P. cerasus*) are natural virus hosts (Richter et al., 1978; Kleinhempel et al., 1980). It is transmitted with the vegetative planting material and nematodes are its vectors in nature.

Sowbane Mosaic Virus (SwMV)

SwMV, first isolated from apple trees and described as ApLtV (Kirkpatrick et al., 1965), was later found identical to SwMV (Bancroft and Tolin, 1967). Also isolated from sour cherry (Šarić, 1971), *Prunus domestica* is one of its hosts (Šutić and Juretić, 1976).

Tobacco Mosaic Virus (TMV)

TMV, first discovered in apple and pear trees (Kirkpatrick and Lindner, 1964) and later confirmed, was also isolated from sweet cherry, sour cherry (Gilmer, 1967), and plum trees (Albrechtová et al., 1974). No damage has been attributed to TMV in woody hosts.

Tobacco Necrosis Virus (TNV)

Isolates of TNV were isolated from apple, pear, and apricot trees (Uyemoto and Gilmer, 1972). Infected woody hosts suffer no damage.

Tobacco Ringspot Virus (TRsV)

TRsV was described as the cause of union incompatibility in apple trees in Canada (Lana et al., 1983). Irregular zig-zag outgrowths, accompanied by tissue necrosis, are formed at the graft union points in infected trees, weakening their mechanical strength, which causes susceptibility to breaking. The leaves are fewer, smaller, and finally become chlorotic.

Tomato Bushy Stunt Virus (TmBuSnV)

TmBuSnV was isolated from apple trees by Allen (1969) and Richter et al. (1977). TmBuSnV-infected trees suffer no economic damage.

Tomato Ringspot Virus (TmRsV)

TmRsV infects all fruit tree species. In apple trees, it causes union necrosis characterized by tissue necrosis and disintegration in the graft union sections. The bark surrounding the graft union becomes abnormally thick and spongy. A transverse black line spreading from the cambium toward the bark is visible on its inner surface. Deep pits and necrotic tissues appear along the grafting structure on the surface of the woody cylinder. The severity of pathological changes depends on rootstock and cv. susceptibility. TmRsV is transmitted with the infected planting material and by *Xiphinema americanum* and *X. rivesi* nematodes in nature. Measures recommended for protection of apple trees include the use of virus-free planting material plus control of both soil nematode vectors and weed hosts of TmRsV. More tolerant/resistant rootstocks and cvs. should be used.

Virus Diseases of Pear (*Pyrus communis*) and Those Assumed Caused by Viruses

Fifteen virus and virus-like diseases have been described in pear and, for most, the causal agents are ill-defined; thus their identity remains uncertain—whether all differ or some are synonymous with separate names. For example, pear ring mosaic and pear ring pattern mosaic were described as two distinct diseases and now, with further research, their identity has shown that both are caused by ApCLsV (Delbos and Dunez, 1988). Likewise, pear vein yellows and pear red mottle are now known as symptom types, both a result of stony pit disease, and probably caused by a virus. Based on symptoms, it is assumed that pear corky pit and pear stony pit are closely related diseases. Future investigations will more fully elucidate the etiology of individual pear diseases and determine their mutual taxonomic relationships.

Diseases of Pear with Characteristic Leaf Symptoms

Pear Ring Disease (PrRDs)/Ring Pattern Mosiac Disease (PrRPtMDs) Caused by apple chlorotic leaf spot virus (ApCLsV)

ApCLsV, a presumed member of the closterovirus group, causes the pear ring disease (PrRDs). Symptomologically, this disease is also described as pear green spot, pear mosaic, and pear ringspot mosaic disease (Németh, 1986). It is widespread in the U.S., Canada, Asia (Isreal and Japan), New Zealand, South Africa, and almost all European countries.

In susceptible cvs., PrRDs can reduce young pear tree growth by 20% and yields by 15% (Kegler, 1977), and it decreases the winter hardiness of susceptible plants and thus is an economically damaging disease. ApCLsV infects all rosaceous fruit trees and probably most rosaceous ornamental trees. Quince, used as a vegetative root stock for pear, plays an important role in the pear ring mosaic development cycle.

Many pear cvs. are symptomless, but Beurré Hardy, Beurré Bosc, Nouveau Poiateau, Marianne, and Williams are susceptible to PrRDs. Symptoms occur in leaves and in some frutis of susceptible cvs. Initially, symptoms that occur in young leaves upon bud opening include mosiac, green ring mosaic, and mosaic ringspot. Leaf midribs grow slowly, bend, and result in a curled appearance. Pale-green and ring spots appear on the surface of the Beurré Hardy cv. fruits and, in the Köstliche von Charneu cv. disease, symptoms occur only on the fruits (Kegler, 1977). Many cvs. have diffuse symptoms of indistinct mottle as numerous pale-green or greenish yellow spots. The symptoms, initally pronoucned in young plants, remain in older plants in a much milder form. Warm and dry weather conditions favor symptom severity.

Pyrus communis cv. Beurré Hardy with variable mosaic symptoms and *Chenopodium quinoa* with local chlorotic spots, rings, and lines are suitable indicator hosts.

Infected pear parent trees and quince clones used as vegetative rootstocks for pear represent natural sources of infection. Grafting of infected plant parts from infected stocks represents the sole means of disease spread. ApCLsV is transmitted mechanically into herbaceous *Chenopodium quinoa* and other plants (see also ApCLsV section), which is important only for experimental purposes.

The production and use of virus-free healthy stocks represents the primary method for control of ApCLsV. Initial healthy planting material can be obtained by incubation of young plants at 37°C for 2 to 3 weeks (Posnette et al., 1962; Campbell, 1962a).

Pear Bud Drop Disease (PrBdDpDs)

PrBdDpDs was discovered and described as a new pear disease in France (Morvan, 1965). The causal agent, still unidentified and presumably a virus, is based on symptoms and budding/grafting transmission. Based on resemblance of symptoms, PrBdDpDs occurs in only a few countries (Bulgaria, Italy, former U.S.S.R.).

Natural infections were discovered in pear cvs. Beurré d'Amanlis, Beurré Hardy, and Comice (Morvan, 1965; Trifonov, 1971). Trifonov (1971) found 18 cvs. tolerant to bud drop disease.

Symptoms of PrBdDpDs occur on leaves and buds, and leaves curl downward and display a characteristic epinasty. Leaf buds formed in the previous year open at the initiation of the vegetation period. Bud drop occurs due to tissue necrosis at the bud base where the corky tissue (periderm) facilitates the process. A few remaining infected buds grow slowly and leaves that develop from them wilt and dry early in the summer. Rosettes form as secondary buds grown around the drop point, resulting in depression of growth and eventual death of fruit trees.

Pyrus communis cv. Beurré Hardy, which exhibits the characteristic bud drop and decline, is the disease indicator plant.

The causal agent of PrBdDpDs is graft transmitted only with infected buds and other plant parts during the grafting-propagation process. Production and use of healthy planting material is the only effective disease control measure.

Pear Necrotic Spot Disease (PrNcsDs)

PrNcsDs was discovered and described as a new pear disease in Japan, but the properties of its causal agent are yet unknown (Kishi et al., 1972; 1976). Many Japanese pear cvs. (*Pyrus serotina* var. *culta*) are susceptible, but some have tolerance to PrNcsDs, which causes economic damage only in Japan.

Pyrus serotina and *P. communis* ssp. *sativus* are natural hosts of this disease whose causal agent has been experimentally transmitted into Chinese pear and hybrid seedlings. PrNcsDs symptoms occur only in leaves as numerous chlorotic spots that quickly changing to red or black. In severe infections, defoliation results in important yield losses. *Pyrus serotina* var. *culta* HN 39 with red and black necrotic spots is a good indicator plant.

Since the PrNcsDs causal agent is transmitted during grafting, pathogen-free, healthy planting material represents the key measure for prevention infection.

DISEASES OF PEAR WITH CHARACTERISTIC FRUIT SYMPTOMS ASSUMED CAUSED BY VIRUSES

Pear Stony Pit Disease (PrStnPiDs)/Pear Vein Yellows Disease

Pear stony pit and pear vein yellows, earlier described as separate and distinct diseases of pear, are now known based on fruit and leaf symptoms, most likely caused by the same presumed virus. PrStnPiDs has also been named the red mottle disease, and the same causal agent probably causes apple Spy epinasty and decline, apple stem pitting, and quince sooty ringspot diseases (Kegler and Schimanski, 1988).

The PrStnPiDs, encountered frequently, is distributed worldwide, probably wherever pear is grown. PrStnPiDs was observed in 70% of the Beurré Bosc cv. trees in Oregon and California (Kristensen, 1963). Studies of several pear cvs. in Romania revealed that PrStnPiDs accounted for 60% of the cases of disease (Minoiu, 1973). Infected trees yield 18 to 94% deformed fruits and, since deformed fruits have no market value, PrStnPiDs is quite damaging in susceptible cvs.

Pear (*Pyrus communis*) and quince (*Cydonia oblonga*) are natural hosts of PrStnPiDs, whose causal agent can be transmitted experimentally also into *Pyrus betulaefolia*, *P. calleryana*, *P. ussuriensis*, and *Pyronia veitchii*.

The symptoms of PrStnPiDs on fruit of susceptible pear cvs. occur early in the season, 10 to 20 days after petal fall, and are first observed as dark-green patches beneath the epidermis (Figures 153 and 154). Individual parenchymal cells become lignified, which causes uneven, irregularly deformed fruit development. Susceptible cvs. have many such cells; even occasionally entire fruits become so lignified they cannot be cut.

Vein yellows occurs along secondary and smaller leaf veins in several pear cvs., in parts or in entire leaves. Occasionally, red mottle also occurs along leaf veins. Disease symptoms most evident in young trees, gradually become milder with time and finally disappear completely. Cracks and necroses of bark differing in severity on stems in some pear cvs. depend on plant susceptibility. PrStnPiDs trees are more sensitive to winter freezing injury; for example, Beurré Bosc, Doyenné du Comice, and Nouveau Poiteau cvs. are highly susceptible, and Köstliche von Charneu and Winter Nelis are highly tolerant pear cvs. (Fridlund, 1976). *Pyrus communis* cv. Beurré Bosc with stony pit and vein yellows, and *Pyronia veitchii* with leaf epinasty and shoot-tip necrosis, are suitable indicator hosts.

Since the causal agent of PrStnPiDs is transmitted only through infected plant parts, the use of healthy stocks represents the only logical plant protection measure. Healthy planting material can be obtained by incubating at 37 to 38°C for 4 to 5 weeks, followed by shoot tip propagation (Kegler and Schimanski, 1988).

A corky pit disease, assumed a strain of PrStnPiDs, was described by Keane and Welsh (1960) in Canada. Although these two diseases have a striking symptomological resemblance, their

FIGURE 153 Beurré Bosc pears misshapen by the stony pit virus. (Courtesy A.L. Jones. Reproduced by permission from APS.)

FIGURE 154 Pear stony pit disease on pear.

relationship is quite unclear (Németh. 1986). The causal agent of the corky pit disease differs from that of the PrStnPiDs because it causes no symptoms in the Beurré Bosc pear cv. The indicator host of the corky pit disease is the Flemish Beauty pear cv., with fruit symptoms of shallow pits or grooves and corky areas under the depressions.

Pear Freckle Pit Disease (PrFrPiDs)

The PrFrPiDs, described in Canada (Wilks and Welsh, 1965), also occurs in the U.S. Symptomologically, it resembles PrFrPiDs except that fruit symptoms appear only 1 month before harvest; no lignification accompanies the pit formation and no bark symptoms occur.

Natural hosts are *Pyrus communis* cvs. Flemish Beauty and Anjou. Anjou is an excellent indicator plant.

Symptoms of PrFrPiDs occur on fruits as numerous dark-green sunken pits beneath which dark green strands form. During disease development, pits become pale green and the strands become darker. Pits are most abundant near the calyx and the peduncle.

PrFrPiDs is transmitted with infected planting material, but possibly also naturally in the Anjou cv. (Parish and Raese, 1986). PrFrPiDs is not economically damaging.

DISEASES OF PEAR ASSUMED CAUSED BY VIRUSES OR VIROIDS WITH CHARACTERISTIC BARK SYMPTOMS

Several virus-like diseases with bark symptoms were described in pear in the U.S. and Europe from 1957 to 1965. These diseases, named after characteristic symptoms, include pear rough bark (PrRgBkDs) and pear bark split (PrBkSplDs) in Denmark (Kristensen and Jörgensen, 1957), pear blister canker (PrBsCnDs) in England (Cropley, 1960), pear bark measles (PrBkMsDs) in the U.S. (Cordy and MacSwan, 1961), and pear bark necrosis (PrBkNcDs) in Germany (Kegler, 1965). Recent evidence suggests a viroid causal agent (Hernandez et al., 1991).

Their causal agents in many cases are still unknown. These diseases commonly cause detrimental pathological alterations only in the bark of susceptible pear cvs., and the symptoms usually appear only in the second year. Primary symptoms include measles and pustules, followed by superficial cracks on the epidermis. Secondary symptoms include scattered variable-size cankers and deep bark splits and necrosis. Severely infected trees usually die rapidly. *Pyrus communis* A 20 is the indicator for the majority of strains.

Due to etiological questions still unanswered, uncertainty exists about differences or similarities of these diseases. Researchers assume that pear blister canker and pear rough bark are caused by the same agent, and that a close relationship exists between pear blister canker and pear bark measles (Németh, 1986). However, symptom variability observed and work on differential host reactions done in different countries are difficult to compare without knowing the causal agent.

In trying to elucidate the etiological nature of this disease group, an important discovery showing research progress suggests the possible viroid nature of the causal agent of pear blister canker made by Hernandez et al. (1991), who isolated a new viroid RNA (PrBsCnVd) from two pear trees. The purified preparations of this viroid replicate in inoculated cucumber (mild or without symptoms) and pear plants. The viroid was discovered by polyacrylamide gel electrophoresis (PAGE) directly in the leaves of infected trees or pear seedlings 2 months after inoculation, highlighting quick disease detection by biochemical and biophysical techniques (Hernandez et al., 1991).

Pear: Natural Alternative Host for Some Viruses

Apple Virus and Virus-like Diseases

Apple stemgrooving virus (ApStmGgV) was isolated from pear in the U.S. (Waterworth, 1965) and then identified (Waterworth and Gilmer, 1969). The ApStmGgV is latent in pear cvs., but causes a severe decline in Packham's Triumph cv. (van Siebert and Engelbrecht, 1983). PrStmPiDs, first described in the U.S. (Millikan et al., 1964), is identical with the ApStmPgD. Apple skin group viroids were discovered in imported pear germplasm (Hurtt et al., 1991). Using various procedures for rapid viroid detection in pear germplasm, with the aid of cRNA probes and pear grafting with cv. Nouveau Poiteau, these authors developed a rapid method for testing germplasm in the quarantine program.

Some Viruses of Herbaceous Hosts

Carnation ringspot virus (Kegler et al., 1977; Richter et al., 1978) and TMV (Kirkpatrick and Lindner, 1964; Uyemoto and Gilmer, 1972) were isolated from pear and, although they have no great importance in pear production, this woody host may be of significant interest in their host range.

Virus Diseases of Quince (*Cydonia oblonga*) and Those of Quince Assumed Caused by Viruses

Quince, the natural host of several virus and virus-like diseases, has susceptibility to viruses similar with apple, also in the Rosaceae. No virus has been described in quince as its primary host, yet some quince virus diseases are identical to those observed in apple and pear. Pear, used as rootstock in quince production, is quite important in the quince virus disease development cycle.

As with all other rosaceous fruit trees, quince is the natural host of ApCLsV. Similar to other pome fruits, quince diseases are known variously as quince sooty ringspot, apple Spy epinasty and decline, and apple/pear stem grooving. Insufficient data are available on this group of diseases, which require further studies on their etiology.

Apple Chlorotic Leaf Spot Virus (ApCLsV) in Quince

Quince Stunt Disease

ApCLsV causes typical quince stunt (Delbos and Dunez, 1988), although other researchers interpreted it as a complex of ApCLsV and quince sooty ringspot disease (Posnette et al., 1962), or possibly as interacting with the apple rubbery wood phytoplasma in this complex (Kegler, 1977).

Quince stunt was described first as a latent pear disease in England (Posnette and Cropley, 1958), and then later discovered in the U.S., Canada, New Zealand, South Africa, and Europe (Németh, 1986).

ApCLsV occurs naturally in quince (*Cydonia oblonga*) and pear (*Pyrus communis*). Symptoms are evident in susceptible quince C7/1 and E cultivars, whereas the A, B, and C clones are tolerant. Susceptible leaves exhibit irregularly shaped yellow spots and curl. Shoots grow slowly and die, depending on degree of plant susceptibility. Susceptible quince cv. buds grafted onto infected rootstocks fail to develop or grow only a few leaflets. Pear is a tolerant disease host. The quince cv. C7/1 is a suitable indicator host, with chlorotic leaf spots and stunt of plants.

The use of healthy parent plants is the only known measure to eliminate infection. Virus-free quince C7/1 planting material was obtained by incubating at 37°C for 11 days.

Quince Ring Pattern Mosaic

ApCLsV causes this disease (Kegler, 1977). Symptoms are evident in some cvs. and E clones, whereas most cvs. remain symptomless. Pale-green and yellow ring and line patterns occur on the leaves of the tested cvs. Quince, apple, pear, and many other rosaceous host plants are natural disease hosts.

Quince Sooty Ringspot Disease (QStyRsDs)

QStyRsDs occurs frequently on quince, but its causal agent has not been isolated; thus, no properties are known. QStyRsDs has been reported in the U.S., Australia, New Zealand, South Africa, and Europe.

Other natural hosts include apple and pear in which this disease has been described under separate names.

Quince cvs. and stocks differ substantially, based on susceptibility to QStyRsDs, but only few display the characteristic sooty ringspot symptoms (Posnette and Cropley, 1963a). Quince cvs. E and C7/1 are as susceptible as clone B. Young leaf epinasty characterizes symptoms in Quince E and C7/1 cvs. Necrosis occurs along smaller leaf veins. The reddening in the leaf cuticle around the veins plus the pale yellow spots give the leaves an appearance of sooty mold.

In mixed infections with the phytoplasma causing apple rubbery wood, the sooty ringspot causal agent produces a rapid decline of quince C/71 shoot. Although normally latent, QStyRsDs is very frequently observed in pear, which is attributed to the use of infected tolerant quince rootstocks for pear stock production. Quince clones used as rootstocks for pear production have been found tolerant to the QStyRsDs causal agent. However, quince clones that are incompatible with pear and the majority of its seedlings are susceptible (Posnette and Cropley, 1963a). The quince cv. C7/1 is a suitable indicator plant, with black pigments in the leaf cuticle, vein clearing and yellowing, and bark necrosis, plus plant decline.

The QStyRsDs causal agent is transmitted with infected planting material; thus, healthy parent plants for production must be the main control measure. Initial healthy planting material can be obtained by incubating infected young plants at 36°C for three weeks.

Quince Fruit Deformaton Disease (QFtDnDs)

The unidentified causal agent of quince fruit deformations resulting from the QFtDnDs is transmitted by grafting, as described for quince. Christoff (1935) ascribed QFtDnDs to the causal agent of PrStnPiDs; Scaramuzzi (1957) reported a possible virus cause of deformation; and several authors reported on quince fruit malformations resulting from infection with the QStyRsDs causal agent (Desvignes, 1971; Refatti and Osler, 1973). The causal agent of QFtDnDs still remains a question.

The QFtDnDs observed in Europe, based on very little spread, is of no great economic importance. Quince is the natural disease host and its C7/1 cv. is a suitable indicator plant. Infected fruits have very uneven growth plus crinkle and depressions of varying shapes and sizes on the surface. The skin in the depressions is greenish to brown. Due to rough hard cell formations in the parenchymal tissue beneath the skin, similar to PrStnPiDs symptoms in pear fruits, the fruits lose market value. Diffuse pale-green spots and deformations on leaves occur in infected trees.

VIRUS DISEASES OF STONE FRUIT TREES

All stone fruit tree species except cherries are natural hosts of the widespread PsNcRsV, PsDV, and PlPV. This group of viruses is the most economically damaging of any group of pathogens in stone fruit tree production and possibly in fruit production in general. However, ApCLsV and TmRsV, which infect all cultivated species in the *Prunus* genus, plus a broad range of other hosts, also cause a serious problem in stone fruit production.

While testing several hundred *Prunus* and *Malus* cvs., Desvignes and Boye (1988) classified the different types of symptoms caused by ApCLsV strains, including fine chlorotic line patterns on peach leaves, chlorotic distortion and puckering on *Prunus* leaves, split bark or rough bark in cherry and prune d'Agen trees, necrotic pitting in apricot, cherry, and plum fruits (pseudopox), discolored or chlorotic rings and flecks on peach fruits, and skin russeting in ring patterns on apple fruits. Also, an incompatibility caused by ApCLsV strains in some *Prunus* graft combinations leads to severe diseases: apricot incompatibility, stem pitting in some plums, etc. TmRsV strains cause prunus necrotic stem pitting in all cultivated stone fruit species, plus peach yellow bud mosaic in *Prunus persica*, *P. amygdalus*, *P. avium*, and *P. armenicaca*.

Based on epidemiological traits, the viruses that are spread naturally during vegetative propagation are most damaging. PlPV is nearly the only pome and stone fruit virus transmitted by several aphid species. PsNcRsV and PsDV are transmitted naturally by pollen, which causes new infections at long distances. ChLRlV naturally spread by walnut pollen (*Juglans regia*) under favorable conditions significantly increases the number of infected trees.

Plum Virus Diseases

Plum Pox Potyvirus (PlPV) (Syn: Sharka Virus)

Properties of Viral Particles

The PlPV filamentous particles are 720×18 nm (Ranković, 1974); some isolates vary from 660 to 770 nm (Kassanis and Šutić, 1965) and from 700 to 800 nm (Cropley, 1968). The single-stranded RNA is 3500 kDa (Dunez and Šutić, 1988) and accounts for 7% of the particle weight (Ranković, 1974). Three polypetides of 43.5, 29, and 27 kDA were separated in the protein capsid (Kerlan and Dunez, 1976; Ranković, 1976).

The TIP of PlPV is 52 to 58°C; the DEP is 1×10^{-3} to 1×10^{-5}; and LIV at 20°C is 24 to 50 hours (Kassanis and Šutić, 1965; Cropley, 1968; Van Oosten, 1972; Krczal and Kunze, 1972). The titres of antisera depend upon the mode of preparation (Ranković, 1974; Dounine and Minoiu, 1968; Schade, 1969; Kerlan and Dunez, 1976).

Šutić et al. (1971) described three PlPV strains as yellow, yellow/necrotic (intermediate), and necrotic. The yellow strain, which causes yellow local spots, develops slowly and causes no leaf fall in differential *Chenopodium foetidum* plants. The necrotic strain that spreads quickly causes rapid tissue necrosis within the local spots and the infected leaves fall off. The yellow/necrotic strain causes yellow spots with more or less pronounced necrosis within the spots and infected leaves do not fall off (or if so, only occassionally, and it occurs late in the season). The yellow strain causes mild, frequently unnoticeable mosaic spots, the necrotic strain severe necrosis, and the yellow/necrotic strain mosaic spots and necrosis on the leaves of systemically infected *Nicotiana clevelandii* plants. Peas *Pisum sativum* cv. Zeiners Grüne Bastard is susceptible to all strains of PlPV, whereas the Petit Provançal, Lincoln, and Delikates cvs., suscpeptible to the necrotic and yellow/necrotic, but resistant to the yellow strain, can be used as "filter" plants for the isolation and differentiation of individual strains in suscpected mixed infections. Marénaud et al. (1976) described two virus ecotypes in France based on virulence: the Southwestern ecotype and the less virulent Southeastern ecotype. In former Yugoslavia, some plum isolates did not infect peach (Ranković and Šutić, 1980, 1986), whereas peach isolates were more infective on plum than any other (Jordović, 1985).

These strains also differ in their biophysical properties. The necrotic strain has a TIP of 57°C, a DEP of 8×10^{-3}, and an LIV of 72 hours. The yellow strain has a TIP of 53°C, a DEP of 4×10^{-3}, and an LIV of 63 hours (Ranković, 1974). Virus strains compared with antisera originating from former Yugoslavia, Holland, and Germany were found serologically identical (Šutić, 1973a), which corroborates the conclusion of Kerlan and Dunez (1979) that limited anitgenic variation exists in PlPV.

Geographic Distribution

The PlPV disease was first discovered in Bulgaria in 1915 to 1918 (Attanasoff, 1932/33), then in 1936 in former Yugoslavia (Josifović, 1937), and in 1941 in Romania (Savulescu and Macovei, 1965). Some 30 years later, the PlPV disease had been found in most European countries: Albania, Austria, Belgium, former Czechoslovakia, France, Germany, Great Britian, Greece, Hungary, Italy, the Netherlands, Poland, Switzerland, Sweden, and the former U.S.S.R. (*Bull. OEPP*, 1974).

Based on its distribution, PlPV is typical for the European continent, particularly the southern and central areas. It is found in Cyprus, Turkey, and Syria (Dunez and Šutić, 1988). Based on symptoms, it was described in apricot trees in New Zealand (Fry and Wood, 1973). With the rapid mode of distribution of planting materials to other continents where significant plum production occurs, the U.S. in particular, they are endangered by the PlPV.

Economic Importance

Economic damages caused by the PlPV are mostly changes in fruits of infected plum cvs., which in susceptible cvs. (e.g., Požegača) are smaller, distorted, of unbalanced chemical composition, and drop prematurely early in the season. All changes result in significant reductions in fruit quality and yield. (See the chapter on "The Influence of Viral Infections on Fruit Tree Development and Yields".)

Researchers and production experts believe that PlPV may be the most damaging virus in fruit production. The extent of damage in any one country primarily depends on the number of susceptible plum cultivars and trees of each still in production. PlPV is a very serious disease in Hungary and former Czechoslovakia (Šutić, 1971).

Former Yugoslavia had the largest number of plum trees in the world, about 84 million, 40% of which were the cv. Požegača, the most susceptible to PlPV. Nearly 20% of all plum trees are infected with PlPV, which dictactes the need for urgent solutions for disease control.

Host Range

PlPV infects numerous hosts, both woody and herbaceous. The woody species include 24 susceptible in the *Prunus* genus. At first believed restricted to only this genus, Šutić et al. (1972a) later showed that *Sorbus domestica* is an experimental host for PlPV, which represents the exception.

The following hosts of PlPV in the *Prunus* genus have been reported: *P. amygdalis, P. armeniaca, P. brigantina, P. cerasifera* (cvs.: Blireiana, Pissardi, and Trailblazer), *P. curdica, P. domestica, P. glandulosa, P. holoserica, P. hortulana, P. insititia, P. italica, P. japonica, P. manshurica, P. maritima, P. mume, P. nigra, P. persica, P. salicina, P. simoni, P. spinosa, P. tomentosa,* and *P. triloba*. The infection remains localized in experimentally infected *P. mahaleb* plants.

Natural hosts among these include: *P. armeniaca* (apricot), *P. cerasifera* (myrobalan), *P. domestica* (European plum), *P. glandulosa* (almond cherry), *P. insititia* (Damson plum), *P. persica* (peach), *P. salicina* (Japanese plum), and *P. spinosa* (sloe plum) (*OEPP Bull.*, 1974; Németh, 1986). Recent investigations suggest the possible presence of PlPV in sweet cherry (*P. arium* cv. Ferrovia) in southern Italy (Crescenzi et al., 1995) as well as in sour cherry (*P. cerasus*) in Moldava (Nemchinov and Hadibi, 1996).

It was long believed that woody plants were the only PlPV hosts, but Šutić (1961) succeeded in transmitting it from plum tree leaf sap (cv. Požegača) into *Nicotiana quadrivalvis* tobacco, thus the first herbaceous host of PlPV with systemic infection, later confirmed by other researchers. Thus, it is one of the herbaceous virus indicator plants. Németh (1963) later discovered that *Chenopodium foetidium* plants also react to PlPV invection with local spots. Kassanis and Šutić (1965) discovered that *Nicotiana clevelandii* tobacco is also a systemic host of PlPV; thus was developed the basic procedure for transmission of PlPV into herbaceous plants.

The discovery of herbaceous hosts for PlPV substantially facilitated this research (*OEPP Bull.*, 1974). About 78 herbaceous plants in nine families were then described as virus hosts by several authors (Bode and Babović, 1969; Pleše et al., 1969; van Oosten, 1970; 1970a; 1971; Šutić, 1971; Kröll, 1973). New herbaceous hosts, such as opium poppy (*Papaver somniferum* cv. Malvo) in the Papaveraceae and chick weed (*Stellaria media*) in the Caryophyllaceae (Šutić, 1977a; 1981), have now been discovered. Other herbaceous hosts worth mentioning include *Lamium album, L. purpureum, Melilotus albus, Vicia sativa, V. cracca, Ranunculus arvensis, Nicotiana acuminata, N. bigelowii, N. excelsior,* and *N. megalosiphon*. Most herbaceous hosts react locally to PlPV infection, with local lesions of differing size, chlorotic halos, yellow-brownish and necrotic tissues. PlPV-infected herbaceous plants react with systemic chlorotic spots, ringspots, mottle, interveinal necrosis, leaf stunt, and distortion (early infected leaves), plus disappearing leaf symptoms (in some cases).

Regardless of intensive investigations, the PlPV has not been found in any naturally infected herbaceous plants in Yugoslavia. However, in some countries, herbaceous plants are listed as potential natural virus hosts, such as *Lamium album, Lupinus albus, Melilotus officinalis, Ranunculus acer, Trifolium incarnatum, T. repens,* and others (Kröll, 1973), which should be studied and confirmed by further research. Hosts recommended as indicators for PlPV include *Prunus domestica* cv. Požegača, with groove-like depressions on fruits (assess fruit symptoms in two harvests); *P. tomentosa*, with chlorotic diffuse mottle on lower leaf surface and chlorosis along lateral leaf veins of seedlings; herbaceous indicators: *Chenopodium foetidum* with yellow and yellow/necrotic spots on leaves caused by respective strains; and *Nicotiana clevelandii* and *N. acuminata*, with systemic infection: mosaic mottle = yellow strain, mosaic mottle with mild necrosis = yellow/necrotic strain, mosaic mottle with severe necrosis = necrotic strain.

Symptoms

PlPV causes symptoms on both leaves and fruits of economically very important hosts, including plum, apricot, peach, and myrobalan plum. Some hosts have symptomless fruits that are considered tolerant. The discovery of symptoms in *P. spinosa* may be epidemiologically important because this woody host may provide a perennial source of PlPV in nature.

Symptoms in P. Domestica

The susceptibility of plum and symptom expression varies, depending on the genotypes, virus strains, and relationships between vectors and virus strains. Most of the 400 plum cutlivars, clones, and types investigated in different countries were susceptible, with symptoms on both leaves and fruits (cvs. Požegača, Italian Prune, Zimmer's Prune, Green Gage, Monarch, Wagenheim, etc.) (*OEPP Bull.*, 1974; Šutić, 1983). Some cvs. listed as tolerant, with symptoms only on leaves, not (or very slightly) fruits (cvs. Stanley, Anna Spät, Prune d'Agen, Čačanska Rana, Large Sugar Prune, etc.) (Németh, 1986; Šutić, 1983). Six cvs., including Jelta boutil covidna and Scoldus No. 1, classify as completely resistant (Šutić and Ranković, 1981; 1983). Symptoms of disease appear on both the leaves and the fruits of susceptible and highly susceptible plum genotypes (Figure 155). The first leaf symptoms consist of vein clearing and chlorotic patterns bordering main and secondary veins. Some of these early symptoms are pale-green areas and oily patches irregularly distributed on the leaves, on which later chlorotic spots, rings, and open rings develop. Leaf symptoms may ameliorate or disappear later in the summer. The margins of chlorotic rings and spots in some cvs. turn reddish-brown during the season, thus making them more conspicuous. Infected leaves of the most susceptible cvs. are stunted and deformed.

The first early fruit symptoms are irregular blue and shallow depressions on the skin. As fruits ripen, deep irregular grooves, rings, and lines appear on their surface, and the tissues beneath

FIGURE 155 Plum pox virus (Šarka) on plum varieties 'Stanley' (left), 'Early Red' (Center), and 'Požegača' (right).

become necrotic and brownish-red (Figures 156 and 157). The fruit flesh is fibrous and gummy. Irregular brown and reddish spots appear on the stone fruit surface. Infected fruits are not tasty, become useless, and drop prematurely (about 3 to 4 weeks before ripening), especially in late cvs. (See the morphological and anatomical alterations in affected plum fruits in the chapter "Influence of Viral Infections on Fruit Tree Development and Yields".)

FIGURE 156 Fruit skin blotches caused by plum pox virus in plum cv. Czar. (Reproduced with permission from APS.)

FIGURE 157 Fruit symptoms caused by plum pox virus in apricot. (Courtesy J.A. Foster. Reproduced with permission from APS.)

Symptoms in P. Cerasifera

In susceptible myrobalan plants, chlorotic spots and mottle appear on leaves, and chlorotic rings and spots of varying size develop on fruits. Leaves and fruits are misshapen in highly susceptible cvs. Myrobalans with colored leaves react severely to virus infections, although some myrobalan seedlings are tolerant (symptomless on leaves and fruits) to PlPV (Marénaud, 1972; Šutić, 1968).

Pelet reported 15 myrobalan type rootstocks of stone fruits as slightly or moderately susceptible to PlPV (*OEPP Bull.*, 1974) with leaf symptoms of chlorotic mottle, spots, and rings surrounded by chlorotic or brownish halos varying in size. Five myrobalans were symptomless.

Symptoms in P. Spinosa

The symptoms on young leaves are irregular chlorotic spots, rings, and bands that fade on older ones and with symptomless fruits. Some resistant clones have been found mostly in *Dasyphylla forma depressa* with small flowers and narrow leaves. In some, the symptoms in PlPV-infected susceptible plum shoots grafted onto resistant clones caused symptoms to disappear, and thus the shoot to recover (Macovei et al., 1971; Šutić, 1965). (Data pertaining to the recovery phenomenon can be found in the section on vegetative protection).

The PlPV causes irregular pollen formation in the plum cv. Požegača and changes its morphological and physiological properties. Pollen germination from infected plants was 1.5 to 5 times lower than for pollen from healthy plants (Pejkić and Šutić, 1971). PlPV-infected Agen plum pollen germinated at 14 to 15% compared with 89% for healthy pollen (Macovei, 1970a,b).

Symptoms gradually disappear (recover) in some infected plants. Among five *P. domestica* (cv. Early Red) and three *P. persica* (cv. Vinogradarska Breskva) seedlings that recovered, the presence of PlPV was proved in only one (Šutić, 1979). The rapidity of the recovery phenomenon may depend on virus strains and host species. In *Stellaria media* infected with the yellow and intermediate PlPV strains, recovery occurred after 6 months infection; but in those infected by the necrotic strain, recovery required 1 year (Šutić, 1980). Plant recovery represents an interesting phenomenon for further theoretical and applied investigations.

Symptoms in P. armeniaca

See "Plum pox virus in apricot."

Symptoms in P. persica

See "Plum pox virus in peach."

Pathogenesis

PlPV is preserved in its natural *Prunus* hosts, and plum is the most important among eight such hosts. *P. spinosa*, a widely spread spontaneous host, seemed to represent an important source of natural infection (Baumann, 1968; Jordović et al., 1971), but investigations 2 decades later failed to confirm this assumption (Ranković and Dulić-Marković, 1991). PlPV is spread naturally through infected planting material and by aphids. Possible transmission through seed and pollen are unconfirmed due to inconsistent data (*see* the chapter on "Transmission Through Infected Seed and Pollen"). PlPV-infected planting material, especially infected suckers, played the key role in its distribution and its introduction into other countries. Up to 100 suckers are formed from roots of plum trees; thus, if the plum tree is infected by PlPV, then all suckers are infected.

Aphids actively transmitt PlPV nonpersistently. *Brachycaudus helichrysi*, the first aphid vector described in Bulgaria (Christoff, 1947), was followed by the discovery of others; for example, *Phorodon humuli* (Vaclav, 1960), *Myzus persicae* (Kassanis and Šutić, 1965), and *Brachycaudus cardui* (Šutić et al., 1972) in former Yugoslavia; and *Aphis cracivora, A. fabae, A. spiraecola*, and *Myzus varians* (Leclant, 1973) in France. Aphids can acquire PlPV within 5 minutes and quickly transmit it to infect healthy plants.

The role of aphids in the epidemiology of plum pox depends on factors such as the source of infection, vector overpopulation, presence of susceptible plants, and remoteness of infection source.

For example, PlPV was transmitted to 100% of healthy plants at a distance of 100 m in a 10-year test, but to only 1.5% at a distance of 500 to 800 m during that test (Jordović, 1968). These data help determine isolation space requirements for establishing new plantations in areas of known PlPV incidence.

Understanding the relationship between individual vectors and virus strains is important for dealing with disease epidemiology. Experimentally, *Myzus persicae* transmitted all virus strains, the necrotic and the intermediate strains at 24.5% efficiency, but the yellow strain at only 8.0% efficiency (Šutić et al., 1976). The role of the vector undoubtedly is influenced by the suitability of individual virus hosts for aphid feeding and multiplication.

PlPV, transmitted mechanically with infected plant sap, is important only for experimental purposes; it is not transmitted through the soil or with tools used for pruning and grafting.

Control

The production of virus-free planting material for the establishment of new plantations requires the primary virus control measure. If virus-tested planting materials are used, they must not contain the PlPV.

PlPV can be eliminated from young infected plants by incubation at 37 to 38°C for 2 months. Plants resulting from grafting of terminal buds of thermotherapy-derived plantlets onto virus-free rootstocks are used satisfactorily as healthy parent material. Suitable previously mentioned woody and herbaceous indicator plants are used for virus detection (*see* the section on "Host Range of Plum Pox Virus"). The enzyme-linked immunosorbent assay (ELISA) is efficiently used for PlPV detection (Clark and Adams, 1977; Adams, 1978; Šutić and Ranković, 1981) and the immunosorbent electron microscopy (ISEM) is quite sensitive (Kerlan et al., 1981).

Although natural virus spread cannot be prevented completely, it can be reduced substantially by certain measures. Of utmost importance, one must plant production material only in noninfected areas with no known adjacent sources of infection because, in an infected area, the plantation is continuously exposed to infections naturally transmitted by aphids. All infected plum, myrobalan, apricot, and other *Prunus* spp. known as hosts of PlPV should be destroyed prior to orchard establishment. This process is especially successful when establishing new plum plantations in slightly infected areas. All trees displaying initial signs of infection should be immediately removed and destroyed from new plantations. The destruction of infected trees is also recommended for older plantations with low incidences of infection. It is best to destroy infected trees with the aid of arboricides, which both ensures tree death plus prevents regeneration from roots (Perišić and Čuturilo, 1960).

Control of aphids to limit virus spread is indispensable in the protection of young nurseries, and also established plum plantations, in less severly infected areas. Quarantine measures may help prevent PlPV spread, since they are obligatory.

Although all these measures are indispensable, they cannot provide absolute protection against PlPV infection under current production practices. Resistant and tolerant plants provide the best solution to this problem.

The testing for resistance against PlPV of 400 plum, apricot, and peach species and cvs. showed that some tolerant plum cvs. and clones, especially early maturing cvs. with large fruits, are particularly important (Šutić and Ranković, 1981). Stanley may be the most important tolerant cv. (Jordović and Janda, 1963; Vaclav, 1965; Šutić, 1977; Festić, 1980) because it is widely used in several countries where PlPV causes serious production problems. Other less susceptible cvs. include Washington, Imperial, Wagenheim, Anna Spät, California, and Rut Gerstetter (Jordović and Janda, 1963; Šutić, 1977).

While testing an international stone fruit tree collection, Šutić and Ranković (1981; 1981a) reported resistance to PlPV in Mirabelca Jaune de Plovdiv, Hybride Christoff No. 1, and the *Prunus cerasifera* cv. Edra Jelta Afisca, and clones Požegača Br and Požegača S. In these trials, they found complete resistance (immunity) in cvs. Jelta Boutil Covidna and Scoldus No. 1 under both natural and experimental inoculation conditions. Ranković (1981) found tolerance in cvs. Opal, symptomless, and home plum selection cvs. Čačanska Lepotica and Čačanska Najbolja, symptomless or rarely with fruit symptoms. The XVII/6 and particularly the XII/40 hybrid, with mild symptoms only on individual leaves and symptomless fruits, may be classified tolerant to PlPV (Ranković and Paunović, 1988). (Some comments on plum resistance can be found in the chapter on "Resistant and Tolerant Fruit Tree cvs.")

Investigations in former Yugoslavia and other countries found a number of plum cultivars resistant or tolerant to the PlPV, some of which should be grown in areas with completely infected plum plantations. Their use depends also on other factors of fruit quality, adaptability to ecological and edaphic conditions, etc. Resistant and tolerant plum cultivars and clones also represent important genetic material in the breeding of new hybrids. Some results obtained in protection of plum by using cross protection and vegetative protection methods, plus the creation of new genetic resistances are discussed in the chapter on "Biological protection."

Prune Dwarf [possible] Ilarvirus (PsDV)

Properties of Viral Particles

The PsDV consists of quasi-isometric particles 19 to 20 nm diameter, and bacilliform particles 20 × 73 nm (Halk and Fulton, 1978). This multicomponent virus has six particle types, with sedimentation coefficients of 75, 81, 85, 99, and 113S. Infectivity is associated with the 99 and 113S particles (Németh, 1986). PsDV contains 14% RNA (Fulton, 1981) and has a 24-kDa protein (Halk and Fulton, 1978).

Since PsDV is labile in undiluted cucumber sap, it loses infectivity in 30 minutes, and in a phosphate buffer in 15 to 18 hours (Fulton, 1970). The TIP of PsDV (in chemically stabilized sap) is 45 to 54°C (Waterworth and Fulton, 1964). PsDV has moderate immunogenicity. It does not react serologically with antisera of PsNcRsV, which it resembles based on other traits (Richter and Kegler, 1967).

Strains

PsDV strains differ symptomologically from the type, Fulton's virus B strain causing different stone fruit diseases that earlier were considered distinct viruses. Thus, PsDV strains now include apricot gummosis virus, cherry chlorotic ringspot virus, cherry chlorotic-necrotic ringspot virus, cherry ring mosaic virus, cherry ring mottle virus, cherry S virus, cherry yellow mosaic virus, cherry yellow mottle virus, cherry (sour) yellows virus, peach stunt virus, and prune dwarf virus apricot strain (Németh, 1986).

Geographic Distribution

PsDV is considered present wherever fruit is grown. It is widespread in cherry and sour cherry plantations in North America, Europe, and is also described in Australia, New Zealand, and Japan. PsDV was first described in Italian prune in New York, Ontario, and British Columbia (Thomas and Hildebrand, 1936). It was also identified in the home plum cultivar Crvena Ranka (Red Early) and several Požegača types (Ranković, 1976a; Jordović, 1955).

Economic Importance

PsDV causes one of the most damaging fruit tree diseases, evident because it infects all economically important stone fruit species, reducing yields and quality. PsDV causes yield losses up to 60% in the Grosse Grüne Reneklode plum and 80% in Italian prune. Sugar and acid contents are reduced substantially in infected fruits (Kegler, 1977). Eight Italian prune cvs. averaged 33 kg from PsDV-infected trees, compared with 67 kg from healthy ones (Blumer, 1957).

Host Range

The most important natural hosts of PsDV in the *Prunus* genus include plum (*Prunus domestica*), cherry (*P. avium*), sour cherry (*P. cerasus*), peach (*P. persica*), apricot (*P. armeniaca*), myrobalan plum (*P. cerasifera*), European bird cherry (*P. padus*), rock cherry (*P. mahaleb*), and Damson plum (*P. insititia*).

PsDV has many experimental hosts, both woody and herbaceous, in 15 dicot families (Fulton, 1957), of which Japanese flowering cherry *P. serrulata* cv. Shirofugen is an important one. Severe necrosis and gum flow characterize symptoms on bud grafts from PsDV-infected plants onto Japanese flowering cherry, thus a good indicator for detection of PsDV and PsNcRsV. *Prunus tomentosa* is an experimental host suitable for comparative studies of PsDV, PlPV, and ApCLsV (Ranković, 1980).

Experimental herbaceous hosts equally suitable as indicator and assay hosts include *Momordica balsaminea* (initial spots and systemic mottle), *Tithonia speciosa* (some isolates, severe chlorotic line patterns), *Sesbania exaltata* (small dark local spots on cotyledons), *Crotolaria spectabilis* (small dark local spots), *Cucurbita maxima* cv. Buttercup (chlorotic interveinal systemically infected leaves with areas becoming pale yellow later in the season, *Cucumis sativus* (small chlorotic initial spots on cotyledons; systemic mosaic sometimes only on a part of the leaf blade). PsDV is probably the first virus from fruit trees transmitted mechanically into herbaceous plant hosts.

Symptoms

Italian and Lombard plum and cv. Emma Heppermann are quite susceptible to PsDV, although dwarfing occurs also in cvs. Giant, Standard, President, Emilie, Tragedy, Green Gage, and Albion (loc. cit. Gilmer et al., 1976).

PsDV markedly reduces leaf size, especially narrowing; thus, the disease known as the plum willow-like leaf. It also reduced dentations and hairs. Infected leaves are wrinkled, notably on midribs, with diffuse mottle and the surface with a slightly compacted, leathery appearance. Leaf margins appear frequently irregular, as if nibbled by insects, more pronounced near the leaf base. Shoot internodes shorten and some branches apparently uninfected maintain a normal appearance. Many flowers have stunted styles plus elongated and irregular calyxes.

PsDV causes similar changes and yield losses in the cv. Požegača plum. Milder symptoms in Lombard plum affect yields much less and several cvs. are tolerant to PsDV.

Pathogenesis

PsDV, preserved in many widespread perennial hosts, is transmitted regularly by grafting (buds, scions), particularly during production of planting materials.

PsDV is spread naturally through infected pollen, particularly in cherry and sour cherry plantations. Natural spread of PsDV in plum trees has not been observed. Spreading of PsDV in young sour cherry plantations is insignificant immediately following planting; then it increases from the 5th to 15th year when all the trees become infected. All factors that affect pollination (man, insects, etc.) also affect transmission of PsDV through the pollen. PsDV transmission occurs through cherry, sour cherry, mahaleb, and myrobalan seed (loc. cit. Gilmer et al., 1976), and especially with seedlings of these plants if infected and if used as rootstocks.

PsDV is transmitted mechanically with infective sap from succulent parts of woody plants into herbaceous hosts, which is useful for further investigations of the virus. Insect transmission of PsDV is unknown.

Control

The use of virus-free planting material is a key control measure and the virus-tested material must not contain PsDV. Mother stock material can be made PsDV-free by exposure to 37°C for 2 to 4 weeks, followed by separate propagation of the buds that grew during this treatment. The seed of myrobalan plants, used to produce rootstock, must be only from healthy plants. These preventive measures can ensure successful disease prevention in plum plantations because natural virus spread seldom occurs in plum trees. (Additional information on PsDV occurrence in stone fruits can be found in the chapter on virus diseases of these fruit trees.)

Prunus Necrotic Ringspot Ilarvirus (PsNcRsV)

Properties of Viral Particles

The PsNcRsV particle is isometric, 23 nm diameter, and multicomponent with sedimentation coefficients of 72, 90, and 95S (Loesch and Fulton, 1975), plus some bacilliform particles (Fulton, 1981). PsNcRsV contains 16% RNA and a major polypeptide coat protein of 25 kDa (Barnett and Fulton, 1969), plus some smaller components. PsNcRsV is unstable in undiluted plant sap, thus it retains infectivity for only a few minutes, but it remains infective several hours in buffered sap. The DEP of PsNcRsV is 5×10^{-1} and the TIP is 55 to 60°C (Fulton, 1970a). PsNcRsV has moderate immunogenicity. Serologically, it is related closely to Danish plum line pattern and distantly to ApMV and rose mosaic virus (RoMV) which cause plum line pattern (Fulton, 1968).

PsNcRsV consists of strains that cause specific symptoms in *Prunus* spp., such as PsNcRsV type, cherry recurrent ringspot virus, cherry stecklenberg virus, cherry virus L, Fulton's cherry viruses E and G, almond calico virus, plum concentric ringspot virus, plum decline virus, plum yellow ringspot virus, and Victoria plum line pattern virus (Németh, 1986).

Geographic Distribution and Economic Importance

The first occurrence of a ringspot disease, possibly caused by PsNcRsV, was reported in plum and peach (Valleau, 1932) and later in sweet and sour cherry (Thomas and Rawlins, 1939; Moore and Keitt, 1944). PsNcRsV is one of the most widely distributed viruses in *Prunus* and *Rosa* species in temperate regions.

PsNcRsV reduces infected tree growth, vigor, and yields expressed as stunt and dwarf. Extent of detrimental effects depends on susceptibility of plum cvs., virulence of strains, and a potential complex of mixed infections usually in combination with PsDV and/or ApCLsV. Yield losses were measured in Oullins Golden Gage, Cambridge Gage, and Marjorle's Seedling cvs. (Posnette and Cropley, 1970). This generally distributed virus especially causes broad damage in fruit production, greater in peach and particularly extensive in sweet and sour cherry. (For more details on the subject, see the descriptions of individual virus diseases in stone fruits.)

Host Range

PsNcRsV has many natural and experimental hosts in 21 families (Fulton, 1957; Kirkpatrick et al., 1967). Natural hosts include nearly all cultivated fruit tree species of Rosaceae, and the *Prunus* species are most numerous, among them sweet cherry (*Prunus avium*), sour cherry (*P. cerasus*), plum (*P. domestica*), peach (*P. persica*), nectarine (*P. persica* cv. *nectarina*), apricot (*P. armeniaca*), myrobalan (*P. cerasifera*), rock cherry (*P. mahaleb*), European bird cherry (*P. padus*), sloe (*P. spinosa*), black cherry (*P. serotina*), apple (*Malus pumila*), and several rose (*Rosa* sp.) species. Hop (*Humulus lupulus*) is a natural host in the Cannabinaceae.

Diagnostic hosts of PsNcRsV include *Cucumis sativus*, with chlorotic primary spots on cotyledons, necrosis or mosaic on permanent leaves; *Momordica balsamina*, with necrotic primary spots and sometimes systemic infection; and *Prunus serrulata* cv. Shirofugen, with necrosis in the bud grafts.

Symptoms

Symptoms depend on stage of development in the disease. The shock stage occurs immediately upon infection; for example, in susceptible plum cvs., the symptoms of ringspots (plum ringspot disease) and lines surrounded by dark necrotic tissues occur on the first emerging leaves in the spring. Later, necrotic tissues tatter and fall out, leaving irregular shot-holes as diagnostic evidence of PsNcRsV. Symptoms fade or fail to appear in leaves formed later in the season, a sign of the latent or chronic stage.

Symptom dynamics and severity vary, depending on plum cvs., virus strains, and environment. Cultivars with apparent symptoms of PsNcRsV infection include Brompton, Burbank, Florentina, and Santa Rosa (Kegler, 1977). Characteristic symptoms ascribed to individual PsNcRsV strains are plum chlorotic ringspot, plum yellow ringspot, stripe mosaic (European), and Victoria plum mosaic (Posnette, 1953; Senevirante and Posnette, 1970). A separate strain of PsNcRsV is a potential cause of plum decline discovered in certain rootstock and scion combinations (Posnette and Cropley, 1970). Most plum cultivars, including Italian prune and Japanese plum (*Prunus salicina*), are tolerant to PsNcRsV.

Pathogenesis

Infected parent trees from which buds and scions are grafted onto infected vegetative rootstocks, plus suckers used as planting material, represent the main sources of infection, and those with latent infections create the highest risk for natural spread of PsNcRsV.

The role of plum seed and plum pollen in natural dissemination is not clear. Traylor et al. (1963) experimentally transmitted PsNcRsV from infected onto healthy plum trees via pollen. Barbara (1988) demonstrated that the natural dissemination of PsNcRsV in plum trees in Spain is less than 2% annually. Digiaro et al. (1991) showed the presence both externally and internally of PsNcRsV and PsDV in pollen of plum and apricot and in ovules of all plum cvs. tested. Possibly important is the greater percentage transmission of PsNcRsV through seed of *P. cerasifera*, whose seedlings are used as rootstocks for plum and apricot (*see* the chapter on "Transmission Through Seed and Pollen").

The *Vasates fockeui* mites and *Longidorus macrosoma* nematodes are potential (not yet confirmed) vectors of PsNcRsV (loc. cit. Fulton, 1970a). PsNcRsV is transmitted mechanically with infective sap, currently important only for experimental purposes.

Control

Based on these facts about the pathogenesis of PsNcRsV, virus-free planting material for establishment of plantations represents the key preventive measure. Infected plants as potential sources of infection should be eliminated immediately from young plantations. The phytosanitary condition of parent trees should be monitored regularly to avoid any potential foreign infections, especially of myrobalan trees used as generative rootstocks for plum and apricot. Healthy planting materials can be generated by exposing the tips of young plants to 38°C for 24 to 32 days, then grafting this virus-free material onto healthy rootstocks.

Plum Line Pattern Viruses

Plum American Line Pattern Virus (PlALnPtV) (assumed member of the Ilarvirus group)

PlALnPtV is quasi-isometric with four particle types, 26 (T), 28 (M), 31 (B), and 33 (B_2) nm in diameter, each with respective sedimentation coefficients of 95, 100, 114, and 125S (Fulton, 1984). The presumed SS-RNA accounts for 17% of viral particle weight. All four types are infective. PlALnPtV is unstable in crude sap, but with antioxidants its DEP is 6.4 to 12.8×10^{-3}, and its TIP is 66°C (Paulsen and Fulton, 1968). It is moderately immunogenic (Fulton, 1984). Some isolates differ, based on host susceptibility.

PlALnPtV seems widespread wherever plum is grown. First observed in the U.S. and Canada (Willison, 1945), it now occurs in India, Israel, New Zealand, Greece, Norway, and Poland (Németh, 1986).

Natural hosts of PlALnPtV are *Prunus* species, including *Prunus domestica*, *P. cerasifera*, *P. insititia*, *P. persica*, *P. salicina*, and *P. serrulata*. Its experimental woody rosaceous hosts are *Chaemomales japonica*, *Malus sylvestris*, *Prunus armeniaca*, *P. amygdalus*, *P. avium*, *P. cerasus*, *P. kansuensis*, and *P. mahaleb*. It has been transmitted mechanically into 78 species in 8 families (Paulsen and Fulton, 1968).

Symptoms of PlALnPtV vary in different plum cvs. Symptoms in leaves are lines, striated spots, and oak-like mosaic, plus small rings and striped chlorosis along the veins. The spots in myrobalan leaves, a tolerant host, are clearly yellow or yellow-green in the spring and pale yellow in the summer. Some cvs. express irregular chlorotic lines along the veins and oak-line patterns on the leaves. A striking chlorotic line pattern mosaic occurs in the Shiro plum.

Plants diagnostic for PlALnPtV include *Vigna cylindrica*, with necrotic local spots, discoloration, and necrosis; *Petunia hybrida*, with necrotic or chlorotic spots, discoloration, and necrosis; and *N. megalosiphon*, with local symptoms of chlorotic or necrotic local spots and ringspots and systemic symptoms of chlorotic mottle and necrosis.

PlALnPtV, preserved in many stone fruit species, is graft-transmitted from infected into healthy plants, the only known mode of transmission. Vectors are unknown. Mechanical transmission of PlALnPtV into herbaceous plants is of only experimental importance, as is transmission by dodder *Cuscuta campestris* from *N. megalosiphon* into petunia.

Infections of PlALnPtV in plum trees can be successfully avoided with the use of virus-free planting material. No known risks of natural infection exist for spread of PlALnPtV in orchards.

Plum European Line Pattern Viruses (PlELnPtV)

ApMV and PsNcRsV cause this type of disease (Senevirante and Posnette, 1970). ApMV may cause line pattern symptoms on different tissues, whereas PsNcRsV causes mosaic symptoms on apple leaves. ApMV, serologically related to PsNcRsV, is a serotype of PlELnPtV. (More details on the properties of these viruses may be found in relevant chapters on apple mosaic virus and Prunus necrotic ringspot virus.)

PlELnPtV, first described in Bulgaria (Attanasoff, 1935), was discovered later in nearly all European countries and is now known in India, Japan, South Africa, and the U.S. Characteristic symptoms in infected plum include leaf line and oak-like patterns associated with chlorotic lines. Severity of discoloration (lighter or darker) depends on virus strains and plum cvs. Leaf symptoms, clearly visible in the spring and early summer, become masked (latent) at higher temperatures later in the season. Highly susceptible cvs. include Grosse Grüne Reinklode, Ruth Gerstetter, Ontario, Wageningen Early, and Magna Glauca (Németh, 1986). European plum line pattern is an economically damaging plum disease.

Plum: Natural Alternative Host of Some Viruses

Apple Chlorotic Leaf Spot Virus (ApCLsV)

ApCLsV, first discovered in stone fruits in Great Britain (*E. Mall. Res. Sta. Ann. Rept.*, 1966; 1967) was found later in the U.S., Canada, and Europe. Research in France showed natural infection of ApCLsV in 35% of plants tested in *Prunus* spp. (Marénaud et al., 1976).

ApCLsV causes two characteristic diseases of plum: plum pseudopox (PlPdPDs) and plum bark split (PlBkSplDs). PlPdPDs is characterized by fruit symptoms of irregular grooves and deformations, particularly in cvs. with blue fruits. ApCLsV-infected fruits do not drop, the primary difference when compared with PlPV-infected fruits. Pseudopox was first described in Great Britain (Posnette and Ellenberger, 1963), then later in several countries in fruits of several cvs. (e.g., Ersinger, Zimmers, Požegača, etc.) in a number of countries. Van Oosten (1970) and Marénaud (1971) reported some ApCLsV strains as potential causal agents of this disease; then Schmid (1980) reported that it represents a genetical-physiological disorder, possibly caused by a/some virus combination(s).

PlBkSplDs was first described in Great Britain and its viral nature established (Posnette, 1953; Posnette and Ellenberger, 1957), and it is now believed to be a separate strain of ApCLsV (Dunez et al., 1976). Brownish-red lateral and irregular spots on the stem and branch bark appear initially, then tissues within these spots become necrotic, followed by severe cracks and splits. The necrotic process spreads to the cambium, causing branch splitting and death. Characteristic stem and branch flattening is clearly a consequence of irregular development caused by ApCLsV. This disease is most damaging in susceptible cvs. Yield losses in infected French Prune d'Ente plum cultivar trees, depending on the virus strains, are 40 to 60% (Marénaud, 1971).

ApCLsV, generally widespread, also induces severe graft incompatibilites in some Prunus combinations, causing serious nursery problems (German et al., 1991). It usually occurs in mixed infections with PsNcRsV, PsDV, and PlPV. Their potential influence on other viruses involved in mixed infections is unclear. ApCLsV was isolated from some plum cultivars and clones resistant or tolerant to PlPV in former Yugoslavia. The cv. Scoldus No. 1 immune to PlPV was heavily infected with ApCLsV. The tolerant cv. Požegača Br and S clones infected with PlPV also contained ApCLsV and PsDV (Šutić and Ranković, 1981). The extent to which such mixed infections influence resistance or tolerance of plum cultivars and clones to PlPV represents an interesting subject for further research.

Tomato Ringspot Virus (TmRsV)

TmRsV and its strains infect many woody, semi-woody, and herbaceous plants in nature, most importantly many in the *Prunus* genus. It causes stem pitting in plum (*Prunus domestica*) and apricot (*P. armeniaca*), and stem pitting and yellow mosaic diseases in peach (*P. persica*). Symptoms in these hosts are similar but differ in severity, depending on cv. susceptibility and strain virulence. Characteristic symptoms occur as enlargements on the lower stem, covered with thicker bark than on healthy trees. Deep or shallow pits and grooves become necrotic in the phloem/cambium at such points. Infected plants have reduced vigor and the less well-developed leaves turn yellow early. Symptoms were observed in plum cvs. Stanley, Italian Prune, Late Smith, and Bradshaw (Németh, 1986).

Raspberry Ringspot and Strawberry Latent Viruses

These viruses were isolated from naturally infected plum (*Prunus domestica*) trees featuring atypical disease symptoms. Such infections occur only rarely in inidividual trees, and thus they are of no economic importance in production.

Latent Viruses

Plum is the natural host of some viruses described in herbaceous plants: carnation ringspot (CaRsV), CMV, SwMV, TMV and TNV. Due to limited or no spread they are of no economic importance, except they should not be allowed in planting materials.

Plum Diseases Assumed Caused by Viruses

Similar to other fruit tree species, a few diseases caused by unidentified pathogens were described in plum. In all cases, they are transmitted through vegetative plant parts by grafting, budding, and chip budding, which is one of the high-risk traits of tree fruit virus diseases. Some were assessed as economically damaging, but fortunately all were discovered by phytosanitary controls.

Plum fruit crinkle (PlFtCrDs) was described in New Zealand (Chamberlain et al., 1959). *Prunus salicina* cv. Sultan has leaf and fruit symptoms that resemble those caused by PlPV in the natural disease host, but none has been identified. *P. salicina* cv. Billington (sensitive) and Purple King (symptomless) are the natural hosts of *plum mottle leaf disease* (PlMtLDs), also described in New Zealand (Wood, 1979). *Plum ochre mosaic* (PlOMDs) was described in former Czechoslovakia (Blattny, 1961). The causal agent of this disease is transmitted by vegetative plant parts and by the aphid *Myzus persicae* as its vector. *Prunus domestica* and *P. insititia* cvs. Krikon and St. Julien d'Orleans are natural hosts of this disease, discovered in France (Desvignes and Savio, 1971). This disease is dangerous when St. Julien seedlings are used as rootstocks. In the *Prunus* genus also, plum is the host of *peach asteroid spot disease* (Wagnon et al., 1963), present in Israel and the U.S. and suspected of causing severe damage to production. *Prunus persica* cv. J.H. Hale is the indicator plant.

Peach leaf necrosis (PhLNDs) was discovered in Germany as a latent disease in plum, wherefrom it was graft transmitted onto *Prunus persica* (Kegler, 1964). *P. persica* seedlings, with dwarfing and brown necrotic spots on the leaves, are effective indicators. Kegler (1964) also described *peach pseudo stunt* (PhPdSnDs) as a latent disease in plum, discovered when peach seedlings were used as indicators (shoot stunting).

The causal agents of all the aforementioned diseases, except plum ochre mosaic, are exclusively graft transmitted through vegetative plant parts; healthy virus-free planting materials for establishment of plantations represents the key, possibly the only, disease control measure.

Virus Diseases of Peach (*Prunus persica*)

Peach Rosette Mosaic Nepovirus (PhRoMV)

Properties of Viral Particles

The PhRoMV isometric particles are 24 to 31 nm in diameter. Purified preparations contain three particle types: T (top), M (middle), and B (bottom), with sedimentation coefficients of 52, 115, and 134S, respectively (Dias and Cation, 1976). T contains no RNA, and thus is noninfective, while M and B are infective containing SS-RNA of 2200 kDa (RNA-1) and 2500 kDa (RNA-2) plus a 57-kDa coat protein subunit (loc. cit. Murant, 1981).

The biophysical properties include a TIP of 58 to 68°C, DEP of 1×10^{-3} to 1×10^{-5}, and LIV at 20°C of 15 to 25 days (Dias, 1975). PhRoMV is moderately immunogenic (Dias, 1975; Ramsdell et al., 1979; Stubbs and Barker, 1985). Isolates from peach and grape are serologically identical in gel-diffusion tests.

Geographic Distribution and Economic Importance

PhRoMV, widespread only in certain areas, has been reported in Michigan and New York in the U.S., in grape in Ontario, Canada, and in Europe observed only in Italy (loc. cit. Klos, 1976).

Although PhRoMV causes yield losses in infected trees, the damage is usually minimal because few trees are infected in any single orchard. However, the risk is high of rapid disease spread among peach and grape, especially when or if either species succeeds the other in a plantation.

Host Range

Peach is the most important natural host of PhRoMV. Some peach cvs. (e.g., South Haven, Hale Haven, J.H. Hale Elberta, and others) are susceptible to PhRoMV; Italian plum and Eickson (*P. salicina* × *P. simonii*) react with symptoms; and some cvs. of Damson, Burbank, Red June, and Abundance are tolerant. Grape is a natural host of PhRoMV, which can be transmitted through its seed (Bovey et al., 1980).

PhRoMV has been discovered in the roots of *Solanum carolinense*, *Rumex crispus*, and *Taraxacum officinale* plants, and may be seed transmitted through dandelion *T. officinale*. It may be transmitted experimentally into a number of herbaceous species, with resultant mild symptoms. Susceptible indicator hosts include *Chenopodium amaranticolor*, with mild local chlorotic spots and systemic leaf mottle; and *C. quinoa*, with mild local chlorotic spots accompanied by epinasty and leaf drop. *Nicotiana glutinosa*, *Petunia hybrida*, and tomato are systemic hosts of PhRoMV.

Symptoms

The growth of early leaves is slow and sometimes they reach only one third of normal size. Leaf blades are wrinkled with scattered chlorotic spots. Leaves formed later remain smaller, but are without spots. Internodes of shoots are shortened which compacts leaves into rosettes. General alterations include delayed foliation, chlorotic mottle, deformed first leaves, and rosette-like shoots and crowns.

Pathogenesis

PhRoMV is preserved in infected peach and grape, is graft transmitted as its main mode of spread and, in the soil, it can be transmitted from infected into healthy plants by the *Xiphinema americanum* nematode, which can allow transmission from herbaceous plants to woody (or visa versa). Infected dandelion seed also plays an active role in the preservation of PhRoMV in nature, and its mechanical transmission from woody into herbaceous plants is solely of experimental interest.

Control

The use of healthy virus-free planting material for establishment of new peach and grapevine plantations is the primary control measure. However, all potential sources of infection must be eliminated from young plantations, and whenever necessary, soil must be disinfested to control nematodes, especially soils recently cleared of infected trees.

Peach Enation Virus (PhEnV)

PhEnV, described in *Prunus persica* cv. Shinensi, its only natural host in Japan (Kishi et al., 1973), is isometric and 33 nm in diameter. Its TIP is 50 to 60°C; its DEP is 1×10^{-3} to 1×10^{-4}; and its LIV is 30 to 40 days. PhEnV has good immunogenicity.

The virus has been transmitted experimentally into *Prunus persica* seedlings and several herbaceous plants, including *Chenopodium amaranticolor*, *C. quinoa*, *C. album*, *G. globosa*, *Nicotiana tabacum*, and *Petunia hybrida*. Small enations along the main vein, usually followed by leaf distortion and shoot internode shortening, occur in PeEnV-infected trees. Interestingly, Jordović (1967) and Ranković (1970) described a virulent PsNcRsV strain from sweet cherry that causes small enations on the lower leaf surface of infected peach seedlings. Indicator plants for PhEnV are *Prunus persica*, with enations on the lower leaf surface; *Petunia hybrida*, with inoculated leaves and local lesions and systemic leaves symptomless later in the season. PhEnV is transmitted only with infected planting material.

Peach Line Pattern and Leaf Curl Virus (PhLnPtLClV)

The PhLnPtLClV is isometric, 33 nm in diameter, and 1800-kDa nucleic acid (Németh et al., 1983; Kerlan et al., 1986). It has strong immunogenicity.

PhLnPtLClV has been reported only in Hungary (Németh et al., 1983). *Prunus persica* and *P. avium* are its natural hosts, whereas it has been experimentally transmitted into many woody hosts, including *Prunus armeniaca*, *P. domestica*, and *P. cerasus*, and herbaceous plants such as *Chenopodium amaranticolor*, *C. quinoa*, *C. sativus*, *Nicotiana clevelandii*, and others (Figure 158).

FIGURE 158 Plum line pattern virus on peach.

Yellow-green discolorations, followed by irregular leaf growth and deformity, characterize this disease in PhLnPtLClV-infected trees.

Peach Latent Mosaic Viroid (PhLtMVd)

Peach latent mosaic disease was first observed in France in peach cvs. introduced from North America (Desvignes, 1980) and its etiology remained unknown, although comparative features led to the belief its causal agent was a virus. Flores et al. (1991) showed the true disease agent is a viroid, thus was named peach latent mosaic viroid. The GF-305 peach seedlings (indicator plants) from which the viroid RNA was isolated were experimentally infected with purified preparations of it. Investigations in progress pertain to the detection of the viroid with the aid both of biological indexing methods on GF-305 peach seedlings and of polyacrylamide gel electrophoresis (PAGE) techniques to identify circular viroid molecules (Flores et al., 1991).

Introduced peach cvs. in France show a 20 to 25% infection, which highlights the substantial spread and damage done by PhLtMVd (Desvignes and Boye, 1978). In peach orchards in Valencia (Spain), infection with PhLtMVd was discovered in 88% of American cvs. Arm King, May Crest, Spring Crest, Star Crest, Primorose, and Queen Crest (Flores et al., 1991), but was not found in Spanish cvs. Marujas, Jerónimos, Calabaceros, Campillo, Paraguayo and Brasileno. PhLtMVd is relatively widespread and can occur in Europe, Japan, and China (Desvignes, 1988).

Peach is the natural and only viroid host. Other *Prunus* spp. are immune. In the GF-305 peach cv., the viroid can be detected in roots, leaves, young shoots, old branches, and bark; thus, its seedlings are excellent viroid indicator plants. Symptom expression depends on the cv. and viroid strain, which can be mild or severe. Commercial cvs. are mostly infected with latent strains.

Leaves in PhLtMVd-infected trees develop more slowly from buds and may display typical mosaic, vein banding, blotch, and calico. The shoots grow slowly with shortened internodes and delayed flowering and fruit ripening. Fruits remain smaller, distorted, and with cracked sutures.

Myzus persicae (Desvignes, 1981) and possibly other aphids transmit PhLtMVd in peach orchards. The viroid has not been detected in seed, and no proof exists of its mechanical transmission by pruning tools. Primary measures for prevention of infection include the use of viroid-free planting material and the removal and destruction of infected trees, especially in young orchards.

Peach: Natural Alternative Host of Some Viruses

Apple Chlorotic Leaf Spot Virus (ApCLsV)

ApCLsV causes rootstock and scion incompatibility. Detrimental effects result also from interspecific parent grafting such as peach and apricot or peach and plum (Marénaud, 1971). An incubation period of two years occurs before evidence of incompatibility appears. ApCLsV also causes incompatibilities of the "latent" death type, characterized by the death eight years after peach was grafted onto plum (Marénaud, 1971). Potential progressive degeneration, both of rootstocks and cvs. propagated for several years, is also atrributed to the wide distribution of ApCLsV in its latent form.

ApCLsV causes narrow, slightly depressed, dark-green flexuous lines and spots on the leaves of susceptible plum cvs., thus also known as peach dark green sunken mottle. Peach cvs. originating in Europe develop dark-green diffused mottle on the leaves, which varies in severity depending on cv. susceptibility.

Often, ApCLsV occurs jointly with PsNcRsV, PsDV, PlPV, and strawberry latent ringspot virus. In mixed infections, ApCLsV usually causes more intense symptoms by these other viruses. ApCLsV is graft transmitted, and therefore, the production and use of virus-free planting material is an essential and successful control measure. The use of virus-free planting material would contribute greatly to increased vitality of peach plants that otherwise die quite prematurely in peach orchards.

Prune Dwarf Virus (PsDV)

PsDV causes the peach stunt disease in the U.S. (Milbrath, 1957, Wagnon et al., 1960), which is identical with the peach Muir dwarf disease (Hutchins et al., 1951). The Muir peach cv. is very sensitive to PsDV expressing symptoms of shortened shoot internodes, thus leaves compacted along lateral branches. Leaves are dark green and erect, with symptoms most obvious in the spring. Severe strains cause marked delays in growth and reduced yields.

Prunus Necrotic Ringspot Virus (PsNcRsV)

PsNcRsV is widespread in peach and other stone fruit species. Juarez et al. (1988) found 14% of peach trees infected in the Murcia area in Spain. The PsNcRsV type strain has been described in peach, plus other variants differing based on their host range and isolate properties. Jordović (1967) and Ranković (1970) described a severe strain of PsNcRsV isolated from sweet cherry that caused enations on leaves of peach seedlings. PsNcRsV is one of the most damaging peach viruses, the extent of damages dependant on virus strains, cvs., environmental conditions, and cultural operations. (For more details, see the chapter on "The Influence of Viral Infections on Fruit Tree Development and Yields.")

The disease caused by PsNcRsV develops in two distinct stages: acute (shock) and chronic stages. Acute stage symptoms are most severe, including delayed bud opening, and death of flower and leaf buds prior to the opening. Branches from the previous year are killed and canker symptoms appear at the nodes. Irregular chlorotic spots and rings, sometimes completely necrotic, form on young leaves in the spring (Figure 159). Recovery occurs during the chronic stage, as evidenced by symptom withdrawal or masking, but the vigor and yields of such plants remain reduced.

FIGURE 159 Prunus necrotic ringspot virus causes enations on peach leaves.

PsNcRsV occurs frequently in mixed infections with PsDV because both viruses are pollen transmitted. Depending on the cv., mixed infections cause more severe symptoms in plant growth and greater yield losses than individually. The symptoms caused by a mixed infection of PsDV and PsNcRsV were described as peach rosette and decline (Stubbs and Smith, 1971; Smith and Challen, 1972). The shortened internode "court noué" disease, also caused by ApCLsV and strawberry latent ringspot virus mixed infection, was described in France (Dunez and Marénaud, 1966). Mixed infections with PsNcRsV and ApMV cause peach-like pattern, economically and symptomologically identical with the European plum line pattern disease.

Plum Pox Virus (PlPV)

Peach, first described as an experimental host (Šutić, 1962; Schuch, 1962), was later labeled as a natural host of PlPV in Hungary (Németh, 1962, 1963), Greece (Demetriades and Catsimbas, 1968), and other European countries (Romania, former Czechoslovakia, France, Germany, Moldavia, Switzerland, and former Yugoslavia). Peach represents one of the main hosts for aphid development throughout the whole vegetation period, especially *Myzus persicae*, which contributed, *inter alia*, to the epidemiological spread of PlPV.

PlPV causes one of the most economically damaging peach diseases, evidenced by fruit drop and marked reduction in the quality of fruits.

The first symptoms of plum pox occur on peach leaves upon bud break in the spring as vein clearing (Figure 160). Infected veins grow slowly, causing changes in leaf shape (i.e., wrinkled and asymmetric). Chlorotic areas spread toward the midrib and form around affected veins. Vein clearing caused by PlPV is specific in peach, and thus of diagnostic importance. Sometimes, ringspots and irregular chlorotic spots resembling those caused on plum leaves also occur on infected peach leaves.

FIGURE 160 Plum pox virus (Šarka)-infected peach leaf; healthy (right).

Symptoms of PlPV on peach fruits are strikingly diagnostic. Several ringspots, frequently overlapping or linked together, appear on the fruit surface during ripening (Figure 161). In cvs. with white mesocarps (flesh), these spots are white; whereas in those with yellow mesocarps, they are greenish or yellowish. Ringspot color is vivid on nectarine fruits. Some peach cvs. symptoms are evident only after the fruit is canned. As with plum, susceptible peach fruits drop prematurely in large numbers when ripening begins and also later; thus, plum pox causes severe economic loss.

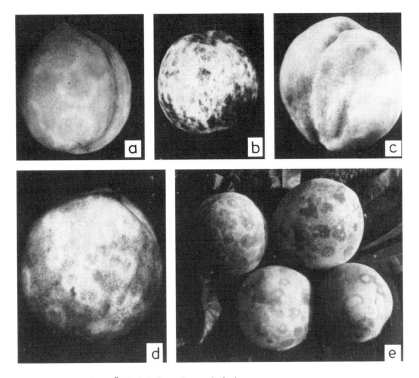

FIGURE 161 Plum pox virus (Šarka)-infected peach fruit.

Mainou and Syrginaidis (1991) studied the susceptibility of 107 peach and nectarine cvs. and clones naturally infected in the field in Greece, and observed symptoms of PlPV infection in all plants tested. However, with a low sensitivity index in 16 cvs., they are considered tolerant. These cvs. have mild symptoms, with only a small percentage of fruits displaying symptoms. They include May Crest, Early Red Free, Early Coronet, Early Red Haven, Cardinal, J.H. Hale, S-570, Fair Havens-1380, Sudanel, Dixi Red, S-1057, Loadel, Ribet, and Silverlode (nectarine).

All measures recommended for protection of plum against PlPV should be used for peach. Spatial isolation of peach from plum and apricot orchards is necessary to reduce the risk of PlPV spread. When selecting cvs. for production, one would select those most tolerant to plum pox. A more permanent solution is to breed peach plants of origins immune to create new, PlPV-resistant variants and cultivars.

Tomato Black Ring Virus (TmBRV)

TmBRV causes peach shoot stunt, a severe disease first observed in Germany (Mischke and Schuch, 1962). Shoot internodes are shortened, leaves turn yellow, and the entire plant grows slowly. The leaves of TmBRV-infected young trees feature chlorotic spots and deformation, and the general appearance of those trees is characterized by bare twigs, rosetted growth, and few fruits.

FIGURE 162 Ringspot on a peach leaf caused by the peach yellow bud mosaic strain of tomato ringspot virus. (Courtesy J.K. Uyemoto. Reproduced with permission from APS.)

Tomato Ringspot Virus (TmRsV)

Strains of TmRsV cause peach stem pitting (PhStmPgDs) and peach yellow bud mosaic (PhYBdMDs), both severe diseases. PhStmPgDs, first observed in New Jersey (Christ, 1960), was considered a consequence of rootstock and scion incompatibility. Then, researchers easily transmitted its causal agent from infected to healthy peach, thereby proving its infective nature (Mircetich et al., 1970, 1971). It is widespread in the U.S. and Europe. Pits, which are clearly evident on the lower (ground) trunk following bark removal, represent the diagnostic symptom. Leaves of TmRsV-infected trees remain smaller, yellow-to-reddish, and fall prematurely.

PhYBdMDs was discovered first in the U.S., where it causes substantial economic damage. Mircetich and Civerolo (1972) established that TmRsV isolates causing PhStmPgDs are serologically related to the TmRsV strain that causes PhYBdMDs. Symptoms on leaves start with characteristic mosaic of irregular chlorotic spots (Figure 162) along the midribs and usually also close to the leaf base. The "yellow bud" stage occurs in the second year. Leaf development from the buds is arrested, or those that do form remain small and pale yellow with a perennial infection. Stunted leaves grow slowly, foliage remains sparse, and rosettes form during the season. In severe infections, flower buds drop, the fruits that do set are much smaller, and yields are reduced markedly. *Prumus persica* cv. Lovell or Elberta seedlings are disease indicator plants.

Strawberry Latent Ringspot Virus (SbLtRsV)

SbLtRsV was observed in peach in Germany (Richter and Kegler, 1967a), later in Italy (Belli et al., 1980), France (Vuittenez and Kuszala, 1968), and Hungary (Németh, 1980). Symptoms in infected trees are mild to severe, depending on the cvs. and virus strains. Flowering and leafing is delayed in SbLtRsV-infected plants, with narrower leaves and shorter shoots early in infection and development later becoming more normal. Few, mostly distorted fruits are formed.

Latent Viruses and Diseases

Latent infections in peach trees can be caused by ArMV, CMV, and sour cherry green ring mottle disease (SChGnRMtDs). No data are available on their influence on peach growth and yields.

Diseases of Peach Assumed Caused by Viruses

Similar to other fruit tree species, several diseases caused by pathogens of yet unknown properties have been described in peach. With more study, the causes of some of these diseases were found to be already known viruses and strains; thus, peach represents an alternate natural host. This group of diseases, *inter alia*, includes peach stunt and Muir dwarf disease syndrome caused by PsDV, peach necrotic ringspot, necrotic leaf spot, chlorotic leaf spotting and blotching, and calico, which are simply variants of the disease syndrome caused by PsNcRsV and PhStmPgDs and PhYBdMDs caused by TmRsV (for more details, see the chapter on "Peach: Alternate Natural Host of Some Viruses").

The causal agents of some separately described diseases have not been identified yet, thus their true viral nature must be elucidated in further etiological research. Some diseases in this group, such as peach mosaic disease, merit special attention.

Peach Mosaic Disease (PhMDs)

Peach mosaic, described in 1931 (Hutchins, 1932; Bodine, 1934), still begs knowledge of its causal agent. Natural infections occur only in the *Prunus* genus, and the causal agent of PhMDs is mite transmitted by *Eriophyes insidiosus*. Leaf symptom resemblances caused some confusion regarding the relatedness of PhMDs and plum pox; however, comparative investigations have ruled out any relatedness (Pine and Šutić, 1968).

PhMDs, first discovered in the U.S. and Mexico (Pine, 1976) and later in Europe (Greece and Italy) and India (Németh, 1986), is economically damaging both in peach and nectarine (*Prunus persica* cv. *nectarine*) production. In several freestone peach cvs., PhMDs results in smaller and severely malformed fruits that lose their market value. Branches are shortened and severely diseased trees remain stunted with a negative impact on yield.

In addition to peach and nectarine, PhMDs has 20 other hosts exclusive to the *Prunus* genus (Pine, 1976), some cultivated such as apricot, plum, almond, Japanese plum and myrobalan. Cherry, sour cherry, and mahaleb are not susceptible.

Chlorotic leaf mottle generally occurs first in the spring or early summer. Chlorotic spots differing in size or shape as blotches, lines, and feather-like patterns occur on leaf veins. Mosaic appears on leaves formed during the entire vegetation period. Severe infections result in smaller, narrower, and wrinkled leaves with irregular margins. Tissues within the spots frequently become necrotic and drop out, resulting in the shot-hole appearance.

Irregular whitish and pink spots occur on the petals. Petal discoloration is accompanied by pronounced flower mottle. In the "stone hardening" stage, the surface of PhMDs-diseased fruits becomes rough and uneven, and they never obtain the size, shape, or quality of healthy ones, maturing several weeks prematurely. Severe symptoms occur in many cvs., especially the freestones.

PhMDs branches have shorter internodes with leaves compacted into rosettes. Severe disease results in markedly reduced growth. Mosaic develops in some cvs. in two stages: the acute (shock) and the chronic. The acute stage may occur first, but usually early in the second year of infection, after which diseased trees usually enter the chronic stage characterized by mild, less pronounced symptoms.

Most freestone cvs. are more susceptible to PhMDs than the clingstones. Highly susceptible peach cvs. include Dixi Red, Early Hale, Early Red, Elberta, J.H. Hale, May Gold, Red Haven, and Sunhigh. Moderately susceptible cvs. are Cardinal, Coronet, Early East, Jerseyland, Red Skin, Rich Haven, and Sun Haven. Andora, Carolyn, Johnson, Springtime, Stuart, and Sun Crest are among those least susceptible (Pine, 1976).

Cultivated plum cvs. are susceptible to PhMDs, with symptoms similar to those in peach, but somewhat milder. Severity of symptoms is maintained consistent year to year. Natural infections of plum are seldom seen, but of apricot actually more frequent than peach. Apricot trees with PhMDs display symptoms very similar to peach, that is, plant vigor and yields are substantially reduced.

PhMDs natural host trees, of which peach is the most frequently infected species, represent permanent sources of infection. Transmission of the causal agent occurs either most importantly through grafts of buds and scions from infected plants or naturally spread with the feeding of *Eriophyes insidiosus* mites. A single mite can transmit the causal agent. Environmental conditions influence mite overpopulation, activity, and, thus, the extent of PhMDs spread in nature. Direction and velocity of air currents by which mites are transported dictate the actual distance of spread. The causal agent of PhMDs has not been transmitted mechanically by infective plant sap or any other means.

The primary disease control measures include use of pathogen-free planting material for establishment of new plantations, elimination of infected trees immediately upon detection, growing tolerant peach cvs. especially in known heavily infected areas, and control of mite vectors.

Peach Asteroid Spot Disease (PhAdsDs)

The causal agent of PhAdsDs is unknown. The disease, discovered in 1938 in California (Cochran and Smith, 1938) and later elsewhere in North America, is believed identical to the peach purple mosaic disease described at the same time in Bulgaria (Christoff, 1958).

Although it reduces peach yields, PhAdsDs is considered economically unimportant because it is so limited geographically. Peach, nectarine, and sweet cherry are potential economically important virus hosts and other known hosts are *Prunus amygdalus*, *P. andersonii*, *P. armeniaca* cv. *ansu*, *P. domestica*, *P. mume*, and *P. salicina*.

Spots in infected peach leaves have asteroid margins. Symptoms that occur both early in the vegetation period and on completely developed leaves become increasingly pronounced during disease development, especially at high temperatures. Infected leaves gradually turn yellow with yellow-green spots, variable in shape and size, distributed over the entire surface. Some chlorotic spots become necrotic and fall out, leaving a hole in the leaf. In severe infections, complete defoliation occurs by summer's end. Although fruits have no external symptoms, if leaves fall early, fruit quality and quantity are reduced.

Symptoms on PhAdsDs sweet cherry leaves are similar to those in peach and nectarine, but vary depending on cv. susceptibility and especially high temperatures. Once initial symptoms appear, diseased leaves age quickly. General yellowing with only small green areas remaining is characteristic. Just as for peach and nectarine, defoliation by summer's end is the final symptom. PhAdsDs also results later in bud and small branch death in sweet cherry. Severely diseased trees die after 2 to 3 years. Once leaves drop, the fruits remain small, mature later, and have poor quality. Susceptible sweet cherry cvs. include Lambert, Bing, Deacon, and Van, whereas Black, Tartarian, Napoleon, and Windsor cvs. are more tolerant (Wadley, 1959).

Symptoms in diseased apricot leaves are similar to those in peach. In addition to leaf spots in some apricot cvs., leaf formation and development are delayed substantially. Other cvs. prematurely defoliate without any symptoms of disease.

PhAdsDs is preserved naturally in peach, sweet cherry, and apricot as the most important hosts. It is graft transmitted, and other modes of transmission remain unknown. PhAdsDs has become established in some wild species of *Prunus*, which suggests some mode of natural spread, although a vector is unknown. The use of virus-free planting material is the key control measure. Elimination of infected trees immediately upon detection in orchards is recommended.

Peach Wart Disease (PhWrDs)

The properties of the PhWrDs causal agent(s) are unknown. First described in 1938 in Idaho (Blodgett, 1939), it was observed later in other states of North America and in Europe (Italy, possibly also France) and Turkey (Németh, 1986).

PhWrDs does not cause economic damage because of its rare occurrence. Infected fruits covered with wart-like formations differ in shape, remain small, and lose all market value. The observed spread of PhWrDs in new plantations implies some risks for production.

Peach is the only natural host known, whereas experimental hosts include *Prunus armeniaca*, *P. avium*, *P. carasus*, *P. domestica*, *P. persica* cv. nectarine, and *P. serrulata*. *P. tomentosa* is a suitable indicator plant.

Wart-like formations on the fruits, thus the name, characterize the key symptoms that occur early-season on young fruits following petal drop as whitish, wrinkled, or pronounced edges of pale brown to vivid red. During development, the fruit skin severs and becomes rough. The mesocarp beneath such warts is hard, tough, gummy, and sometimes compacted, forming gummous plugs. Wart formation may also occur on their fruits as a trait of some peach cvs., or caused by insect damage, and should be distinguished from the infective PhWrDs. Elberta, Fireglow, J.H. Hale, Prairie Gold, and Rio Oso Gem cvs. are very susceptible to PhWrDs, whereas Clingstone peach cvs. display only mild fruit symptoms.

The causal agent of PhWrDs is graft transmitted from infected peach trees through buds and scions. No other mode of transmission is known, although disease spread in plantations has been observed. Infections resulting in PhWrDs can be avoided by using healthy planting material. However, it is essential to remove all PhWrDs trees as soon as detected. The Shiro plum is a suitable virus indicator plant (Lazar and Fridlund, 1967).

Peach Stubby Twig Disease (PhSbTwDs)

PhSbTwDs was discovered in peach and nectarine orchards in California (Wagnon et al., 1958), but no causal agent was found. PhSbTwDs results in yield reductions, and thus is considered an economically important disease. *Prunus persica* is the natural host and *P. persica* cv. Elberta seedlings a suitable indicator host. Leaf symptoms vary, from pale and dark green areas of marbled appearance initially to leaves that are narrow, malformed, and have irregular margins, and finally to reddish-brown spots evident on apical leaves at season's end. Branches bearing such leaves are stubby, thickened, and brittle, and lateral buds do not form or die quickly. The causal agent of PhSbTwDs can be inactivated in the buds by incubation at 38°C for 3 to 5 weeks (Németh, 1986).

Peach Bark and Wood Grooving Disease (PhBkWdGgDs)

PhBkWdGgDs was described in Michigan (Rosenberger and Jones, 1976), but no properties have been assigned its causal agent. Symptoms occur in trees 2 (or more) years old as elongated grooves in the bark and flattening of some branches. Diseased tree vigor is reduced.

Peach Oil Blotch Disease (PhOlBhDs)

The causal agent of PhOlBhDs is unknown and it has been described only in Japan (Kishi et al., 1973a). No information is available regarding the damage it can cause. Oil blotches on leaves characterize this disease.

Peach Yellow Mosaic Disease (PhYMDs)

The properties of this causal agent of PhYMDs described in Japan (Kishi et al., 1973a) are unknown. *Prunus persica* is its natural disease host and its seedlings are suitable indicator plants. First symptoms on infected leaves are yellow mosaic spots and malformations that become masked at higher temperatures during the season.

Peach Seedling Chlorosis Disease (PhSgCsDs)

PhSgCsDs was dsecribed in New Zealand (Fry and Wood, 1973). Symptoms are visible on the cv. Golden Queen peach seedlings, whereas sweet cherry and plum infections remain symptomless.

Virus Diseases of Apricot (*Prunus armeniaca*)

No virus species has been described in apricot as its primary host. However, it serves regularly as an alternate natural host of viruses infective in other rosaceous fruit trees, such as ApCLsV, PsDV, PsNcRsV, and PlPV. Apricot is the host of many "virus" diseases whose causal agents have yet to be isolated for study. Apricot was the first transgenic rosaceous fruit to receive the coat protein gene of PlPV via genetic engineering technology. The possibilities improve the chances of creating a different type of genetic resistance in fruit and other woody species. (For more details, see the chpater on creating of genetic resistance.)

Apple Chlorotic Leaf Spot Virus (ApCLsV)

ApCLsV, which is latent in apricot, had spread for decades unnoticed and naturally through infected planting material. For example, seedlings of several apricot cultivars in France were highly susceptible to ApCLsV; thus, new tolerant rootstocks of the Rein Claude type plum were introduced into apricot production. Unfortunately, these rootstocks, originating mostly from infected suckers, contributed greatly to further spread of ApCLsV and apricot decline (Marénaud, 1971). Di Terlizzi et al. (1991) in Italy discovered a 53% incidence of ApCLsV in 370 trees tested in 15 apricot cvs.

ApCLsV is one of the causal agents of apricot incompatibility and decline. Incompatibility is induced by grafting healthy cvs. onto infected rootstocks. This type of incompatibility is due to the hypersensitive reaction of ApCLsV with the graft union points resulting in tissue necroses. Research has shown that hypersensitive reactions partially encompass the tissues, since the virus is transmitted through graft union points onto the whole plant. If both rootstock and cv. are infected with ApCLsV at time of grafting, graft union incompatibility is reduced or even unobserved the first year. This delay phenomenon results from cross protection. This type of incompatibility has also been experimentally confirmed by grafting an infected apricot cv. clone A 238 onto randomly selected A 843 apricot seedlings. Incompatibility is most pronounced when *Prunus domestica*, *P. persica*, and *P. armeniaca* are used as rootstocks for apricot production. When grafting onto *P. cerasifera* rootstocks, incompatibility is delayed (Delbos and Dunez, 1988).

In infected apricot, especially the susceptible Luiset cv., ApCLsV causes internode shortening of shoots, resulting in leaves compacted into a rosette, thus the disease known as apricot Luiset rosette (Morvan and Castelain, 1967; Bernhard and Dunez, 1971). Bud death and local gummosis also result in susceptible cultivars and clones when an ApCLsV-infected partner is used for grafting.

ApCLsV causes two fruit symptoms: apricot pseudopox (or varmuela) and apricot fruit blotch (butterature). Apricot pseudopox was first described in Spain in Bulida (Benlloch, 1964; Pena-Iglesias and Ayuso Conzales, 1975), a highly susceptible cv. nearly completely infected. Depressions and deformations followed by deep red or brownish-red discolorations occur on ApCLsV-infected susceptible cv. fruits. Pseudopox is economically damaging to susceptible apricot cvs. Characteristic symptoms on apricot fruit are brown necrotic blotches with very dark-brown necrotic mesocarp tissues beneath, whereas holes and cracks appear on abnormally lignified stones. Pathological changes reduce fruit market value.

ApCLsV is graft transmitted in infected vegetative material used for general reproduction. No other mode of transmission is known. The production and use of virus-free planting material for estblishment of new plantations represents an effective and successful control measure.

Prune Dwarf Virus (PsDV)

PsDV occurs regularly in most apricot crops. In Apulia, Italy, PsDV was observed in 5% of 350 trees tested in 15 apricot cvs. (Di Terlizzi et al., 1991). PsDV occurs naturally in mixed infections with one or more of ApCLsV, PsNcRsV, PlPV, and ApMV.

Fridlund and Bottorff (1968) found that PsDV apricot strain causes apricot gummosis, observed first in 1947 in Washington state (U.S.). PsDV may cause severe damage locally with up to 50% incidence in some orchards.

Necrotic spots occur first in PsDV-infected apricot leaves, later yellowing, and then fruit drop prior to ripening. Gum flow along affected stems and most branches is a diagnostic symptom. Dark-brown necrotic zones in which gum flows are formed occur first on smaller branches. Diseased branches gradually die, then entire trees some years later. PsDV, causal agent of apricot gummosis, is both graft and seed transmitted (Fridlund, 1966, reported the seed transmission of PsDV through apricot). Digiaro et al. (1991) discovered PsDV in both apricot pollen grains, both externally and internally, and ovules. (More details on the properties of this virus and control measures can be found in the chapter on prune dwarf virus.)

Prunus Necrotic Ringspot Nepovirus (PsNcRsV)

Apricot is a natural, albeit less frequent host of PsNcRsV. The presence locally of PsNcRsV in apricot may be quite important; for example, in Apulia, Italy, Digiaro et al. (1991) found a 39% incidence in apricot.

Based on symptoms, two disease types occur in leaves of apricot infected with strains of PsNcRsV: apricot line pattern and apricot ringspot diseases. Line pattern is characterizeed by pale-yellow or yellow-green bands, spots, and ringspots on new leaves in the spring. Leaves that develop later are usually symptomless. Severe symptoms that characterize the shock stage include chlorotic spots and ringspots which gradually become necrotic and necrotic tissues fall out, leaving a shot-hole appearance. Susceptible cvs. frequently die in this stage of disease. Infected leaves are usually symptomless in the chronic stage. Apricot ringspot is also characterized by gum flow on stems and branches, frequent branch death, and reduced fruit yields.

Mixed infections of PsNcRsV with other viruses generally cause very severe growth alterations and reduced yields. Di Terlizzi et al. (1991a) reported that complex infections of apricot with PsNcRsV + PsDV + ApCLsV caused decline, low yields, and unusual fruit disorders in the susceptible Titynthos cv.

PsNcRsV is mostly graft transmitted. Although transmitted through pollen and seed of stone fruits, the importance is unknown for these modes of transmission in apricot. PsNcRsV has been observed in pollen grains and ovules of tested apricot cvs. (Digiaro et al., 1991), which suggests the possibilties of such natural virus spread. (More details on the properties of this virus and infection control measures can be found in the chapter on prunus necrotic ringspot virus.)

Plum Pox Potyvirus (PlPV)

PlPV, discovered in apricot in Europe, is especially widespread in the Mediterranean where rosaceous stone fruits are grown intensively. PlPV infections occur in regions where infected plum and peach are also grown. PlPV generally is more rare in apricot than in either plum or peach, although the number of locally infected apricot trees may be quite high. PlPV is considered one of the most dangerous apricot pathogens in Greece (Karayiannis, 1991).

Initial symptoms of PlPV infection that occur on newly emerging apricot leaves in the spring are pale chlorotic areas surrounding leaf veins, later merging into bands along the veins (Figure 163). Then later, ringspots and irregular chlorotic spots differing in size and with diffuse margins appear on leaves. During the summer, symptoms of disease become milder and less noticeable. Symptoms of PlPV on young leaves become more severe in autumn, enabling easy detection. A high percentage (about 62%) of PlPV-inoculated apricot seedlings display characteristic plum pox symptoms on leaves within one month of inoculation and thus are recommended as indicator plants for rapid detection and diagnosis in greenhouses (Šutić, 1964).

FIGURE 163 Plum pox virus (Šarka)-infected plum leaf.

PlPV causes severe alterations in apricot fruits very similar to, and dependant on cv. suscpetiblity, those in infected plum (Figure 164). Chlorotic roundish ringspots usually appear on the fruit surface at the time of ripening. Groove-like depressions usually form below these spots as a result of uneven development. Such alterations spread further into the mesocarp (flesh) and along the stone border, forming depressed dark-brown spots. External fruit symptoms seldom occur in some cvs., but numerous whitish ringspots and mottles may be observed on the stones. The quality of PlPV-infected fruit is reduced and significant premature fruit drop occurs at fruit ripening. These changes caused by PlPV represent an important risk in apricot production.

Most apricot cvs. tested are susceptible to PlPV. Under natural conditions of PlPV infection, only 5 of 34 cvs. tested in Bulgaria (Christoff, 1958) remained symptomless (Bademova, Malatia, Paviot, Stella, and Stark Early Orange). Preliminary investigations in France showed that the following apricot cvs., mostly from North America, inoculated either by chip budding or by aphid transmission, displayed high degrees of resistance to PlPV: Harcot, Stella, Stark Early Orange, and Henderson (Dosba et al., 1991). Syrgianidis (1980), during a 20 cv. test under natural conditions of infection by PlPV, found two symptomless ones—Sunglo and Veecot. Research in Yugoslavia on many apricot cvs. under conditions of both natural and artificial PlPV infections revealed the Hybrid Banaesa 33/13 as resistant (Šutić and Ranković, 1981). Continuing these investigations, Ranković et al. (1991) added the following cvs. as resistant to PlPV: Alfred, Blenril, Farmingdale, Riland, Krupna Skopljanka, and Selection Čačak RS. The susceptibility/resistance of individual apricot cvs. may differ from country to country because of various virus strains; therefore, such investigations require international cooperative testing.

All measures recommended for protection of plum against PlPV also apply to apricot production. When establishing new plantations, special attention should be paid to spatial isolation of apricot

FIGURE 164 Plum pox virus (Šarka)-infected apricot pit; healthy (left).

from plum and peach, which are natural hosts of PlPV. Tolerant and resistant cvs. should be used, while bearing in mind that such cultivars can react differently to PlPV strains in other countries.

Tomato Ringspot Virus (TmRsV) in Apricot

Stem pitting in apricot, as for peach, is caused by a strain of TmRsV (Mircetich and Moller, 1977). Characteristic symptoms include lower trunk thickening, usually accompanied by length-wise bark cracks. The bark is thickened, spongy, and sporadically necrotic. Pits and necrosis of some phloem tissues occur in the woody cylinder. This virus causes development disorders, and discolorations and early decline of leaves.

Strawberry Latent Ringspot (SbLtRsV) Plus Cucumber Green Mottle Mosaic Virus (CGnMtMV (mixed infection)

Mixed infections with these two viruses cause the apricot bare twig and unfruitfulness disease described in former Czechoslovakia, where it caused high losses in apricot orchards in Moravia (Blattny and Janečkova, 1980; Cech et al., 1980).

Diseases of Apricot Assumed Caused by Viruses

Apricot Ring Pox Disease (AtRPDs)

AtRPDs, discovered in Colorado, later was observed in Washington, British Columbia, Utah, and California (loc. cit. Hansen et al., 1976). Additionally, it is also suspected in Italy (Ragozzino et al., 1991). It was described first as apricot ringspot (Bodine and Kreutzer, 1942), then as apricot ring pox (Reeves, 1943).

The causal agent of AtRPDs has not been isolated from infected plants; thus, no properties are known. Based on symptoms on apricot fruits, this disease appears quite similar to the PlPV disease, but it has been proven that no relationship exists between these two causal agents (Pine and Šutić, 1968). However, a possible identity exists for the causal agents of AtRPDs and cherry twisted leaf disease because certain specific symptoms of these diseases occur in cross-inoculated Bingsweet cherry and Tilton apricot plants (Hansen et al., 1976). Two disease causal agent strains have been described, based on characteristic reactions in cvs. Tilton, Wenatchee, Royal, and

Perfection, that is, the common strain that causes superficial fruit symptoms, but none on the seed; and the pit pox strain, which penetrates the fleshy fruit, nearly always reaching the seed. Due to detrimental fruit alterations, particularly in susceptible cultivars, AtRPDs represents an economically important disease in all areas where it occurs.

In addition to apricot, other natural AtRPDs hosts include sweet cherry, hybrid plums (*Prunus salicina* × *P. simonii*), and *P. virginiana* cv. *demissa*. The virus has been transmitted experimentally into some almond, peach, nectarine, Japanese plum, and sour cherry cvs. Resistant plants include Abundance plum (*Prunus salicina*), Shiro plum, Italian plum (*P. domestica*), American plum (*P. americana*), and *P. emarginata* (loc. cit. Hansen et al., 1976).

AtRPDs symptoms occur on leaves, fruits, and bark of annual branches. Leaves formed early in the season usually remain symptomless; whereas later, irregularly distributed dark spots, vein banding, chlorotic spots, stripes, and ringspots occur. Symptoms of disease are quite pronounced by mid-summer at lower temperatures. During development of disease, the tissues within the spots become necrotic and fall out, resulting in the shot-hole symptom. Reddish spots, sometimes becoming necrotic, appear on annual branches of susceptible cvs.

Discolored or necrotic sections occur on the fruit surface, beneath the skin, and in the mesocarp. They increase, merge together, and form reddish-violet ring- or semi-ring-shaped spots. During ripening, the spots are clearly surrounded by a border of concentric lines with cracked margins. Some fruits discolor and necrosis spreads deep into the mesocarp. The incidence of fruit symptoms in AtRPDs-infected trees may vary from year to year. Fruit drop has been recorded.

The Tilton and Wenatchee Moorpark cvs. are highly susceptible, and the Royal (Blenheim), Reliable, Sunglo, and Blenril cvs. are mostly tolerant to AtRPDs.

AtRPDs is naturally transmitted only through the grafts of buds and scions from infected trees onto healthy rootstocks. The common strain transmits much more easily than the pit pox strain. Incubation lasts 2 years, but symptoms can occur the year following grafting. Although assumed to be vector transmitted, none are known. Production of virus-free planting material and its use for establishment of new plantations represents the main disease control measure. All symptom-bearing trees should be eliminated immediately from orchards to reduce the potential for natural spread of the AtRPDs causal agent.

Some Other Diseases of Apricot

Apricot asteroid spot disease (AtAdsDs), first described in the U.S. (Cochran and Smith, 1938), is caused by the same agent as the *cherry Utah Dixie rusty mottle disease*, to which Tilton and Royal cvs. are highly susceptible (Williams et al., 1976). *Apricot mosaic disease*, first observed in the U.S. (Bodine, 1934), has comparatively limited distribution (India, Mexico, and South Africa) and is a causal agent identical to that which causes the *peach mosaic disease*. *Apricot pucker leaf disease*, also originating from the U.S. (Utah) (Wadley, 1966), is quite limited in distribution and has no great economic importance. Susceptible cvs. include Moorpark, Perfected, Riland, Tilton, and Sunglo. Based on specific symptoms in their only natural host, the Moorpark apricot cv., *apricot Moorpark mottle* (Chamberlain et al., 1954), and *apricot chlorotic leaf mottle* (Wood, 1975) diseases described in New Zealand are of limited distribution and the disease(s) has no great economic importance.

All apricot diseases, although lacking published properties of their causal agents and thus no identification, have the common trait of being transmitted only be grafting and budding. Therefore, establishment of plantations should use only healthy planting material as the primary control measure.

Virus Diseases of Sweet Cherry (*Prunus avium*)

Cherry Leaf Roll Nepovirus (ChLrV)

Properties of Viral Particles

The ChLrV isometric particle is 28 nm diameter, consisting of three different particle types: an RNA-free (T) component and two nucleoprotein components (M and B) with sedimentation coefficients of 52S (T), 115S (M), and 128S (B), and MW of 3240 kDa (T), 5530 kDa (M), and 6060 kDa (B) (Jones, 1985). The single-stranded RNA consists of two genomic species RNA-1 and RNA-2, separately located in the B and M particles, of 2400 kDa (RNA-1) and 2100 kDa (RNA-2). All particle types contain a single 54-kDa coat protein (Murant, 1981). M and B particles are not infective alone, but only when simultaneously present (Jones and Duncan, 1980).

The TIP of ChLrV is 50 to 60°C; the DEP is 1×10^{-3} to 1×10^{-5}; and LIV at 20°C is 4 to 16 days (Cropley, 1961; Walkey et al., 1973). ChLrV is moderately immunogenic in rabbit (Jones, 1985). Several strains have been described, based on reactions in natural and experimental hosts (e.g., type (cherry), elm mosaic, rhubarb, golden elderberry, red elder ringspot, dogwood ringspot, birch, walnut, walnut yellow vein, blackberry, and red raspberry strains) (Jones, 1985). Jones (1976) suggested that six strains might be classified serologically into two groups: one comprising strains from the U.S. and another strain from Europe.

Geographic Distribution and Economic Importance

ChLrV is widespread in North America, Europe (several countries), Turkey, the former U.S.S.R., and New Zealand (Jones, 1985).

ChLrV causes tree decline, especially in susceptible cvs., usually 5 years after the first symptoms appear. In mixed infections, most frequently with the PsNcRsV, bud necrosis occurs after bud opening. Damage in individual orchards depends on the incidence of disease.

Host Range

Sweet cherry and sour cherry (*Prunus cerasus*) are the main natural hosts of ChLrV, and other hosts are in the following genera: *Betula, Cornus, Juglans, Ligustrum, Olea, Ptelea, Rheum, Sambucus*, and *Ulmus* (Cooper, 1979). ChLrV was transmitted experimentally into many herbaceous dicots in over 36 families (Schmelzer, 1966; Hansen and Stace-Smith, 1971; Horvath, 1979). The monocots *Tinatia erecta* and *Cammelina* spp. are also experimental hosts of ChLrV (Schmelzer, 1966; Horvath, 1979). Herbaceous indicator hosts include *Chenopodium murale* and *C. quinoa*, with chlorotic and necrotic local lesions and systemic pattern and tip necrosis; *Cucumis sativus*, with chlorotic lesions in inoculated leaves followed by chlorotic rings and line pattern symptoms in systemic leaves; and *Nicotiana tabacum* cv. White Burley, with necrotic primary spots and systemic chlorotic and necrotic ringspots.

Symptoms

ChLrV causes similar symptoms in sweet and sour cherry. Infected trees express delayed development in the spring, shoots grow slowly, and leaves are compacted into rosettes. Leaf margins twist and change colors, depending on cherry cv. susceptibility (Figure 165). The leaves of Early Rivers cv. become reddish with pale-green spots, accompanied by enations in Bing cv. leaves. Cracks and gum flows appear on infected susceptible trees.

ChLrV causes ringspots in *Sambucus racemosa* and leaf yellow net in *S. nigra* leaves. Leaf mosaic is the main symptom on infected elm, whereas leaf patterns or vein yellowing occurs in *Rubus* spp. and *Betula* spp. plants. ChLrV also causes blackline at the scion/stock union points of *Juglans regia* grafted onto *J. nigra*, and *J. hindsii* grafted onto cv. Paradox, plus walnut yellow mosaic and ringspot symptoms. Symptomless infections occur in naturally infected olive, rhubarb, *Rumex obtusifolius*, and others. The seedlings of *Prunus avium* × *P. pseudocerasus* rootstocks have virus resistance based on hypersensitive symptoms (Cropley, 1968b).

Pathogenesis

ChLrV is preserved permanently in its natural hosts and is graft transmitted by buds or scions; it is also spread through vegetative planting material if the mother plants are propagated from ChLrV-infected ones; and is spread by seed and pollen of various hosts (Jones, 1985). ChLrV has been detected in the seed (0.5 to 35%) of American elm, rhubarb, *Sambucus* spp., and walnut natural hosts, and elm mosaic and birch strains are transmitted through both pollen and ovule to the seed.

FIGURE 165 Cherry leafroll virus in Napoleon cherry.

Walnut strains are spread in orchards via pollen from infected trees. In addition to woody plants, ChLrV has been found in seed, up to 100%, of several herbaceous hosts. Transmission of some cherry isolates by nematodes (*Xiphinema coxi*, *X. diversicaudatum*, *X. vuittenezi*) has not been confirmed. Mechanical transmission of ChLrV with infective plant sap is experimentally important.

Control

The use of healthy planting material represents the primary preventive measure for virus control in sweet cherry, sour cherry, and other woody hosts. Infected trees represent sources from which ChLrV can be transmitted by pollen; therefore, their immediate elimination from young plantations upon detection is recommended.

Cherry Rasp Leaf [probable] Nepoviruses (ChRpLV)

ChRpLV causes characteristic rasp-like enations on the underside of cherry leaves (Figure 166). *American cherry rasp leaf* and *European cherry rasp leaf* are distinguished, based on the strain of ChRpLV.

Cherry American Leaf Rasp Disease (ChARpLDs)

This disease is caused by ChRpLV, consisting of isometric particles of 30-nm diameter. Three particle types include empty protein shells without RNA (T) and two nucleoproteins (M and B) with sedimentation coefficients of 56S (T), 96S (M), and 128S (B) (Murant, 1981). Viral RNA is single stranded. The M and B components each contain one RNA of 1500 kDa and 2000 kDa, respectively. The virus capsid consists of two polypeptides of 24 kDa and 22 kDa (Stace-Smith and Hansen, 1976). The TIP of ChRpLV is 58°C; the DEP is 1×10^{-4}; and the LIV at 4°C is 7 days. ChRpLV has moderate immunogenicity. Only slight symptomological differences are observed among virus isolates in herbaceous hosts.

ChRpLV, discovered in North America (Nyland, 1976), has been isolated in nature from individual detached or grouped cherry, mahaleb, peach, and apple trees. Symptomless natural infections

FIGURE 166   Cherry leaf with enation and distortion caused by cherry rasp leaf virus. (Reproduced with permission from APS.)

have been detected in balsam root (*Balsamorhiza sagitata*), dandelion (*Taraxacum officinale*), and plantain (*Plantago major*) weeds in orchards. ChRpLV has 22 experimental herbaceous hosts, including *Cucumis sativus* with chlorotic primary spots on coytledons and mild systemic mottling; *Chenopodium amaranticolor* with systemic leaf mottle; and others used as assay plants.

ChRpLV causes two main diseases, the cherry rasp leaf disease and the flat apple fruit disease, in some cultivars. It also usually causes systemic infection in sweet cherry, mahaleb, and peach plants. Rasp enations of differing size and shape occur on lower leaf surfaces of sweet cherry; similar symptoms can also be observed on apple leaves. ChRpLV causes peach and sweet cherry stunt and decline.

ChRpLV is primarily graft transmitted with buds and scions from infected trees. *Xiphinema americanum* nematodes are a soil vector. ChRpLV is transmitted through *Chenopodium quinoa* and *Taraxacum officinale* seed at 10 to 20% (Stace-Smith and Hansen, 1976).

Cherry European Rasp Leaf Disease (ChERpLDs)

Potential causal agents of the ChERpLDs include raspberry ringspot (RbRsV) and ArMV, as well as the complexes RbRsV + ChLrV, RbRsV + PsDV, ArMV + PsDV, and PsDV + strawberry latent ringspot virus (SbLtRsV) (Németh, 1986). For disease characterization, it is important to note that PsDV is an ilarvirus, whereas all others are nepoviruses.

Well-known diseases caused by these viruses are Pfeffinger disease (RbRsV), Eckelrader disease (RbRsV + ChLrV), cherry rosette disease (RbRsV + ChLrV), and cherry rasp leaf and decline (PsDV + SbLtRsV). Reports on individual disease types exist in most European countries. Hungarian cherry rasp leaf is probably caused by a PsDV strain (Németh and Kegler, 1960; Kegler, 1968).

Pfeffinger disease, which has destroyed cherry orchards in Switzerland and other countries, is considered the most dangerous disease in this group. Symptoms that occur on leaves of infected trees are of large, yellow-green leaf flecks. Leaves appearing normal early in the season, later become quite narrow, slightly wavy, and deformed. One half of the leaf is usually narrower, thus bent and assymetrical.

Enations on the lower surface of leaves are diagnostic secondary symptoms of ChERpLDs. Formed starting from the midrib, enations gradually appear along lateral veins toward the margins. Affected leaf blades are brittle and break easily. Foliation is delayed in infected trees. Shortened internodes form rosettes of compacted buds, some of which fail to develop. Shoots grow slowly and some branches die.

Fruit development is highly uneven, with short peduncles and prolonged ripening. Yields of ChERpLDs-infected trees are reduced substantially. In nursery trees, numerous buds fail to develop following bud break or, if so, then they are stunted and leaves fall prematurely.

Similar more severe symptoms on leaves and the rosette formation occur on trees affected by the Eckelrader disease. Symptoms of diseases in this group vary based on cv. susceptibility and viral strains. However, common features include enations on the lower leaf surface, severe leaf deformations, delayed growth and development, and reduced yields.

The use of virus-free planting material for establishment of orchards is the primary control measure for ChRpLV. Elimination of weeds known as hosts of ChRpLV is also recommended for both European and American types. The control of nematode vectors of nepoviruses is feasible primarily on smaller acreages, and especially for the production of planting materials. Destruction of trees infected with ChRpLV reduces the possibility of virus transmission through the pollen in orchards.

Cherry Epirus Virus (ChEpV)

ChEpV, isolated recently from cherry trees displaying rasp leaf symptoms in Epirus, Greece (Avgelis et al., 1988; Marina and Avgelis, 1991), is a new multicomponent, bacilliform particle of 18 nm × 26, 34, or 44 nm. The 26-nm particles are most abundant and 44-nm particles are least abundant. ChEpV has single-stranded RNA with three bands of 780, 340, and 310 kDa as resolved in 1% agarose gels, plus a coat protein of one single band (Marina and Avgelis, 1991). The TIP of ChEpV is 50 to 55°C; the DEP is 1×10^{-4}; and the LIV is over 15 days (Avgelis et al., 1988). ChEpV is serologically distinct from some tested multicomponent viruses.

The cherry trees infected by ChEpV had severe rasp symptoms with numerous enations on the lower leaf surface. ChEpV was transmitted into many hosts with the aid of infective plant sap, but not by aphids. Seed transmission occcured in three mechanically inoculated *Vicia faba* cvs.: Aquadulce var. major, Tanagra, and Policarpi var. minor (Avgelis, 1991).

ChEpV resembles the Ourmia melon virus (OuMnV), but differs serologically; Marina and Avgelis (1991) described it as the second member of the Ourmiaviruses, a newly proposed virus group.

Sweet Cherry: Natural Alternative Host of Some Viruses

Prune Dwarf Virus (PsDV)

PsDV is widespread in sweet cherry. Alone, it causes yield reductions of 35%, and greater reductions in infections mixed with PsNcRsV (Kunze, 1988). The strains of PsDV earlier described as separate viruses cause varying symptoms on cherry. The type symptoms include necrotic leaf mottle, chlorotic necrotic ringspot, chlorotic ringspot, ring mosaic, ring mottle, yellow mosaic, and yellow mottle, most difficult to distinguish. Symptoms may appear for a short period, then disappear. Symptoms in some susceptible cvs. (Kassins Frühe, Schneiders Späte Knorpel, Weisse Spanische) occur on just a few leaves. Also, some symptoms of chlorotic ringspots may recur for several years, while in other cvs. (Bing and Maibigarreau) they occur only the year following infection. Similar symptoms of pale-green diffuse rings may be caused by ChLrV and Pfefflinger disease. The susceptibility of various cherry cvs. also influences symptom variability.

PsDV is graft transmitted with buds and scions from infected trees. PsDV is transmitted naturally with pollen and seed. Pollination with pollen from SChYV-infected trees reduces fruit set by 10 to 75% (Way and Gilmer, 1963). Infection of 15 to 30% incidence was found in seedlings from commercial mahaleb and mazzard seed lots (Gilmer and Kamalsky, 1962). Infections of rootstocks originating from seed undoubtedly contribute to the wide distribution of PsDV in cherry.

Healthy parent cultivated cherry and virus-free rootstock seedlings must be used to produce virus-free planting material. Parent trees for seed production should be protected against PsDV-infected pollen by spatial isolation from PsDV-infected trees. Orchards 1 to 5 years old should be inspected regularly, 3 to 4 weeks after petal drop, and all infected trees must be eliminated immediately and replaced with healthy stock. Infected trees should also be eliminated from 6- to

FIGURE 167a Prunus necrotic ringspot virus causes enations on cherry leaf.

10-year-old orchards, although costs of replacement may not be economically justified. Removing infected trees from orchards over 10 years old is probably uneconomical. The fill-in of empty spaces in mature orchards gains little because older trees are sources of spread of PsDV to young trees as soon as they flower (Gilmer et al., 1976).

Prunus Necrotic Ringspot Virus (PsNcRsV)

PsNcRsV is widespread in all temperate areas and any region where *Prunus* and *Rosa* genera grow. PsNcRsV strains cause many diseases of differing symptoms (e.g., cherry line pattern, cherry necrotic ringspot, cherry tatter leaf, cherry lace leaf, cherry rugose mosaic, and stecklenberger diseases). Its mixed infection with other viruses usually occurs as necrosis; for example, cherry fruit necrosis (ApCLsV + PsNcRsV) and cherry necrotic line pattern (ApMV + PsNcRsV). Individual disease symptom types vary, depending on the strains of PsNcRsV, cv. susceptibility, and environmental conditions. Some symptoms gradually decrease during disease development, often making it difficult to detect with certainty.

The initial symptoms in the spring are dark-brown necrotic spots and lines on the young, partially developed leaves (Figures 167a and b). During leaf expansion, these necrotic spots drop out, resulting in the shot-hole effect. Similarly, necrosis, described as necrotic striped mosaic or necrotic ringspot, can also occur along leaf line patterns and ringspots. Enations can accompany these symptoms, especially in young leaves of Bing cv. in nurseries (Figure 168).

After PsNcRsV infection during the acute stage of disease, necrosis and decline of all young leaves plus shoot tip death occur. Symptoms do not form or are much less severe on leaves formed late in the season. Latent or mild symptoms characterize the plant recovery stage in later years.

FIGURE 167b Necrotic spots and shot-holes caused by Prunus necrotic ringspot virus which appear on cherry leaves during first year of infection. (Reproduced with permission from APS.)

FIGURE 168 Fruit marking on Rainier sweet cherry associated with some Prunus necrotic ringspot virus strains. (Reproduced with permission from APS.)

Cherry cvs. highly susceptible to PsNcRsV are Germesdorfer, Schneiders Späte Knorpel, and Weisse Spanische; the Bing cherry cv. is symptomless following the annual actue stage (Kegler, 1977).

PsNcRsV is transmitted through infected planting material and spread in the orchards with the pollen from infected trees, which is epidemiologically important. (For more details, see the section

Virus Diseases of Fruit Trees

FIGURE 169 Deformed leaves and necrotic fruits caused by tomato bushy stunt virus on Sam sweet cherry; healthy (right). (Courtesy J.K. Ukemoto. Reproduced with permission from APS.)

on transmission through infected seed and pollen. Concerning virus control, see the section on *Prunus* necrotic ringspot virus).

Tomato Bushy Stunt Virus (TmBuSnV)

TmBuSnV has natural hosts both herbaceous and woody. (For more details, see the section on tomato bushy stunt virus.) Woody virus hosts are sweet cherry, apple, and grapevine. Vein necrosis occurs on TmBuSnV-infected cherry leaves and causes uneven leaf development and malformations (Figure 169). The shortened internodes compact leaves into rosettes. The diseases caused by PsNcRsV are cherry fruit pitting (Davidson and Allen, 1976) and cherry detrimental canker (Albrechtová et al., 1975).

Tomato Ringspot Virus (TmRsV)

TmRsV, first isolated from cherry in Oregon (Milbrath and Reynolds, 1961), was found to cause cherry "'Eola" rasp leaf disease. Later, a strain of TmRsV was shown to cause peach yellow bud mosaic in California. Chlorotic spots near veins and enations along lower midribs occur in newly emerging leaves (in the shock stage). During disease development, some spurs, shoots, and branches die. Deformation symptoms appear at the stem base and gradually spread toward the crown.

TmRsV strains cause cherry stem pitting with symptoms resembling those in other stone fruits. Symptomological changes occurring near the stem base are seen outside by a pronounced thickening beneath the bark. Pit depressions of varying size are seen upon bark removal. Infected leaves are chlorotic and rolled. Shoot growth slows and terminal shoots die.

Apple Chlorotic Leaf Spot Virus (ApCLsV)

ApCLsV naturally infects cherry (Boxus, 1966; Gilmer, 1967) as a latent virus. It may cause split bark or rough bark in cherry (Desvignes and Boye, 1988).

Some Viruses of Herbaceous Hosts

TMV (Gilmer, 1967) and CMV (Willison and Weintraub, 1957) cause latent infections in cherry. In sweet cherry, Pfeilstetter et al. (1991) identified petunia asteroid mosaic (PnAdMV) and carnation Italian ringspot viruses (CaIRsV) as the cause of cherry twig necrosis in Bulgaria and Germany. Symptomologically, the disease is characterized by necrotic spots on leaf veins and fruits, plus necrotic zones at branch ends.

Other Viruses and Virus Complexes

More information on these viruses, as well as on the diseases they cause, can be found in the section on cherry European rasp leaf viruses.

Diseases of Cherry Assumed Caused by Viruses

Cherry Green Ring Mottle Disease (ChGnRMtDs)

ChGnRMtDs is widely distributed (first discovered in Michigan, then in other states) (Parker et al., 1976) and now occurs in Australia, Europe, New Zealand, and Japan (Desvignes, 1988a).

It occurs naturally in sweet cherry, sour cherry, and flowering cherry (*Prunus serrulata*). Frequently affected sweet cherry cvs. include Bing, Lambert, Black Republican, Napoleon, and Deacon. Nectarine and apricot are naturally tolerant. Hale Haven, Sun Haven, Rich Haven, Suncling, and Rio Oso Gem are generally infected peach cvs. (loc. cit. Parker et al., 1976). *P. tomentosa* is susceptible to this virus, which has been transmitted experimentally into mahaleb cherry (*P. mahaleb*) plants.

ChGnRMtDs in sweet cherry is symptomless. Pronounced symptoms on leaves and fruits of Montmorency sour cherry include yellow and green mottle on leaves 4 to 6 weeks after petal fall, when symptoms develop for 2 to 3 weeks then leaf fall begins. Leaf tissues along the midrib and most lateral veins grow slowly and chlorosis occurs in the narrowed sections between leaf veins.

Necrotic depressions or ringspots on the fruit epidermis with necrotic points in the mesocarp result from the ChGnRMtDs and quality of infected fruit is substantially reduced. Where symptoms are milder, the necrotic points in the mesocarp do not occur.

Some diseases are characterized by leaf epinasty on flowering cherry *P. serrulata* cv. Kwanzan and cv. Shirofugen. Interveinal leaf tissue necrosis causes uneven growth, leaf curl, and shortened internodes; shoot symptoms in flowering cherry are more severe at 18°C than at 22 to 26°C (Parker et al., 1976). During high summer temperatures, yellow and green mottle occurs neither on leaves nor on stems of Montmorency sour cherry.

The causal agent of ChGnRMtDs is graft transmitted, especially in sweet cherry, peach, and apricot trees with latent symptoms. No other mode of transmission in cherry is known.

Healthy planting material represents the key preventive measure. Elimination of ChGnRMtDs trees helps prevent spread in orchards. Since ChGnRMtDs is not seed transmitted, seedlings may be used in nursery production.

Cherry Rusty Mottle Diseases (ChRyMtDs)

ChRyMtDs, described in the U.S. (Reeves, 1940) and later in England (Posnette, 1951), were named cherry American rusty mottle (ChARyMtDs) and cherry European rusty mottle (ChERyMtDs), whose causal agents have been neither isolated nor identified; thus, etiological relations are unknown.

Wadley and Nyland (1976) classified the American ChRyMtDs type into cherry rusty mottle, mild rusty mottle, necrotic rusty mottle, and Lambert mottle. Cherry rusty mottle and necrotic rusty mottle (syn: cherry bark blister) have been separately described based on resemblance of traits.

ChRyMtDs is widespread in the U.S., the severe type in Washington, Idaho, and Montana, and the mild type in Oregon. Necrotic rusty mottle has also been observed in North America and Europe.

Sweet cherry and sour cherry are natural hosts of ChRyMtDs, which has been transmitted experimentally onto other *Prunus* spp. some of which are tolerant. Lambert, Seneca, Sam, Hudson, and Bing sweet cherry cvs. are quite susceptible to ChRyMtDs, while others (Napoleon, Black Republican, Van, Windsor and Lyons) are moderately susceptible. Finally, Black Tartarian, Burbank, Orb, Schmidt, and others are symptomless or have mild symptoms (loc. cit. Wadley and Nyland, 1976). Significant damage occurred in Frogmore, Noble, Lambert, and Bing cvs. in Great Britain (Posnette and Cropley, 1964).

FIGURE 170   Cherry rusty mottle virus-infected cherry leaves.

Symptoms of disease vary depending on cherry cv., disease severity, and environment during the initiation of vegetative growth (Figure 170). Necrotic rusty mottle symptoms occur earlier and are more severe in a cold, late spring. Bud break in spring is delayed and leaf symptoms that occur 3 to 5 weeks after bud opening are large, dark-brown necroses with irregular margins resembling cherry necrotic ringspot disease (ChNcRsDs); 2 to 3 weeks later, entire leaves turn yellow and the most damaged ones drop; and chlorosis and rusty ringspots appear on any remaining leaves during fruit ripening. Autumn changes in color occur earlier in such leaves. Detection in autumn is easy with dark-green and striped spots. Finally, buds, shoots, and branch parts die, all of which cause substantial yield losses, especially in susceptible cvs.

The development of cherry rusty and mild rusty mottle is similar to that of necrotic rusty mottle. Initial symptoms are chlorotic mottle in young leaves near the shoot base. Symptoms in leaves are not altered much during the season, except for a bronze appearance during the summer. Autumn colors appear earlier in infected leaves. Infected plants lose 30 to 70% of leaves at ripening and fruits remain smaller and tasteless, and mature much later. These alterations cause severe damage that is correlated with incidence of infected trees.

Cherry European rusty mottle, widely distributed in England, affects many cvs., some of which are tolerant. The disease causes reduction of growth of Mazzard F 12/1 cherry rootstock by 23% and of yield by 25% in some sweet cherry cvs. (Posnette and Cropley, 1956a; 1960). Primary symptoms on older leaves during July are clearing and yellowing along smaller veins, which then become pale green and in late August a rusty-red mottle, or a dark-red mottle with yellow pigmentation occurs along the yellow veinlets. Sweet cherry cvs. Lambert, Bing, Sam, and Mazzard F 12/1 are suitable indicator plants (Blattný and Janečkova, 1978).

The cherry rusty leaf mottle agent is graft transmitted with buds and scions used from infected trees. A slow spread of disease observed in the orchards still remains puzzling.

The use of healthy planting material represents the only successful measure to help prevent disease spread. Elimination of infected trees from young plantations also helps during a few years after planting. The cherry necrotic rusty mottle agent can be inactivated by incubating Lambert cv. scions in hot water at 50°C for 10 to 13 minutes or at 52°C for 5 minutes (Nyland, 1959). The

European disease type can be eliminated from terminal buds by incubating infected shoots to 36°C for 26 days (loc. cit. Németh, 1986).

Cherry Little Cherry Disease (ChLlChDs)

The causal agent of ChLlChDs has not yet been determined, but is a suspected virus or phytoplasma-like organism. Filamentous virus-like particles were observed in phloem of infected sweet cherry leaves and petioles (Raine et al., 1979; Ragetli et al., 1982), but the infectivity of these particles is still unproven, and thus, disease etiology is unknown.

ChLlChDs, observed in Europe, Australia, Canada, Japan, New Zealand, and the U.S. (Welsh and Cheney, 1976), are economically damaging because infected fruits are 25 to 60% smaller than the healthy ones, ripen later, and have poor color. Natural hosts include sweet cherry and sour cherry, plus some ornamental and wild *Prunus* species (Welsh and Cheney, 1976). Plum (*Prunus domestica*), peach (*P. persica*), and apricot (*P. armeniaca*) are not suceptible to ChLlChDs. Flowering Japanese cherry as a tolerant host represents an important latent source of infection.

Symptoms of ChLlChDs occur in sweet, sour, and duke cherries, especially in the fruits and leaves of many cvs. Plant vitality is greatly reduced in perennial infections. Initially, fruits develop normally, then later grow slowly and at harvest are only one half or two thirds of normal size, heart shaped, of poor color, and never completely ripen (Figure 171). Sweet cherry cvs. with highly susceptible fruits include Early Rivers, Hedelfinger, and Waterloo, and the tolerant ones are Ember, Florence, and Merton Bigarreau (Kegler, 1977). The vitality and yields of infected trees decrease progressively from year to year. Cvs. that display leaf symptoms are Lambert, Sam, Star, Vana, and F 12/1.

FIGURE 171 Small, pointed Bing sweet cherry with little cherry disease; healthy (right). (Courtesy G.I. Mink. Reproduced with permission from APS.)

Infected sweet and sour cherry trees represent important sources of infection and the causal agent of ChLlChDs is regularly graft transmitted with buds and scions. The *Macrosteles fascifrons*, *Scaphytopius acutus*, and *Psamotettix lividellus* leafhoppers are well-known vectors of the causal agent (Wilde, 1960). No other mode of transmission is known, although disease spread is rapid in sweet cherry orchards. Healthy planting materials represent the main preventive measure. Healthy buds can be obtained by incubating the shoots at 37°C for 3 weeks. Rootstocks used for stocks production must also be healthy. Control of leafhopper vectors is a necessary preventive measure. Destruction of ChLlChDs-infected trees in orchards, plus elimination of all nearby flowering cherries, is recommended.

Cherry Twisted Leaf Disease (ChTdLDs)

The etiology of ChTdLDs is insufficiently studied, but probably has the same causal agent as apricot ring pox disease (Hansen et al., 1976; Lott and Keane, 1960). The area of its distribution is limited to Canada, a few European countries, and the U.S.

Cherry (*Prunus avium*) and *P. virginiana* var. *demissa* plants are natural hosts of ChTdLDs (Lott and Keane, 1960). ChTdLDs sweet cherry cvs. differ in reaction from most susceptible Bing, Lamida, Long Stem Bing, and Rainier cvs. (Hansen and Cheney, 1976), to several Lambert and Napoleon cvs., to tolerant Black Tartarian, Deacon, Sam, Start, and Sue (Reeves and Cheney, 1956; Lott and Keane, 1960). Symptomless hosts experimentally infected are *Prunus dulcis*, *P. salicina*, *P. mahaleb*, *P. cerasus*, and *P. persica*.

Symptoms that vary primarily depending on cv. susceptibility include an abrupt characteristic kink in the midrib and petiole, twisted leaves curled downward, and asymmetrical and necrotic margins along veins. Internodes become shortened but if present, fruit symptoms have no diagnostic importance. *P. avium* cv. Bing is a suitable indicator plant.

The causal agent of ChTdLDs is easily graft transmitted by budding. The use of healthy propagating material is the only recommended preventive measure. Elimination of ChTdLDs trees immediately upon detection in orchards is recommended.

Cherry Leaf Mottle Disease (ChLMtDs)

ChLMtDs, first observed in Oregon, is known in other states (Cheney and Parish, 1976). Economic damage varies with region, depending on infections in sweet cherry ranging from severe to mild.

Sweet cherry is the natural disease host of ChLMtDs and most peach cvs. are symptomless natural hosts. The ChLMtDs occurs naturally in apricot and some ornamental flowering cherries (Cheney and Parish, 1976). Several *Prunus* species that react as tolerant hosts can be infected experimentally. Sweet cherry cvs. have different degrees of susceptibility, from severe to mild to symptomless.

Symptom severity depends on sweet cherry cv. susceptibility. Irregular chlorotic mottle and deformation of terminal leaves are characteristic ChLMtDs symptoms, including shortened internodes forming rosettes and smaller, later ripening fruit in susceptible cvs.

Causal agents of ChLMtDs are graft transmitted by budding and spread naturally by *Eriophyes inaequalis* leaf mites living on bitter cherry (*Prunus emarginata*) in North America. Healthy planting material and removal of infected ChLMtDs trees from orchards are the two key measures for control.

Cherry Spur Cherry Disease (ChSprChDs)

ChSprChDs, discovered in North America (Blodgett and Aichele, 1976), has limited distribution. Sweet cherry is its natural host and apricot (cvs. Tilton and Riland), plus other *Prunus* species are experimental hosts. Severe symptoms of ChSprChDs were observed in the Badacsonyi Óriás, Bing, Van, and Sparkle sweet cherry cvs. (Németh, 1986).

Leaf epinasty, internode shortening, and bark necrosis in young shoots are diagnostic symptoms. Apical shoots are thicker, rougher, and brittle. Fruits are symptomless. ChSprChDs is graft transmitted by budding; thus, the use of healthy planting material for establishment of orchards is the primary measure for control.

Cherry Black Canker Disease (ChBCnDs)

ChBCnDs, of limited distribution, was discovered in the U.S. (Zeller et al., 1947; 1951). Sweet cherry is its only natural host. Susceptible cultivars include Napoleon, Bing, Republican, and Deacon (Cheney et al., 1976). ChBCnDs is characterized by rough bark cankers on branches of diseased trees, which develop initially as blisters that split longitudinally to form wounds that vary in shape and size. During disease development, the canker may girdle the branches, killing the section. No leaf and fruit symptoms have been observed. ChBCnDs is graft transmitted by budding; thus, establishment of new orchards requires healthy planting material.

Virus Diseases of Sour Cherry (*Prunus cerasus*)

Prune Dwarf Virus (PsDV)

PsDV infects several fruit tree species of *Prunus*, sour and sweet cherry most frequently. In sour cherry, PsDV causes sour cherry yellows (often found in the literature as SChYsV) and its strains cause sour cherry chlorotic ringspot, sour cherry chlorotic-necrotic ringspot, sour cherry ring mosaic, and sour cherry ringspot. Sour cherry yellows, found worldwide, is spread mostly by exchanges of PsDV-infected planting materials. Described as highly damaging in the U.S., it caused reductions in yield of 50% or more (Klos and Parker, 1960; Davidson and George, 1965). The disease is also found in South Africa and Europe wherever sour cherry grows. The amount of damage by yellows disease depends on sour cherry cvs. Kegler et al. (1972) recorded reductions in yield up to 62% in Montmorency cv., and 94 to 96% in Schattenmorelle cv.

The Montmorency cv. with pronounced symptoms is highly susceptible to PsDV. Initial symptoms occur 1 month after flowering on older leaves, usually somewhat larger than healthy ones with yellow-green ringspots. The yellows engulfs the entire leaf and the green remains immediately around the midrib and veins radiating from it. Infected leaves usually drop by August 1; thus, trees appear to recover. A second defoliation occurs again 1 month later when the trees have lost nearly half of their leaves.

PsDV-infected shoots grow slowly and only a few leaf and flower buds form on terminal shoots, which results in substantial reductions in productivity. Pollen germinability is reduced, leading to poor pollination and flowering, and reduced fruit yields. More tolerant sour cherry cvs. have less pronounced yellows symptoms, but yet characteristic and chlorotic ringspots on the leaves, shortened shoots, and reduced tree vigor.

PsDV is graft transmitted with infected plant material. Pollen from PsDV-infected plants is important in natural virus transmission (George and Davidson, 1963; 1964) into sour cherry seed (20 to 70%) (Gilmer and Way, 1960). PsDV is transmitted with commercial mahaleb and sweet cherry seed from which seedlings expressed 15 to 30% incidence of disease. Such seed infections contribute greatly to the widespread nature of PsDV in sour and sweet cherry. The rate of natural spread in sour cherry depends on tree age. In young orchards, 2 to 4 years after establishment, disease spreads slowly, but increases rapidly in 5- to 15-year-old trees until nearly all become infected.

Healthy planting material is the primary yellows control measure. Also, phytosanitary virus-free scions and rootstocks are essential to reduce the spread of PsDV. Mother trees used for seed production must be virus-free. Control of spread of PsDV in orchards is the same as recommended for sweet cherry.

Prunus Necrotic Ringspot Virus (PsNcRsV)

PsNcRsV, generally widespread in stone fruits, has many strains that cause diseases which differ in symptoms on different tree species; for example, a separate strain causing sour cherry Stecklenberger disease described in sour cherry is most closely related to Fulton's type strains associated with the E virus. It occurs in the U.S. and in Europe. Stecklenberger is the most damaging disease of sour cherry in Europe. Extent of damage depends on susceptibility of sour cherry cv., stage of disease development, age at which infection takes place, and environmental conditions. In susceptible cvs., sour cherry in the shock stage of disease yields only 30% of normal (Baumann, 1967; Kunze, 1969). Kegler at al. (1972) recorded only 10 to 30% of normal yield in trees over 7 years old, some killed. For the Montmorency sour cherry cultivar, Németh (1972a) showed an approximate 10% reduction of successful bud grafts, a 33% reduction of infection scions grafted to healthy rootstocks, and a 44.9% reduction of successful grafts when both scions and rootstocks were PsNcRsV infected.

Symptom development differs at each disease stage; for example, darkening and dessication of flower and leaf buds on branches in the acute stage (spring) with pale-green and brown necrotic ringspots on older leaves later in the season when the necrotic spots break, fall out, and leaves take on the diagnostic shot-hole appearance. The chronic stage leaves are characterized by intensive green with enations formed primarily on the lower surface of apical shoot leaves. After a long chronic stage, severe symptoms may once again occur, reducing plant vitality and leading to some death. A mean daily temperatures above 20°C can mask symptoms (Kegler, 1977).

Sour cherry cvs. differ in their susceptibility to PsNcRsV (Németh, 1986). The most susceptible are Schattenmorelle, Diemitzer, Heiman 29, Königin Hortensia, Ostheimer, Spanyol, Üvegmeggy, and Cigánymeggy. The most susceptible rootstock is wild sour cherry, whereas mahaleb is a frequent latent carrier.

In pathogenesis, PsNcRsV is transmitted through sweet and sour cherry pollen and seed. PsNcRsV-infected Schattenmorelle sour cherry stones resulted in 36% incidence, and 33% incidence for cv. Koroser. PsNcRsV infection of cherry seedlings in nurseries was frequently 5%. PsNcRsV is widespread among mahaleb seedlings that are symptomless (Baumann, 1967).

The use of virus-free planting material for the establishment of orchards is the primary measure of plant protection. The phytosanitary virus-free condition of mother trees for scion production should be inspected regularly and all flower buds should be removed to avoid pollen transmission of PsNcRsV. Rootstock seedlings should be produced only from healthy seed. Infected trees that represent sources of PsNcRsV should be eliminated from young orchards. The application of optimal cultural operations in older orchards can partially protect plant yields.

Some Viruses and Virus Complexes in Sour Cherry

Sour cherry is the natural host of ChLrV and PsNcRsV, with symptoms similar to those in sweet cherry. Complexes of infection with AChLsV and PsNcRsV that may cause the disease sour cherry fruit necrosis were identified in sour cherry trees (Baumann, 1971). Complexes of infection with ChLrV and PsNcRsV cause line mosaic symptoms in sour and sweet cherry plants, etc.

Diseases of Sour Cherry Assumed Caused by Viruses

Sour Cherry Green Ring Mottle Disease (SChGnRMtDs)

SChGnRMtDs, first observed in the U.S., was later described in South Africa and Europe. Its causal agent is the same one that causes the sweet cherry green ring mottle disease. Characteristic symptoms include green ring mottle and vein spot caused by the same strain of causal agent. Separate strains cause severe and mild symptoms (Milbrath, 1966).

SChGnRMtDs is widespread among both sour and sweet cherry. Cell necrosis appears as variously shaped spots on the fruit surface. These detrimental alterations damage sour cherry production.

Sour cherry cvs. differ in susceptibility to SChGnRMtDs, with Montmorency especially susceptible; most sour and sweet cherry cvs. are tolerant and act as latent carriers.

Initial symptoms occur on leaves of lower or inner shoots in the crown of infected Montmorency sour cherry, becoming apparent 1 month after flowering, first as individual pale-green stained spots scattered among dark green ringspots and line patterns over the entire leaf surface. These pale-green spots gradually enlarge and turn yellow and are clearly distinguishable from other dark-green spots. Affected leaves drop after 2 weeks in the acute disease stage; then a seeming recovery of plants marks the beginning of the chronic disease stage.

If severe symptoms occur once more on leaves in the autumn, it is a sign of a second acute stage. In some cvs. (e.g., Doppete and Natta), severe disease is characterized by pale-green and chlorotic spots along the midrib and other veins spreading therefrom. This represents a highly characteristic symptom of diagnostic importance. The shoots nearly cease growth.

Measures as recommended for the control of SChGnRMtDs in sweet cherry plants should be applied also for sour cherry.

Some Other Sour Cherry Diseases

Sour cherry is the natural host of cherry American rusty mottle (mild rusty mottle). Sour cherry line pattern disease with leaf symtpoms of light-green and diffused rings, spots, and vein banding was described in former Czechoslovakia (Paulechová, 1968). Several diseases were observed in the Montmorency sour cherry cv. in the U.S.; that is, sour cherry pink fruit with smaller fruits, pinkish-yellow and necrosis in mesocarps (Reeves, 1943), sour cherry bark splitting brown stripes, bark splitting, and gum exudation on shoots and branches (Cameron, 1954), and sour cherry gummosis with gum pockets, solidified gum, and exudate on shoots, limbs, and trunk (Blodgett et al., 1964).

VIRUS DISEASES OF ALMOND (*PRUNUS COMMUNIS*)

Prunus Necrotic Ringspot Virus (PsNcRsV)

PsNcRsV is probably the most widespread and the most damaging virus in almond. PsNcRsV strains cause almond calico, almond line pattern, and almond necrotic ringspot diseases.

Almond calico disease (AdCaDs), first observed in the U.S. (Thomas and Rawlins, 1939) and identified later (Nyland and Lowe, 1964), may be identical or closely related to almond bud failure and peach mule's ear diseases (Nyland, 1976a). Chlorotic spots occur first in infected leaves, then gradually merge, forming pale yellow or white areas especially evident in early spring. Symptom severity depends on the susceptibility of almond cvs. and seedlings, but also varies year to year. Lateral shoot buds frequently fail to develop, as observed especially in Peerless, Nonpareil, and Drake cvs. (Nyland et al., 1976).

Almond line pattern disease (AdLnPtDs) occurs in nearly all European countries where almond is grown. PsNcRsV is found alone or in complexes with other viruses. Line pattern symptoms in Bulgaria may be caused by either PsNcRsV or ApMV (Topchiiska and Topchiiski, 1976a; Topchiiska, 1977). This disease, although sporadic, occurs in many trees, so its damage depends on the incidence of infection in orchards.

Almond necrotic ringspot disease (AdNcRsDs) is reported in the U.S., South Africa, and Mediterranean countries. Symptoms and disease development reveal its similarity to disease caused by PsNcRsV in other stone fruit species. Severe symptoms characterized by necrotic ringspots in leaves and potential bud and bark necrosis occur in the shock stage. In the mild chronic stage, chlorotic spots form in leaves.

PsNcRsV is pollen and seed transmitted in almond. PsNcRsV has been observed on the pollen surface (Cole et al., 1982), which highlights the potential role of bee pollinators in flower infection. Some studies showed 9% infected seedlings from the seed from buds and PsNcRsV-infected trees (Barba et al., 1986). All the measures recommended to protect sweet cherry and other stone fruits should be used for almond. PsNcRsV can be eliminated from mother (donor) plants and virus-free almond plants recovered with the use of an *in vitro* shoot-tip grafting method (Juarez et al., 1991).

Prune Dwarf Virus (PsDV)

Almond is the natural host of PsDV. Pleše (1972) isolated PsDV from 12 of 30 trees from different regions, which shows that the extent of almond infection can be quite important. PsDV-infected trees are tolerant (symptomless), a possible reason for difficulty in controlling it. Using the *in vitro* shoot-tip grafting method, Juarez et al. (1991) eliminated PsDV and regenerated virus-free almond plants. Because it is spreading, PsDV in almond plants should be expected in complexes of infections with other stone fruit viruses.

Plum Pox Virus (PlPV)

Although not a natural host, almond has been infected experimentally with PlPV (*OEPP Bull.*, 1974). Almond has some advantages as an experimental host, and thus as an indicator plant to detect PlPV in woody hosts (Šutić, 1983). Almond may be an occasional natural host of PlPV (Dunez and Šutić, 1988).

Apple Chlorotic Leaf Spot Virus (ApCLsV)

Latent infections with ApCLsV, which is widespread among different fruit tree species, have been identified in almond. No data exist on possible damage caused by ApCLsV in almond, but its presence should be monitored in reproducing planting materials.

Tomato Ringspot Virus (TmRsV)

TmRsV causes yellow bud mosaic and stem pitting in almond. Almond yellow bud mosaic is identical to peach yellow bud mosaic caused by TmRsV (more details on disease characteristics can be found in the section on peach virus diseases). The extent of damage in almond caused by TmRsV, which was serious in California, depends on symptom severity in infected trees (Karle and Nyland, 1959).

Other Viruses from Herbaceous Hosts in Almond

CMV was isolated from almond in latent form (Topchiiska and Topchiiski, 1976). TNV infects almond and many other cultivated and wild plants, usually causing local and necrotic symptoms (Uyemoto, 1988). TmBRV was isolated from almond trees with narrow leaves and abundant enations on the lower leaf surface (Mischke and Bercks, 1965).

Almond bud failure disease (AdBdFaDs) in the U.S. lacks a descriptive etiology. In almond, it occurs jointly with the calico strain of PsNcRsV, but the interaction of this natural relationship is unknown. The causal agent of bud failure may be identical to that of peach willow twig disease, although not proven (Nyland, 1976a).

In nature, almond bud failure occurs in several almond, peach, and nectarine cvs. Pear, sour cherry, Shiro plum, and mahaleb seedlings are also carriers (loc. cit. Nyland, 1976a).

In almond, bud failure is characterized by flower and leaf bud death. It occurs regularly, accompanied by the calico strain of PsNcRsV. Many buds fail to form in leaf axils, particularly in shoots of slow growth. Dead lateral buds can remain on the shoots for several months. Terminal buds frequently fail to develop, resulting in lateral shoots of bushy apperance.

The bud failure disease agent is graft transmitted through buds and scions; thus, infected mother stocks represent the primary mode of spread. Only healthy planting materials for establishment of orchards are recommended for protection of almond.

Almond Mosaic Disease (AdMDs) (Figure 172) is identical to peach mosaic, and is thus believed caused by the same unknown causal agent. Natural hosts are several species of *Prunus*, in addition to almond, economically important species include peach and apricot. (For disease characteristics and its control, see the chapter on peach mosaic disease.)

VIRUS DISEASES OF SHELL FRUITS

Virus Diseases of Walnut (*Juglans regia*)

Cherry Leaf Roll Virus (ChLrV) in Walnut

Walnut, in addition to sweet cherry, is an economically important natural host of ChLrV. This disease was observed in southern Italy (Savino et al., 1976) with two disease types: walnut ringspot and walnut yellow mosaic. These isolates of ChLrV clearly differ, thus their assignment as separate strains of ChLrV. ChLrV was later discovered in the U.K. (Cooper, 1979; 1980), former Czechoslovakia (Novák and Lanzová, 1981), Hungary (Németh et al., 1982), France (Delbos et al., 1983), and Bulgaria (Topchiiska, 1988).

In the U.S., ChLrV was isolated and identified as causing blackline disease in English walnut grafted onto Paradox rootstock and in *Juglans hindsii* (Mircetich et al., 1980; Mircetich and Rowhani, 1984), thus explaining the etiology of the problem described earlier in the U.S. (Miller et al., 1958; Serr and Forde, 1959). The blackline disease substantially reduces walnut yields in

FIGURE 172   Almond mosaic disease in almond leaves.

California. ChLrV damages walnut by reducing growth and yields, and causing a decline of infected trees.

Virus isolates from different species and even from the same represent ChLrV, although some may differ (*see* the chapter on cherry leaf roll virus). For example, among the isolates obtained from walnut with yellow mosaic, some differ in biophysical traits (Németh, 1986) but remain within the limits of natural variation.

ChLrV causes leaf patterns and blackline at the graft union points in infected walnut trees. Leaf patterns include yellow mosaic and ringspot, which are considered separate diseases. Yellow mosaic has chrome-yellow spots and mottle on leaves and green fruits. Leaves with many spots become chlorotic or turn yellow. Ringspot disease has ringspots between the veins and along leaf margins that sometimes form a line mosaic on the leaves of some branches. Secondary symptoms are leaf yellowing, vein necrosis, deformation of the stone shell, and shrinkage and browning of the nut kernel.

Leaf yellowing, wilt, and early defoliation are blackline symptoms; terminal shoots develop poorly and die when new suckers appear from the rootstock. The narrow blackline at the scion and rootstock union point is the characteristic diagnostic symptom. This initial broken blackline gradually spreads, then after girdling the whole trunk, results in rapid death of the English scion in the following years. Blackline disease occurs when the English scion (*Juglans regia*) is grafted onto rootstocks of *Juglans hindsii*, *J. californica*, *J. nigra*, and their hybrids with *J. regia* (Németh, 1986).

ChLrV is graft transmitted when scions from infected trees are used. In nature, ChLrV is pollen and seed transmitted in walnut. Kölber and Németh (1983) report 8 to 54% incidence of ChLrV in germinating seed. Seedlings from naturally infected walnut seed resulted in 3 to 5% incidence of infected plants (Mircetich et al., 1982). ChLrV is transmitted in orchards with pollen from infected to healthy trees of the same cv. or to those flowering simultaneously with infected trees. In 3 subsequent years, ChLrV occurred in 20 to 36% plants annually in California orchards (Mircetich et al., 1980).

Virus-free planting material—both scions and walnut seedlings—should be used for establishment of orchards. ChLrV-infected trees that provide sources of infected pollen for all simultaneously flowering trees should be eliminated immediately. Seed infections can be reduced by storing seed at 4 to 5°C prior to germination (Quacquarelli and Savino, 1977; Cooper, 1980).

Walnut Line Pattern Disease (WtLnPtDs)

WtLnPtDs, first described in Bulgaria (Christoff, 1958), was later described in Italy, but its relatedness with the line pattern mosaic in Bulgaria was not established (Savino et al., 1976). Line pattern walnut mosaic symptoms in former Yugoslavia were first observed in 1953; but because the causal agent was not studied, its relationship with other walnut viruses could not be confirmed (Šutić, 1953).

The sporadic occurrence of walnut line pattern mosaic represents no serious danger to walnut production. Characteristic pale-green or yellow spots, plus broad lines and stripes, usually spread around and along leaf veins. These spots usually appear only on the leaves of some branches; thus, the disease is difficult to detect. Because it is graft transmitted, healthy seedlings must be used to establish new orchards—which represents the only possible control measure.

Walnut Mosaic Disease (WtMDs)

Walnut mosaic, described in Bulgaria (Christoff, 1937), is particularly important on fruits and causes economic damage in walnut production, but its causal agent is still unknown.

Walnut is the natural host and apricot and peach seedlings are experimental hosts of WtMDs (Christoff, 1976). Small, usually roundish, dark-grey or yellow-green spots occur on the leaves of infected walnut trees, especially evident in May/June, with individual spots gradually merging together to become more visible. Cell necroses also occur within the mosaic spots, which creates some leaf distortion.

Small irregular pit and groove-like depressions appear on the fruit stone shell, resembling the symptoms on plum from PlPV infection. Infected fruits are altered in shape and smaller. In some cvs., nut kernels grow slowly and dry early, and fruits drop prematurely. The disease is graft transmitted. Successful prevention results primarily for the use of healthy seedlings.

VIRUS DISEASES OF HAZELNUT (*CORYLUS AVELANA*)

Hazelnut Mosaic Diseases (HtMDs)

Symptoms of HtMDs were first described in Bulgaria by Atanasoff (1935a), who did not identify the causal agent. Cooper (1979) found ApMV in mosaic-affected hazelnut plants. Apple Tulare mosaic virus (ApTuMV) was isolated from infected hazelnut in France (Cardin and Marénaud, 1975; Marénaud and Germain, 1975). (More details about the features of these viruses can be found in the respective chapters.) Natural infections are caused by individual viruses or virus complexes, which thus determine symptom severity and/or extent of damage. These viruses are transmitted through infected planting material. ApTuMV is seed transmitted through infected hazelnut, but was not confirmed for Tulare ApTuMV (Cameron and Thompson, 1986).

Hazelnut Line Pattern Disease (HtLnPtDs)

Symptoms of HtLnPtDs were described in Bulgaria simultaneously with mosaic symptoms of HtMDs (Atanasoff, 1935a), and also later in hazelnut in Italy (Scaramuzzi and Cifferi, 1957). The causal agent is unknown; thus establishing any relationship with other viruses is impossible.

HtLnPtDs occurs naturally in *Corylus avellana* and *C. maxima*. Mosaic chlorotic mottle in lines and stripes occur along the veins of infected leaves, the characteristic symptom for which the disease was named. Insisting on healthy planting material is the key prerequisite for establishment of healthy production orchards.

Hazelnut Dieback Disease (HtDkDs)

HtDkDs is characterized by tree and shoot death in infected plants and is of great economic importance in large-scale production. Described first in Italy (Corte and Pesante, 1963), it is believed widespread in other southern European countries. Its causal agent is unknown.

Hazelnut is the natural host of this disease; the causal agent has been transmitted mechanically into *Cucumis sativus*, *Cucurbita maxima*, *Datura stramonium*, *Lycopersicon lycopersicum*, *Nicotiana tabacum*, and *Petunia hybrida*. Symptoms in HtDkDs hazelnut during vegetation include defoliation, reduced growth, dessication, and death of shoots or entire trees resulting in important production losses. The use of healthy planting material provides full protection in hazelnut orchards.

VIRUS DISEASES OF CHESTNUT (*CASTANEA SATIVA*)

Chestnut Mosaic [probable] Tobamovirus (CtMV)

The chestnut mosaic symptoms observed in France are caused by CtMV, with rod-shaped particles 300×20 nm (Desvignes, 1991). It was introduced with cvs. or hybrids of *Castanea crenata*, which are sometimes symptomless. CtMV resembles the mosaics also described in Japan, Hungary, and Italy (loc. cit. Desvignes, 1991).

The latent and severe virus strains that have been described, vary because CtMV can occur as a latent infection, but also severe infections can become latent. Cross protection phenomena have been demonstrated between latent and severe strains.

Mosaic spots and yellow mottle of semi circular rings occur on CtMV-infected leaves of susceptible cultivars, with leaf epinasty and young shoot necrosis also observed. CtMV is spread in nature, but no vectors are known. Seedlings and cuttings of susceptible Maraval and Marigoule cvs. are suitable glasshouse indicator plants. Healthy clones should be selected for chestnut production (Desvignes, 1991).

Chestnut Line Pattern Mosaic Disease (CtLnPtMDs)

CtLnPtMDs was described in Italy (Gualaccini, 1959), but its causal agent remains unknown. It has a limited spread and is of no great economic importance.

Chestnut is the natural host of CtLnPtMDs. Characteristic pale-green or yellow stripes and zigzag lines occur along the veins on both sides of the midrib. Chlorotic lines merge to form oak-leaf patterns close to the leaf base. Small, pale-green ringspots can also occur on the leaf surface. Disorders in leaf tissue development and partial leaf blade wrinkling are observed in the interveinal leaf parts. The use of healthy planting material is the only successful measure for protection of chestnut against CtLnPtMDs.

VIRUS DISEASES OF SOME TROPICAL AND SUBTROPICAL FRUIT CROPS

VIRUS DISEASES OF AVOCADO (*PERSEA AMERICANA*)

Avocado Sunblotch Viroid (AvSbcVd)

Properties
The AvSbcVd consists of SS-RNA (80 kDa) and 247 nucleotides per molecule (Dale et al., 1982). Smaller than the potato spindle tuber and chrysanthemum stunt viroids with which it has 18% sequence homology, it resembles the smallest RNA of the cadang-cadang viroid (Dale et al., 1982). AvSbcVd has high thermal stability and is susceptible to low RNase concentrations (Desjardins et al., 1980).

Geographic Distribution

AvSbcVd, observed first in California in 1928 and later in Florida (Zentmyer, 1953), is well known in countries where avocado (*Persea americana*) is grown (e.g., Australia, Israel, Peru, South Africa, the U.S., and Venezuela (Dale et al., 1982)). Since beginning the cultivation of avocado in southern Europe, regular inspections aimed at the detection of viroids are recommended. Alterations in color, structure, and appearance of infected fruits substantially reduce their market value.

Host Range and Symptoms

AvSbcVd infects only a few plants in the family Lauraceae: avocado cinnamon (*Cinnamonum camphora*), *C. zeylanicum*, and *Ocotea bullata*.

Different symptoms occur on AvSbcVd-infected avocado stems, and symptomless infections are frequent. Diagnostic symptoms are yellow streaks on green stems and branches, and yellow-red streaks on the fruits (Figure 173). AvSbcVd-infected fruits remain green up to maturity, but with yellow-orange color along longitudinal grooves and with streaks on their surface. The streaks usually become bright red on the fruits and then turn black or purple at maturity (Zentmyer, 1953). Vein chlorosis and slight deformations on leaves are less diagnostic symptoms.

Pathogenesis

AvSbcVd survives naturally in infected avocado, either with symptoms of disease or symptomless. AvSbcVd is 80 to 100% seed transmitted through seed from infected symptomless plants; infected seedlings from such seed also remain symptomless. Transmission of AvSbcVd through the seed from symptom-bearing infected plants is under 5% and their resultant seedlings are symptom-

FIGURE 173 Symptoms of sunblotch on avocado leaves. (Courtesy H.D. Ohr. Reproduced with permission from APS.)

bearing (Wallace and Drake, 1962). AvSbcVd also is easily graft transmitted, an important consideration in the production of planting materials.

The AvSbcVd was transmitted mechanically from avocado to avocado by a razor-slash technique that was also successful in transmitting it from infected fruit extracts into healthy avocado plants (Desjardins et al., 1980). Desjardins et al. (1979) experimentally transmitted the AvSbcVd through pollen with bees as pollinators.

Control

Based on the economic importance of avocado sunblotch disease detection in symptomless trees, AvSbcVd is of particular importance. Detection can be done by graft transmission to healthy avocado seedlings or with the use of rapid laboratory methods such as PAGE and molecular hybridization using specific complementary DNA (Palukaitis et al., 1981). Healthy parent trees for seed, rootstocks, and budwood sources represent the most important requirement to establish viroid-free nurseries.

VIRUS DISEASES OF BANANA (*MUSA PARADISIACA*)

Banana Bunchy Top Virus (BaByTpV)

The isometric BaByTpV particle is 30 nm diameter; and although previously thought to be a possible luteovirus (Agrios, 1988), it is now known that BaByTpV has a ssDNA with a multicomponent genome, six of which have been cloned and sequenced (J. Dale and R. Harding, 1994; Brisbane, personal communication). It occurs in the most well-known banana production regions of Australia, Asia, the Pacific Islands, and Africa.

BaByTpV endangers banana production. In N.S. Wales in 1922, over 90% of banana acreages were excluded from banana production; thus, in 1927, the Ministry of Agriculture ordered destruction of all diseased plantations and allowed replanting with virus-free suckers in 1928 (Cann, 1952; Broadbent, 1964).

In the field, BaByTpV-infected trees are diagnosed based on symptoms in leaves and their arrangement along the stem. Irregular dark-green streaks develop along leaf veins. The leaves in the apical plant part are bunched as rosettes. BaByTpV-infected plants grow slowly and fail to develop fruits.

BaByTpV is preserved permanently in infected perennial banana. It is transmitted naturally with infected suckers used as planting material, and also aphid-transmitted semipersistently by *Pentalonia nigronevosa* (Maggee, 1940).

Only virus-free suckers should be used to establish new plantations. Disease spreading by aphids in the field can be prevented or limited with preventive chemical insecticides. BaByTpV-infected trees should be destroyed immediately upon detection because the aphid vectors pose a continuing threat in the presence of perennial sources of infection. Quarantine regulations pertaining to seed commercialization and introductions of BaByTpV-infected plant materials should be strictly respected.

Banana Bract Mosaic Potyvirus (BaBtMV)

Properties of Viral Particles

BaBtMV is a flexuous rod, 660 to 760 × 12 nm (Espino et al., 1990), about 10 kb, and the partially purified viral particles consist of positive-sense ssRNA (Bateson, 1995). BaBtMV occurs in low concentration in the banana plant and has been partially purified with difficulty.

The coat protein is a single polyprotein (about 36 kDa). A partial sequence of the amino acids has been determined for the coat protein (Bateson, 1995). Further progress had been made in sequencing and cloning several genes of BaBtMV (Rodoni, 1999). The BaBtMV is highly antigenic (Espino et al., 1990).

Geographic Distribution

The BaBtMV disease, observed as early as 1979, has been reported in the Philippines (Magnaye and Espino, 1990). Similar symtpoms have been observed in the Tamil Nadu region of India (INIBAP, 1992).

Host Range

Symptoms, which occur at all stages of plant development, have been noted in cvs. Canara, Pisang pulot, Madurangga, Cardaba, Abuhon, Saba, and Cavendish. The cv. K. Thong Det 52 exhibits some tolerance or resistance.

Symptoms

Characteristic symptoms in banana attributed to BaBtMV are broad green or brown spindle streaks and discontinuous stripes irregularly distributed along the petioles; mosaic streaks in the leaf blade, which may be absent; and dark, broad purple streaks on the bracts of inflorescences and the fruit bunch (see Figures 174 and 175).

FIGURE 174 Mosaic and streaks on petioles of banana bract mosaic virus-infected banana. (Courtesy J.L. Dale.)

Pathogenesis

BaBtMV has been experimentally transmitted nonpersistently by two aphid species: *Rhopalosiphum maidis* and *Aphis gossypii* (Magnaye and Espino, 1990). The BaBtMV causes both apparent losses of yield and smaller/fewer fruits per hand. It can be distributed widely in the Philippines and is of possible significance in India.

Control

The control of BaBtMV will be difficult due to the systemic infection of all shoots, unique cultural practices, the standard vegetative propagation system, and the lack of progress in traditional plant breeding to incorporate plant genetic resistance. However, the prospect is bright through the use of biotechnology of incorporating viral resistance by incorporating the coat protein gene of BaBtMV into banana (J. Dale, personal communication, 1994; Fauquet and Beachy, 1993).

FIGURE 175 Purple streaks on flower bracts of banana infected with banana bract mosaic virus. (Courtesy J.L. Dale.)

Cucumber Mosaic Cucumovirus (CMV)

Banana is one of many natural hosts of CMV. Some CMV strains cause a banana mosaic disease with significant damage in certain areas (e.g., Honduras) (Adam, 1962). Vegetative banana reproduction allows the easiest, most dangerous mode of virus spread when phytosanitary sucker control is not abided by. Aphid vectors also play an active role in its spread.

Disease control can be effective by applying measures similar to those recommended for BaBtTpV. Extensive, restrictive protection programs in Honduras markedly reduced disease and prevented virus introduction into newly established farms (Adam, 1962; Broadbent, 1964).

VIRUS DISEASES OF CITRUS (*CITRUS* SPP.)

Citrus Leaf Rugose Ilarvirus (CsLRgV)

Properties of Viral Particles

CsLRgV is isometric, with four particle sizes (nucleoproteins) of 32 (NP1), 31 (NP2), 26 (NP3), and 25 (NP4) nm in diameter; respective 105, 98, 89, and 79S sedimentation coefficients; and respective SS-RNAs of 1100, 1000, 700, and 300 kDa (Garnsey and Gonsalves, 1976). The coat protein is a single polypeptide species of 26 kDa.

Neither the individual RNAs of the multicomponent CsLRgV nor a mixture of the three largest RNAs are infective. However, their infectivity is activated by the addition of either the smallest RNA or the coat protein (Gonsalves and Garnsey, 1975). CsLRgV infectivity can also be activated by the addition of the smallest RNA or the coat protein from AMV or CsVgV, or the coat protein from TSkV. Also, vice versa, the addition of the coat protein or the smallest RNA of CsLRgV is able to activate the largest RNAs of AMV or CsVgV (Gonsalves and Garnsey, 1975; 1975a).

The CsLRgV is more stable in crude sap than other ilarviruses. The TIP of CsLRgV in citrus leaf sap is 60 to 65°C and LIV at 25°C is 24 to 48 hours (Garnsey and Gonsalves, 1976). CsLRgV has moderate immunogenicity. The two described strains differ from the type strain and do not cause stunt of Duncan grapefruit (*Citrus paradisi*).

Geographic Distribution

CsLRgV was discovered only in a single area of Florida (Fulton, 1981).

Host Range, Symptoms, and Pathogenesis

Natural hosts of CsLRgV are only in the *Citrus* genus, but it has been transmitted mechanically into several citrus and herbaceous plants. Characteristic symptoms for individual host species include variable leaf rugosity in Mexican lime (*Citrus aurantifolia*), variable leaf fleck in Eureka lemon (*C. limon*), and severe stunt of young Duncan grapefruit (*C. paradisi*) plants (Garnsey and Gonsalves, 1976). Experimental hosts that display local lesions include *Phaseolus vulgaris* (cv. Red Kidney), *Vigna unguiculata* (Early Ramstorn cowpea), and *Crotalaria spectabilis*. Animal vectors of CsLRgV are unknown, but it can be transmitted experimentally with contaminated cutting tools. Due to its isolated occurrence, the control of CsLRgV has received little attention.

Citrus Tristeza Closterovirus (CsTzV)

Properties of Viral Particles

CsTzV has filamentous thread-like particles of 2000 × 10 to 12 nm, with a sedimentation coefficient of 140 ± 10S (Price, 1970). Its single-stranded RNA is 6400 to 6900 kDa and its single polypeptide coat protein is 25 kDa (Bar-Joseoh, 1988). CsTzV has not been transmitted mechanically into herbaceous plants.

Severe natural strains of CsTzV, described based on virulence and aphid transmissibility, include seedling yellows (SY, accepted as a symptom of severe CsTzV isolates), grapefruit stem pitting, lime die-bark, and Ellendale mandarin decline (Price, 1970). Serological studies show that CsTzV strains do not differ based on antigenic properties (Bar-Joseph, 1988).

Geographic Distribution and Economic Importance

CsTzV occurs wherever citrus is grown. Although not widespread in Mediterranean countries, it has been established experimentally in Italy (Russo, 1956), Israel (Reichert et al., 1956), and Spain (Bar-Joseph, 1988).

CsTzV affects all *Citrus* species, and especially sweet orange (*Citrus sinensis*), grapefruit (*C. paradisi*), and lime (*C. aurantifolia*). Yield losses depend on the scion-rootstock combination, virus strain, and vector activity. The greatest losses are found in sweet orange (*C. sinensis*) grafted onto sour orange rootstocks. An estimated 15 million trees have died in South America; for example, about 75% (6 million) of the trees in 12 years in the state of Sao Paolo, Brazil (Bennett and Costa, 1949). In the 15 years following discovery in California, nearly 400,000 citrus trees were destroyed. Worldwide, 40 million citrus trees were lost due to CsTzV during the last 50 years (Bar-Joseph, 1988).

Host Range

Natural hosts of CsTzV are mostly in the family Rutaceae. It infects nearly all citrus species, and of particular importance are sweet orange (*Citrus sinensis*), lemon (*C. limon*), mandarin (*C. reticulata*), grapefruit (*C. paradisi*), as well as others, such as *C. aurantium, C. aurantifolia, C. excelsa, C. jambhiri, C. limonia, C. macrophylla, C. medica, C. mitis, C. pennivesiculata*, and *C. reshni*. Trifoliate orange (*Poncirus trifoliata*), sweet lime (*C. limetta*), *C. volkameriana*, and *Severenia buxifolia* are resistant to CsTzV (Kegler, 1977). Natural hosts in genera are *Aeglopsis chevalieri, Afraegle paniculata*, and *Pamburus missiones*. Some plants in the *Passiflora* genus are the only hosts of CsTzV outside the Rutaceae family (Costa, 1949).

Symptoms

Characteristic symptoms of CsTzV infection vary, depending on the host species, rootstock combinations, and virus strains. Characteristic symptoms of the drastic tristeza disease were observed first in Argentina (1930, 1931); then Brazil (about 1937), named after extensive damage occurred (*tristeza* means sadness in Spanish); then in California (1939) where the disease was described as quick decline or bud union decline, with worldwide distribution; tristeza occurs in all citrus, especially sweet orange, grapefruit, and lime.

Typical tristeza symptoms appear in sweet orange and other citruses grafted onto sour orange rootstocks. Disease development begins with cytopathological effects in the phloem cells in which virus particles and virus aggregates accumulate in a large number. Characteristic overgrowths (pathological meristemization) and honeycomb pitting just below the union occur in the bud union points. The meristematic (callus) tissue becomes necrotic, sieve tubes clog, and wood ray cells grow slowly. Root supplies of mineral elements are restricted and, with less starch, rootlets dessicate, which results in reduced starch accumulation and root decline; plant nutrition and water supplies are seriously disturbed, resulting in leaf yellowing and twisting. Finally, trees collapse, with a sudden wilt and drying of leaves, plus branch and whole-tree dieback.

Several citrus species (especially grapefruit and lime), when infected with CsTzV, show stem pitting (stem pitting and bushy stunt of grapefruit, lime die-back). Stem pitting results from disorders in cambial tissues characterized by elongated stem grooves and depressions. Final symptoms include reduced plant growth, decreased branch development, and small, deformed fruits.

Seedling yellows represents a special tristeza disease symptom occurring in the seedlings of some citrus species (grapefruit, sour orange, and lemon) when infected with the seedling yellows strain of CsTzV. Those seedling leaves grow slowly and remain small and yellow. This symptom is believed to result from the degeneration of cells adjacent the phloem sieve tubes, which are mildly affected in the leaves and stems (Agrios, 1988).

Diagnostic plants with varying symptoms include *Citrus aurantifolia* seedlings with vein clearing, yellow spots on leaves, and stem pitting; *C. sinensis* grafted onto *C. aurantium*, with quick decline or overgrowths at the graft union point with a honeycomb pitting below the union; and *C. limon* seedlings (cv. Eureka and Lisbon); whereas *C. aurantium* and *C. paradisi* are diagnostic for CsTzV seedling yellows strains with chlorotic leaves, slightly twisted and smaller, growth ceases, and deformation caused by virulent isolates.

Several laboratory methods can be used effectively for quick virus diagnosis: for example, sodium dodecylsulphate (SDS) immunodiffusion (Gonsalves et al., 1978), ISEM, ELISA, radio-immunoassay (RIA) (Garnsey et al., 1981), double-stranded RNA profiles (Dodds and Bar-Joseph, 1983), and nucleic acid hybridization with cloned cDNA (Rosner et al., 1983).

Pathogenesis

The CsTzV preserved in infected citrus trees represents a perennial source of infection. Primary infections occur in orchards when CsTzV-infected propagative materials—both scion and rootstock—are used for their establishment. CsTzV spreads naturally when susceptible citrus species and cvs. are grown where aphid vectors predominate.

Tropical citrus aphids *Toxoptera (Aphis) citricida*, *Aphis gossypii*, and *A. citricola*, the most active, transmit CsTzV semipersistently; also, a high level of transmission of some CsTzV strains occurs with *A. gossypii* (Bar-Joseph and Lobenstein, 1973; Roistacher et al., 1980). CsTzV is neither seed transmitted nor mechanically transmitted with the sap of infected plants.

Control

CsTzV control in established citrus orchards with susceptible scion rootsock combinations is nearly impossible. The only economically justified measure is to remove infected trees and replant with resistant rootstock/scion combinations.

New plantations should be established with healthy, virus-free planting material and tolerant scion/rootstock combinations. Recommended rootstocks are sweet orange, rough lemon, Cleopatra mandarin, and Troyer citrange. It is not advisable to use CsTzV tolerant rootstocks if they are susceptible to other pathogens (Agrios, 1988).

The cross protection method is effective in protecting commercial orchards against possible subsequent infections with severe CsTzV strains transmitted by aphid vectors in nature (Costa and Muller, 1980).

Citrus Variegation Ilarvirus (CsVgV)

Properties of Viral Particles

CsVgV isometric particles are 30 nm diameter, of multicomponent structure with particles of several different sizes. The zones with fastest sedimentation contain particles of 28, 31, and 33 nm diameter, but too few particles accumulated in the fourth (slowest top) zone to measure (Fulton, 1981). Individual particles are not infective. A mixture of three of the largest particles allowed the greatest infectivity, whereas a mixture of any two particle types were less infective.

Individual CsVgV particles have sedimentation coefficients of 79, 83, 93, and 110S—smallest to largest (Corbett and Grant, 1967; Gonsalves and Garnsey, 1976). Comparatively unstable, CsVgV has an LIV in plant sap at 20 to 22°C of 8 hours and a TIP of 55 to 60°C (Grant and Corbett, 1964).

CsVgV consists of 5 RNAs. A mixture of the three largest RNA species is not infective, but infectivity follows the addition of either the smallest RNA or the coat protein (protein activation) (Gonsalves and Garnsey, 1975). Characteristically, a mixture of these RNAs can be activated by the addition of the smallest RNA or coat protein from CsLrV or AMV, and by the addition of the coat protein from TSkV (Fulton, 1981).

CsVgV is serologically related to CsLRgV from which it differs based on other traits. Its characteristic strain causing citrus crinkly leaf disease (CsCrLV) was earlier considered a separate virus species (Yot-Dauthy and Bove, 1968).

Geographic Distribution

CsVgV occurs worldwide wherever citrus is grown, but it is economically important only in areas where severe strains dominate.

Host Plants and Symptoms

CsVgV naturally infects only citrus species in which the type isolate causes chlorotic spots and crinkle on leaves and fruit deformations (Figure 176). Infections by the severe CsCrLV isolates cause warping and pocketing of leaves, whereas CsVgV causes severe distortions and variegation in Etrog citron (*C. medica*), but not rugosity in Mexican lime (*C. aurantifolia*), thus differing from CsLRgV (Garnsey, and Gonsalves, 1976).

Lemon and sour orange seedlings and ELISA tests are used for virus diagnosis (Garnsey et al., 1983).

Pathogenesis and Control

CsVgV is transmitted and spread naturally with infected planting material. CsCrLV isolates are also seed transmitted, but at a low incidence. CsVgV is transmitted mechanically with infective plant sap, an important prerequisite for experimental work. No known natural vectors have been reported (Klotz, 1961). Only healthy, virus-free planting material obtained by thermotherapy, nuclear embryology, or shoot-tip grafting should be used for establishment of orchards (Cambra and Navarro, 1988).

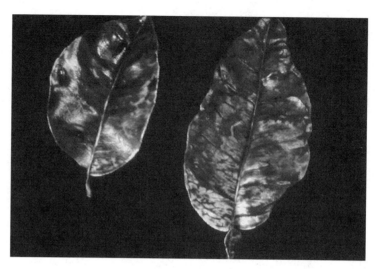

FIGURE 176 Chlorosis caused by citrus variegation virus on 'Duncan' grapefruit leaves. (Reproduced with permission from APS.)

Citrus Exocortis Viroid (CsExVd)

Properties

The CsExVd consists of only ribonucleic acid without a protein coat. The RNA is single stranded, covalently closed and circular, and with a known sequence of 371 nucleotide residues (Gross et al., 1982). High sequence homology (60 to 73%) exists between CsExVd and PSlTbVd and CmSnVd (Bové, 1988). CsExVd is very stable with a TIP (in *Gynura aurantiaca* sap) of 90 to 100°C. Many strains differ based on symptom severity, incubation period and the degree of stunt of infected plants. Mild strains occur on leaves and cause stunt, whereas virulent strains cause bark scaling (loc. cit. Semancik, 1980).

Geographic Distribution and Economic Importance

CsExVd is widespread in areas where citrus is grown on susceptible rootstocks, for example, South America (Brazil and Argentina), Australia, and the Mediterranean and somewhat less in North America and northern Africa. Following an inspection of citrus orchards in several Mediterranean countries, Reichert (1957) found CsExVd only in citruses grafted onto hypersensitive *Poncirus trifoliata* rootstock, thus being of substantial importance only in former Yugoslavia. Bark scaling causes economic damage by reduction of yields up to 40%, especially when grafted onto susceptible rootstocks (Agrios, 1988).

Host Range

Natural hosts of CsExVd are in the *Citrus* genus: sweet orange (*Citrus sinensis*), lemon (*C. limon*), mandarin (*C. reticulata*), grapefruit (*C. paradisi*), sour orange (*C. aurantium*), lime (*C. aurantifolia*), trifoliate orange (*Poncirus trifoliate*), *C. limonia*, *C. medica*, *C. jambhiri*, *C. latifolia*, and *C. paradisi* × *C. sinensis*. Experimental hosts include some species in the Solanaceae (*Solanum tuberosum, Lycopersicon lycopersicum, Petunia hybrida*) and Compositeae (*Gynura aurantiaca, G. saramentosa*) families. *L. lycopersicum* and *G. aurantiaca* are diagnostic, in which viroids cause leaf epinasty, vein cracking, and darkening on the lower leaf surface.

Symptoms

Symptoms of CsExVd depend on strain virulence and rootstock susceptibility. Virulent strains cause severe symptoms in plants grafted onto susceptible *P. trifoliata* rootstock. Main stem disorders are characterized by the cracking, scaling, anddropping of individual bark parts; shoots are stunted

FIGURE 177 Etrog citron inoculated with citrus exocortis viroid showing leaf epinasty in severe (left) and mild (center) reactions; healthy (right). (Reproduced with permission from APS.)

with shortened internodes; and leaves become chlorotic (Figures 177 and 178). All pathological disorders noted occur in rapid sequence, resulting in necrosis and death of rootstock tissues. Less virulent strains cause milder symptoms; and *C. limonia* rootstock has the ability to condition milder symptoms. Some citrus species (sweet orange, sour orange, rough lemon, and grapefruit) are tolerant to CsExVd and no bark scaling occurs.

Pathogenesis and Control

CsExVd, permanently present in infected citrus and herbaceous hosts, is transmitted through infected planting material as the primary mode of spread. It is also transmitted mechanially, both with artificial inoculations and infected cutting tools and other equipment (Garnsey and Jones, 1967). CsExVd retains infectivity on knife blades up to 8 days. No natural vectors of CsExVd are known, although numerous insects have been tested (Laird et al., 1969).

Healthy, viroid-free parent planting material is necessary to prevent CsExVd problems. The phytosanitary condition of the parent material should be checked regularly by grafting onto susceptible *P. trifoliata* rootstocks or other indicator plants. Viroid-resistant or -tolerant rootstocks should be used in production. All tools should be disinfested between cuts of different plants by dipping into 10 to 15% NaOCl (Bové, 1988).

Satsuma Dwarf Virus (SaDV)

Properties of Viral Particles

SaDV is isometric, 26 nm diameter, and upon centrifugation, separates into three components: top (T) in low quantities, middle (M), and bottom (B), with 119S (M) and 129S (B) sedimentation coefficients. T particles probably contain no RNA, whereas M and B particles contain RNA and are particularly infective as a mixture (Usugi and Saito, 1979). The RNA is probably single-stranded and accounts for about 33% (M) and 37% (B) particle weight. Purified preparations contain proteins of 21 and 42 kDa.

The TIP of SaDV in *Physalis floridana* sap ranges from 50 to 55°C; the DEP is 1.2 to 2.5 × 10^{-3}; and the LIV at 24°C is 8 to 12 days (Usagi and Saito, 1979). SaDV is highly antigenic (Bové, 1988a), and strains differ, based on symptomlological reactions of indicator hosts.

FIGURE 178 Bark scaling caused by citrus exocortis viroid on the trifoliolate orange rootstock of a lemon tree. (Reproduced with permission from APS.)

Geographic Distribution and Economic Importance

SaDV, widespread in Satsuma orange in Japan (Yamada and Sawamura, 1952), occurs in Turkey and possibly in former Yugoslavia and the former U.S.S.R. (Bové, 1988). It causes a marked stunt of stems, leaves, shoots, and fruits, leading to severe yield losses.

Host Range

SaDV infects citruses in nature and can be transmitted mechanically into eight herbaceous dicot families (Usagi and Saito, 1929). Included among susceptible hosts are citranges, *Citrus excelsa*, *C. junos*, clementine, grapefruit, lemon, mandarins, natsudaidai, *Poncirus trifoliata*, rough lemon, satsuma, sour orange, sweet orange, tangelos, and small acid lime (Bové, 1988). Etrog citron, Hassaku, Tahiti lime, and Orlanda tangelo are symptomless hosts. Susceptible herbaceous plants, *inter alia*, include cowpea, French bean, sesame, and tobacco, whereas *Chenopodium amaranticolor*, cucumber, and *Nicotiana glutinosa* are symptomless virus hosts.

Symptoms and Pathogenesis

General symptoms in citrus are stem stunt plus leaf and fruit deformations. Leaves developed in the spring are spoon- or inverted boat-shaped (Figure 179) (Tanaka, 1980). The fruits are underdeveloped with a thick peel. Indicator plants include *Chenopodium quinoa*, with only local lesions; *Sesamum indicum*, with local lesions in inoculated leaves and vein clearing, necrosis, malformation in systemically infected leaves; and cowpea cv. Blackeye, with local lesions in inoculated leaves, vein clearing, and mottling in systemically infected leaves and necrotic streaks in petioles and stems.

Virus Diseases of Fruit Trees

FIGURE 179 Boat-shaped leaves of satsuma mandarin infected with satsuma dwarf virus. (Reproduced with permission from APS.)

SaDV, which remains infective in natural hosts, is transmitted and spread through infected planting materials. Virus vectors of SaDV are unknown and its mechanical transmission is of experimental importance only.

Diseases of Citrus Assumed Caused by Viruses

Several diseases of unknown etiology occur in different citrus species. Infections of plants affected by these diseases are systemic and graft transmitted, suggesting they may be caused by viruses. Susceptible plants display various symptoms or remain symptomless. A common and generally valid measure for the control of these diseases is to use healthy propagative budwood. In commercial production, priority should be given to the use of resistant or tolerant cvs. and rootstocks. The most important information on this group of diseases (Bové, 1988) will also relate to this control measure.

Citrus Cachexia-Xyloporosis Disease (CsCxDs)

CsCxDs is widespread wherever citrus is grown, but is most widespread in Mediterranean countries (Reichert, 1957). Susceptible hosts include *Citrus macrophylla*, clementine, kumquat, mandarin, Rangpur lime, Satsuma, sweet lime, and tangelos. Symptomless hosts include sweet orange, grapefruit, lemons, and acid limes (Bové, 1988). Tolerant rootstocks are *Poncirus trifoliata*, citranges, citrumelos, rough lemon, and sour orange.

Characteristic symptoms of CsCxDs include bark pegging and pitting on stems, branches, and shoots. Pegs form on the bark that project into pits in the wood below the bark. The pegs are located on the inner part of the bark and connected into the pits. These pegs, which penetrate deep into the cambium, crush the cells in the phloem region and the rings of xylem, plus accompanying gum deposits. If the scion is susceptible, characteristic symptomological alterations occur above the bud union, or below this point if the rootstocks are susceptible. Specific symptoms may differ substantially in different host species (Childs, 1980), which has diagnostic importance. Because it is highly susceptible, the Parsons special mandarin is a suitable cachexia disease indicator plant.

In addition to spread by infected planting material, the causal agent of CsCxDs is also transmitted mechanically with pruning knives, which can be prevented by disinfestation of cutting knives in sodium hypochlorite (NaOCl) solutions (Roistacher et al., 1980a).

Citrus Concave Gum Disease (CsCgDs), Citrus Blind Pocket Disease

CsCgDs is generally widespread, especially in Mediterranean countries. Susceptible hosts include sweet orange, mandarin, clementine, tangelo, and grapefruit. Symptomless hosts include sour orange, sweet lime, *Poncirus trifoliata*, citrons, lemons, rough lemon, and acid limes (Bové, 1988).

Characteristic symptoms of CsCgDs are concavities occurring on stems and branches of infected citrus. These concavities may be large (wide) and in some cvs. (Washington navel orange and Orlando tangelo) accompanied by the formation of gum deposits, thus the name "concave gum" disease. Another symptom is narrow, pronounced depressed concavities/deep invaginations with adjoining vertical sides, and thus the name "blind pocket" disease. Individual or combined symptoms of vein flecking and oak-leaf patterns may occur on young leaves, although not diagnostic because they have been seen in some other citrus diseases (Klotz, 1980).

In nature, the causal agents of CsCgDs are spread with infected planting material. Transmission by bark inoculation is suitable for indicator plant (mandarin and sweet orange seedlings) indexing in glasshouses and for other experimental purposes.

Citrus Cristacortis Disease (CsCtDs)

The CsCtDs is comparatively more restricted than other citrus diseases, having been observed in some Mediterranean countries, in the Near and Middle East, and in northern Yemen. It is not considered economically important because it does not greatly influence yields. Susceptible hosts include *Citrus pectinifera*, clementine, grapefruit, mandarin, rough lemon, satsuma, siamelos, sour orange, sweet lime, sweet orange, tangelos, and tangers. Symptomless hosts are bergamot, citranges, citrons, *Citrus hystrix*, *C. volkameriana*, lemons, *Poncirus trifoliata*, and acid limes (Bové, 1988).

Stem pitting with pegs on the cambial side of the bark occur in the woody parts of CsCtDs stems, branches, and shoots (Figure 180). Gums form on the bottom of the pits and at the top of the pegs in some hosts (tangelos). During radial growth, visible depressions disappear from the woody parts, but new ones form later. Symptoms of vein flecking and oakleaf patterns on young diseased leaves are quite similar to those observed in other citrus diseases in this group (Vogel, 1980). Orlando, Webber, or Williams tangelos grafted onto sour orange are suitable indicator plants (e.g., with tangelo scion and sour orange rootstock, stem pitting usually appears 1 year after inoculation).

The causal agent is transmitted with infected planting material. It can be transmitted experimentally through the pollen, but not in nature (Vogel and Bové, 1980).

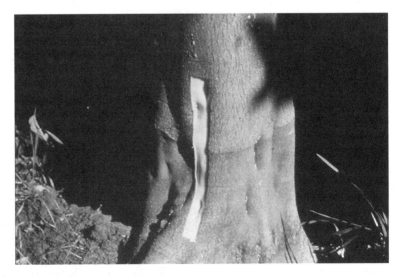

FIGURE 180   Stem pitting of sweet orange on sour orange rootstock infected with cristacortis. (Reproduced with permission from APS.)

Citrus Gummy Bark Disease (CsGmBkDs)

CsGmBkDs has been recorded in the Near and Middle East and in Greece. It is not considered economically important, although it causes some stunt of infected plants. Only sweet orange and rough lemon have symptoms of disease, and all other hosts remain symptomless (Bové, 1988).

The formation of gum on the bark near the bud union point, accompanied by discoloration spreading along affected plant parts, is characteristic. Gum impregnation of bark accompanies the occurrence of stem pitting symptoms. Affected bark parts become detached and fall off above the union line, revealing streaks of reddish-brown gum impregnated tissues (Nour Eldin, 1980). This symptom is diagnostic.

In nature, the disease is graft transmitted by bud propagation. Establishment of orchards with healthy planting material represents the primary disease control measure.

Citrus Impietrature Disease (CsImDs)

CsImDs has been observed in the Mediterranean, the Near East (Cyprus), the Middle East, and in South Africa, Texas, and Venezuela. Fruits drop prematurely—thus a complete loss in their market value. Main disease hosts include bergamont, *Citrus volkameriana*, clementine, grapefruit, lemon, rough lemon, sour orange, sweet orange, and tangelo. Symptomless hosts are Chinotto, citron, and kumquat (Bové, 1988).

Specific CsImDs symptoms of diagnostic importance occur on the fruits as gum pockets in the albedo (albedo gumming), hard and stone-like consistency, and possible gum impregnation of the axis. Such fruits remain small, lose market value, and frequently drop prematurely. Striking symptoms on young leaves appear as vein flecking and oak patterns, which might be confused with other diseases in this group. Three-year-old *Citrus volkameriana* lemon seedlings with typical albedo gumming and fruit hardening symptoms are suitable indicator plants.

CsImDs is spread by budwood propagation and experimentally by graft inoculation. The use of healthy planting material is the key disease control measure.

Citrus Psorosis Disease (CsPsDs)

The CsPsDs, first observed in Florida, is well-known in all areas where citrus is grown. It represents the main cause of tree decline and reduced yields in orchards that were established with propagative material previously not subject to a strict phytosanitary monitoring. Characteristic symptoms of CsPsDs are also known as scaly bark or scaly bark psorosis. The two disease types described are psorosis A (more frequently occurring in production) and psorosis B (less frequent, but more severe and dangerous) (Fawcett, 1932).

Susceptible hosts that display scaly bark symptoms are grapefruit, mandarin, sweet orange, and tangelo. These hosts may also display the young leaf symptoms frequently evident also in citron, lemon, acid and sweet limes, pummelo, rough lemon, and sour orange (Bové, 1988).

Scaly bark on trunks and limbs is a characteristic symptom of CsPsDs, occurring in orchards less than 6 years old, but often in trees 12 to 15 years old (Bové, 1988). Some strains (e.g., California strain 339) of psorosis causal agents, under certain conditions, cause severe bark scaling. The vein fleck described on young leaves affected by psorosis A appear during growth flushes (Fawcett, 1933).

Flecking (small, elongated cleared patches around the veinlets) may partially complete or encompass the leaf blades. Joining of small flecks results in the formation of zonate and oak patterns around the midrib. However, this symptom is not diagnostic because it also occurs in citruses affected by other diseases in this group (Roistacher, 1980b).

Young sweet orange, sweet tangor, or mandarin seedlings are suitable for CsPsDs indexing (leaves: vein flecking and/or oak-leaf pattern and newly developed axillary shoots). Sweet orange young seedling inoculations can be used for the differentiation of normal non-lesion bark and lesion bark isolates causing psorosis A and psorosis B symptoms.

Because CsPsDs spreads by bud propagation, the use of virus-free healthy buds represents the primary control measure. Transmission of the causal agent can be done experimentally by inoculations with the use of infected tree bark.

Citrus Ringspot Disease (CsRsDs)

CsRsDs, widespread in Mediterranean and African countries, causes economic damage only in mixed infections with severe scaly bark disease. Susceptible hosts (with characteristic symptoms) include numerous citrus species and hybrids. However, some infected trees in the orchards remain symptomless (Moreno and Navarro, 1988).

CsRsDs is easily recognized, based on symptoms in older leaves of yellow blotches, chlorotic vein banding, and chlorotic rings—thus its name. Similar symptoms can be observed in earlier branch, thorn, and fruit development stages. Scaly bark can also occur in some cases. Grapefruit, sweet orange (Figure 181), and mandarin seedlings reacting with leaf spot, fleck, and chlorotic rings (specific symptoms) are suitable indicator plants. Some isolates can also cause necrotic shock reactions. Local chlorotic-necrotic lesions occur on inoculated herbaceous indicator *Chenopodium quinoa* leaves.

FIGURE 181 Citrus ringspot in mature leaves of sweet orange. (Reproduced with permission from APS.)

In nature, CsRsDs spreads with infected propagative material. Although circumstantial evidence of natural disease spread exists, the exact mode is unknown (Timmer and Garnsey, 1980). Mechanical transmission into herbaceous plants (Garnsey and Timmer, 1980) is of importance only for experimental work. Disease control is ensured with the use of healthy planting material for establishment of orchards and, if quarantaine measures are followed, to prevent potential natural spread of some isolates (Moreno and Navarro, 1988).

Citrus Vein Enation Disease (CsVEnDs) and Citrus Woody Gall Disease

CsVEnDs was first described as citrus vein enation disease (Wallace and Drake, 1953) and then later as citrus woody gall disease (Fraser, 1958). It is now established that these are synonyms for a disease with the same causal agent (Wallace and Drake, 1960). Although widespread in areas where citrus is grown, in Europe it has been reported only in Italy (Ballester et al., 1979). Its distribution is connected with the use of rough lemon-susceptible rootstocks.

Characteristic symptoms of CsVEnDs include irregular tumors along leaf veins on the lower surface, with corresponding depressions on the upper surface. Under natural conditions, these symptoms occur in Mexican lime and sour orange, which serve as diagnostic plants. Tumors appearing on trunks, limbs, or roots form first as small swellings from which irregularly shaped woody galls develop later, differing in size. The enation/gall symptoms have been observed in rough lemon, *Citrus volkameriana*, Rangpur lime, and Mexican lime, whereas some hosts remain symptomless (Navarro and Pino, 1988).

In nature, the causal agent of CsVEnDs is transmitted with infected propagative material. Aphids *Myzus persicae*, *Toxoptera citricidus*, and *Aphis gossypii* also transmit the causal agent, which contribute to its spread. It may be experimentally graft transmitted and dodder transmitted. The use of healthy budwood and tolerant rootstocks represents the primary disease control measure. Thermotherapy and shoot-tip grafting can be used successfully to obtain healthy plants for production (Navarro and Pino, 1988).

VIRUS DISEASES OF FIG (*FICUS CARICA*)

Fig Mosaic Disease (FMDs)

FMDs is the most widespread and most frequent disease of fig (*Ficus carica*). The causal agent of FMDs is unknown. In parenchymal cells of diseased fig leaves, broad polymorphous particles not present in healthy plant leaves were discovered with the aid of the electron microscope: isometric particles of 120 to 150 nm in diameter and elongated particles of 200 nm. Some particles are associated with membranes (Plavšić and Miličić, 1980).

FMDs, observed in all warmer climatic areas where fig is grown, occurs in North America, Australia, Africa (Egypt), Asia (China), and Europe, and is especially widespread in Mediterranean countries.

No FMDs loss data are available because it is difficult to find enough non-infected trees for the comparative investigations required. Producers have simply accepted both current yields and fruit quality. FMDs reduces yields significantly and lowers fruit quality in particular.

The severity of FMDs is clearly evident in symptoms on leaves and fruits, and dependant on cultivar susceptibility. Primary symptoms on leaves are mosaic, chlorotic fleck, and veinal chlorosis (Figure 182). Tissue necrosis may also occur within the flecks, starting from the center. Veinal chlorosis affects tissues along and around the veins, also spreading into the interveinal tissues. These symptom types merge, and thus are difficult to distinguish. High temperatures during the vegetation period alter symptoms.

In cvs. susceptible to FMDs, leaves remain smaller and of irregular shape. Individual shoots grow slowly, internodes are shortened, and the leaves along the shoots are compacted into rosettes. Roundish, elongated pale yellow ring spots appear on the fruits. In cvs. Crnica hercegovačka and Della Madonna, the pale-yellow spots disappear during maturation; but in other cvs., these spots become dark blotches, reducing the market value of ripe fruits. Flecks spread and form grooved depressions on fruits of the susceptible cvs. Zamorčica and Vodnjača, and premature fruit drop occurs in the plants with severe disease symptoms.

FMDs affects at least 16 *Ficus* species, including *F. altissima*, *F. carica*, *F. carica* cv. *sylvestris*, *F. krishna*, and *F. tsida*. Research in former Yugoslavia shows all trees tested are infected and fig cvs. are divided into two groups based on symptomology (Omčikus and Blagojević, 1957). One group is cvs. with obvious leaf and fruit symtpoms; for example, Zamorčia (Tenica), Petrovača Bela, Grčka Bela, Kolačuša, Crna, Grčka Crna, Dugošija, Sitnica, and others. The other group is cvs. with symptoms on leaves, but not on fruits, and includes Šapanka, Petrovača Crna, Gračanka, and Sultanija Bela. Zamorčica cv. is excellent for drying, but its fruits are highly disfigured by the mosaic disease.

The causal agent of mosaic is permanently preserved in infected fig trees. It is spread naturally with infected cuttings and suckers by which fig is nearly exclusively propagated. It is graft transmissible, an important consideration both for production and investigations of disease etiology.

FIGURE 182 Fig mosaic virus in fig leaves.

The *Aceria ficus* mite also contibutes to virus spread in the field (Flock and Wallace, 1955; Proeseler, 1972). The omnipresence of mosaic in fig makes it difficult to select healthy stock as mother trees in nature. Therefore, thermotherapy might be the method of choice, but it has not yielded satisfactory practical results (Šutić and Blagojević, 1964). However, recent research suggests possible success soon in the production of healthy fig stocks.

Virus Diseases of Olive (*Olea europea*)

Olive Latent Ringspot Nepovirus (OlLtRsV)

Properties of Viral Particles

The OlLtRsV isometric particles are 28 nm diameter. Purified preparations yield three particle types (T, M, B), with sedimentation coefficients of 52S (T), 97S (M), and 132S (B) (Gallitelli et al., 1985). The single-stranded RNA consists of species RNA-1 from particle B and RNA-2 from both particles M and B (Savino et al., 1983). Both RNA species are necessary to cause infection. The TIP of OlLtRsV in *Chenopodium quinoa* sap is 60°C; the DEP is 1×10^{-3}; and the LIV at 20°C is 15 days. OlLtRsV has good immunogenicity. No strains have been described.

Geographic Distribution

OlLtRsV has been reported only in central Italy (Gallitelli et al., 1985).

Host Range and Symptoms

Olive is the natural host of OlLtRsV, which may be transmitted mechanically by sap into seven herbaceous species, of which five are dicots. (Savino et al., 1983).

OlLtRsV-infected olive trees are symptomless. Experimentally infected *Chenopodium quinoa*, *C. amaranticolor*, and *Gomphrena globosa* express symptoms of local lesions and systemic mosaic that make them suitable for disease diagnosis. OlLtRsV, being latent, is readily preserved in infected olive trees and transmitted with the propagative material. No other mode of spread of OlLtRsV is known.

Olive Latent Viruses 1 and 2

Two species of latent viruses have been isolated from symptomless olive trees: olive latent virus 1 with isometric-shaped particles and olive latent virus 2 with quasi-spherical to bacilliform particles (Savino et al., 1983). Both virus species are transmitted mechanically with infective plant sap. Olive latent virus 1, in contrast to other olive viruses, causes mild fasciation and apical bifurcation of twigs and leaves in cv. Paesana (Martelli, 1988).

Olive: Natural Alternative Host of Some Viruses

Olive is the symptomless host of nepoviruses ArMV, ChLrV, SbLtRsV, and CMV (Gallitelli et al., 1985). All are transmitted with infective plant sap and have been isolated from symptomless olive trees—except for SbLtRsV, which causes leaf and fruit malformations in the Ascolona cultivar (Martelli, 1981; 1988).

VIRUS DISEASES OF PALM (*PALMACEAE*)

Coconut Cadang-Cadang Viroid (CcCcgVd)

Properties

The smallest known viroid is CcCcgVd, with a circular or linear single-stranded RNA occurring in monomeric and dimeric forms of 246 nucleotides, (Randles and Imperial, 1984). According to Haseloff et al. (1982), all sequenced isolates of CcCcgVd have the same minimal nucleotide sequence. It is believed to cause tinangaja disease (i.e., a CcCcgVd strain of equal size that has a high nucleotide sequence homology with CcCcgVd) (Randles and Imperial, 1984).

Geographic Distribution and Economic Importance

Coconut cadang-cadang disease is widespread in the central Phillipines and the tinangaja disease only in Guam (Marianas Islands) (Randles and Imperial, 1984). Cadang-cadang disease is dramatic, causing massive coconut tree death and drastically endangering production. The number of CcCcgVd-infected trees has increased steadily since discovery. De Leon, in 1952, reported that it destroyed more than 1.5 million trees in the Biocol area; and by 1958, this number had increased to nearly 8 million—or almost half of the total number of coconut trees in the area (Price, 1958; Corbett, 1964). The latest report indicates that 30 million palm trees have been lost since the discovery of CcCcgVd (Zelazny et al., 1982).

Host Range and Symptoms

CcCcgVd hosts belong only to the Palmae family. Coconut (*Cocos nucifera*) is the best-known host of CcCcgVd.

Characteristic symptoms occur in all organs of infected plants and CcCcgVd is also present in roots. A yellow-bronze coloration appears on the lower two-thirds of fronds, and small chlorotic spots gradually coalesce and form chlorotic mottle on the third and fourth fronds below the spear leaf. Nuts grow slowly and are visibly deformed. Inflorescences with necrosis are unfertile; fail to produce nuts, and finally with crown volume reduction, the trees die usually 8 to 16 years after first symptom appearance (Zelazny and Niven, 1980). In tinangaja disease, the nuts are smaller, elongated, and mummified (Boccardo et al., 1981).

In natural infections, CcCcgVd causes similar decoloration in young oil palm (*Elaeis guineensis*) fronds and inflorescences fail to form. Suitable diagnostic plants are coconut (*Cocos nucifera*) seedlings and some plants of other species (Randles and Imperial, 1984).

Pathogenesis

The natural mode of CcCcgVd pathogenesis is unknown: CcCcgVd is seed transmitted into one infected seedling of 320 tested samples; however, no natural vectors are known (Randles and Imperial, 1984). It is transmitted mechanically by a special procedure (Randles et al., 1977) that is highly important for the investigation of CcCcgVd, its hosts, and their mutual relations.

VIRUS DISEASES OF PAPAYA (*CAPRICA PAPAYA*)

Compendium of Tropical Fruit Diseases by Ploetz (1994) is a good companion book to this section.

Papaya Mosaic Potexvirus (PaMV)

Properties of Viral Particles

PaMV is a filamentous particle of 530 nm (de Bokx, 1965) with a 118.7S sedimentation coefficient and MW of 3140 kDa. The single-stranded RNA of 220 kDa and a sedimentation coefficient of 31.8S accounts for 7% of the particle weight; the capsid is composed of homologous coat protein subunits of 19 kDa (loc. cit. Purcifull and Hiebert, 1971). PaMV is stable, with a TIP in papaya sap of 73 to 76°C, a DEP of 1×10^{-4}, and an LIV at 24°C of up to 6 months (Conover, 1964). PaMV has strong immunogenicity (de Bokx, 1965).

Geographic Distribution

PaMV is widespread in Florida and Venezuela, but is of little economic importance (Purcifull and Hiebert, 1971).

Host Range and Symptoms

Papaya is the only natural host of PaMV, which causes mosaic and retarded growth of leaves. It can be transmitted experimentally to 17 species in 9 dicot families (Purcifull and Hiebert, 1971). Susceptible herbaceous plants include *Antirrhinum majus*, *Cassia occidentalis*, *Chenopodium amaranticolor*, *C. quinoa*, *Cucumis sativus*, *Glycine max.*, *Gomphrema globosa*, *Nicotiana benthamiana*, *Phaseolus vulgaris*, *Vigna sinensis*, *Vinca rosea*, and *Zinnia elegans* (loc. cit. Purcifull and Edwardson, 1981). Diagnostic plants with local lesions on inoculated leaves include *Cassia occidentalis*, *Chenopodium amaranticolor*, and *G. globosa*.

Pathogenesis

PaMV is transmitted readily with infective plant sap. No vectors in nature are known: PaMV is not seed transmitted and no other mode of transmission is known (Purcifull and Hiebert, 1971).

Papaya Ringspot Potyvirus (PaRsV)

Properties of Viral Particles

PaRsV is a filamentous particle, 780 × 12 nm (Purcifull et al., 1984), with single-stranded RNA (De La Rose and Lastra, 1983) and coat protein subunits of 36 kDa. PaRsV consists of two strains: PaRsV-Pa infects both papaya and cucurbits but the former WmMV1 (watermelon mosaic virus 1), now known as the Wm strain of PaRsV (PaRsV-Wm), infects only watermelon and cucurbits, but not papaya.

Pa and Wm strains differ in symptomological reactions in individual hosts (Conover, 1964; Gonsalves and Ishii, 1980; Russo et al., 1979). PaRsV differs from WmMV1 in biophysical properties. The TIP of PaRsV in papaya sap is 54 to 56°C; the DEP is 1×10^{-3}, and LIV at 24°C is 8 hours (Conover, 1964a). The TIP of WmMV1 in squash sap is 60°C; the DEP is 5×10^{-4}; and LIV is 40 to 60 days (Webb and Scott, 1965).

Geographic Distribution

PaRsV-Pa and -Wm are widespread wherever papaya and cucumber, their main hosts, are grown. PaRsV-Pa has been recorded in Africa, Caribbean countries, India, Okinawa, South Africa, and Taiwan; and PaRsV-Wm has been recorded in cucurbits in Australia, Caribbean countries, Europe (France, Italy), India, Mexico, Middle East countries, South America, and the U.S. (Purcifull et al., 1984).

Host Range and Symptoms

PaRsV-Pa naturally infects papaya and cucurbits and may be transmitted experimentally to 15 dicot plants in the Caricaceae, Chenopodiaceae, and Cucurbitaceae families (Jensen, 1949; Conover, 1964a), whereas PaRsV-Wm infects 38 species in 11 genera of the Cucurbitaceae and two genera of the Chenopodiaceae.

FIGURE 183   Pawpaw (papaya) leaves naturally infected by papaya ringspot virus (Qld., Australia).

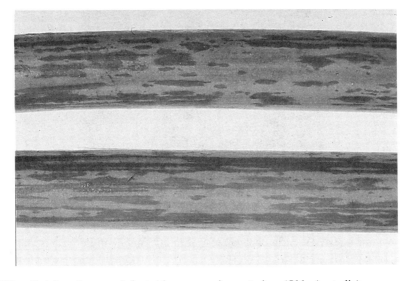

FIGURE 184   Petioles of papaya infected by papaya ringspot virus (Qld., Australia).

FIGURE 185 Pawpaw (papaya) infected by papaya ringspot virus (Qld., Australia).

FIGURE 186 Pawpaw (papaya) infected by papaya ringspot virus (Qld., Australia).

Main symptoms of PaRsV-Pa in papaya are leaf mottle and distortion, rings and spots on the surface of misshapen fruits, and streaks on stems and petioles. Quality and yield of infected plants are markedly reduced. Cucurbit plants infected with PaRsV-Wm display chorlotic mottle plus leaf and fruit malformation symptoms.

FIGURE 187 Shoestring mosaic of papaya foliage caused by papaya ringspot virus. (Courtesy D. Gonsalves. Reproduced with permission from APS.)

Papaya seedlings not susceptible to PaRsV-Wm provide suitable hosts for PaRsV-Pa diagnosis. Young inoculated *Cucurbita pepo*, *Cucumis metuliferus*, and *Luffa acutangula* plants are used to characterize PaRsV-Pa and WmMV1, whereas *Nicotiana benthaminana*, which is susceptible to neither one, may be systemically infected with WmMV2 (Purcifull et al., 1984). Serological methods are also used for plant indexing, including SDS immunodiffusion, ELISA, and ISEM (Purcifull and Hiebert, 1979).

Pathogenesis

PaRsV-Pa is naturally preserved in infected perennial plants and PaRsV-Wm in continuously grown various cucurbit species. In the field, PaRsV-Pa and -Wm are transmitted by numerous aphid species nonpersistently. PaRsV-Pa may be transmitted by 21 species in 11 genera, including *Myzus persicae* and *Aphis gossypii*. PaRsV-Wm is transmitted by 21 species in 15 genera, including *M. persicae*, *Aulacorthum solani*, *Aphis craccivora*, and *Macrosiphum euphorbiae* (Purcifull et al., 1984).

PaRsV strains are transmitted easily with infective plant sap, which may be as important for field spread as it is for experimental purposes. Neither transmission of PaRsV-Pa through papaya seed, nor PaRsV-Wm through cucurbit seed, has been confirmed.

Control

Papaya has been engineered for resistance to PaRsV by cloning the coat protein gene into papaya (Gonsalves 1998). This case study gives an account of the work history, how the papaya was commercialized, and provides brief account of the technology transfer program to developing countries. The first commercial fruit was harvested in 1998 in Hawaii, and will be marketed under the label "Rainbow." [One of us (Ford) had the privilege of seeing the test field and eating several of the 'Rainbow' fruits.] This resistance should now be made widely available to papaya growers because PaRsV causes disasterous losses in all papaya growing areas.

Papaya: Natural Alternative Host of Some Viruses

Papaya is also the alternate natural host of viruses described in other plants (TRsV, TmSpoWV, and one or more rhabdoviruses) (loc. cit. Purcifull et al., 1984), all of which are in other groups which, based on morphological traits, are easily distinguishable from PaMV and PaRsV.

REFERENCES

Abel, P.P. et al. 1986. Delay of disease development in transgenic plants that express the tobacco virus coat protein. *Science* 232:738–743.
Adam, A.V. 1962. An effective program for the control of banana mosaic. *Plant Dis. Reptr.* 46:366–370.
Adams, A.N. 1978. The detection of plum pox virus in *Prunus* species by enzyme-linked immunosorbent assay (ELISA). *Ann. Appl. Biol.* 90:215–251.
Agrios, G.N. 1988. *Plant Pathology*, 3rd edition, Academic Press, New York, 803 pp.
Albrechtová, L. et al. 1974. Dúkaz virus mozaiky tabaku v listech svestky (*Prunus domestica* L.) v CSSR. *Ochr. Rostl.* 47:113–116.
Albrechtová, L. et al. 1975. Nachweis des Tomatenzwergbusch-Virus (Tomato bushy stunt virus) in Süssikirschen, die mit virösen Zweigkrebs befallen waren. *Phytopathol. Z.* 82:25–34.
Allen, W.R. 1969. Occurrence and seed transmission of tomato bushy stunt virus in apple. *Can. J. Plant Sci.* 49:797–799.
Atanasoff, D. 1932. Sarka po slivite, edna nove virusna bolest. *Jb. Univ. Sofia. Agronom. Fak.* 11:49–70.
Atanasoff, D. 1932–33. Plum pox: a new virus disease. *Ann. Rept. Univ. Sofia, Fac. Agronom. et Sylvic.* 11:49–69.
Atanasoff, D. 1935. Mosaic of stone fruits. *Phytopathol. Z.* 8:259–284.
Atanasoff, D. 1935a. Old and new virus diseases of trees and shrubs. *Phytopathol. Z.* 8:197–223.
Avgelis, A. et al. 1988. Epirus cherry virus, a new virus isolated from cherry in Greece (Abstr.). *XIVth Int. Symp. Fruit Tree Virus Dis.*, 12–18 June, Thessaloniki, Greece.
Avgelis, A. 1991. Seed transmission of Epirus cherry virus in broad bean (Abstr.). *XVth Int. Symp. Virus and Virus Dis. Temp. Fruit Crops*, 8–13 July, Vienna, Austria.
Ballester, J.F. et al. 1979. Estudios sobre el "vein enation woody gall" de los agrios en Espana. *Anales Inst. Nacl. Invest. Agrarias, Prot. Vegetal* 12:127–138.
Bancroft, J.B. and Tolin, S.A. 1967. Apple latent virus 2 is sowbane mosaic virus. *Phytopathology* 57:639–640.
Barba, M.P. et al. 1986. Meaning of seeds in the epidemiology of almond viruses. *Acta Hort.* 193.
Barbara, D.J. 1988. *Prunus* necrotic ringspot virus (PNRSV). In: Smith, I.M. et al. Eds. *European Handb. Pl. Dis.*, Blackwell Sci. Publ., London, p. 17–19.
Bar-Joseph, M. and Loebenstein, G. 1973. Effects of strain, source plant and temperature on the transmissibility of citrus tristeza virus by the melon aphid. *Phytopathology* 63:716–720.
Bar-Joseph, M. 1988. Citrus tristeza virus (CTV). In: Smith, I.M. et al., Eds. *European Handb. Pl. Dis.*, Blackwell Sci. Publ., London, p. 9–11.
Barnett, O.W. and Fulton, R.W. 1969. Some chemical properties of *Prunus* necrotic ringspot and Tulare apple mosaic viruses. *Virology* 39:556–561.
Bateson, M.B. 1995. Characterization of Banana Bract Mosaic Virus Through Polymerase Chain Reaction Amplification Using Degenerate Primers. Ph.D. dissertation, Queensland University of Technology, Brisbane, Australia.
Baumann, G. 1967. Die Stecklenberger Krankheit der Sauerkirschen. *Erwerbsobstbau* 11:1–3.
Baumann, G. 1968. Steinobstviren in Ziergehölzen der Gattung Prunus.-7. *Eur. Symp. Über Viruskrankheiten der Obstbäume, Aschersleben, 1967. Tagungsberichte, Berlin* 97:211–220.
Baumann, G. 1971. Untersuchungen über die Ockst. *Erwerbsobstbau* 13:203–205.
Belli, G. et al. 1980. Properties of a strain of strawberry latent ringspot associated with a rosetting of peach in northern Italy. *Acta Phytopathol. Acad. Sci. Hung.* 15:113–117.
Benlloch, M. 1964. Notas sobra la enfermedad denominada "Viruela" del albaricoquero. *Bol. Pat. Veg. Ent. Agric.* 27:1–14.
Bennett, C.W. and Costa, A.S. 1949. Tristeza disease of citrus. *J. Agric. Res.* 78:207–283.
Bernhard, R. and Dunez, J. 1971. Chlorotic leaf spot virus. Contaminations of different Prunus species. Populations of viruses. *Ann. Phytopathol., No. hors série*, 317–336.
Blattný, C. 1961. Studie virus a z viros podezrelých onemocreni ovocných srevin v ČSSR. *Sbornik Vysoke skoly chem. Techn Praze* 5:205–286.
Blattný, C. and Heger, M. 1965. Some remarks on the economical importance of Sarka disease in Czechoslovakia. *Zaštita bilja* 16:85–88, 417–418.
Blattný, C. and Janečková, M. 1978. Reakce dřevitýah indikátorů na infekci zhonbnoro rakovinu crešně. *Sbornik UVTIZ-Ochrana Rostlin* 14:13–15.

Blattný, C. and Janečková, M. 1980. Apricot bare twig and unfruitfulness. *Acta Phytopathol. Acad. Sci. Hung.* 15:383–390.

Blodgett, E.C. 1939. Some obscure peach diseases in Idaho. *Plant Dis. Reptr.* 23:216–218.

Blodgett, E.C., et al. 1964. Virus gummosis of Montmorency cherry. *Plant Dis. Reptr.* 48:277–279.

Blodgett, E.C. and Aichele, D. 1976. Spur cherry. In: *Virus Diseases and Noninfectious Disorders of Stone Fruits in North America. USDA Handb.* 437:252–253.

Blumer, S. 1957. Beobachtungen uber die Weidenblättrigkeit der Fellenbergzwetsche (Prune dwarf). Schweiz. *Zeitschrift fur Obst - und Weinbau* 66:531–537.

Boccardo, G. et al. 1981. Tinangaja and bristle top, coconut diseases of uncertain etiology in Guam, and their relationship to cadang-cadang disease of coconut in the Philippines. *Phytopathology* 71:1104–1107.

Bode, O. and Babović, M. 1969. Untersuchungen über die Eigenschaften auf den Wirtspflanzenkreis des Sharka-virus. *Jber. biol. Budesanst. Landw. Forstwirtsch., Berlin-Dahlem*, 1968:758.

Bodine, E.W. 1934. Occurrence of peach mosaic in Colorado. *Plant Dis. Reptr.* 18:123.

Bodine, E.W. 1936. Peach mosaic disease in Colorado. *Colo. Expt. Sta. Bul.* 421:11.

Bodine, E.W. and Kreutzer, W.A. 1942. Ring spot of apricot. *Phytopathology* 32:333–335.

Bové, J.M. 1988. Citrus exocortis viroid. In: Smith, I.M. et al., Eds. *European Handb. Pl. Dis.*, Blackwell Sci. Publ., London, p. 56.

Bové, J.M. 1988a. Satsuma dwarf virus (SDV). In: Smith, I.M. et al., Eds. *European Handb. Pl. Dis.*, Blackwell Sci. Publ., London, p. 29–30.

Bovey, R., et al. 1980. Maladies à virus et affections similaires de la vigne. *Atlas en Couleurs des Symptoms.*

Boxus, P. 1966a. Étude de deux souches de virus isolées mécaniquement d'arbres fruitièrs. *Parasitica* 22:216–220.

Boxus, P. 1969. Viroses des arbres fruitiers. *Fruit Belge* 331:3–16.

Bradford, F.C. and Joley, L. 1933. Infectious variegation in the apple. *J. Agric. Res.* 46:901–908.

Broadbent, L. 1964. Control of plant virus diseases. In: Corbett, M.K. and Sisler, H.P., Eds. *Plant Virology.* University of Florida Press, Gainesville, p. 330–364.

Bulletin OEPP. 1974. Revue de recherche et de technologie phytosanitaire. *Progrès réalisée dans la connaissance de la sharka*, 4(1):125.

Cambra, M. and Navarro, L. 1988. Citrus variegation virus (CVV). In: Smith, I.M. et al. Eds. *European Handb. Pl. Dis.*, Blackwell Sci. Publ., London, p. 15.

Cameron, H.R. 1954. A bark splitting virus disease of Montmorency sour cherry. (Abstr.) *Phytopathology* 44:484.

Cameron, H.R. and Thomson, M. 1986. Seed transmission of apple mosaic virus in hazelnut. *Acta Hort.* 193:131.

Campbell, A.I. 1962. The effect of some apple viruses on the growth of *Malus* species and varieties. *J. Hort. Sci.* 37:239–246.

Campbell, A.I. 1962a. Apple virus inactivation by heat therapy and tip propagation. *Nature* 195:520.

Campbell, A.I. and Best, M.W. 1964. The effect of heat therapy on several apple viruses. *Rep. Long Ashton Res. Sta.* 1963:65.

Campbell, A.I. 1968. Heat sensitivity of some apple viruses. *Tagungsber. DAL DDR, Berlin* 97:311–316.

Campbell, A.I. and Bould, C. 1970. Virus, fertilizer and rootstock effects on the nutrition of young apple trees. *J. Hort. Sci.* 45:287–294.

Cann, H.J. 1952. Bunchy top disease of bananas "at all time low" in N.S.W. *Agric. Gaz. N.S.W.* 63:73–76.

Canova, A. 1963. Apple ring spot. In: *Virus Diseases of Apples and Pears. Tech. Commun. Bur. Hort. E. Malling* 30:47–49.

Cardin, L. and Marénaud, C. 1975. Relations entre le virus de la mosäique du noisetier et le tulare apple mosaic virus. *Ann. Phytopathol.* 7:159–166.

Cation, D. 1949. Transmission of cherry yellows virus complex through seeds. *Phytopathology* 39:37–40.

Cation, D. 1952. Further studies on transmission of ringspot and cherry yellows viruses through seeds. *Phytopathology* 42:4.

Cech, M., et al. 1980. Strawberry latent ringspot and cucumber green mottle mosaic viruses in apricot with bare twig and unfruitfulness disease syndrome. *Acta Phytopathol. Acad. Sci. Hung.* 15:391–396.

Chairez, R. and Lister, R.M. 1973. Soluble antigens associated with infections with apple chlorotic leaf spot. *Virology* 54:506–514.

Chamberlain, E.E. et al. 1954. Moorpark mottle, a virus disease of apricots in New Zealand. *N.Z. J. Sci. Techn.* 35:471–477.

Chamberlain, E.E. et al. 1959. Two diseases of plum, causing distortion and internal necrosis of fruits. *N.Z. J. Agric. Res.* 2:174–180.
Cheney, P.W. et al. 1967. Graft transmission of the flat apple disease. *Plant Dis. Reptr.* 51:509–510.
Cheney, P.W. et al. 1976. Black canker. In: *Virus Diseases and Noninfections Disorders of Stone Fruits in North America.* ARS, *USDA Agric. Handbook,* N.437:209–210.
Cheney, P.W. and Parish, C.L. 1976. Cherry mottle leaf. In: *Virus Diseases and Noninfectious Disorders of Stone Fruits in North America. USDA Agric. Handb.* 437:216–218.
Childs, J.F.L. 1980. Cachexia-Xyloporosis. In: Bové J. and Vogel R., Eds. *Descr. Illus. Virus Virus-like Dis. Citrus* SETCO-IRFA, Paris.
Christ, E.G. 1960. New peach tree problem. Stem pitting. *Hort. News N.J. Hort. Soc.* 41:4006.
Christoff, A. 1935. Mosaikfleckigkeit, Chlorose und Stippenfleckigkeit bei Apfeln, Birnen und Quitten. *Phytopathol. Z.* 8:285–296.
Christoff, A. 1937. Virusnite bolesti po owotschniti dirweta. *Min zemed. dirsch. sl. zascht. rast.* 29:1–23.
Christoff, A. 1947. Sharkata po slivite. *Izvest. na Kam. na nar. kultura Ser. Biol. Zemed. Lesovod.* 1(60):261–296.
Christoff, A. 1958. Die Obstvirosen in Bulgarien. *Phytopathol. Z.* 31:381–436.
Christoff, A. 1976. Mozaika po oreha. *Gradinarska i lozarska nauka* 12:2, 50–55.
Clark, M.F. and Adams, A.N. 1977. Characteristics of the microplate method of enzyme-linked immunosorbent assay for detection of plant viruses. *J. Gen. Virol.* 34:475–483.
Cochran, L.C. and Smith, C.O. 1938. Asteroid spot, a new virosis of peach. *Phytopathology* 28:278–281.
Cole, A. et al. 1982. Location of *Prunus* necrotic ringspot virus on pollen grains from infected almond and cherry trees. *Phytopathology* 72:1542–1545.
Coman, T. and Cociu, V. 1976. Transmission de la Sharka par le pollen et par les graines. *Bull. d'inform. Sharka* No. 2:15–21.
Conover, R.A. 1964. Mild mosaic and faint mottle ringspot, two papaya virus diseases of minor importance in Florida. *Proc. Fla. State Hort. Soc.* 77:444–448.
Conover, R.A. 1964a. Distortion ringspot, a severe virus disease of papaya in Florida. *Proc. Fla. State Hort. Soc.* 77:440–444.
Cooper, J.I. 1979. Virus Diseases of Trees and Shrubs. Institute of Terrestrial Biology, Cambridge, p. 74.
Cooper, J.I. 1988. Cherry leaf roll virus (CLRV). In: Smith, I.M. et al., Eds. *European Handb. Pl. Dis.,* Blackwell Sci. Publ., London, p. 26.
Cooper, J.I. 1980. The prevalence of cherry leaf roll virus in *Juglans regia* in the United Kingdom. *Acta Phytopathol. Acad. Sci. Hung.* 15:139–145.
Corbett, M.K. 1964. Introduction. In: Corbett, M.K. and Sisler, H.D., Eds. *Plant Virology,* University of Florida Press, Gainesville, p. 1–16.
Corbett, M.K. and Grant, T.J. 1967. Purification of citrus variegation virus. *Phytopathology* 57:137–143.
Cordy, C.B. and MacSwan, I.C. 1961. Some evidence that bark measles is seed borne. *Plant Dis. Reptr.* 45:891.
Corte, A. and Pesante, A. 1963. Rapporti tra il "seccume" del nocciolo gentile delle langhe langhe e una virosi accerata sperimentalmente. *Riv. Pat. Veg.* 3:3–14.
Costa, A.S. and Muller, G.W. 1980. Tristeza control by cross protection. *Plant Dis. Reptr.* 64:538–541.
Crescenzi, A. et al. 1995. Plum pox virus (PPV) in sweet cherry. *Acta Hort.* 386. Fruit Tree Virus Diseases XVI.
Cropley, R. 1960. Pear blister canker: a virus disease. *Ann. Rept. East Malling Res. Sta. 1959,* A:43–159.
Cropley, R. 1961. Cherry leaf roll. *Ann. Appl. Biol.* 49:542–549.
Cropley, R. 1963. Apple star crack. In: *Virus Diseases of Apples and Pears. Techn. Commun. Bur. Hort. E. Malling* 30:47–49.
Cropley, R. 1968. The identification of plum pox (Sharka) virus in England. *Pl. Pathol.* 17:66–70.
Cropley, R. 1968a. Varietal reactions to viruses causing star crack and russet rings on apple fruits. *J. Hort. Sci.* 43:157–165.
Cropley, R. 1968b. Testing cherry rootstocks for resistance to infection by raspberry ringspot and cherry leaf roll viruses. *Rept. East Malling Res. Stn.* 1967:141–143.
Cuozzo, M., et al. 1988. Viral protection in transgenic tobacco plants expressing the cucumber mosaic virus coat protein or its antigens RNA. *Biotech.* 6:549.
Cutting, V.C. and Montgomery, H.B.S. 1973. More and better Fruit with EMLA. *Rept. Malling Res. Stn. and Long Ashton Res. Stn.,* p. 1–29.

Da Câmara Machado, A. et al. 1991. Coat protein-mediated protection against plum pox virus (Abstr.). *XVth Int. Symp. in Virus and Virus Diseases of Temperate Fruit Crops.* 8–13 July, Vienna, Austria.
Dale, J.L. et al. 1982. Avocado sunblotch viroid. *CMI/AAB Descriptions of Plant Viruses*, No. 254.
Dale, J.L. 1994. Banana Virus Research at the Centre for Molecular Biotechnology. Queensland University of Technology, Brisbane, Australia.
Davidson, T.R. and George, J.A. 1965. Effects of necrotic ringspot and sour cherry yellows on the growth and yield of young sour cherry trees. *Can. J. Pl. Sci.* 45:525–535.
Davidson, T.R. and Allen, W.R. 1976. Fruit pitting of sweet cherry. In: *Virus Diseases and Noninfectious Disorders of Stone Fruits in North America. USDA Handb.* 437:227–230.
De Bokx, J.A. 1965. Hosts and electron microscopy of two papaya viruses. *Plant Dis. Reptr.* 49:742–746.
De La Rosa, and Lastra, R. 1983. Purification and partial characterization of papaya ringspot virus. *Phytopathol. Z.* 106:329–336.
Delbos, R. et al. 1983. Virus infection of walnuts in France. *Acta Hort.* 130:123–131.
Delbos, R.P. and Dunez, J. 1988. Apple chlorotic leaf spot virus (ACLSV). In: Smith, I.M. et al., Eds. *European handb. Pl. Dis.*, Blackwell Sci. Publ., London, p. 5–7.
De Leon, D. 1952. Cadang-cadang disease of coconuts in the Philippines. *Foreign Agric.* 16:103. (Absts. in *RAM* (1954) 33:152).
Demetriades, S.D. and Catsimbas, C. 1968. Attaques et nouveaux ennemis signals (Sharka). *Bull. Phytosan. FAO* 16(1):10–11.
De Sequeira, O.A. 1967. Purification and serology of an apple mosaic virus. *Virology* 31;314.
Desjardins, P.R. et al. 1979. Pollen transmission of avocado sunblotch virus experimentally demonstrated. *Calif. Agric.* 33(10):14–15.
Desjardins, P.R. et al. 1980. Infectivity studies of avocado sunblotch disease causal agent possibly a viroid rather than a virus. *Plant Dis.* 64:313–315.
Desjardins, P.E. 1980a. Infectious variegation crinkly leaf. In: Bové, J. and Fogel, R., Eds. *Descr. Illus. Virus Virus-like Dis. Citrus.* SETCO-IRFA, Paris.
Desvignes, J.C. 1971. Observations on some pear virus diseases, vein yellows, quince sooty ringspot, ring pattern mosaic. *Ann. Phytopathol. No. hors série* 295–304.
Desvignes, J.C. and Savio, A. 1971. A virus disease causing necrosis on *Prunus insititia* L. var. Krikom. *Ann. Phytopathol. No. hors série*, 75–84.
Desvignes, J.C. and Bovey, R. 1978. Trois Nouvelles Maladies de Dégénérescence du Pêcher Déectés en France. C.T.I.F.L. Documents No. 59 - IVᵉtrim, 127–133.
Desvignes, J.C. 1980. Different symptoms of the peach latent mosaic. *Acta Phytopathol. Acad. Sci. Hung.* 15:83–190.
Desvignes, J.C. 1981. Epidémiologie de la mosäique latente du pêcher (PLMV), Ier *Colloque sur les Recherches Frutières*, Bordeaux, 263–276.
Desvignes, J.C. and Boye, R. 1988. Different disease caused by the chlorotic leaf spot virus on the fruit trees. (Abstr.) *XIVth Int. Symp. on Fruit Tree Virus Diseases*, 12–18 June, Thessaloniki, Greece.
Desvignes, J.C. 1988. Peach latent mosaic disease. In: Smith, I.M. et al., Eds. *European Handb. Pl. Dis.*, Blackwell Sci. Publ., London, p. 111.
Desvignes, J.C. 1988a. Cherry green ring mottle disease. In: Smith, I.M. et al., Eds. *European Handb. Pl. Dis.*, Blackwell Sci. Publ., London, p. 103–104.
Desvignes, J.C. 1991. Characterization of the chestnut mosaic (Abstr.). *XVth Int. Symp. on Virus and Virus Dis. Temp. Fruit Crops*, 8–13 July, Vienna, Austria.
Desvignes, J.C. et al. 1991. Quick detection of the principal apple and pear virus diseases (Abstr.). *XVth Int. Symp. Virus and Virus Dis. Temp. Fruit Crops*, 8–13 July, Vienna, Austria.
Detienne, G. et al., 1980. Use and versatility of the immunoenzymatic ELISA procedure in the detection of different strains of apple chlorotic leaf spot virus. *Acta Phytopathol. Akad. Sci. Hung.* 15:39–45.
Dias, H.F. 1975. Peach rosette mosaic virus. *CMI/AAB Descriptions of Plant Viruses*, No. 150.
Dias, H.F. and Cation, D. 1976. The characterization of a virus responsible for peach rosette mosaic and grape decline in Michigan. *Can. J. Bot.* 54:1228–1239.
Digiaro, M. et al. 1991. Ilaraviruses in apricot and plum pollen (Abstr.). *XVth Int. Symp. Virus and Virus Dis. Temp. Fruit Crops*, 8–13 July, Vienna, Austria.
Di Terlizzi, B. et al. 1991. Viruses of peach, plum and apricot in Apulia (Abstr.). *XVth Int. Symp. Virus and Virus Dis. Temp. Fruit Crops*, 8–13 July, Vienna, Austria.

Di Terlizzi, B. et al. 1991a. Unusual fruit diseases and disorders in Tirynthos apricot (Abstr.). *XVth Int. Symp. Virus and Virus Dis. Temp. Fruit Crops*, 8–13 July, Vienna, Austria.

Dodds, J.A. and Bar-Joseph, M. 1983. Double stranded RNA from plants infected with Closteroviruses. *Phytopathology* 73:419–423.

Dosba, F. et al. 1991. Breeding for resistance to sharka in apricot trees (Abstr.). *XVth Int. Symp. Virus and Virus Dis. Temp. Fruit Crops*, 8–13 July, Vienna, Austria.

Dounine, S.M. and Minoiu, N. 1968. On serodiagnosis of plum pox virus. *J. Europ. Symp. uber Viruskrankheiten der Obstbäume, Aschersleben, Tagungsberichte* 97:259–267.

Dunez, J. and Marénaud, C. 1966. Identification des constituans d'un complex viral, de Court-Noué du pêcher. *Ann. Epiphytes* 17, H.S., 15–30.

Dunez, J. et al. 1972. Variability of symptoms induced by the apple chlorotic leaf spot (CLSV)—a type CLSV probably responsible for bark split disease of prune trees. *Plant Dis. Reptr.* 56:293–295.

Dunez, J. et al. 1976. Myrobalan latent ringspot virus. *CMI/AAB Descriptions of Plant Viruses*, No. 160.

Dunez, J. et al. 1982. Detection of virus and virus-like diseases of fruit trees. *Acta Hort.* 130:319–330.

Dunez, J. and Šutić, D. 1988. Plum pox virus (PPV). In: Smith, I.M. et al., Eds. *European Handb. Pl. Dis.*, Blackwell Sci. Publ., London, p. 44–46.

Espino, T.M. et al. 1990. Banana bract mosaic, a new disease of banana. II. Isolation and purification for monoclonal antibody production. *The Philippine Agr.* 73(1):61–68.

Fauquet, C.M. and Beachy, R.N. 1993. Status of coat protein-mediated resistance and its potential application for banana viruses. *Proc. Workshop Biotech. Appl. Banana/Plantain Impvt., INIBAP,* Montpellier, France, 1992, p. 69–84.

Fawcett, H.S. 1932. New angles on treatment of bark diseases of citrus. *Calif. Citrograph* 17:406–408.

Fawcett, H.S. 1933. New symptoms of psorosis, indicating a virus disease of citrus. *Phytopathology* 23:930.

Festić, H. 1980. Susceptibility of Stanley plum variety to Sharka virus. *Acta Hort. Fruit Tree Virus Diseases* 94:309–313.

Flock, R.A. and Wallace, J.M. 1955. Transmission of fig mosaic by the eriophyid mite *Aceria ficus*. *Phytopathology* 45:52–54.

Flores, R.C. et al. 1991. Studies on detection, transmission and distribution of peach latent mosaic viroid in peach trees. (Abstr.). *XVth Int. Symp. Virus and Virus Diseases of Temperate Fruit Crops*, 8–13 July, Vienna, Austria.

Fraser, L. 1958. Virus diseases in Australia. *Proc. Linnean Soc. NSW* 83:9–19.

Fridlund, P.R. 1966. Transmission and lack of transmission of seven viruses through *Prunus* seed. *Plant. Dis. Reptr.* 50:902–904.

Fridlund, P.R. and Bottorff, F.O. 1968. Virus gummosis of apricot caused by some prune dwarf virus strains. *Phytopathology* 58:119–120.

Fridlund, P.R. 1976. Pear vein yellows virus symptoms in greenhouse-grown pear cultivars. *Plant Dis. Reptr.* 60:891–894.

Fritzsche, R. and Kegler, H. 1968. Nematoden als Vektoren von Viruskrankheiten der Obstgehölze. Tagungsber. *DAL DDR Berlin* 97:289–295.

Fry, P.R. and Wood, G.A. 1973. Further viruses of *Prunus* in New Zealand. *N.Z. J. Agric. Res.* 16:131–142.

Fuchs, E. 1980. Serological detection of apple chlorotic leaf spot virus (CLSV) and apple stem grooving virus (RGV) in apple trees. *Acta Phytopathol. Acad. Sci. Hung.* 15(1–4):69–73.

Fulton, R.W. 1957. Comparative host ranges of certain mechanically transmitted viruses of *Prunus*. *Phytopathology* 47:215–220.

Fulton, R.W. 1965. A comparison of two viruses associated with plum line pattern and apple mosaic. *Zaštita bilja* 16:427–430.

Fulton, R.W. 1967. Purification and some properties of tobacco streak and Tulare apple mosaic viruses. *Virology* 32:153–162.

Fulton, R.W. 1968. Serology of viruses causing cherry necrotic ringspot, plum line pattern, rose mosaic and apple mosaic. *Phytopathology* 58:635–638.

Fulton, R.W. 1970. Prune dwarf virus. *CMI/AAB Descriptions of Plant Viruses*, No. 19.

Fulton, R.W. 1970a. *Prunus* necrotic ringspot virus. *CMI/AAB Descriptions of Plant Viruses*, No. 5.

Fulton, R.W. 1971. Tulare apple mosaic virus. *CMI/AAB Descriptions of Plant Viruses*, No. 42.

Fulton, R.W. 1972. Apple mosaic virus. *CMI/AAB Descriptions of Plant Viruses*, No. 83.

Fulton, R.W. 1981. Ilarviruses. In: Kurstak, E., Ed. *Hand. Pl. Virus Infec. Comp. Diag.*, Elsevier/North-Holland Biomedical Press, p. 373–413.

Fulton, R.W. 1984. American plum line pattern virus. *CMI/AAB Descriptions of Plant Viruses*, No. 280.

Gallitelli, D. et al. 1985. Olive latent ringspot virus. *CMI/AAB Descriptions of Plant Viruses*, No. 301.

Garnsey, S.M. and Jones, J.W. 1967. Mechanical transmission of exocortis virus with contaminated budding tools. *Pl. Dis. Reptr.* 51:410–413.

Garnsey, S.M. and Gonsalves, D. 1976. Citrus leaf rugose virus. *CMI/AAB Descriptions of Plant Viruses*, No. 164.

Garnsey, S.M. and Timmer, L.W. 1980. Mechanical transmission of citrus ringspot virus isolates from Florida, Texas and California. In: *Proc. 8th Conf. Int. Org. Citrus Virologists*, University of California, Riverside, p. 174–179.

Garnsey, S.M. et al. 1981. Applications of serological indexing to develop control strategies for citrus tristeza virus. In: *Proc. Int. Soc. Citrusculture*.

Garnsey, S.M. et al. 1983. A mild isolate of citrus variegation virus found in Florida citrus. In: *Proc. 9th Conf. Int. Org. Citrus Virol.*, p. 188–195.

George, J.A. and Davidson, T.R. 1963. Pollen transmission of necrotic ring spot and sour cherry yellows viruses from tree to tree. *Can. J. Plant Sci.* 43:273–288.

George, J.A. and Davidson, T.R. 1964. Further evidence of pollen transmission of necrotic ring spot and sour cherry yellows viruses in sour cherry. *Can. J. Plant Sci.* 44:383–384.

George, J.A. and Davidson, T.R. 1966. Virus assay of seeds from selected Montmorency cherry trees. *Can. J. Plant. Sci.* 46:501–505.

German, S. et al. 1991. Genomic organization of apple chlorotic leaf spot closterovirus (ACLSV). *XVth Int. Symp. Virus and Virus Dis. Temp. Fruit Crops*, 8–13 July, Vienna, Austria.

Gilmer, R.M. 1958. Two viruses that induce mosaic of apple. *Phytopathology* 48:432–434.

Gilmer, R.M. and Way, R.D. 1960. Pollen transmission of necrotic ring spot and prune dwarf viruses in sour cherry. *Phytopathology* 50:624–625.

Gilmer, R.M. 1961. A possible history of sour cherry yellows. *Phytopathology* 51:265–266.

Gilmer, R.M. and Kamalsky, L.R. 1962. The incidence of necrotic ringspot and sour cherry yellows viruses in commercial mazzard and mahaleb cherry rootstocks. *Plant Dis. Reptr.* 46:583–585.

Gilmer, R.M. 1967. Apple chlorotic leaf spot and tobacco mosaic viruses in cherry. *Plant Dis. Reptr.* 51:823–826.

Gilmer, M.R. et al. 1976. Prune dwarf. In: *Virus Dis. Noninfec. Disorders Stone Fruits North America*, USDA Handb. 437:179–190.

Gonsalves, D. and Garnsey, S.M. 1975. Nucleic acid components of citrus variegation virus and their activation by coat protein. *Virology* 67:311–318.

Gonsalves, D. and Garnsey, S.M. 1975a. Infectivity of heterologous RNA-protein mixtures from alfalfa mosaic, citrus leaf rugose, citrus variegation, and tobacco streak viruses. *Virology* 67:319–326.

Gonsalves, D. and Garnsey, S.M. 1976. Association of particle size with sedimentation velocity of the nucleoprotein components of citrus variegation and citrus leaf rugose viruses. *Proc. 7th Conf. Int. Org. Citrus Virol.*, p. 109–115.

Gonsalves, D. et al. 1978. Purification and serology of citrus tristeza virus. *Phytopathology* 68:553–559.

Gonsalves, D. and Ishii, M. 1980. Purification and serology of papaya ringspot virus. *Phytopathology* 70:1028–1032.

Gonsalves, D. 1998. Control of papaya ringspot virus: a case study. *Annu. Rev. Phytopathol.* 36:415–437.

Gualaccini, F. 1959. Una virosi nuova del Castagno. *Boll. Staz. Pat. Veget.* 3:67–75.

Grant, T.J. and Corbett, M.K. 1964. Properties of citrus variegation virus. *Phytopathology* 54:946–948.

Gross, H.J. et al. 1982. Nucleotide sequence and secondary structure of citrus exocortis and chrysanthemum stunt viroid. *Eur. J. Biochem.* 121:249–257.

Hadidi, A. et al. 1990. Homology of the agent associated with dapple apple disease to apple scar skin viroid and molecular detection of these viroids. *Phytopathology* 80:263–268.

Hadidi, A. 1991. Detection of fruit crop viroids by reverse transcription polymerase chain reaction (Abstr.), *XVth Int. Symp. Virus and Virus Dis. Temp. Fruit Crops*, 8–13 July, Vienna, Austria.

Hadidi, A. and Parish, C.L. 1991. Apple scar skin and dapple apple viroids are seed borne (Abstr.). *XVth Int. Symp. Virus and Virus Dis. Temp. Fruit Crops*, 8–13 July, Vienna, Austria.

Halk, E.L. and Fulton, R.W. 1978. Stabilization and particle morphology of prune dwarf virus. *Virology* 91:434–443.

Hansen, A.J. and Stace-Smith, R. 1971. Properties of a virus isolated from golden elderberry. *Phytopathology* 61:1222–1229.

Hansen, J.A. and Cheney, W.P. 1976. Cherry twisted leaf. In: *Virus Diseases and Noninfectious Disorders of Stone Fruits in North America. USDA Agr. Handb.* 437:222–225.

Hansen, A.J. et al. 1976. Apricot ring pox. In: *Virus Diseases Noninfectious Disorders of Stone Fruits in North America. USDA Handb.* 437:45–48.

Haseloff, et al. 1982. Viroid RNAs of cadang-cadang disease of coconnuts. *Nature (London)* 299: 316–321.

Hernandez, C. et al. 1991. Evidence supporting a viroid etiology for pear blister canker disease (Abstr.). *XVth Int. Symp. Virus and Virus Dis. Temp. Fruit Crops*, 8–13 July, Vienna, Austria.

Hickey, K.D. and Shear, G.M. 1975. Growth response to latent infection of apple trees on seedling rootstocks. *Acta Hort.* 44:237–244.

Holmes, F.O. 1960. Rose mosaic cured by heat treatments. *Plant Dis. Reptr.* 44:46–47.

Horvath, J. 1979. New artificial hosts and non-hosts of plant viruses and their role in the identification and separation of viruses. XIII. Nepovirus group (CLRV sub-groups): cherry leaf roll virus. *Acta Phytopathol. Acad. Sci. Hung.* 14:319–326.

Howell, W.E. and Mink, G.I. 1991. Rapid biological detection of apple scar skin and dapple apple viroids (Abstr.). *XVth Int. Symp. on Virus and Virus Dis. Temp. Fruit Crops*, 8–13 July, Vienna, Austria.

Hurtt, S. et al. 1991. Early detection of apple scar skin group viroids from imported pear germplasm (Abstr.). *XVth Int. Symp. on Virus and Virus Dis. Temp. Fruit Crops*, 8–13 July, Vienna, Austria.

Hutchins, L.M. 1932. Peach mosaic—a new virus disease. *Science* 76:123.

Hutchins, L.M. et al. 1951. Muir peach dwarf. In: *Virus Dis. Other Disorders Virus-like Symptoms of Stone Fruits in North America. USDA Handb.* 10:62–70.

INIBAP. 1992. Annual Report. International Network Impr. Banana and Plantain, Montpellier, France.

Jang, X. et al. 1991. Nucleotide sequence of apple scar viroid reverse transcribed in host extracts and amplified by the polymerase chain reaction (Abstr.). *XVth Int. Symp. Virus and Virus Dis. Temp. Fruit Crops*, 8–13 July, Vienna, Austria.

Jensen, D.D. 1949. Papaya virus diseases with special reference to papaya ringspot. *Phytopathology* 39:191–211.

Jones, A.L. and Aldwinckle, H.S. 1990. *Compendium of Apple and Pear Diseases*. APS Press, St. Paul, MN, 100 pp.

Jones, A.T. 1976. Serological specificity of isolates of cherry leaf roll virus from different natural hosts. *Poljopr. Znan. Smotra* 39:527–532.

Jones, A.T. and Duncan, G.H. 1980. The distribution of some genetic determinants in two nucleoprotein particles of cherry leaf roll virus. *J. Gen. Virol.* 50:269–273.

Jones, A.T. 1985. Cherry leaf roll virus. *CMI/AAB Descriptions of Plant Viruses*, No. 306.

Jones, A.T. et al. 1988. Recent research on the control of virus infections in raspberry in Britain (Abstr.). *XIVth Int. Symp. Fruit Tree Virus Dis.*, 12–18 June, Thessaloniki, Greece.

Jordović, M. 1955. Kržljavost šljive virozna bolest u našoj zemlji. *Zaštita bilja* 30:61–62, Beograd.

Jordović, M. and Nikšić, M. 1957. Uticaj šarke šljive (*Prunus* virus) na prinose i hemijsko-tehnološka svojstva plodova požegače. *Archiv za poljoprivredne nauke 1*, 10(28):86–95.

Jordović, M. 1963. Investigation of the spread and some factors of spreading plum pox virus disease. *Phytopathol. Mediter.* 2(3):167–170.

Jordović, M. and Janda, L. 1963. Morphological, anatomical and chemical changes on the fruits of some plum varieties infected by virus plum pox disease. *Zaštita bilja* 14(16):653–670.

Jordović, M. and Janda, L. 1963a. Morfološko-anatomske i hemijske promene na plodovima nekih sorti šljive zaraženih virusom šarke. *Zaštita bilja* 76:653–670.

Jordović, M. 1967. Simptomi virozne enacije na lišću breskve. *Zaštita bilja* 18(93–95):143–145, Beograd.

Jordović, M. 1968. Effect of sources of infection on epidemiology of Sharka (plum pox) virus disease. *Tagungsber. DAL/DDR Berlin* 97:301–308.

Jordović, M. et al. 1971. *Prunus spinosa* L. as host of sharka virus. Isolation and some properties of the virus. *Ann. Phytopathol. H.S.* 179–184.

Jordović, M. 1985. Prilog proučavanju šarke šljive i breskve. *Zaštita bilja* 172:155–159.

Josifovič. M. 1937. Mozaik šljive. *Archiv. Minist. poljoprivrede* 4(7):131–143.

Juárez, J. et al. 1988. Recovery of virus-free peach trees from selected clones by shoot-tip grafting in vitro (Abstr.). *XIVth Int. Symp. Fruit Tree Virus Dis.*, 12–18 June, Thessaloniki, Greece.

Juárez, J. et al. 1991. Recovery of virus-free almond plants by shoot-tip grafting in vitro (Abstr.). *XVth Int. Symp. on Virus and Virus Dis. Temp. Fruit Crops*, 8–13 July, Vienna, Austria.

Kalashjan, J.A. et al. 1988. Cytopathology of apple stem grooving virus and the methods of its diagnostics (Abstr.). *XIVth Int. Symp. Fruit Tree Virus Dis.*, 12–18 June, Thessaloniki, Greece, p. 30.

Karayiannis, I. 1991. Susceptibility of apricot cultivars to plum pox virus natural infection. *XVth Int. Symp. Virus and Virus Dis. Temp. Crops* (Abstr.), 8–13 July, Vienna, Austria.

Karle, H.P. and Nyland, G. 1959. Yellow bud mosaic virus in almond. *Plant Dis. Reptr.* 43:520–521.

Kassanis, B. and Šutić, D. 1965. Some results of recent investigations of sharka (plum pox) virus disease. *Zaštita bilja* 16:335–340.

Katwijk, van W. and Meijneke, C.A.R. 1963. Apple rough skin. In: *Virus Diseases of Apples and Pears. Techn. Commun. Bur. Hort. E. Malling* 30:43–44.

Keane, F.W.L. and Welsh, M.F. 1960. Transmissible corky pit of Flemish Beauty pear. *Plant Dis. Reptr.* 44:636–638.

Kegler, H. 1962. Chenopodium-Arten als Test-und Wirtspflanzen für Kirschenviren. *Phytopathol. Z.* 45:248–259.

Kegler, H. 1964. Latente Steinobstviren. *Mitt. Biol. Bundesanst. Land-Forstwirtsch. Berlin-Dahlen*, 115:128–136.

Kegler, H. 1965. Bark split and decline in "Beurre Hardy" pear trees. *Zaštita bilja* 16:311–316, Beograd.

Kegler, H. 1968. Obstgehölze und Beerenobst. In: Klinkowski, M. et al., Eds. *Pflanzliche Virologie*. Vol. II/1, Akademie-Verlag, Berlin, 208–379.

Kegler, H. et al. 1969. Nachweis des Tabaknekrosenvirus (Tobacco necrosis virus) in Obstgehölzen. *Phytopathol. Z.* 65:21–42.

Kegler, H. and Schade, C. 1971. Plum pox virus. *CMI/AAB Description of Plant Viruses*, No. 70.

Kegler, H. et al. 1972. Einfluss von Viren auf Ertrag und Wuchs der Sauerkirschensorte "Schattenmorelle." *Arch. Gartenbau* 20L479–487.

Kegler, H. 1977. Obstgehölze. In: Klinkowski, M. et al., Eds. *Pflanzliche Virologie*, Vol. 3, 3rd edition, Akademie-Verlag, Berlin, p. 139–315.

Kegler, H. 1977a. Die Tristeza-Krankheit bei Citrus. In: Klinkowski, M. et al., Eds. *Pflanzliche Virologie*, Vol. 3, 3rd edition, Akademie-Verlag, Berlin, p. 301–303.

Kegler, H. 1977b. Das Apfelmosaik. In: Klinkowski, M. et al., Eds. *Pflanzliche Virologie*, Vol. 3, 3rd edition, Akademie-Verlag, Berlin, p. 140–142.

Kegler, H. et al. 1977. Isolierung und Identifizierung eines Virus von Birnen mit Steinfrüchtigkeit (pear stony pit). *Arch. Phytopathol. Pflanzenschutz.*, Berlin 13:297–310.

Kegler, H. et al. 1979. Untersuchungen uber Wechselbeziehungen verschiedener Kern - und Steinobstvirosen. *Archiv. fur Gartenbau* 27:325–336.

Kegler, H. and Schimanski, H.H. 1988. Pear stony pit disease/pear vein yellows disease. In: Smith, I.M. et al., Eds. *European Handb. Pl. Dis.*, Blackwell Sci. Publ., London, p. 111–112.

Kegler, H. and Gruntzig, Maria. 1991. The hypersensitivity of the plum hybrid K4 and its progenies to plum pox virus (Abstr.). *XVth Int. Symp. Virus and Virus Dis. Temp. Fruit Crops*, 8–13 July, Vienna, Austria.

Kerlan, C. and Dunez, J. 1976. Some properties of plum pox virus and its nucleic acid protein components. *Acta Hort.* 67:185–191.

Kerlan, C. and Dunez, J. 1979. Différenciation biologiqie et sérologique de souches de virus de la sharka. *Ann. Phytopath.* 11:241–250.

Kerlan, C. et al. 1980. Preliminary studies on the antagonism between strains of plum pox virus. *Acta Phytopathol. Acad. Sci. Hung.* 15:57–67.

Kerlan, C. et al. 1981. Immunosorbent electron microscopy for detecting apple chlorotic leaf spot and plum pox viruses. *Phytopathology* 71:400–404.

Kerlan, C. et al. 1986. Further information about a new virus isolated from peach in Hungary, *Acta Hort.* 193.

Kirkpatrick, H.C. and Lindner, R.C. 1964. Recovery of tobacco mosaic virus from apple. *Plant. Dis. Reptr.* 48:855–857.

Kirkpatrick, H.C. et al. 1965. Purification of a second mechanically transmissible virus latent in apple. *Phytopathology* 55:386–290.

Kirkpatrick, H.C. et al. 1967. A selective host range for certain mechanically transmitted stone fruit viruses. *Plant Dis. Reptr.* 51:786–790.

Kishi, K. et al. 1972. Studies on pear necrotic spot. *Bull. Hort. Res. Stn. Japan (Hiratsuka)* A-11:139–150.

Kishi, K. et al. 1973. Studies on the virus diseases of stone fruits. IX. A new virus disease, pear enation. *Ann. Phytopathol. Soc. Japan* 39:297–304.

Kishi, K. et al. 1973a. Studies on virus diseases of stone fruits. IV. New virus diseases of peach, yellow mosaic, oil blotch and star mosaic. *Bull. Hort. Res. Stn., Japan* A–12:197–208.

Kishi, K. et al. 1976. Pear necrotic spot, a new virus disease in Japan. *Acta Hort.* 67:269–273.

Kleinhempel, H. et al. 1980. Investigation of carnation ringspot virus in fruit trees. *Acta Phytopathol. Acad. Sci. Hung.* 15:107–111.

Klos, E.J. and Parker, K.G. 1960. Yields of sour cherry affected by ringspot and yellows viruses. *Phytopathology* 50:412–415.

Klos, E.J. 1976. Rosette mosaic. In: *Virus Diseases and Noninfectious Disorders of Stone Fruits in North America. USDA Handb.* 347:135–138.

Klotz, J.L. 1961. *Color Handbook of Citrus Diseases.* Citrus Exp. Sta. Riverside, CA. p. 74.

Klotz, L.J. 1980. Concave gum blind pocket. In: Bove, J. and Vogel, R., Eds. *Descr. Illus. Virus Virus-like Dis. Citrus.* SETCO-IRFA, Paris.

Koganezawa, H. et al. 1982. Viroid-like RNA associated with apple scar skin (or dapple apple). *Acta Hort.* 130:193–197.

Kölber, M. and Németh, M. 1983. Screening of fruit seed lots with ELISA technique. *Proc. Int. Conf. Integr. Plant Prot.*, Budapest, 4th-9th July, 2:50–55.

Krczal, H. and Kunze, L. 1972. Untersuchungen zur Übertragung des Scharkavirus durch Blattläuse. *Mitt. Biol. Bundesanst. Land - u. Forstwirtsch., Berlin-Dahlem* 144:71–83.

Kristensen, H.R. and Jörgensen, H.A. 1957. Plantesygdomme i Danmark 1955. Arsover sigt samlet ved Statens plante-pathologiske Forsøg., *T. Planteavl* 61:617.

Kristensen, H.R. and Thomsen, A. 1962. Apple mosaic virus host plants and strains, *Proc. 5th Eur. Symp. Fruit Tree Virus Diseases,* 21–26, Bologna.

Kristensen, H.R. 1963. Apple green crinckle. In: *Virus Diseases of Apples and Pears. Techn. Commun. Bur. Hort. E. Malling* 30:33–34.

Kristensen, H.R. 1963a. Stony pit of pear. In: *Virus Diseases of Apples and Pears. Techn. Commun. Bur. Hort. E. Malling* 30:99–101.

Kröll, J. 1973. Natural and experimental host plants of the Sharka virus of plum trees. *Plant Virology* 397–402, Prague.

Kunze, L. 1969. Der Einfluss der Stecklenberger Krankheit auf den Ertrag von Sauerkirschen. *Erverbsobstbau* 11:1–3.

Kunze, L. and Krczal, H. 1971. Transmission of Sharka virus by aphids. *Ann. Phytopathol. No. hors-serie* 255–260.

Kunze, L. 1988. Prune dwarf virus (PDV). In: Smith, I.M. et al., Eds. *European Handb. Pl. Dis.*, Blackwell Sci. Publ., London, p. 15–17.

Laimer da Câmara Machado, M. et al. 1991. Transformation and regeneration of plants of *Prunus armeniaca* with the coat protein gene of plum pox virus (Abstr.). *XVth Int. Symp. Virus Virus Dis. Temp. Fruit Crops,* 8–13 July, Vienna, Austria.

Laird, E.F. et al. 1969. Attempts to transmit citrus exocortis virus by insects. *Plant Dis. Reptr.* 53:850–851.

Lana, A.F. et al. 1983. Association of tobacco ringspot virus with a union incompatibility of apple. *Phytopathol. Z.* 106:141–148.

Lazar, A.C. and Fridlund, P.R. 1967. Shiroplum: a rapid greenhouse indicator for the peach wart virus. *Plant Dis. Reptr.* 51:831–833.

Leclant, F. 1973. Ecological aspects of Sharka (plum pox) virus transmission in southern France: Additional aphid vectors. *Ann. Phytopathol.* 54:431–439.

Lister, R.M. et al. 1965. Some sap-transmissible viruses from apple. *Phytopathology* 55:859–870.

Lister, R.M. 1970. Apple chlorotic leaf spot virus. *CMI/AAB Descriptions of Plant Viruses,* No. 30.

Lister, R.M. 1970a. Apple stem grooving virus. *CMI/AAB Descriptions of Plant Viruses,* No. 31.

Lister, R.M. and Bar-Joseph, M. 1981. Closteroviruses. In: Kurstak, E., Ed., *Handb. Pl. Virus Infec. Comp. Diag.*, Elsevier/North-Holland Biomedical Press, p. 800–844.

Loesch, L.S. and Fulton, R.W. 1975. Prunus necrotic ringspot virus as a multicomponent system. *Virology* 68:71–78.

Loesch-Fries, L.S. et al. 1987. Expression of alfalfa mosaic virus RNA 4 in transgenic plants confers resistance. *EMBO J.* 6:1845.

Lott, B. and Keane, F.W.L. 1960. The host range of the virus of twisted leaf of cherry. *Plant Dis. Reptr.* 44:240–242.

Luckwill, L.C. and Campbell, A.L. 1959. *Malus platycarpa* as an apple virus indicator. *J. Hort. Sci.* 34:248–254.

Luckwill, L.C. and Campbell, A.L. 1963. Apple chlorotic leaf spot. In: *Virus Diseases of Apples and Pears. Tech. Commun. Bur. Hort. E. Malling* 30:7–9.

Luckwill, L.C. and Campbell, A.L. 1963a. Platycarpa scaby bark. In: *Virus Diseases of Apples and Pears. Techn. Commun. Bur. Hort. E. Malling* 30:61.

Luckwill, L.C. and Campbell, A.L. 1963b. Platycarpa dwarf. in: *Virus Diseases of Apples and Pears. Techn. Commun. Bur. Hort. E. Malling* 30:59.

Macovei, A. 1970. Date privid influenta viruslui plum pox asupra morfologiei si germinatiei polenuli di prun. *St. si Cerc. de Biologie, S.B. Tom* 22(4):325–328.

Macovei, A. 1970a. The results concerning influence of plum pox virus on plum pollen morphology and germination. *Stud. si cere Biol. Ser. Bot.* 22(4):325–328.

Macovei, A. et al. 1971. New developments of the study in influence of *Prunus spinosa* L. on the appearance of Sharka symptoms. *Ann. Phytopathol. H.S.*, 171–178.

MAFF (Ministry of Agriculture, Fisheries and Food). 1978. Plum pox disease. Advisory Leaflet 611.

Magee, J.P. 1940. Transmission studies on the banana bunchy-top virus. *J. Austr. Inst. Agric. Sci.* 6:109–110.

Magnaye, L.V. and Espino, R.R.C. 1990. Note: Banana bract mosaic, a new disease of banana. I. Symptomatology. *The Philippine Agr.* 73(1):55–59.

Mainou, A.C. and Syrginaidis, G.D. 1991. Evaluation of peach and nectarine varieties according to their resistance to Sharka (plum pox) virus (Abstr.). *XVth Int. Symp. Virus and Virus Dis. Temp. Fruit Crops*, 8–13 July, Vienna, Austria.

Marénaud, C. 1971. Contribution de l'Étude d'un Virus du Type Chlorotic Leaf Spot des Arbres Frutiers à Noyau. Thèse (Docteur-Ingenieur) à l'Université de Bordeaux I, p. 1–176.

Marénaud, C. 1972. Observations relatives au déterminisme génétique de résistance aux virus chez les *Prunus*. *C.R. 3ᵉ Congr. Union Phytopathol. Medit.*, Oeiras, 1972, 475–476.

Marénaud, C. and Germain, E. 1975. La mosaïque du noisetier (Hazelnut mosaic). *Ann. Phytopathol.* 7:133–146.

Marénaud, C. et al. 1976. Identification and comparison of different strains of apple chlorotic leaf spot virus and possibilities of interactions. *Acta Hort.* 67:219–226.

Marénaud, C. et al. 1976. Observations sur la diffusion et la détection du virus de la Sharka. *Phytoma* 280:20–24.

Marina, B. and Avgelis, A. 1991. Epirus cherry virus: a new progress report (Abstr.). *XVth Int. Symp. Virus and Virus Dis. Temp. Fruit Crops*, 8–13 July, Vienna, Austria.

Martelli, G.P. 1981. Le virosi dell'olivo esistono? *Inform. Fitopatol.* 31(L2):97–100.

Martelli, G.P. 1988. Olive latent virus 1 and olive latent virus 2. In: Smith, I.M. et al., Eds. *European Handb. Pl. Dis.*, Blackwell Sci. Publ., London, p. 96.

McGrum, C.R. et al. 1963. Dapple apple. In: *Virus Diseases of Apples and Pears. Techn. Commun. Bur. Hort. E. Malling* 30:29–30.

Meijneke, C.A.R. et al. 1963. The economic importance of virus diseases of apples and pears. In: *Techn. Commun. Bur. Hort. Plant Crops* 30:1–4.

Meijneke, C.A.R. et al. 1975. Growth, yield and fruit quality of virus-infected and virus-free Golden Delicious apple trees. *Acta Hort.* 44:209–212.

Milbrath, J.A. 1957. The relationship of peach ringspot virus to sour cherry yellows, prune dwarf and peach stunt. *Phytopathology* 47:529.

Milbrath, J.A. and Reynolds, J.E. 1961. Tomato ringspot virus isolated from "Eola" rasp leaf of cherry in Oregon. *Plant Dis. Reptr.* 45:520–521.

Milbrath, J.A. 1966. Severe fruit necrosis of sour cherry caused by strains of green ring mottle virus. *Plant Dis. Reptr.* 50:59–62.

Miller, P.W., et al. 1958: Blackline and Rootrots of Persian Walnuts in Oregon. Agric. Exp. Sta. Oregon State College, Corvallis, Oregon, Misc. Paper 55:31.

Millikan, A.A. et al. 1964. A wood-pitting symptom in pear; its occurrence, distribution and association with certain pear virus diseases (Abstr.). *Phytopathology* 54:1435.

Millikan, F.D. 1963. Apple scar skin. In: *Virus Diseases of Apples and Pears. Techn. Commun. Bur. Hort. E. Malling* 30:51–52.

Mink, G.I. et al. 1963. Properties of Tulare apple mosaic virus. *Phytopathology* 53:973–978.

Mink, G.I. and Aichele, M.D. 1984. Detection of *Prunus* necrotic ringspot and prune dwarf viruses in *Prunus* seed and seedlings by enzyme-linked immunosorbent assay. *Plant Dis.* 68:378–381.
Minoiu, N. 1973. Cercetari asupra virozelor parului. *An. I.C.P.P.*, IX, 57-72, Bucuresti.
Mircetich, S.M. et al. 1970. Peach stem pitting: transmission and natural spread. *Phytopathology* 60:1320–1334.
Mircetich, S.M. et al. 1971. Relative differential efficiency of buds and root chips in transmitting the causal agent of peach stem pitting and incidence of necrotic ringspot virus pitted trees. *Phytopathology* 61:1270–1276.
Mircetich, S. M. and Civerolo, E.L. 1972. Relationship between stem pitting in peach and other *Prunus* species. *Phytopathology* 62:1294–1302.
Mircetich, S.M. and Moller, J.W. 1977. *Prunus* stem pitting. *EPPO Bull.* 7(1):29–36.
Mircetich, S.M. et al. 1980. Natural spread, graft transmission and possible etiology of walnut blackline disease. *Phytopathology* 70:962–968.
Mircetich, S.M. et al. 1982. Seed and pollen transmission of cherry leaf roll virus (CLRV-W) the causal agent of the blackline (BL) of English walnut trees. (Abstr.) *Phytopathology* 72:988.
Mircetich, S.M. and Rowhani, A. 1984. The relationship of cherry leaf roll virus and the blackline disease of English walnut trees. *Phytopathology* 74:423–428.
Mirkov, T.E. et al. 1997. Transgenic virus resistant surgarcane. Proc. ISSCT 5th Pathol. and 2nd Mol. Biol. Workshop.
Mischke, W. and Schuch, K. 1962. Untersuchungen über eine viröse Triebstauchung des Pfirsichs. *Phytopathol. Z.* 44:4006.
Mischke, W. and Bercks, R. 1965. Kurze Mitteilung über ein Vorkommen des Tomaten Schwarzringflecken-Virus (tomato black ring virus) an Mandelbäumen (*Prunus amygdalus* Batsch.). *Nachrichtenbl. dt. Pflanzenschutzd. (Braunschweig)* 17:186–187.
Moore, D.J. and Keitt, G.W. 1944. Host range studies of virus disease necrotic ringspot and yellows of sour cherry. (Abstr.) *Phytopathology* 34:1009.
Moreno, P. and Navarro, L. 1988. Citrus ringspot disease. In: Smith, I.M. et al., Eds. *European Handb. Pl. Dis.*, Blackwell Sci. Publ., London, p. 108.
Morvan, G. 1965. La chute des bourgeons du poirier une nouvelle maladie à virus. *Ann. Epiphyt.* 16:153–162.
Morvan, G. and Castelain, C. 1967. Une affection virale distincte de l'enroulement chlorotique: la rosette d'abricotier var. Luizet. *Ann. Epiphyt.* 18:205–216.
Müller, G.W. and Costa, A.S. 1968. Further evidence on protective interference in *Citrus tristeza*. In: Childs, J.F.L., Ed., *Proc. 4th Cont. Int. Organ. Citrus Virol.*, University of Florida Press, Gainesville, p. 71–82.
Murant, A.F. 1981. Nepoviruses. In: Kurstak, E., Ed. *Handb. Pl. Virus Infec. Comp. Diag.*, Elsevier/North-Holland Biomedical Press, p. 197–238.
Navarro, L. et al. 1982. Shoot tip grafting in vitro for elimination of viruses in peach plants. *Acta Hort.* 130:185–192.
Navarro, L. and Pino. 1988. Citrus vein enation disease (Wallace and Drake, 1953) and *Citrus* woody gall disease (Frazer, 1958). In: Smith, I.M. et al., Eds., *European Handb. Pl. Dis.*, Blackwell Sci. Publ., London, p. 108–109.
Nemchinov, L. and Hadidi, A. 1996. Characterization of the sour cherry strain of plum pox virus. Phytopathology 86: 575–580.
Németh, M. and Kegler, H. 1960. Untersuchungen über das Vorkommen von Kirschenvirosen in Ungarn. *Nachrichtenbl. Dt. Pflanzenschutzd. (Berlin)* 14:110–143.
Németh, M. 1962. Obstvirosen und ihre Bekämpfung in der Ungarischen Volksrepublik. *Arch. Gartenbau (Berlin)* 10(2):99–112.
Németh, M. 1963. Field and greenhouse experiments with plum pox virus. *Phytopath. Médit.* 2(3):162–166.
Németh, M. 1972. Interferencia vizsgálotok a csonthéjas gyümöles fák gyürüsfoltosság (ringspot) virus aival. *Növényvédelem* 8(2):64–71.
Németh, M. 1972a. A gyürüsfoltosság vírusok faiskolai kártétele. *Kertgazdaság* 4(3):31–38.
Németh, M. 1980. Die Wirkung in Ungaren vorkommender Pfirsichviren auf das Wachstum von GF 305 - Pfirsichsämlingen beim Komplexinfektionen. Einfluss der Viren auf den Verwachsungerfolg der Okulationen und auf die vegetative Entwicklung der Bäume bei zwei Pfiersichsorten in der Baumschule. *Tagungsber. DAL DDR, Berlin* 184, 403–408.
Németh, M., et al. 1982. A dió cherry leaf roll virus fertozöttsége. I. A virus azonositása és elofordulása Magyarorzágon (cherry leaf roll virus in *Juglans regia*. I. Identification and distribution of the virus in Hungary). *Növényvédelem* 18(1):1–10.

Németh, M. and Köbler, M. 1983. Additional evidence on seed transmission of plum pox virus in apricot, peach and plum proved by ELISA. *Acta Hort.* 130:293–300.

Németh, M., et al. 1983. A new virus isolated from peach in Hungary. *Acta Hort.* 130:33–45.

Németh, M. 1986. *Virus Mycoplasma and Rickettsia Disease of Fruit Trees.* Académiai Kiadó, Budapest.

Nour-Eldin, F. 1980. Gummy bark. In: Bové, J. and Vogel, R., Eds. *Descr. Illus. Virus Virus-like Dis. Citrus*, SETCO-IRFA, Paris.

Novák, J.B. and Lanzová, J. 1981. Virus svinutky třešně z ořešáku vlašského (J. regia) v Čechoslovensku. *Sbor. UVTIZ - Ochr. Rostl.* 17(1):1–18.

Nyland, G. 1959. Hot water treatment of Lambert cherry bud sticks infected with necrotic rusty mottle virus. *Phytopathology* 49:157–158.

Nyland, G. and Lowe, S.K. 1964. The relation of cherry rugose mosaic and almond calico viruses to *Prunus* ringspot virus. (Abstr.) *Phytopathology* 54:1435–1436.

Nyland, G. 1976. Cherry rasp leaf. In: *Virus Dis. Noninfec. Disorders of Stone Fruits in North America. USDA Handb.* 437:219–221.

Nyland, G. 1976a. Almond virus bud failure. In: *Virus Dis. Noninfec. Disorders of Stone Fruits in North America. USDA Handb.* 437:33–42.

Ogawa, J.M. et al. 1995. *Compendium of Stone Fruit Diseases.* APS Press, St. Paul, MN, 90 pp.

OEPP 1974. *Eur. Pl. Prot. Organ. Bull.* 101.

Omčikus, Č. and Blagojević, M. 1957. Neka zapažanja o simptomima mozaika na raznim sortama smokve u Hercegovini. *Zaštita bilja (Plant Protection)* 44:27–34.

Palukaitis, P., et al. 1981. Rapid indexing of the sun blotch disease of avocado using a complementary DNA probe to avocado sun blotch viroid. *Ann. Appl. Biol.* 98:439–449.

Parish, C.L. 1977. A comparison between the causal agents of cherry rasp leaf and flat apple disease. *Acta Hort.* 67:199–202.

Parish, C.L. and Raese, J.T. 1986. Freckle pit and/or an alfalfa greening-like disorder of "Anjou" pears: possible transmission and natural spread. *Acta Hort.* 193.

Parker, K.G., et al. 1976. Green ring mottle. In: *Virus Dis. Noninfec. Disorders of Stone Fruits in North America. USDA Handb.* 437:193–199.

Paulechová, K. 1968. Untersuchengen zur Identifizierung des Bandmosaiks der Sauerkirsche. *Tagungsber. DAL DDR Berlin* 97:165–172.

Paulesen, A.Q. and Fulton, R.W. 1968. Hosts and properties of a plum line pattern virus. *Phytopathology* 58:766–772.

Pejkić, B. and Šutić, D. 1971. Morfologie et germination due pollen du prunier (*Prunus domestica* var. Požegača) atteint par le virus de la Sharka. *Ann. Phytpathol.* H.S. 235–244.

Pena-Iglesias, A. and Ayuso Conzalez, P. 1975. Preliminary identification of the viruses producing Spanish apricot pseudopox (Viruela) and apricot mosaic diseases. *Acta Hort.* 44:255–266.

Perišić, M. and Čuturilo, S. 1960. Uništavanje stabala šljive Požegače obolelih od "šarke" primenom arboricida. *Zaštita bilja Beograd* 62:23–33.

Pfeilstetter, E. et al. 1991. Occurrence of petunia asteroid mosaic (PAMV) and carnation Italian ringspot (CIRV) viruses in cherry orchards in northern Bavaria (Abstr.). *XVth Int. Symp. Virus and Virus Dis. Temp. Fruit Crops*, 8–14 July, Vienna, Austria.

Pine, T.S. 1964. Influence of necrotic ringspot virus on growth and yield of peach trees. *Phytopathology* 54:604–605.

Pine, T.S. and Šutić, D. 1968. Šarka (plum pox) disease in reference to virus diseases of *Prunus* in the United States. *Plant Dis. Reptr.* 52:250–252.

Pine, T.S. 1976. Peach mosaic. In: *Virus Dis. Noninfec. Disorders Stone Fruits North America. USDA Handb.* 437:61–70.

Plavšić, B. and Milišić, D. 1980. Intracellular changes in trees infected with fig mosaic. *Acta Hort., Virus diseases of ornamentals.* 110:281–286.

Pleše, N. et al. 1969. Novi domadari i intracelularne inkluzije virusa šarke. *Zaštita bilja* 104:143–150.

Pleše, N. 1972. Virus klorotične prstenaste pjegavošti trešnje na bademu. *Acta Bot. Crost.* 31:21–27.

Pleše, N. et al. 1975. Pathological anatomy of trees affected with apple stem grooving virus. *Phytopathol. Z.* 82:315–325.

Ploetz, R.C. et al. 1994. *Compendium of Tropical Fruit Diseases.* APS Press, St. Paul, MN. 88 pp.

Poggi-Pollini, C. et al. 1983. Sulla frequenza dei virus che si transmettono per polline in semenziali di alcune drupacee. *Inf. Fitopat.* 33(4):75–78.

Polak, J. 1988. Apple stem grooving. In: Smith, I.M. et al., Eds. *European Handbook of Plant Disease*, Blackwell Sci. Publ., London, p. 89.

Posnette, A.F. 1951. Virus diseases of sweet cherries. *Annu. Rep. East Malling Res. Sta. 1950*, A 34:209–210.

Posnette, A.F. 1953. Virus diseases of plums. *Annu. Rep. East Malling Res. Sta. 1952*, 131.

Posnette, A.F. and Cropley, R. 1956. Apple mosaic virus. Host reaction and strain interference. *J. Hort. Sci.* 31:119–133.

Posnette, A.F. and Cropley, R. 1956a. Virus diseases of cherry trees in England. II. The supression of growth caused by some viruses. *J. Hort. Sci.* 31:298–302.

Posnette, A.F. and Ellenberger, C.E. 1957. Bark split—a virus disease of plums. *Ann. Appl. Biol.* 45:573–579.

Posnette, A.F. and Cropley, R. 1958. Quince indicators for pear viruses. *J. Hort. Sci.* 33:289–291.

Posnette, A.F. and Cropley, R. 1959. Transmission of virus causing star crack of apples. *J. Hort. Sci.* 34:126–129.

Posnette, A.F. and Cropley, R. 1959a. The reduction in yield caused by apple mosaic. *Annu. Rep. East Malling Res. Sta. 1958*, 89–90.

Posnette, A.F. and Cropley, R. 1960. Virus diseases of cherry trees in England. III. Crop reduction caused by some viruses. *Annu. Rep. East Malling Sta. 1959*, 92–95.

Posnette, A.F. et al. 1962. Heat inactivation of some apple and pear viruses. *Annu. Rep. E. Malling Res. Sta. 1961*, 94–96.

Posnette, A.F. 1963. Virus diseases of apples and pears. *Commun. Bur. Hort., E. Malling, Techn. Comm.* No. 30.

Posnette, A.F. 1963a. Apple mosaic. In: Virus Diseases of Apples and Pears. *Commun. Bur. Hort., E. Malling, Techn. Comm.* No. 30, 19–21.

Posnette, A.F. and Cropley, R. 1963. Apple stem pitting. In: Virus Diseases of Apples and Pears. *Commun. Bur. Hort. E. Malling, Techn. Comm.* 30:77.

Posnette, F.A. and Cropley, R. 1963a. Quince sooty ringspot. In: Virus Diseases of apples and pears. *Commun. Bur. Hort. E. Malling.* 30:111–112.

Posnette, A.F. and Ellenberger, C.E. 1963. A disease resembling plum pox in England. *Plant Pathol.* 12:115–117.

Posnette, A.F. and Cropley, R. 1964. Necrotic rusty mottle virus disease of sweet cherry in Britain. *Plant Pathol.* 13:20–22.

Posnette, A.F. and Cropley, R. 1965. The growth of apple trees with and without latent virus infection. *Annu. Rep. East Malling Res. Stn. 1964*, 150–151.

Posnette, A.F. et al. 1968. The incidence of virus diseases in England sweet cherry orchards and their effect on yield. *Ann. Appl. Biol.* 61:351–360.

Posnette, A.F. and Cropley, R. 1970. Decline and other effects of five virus infections on three varieties of plum (*Prunus domestica* L.). *Ann. Appl. Biol.* 65:111–114.

Post, J.J. et al. 1979. Verchillen in virusfrije en niet-virusfrije. *Conf. Doyenné du Comice. Fruitt.* 70:462–464.

Price, W.C. 1970. Citrus tristeza virus. *CMI/AAB Descriptions of Plant Viruses*, No. 33.

Proeseler, G. 1958. Übertragunsversuche mit dem latenten Prunus Virus und der Gallmilbe Vasates fockeui Nol. *Phytopathol. Z.*, 63:1–9.

Proeseler, G. 1968. Übertragung eines latenten *Prunus*—Virus durch die Gallmilbe Vasates fockeui Nal. *Tagunsber. Dt. Akad. Landirtschaftsuiss.* No. 97, 297–300.

Proeseler, G. 1972. Beziehungen zwishen Virus, Vektor und Wirtpflanze am Beispiel des Fiegen-Mosaik Virus und Aceria ficus (Eriophyidae). *Acta Phytopathol. Acad. Sci. Hung.* 7:297–300.

Purcifull, D.E. and Hiebert, E. 1971. Papaya mosaic virus. *CMI/AAB Descriptions of Plant Viruses*, No. 56.

Purcifull, D.E., and Hiebert, E. 1979. Serological distinction of watermelon mosaic virus isolates. *Phytopathology* 69:112–116.

Purcifull, D.E. and Edwardson, J.R. 1981. Potexviruses. In: Kurstak, E., Ed., *Handb. Pl. Virus Infec. Comp. Diag.*, Elsevier/North-Holland Biomedical Press, p. 627–692.

Purcifull, D. et al. 1984. Papaya ringspot virus. *CMI/AAB Descriptions of Plant Viruses*, No. 292.

Quacquarelli, A. and Savino, V. 1977. Cherry leaf roll virus in walnut. II. Distribution in Apulia and transmission through seed. *Phytopathol. Medit.* 16(2–3):154–156.

Ragozzino, A. et al. 1991. Apricot ring-pox-like disease and related disorders of apricot fruits in Gampania (Abstr.). *XVth Int. Symp. Virus and Virus Dis. Temp. Fruit Crops*, 8–13 July, Vienna, Austria.

Ragetli, H.W.J. et al. 1982. Isolation and properties of filamentous virus-like particles associated with little cherry diseases in *Prunus avium*. *Can. J. Bot.* 60:1235–1248.

Raine, J. et al. 1979. Hexagonal tubules in phloem cells of little cherry infected trees. *J. Ultrastr. Res.* 67:109–116.

Ramsdell, D.C. et al. 1979. A comparison between enzyme-linked immuno-sorbent assay (ELISA) and *Chenopodium quinoa* for detection of peach rosette mosaic virus in "Concord" grapevines. *Plant Dis. Reptr.* 63:74–78.

Randles, J.W. and Imperial, J.S. 1984. Coconut Cadang-Cadang Viroid. *CMI/AAB Descriptions of Plant Viruses*, No. 287.

Randles, J.W. et al. 1977. Transmission of the RNA species associated with cadang-cadang of coconut palm, and the insensitivity of the disease to antibiotics. *Phytopathology* 67:1211–1216.

Ranković, M. 1970. Proučavanje i identifikacija virusa prouzrokovača enacija na lišću breskve. *Zaštita bilja* 110–111:335–356.

Ranković, M. 1974. Prečišćavanje virusa šarke šljive i neke njegove osobine. Doktorska disrtacija, Poljoprivredni fakultet, Zemun.

Ranković, M. 1976. Protein fragments of Sharka virus and its use in agar gel-diffusion test. *III Kongres Mikrobiologa Jugoslavije, Bled*, 178–179.

Ranković, M. 1976a. The most frequent viruses among various plum cultivars in Yugoslavia. *Mitt. Biol. Bundesanst. Land-und Forstwirtsch. Berlin-Dahlem, H.* 170:51–55.

Ranković, M. and Šutić, D. 1980. Investigation of peach as a host of Sharka virus. *Acta Phytopathol. Acad. Sci. Hung.* 15(1–4):201–205.

Ranković, M. 1980. Use of *Prunus tomentosa* for the detection and differentiation of Sharka and other viruses of plum. *Acta Hort. Fruit Tree Virus Diseases* 94:303–308.

Ranković, M. 1981. The degree of susceptibility of some plum cultivars and hybrids to Sharka (plum pox) virus (manuscript).

Ranković, M. 1983. The degree of sensitivity of some plum cultivars and hybrids to Sharka (plum pox) virus disease. *Acta Hort.* 130:93–98.

Ranković, M. and Šutić, D. 1986. Resistance of some peach cultivars and variable pathogenicity of the Sharka (Plum pox) virus. *Acta Hort.* 193:193–199.

Ranković, M. and Paunović, S. 1988. Further studies of the resistance of plums to Sharka (plum pox) virus. *Acta Hort. (Fruit Tree Virus Diseases)* 235:283–290.

Ranković, M. and Dulić-Marković, J. 1991. Evaluation of *Prunus spinosa* L. as host of sharka and other viruses (Abstr.) *XVth Int. Symp. Virus and Virus Diseases of Temperate Fruit Crops*, 8–13 July, Vienna, Austria.

Ranković, M. et al. 1991. Selection of *Prunus* sp. and Sharka virus. Project 40-ARS-138-IB-158 (PP759), Annual report 1990/91.

Ravelonandro, M. et al. 1991. Transgenic tobacco plants that contain the plum pox virus (PPV) coat protein gene (Abstr.). *XVth Int. Symp. on Virus and Virus Diseases of Temperate Fruit Crops*, 8–13 July, Vienna, Austria.

Reeves, E.L. 1940. Rusty mottle, a new virosis of cherry. *Phytopathology* 30:789.

Reeves, E.L. 1943. Virus diseases of fruit trees in Washington. *Wash. State Dept. Agric. Bull.* 1, 25 p.

Reeves, E.L. and Cheney, P.W. 1956. A new form of the twisted leaf virus disease of cherries (Abstr.). *Phytopathology* 46:639.

Reichert, I.A. et al. 1956. Transmission experiments on the tristeza and xyloporosis diseases of citrus. *Ktavim* 6:69–75.

Reichert, I. 1957. A survey of citrus virus diseases in the Mediterranean area. *Proc. Conf. Citrus Virus Dis.*, Univ. Calif., Riverside, Nov. 18–28, p. 23–38.

Refatti, E. and Osler, R. 1973. Anomalie dei frutti e alterazioni di colore delle foglie di cologno causate dal virus della annueatura fuligginosa. *Riv. Pat. Veg. (Pavia) Ser. IV*, 9:45–61.

Refatti, E. and Osler, R. 1975. Possible relationships among some fruit viruses detected in graft transmission trials. *Acta Hort.* 44:201–208.

Richter, J. and Kegler, H. 1967. Unteruschungen über Ringfleckenkrankheit der Kirsche. III Serologische Untersuchungen. *Phytopathol. Z.* 60:262–272.

Richter, J. and Kegler, H. 1967a. Isolierung des latenten Erdbeerringfleckenvirus (strawberry latent ringspot virus) aus stauchekranken Pfirsichbäumen, *Phytopathol. Z.* 58:298–301.

Richter, J. et al. 1977. Nachweis des Tomatenzwergbusch-virus (Tomato bushy stunt virus) in Obstgehölzen. *Arch. Phytopathol. Pflanzensuchtz, Berlin* 13:367–368.

Richter, J. et al. 1978. Identifizierung eines virus von Birnen mit Steinfrüchtigkeit als Nelkenringflecken-Virus (Carnation ringspot virus). *Arch. Phytopathol. Pflanzenschuchtz, Berlin* 14:411–412.

Rodoni, B. 1999. Characterisation and Control of Banana Bract Mosaic Potyvirus. Ph.D. dissertation, Queensland University of Technology, Brisbane, Australia.

Roistacher, C.N.E. et al. 1980. Transmission of citrus tristeza virus by *Aphis gossypii* reflecting changes in virus transmissibility in California. In: Calavan, et al. Eds., *Proc. 8th Conf. Int. Organ. Citrus Virol. IOCV,* Univ. Calif., Riverside, p. 76–82.

Roistacher, C.N.E. et al. 1980a. Transmissibility of cachexia, sweet mottle, psorosis, tatterleaf and infectious variegation viruses on knife blades and its preservation. In: Calavan, et al., Eds., *Proc. 8th Conf. Int. Organ. Citrus Virol.*, Univ. Calif., Riverside, p. 225–229.

Roistacher, C.N.E. 1980b. Psorosis A. In: Bové J. and Vogel R., Eds., *Descr. Illus. Virus Virus-like Dis. Citrus.* SETCO-IRFA, Paris.

Rosenberger, D.A. and Jones, A.L. 1976. A graft-transmissible agent associated with bark and woodgrooving disease of peach and nectarine. *Phytopathology* 66:729–730.

Rosner, A. et al. 1983. Molecular cloning of complementary DNA sequences of citrus tristeza virus RNA. *J. Gen. Virol.* 64:1757–1763.

Russo, F. 1956. La presenza del virus della tristeza su limone "Dwarf Meyer" e mandarino "Satsuma" riscontrata in Sicilia, *Riv. Agrumicolt.* 1:281–289.

Russo, M. et al. 1979. Comparative studies on Mediterranean isolates of watermelon mosaic virus. *Phytopathol. Medit.* 18:94–101.

Šarić, A. 1971. The occurrence of sowbane mosaic in sour cherry. *Ann. Phytopathol.* 71:155–158.

Saunier, R. 1970. Modification du comportement de deux cultivars de pêcher par élimination d une viruse du type ringspot. Mémoire présenté pour l'obtention du titre d'Ingenieur DPE Specialité, Agriculture.

Savino, V. et al. 1976. Occurrence of two sap-transmissible viruses in walnut. *Mitt. Biol. Budasanst. Land-Forstwirsch. Berlin-Dahlem* 170:23–27.

Savino, V. et al. 1983. Olive latent ringspot virus, a newly recognized virus infecting olive in Italy. *Ann. Appl. Biol.* 103:243–249.

Savulescu, A. and Macovei, A. 1965. Studies on the Sharka (plum pox) virus and related pattern line virus. *Zastita bilja* 16:357–366.

Scaramuzzi, G. 1957. Una virosi con deformazione, maculatura verde e suberosi interna dei frutti di cotogno (*Cydonia oblonga*). *Phytopathol. Z.* 30:259–274.

Scaramuzzi, G. et Ciferri, R. 1957: Una nuova virosi: la "maculatura lineare" del nocciolo. *Ann. Sper.* N.S.:11–21.

Schade, C. 1969. Eigenschaften und Serologie des Scharkavirus der Pflaume (Abstr.). *Zbl. Bakteriol. Parasitennde Infekt. Krankh. Hyg. II*, 123:219–304.

Schmelzer, K. 1966. Untersuchungen an Viren der Zier-und Wildgehölze. *Phytopathol. Z.* 55:317–351.

Schmelzer, K. 1980. Übertragungsmöglichkeiten. In: Klinkowski, M., et al. Eds. *Pflanzliche Virologie*, Akademie-Verlag, Berlin. Bd. 1, 3. Aufl.:111–133.

Schmid, G. 1980. Transmission experiments with pseudo-pox and similar disorders of plum. *Acta Hort. Fruit Virus Disease* 94:159–165.

Schuch, K. 1962. Untersuchungen über die Pockenkrankheit der Zwetsche. *Z. Pflanzenkrankh. Pflanzenschutz* 69:137–142.

Semancik, J.S. 1980. Citrus exocortis viroid. *CMI/AAB Descriptions of Plant Viruses*, No. 226.

Senevirante, S.N. and Posnette, A.F. 1970. Identification of viruses isolated from plum trees affected by decline, line pattern and ringspot diseases. *Ann. Appl. Biol.* 65:115–125.

Serr, E.F. and Forde, H.I. 1959: Blackline, a delayed failure at the union of *Juglans regia* trees propagated on *Juglans* species. *Proc. Am. Soc. Hort.* 74:220–231.

Smith, G.R. and Harding, R.M. 1999. Genetic engineering for virus resistance in surgarcane. In Rao, G.P. et al. *Current Trends in Sugarcane Pathology. III. Virus and Phytoplasma Diseases*. Oxford Press (in press).

Smith, P.R. and Challen, D.I. 1972. Aetiology of the rosette and decline disease of peach and interactions between *Prunus* necrotic ringspot, prune dwarf and dark green sunken mottle viruses. *Austr. J. Agric. Res.* 23:1027–1034.

Smith, I.M. et al., Eds. 1988. *European Handb. Pl. Dis.* 1988. Blackwell Sci. Publ., p. 583.

Stace-Smith, R. and Hansen, A.J. 1976. Some properties of cherry rasp leaf virus. *Acta Hort.* 67:193–197.

Stubbs, L.L. and Smith, R.P. 1971. The association of prunus necrotic ringspot, prune dwarf and dark green sunken mottle viruses in rosetting and decline of peach. *Austr. J. Agric. Res.* 22:771–781.

Stubbs, L.L. and Barker, D. 1985. Rapid sample analysis with a simplified ELISA. *Phytopathology* 75:492–495.

Šutić, D. 1953. Some observations on walnut virus diseases in Yugoslavia (personal communication, nonpublished data).

Šutić, D. 1961. Assay of the transmission of Sharka virus disease by sap inoculation to herbaceous plants. *T. Planteavl.* 65 (Saernummer):138–146.

Šutić, D. 1962. The peach seedlings as test plants in the experiments of Sharka (plum pox) virus transmission. *Phytopathol. Medit.* 2(3):171–174.

Šutić, D. 1964. Sejanci kajsije kao indikatori virusa šarke. *Zaštita bilja* 77:87–91.

Šutić, D. and Blagojević, M. 1964. Application of heat-therapy for the inactivation of fig mosaic virus. *Phytopathol. Medit.* III:32–34.

Šutić, D. 1965. Vegetative effect of some plants on the curing of plum infected with Sharka (plum pox) virus. *Zaštita bilja* 85–88:347–351.

Šutić, D. 1968. Some aspects in the control of Sharka virus disease. *Acta Hort., Bucharest* 2(10):467–478.

Šutić, D. et al. 1971. Comparative studies on some Sharka virus isolates. *Ann. Phytopathol. No. hors série*, 185–194.

Šutić, D. 1971. État de recherches sur le virus de la Sharka. *Ann. Phytopathol. H.S.*, 161–170.

Šutić, D. et al. 1972. *Brachycaudus cardui* L., novi vektor virusa šarke u Jugoslaviji. *Jugoslovensko vocárstvo* 21–22:791–795.

Šutić, D. et al. 1972a. Šarka (Plum pox) virus disease. Project No. E30-CR-19, PL 480, Fin. Techn. Report.

Šutić, D. 1973. The susceptibility of some Fabaceae plants to the Šarka (plum pox) virus. *Phytopathol. Z.* 78:245–252.

Šutić, D. 1973a. Neke serološke osobine sojeva virusa šarke. *Mikrobiologija* 10(2):199–206.

Šutić, D. 1975. Is the gall mite *Eriophyes phloecoptes nalepa* a vector of Sharka virus? *Bull. Infor. Sharka*, No. 1, 10–11.

Šutić, D. 1975a. Protection with some strains of Šarka virus. *Acta Hort. XI Int. Symp. Fruit Tree Virus Dis.* 44:165–168.

Šutić, D. 1975b. La greffe comme technique de protection des plantes. *Semaine d étude Agriculture et Hygiène des plantes.* Faculté des science agronomique de l'État et Centre de Recherches Agronomiques. Gembloux.

Šutić, D. 1976. Intérêt de *Prunus spinosa* comme porte-greffe de prunier. *Bull. Info. Sharka*, No. 1, 8–9.

Šutić, D. and Juretić, N. 1976. Occurrence of sowbane mosaic virus in plum trees. *Mitt. Bundesanst. Land-Forstwirtsch.* 170:43–46.

Šutić, D. et al. 1976. Transmissibility of some Sharka virus strains by *Myzus persicae* Sulz, depending on various infection sources. *Acta Hort. Fruit Tree Virus Dis.* 67:171–177.

Šutić, D. 1977. Sensibilité de certains cultivars du prunier à l'infection naturelle par le virus de la sharka. *Travaux dédiés à G. Viennot-Bourgin*, 361–443.

Šutić, D. 1977a. Hôtes herbacés du virus de la Sharka parmi les plantes de la famille des Papavéracées. *C.R. Acad. Agr. France* 6:440–443.

Šutić, D. 1979. Recovery of some *Prunus* sp. plants infected with Sharka virus. *Phytopathol. Medit.* 18:211–212.

Šutić, D. 1980. Biljni virusi. Nolit, Beograd.

Šutić, D. and Ranković, M. 1981. Resistance of some plum cultivars and individual trees to plum pox (Sharka) virus. *Agronomie* 1(8):617–622.

Šutić, D. and Ranković, M. 1981a. Otpornost nekih kultivara šljive, breskve i kajsije prema virusu šarke. *IV kongres mikrobiologa Jugoslavije, Beograd*, Sep. 22–25.

Šutić, D. 1981. Mišjakinja (*Stellaria media* Vill.) domaćin i virusa šarke šljive i virusa chlorotične lisne pegavosti jabuke. *IV kongres mikrobiologa Beograd Jugoslavije*, 22–25 Sept., p. 224–225.

Šutić, D. 1983. Viroze biljaka, Nolit, Beograd.

Šutić, D. and Ranković, M. 1983. Sensitivity of some stone fruit species to Sharka (plum pox virus) disease. *Zaštita bilja* 34:241–348.

Šutić, D. and Sinclair, J.B. 1991. *Anatomy and Physiology of Diseased Plants.* CRC Press LLC, Boca Raton, FL.

Sweet, J.B. and Campbell, A.I. 1976. Pome fruit virus infection of some woody ornamental and indigenous species of Rosaceae. *J. Hort. Sci.* 51:91–97.

Sweet, J.B. et al. 1978. Improving the quality and the virus status of hardy ornamental trees and shrubs. *Rept. Long Ashton Res. Stn. 1977*, p. 36–37.

Sweet, J.B. 1980. Fruit tree infections of woody exotic and indigenous plants in Britain. *Acta Phytopathol. Acad. Sci. Hung.* 15:231–238.

Syrgianidis, G.D. 1980. Selection of the apricot varieties resistant to Sharka virus. *Acta Phytopathol. Acad. Sci. Hung.* 15(1–4):85–87.

Szirmai, J. 1961. Report of fruit tree virus diseases in Hungary. *T. Plantearl.* 65 (Saernummer):220–229.

Tanaka, H. 1980. Satsuma dwarf. In: Bové, J. and Vogel, R., Eds. *Descr. Illus. Virus Virus-like Dis. Citrus.* SETCO-IRFA, Paris.

Thomas, H.E. and Hildebrand, E.M. 1936. A virus disease of prune. *Phytopathology* 26:1145–1148.

Thomas, H.E. and Rawlins, T.E. 1939. Some mosaic diseases of *Prunus* species. *Hilgardia* 12:623–644.

Thomsen, A. 1975. Cross-protection experiment with apple mosaic virus. *Acta Hort.* 44:119–122.

Timmer, L.W. and Garnsey, S.M. 1980. Natural spread of citrus ringspot virus in Texas and its association with psorosis-like diseases in Florida and Texas. In: Calavan, et al., Eds. *Proc. 8th Conf. Int. Organ. Citrus Virol.*, Univ. Calif., Riverside, p. 167–173.

Topchiiska, M. and Topchiiski, I. 1976. Isolation of cucumber mosaic virus from almond (*Prunus amydalus* Stocks). *Hort. Vitic. Sci. (Sofia)* 13(3):44–54.

Topchiiska, M. and Topchiiski, I. 1976a. Studies on the properties of two strains of the necrotic ringspot virus (NRV) isolated from almond trees (*Prunus amygdalus* stocks) grown in Bulgaria. *Hort. Vitic. Sci. (Sofia)*, 13(2):57–68.

Topchiiska, M. 1977. Preliminary studies on a strain of apple mosaic virus isolated from almond. *Hort. Vitic. Sci. (Sofia)* 14(1):37–43.

Topchiiska, M. 1983. Effect of *Prunus* necrotic ringspot virus (PNRV) and prune dwarf virus (PDV) on some biological properties of peach. *Acta Hort.* 130:307–312.

Topchiiska, M. 1988. Detection of cherry leaf roll virus - CLRV in *Juglans regis* L. by enzyme-linked immunosorbent assay (ELISA) in Bulgaria (Abstr.). *XIVth Int. Symp. Fruit Tree Virus Dis.*, 12-18 June, Thessaloniki, Greece.

Traylor, J.A. et al. 1963. Studies on the passage of prunus ringspot virus complex through plum seed (Abstr.). *Phytopathology* 53:1143.

Trifonov, D. 1971. Bud drop of pear. *Ann. Phytopathol.* 3:285–294.

Turner, N.W. et al. 1987. Expression of alfalfa mosaic virus coat protein gene confers cross protection in transgenic tobacco and tomato plants. *EMBO J.* 6:1181.

Usugi, T. and Saito, Y. 1979. Satsuma dwarf virus. *CMI/AAB Descriptions of Plant Viruses*, No. 208.

Uyemoto, J.K. and Gilmer, R.M. 1972. Properties of tobacco necrosis virus strains isolated from apple. *Phytopathology* 62:478–481.

Uyemoto, J.K. 1988. Tobacco necrosis virus (TNV). In: Smith, I.M., et al., Eds. *European Handbook of Plant Diseases*, Blackwell Sci. Publ., London, p. 80–81.

Vaclav, V. 1960. *Ispitivanje šarke šljive, Sarajevo*, 199–206.

Vaclav, V. 1965. Rezultati ispitivanja uticaja šarke šljive. *Radovi Poljoprivrednog faulteta, XIV*, 16:49–91, Sarajevo.

Valleau, W.D. 1932. A virus disease of plum and peach. *Ky. Agr. Exp. Sta. Bull.* 327:89–103.

Van Dun, C.M.P. et al. 1987. Expression of alfalfa mosaic virus and tobacco rattle virus coat protein genes in transgenic tobacco plants. *Virology* 159:299.

Van der Meer, F.A. 1975. Plant species outside the genus *Malus* as indicators for latent viruses of apple. *Acta Hort.* 44:213–220.

Van der Meer, F.A. 1976. Observations on apple stem grooving virus. *Acta Hort.* 67:293–304.

Van Oosten, H.J. 1970. Herbaceous host plants for the Sharka (plum pox) virus. *Neth. J. Pl. Pathol.* 76:253–260.

Van Oosten, H.J. 1970a. The isolation of Sharka (plum pox) virus from leaves and fruits of plum with herbaceous plants. *Neth. J. Pl. Pathol.* 76:99–103.

Van Oosten, H.J. 1971. Further information about herbaceous host range of Sharka (plum pox) virus. *Ann. Phytopathol. H.S.* 195–201.

Van Oosten, H.J. 1972. Purification of plum pox virus with the use of Triton X-100. *Neth. J. Pl. Pathol.* 78:33–44.

Van Oosten, H.J. 1976. The importance of virus-free apple trees for fruit growing. *Gewasbescherming* 7:1–15.

Van Siebert, Z. and Engelbrecht, D.J. 1983. Association of apple stem grooving virus with a decline of Packham's Triumph pear on seedling rootstock. *Acta Hort.* 130:47–52.

Vogel, R. 1980. Cristacortis. In: Bové, J.M. and Vogel, R., Eds. *Description and Illustration of Virus and Virus-like Diseases of Citrus.* SETCO-IRFA, Paris.

Vogel, R. and Bove, J.M. 1980. Pollen transmission to citrus of the agent inducing cristacortis stem pitting and psorosis young leaf symptoms. In: Calavan, et al. Eds. *Proc. 8th Conf. Int. Organ. Citrus Virol.*, Univ. Calif., Riverside, p. 188–190.

Vuittenez, A. and Kuszala, J. 1968. Propriétés d'un virus de type "ringspot" non labile associé à la maladie du Court-Noué du Pêcher. *Ann. Epiphytes* 19:No. hors série, 269–274.

Wadley, B.N. 1959. Rusty mottle virus complex in Utah (Abstr.). *Phytopathology* 49:114.

Wadley, B.N. 1966. Apricot leaf pucker, a virus-induced disorder (Abstr.). *Phytopathology* 56:152.

Wadley, N. and Nyland, G. 1976. Rusty mottle groups. In: *Virus Dis. Noninfec. Disorders Stone Fruits North America. USDA Handb.* 437:242–249.

Wagnon, H.K. et al. 1958. Stubby twig, a new virus disease of peach and nectarine in California. *Phytopathology* 48:465–468.

Wagnon, H.K. et al. 1960. Observations on the natural spread of the so-called peach stunt virus in California peach orchard. *Plant Dis. Reptr.* 44:488–490.

Wagnon, H.K. et al. 1963. The natural occurrence of peach asteroid spot virus in a native *Prunus* (*P. Andersonii* L.) in California and Nevada. *Phytopathol. Medit.* 2:196–198.

Walkey, D.G.A. et al. 1973. Serological, physical and chemical properties of strains of cherry leaf roll virus. *Phytopathology* 63:566–571.

Wallace, J.N. and Drake, R.J. 1953. A virus-induced vein enation in citrus. *Citrus Leaves* 33:22–24.

Wallace, J.M. and Drake, R.J. 1960. Woody galls on citrus associated with vein enation virus infection. *Plant Dis. Reptr.* 44:580–584.

Wallace, J.M. and Drake, R.J. 1962. A high rate of seed transmission of avocado sun-blotch virus from symptomless trees and the origin of such trees. *Phytopathology* 52:237–241.

Waterworth, H.E. 1965. Some properties of an unidentified virus latent in pear introduction (Abstr.). *Phytopathology* 55:1081.

Waterworth, H.E. and Fulton, R.W. 1964. Variation among isolates of necrotic ringspot and prune dwarf viruses from sour cherry. *Phytopathology* 54:1155–1160.

Waterworth, H.E. and Gilmer, R.M. 1969. Dark green epinasty of *Chenopodium quinoa*, a syndrome induced by a virus latent in apple and pear. *Phytopathology* 59:334–338.

Way, R.D. and Gilmer, R.M. 1963. Reductions in fruit sets on cherry trees pollinated with pollen from trees with sour cherry yellows. *Phytopathology* 53:399–401.

Webb, R.E. and Scott, H.A. 1965. Isolation and identification of watermelon mosaic viruses 1 and 2. *Phytopathology* 55:895–900.

Welsh, M.F. and Keane, F.W.L. 1963. Leaf pucker and associated fruit disorders. *Techn. Comm. Commonw. Bur. Hort. Plantn. Crops* 30:11–15.

Welsh, M.F. and Nyland, G. 1965. Elimination and separation of viruses in apple clones by exposure to dry heat. *Can. J. Plant Sci.* 45:443–454.

Welsh, M.F. and May, J. 1973. Suppression of the symptoms of apple leaf pucker disease in an occurrence pattern that indicates virus infection. *Can. J. Plant Sci.* 53(1):137–145.

Welsh, M.F. and Cheney, P.W. 1976. Little cherry. In: *Virus Dis. Noninfec. Disorders Stone Fruits North America. USDA Handb.* 437:231–237.

Williams, H.E. et al. 1976. Asteroid spot. In: *Virus Dis. Noninfec. Disorders Stone Fruits North America. USDA Handb.* 437:50–55.

Wilde, W.H.A. 1960. Insect transmission of the virus causing little cherry disease. *Can. J. Plant Sci.* 40:707–712.

Willison, R.S. 1945. A line-pattern virosis of Shiro plum. *Phytopathology* 35:991–1001.

Willison, R.S. and Weintraub, M. 1957. Properties of a strain of cucumber mosaic virus isolated from *Prunus* hosts. *Can. J. Bot.* 35:763–771.

Wilks, J.M. and Welsh, M.F. 1965. Freckle pit-a virus disease of "Anjou" pears. *Can. Plant Dis. Survey* 45:90–91.

Wilson, N.S. et al. 1955. An eriophyid mite vector of the peach-mosaic virus. *Plant Dis. Reptr.* 39:889–892.

Wood, G.A. 1975. Some bud transmissible disorders in Prunus. *N.Z. J. Agr. Res.* 18:255–259.

Wood, G.A. 1979. Virus and virus-like diseases of plum and stone fruits in New Zealand. *DSIR Bull. Wellington* 226, 87.

Yamada, S. and Sawamura, K. 1952. Studies on the dwarf disease of satsuma orange. Citrus unshin (preliminary report). *Bull. Hort. Div. Tokai-kinki Agric. Expt. Stn.* 1,61–71.

Yarwood, C.E. 1955. Mechanical transmission of an apple mosaic virus. *Hilgardia* 23:613–628.

Yoshikawa, N. and Takahashi, T. 1988. Properties of apple stem grooving and apple chlorotic leaf spot viruses (Abstr.). *XVth Int. Symp. Fruit Tree Virus Dis.*, 29, 12–18 June, Thessaloniki, Greece.

Yot-Danthy, D. and Bové, J.M. 1968. Purification and characterization of citrus crinkly leaf virus. *Proc. 4th Conf. Int. Organ. Citrus Virol.*, p. 255–263.

Zawadzka, B. 1965. Observations and preliminary experiments on fruit tree virus diseases in Poland. *Zaštita bilja* 16(85–88):513–516.

Zelazny, B. and Niven, B.S. 1980. Duration of the stages of cadang-cadang disease of coconut palm. *Plant Dis.* 64:841–842.

Zelazny, B. et al. 1982. The viroid nature of the cadang-cadang disease of coconut palm. *Scientia Filipinas* 2:45.

Zeller, S.M. et al. 1947. Black canker of cherry (Abstr.). *Phytopathology* 37:366.

Zeller, S.M. et al. 1951. Black canker. In: *Virus Dis. Other Disorders Virus-like Symp. Stone Fruits North America. USDA Handb.* 10:137–138.

Zentmyer, G.A. 1953. Diseases of the avocado. In: *Plant Diseases, The Yearbook of Agriculture.* USDA, Washington, D.C. p. 875–880.

6 Virus Diseases of Small Fruits

Small fruit crops include the genera *Fragaria*, (Rosaceae), *Rubus* (Rosaceae), *Ribes* (Saxifragaceae), *Vaccinium* (Ericaceae), and *Vitis* (Vitaceae). The *Fragaria* genus includes various cultivated (*F. ananassa*), wild (*F. vesca*), and other strawberry species. The cultivated genus *Rubus* consists of two species: raspberries (red raspberry, *R. idaeus*) and blackberries (blackberry, *R. occidentalis*). The genus *Ribes* consists of cultivated species including black currant (*R. nigrum*), red currant (*R. grossularia*), and gooseberry (*R. rubrum*). Species of blueberries and cranberries are economically important in the *Vaccinium* genus, whereas the *Vitis* genus includes numerous species among which grapevine (*Vitis vinifera*) represents the most economically important one. Three *Compendia of Diseases* do an excellent job of documenting all diseases in strawberries (Maas, 1984), blueberries and cranberries (Caruso and Ramsdell 1995), and raspberries and blackberries (Ellis et al. 1991).

The many viruses that affect small fruits represent a serious and permanent threat to profitable large-scale production. According to Agrios (1988), strawberries can be infected naturally by at least 20 viruses, raspberries and blackberries by about 15, and each of the other fruit species by about 6 to 10 viruses. Some viruses (e.g., strawberry crinkle, mottle, and mild yellow edge viruses, raspberry bushy dwarf, vein chlorosis, and ringspot viruses, and others) cause severe symptoms and losses in yield or marketability. Complex infections that occur regularly in nature may eliminate yields and cause massive plant decline. Producers are well aware that virus diseases represent a major limiting factor in the commercial production of these plants.

It should be pointed out that virus infections represent a particular danger for small fruits as a result of their biological and phytotechnical traits. First, they are usually perpetuated by vegetative propagation that permits the accumulation as a reservoir to allow further spread of viruses. Second, many viruses affecting small fruits are transmitted by aphids or nematodes, which both increases the rate of spread of virus in plantations and complicates their epidemiology and control. Detrimental effects of viral infections also result from the fact that their hosts are perennials in which the viruses accumulate through the years, causing increasing degeneration, sterility, and plant decline.

Control of virus diseases in small fruit is very difficult because of their mode of spread and the pronounced activity both of aerial and soil vectors in the transmission of some deleterious viruses. The production and use of virus-free planting material (runners, cuttings, stocks) represents a permanent and indispensable measure that must be strictly adhered to in the course of establishment of plantations. Virus-free plants (strawberry crinkle, mild yellow edge, raspberry vein chlorosis, and other viruses) are successfully obtained by heat treatment and/or meristem-tip culture, followed by repeated virus indexing of plants produced in order to determine whether such treatment has been successful.

New plantations should be established on soils uncontaminated with vector nematodes and isolated spatially from existing sources of infection. Plantations, especially young ones, should be sprayed with aphicides to reduce the populations of aphids known as aerial vectors of a number of destructive viruses.

Regardless of the many possible measures for virus-free plant production and virus control, the selection of resistant plants represents the best solution for the control of plant virus diseases.

In this respect, modern molecular biology research methods should help enable the accurate identification and the incorporation of plant genes into small fruit cvs. that confer resistance to virus vectors and tolerance or resistance to viruses, as well as the incorporation of portions of viral genes into these plants to confer resistance or reduce disease damage.

VIRUS DISEASES OF STRAWBERRY (*Fragaria chiloensis*)

Strawberry Crinkle Rhabdovirus (SbCrV)

Properties of Viral Particles

SbCrV, a bacilliform virus, features enveloped particles of 190 to 280 × 69 ± 6 nm (Sylvester et al., 1976) of insufficiently known biophysical and biochemical properties. Several strains have been described, based on symptom severity in host plants. The type strain causes classical symptoms, whereas infections with mild strains result in symptomless strawberry cvs. the *Fragaria vesca* indicator plant. The following strains are listed according to increasing virulence: SbCrV latent A (mild form), SbCrV latent A and B, SbCrV lesion A, SbCrV lesion B, and SbCrV vein chlorosis (loc. cit. Frazier and Mellor, 1970).

Geographic Distribution and Economic Importance

The widespread strawberry crinkle virus occurs wherever cultivated strawberries are attacked by the *Chaetosiphon* sp. of aphids known as virus vectors. It has been observed in North and South America, Europe, the British Isles, New Zealand, and Australia. Some primarily latent strains, first discovered as endemic, were eventually distributed throughout the world by the systemically infected *Fragaria vesca* clones, which are the designated indicator plants.

Strawberry crinkle is one of the most detrimental strawberry diseases. Serious damage occurs in production areas where infection with both virulent and symptomless strains occur. Some findings show that virulent strains severely reduce plant vigor and productivity, whereas symptomless strains (i.e., latent A) reduce plant vigor, runner formation, fruit size, and yields only in some cultivars (Freeman and Mellor, 1962; McGrew and Scott, 1964). Investigations in Yugoslavia have shown that SbCrV reduced strawberry yields by 45% in 1963, and by as much as 63% in 1964 (Babović, 1965), and the disease also caused fruit quality changes. Sugar content of fruits in infected plants was reduced by 11%, dry matter by 17%, and total acid content increased by 29%.

Mixed infections of SbCrV with strawberry mottle, vein banding, or mild yellow edge viruses in nature are particularly damaging in strawberry production.

Host Range

Virus hosts are species belonging only to the genus *Fragaria*, among which none is known to be immune. They include cvs. *Fragaria chiloensis* var. *ananassa*, as well as the wild species *F. vesca*, *F. virginiana*, and *F. chiloensis* (Sylvester et al., 1976).

Symptoms

Disease symptoms depend on the virulence of virus strains and the sensitivity of strawberry cvs. Virulent strains of SbCrV cause variable symptoms in susceptible cvs. For example, Marshall displays distinct symptoms, whereas Shasta remains symptomless. Mild strains of SbCrV are symptomless in all cvs. (Frazier and Mellor, 1970).

In susceptible cvs., symptoms of disease occur on the leaves early in the season as small scattered dotted spots, then develop into large chlorotic spots. The leaf tissue within these spots and in some veins frequently becomes necrotic. Disease symptoms are most pronounced from late April through June. During the summer months of July and August with normally high temperatures, the symptoms become latent. SbCrV disease symptoms once again become clearly evident on newly formed leaves in the months of September and October with the advent of lower temperatures. In the later stages of SbCrV infection, the diseased leaf blades become smaller with irregularly shaped, uneven margins.

Indicator plants of *Fragaria vesca* (clones UC4, UC5, UC6) or *F. vesca* var. *semperflorens* (alpine seedlings) each react to the infection with different symtpoms (Krczal, 1988). Initial symptoms appear as small chlorotic spots on indicator plant leaves (resembling those caused by strawberry mottle or mild yellow edge viruses) and later develop the more characteristic angular epinasty of leaflets, petiole lesions, and petal streaks, which are diagnostic for SbCrV disease detection.

Pathogenesis

SbCrV is preserved permanently in systemically infected strawberry plants, wherefrom it is spread by infected runners as vegetatively propagated plant parts, that is, the main mode of natural transmission and spread of SbCrV.

Chaetosiphon fragaefolii and the dark strawberry *C. jacobi* aphids are active transmitters of SbCrV in a persistent manner (loc. cit. Sylvester et al., 1976). Relations between vectors and viruses may vary, primarily depending on virus strains. According to Krczal (1988), *C. fragaefolii* acquires SbCrV and transmits it within 24 hours; its latent period in the vector is about 2 weeks; and it multiplies in *C. jacobi* aphids.

SbCrV is transmitted by grafting, which is generally used for indexing strawberry plants. It was transferred experimentally by needle inoculation from aphid to aphid (Sylvester et al., 1976), as well as by injection with aphid vector extracts into the non-vector bean aphid *Megoura viciae* (Adams and Barbara, 1988). SbCrV has not yet been transmitted mechanically by infective plant sap, and other modes of transmission are unknown.

Control

The use of virus-free runners represents the main prevention measure in the establishment of strawberry plantations. Control of strawberry aphids also plays an important role in the prevention of virus spread in the fields. All infected plants from which aphids transmit the virus into healthy plants should be rogued promptly from young plantations. All necessary measures for isolation of plantations from potential sources of infection should be applied. Plants can be rendered virus-free by heat therapy (Posnette and Cropley, 1958) and by meristem culture (Miller and Belkengren, 1962).

Strawberry Mild Yellow Edge [possible] Luteovirus (SbMdYEgV)

Properties of Viral Particles

The SbMdYEgV isometric viral particles are about 23 to 28 nm in diameter. SbMdYEgV consists of several strains differing in virulence. A combination of strawberry mottle virus and/or SbCrV causes diseases known as strawberry yellows edge or xanthosis in North America and strawberry yellows in England. Yellows was the first viral disease recognized in strawberries; Horn (1922) described it in California and Harris (1933) named it "yellow edge" in England.

Geographic Distribution and Economic Importance

SbMdYEgV probably occurs worldwide; the yellows complex has been reported in North America, Europe, Rhodesia, Australia, and New Zealand (Mellor and Frazier, 1970). SbMdYEgV frequently infects strawberries in nature, depending primarily on sources of infection and activity of the vectors. While investigating the distribution of strawberry diseases in some prefectures of Japan, Yoshikawa and Inouye (1988) detected the presence of this virus in 66% of plants tested. Infection complexes including SbMdYEgV cause severe damage in strawberry production and, in susceptible cultivars, yield reductions range from 25% (Krczal, 1988a) to 75% (Stitt and Breakey, 1952).

Host Range

Natural virus hosts consist of species only in the genus *Fragaria*. Susceptible hosts displaying symptoms of infection are the wild species *Fragaria vesca*, *F. virginiana*, and some *F. chiloensis* clones. *F. ovalis* ranks among symptomless hosts (Mellor and Frazier, 1970). Martin et al. (1988) experimentally transmitted the virus to *Rubus rosifolius*.

Symptoms

SbMdYEgV alone is symptomless in most commercial cvs. while its reaction in complex infections depends on the companion virus species and cultivar susceptibility. Characteristic yellowing starts from leaf blade margins and spreads gradually toward the mid-section between leaf veins. Shortened and thickened leaf petioles plus curved midribs create a plant with a rosette-like appearance. Reddening appears on these leaves early in the autumn. Symptoms become latent during high summer temperatures with decreased soil moisture.

Royal Sovereign, Tardive de Léopold, and Marshall cvs. are extremely susceptible to yellows infections, whereas those of Huxley, Madame Lefebre, Premier, Senator Dunlap, Northwest, and most of the Ettesburg selections are symptomless carriers (Mellor and Frazier, 1970). Kegler and Kleinhempel (1977) show cvs. Deutsch Everin, Madame Moutôt, Marshall, and Royal Sovereign as highly susceptible, and Brandenburg among others as tolerant.

Fragaria vesca (clones UC1, UC4, UC5) or *F. vesca* var. *semperflorens* (alpine seedlings) are virus indicator plants. Symptoms of single infections by SbMdYEgV include chlorotic spots and chlorosis on young leaves, chlorotic zones along leaf margins, and decline of older leaves. The severity of symptoms increases in complex infections. Spiegel and Martin (1991) described the procedure for virus detection in strawberry plantlets grown *in vitro*. Particular attention should be given to the possibility of using cDNA clones to detect viruses in strawberry and *Rubus* tissues under laboratory conditions (Martin et al., 1988).

Pathogenesis

SbMdYEgV systemically infects strawberry plants, and thus is introduced into new plantations with infected runners. In nature, the aphids *Chaetosiphon fragaefolii*, *C. thomasi*, and *C. jacobi* transmit SbMdYEgV in a persistent manner (Mellor and Frazier, 1970; Krczal, 1980). It is also transmitted by infective plant sap (Miller, 1951; Liu, 1958). Martin et al. (1988), preparing cDNA clones, succeeded in mechanically inoculating *Rubus rosifolius* plants. The virus was not transmitted by dodder, seed, or pollen (Mellor and Frazier, 1970).

Control

Only virus-free propagative materials should be used for establishment of plantations. Chemical control of vectors reduces virus spread in the field and rogueing to eliminate infected plants (in younger plantations) reduces the sources of infection. Virus-free material can be obtained by thermotherapy and meristem-tip culture (Krczal, 1988a).

Strawberry Mottle Virus (SbMtV)

Properties of Viral Particles

Knowledge of the properties of SbMtV, the suspected causal agent of strawberry mottle disease, is insufficient partly because conventional methods for virus purification in strawberry extracts have not been successful. More recent results of pathogen agent transmission directly from infected strawberry plants into experimental plants of *Chenopodium quinoa*, and *Physalis floridana* allow for epidemiological studies of strawberry mottle disease.

Polák and Bezpalcová (1988) discovered some isometric virus-like particles 20 nm in diameter in ultrathin sections of infected strawberry and *C. quinoa* leaves. Polák and Jokeš (1991), who also searched for particles of the causal agent in ultrathin sections of systemic *C. quinoa* leaf tissues from strawberry mottle diseased plants, detected small irregularly shaped inclusions containing very small virus-free isometric particles, 10 to 11 nm in diameter. Leone et al. (1991) studied purification of viral particles from infected *C. quinoa* and *Physalis floridana* plants by combining alternative purification methods with conventional centrifugation steps. Such efforts undoubtedly will gradually result in new knowledge on this virus as the causal agent of strawberry mottle disease.

Geographic Distribution and Economic Importance

Strawberry mottle, considered the most widespread strawberry virus disease, has a very high incidence. Yoshikawa and Inouye (1988) discovered SbMtV in 60% of indexed plants in some parts of Japan. SbMtV reduces strawberry yields by 25 to 30% (Freeman and Mellor, 1962; Horn and

Carver, 1962). In complexes of infections with crinkle, vein banding, and mild yellow edge viruses, yields are reduced even more.

Host Range

Main virus-susceptible hosts include all *Fragaria* species, but symptoms of disease occur only in *Fragaria vesca* (Mellor and Frazier, 1970a). SbMtV was transmitted experimentally into herbaceous plants by mechanical inoculation of infective plant sap or by the aphid vector *Chaetosiphon jacobi* (Frazier, 1968). Infections were reproduced in *Chenopodium amaranticolor* and *C. quinoa* plants, but recurrent strawberry infections did not succeed. Some authors successfully transmitted SbMtV into *C. quinoa* plants, which reacted with systemic symptoms (Polák and Bezpalcová, 1988; Polák and Jokeš, 1991). Leone et al. (1991) used experimentally infected *C. quinoa* and *Physallis floridana* to study purification of SbMtV and SbCrV.

Symptoms

All commercial strawberry cvs. infected with SbMtV, although symptomless, have significantly reduced vigor and yield, but such symptoms are not characteristic or unique because such reductions also occur in strawberry plants infected with other viruses.

Virus strains and variants cause various symptoms in *Fragaria vesca* indicator plants, ranging from mild mottle to severe degeneration, *inter alia*, including mottle, downward leaf curl, vein chlorosis, shortened leaf petioles, leaf deformation, and crown proliferation. As already stressed, experimentally infected *Chenopodium quinoa* and *Physalis floridana* plants react with systemic symptoms of infection.

Pathogensis

SbMtV is permanently preserved in and spread by systemically infected propagative plant material. Prior to the production of virus-free stock by thermotherapy, most old strawberry cvs. were infected with SbMtV.

Aphid vectors that transmit SbMtV semipersistently include, in order of efficiency: *Chaetosiphon fragaefolii*, *C. thomasi*, and *C. minor*. *C. jacobi*, probably a vector of SbMtV to wild strawberries, has not been discovered yet in fields of strawberry cvs. (Mellor and Frazier, 1970a); also less efficient vectors, but important are *Acyrthosiphon pelargonii* var. *rogersii*, *A. porosum*, *Amphorophora rubis*, *Aphis gossypii*, *Chaetosiphon tetrarhodus*, *Myzaphis rasarum*, *Myzus ascalonicus*, and *M. ornatus*.

SbMtV has been transmitted by infective plant sap, which is experimentally important. Successful transmission was performed by the dodders *Cuscuta gronovii* and *C. subinclusa* (Smith and Moore, 1952). No data are available on other modes of transmission.

Control

Strawberry plantations should be established with virus-free plant stocks. They should be located for adequate isolation from potential sources of infection. Chemical control of vectors features a certain importance in practice in view of their number and vector activity. Virus-free plants can be obtained by thermal treatment and/or meristem-tip culture (Mellor and Frazier, 1970a).

Strawberry Vein Banding Caulimovirus (SbVBdV)

Properties of Viral Particles

Isometric-shaped viral particles of SbVBdV are 40 to 50 nm in diameter and, similar to other viruses in this group, probably contain double-stranded DNA (Frazier and Converse, 1980). The sedimentation coefficient of particles determined by co-sedimentation with cauliflower mosaic virus (CfMV) is $200 \pm 10S$ (Morris et al., 1980). An antiserum specific for SbVBdV has not been prepared; however, it can be detected directly by enzyme-linked immunosorbent assay (ELISA) and serological specific electron microscopy (SSEM) using CfMV antisera (Morris et al., 1980). Several strains differ, based on symptoms and transmissibility of SbVBdV by aphids (Frazier and

Posnette, 1958; Frazier, 1960). The leaf curl strain, vein banding strain, and Erdebeere necrose virus, a possible strain, differ according to their incubation periods in indicator plants (loc. cit. Frazier and Converse, 1980).

Geographic Distribution and Economic Importance

SbVBdV occurs in eastern and western North America, Australia, and Brazil, where it was probably imported with infected plant material (Frazier and Converse, 1980). It may be present in Turkey, Ireland, and the former U.S.S.R. (EPPO, 1974). According to Miller and Frazier (1970), disease caused by SbVBdV ranks among the less damaging viral diseases due to its sporadic occurrence and low incidence. As with other viruses, disease damage includes reduced runner production, vigor, and yields (Freeman and Mellor, 1962). Complex infections of SbVBdV with other viruses result in important losses.

Host Range

Natural hosts of SbVBdV are plants in the genus *Fragaria*, including, *inter alia*, *Fragaria chiloensis*, *F. vesca*, and *F. virginiana*, as well as cultivated *F. ananassa* plants (Frazier and Converse, 1980). *Sanguisorba minor* is the experimental virus host (Morris et al., 1980).

Symptoms

Vein banding occurring in tissues along primary and secondary leaf veins is the primary disease symptom. This type of symptom may be more or less pronounced, depending mostly on strain virulence. Accompanying symptoms of yellows, crinkle, and leaf curl appear where multiple virus infections occur.

Fragaria vesca (clone UC6) and *F. virginiana* (clone UC12) are suitable diagnostic plants. Initial symptoms in these plants include various forms of mild leaf curl, accompanied by chlorotic banding along the main veins (of diagnostic importance), followed by stunting and necrosis of older leaves.

Pathogenesis

Initial infections in new strawberry plantations originate from SbVBdV-diseased planting material, particularly when the state of infection is not strictly controlled. Numerous vector aphids transmit SbVBdV nonpersistently, thus spreading it within already established plantations. The most efficient vectors include *Chaetosiphon fragaefolii*, *C. jacobi*, and *C. thomasi*; other active vectors are *Amphorophora rubi*, *Aphis rubifolii*, *Aulacorthum solani*, *Macrosiphum pelargonii*, *M. rosae*, *Myzus ascalonicus*, *M. persicae*, and *Chaetosiphon tetrarhodus* (Frazier, 1955; Mellor and Forbes, 1960). Some vectors, such as *C. fragaefolii*, *C. thomasi*, and *C. ornatus* transmit only some, but not all virus strains. The virus is also transmitted by grafting and by dodder (*Cuscuta subinclusa*) (Frazier, 1955), but neither mechanically with infected plant sap nor other modes of transmission.

Control

General measures recommended for viral control of other viruses are effective if applied to SbVBdV. The use of healthy planting material for establishment of plantations represents a mandatory control measure. Virus-free mother plants can be obtained by thermal treatment for 10 days at 42°C (Bolton, 1967) or by tissue culture of runner tips (Miller and Belkengren, 1963).

STRAWBERRY NEMATODE-BORNE VIRUSES

Strawberry Latent Ringspot Neopvirus (SbLtRsV)

Properties of Viral Particles

SbLtRsV is an isometric-shaped virus, 28 nm (Dunez, 1988) and 30 nm (Murant, 1974) in diameter. Virus sediments contain three groups of particles that sediment at 58S (T), 94S (M), and 126S

(B) (Murant, 1981). The SbLtRsV genome is composed of two RNA species of 2600 kDa (RNA_1) and 1600 kDa (RNA_2). Murant (1981) found two protein molecules of 44 kDa and 29 kDa, but Dunez (1988) found only a single undistinguishable nucleoprotein containing one RNA or two RNA_2 molecules.

The TIP of SbLtRsV is 54 and 56°C; the DEP (in cucumber sap) is 1×10^{-3} to 1×10^{-5}; and the LIV is 50 days or 13 months at 4°C (Lister, 1970). SbLtRsV displays good immunogenic activity (Murant, 1974). The Hampshire strawberry isolate of SbLtRsV is the type strain (Lister, 1964), and other strains that differ in virulence do not differ in biological traits or serological activity.

Geographic Distribution and Economic Importance

SbLtRsV has been isolated from numerous different plant samples in Europe, suggesting that it is a common pathogen there. It was found only once in Canada where its natural spread is unknown (Murant, 1974). Based on the number of hosts that SbLtRsV infects, it is considered similar to arabis mosaic virus with which it shares common vectors.

Crop losses in commercial strawberry production depend on disease severity. Otherwise, it is generally believed that SbLtRsV is less detrimental except when occurring in complex infections with other viruses, such as prune dwarf virus (Dunez, 1988). Some SbLtRsV isolates cause severe peach tree degeneration (Scotto la Massese et al., 1973).

Host Range

SbLtRsV naturally infects a large number of hosts among cultivated and wild plants. The number of its experimentally infected hosts is also large; in fact, 127 plant species belonging to 27 families, most of which are tolerant to the virus in nature, were infected by mechanical inoculations (Schmelzer, 1969).

Natural hosts among small fruit species include strawberry, raspberry, blackberry, black currant, red currant, and grapevine. The virus was isolated and identified from infected cherry, olive, peach, and plum trees. Its natural hosts also comprise other woody plants: *Robinia pseudoacacia*, *Euonymus europaeus*, *Aesculus carnea*, and rose. Natural infections were also discovered in asparagus, rhubarb (rhubarb virus 5), and elderberry. *Chenopodium amaranticolor*, *C. murale*, *C. quinoa*, *Datura stramonium*, *Nicotiana tabacum*, and *Lycopersicon esculentum* as experimental herbaceous plants are easily infected mechanically (Murant, 1981).

Symptoms

In individual hosts and their cvs., SbLtRsV causes various symptoms, including leaf mottle and decline in strawberry and raspberry; reduction of growth, rosette, and dieback in peach; mosaic in *R. pseudoacacia*; yellow mottle in *E. europaeus*; line pattern in *A. carnea*; mottle, spots, and occasional stunt in rose, etc.

Several herbaceous plant species used as indicators important for diagnosis of SbLtRsV include *Chenopodium amaranticolor*, *C. murale*, and *C. quinoa*, with chlorotic or necrotic local spots, systemic chlorosis, necrosis, and mild chlorotic mottle; and *Cucumis sativus*, with or without local spots, systemic chlorosis, or necrosis between leaf veins. Tobacco (*Nicotiana tabacum*) and petunia (*Petunia hybrida*) are tolerant to the virus (Murant, 1974).

Pathogenesis

SbLtRsV is preserved systemically in numerous herbaceous and woody hosts in nature. In the soil, it is transmitted by the nematodes *Xiphinema diversicaudatum* and *X. coxi* (Lister, 1964; loc. cit. Murant, 1974). SbLtRsV remains infective without its plant host up to 80 days in *X. diversicaudatum* (Taylor and Thomas, 1968). Both larval and adult stages of the nematode transmit SbLtRsV, which is also transmitted by the seed of several plant species, frequently above a 70% rate. SbLtRsV has been detected in seed of raspberry, celery, *Stellaria media*, *Mentha arvensis*, *Lamium amplexicaule*, and *Chenopodium quinoa* (Murant, 1974). It can be transmitted by dodder *Cuscuta californica* or *C. subinclusa* (Schmelzer, 1969) and mechanically for experimental purposes.

Control

Strawberry and plantations of other susceptible plants should be established only with healthy planting material, and infected plants should be rogued immediately from young plantations. Only fields free of vector nematodes should be used for the establishment of plantations. If soil nematodes are known or suspected, the soils should be treated with nematicides down to a depth of 50 cm.

Strawberry: Natural Alternative Host for Some Nematode-Borne Viruses

Arabis Mosaic Virus (ArMV)

ArMV was first described in Arabis (rock-cross) plants (Smith and Markham, 1944) and later in several herbaceous and woody hosts in nature. The co-infection of ArMV with other viruses in strawberry was first detected by Cadman (1960) from diseased strawberries, commonly known as "strawberry mosaic."

Strawberry, a natural host of ArMV, was discovered diseased in southern England, Scotland, Ireland, and Germany, and probably exists in other parts of Europe (loc. cit. Lister, 1970). Lister (1970) showed that several tested cvs. and *Fragaria* species are susceptible, and detrimental effects can cause unprofitable commercial production in 1 to 2 years.

Disease symptoms of ArMV infection vary significantly, depending on cv. susceptibility and strain severity. Symptoms include leaf deformations (twisted, cupped, crinkled), interveinal chlorosis, irregular yellow blotches on leaves or leaf edges, and general chlorosis, less obvious in summer. Plant stunting, severely reduced yields, and plant death 1 to 2 years after infection are general disease symptoms (Lister, 1970). *Fragaria vesca* is a susceptible symptomless host, and *Chenopodium amaranticolor*, *C. quinoa*, *Nicotiana tabacum* (inoculated leaves: chlorotic, necrotic lesions; other leaves: symptoms of systemic infections) are suitable indicator plants. Virus vectors are the nematodes *Xiphinema diversicaudatum* and *X. coxi*. ArMV frequently occurs jointly with SbLtRsV in strawberry. Although serologically different, they have common vectors and cause similar symptoms in indicator plants. General control measures recommended for SbLtRsV are applicable for ArMV.

Raspberry Ringspot Virus (RbRsV)

RbRsV, first isolated from raspberry in Scotland (Cadman, 1956), was later found in strawberry (Lister, 1958). Among numerous hosts, it is frequently observed in small fruits and fruit trees. Several strawberry cvs. and *Fragaria* species are susceptible to RbRsV (Lister, 1970). Infections of RbRsV causing economic damage occur in regions in which susceptible cvs. are grown near RbRsV-infected plants and populations of the nematode vectors *Longidorus elongatus* or *L. macrosoma*.

Symptoms of leaf chlorosis, chlorotic spots, rings, and line patterns can occur in infected leaves of some strawberry cvs. Progressive stunting and potential plant death are general symptoms in all cvs. (Lister, 1970). *Chenopodium amaranticolor*, displaying local lesions without systemic infection, is the experimental assay plant. RbRsV is transmitted mechanically from strawberry to herbaceous plants, but not vice versa. It can be transmitted, sometimes in high percentages, through strawberry and raspberry seed. Measures similar to those recommended for other strawberry nepoviruses should be applied also for the control of RbRsV.

Tomato Black Ring Virus (TmBRV)

TmBRV, first discovered as a dangerous pathogen in Scotland (Lister, 1960; 1963), was reported later in other European countries. It frequently occurs co-infecting with raspberry ringspot virus (RbRsV), which it resembles based on both incidence and economic damage in strawberry production. In nature, TmBRV infects numerous monocotyledonous and dicotyledonous plants (Murant, 1970). Several strawberry cvs. and *Fragaria* species that display symptoms similar to those caused by RbRsV are susceptible to TmBRV. The Huxley strawberry cv. is immune to TmBRV, but susceptible to RbRsV (Lister, 1970). Virus vectors are the nematodes *Longidorus attenuatus* and *L. elongatus*.

Tomato Ringspot Virus (TmRsV)

TmRsV was first identified in naturally infected strawberry plants in California (Frazier et al., 1961). In view of its frequency of occurrence in other hosts and the natural movement and spread of its vector (*Xiphinema americanum*), TmRsV probably infects cultivated strawberry in nature, but that has not been proven (Frazier and Mellor, 1970a). The host range is broad among woody and herbaceous plants. Most commercial strawberry cvs., most *Fragaria chiloensis* clones, all tested *F. vesca* indicator plants, plus the *F. virginiana* M1 clone are susceptible in experimental infections. These infections have proven lethal for several cvs., which highlights the potential dangers of TmRsV.

Strawberry: Natural Alternative Host for Some Viruses

Strawberry plants serve as alternate natural hosts for a number of viruses. One group comprises viruses that probably occur incidentally as pathogens in strawberry plants. Although of only minor economic importance, their presence must not be neglected; for example, TMV (Cornuet and Morand, 1960), TNV (Fulton, 1952), and TSkV, causal agent of necrotic shock in strawberry (Stace-Smith and Frazier, 1971).

A second group includes nematode-borne viruses of economic importance in strawberry production, such as ArMV, RbRsV, and TmBRV (*see* the chapter section on "Strawberry: Alternate Natural Host for Some Nematode-Borne Viruses").

Diseases of Strawberry Assumed Caused by Viruses

A number of diseases have been described, the causal agents of which have yet to be identified. These diseases were described in the initial reports of strawberry viruses and were named after characteristic symptoms or hosts. (See Figures 188 and 189.) Their etiological origin was usually defined, based on systemic infection, transmissibility from plant to plant by mechanical transmission, and on experimental biophysical traits in plant sap. One group of diseases is that transmitted by grafting and vector aphids, particularly important for perpetuation and spread of diseases.

FIGURE 188 Strawberry leafroll in a commercial planting. (Courtesy USDA. Reproduced with permission from APS.)

FIGURE 189 Arabis mosaic virus in strawberry.

Strawberry Chlorotic Fleck Disease (SbCFkDs)

The SbCFkDs disease markedly reduced yields of Headliner cv. plants and fruits (Horn and Carver, 1962). SbCFkDs spreads only locally and is of no great economic importance. No symptoms occur in diseased strawberry cvs. The disease agent is transmitted by grafting and by cotton aphid *Aphis gosypii*. Morphological, serological, and other traits of the causal agent transmitted by grafting are unknown (Horn and Carver, 1970).

Strawberry Latent C Disease (SbLt"C"Ds)

The SbLt"C"Ds disease described in a clone of East Malling *Fragaria vesca* (Harris and King, 1942) also spreads in the Howard 17 (Premier) cv., where all indexed plants represented disease carriers. Both single and/or complex infections always displayed detrimental effects. The diseased commercial cvs. have only erratic, nondiagnostic symptoms. Obvious symptoms of epinasty of newly developed leaves and petioles and smaller leaves without epinasty when formed later occur in some *F. vesca* clones. The causal agent was graft transmitted by strawberry aphids and by dodders *Cuscuta subinclusa* and *C. campestris* (McGrew, 1970), but no other information exists on the causal agent of the SbLt"C"Ds.

Strawberry Stunt Disease (SbSnDs)

The SbSnDs disease was described in fruiting fields of strawberries in western Oregon and Washington (Zeller and Weaver, 1941). Its symptoms have been recorded in commercial strawberry cvs.; characteristic symptoms in the Marshall cv. include severe stunt, erect petioles, leaflets erectly folded along midveins, mature leaves made papery which are smaller and rattle, seedy and hard fruits (Frazier, 1970). Frazier (1970) also determined the causal agent of the SbSnDs disease to be graft transmitted and showed that the strawberry aphid *Chaetosiphon fragaefolii* transmitts, but its etiology remains unknown.

The second group comprises diseases for which the causal agents can be graft transmitted, but not vectors. The causal agents of strawberry band mosaic, strawberry leaf roll, strawberry lethal decline, strawberry pallidosis, strawberry feather leaf, and strawberry necrosis are transmitted both by grafting and mechanically (Frazier, 1970).

Further research should enable the determination of the causal agents and their true etiological nature. Symptoms used as a diagnostic criterion are insufficient because one virus species and its strains may cause different symptoms in individual hosts and, on the other hand, several viruses may cause similar symptoms in the same host. Smith (1957) cited strawberry leaf roll as a synonymous symptom of strawberry leaf curl described by Prentice (1952) and synonymized Prentice's leaf curl (virus 5) with SbVBdV. Necrotic shock in strawberry was described as a specific disease (Frazier and Stace-Smith, 1970), and then the symptom caused by TSkV in strawberry was described under the same name (Stace-Smith and Frazier, 1971).

Transmissibility by vectors is quite consistent, but only in investigations relating to the vectors, their strains, and hosts due to potential specific relationships in such a complex. Mixed infections that occur in nature further complicate the identification of viruses and should be studied separately.

New improved methods for virus detection—both in cultivated and wild strawberries—developed recently will contribute greatly to the elucidation of the etiological nature of numerous diseases in strawberry and the definition of their relationships with other such virus diseases.

VIRUS DISEASES OF RASPBERRY (*Rubus Idaeus* and *R. occidentalis*)

Raspberry Aphid-Borne Viruses

Raspberry Vein Chlorosis Rhabdovirus (RbVCsV)

Properties of Viral Particles
RbVCsV is a bacilliform particle rounded at both ends, about 430 to 560 × 65 to 91 nm, or shorter (i.e., rounded only at one end) (Jones et al., 1977). The size was calculated by electron microscopy of fixed tissue in thin sections of infected raspberry leaves (Stace-Smith and Lo, 1973; Jones et al., 1974) and of viruliferous aphids (Murant and Roberts, 1980). RbVCsV has not been purified or mechanically transmitted. RbVCsV is highly stable in raspberry, in which it is not inactivated by exposing infected plants to temperatures of 37°C for several weeks (Jordović, 1963). Mild, moderate, and severe vein chlorosis, symptoms of disease, presumably caused by separate strains, have been described (Cadman, 1952). RbVCsV was discovered in raspberry plants displaying a vein banding mosaic (Putz and Meignoz, 1972), but it is suspected as only one of the causal agents of this disease, believed to include black raspberry and *Rubus* yellow net viruses.

Geographic Distribution and Economic Importance
RbVCsV, first discovered in Scotland (Cadman, 1952), was later detected in Canada, continental Europe, the former U.S.S.R., and New Zealand (loc. cit. Jones et al., 1987). RbVCsV causes various pathological changes; although not lethal, substantial reduction of raspberry vigor occurs. Significant berry weight reduction, thinner canes, earlier maturation, increased pollen abortion, and retarded embryo sac development were observed in some cvs. (loc. cit. Jones et al., 1987). Detrimental effects are more pronounced in infection complexes.

Host Range and Symptoms
Natural infection by RbVCsV were discovered only in red raspberry (*Rubus idaeus* and *R. idaeus* var. *strigosus*) plants, but has been graft transmitted onto Loganberry (*R. loganobaccus*) and by aphids into Apline strawberry (*Fragaria vesca* var. *semperflorens*) plants (Stace-Smith, 1961).

Many raspberry cvs. are susceptible to both natural and experimental infections of RbVCsV; for example, Golden Queen, Great American, La France, Lloyd George, Malling Exploit, Malling Jewel, Malling Promise, Norfolk Giant, and Valjevka, among others (Converse, 1987).

Symptoms of RbVCsV-infected raspberry vary, depending on host susceptibility, virus strain, and growing conditions. Yellow-net patterns occurring along smaller leaf veins characterize this disease. In tolerant cvs., vein chlorosis occurs in patches; while in susceptible cvs., leaf deformity and epinasty can occur (Jordović, 1963). In least susceptible cvs., no symptoms of disease occurs. Based on symptomology, Lloyd George, Malling Delight, Norfolk Giant, and Washington cvs. represent suitable indicator plants (Jones et al., 1977).

Pathogenesis

Infected raspberry plants are natural perennial sources of inoculum for further systemic infection. RbVCsV is introduced into young plantations, primarily via infected nursery stocks. Jones et al. (1987) reported on the wide dissemination of RbVCsV in infected planting material in continental Europe where many stocks of the older and most popular cvs. are 100% infected.

Aphis idaei is a vector of RbVCsV in fields, requiring at least 24 hours for the acquisition of RbVCsV, which they retain for at least 24 hours (Stace-Smith, 1961). The greatest incidence of transmission of RbVCsV (46%) followed acquisition and inoculation access periods of 7 and 30 days, respectively (Jordović, 1963). RbVCsV transmission by grafting is experimentally important; otherwise, RbVCsV is not transmitted mechanically or through seed of red raspberry (Jordović, 1963).

Control

For control of RbVCsV, it is important to use virus-free planting material for establishment of raspberry plantations. Uniformly RbVCsV-infected stocks of some cvs. represent serious future disease pressure. Potential sources of infection by RbVCsV are reduced by rogueing infected plants from younger plantations. Vector control can help reduce infection of healthy planting material, but such measures in the field, when favorable conditions for vector development exist, are quite inefficient.

Some North American cvs. (e.g., Guthburt, Lathan, and Viking) seem immune to inoculation and field infection (loc. cit. Jones et al., 1987). However, existing sources of RbVCsV resistance, possibly from other *Rubus* species, might eventually be incorporated into raspberry cultivars of importance for large-scale production in the future.

Raspberry: Natural Alternative Host for Some Aphid-Borne Viruses

Black Raspberry Necrosis Virus (BRbNV)

Raspberry ranks among the most important natural hosts of BRbNV. (See the description of BRbNV in "Virus Diseases of Black Raspberry.")

Rubus Yellow Net Virus (RsYNtV)

RsYNtV, first isolated from naturally infected Himalaya blackberry (*Rubus procerus*) plants, was later described as one of the viruses in the raspberry mosaic disease complex (Stace-Smith, 1956). As a component of this disease, RsYNtV is believed to be spread worldwide wherever raspberries are grown. No yield loss data are available on the damage caused by infection with RsYNtV alone. Important damage is expected in complex infections with other viruses (e.g., black raspberry necrosis virus).

Net-like chlorosis along the veins, but without important changes in the shape, size, and vigor, characterizes RsYNtV-infected red raspberry leaves (Figure 190) (Stace-Smith, 1955). *Amphorophora agathonica* in North America and *A. idaei* in Europe are well-known vectors of RsYNtV. Raspberry cvs. from North America with the aphid vector immunity genes ensure comparatively little damage from RsYNtV (Stace-Smith and Jones, 1987).

Cucumber Mosaic Virus (CMV) in raspberry

CMV in raspberry (*Rubus idaeus* cv. Lloyd George) was first discovered in Scotland (Harrison, 1958) and later in the Soviet Far East (Gordejchuk et al., 1977). CMV has a large host range, among which raspberry is infrequent. Pale-green blotches that do not influence infected plant vigor occur on red raspberry (cv. Lloyd George) leaves (Harrison, 1958). Gordejchuk et al. (1977) observed small leaves with mild chlorotic mottle in CMV-infected raspberry cv. Visluha. In nature, CMV is easily transmitted mechanically by infected plant sap, the seed of some hosts, and numerous aphid vectors, which is highly important for the pathogenesis of the disease caused in its hosts (see the description of this virus in its main host).

FIGURE 190   Raspberry yellow net virus in raspberry leaves.

Raspberry Nematode-Borne Viruses

Raspberry Ringspot Nepovirus (RbRsV)

Properties of Viral Particles

RbRsV isometric, 20 nm in diameter, with particles of equal size that sediment in three components—Top, Middle, and Bottom—with sedimentation coefficients of 50, 92, and 130S, respectively (Murant, 1987). The coat protein consists of a single species of 54 kDa (Mayo et al., 1971).

The viral genome is composed of two single-stranded RNA species (RNA_1 = 2400 kDa and RNA_2 = 1400 kDa); the M component contains one RNA_2 molecule, while the B component has one RNA_1 and two RNA_2 molecules (Dunez, 1988a). The virus infects only when both RNA species are present (Murant, 1987).

The TIP of RbRsV (in *Nicotiana rustica* sap) is 65 to 70°C; the DEP 1×10^{-3} to 1×10^{-4}; and infectivity at room temperature is maintained for 2 to 3 weeks (Harrison, 1958a). RbRsV is strongly immunogenic.

Three strains of RbRsV have been described. The Scottish and the English strains differ, based on antigenic properties and vector specificity; and the Lloyd George yellow blotch strain resembles the Scottish strain serologically, plus it infects raspberry cvs. that are immune to the Scottish strain (Murant, 1987).

Geographic Distribution and Economic Importance

RbRsV was observed first in Scotland, then later in other parts of Great Britain, continental Europe, the former U.S.S.R., and Turkey (Murant, 1987). RbRsV causes plant decline in some raspberry cvs. and the resultant damage in production is directly correlated with the number of dead plants.

Host Range

RbRsV has a broad range of natural hosts among monocotyledonous and dicotyledonous plants. Perennial hosts among both small fruits and fruit trees play an important role in providing a reservoir of RbRsV inoculum (Converse, 1987). Species naturally susceptible to RbRsV are raspberry (*Rubus idaeus*), blackberry (*R. procerus* cv. Himalaya Giant and *R. sachaliensis*), *Fragaria* (strawberry), *Ribes* species (black currant, gooseberry, red currant), sweet cherry (*Prunus avium*), peach

FIGURE 191 Raspberry ringspot virus in raspberry leaf.

(*P. persica*), and grapevine (*Vitis vinifera*). *Beta vulgaris* subsp. *saccharifera, Capsella bursa pastoris, Cerrestium vulgatum, Myosotis arvensis, Polygonum convulvus, Spergula arvensis, Stellaria media, Veronica agrestis,* and *V. persica* are among important naturally susceptible herbaceous hosts. Experimental diagnostic plants are *Chenopodium amaranticolor* (chlorotic necrotic local spots), *C. quinoa* (chlorotic necrotic local spots and systemic chlorotic spots), *Nicotiana clevelandii* (local necrotic spots and rings; systemic veinal necrosis), among others.

Symptoms

Pale-green irregular spots or ringspots occur initially in the spring following the year of infection on raspberry leaves (Figure 191). Less obvious, sometimes even disappearing during the summer, these symptoms reappear on new leaves in the autumn. Leaf curl characterizes some cvs. (e.g., Norfolk Giant). In some susceptible cvs. (Malling Enterprise, Malling Jewel, Malling Notable, etc.), diseased canes are stunted and brittle, and such plants frequently die two to three years after symptom occurrence. Infected plants of other cvs. that are frequently symptomless or display few symptoms remain less vigorous, but do not die.

Pathogenesis

In nature, RbRsV is preserved in perennial woody and herbaceous plants, and in infected seed of many hosts. Soil-borne nematode vectors have specific relationships in the transmission of some strains of RbRsV. *Longidorus elongatus* is a common vector of the Scottish strain (Taylor, 1962), whereas *L. macrosoma* and *L. attenuatus* are the natural vectors of the English or continental European strains (Taylor and Murant, 1969).

RbRsV, seed borne in many hosts, results in infected progeny seedlings usually without symptoms. RbRsV is transmitted by raspberry (up to 20%) and strawberry (up to 40%) seed from either male or female plants. Infected seed represents an important reservoir for virus transmission from year to year. RbRsV can be graft transmitted or mechanically transmitted by infective plant sap which has enabled detailed studies of the virus and experimental host susceptibility.

Control

Main control measures include the use of virus-free planting material for establishment of plantations, control of nematodes by nematicide application, or crop rotation in combination with plants that are immune or are unsuitable vector hosts for supporting populations. The use of resistant raspberry cvs. is vitally important because this nematode- and seed-borne virus is difficult to control in production.

Raspberry: Natural Alternative Host for Some Nematode-Borne Viruses

Arabis Mosaic Virus (ArMV)

ArMV has been observed in raspberry plants in the British Isles, continental Europe (France, Germany), and in the Soviet Far East. It is related closely to the sap-transmissible, soil-borne virus that causes the raspberry yellow dwarf disease (Cadman, 1960). ArMV is economically damaging only where large numbers of plants are infected.

Symptoms of disease vary, depending on raspberry cv. and virus strain. The leaves of cvs. Malling Exploit, Malling Promise, and Malling Admiral exhibit yellow specks and vein yellowing, or yellow net clearly evident on lower leaves. The canes are stunted and produce few or no fruits. Some raspberry cvs., such as Burnetholm, Malling Enterprise, Malling Leo, Malling Notable, and Malling Orion, resist infection by ArMV (common strain) (Murant, 1987a).

ArMV is graft transmitted, mechanically transmitted by infective plant sap, and seed transmitted in several plants. Its primary natural vector is *Xiphinema diversicaudatum*. Primary virus control measures include the use of virus-free planting stock material, chemical control of vectors, and resistant cvs. in raspberry production.

Cherry Leafroll Virus (ChLrV)

ChLrV was discovered in red raspberry plantations in New Zealand with an incidence of infection within plantations of from one to two plants to over 70% (Jones and Wood, 1978). Infected fruiting raspberry canes display severe leaf symptoms and decreased vigor. Spread of ChLrV can be restricted by planting healthy material for establishment of the raspberry plantation.

Strawberry Latent Ringspot Virus (SbLtRsV)

SbLtRsV has been observed in raspberry in the British Isles, France (Putz and Stocky, 1970), and Italy (Vegetti et al., 1978). It infects raspberry alone or co-infects with ArMV. The raspberry cvs. Lloyd George, Malling Delight, and Norfolk Giant are resistant to the SbLtRsV common strain.

Tomato Black Ring Virus (TmBRV)

Raspberry infected with TmBRV has occurred in Scotland and other parts of Great Britain and only seldom in the former U.S.S.R. (Murant, 1987). Symptom severity varies in individual raspberry cvs. Severe symptoms occurred in the Seedling V cv. (Harrison, 1958), but most other cvs. express comparatively mild symptoms. Ringspots or mild chlorotic mottle first occurs on the leaves of the cv. Malling Exploit, which are later accompanied by cane stunt, fruit deformation, and reduced yields (Taylor et al., 1965). The Norfolk Giant cv. displays only leaf curl symptoms and seems less affected by TmBRV. Obvious symptoms of disease appear in the Burnetholm, Malling Delight, Malling Leo, and other cvs., while Lloyd George, Malling Admiral, Malling Enterprise, Malling Jewel, and other cvs. resist infection by TmBRV (Murant, 1987).

Longidorus elongatus is the common vector of the Scottish strain. TmBRV is transmitted by the seed of numerous hosts, including raspberry and strawberry, and can be transmitted experimentally through grafts and by mechanical inoculations of infected plant sap. The same measures of control as recommended for RbRsV should be adhered to for TmBRV.

Tomato Ringspot Virus (TmRsV)

TmRsV occurs in raspberry in North American and European countries. It is the most common virus occurring in raspberry production in Oregon and Washington (Converse et al., 1970).

Characteristic symptoms of the shock reaction develop the year following infection. Chlorotic ringspots, line patterns, and mild vein chlorosis occur on the new leaves developing in the spring, whereas lower leaves that developed during warm-weather conditions remain symptomless (see Figures 192 and 193).

TmRsV reduces drupelet set in cv. Fairview, but not in others (Glen Clova, Malling Jewel, etc.) (Daubeney et al., 1975). TmRsV-infected plant yields were reduced the third cropping year in cvs. Lloyd George, Lathan, Glen Clova, and others, but no significant reductions occurred in other cvs. such as Canby, Malling Jewel, etc. (Freeman et al., 1975).

The nematodes *Xiphinema americanum* and *X. rivesi* are the vectors of TmRsV, which is otherwise transmitted by infective plant seed in low percentages, and experimentally by mechanical sap inoculation from raspberry onto herbaceous indicator plants. Virus control mandates the use

FIGURE 192 Red raspberry plant infected with tomato ringspot virus with typical ring spots and line patterns. (Courtesy F.D. McElroy. Reproduced with permission from APS.)

FIGURE 193 Crumbly fruit (bottom) of red raspberry infected with tomato ringspot virus; healthy (top). (Courtesy F.D. McElroy. Reproduced with permission from APS.)

of virus-free planting material, nematicide treatment where necessary, and the use of resistant or tolerant raspberry cultivars.

(More detailed descriptions of the traits of this group of viruses can be found in their descriptions in main hosts.)

RASPBERRY POLLEN-BORNE VIRUSES

Raspberry Bushy Dwarf Virus (RbBuDV)

Properties of Viral Particles

RbBuDV is isometric, 33 nm in diameter, and consists of a single particle with a sedimentation coefficient of 115S. The protein coat consists of a single major protein of 29 kDa and three single-stranded RNA species of 2200, 900, and 400 kDa (Murant, 1987b). RbBuDV has a bipartite RNA genome, and RNA_3 is a subgenomic fragment of RNA_2 (Natsuaki et al., 1991). Although it resembles the particle morphology, total genome size and the pollen transmission characteristics of Ilarviruses, RbBuDV is classified differently.

The TIP in *Chenopodium quinoa* sap for the typical isolate from raspberry is 65°C; the DEP is 1×10^{-4}; and the LIV at 22°C is 4 days (Barnett and Murant, 1970). RbBuDV is moderately immunogenic.

All investigated red raspberry isolates originating from various geographical regions display no serological differences. Some *Rubus occidentalis* isolates differed only slightly in serological test and *in vitro* properties from *R. idaeus* isolates (Murant and Jones, 1976). Strains (isolates) of RbBuDV have been shown capable of infecting raspberry cvs. that contain the Bu gene, which accords resistance of raspberry to the type isolate (Barbara, 1988).

Geographic Distribution and Economic Importance

The disease caused by RbBuDV was first described in red raspberry cv. Lloyd George in Great Britain (Cadman and Harris, 1951), then in North America, New Zealand, Australia, South Africa, and Chile. It has been distributed or spread worldwide wherever raspberry cvs. are grown (loc. cit. Murant, 1987b). Damage caused by RbBuDV varies, depending mainly on the level of cv. susceptibility. Significant yield reductions and reductions in cane height and diameter occurred in Canby and Lloyd George cvs., while the Meeker and Creston cvs. were comparatively tolerant to RbBuDV (Daubney et al., 1982). RbBuDV contributes significantly to damage in mixed viral infections (i.e., one of the causes of bushy dwarf disease), resulting in plant decline in the cv. Lloyd George.

Host Range

Natural hosts of RbBuDV include plants belonging to the genus *Rubus*: *Rubus idaeobatus* (blackberries), *R. idaeus* (red raspberry), *R. occidentalis* (black raspberry), *R. phoenicolasius* (wineberry), *R. sacchaliensis*, and *R. vulgatus* ssp. *buschii*. RbBuDV has been transmitted experimentally into 53 plant species in 12 dicotyledonous families (Barnett and Murant, 1971). This group of hosts comprises *Fragaria vesca* (strawberry), *Rubus* ssp. (*R. henryi*, *R. laciniatus*, *R. procerus*, and others), *Cydonia oblonga*, plus herbaceous hosts. *Chenopodium amaranticolor*, *C. murale*, *C. quinoa*, *Cucumis sativus*, *Phaseolus vulgaris* cv. Prince, etc. (Murant, 1987b).

Symptoms

Disease symptoms vary, depending both on the susceptibility of individual cultivars and on whether infections occur alone or in complex with other viruses and/or on environmental influences (see Figure 194). Plants of susceptible raspberry cvs. infected with RbBuDV alone are symptomless, but mixed infections cause severe and characteristic symptoms in these same cvs.; for example, the susceptible Lloyd George cv. displays clear bushy dwarf symptoms when infected with RbBuDV along with BRbNV (Cadman and Harris, 1951). Characteristic symptoms of both diseases include shorter canes that develop slowly in the spring, chlorotic and red shoots, downcurled leaves, and plants tending to form fruits in the autumn.

FIGURE 194 Ring spots and line patterns on a leaf of 'Meeker' red raspberry infected with raspberry bushy dwarf virus. (Courtesy H.A. Daubeny, Agriculture Canada. Reproduced with permission from APS.)

Drupelet abortion resulting in crumbly fruit occurs in some RbBuDV-infected cvs. (Daubeney et al., 1975). That these symptoms are not stable year by year suggests an influence of environmental conditions. Infections associated with RbBuDV cause yellows in some cvs. (Jones et al., 1982), a symptom observed in several raspberry cvs., such as Great American, Malling Landmark, Norfolk Giant, and others (Cadman, 1952). Vein chlorosis on lower leaves is the initial yellows symptom, which gradually extends partially or completely encompassing the leaf blades. For diagnostic purposes, it is very important to distinguish between infectious yellows disease and physiological yellows disorders.

Characteristic symptoms in some diagnostic plants include transient local lesions, systemic chlorotic rings and line patterns (*Chenopodium amarnaticolor*), local necrotic rings without systemic infection (*C. murale*), symptomless systemic infection (*Nicotiana clevelandii*), etc. *Chenopodium quinoa* is a suitable assay plant (Murant, 1976).

Pathogenesis

RbBuDV is transmitted through seed up to 77% and pollen of infected raspberry, by *Rubus phoenicolasius* seed up to 15%, and by strawberry seed up to 2% (loc. cit. Murant, 1976). In nature, RbBuDV is spread by plant pollination, seemingly the main mode of field dissemination. Propagation of infected planting materials plays an active role in the infection of new raspberry plantations.

RbBuDV is experimentally graft transmitted and mechanically transmitted with infected raspberry sap. Natural vectors of RbBuDV are unknown.

Control

New raspberry plantations should be established with virus-free planting material and with spatial isolation from infected cultivated and wild *Rubus* plants. The flowers should be regularly eliminated from the plants in cane nurseries for the purpose of avoiding infections by pollen. Healthy plants can be obtained through continuous heat treatment of several weeks at 36°C and propagation from shoot tips obtained following such treatment (loc. cit. Murant, 1987b). The use of resistant cultivars represents the most successful method of virus control (Converse, 1987).

DISEASES OF RASPBERRY ASSUMED CAUSED BY VIRUSES

Raspberry Leaf Mottle Disease/(RbLMtDs) and Raspberry Leaf Spot Disease (RbLsDs)

RbLMtDs and RbLsDs are described as distinct raspberry diseases, otherwise with a close resemblance based on symptoms in infected plants, vector relationships, and reactions to thermotherapy.

Their symptomological reaction in some raspberry cvs. has been classified as raspberry mosaic 2, raspberry chlorotic spot, and leaf spot mosaic.

No causal agent of either disease has been determined; if viruses, they have been neither purified nor detected by electron microscopy in infected raspberry plants (Jones, 1988). Although unidentified, circumstantial evidence points to the causal agents as viruses in their characteristic production of disease as a systemic infection, by their vector transmissibility, and by their responses to thermotherapy.

RbLMtDs and RbLsDs, as described in several countries (Europe, New Zealand, Australia and North America), are not always clearly distinguishable (loc. cit. Jones, 1987). Both diseases have been discovered to occur separately in nature, but frequently also simultaneously in wild European raspberry, cultivated red raspberry, several cultivated blackberries, and in *Rubus gracilis* and *R. occidentalis*. All investigated, experimentally infected *Rubus* species are susceptible to the causal agent of these diseases (Jones and Jennings, 1980). Symptoms in susceptible plants infected with the agents of one and/or other disease are difficult to distinguish.

Most red raspberry cultivars are symptomless, while a few susceptible to infection react with severe symptoms, frequently resulting in death (Jones and Murant, 1975; Jones and Jennings, 1980). Plant death can occur quickly, or even two to three years after infection. Symptomless latent infections with RbLMtDs and RbLsDs plus other viruses are believed to cause degeneration of vigor in some raspberry cvs. (Jones, 1980; 1981).

Causal agents of RbLMtDs and RbLsDs are probably transmitted semipersistently by the aphid *Amphorophora idaei*. The incidence and rate of spread of infection in the field depends on aphid dynamics. Both agents of disease are graft transmissible, but not mechanically by infective plant sap inoculations.

Establishment of plantations with healthy planting material and adequate spatial isolation from existing known diseased plants are the best ways to prevent disease occurrence. Virus-free planting materials have been obtained by heat treatment, preferably in combination with meristem-tip culture (loc. cit. Jones, 1987). Disease control can be enhanced by producing cvs. that resist vector colonization (Jones, 1979).

Raspberry Leaf Curl Disease (RbLCuDs)

RbLCuDs is a characteristic virus disease of *Rubus* plants recognized in North America. In the U.S. and Canada, it occurs in nearly all regions where raspberries are grown (Stace-Smith and Converse, 1987). A symptom caused by RbRsV and TmBRV in some raspberry cvs. has been described as leaf curl in Europe and the former U.S.S.R. (Murant, 1987). The causal agent of RbLCuDs, an assumed virus, lacks data on morphological, serological, and physical/chemical traits.

Natural hosts of RbLCuDs belong only to the *Rubus* genus: that is, *R. idaeus*, *R. idaeus* var. *strigosus*, *R. occidentalis*, etc. Experimentally infected hosts are in the genera *Fragaria* and *Rubus*. All tested red raspberry cvs. are susceptible (Stace-Smith, 1962). Plants remain symptomless or affected only by slight downcurling of apical leaves in the first year of infection. The next year, canes curl, leaves turn yellow, and shoots proliferate, resulting in rosettes. Diseased plants are stunted and frequently winter-killed. Fruits from infected plants are small and crumbly, with yield reduced by 40% (loc. cit. Stace-Smith and Converse, 1987).

Aphis rubicola is a natural vector of the causal agent of RbLCuDs and plays an important role in field spread. Red raspberry cvs. differ in their resistance to *A. rubicola* colonizations, but cvs. immune to colonization are not known (Stace-Smith and Converse, 1987). Infections may be experimentally graft transmitted, but not by mechanical transmission of infective sap plant. Prevention of disease occurrence and control measures include the use of healthy plants in propagation, treatment with aphicides to reduce the vector population, and the use of raspberry cvs. resistant to colonization by *A. rubicola*.

Other viruses (e.g., ApMV, Figure 195) cause noticeable diseases in red raspberry.

FIGURE 195 Brilliant yellow blotches and ring patterns in leaf of 'Schoenemann' red raspberry naturally infected with apple mosaic virus. (Reprinted with permission from Baumann et al., 1987. Reproduced with permission from APS.)

VIRUS DISEASES OF BLACK RASPBERRY (*Rubus occidentalis*)

Black Raspberry Necrosis Virus (BRbNV)

Properties of Viral Particles

BRbNV occurs in low concentrations in raspberry (natural host) and *Chenopodium quinoa* (experimental host) plants, which complicates its purification and characterization. In partially purified preparations, a few virus-like particles found were 24 to 30 nm in diameter (Jones and Murant, 1972; Murant et al., 1976). Similar particles were observed by electron microscopy in ultrathin sections of other plants, (e.g., *Rubus henryi, R. occidentalis* and *C. quinoa*) infected with the virus (loc. cit. Stace-Smith and Jones, 1987a). Later data showed isometric particles of 28-nm diameter, with the infective component sedimenting at 130S (Jones, 1988a).

The TIP of BRbNV during the winter in *Chenopodium quinoa* sap is 50 to 52°C; the DEP is 1×10^{-1} to 1×10^{-2}; and the LIV at 18°C is 6 to 24 days (Jones and Murant, 1972). BRbNV has poor immunogenic activity. BRbNV causes systemic necrosis in black raspberry similar to the agents of RbLMt and RbLs diseases whose properties remain undefined.

Geographic Distribution and Economic Importance

BRbNV was first described in British Columbia as the causal agent of severe tip necrosis in black raspberry seedlings or the causal agent of a latent and mild disease in red raspberry cvs. (Stace-Smith, 1955a). In North America, it was consistently discovered jointly with Rubus yellow net virus (RsYNtV) in the raspberry mosaic complex (Stace-Smith, 1956). Among other aphid-borne viruses, a virus identical to BRbNV was also described in Europe (Jones and Roberts, 1977).

BRbNV ranks in the group of the commonest viruses affecting black raspberry and raspberry crops. The reactions of black raspberry cvs. to BRbNV vary, but the most tolerant cvs. can be seriously affected (Stace-Smith and Jones, 1987). In Great Britain, BRbNV is one of the first aphid-borne viruses occurring in newly established plantations that can completely infect the Malling Jewel cv. within its first fruiting year (Jones, 1976 1979).

All red raspberry cvs. tested are susceptible to BRbNV, and its damage depends on the duration of infection and cv. susceptiblity. In North America, for example, infections with BRbNV did not influence yields of some cvs.; but in others, yields were reduced by 30% in the first cropping year

(Stace-Smith and Jones, 1987). In Great Britain, BRbNV infections of some cvs. such as Malling Jewel combine with other virus infections and cause degeneration of plantations (Cadman, 1961; Jones, 1981).

Host Range

Among the natural hosts of BRbNV in the genus *Rubus*, economically important ones include raspberry (*Rubus idaeus*), red raspberry (*R. idaeus* var. *strigosus*), and black raspberry (*R. occidentalis*). Other natural virus hosts are *R. fruticosus*, *R. leucodermis*, *R. procerus*, Thornless Young derivative of *R. lasiocarpus* var. *rosifolius*, *R. allegheniensis* cv. Darrow, *R. phoenicolasius*, and *R. loganobaccus* (loc. cit. Stace-Smith and Jones, 1987). The virus has been graft transmitted experimentally or aphid transmitted onto *R. albescens*, *R. henryi*, *R. laciniatus*, and *R. molaccanus*. Herbaceous hosts infected by raspberry sap include *Chenopodium amaranticolor*, *C. quinoa*, *C. murale*, *Petunia hybrida*, *Spinacia oleracea*, *Gomphrena globosa*, and *N. debneyi* (Jones and Murant, 1972; Murant et al., 1976).

Symptoms

Disease symptoms vary, depending mainly on host species and/or environmental conditions. The first symptom to appear in black raspberry plants following aphid inoculation of BRbNV is bending of shoot tips, which later curl down, become brittle, and leaves below the apex develop only partially and gradually wilt. Further disease development includes necrosis of petioles, midribs, infolding leaves, and stem tips. These characteristic symptoms of shock reactions result in plant wilt and death. Mottle varying in intensity and severity develops on the leaves of surviving plants.

Most BRbNV-infected red raspberry cvs. display no symptoms. Other cvs. (e.g., Malling Admiral, Malling Orion, Taylor, and Washington) exhibit chlorotic spots and mottle of leaf veins (Jones and Jennings, 1980). Infected plants of susceptible cvs. produce smaller fruits, thinner and shorter canes, resulting in reduced yields.

Depending on test conditions, symptoms such as chlorotic mottle and spots vary in severity, and apical necrosis with leaf deformations and epinasty occur on the main *Rubus* spp. hosts.

Experimental herbaceous *Chenopodium murale* hosts display necrotic local lesions without systemic infection. Local chlorotic and necrotic spots develop on inoculated *C. amaranticolor* leaves, followed by systemic symptoms. Systemic chlorosis and necrosis occur in the winter of BRbNV-infected *Spinacea oleracea* (Jones and Murant, 1972; Murant et al., 1976).

Pathogenesis

Natural sources of infection in which BRbNV is preserved from year to year are infected cultivated and wild plants of the *Rubus* genus; all clones of some old red raspberry cvs. are infected with BRbNV, which enabled its spread into several parts of the world where raspberries are grown (Stace-Smith and Jones, 1987).

Aphids transmit BRbNV semipersistently in nature. In North America, *Amphorophora agathonica* is the primary vector of BRbNV, and *A. sensoriata*, *A. rubicumberlandii*, and *A. illinoia* (*Mesonaphis*) *rubicola* are less important. In Europe, *A. idaei* is the primary vector of BRbNV, which is also experimentally transmitted from red raspberry into *Chenopodium quinoa* seedlings by *Aulacorthum solani*, *M. euphorbiae*, and *A. idaei*; but BRbNV could not be transmitted by these aphids from *C. quinoa* plants (loc. cit. Stace-Smith and Jones, 1987).

BRbNV can be transmitted mechanically by infective sap from red raspberry and black raspberry into *Chenopodium quinoa*. Such transmission from red raspberry depends on environmental conditions during the vegetation period. It transmits more readily in spring and autumn than in summer, and transmission from glasshouse-grown raspberry is difficult throughout the year (Jones and Murant, 1972; Jones and Roberts, 1977; Jones and Jennings, 1980). BRbNV can be graft transmitted from *Rubus* to *Rubus*, but not from infected herbaceous hosts into *Rubus* (Murant et al., 1976). BRbNV is not transmitted through seed of red raspberry (Jones and Murant, 1972).

Control

Growing of cvs. that resist aphid vector colonization represents the primary control measure. In North America, several resistant cvs. maintain such resistance for several years (Converse et al., 1971). In Great Britain, with four *Amphorophora idaei* biotypes, extensive studies of incorporating resistance into each biotype against them are well underway. Full resistance would represent the most efficient preventive measure against BRbNV spread, although incomplete resistance can also be useful under certain conditions.

BRbNV isolates differ in sensitivity to heat treatment, but most can be eliminated from *Rubus* spp. by treatment at 32 to 37°C for 1 to 4 weeks (Chambers, 1954; Converse, 1963), plus by excising and rooting tip cuttings during heat treatment (Bolton and Turner, 1962).

Black Raspberry Latent Virus (BRbLtV)

Properties of Viral Particles

BRbLtV particles are isometric, 26 nm in diameter, and sediment as Top, Middle, and Bottom, with respective sedimentation coefficients of 81, 89, and 98S at pH 5, and 78, 88, and 93S at pH 7 (Lister and Converse, 1972). Chemical composition is still unknown and only the B particles appear infective. The BRbLtV, according to the same authors, has a TIP in *Chenopodium quinoa* sap of 46 to 49°C, a DEP of 1×10^{-4}, and the LIV is a few hours at 20°C to 3 to 5 days at 4°C. BRbLtV has poor immunogenic activity. No significant differences exist among virus isolates and the type isolate obtained from black raspberry cv. New Logan (Converse, et al., 1966; Converse, 1967).

Geographic Distribution

BRbLtV, first discovered in Maryland, was later found in most cvs. in the eastern U.S. (Converse and Lister, 1970). No precise data on yield losses caused by BRbLtV are available.

Host Range, Symptoms, and Pathogenesis

Raspberry cvs. Black Hawk, Bristol, Cumberland, Morrison, and New Logan, plus some red raspberry cvs., are natural BRbLtV hosts, but with no symptoms. BRbLtV has been transmitted experimentally from infected raspberry into some herbaceous plants, *inter alia*, including *Amaranthus caudatus, Mormordica balsamina, Nicotiana glutinosa, Phaseolus vulgaris, Spinacea oleracea*, etc. (Converse and Lister, 1970).

Diagnostic plants include *Chenopodium quinoa* (local vein chlorosis followed by necrotic, chlorotic lesions, and severe systemic infection symptoms) *Nicotiana tabacum* cvs. White Burley and Turkish (local chlorotic, necrotic ringspots and line patterns), *Gomphrena globosa* (small red rings and spots), etc. (Lister and Converse, 1972).

BRbLtV is seed borne in raspberry and seed and pollen borne in black raspberry. It was seed transmitted up to 10% in New Logan and Cumberland black raspberry cvs. (Lister and Converse, 1972). The virus is also transmitted into *Chenopodium quinoa* and other herbaceous hosts via the sap from flower parts and very young infected black raspberry plants. No vectors are known, and thus, transmission of BRbLtV by pollen may be most important.

Black Raspberry: Natural Alternative Host for Some Viruses

Raspberry Bushy Dwarf Virus (RbBuDV)

Black raspberry is the natural host frequently infected with RbBuDV in the U.S. (Converse, 1973; Murant and Jones, 1976). Black raspberry cvs. susceptible to RbBuDV include Munger, New Logan, and Plum Farmer, and those not found naturally infected in the field are Cumberland, Black Hawk, and Bristol (Murant, 1987). According to Converse (1973), field-infected plants of susceptible cvs., plus graft-inoculated cv. Munger plants, remained symptomless. Seedlings infected through the seed also remain symptomless. Virus-free plants can be obtained by thermotherapy (*see* the description of the raspberry bushy dwarf virus).

Rubus Yellow Net Virus (RsYNtV)

In addition to red raspberry (*Rubus idaeus*), black raspberry (*R. occidentalis*) is the the only natural host of RsYNtV. Net-like chlorosis is diagnostic of infection in black raspberry seedlings after inoculation of RsYNtV by viruliferous aphids. Spots of net-like chlorosis on the fourth and fifth leaves below the apices are the initial symptoms of disease. RsYNtV vein chlorosis spreads progressively to young leaves, and affected plant parts develop slowly. Older leaves below the first infected leaf remain unaltered in appearance, but net-like chlorosis spreads toward shoot apices. Finally, affected leaves become chlorotic, stunted, and they cup downward.

Raspberry Leaf Mottle Disease (RbLMtDs)

Black raspberry, *Rubus occidentalis*, is a natural host of the agent causing RbLMtDs in raspberry. Following graft inoculation, symptoms of mosaic and apical necrosis occur in infected black raspberry plants (Jones and Jennings, 1980).

Raspberry Leaf Curl Disease (RbLCuDs)

The symptoms of RbLCuDs in black raspberries are similar to those in red raspberries. Infected leaves are small, firm, and almost circular. In chronic infections, young canes become stiff, brittle, and frequently branchless (Stace-Smith and Converse, 1987).

Tomato Streak Virus (TmSkV)

As opposed to other *Rubus* hosts, TmSkV is prevalent in black raspberry cvs. Its high incidence is believed a result of establishment of plantations with infected stock. TmSkV-infected black raspberry plants, although systemically invaded, are symptomless and therefore should be detected by an index procedure and then eliminated from young plantations. TmSkV is transmitted through the seed (about 6%) and via pollen originating from infected black raspberry (Converse and Lister, 1969).

Tobacco Ringspot Virus (TRsV)

TRsV has been recorded as naturally infecting wild blackberry (Figure 196).

A Disease of Black Raspberry Assumed Caused by a Virus

Black Raspberry Streak Disease (BRbSkDs)

BRbSkDs is one of the oldest virus diseases described in black raspberry (Wilcox, 1923). It was long believed to cause serious yield reductions. However, more recent information shows it is of lesser economic importance and, thus, now classified as a minor disease (Stace-Smith, 1987).

Black raspberry, *Rubus occidentalis*, is a natural host in which the presence of BRbSkDs has been proved. BRbSkDs was named for the characterisitc faint blue or gray streaks occurring on and under the surface bloom of young canes and sometimes also on fruiting canes (Converse, 1970). Streak symptoms that develop during warm but not cool weather have a dependence on the environment which results in unreliable disease identification (Stace-Smith, 1987).

BRbSkDs is experimentally transmitted by dodder, *Cuscuta subinclusa*, and grafting. Disease etiology has not been studied, but based on mode of transmission, it is presumably of viral origin (Converse, 1970).

Virus and Virus-Like Diseases of Some Other *Rubus* spp.

Bramble Yellow Mosaic Virus (BmYMV)

BmYMV is a filamentous particle of 750 nm (Engelbrecht, 1987). It has a TIP in *Chenopodium murale* sap of 50°C, a DEP of 1×10^{-3} and an LIV at 20°C of 8 days.

BmYMV was isolated in the Cape Province of South Africa from wild trailing blackberry, *Rubus rigidus*, its only known natural host (Engelbrecht and van der Walt, 1974). Early in the

FIGURE 196 Mottling and leaf distortion in wild blackberry, caused by tobacco ringspot virus. (Courtesy G.V. Gooding, Jr.; reprinted from Stace-Smith, 1987b. Reproduced with permission from APS.)

spring, leaves of young canes are usually symptomless, then yellow mosaic and line patterns develop on the leaves, which turn pale and whitish by the end of the season.

BmYMV has been transmitted experimentally into herbaceous *Chenopodium murale*, *C. quinoa*, *Gomphrena globosa*, and *Nicotiana tabacum* cv. White Burley plants. *C. murale* is a suitable assay plant (chlorotic lesions and later systemic yellow necrotic rings develop on inoculated leaves). BmYMV spreads easily by mechanical inoculation of sap from infected blackberry and strawberry plants, and it can be transmitted through infected *C. murale* seed at 86 to 100% of seedlings (Engelbrecht, 1987).

Wineberry Latent Potexvirus (WbLtV)

WbLtV is a filamentous particle 510×12 nm that contains RNA of 2700 kDa and two proteins, the major one of 31 kDa, and the minor one of 28 kDa (loc. cit. Jones, 1985). The TIP of WbLtV in *Chenopodium quinoa* sap is 65 to 75°C; the DIP is 1×10^{-3} to 1×10^{-4}; and the LIV at 18°C is 8 to 16 days (Jones, 1987). WbLtV has poor immunogenic activity.

WbLtV was isolated from *Rubus phoenicolasius* plants imported from the U.S. into Scotland. Naturally infected *R. phoenicolasius* and experimentally graft-infected *Rubus* species and cvs. express no symptoms of disease. WbLtV has been transmitted by mechanical inoculations into herbaceous plants in five families, among which mild systemic infection occurs only in a few *Chenopodium* species. Diagnostic plants include *Chenopodium album*, *C. amaranticolor*, *C. murale*, and *C. quinoa* (which produce small necrotic local lesions and with symptomless systemic infection in *C. album* and *C. quinoa*), *C. ambrosoides* and *Gomphrena globosa* (which produce local red rings, and with symptomless systemic infection in *C. ambrosoides*) (Jones, 1985). Since WbLtV has limited spread, its infections are probably of negligible economic importance.

Blackberry Calico Disease (BbCaDs)

No other information is available except that the causal agent of BbCaDs is graft transmissible. According to Converse (1987), BbCaDs occurs in California and Oregon and probably in the U.S. wherever Pacific trailing blackberry cultivars such as Boysen, Logan, Thornless Logan (Loganberry), and Marion are grown. Universally infected Thornless Logan and Chehalom blackberry cvs. generally have low yields.

Cultivated blackberry cvs. are natural hosts of the BbCaDs, which was also discovered in *Rubus ursinus*, wild Pacific coast trailing blackberry (Converse, 1987). Chlorotic zones with occasional blotches or rings and line patterns develop in floricane leaves. Chlorotic blotches feature white margins and occasional red coloration.

The mode of natural disease spread observed in some cvs. has yet to be clarified. BbCaDs is transmissible by approach graft into several blackberry cvs. and *R. ulmifolius* var. *inermis*. No information is available on natural vectors. The use of positively disease-free Pacific coast trailing blackberry cvs. is the main control measure to protect against BbCaDs.

Thimbleberry Ringspot Disease (TbRsDs)

TbRsDs was discovered in naturally infected thimbleberry, *Rubus parviflorus*, plants. The characteristics of the causal agent of TbRsDs are unknown, other than the isometric particles of 25 nm in diameter observed by electron microscopy in leaf sections of thimbleberry, but it is assumed virus-like, similar to most aphid-transmissible viruses that infect plants in the genus *Rubus* (Stace-Smith, 1987a).

Thimbleberry is a wild plant and the only natural host of TbRsDs. Characteristic symptoms on leaves are irregular ringspot and oak-leaf patterns. In some cases, only patches of net-like vein chlorosis appear on the leaves. Symptoms of TbRsDs are more obvious on completely developed leaves. Leaves with severe mottle are malformed and the canes on which they develop remain stunted.

TbRsDs has been transmitted experimentally into the cultivated *Rubus* species red raspberry (*R. idaeus* cv. Washington), black raspberry (*R. occidentalis* cv. Munger), and *R. henryi* (Stace-Smith, 1987a). Natural vectors of the TbRsDs agent are the aphids *Illionia maxima*, *I. davidsonii*, and *Amphorophora parviflora* (Stace-Smith, 1958). The presence of TbRsDs, as verified by graft or aphid transmissions, has been recorded in *R. occidentalis* and *R. henryi* plants. TbRsDs is generally sporadic in cultivated commercial *Rubus* species, thus control measures have not been designed.

VIRUS DISEASES OF *Ribes* spp.

No virus has been described in *Ribes* crops (gooseberries and currants) as its primary natural host. *Ribes* plants are natural alternative hosts for several viruses and/or are affected by virus-like diseases possibly caused by viral agents. Most of these viruses infect more than one plant species in nature. Thus, nematode-borne ArMV and RbRsV infect gooseberry (*Ribes grossularia*), black currant (*R. nigrum*), and red currant (*R. rubrum*). Aphid-borne CMV and nematode-borne SbLtRsV and TmRsV infect black currant and red currant, whereas red currant plants are natural hosts of tobacco rattle virus.

Virus and virus-like diseases are usually considered of less economic importance in the production of *Ribes* small fruits. Greater damages can be expected when these crops are grown more widely and/or on greater acreages.

Virus Diseases of Gooseberry (*Ribes* sp.)

Gooseberry: Natural Alternative Host for Some Viruses

Arabis Mosaic Virus (ArMV)

Natural infections of gooseberry with ArMV were recorded only in East Germany (Kleinhempel, 1972) and, due to its limited spread, its economic importance is negligible. Since ArMV-infected plants remain symptomless (Kleinhempel, 1972), the virus is detectable by mechanical inoculations of *Chenopodium quinoa*, which serve as suitable assays for indexing and the selection of healthy gooseberry stocks. ArMV is identified with the aid of *C. quinoa* indicator plants and serological analyses (van der Meer, 1987).

Raspberry Ringspot Virus (RbRsV)

RbRsV has been recorded in naturally infected gooseberry only in the Netherlands (Houtman, 1951), and thus seems of minor concern and of little economic importance. Individual cvs. differ in symptomological reactions. RbRsV-infected plants of the cv. Whynham Industry express leaf mosaic, smaller and malformed fruits (gooseberry deterioration), and later maturity, whereas RbRsV infected cv. Whitesmith remains symptomless (van der Meer, 1987a). Reactions of *Chenopodium quinoa* plants and serological tests are used for virus detection and identification.

Diseases of Gooseberry Assumed Caused by Viruses

Gooseberry Vein Banding Disease (GbVBaDs)

GbVBaDs, first observed in Czechoslovakia, was found later throughout Europe; nearly all plantations of old cvs. in England are diseased (loc. cit. Adams and Posnette, 1987). The extent of damage depends mostly on differences in cv. susceptibility.

Plants in the genus *Ribes* are natural hosts of GbVBaDs, and all gooseberry cvs. appear susceptible. In the spring, all leaf veins express the vein banding symptom, which later in the season is restricted to smaller veins being expressed only partially in primary veins.

No causal agent of GbVBaDs has been isolated or identified. The causal agent of GbVBaDs is transmitted by chip budding and patch grafting, but not by mechanical inocualtions from diseased infected plants. The aphids *Aphis grossulariae, A. schneideri, Hyperomyzus pallidus, Nasonovia ribes nigri, Cryptomyzus ribis, Hyperomyzus lactucae*, and *Myzus persicae* transmit the causal agent in a semipersistent manner (Karl and Kleinhempel, 1969). Healthy planting material is highly recommended for control of GbVBaDs because reinfections occur less frequently when small, isolated fields are planted.

Gooseberry Mosaic Disease (GbMDs)

GbMDs was first described in the cv. Lady Delamare in Poland (Basak and Meskiewicz, 1980). Leaf bright yellow mosaic and vein yellowing characteristize symptoms in GbMDs-infected plants. The causal agent has been experimentally graft transmitted into other gooseberry and black currant cvs. Although etiologically undefined, GbMDs is believed to differ based on symptomology and host range from GbVBaDs.

VIRUS DISEASES OF BLACK CURRANT (*RIBES* SP.)

Aphid-Borne Cucumber Mosaic Virus (CMV)

CMV infections have been discovered in some black currant plantations and nurseries in England and Wales, in individual plants in Germany, and in the cv. Primorskij Champion in the Soviet Far East (loc. cit. Adams and Thresh, 1987). CMV-infected plants are stunted and yield less, but due to limited spread, CMV causes little economic damage.

The most important commercial British black currant cv. is susceptible to CMV. The Amos Black cv. is most susceptible, and is thus also used as a suitable indicator plant (Adams and Thresh, 1987). Symptoms of CMV infection in inoculated shoots appear the year following infection. Scattered discolorations that become pale chlorotic spots (i.e., green mottle of black currant) on the surface and/or along midveins characterize symptoms in infected leaves that are most distinct in developed leaves and less so in younger and older leaves.

CMV can be both graft and sap transmitted from infected *Ribes* to herbaceous hosts and occasionally to black currant seedlings. A CMV isolate from black currant was transmitted by the aphids *Myzus persicae, Aphis grossulariae, A. schneideri, Cryptomyzus ribis, Hyperomyzus lactucae*, and *Nasonovia ribes nigri* to other *Ribes* crops (Adams and Posnette, 1987). CMV spreads slowly in black currant plantations. Elimination of CMV-infected mother plants used for stock production is recommended for restriction of spread.

Nematode-Borne Viruses

Black currant serves as a natural host of several widespread nematode-borne viruses from different hosts in nature, including ArMV, RbRsV, SbLtRsV, TmRsV, and TmBRV. Commonly, these viruses are discovered in small numbers, frequently only in one plant and in restricted areas (Converse, 1987). For example, ArMV was discovered in one black currant nursery in England (Thresh, 1966), in one black currant nursery in France (Putz and Stocky, 1971), and in the Soviet Far East (Gordejchuk et al., 1977). SbLtRsV was isolated from one black currant cv. Baldwin bush in Scotland (Lister, 1964), and both RbRsV and TmRsV were isolated from a few black currant cvs. in the Soviet Far East (Gordejchuk et al., 1977). TmBRV was found in a few bushes of different black currant cvs. in one region in Finland (loc. cit. Adams and Thresh, 1987a). Due to the limited occurrence and spread of these several viruses, the diseases caused create no great economic impact in black currant production.

These viruses are commonly graft transmissible among woody plants and mechanically transmissible by inoculation of infective plant sap into herbaceous hosts. In black currant plantations, these viruses are spread by movement of certain nematode species and by propagation of infected cuttings. Nematicide treatments and the use of healthy cuttings in propagation represent the main recommended control measures.

(More details on the traits of these viruses can be found in their descriptions on main hosts.)

Diseases of Black Currant Assumed Caused by Viruses

Black Currant Vein Clearing Disease (BCuVClDs) and
Black Currant Vein Net Disease

BCuVClDs, rarely observed in black currant in Britain (Thresh, 1966) and continental Europe (Baumann, 1974; Putz, 1972), is not believed economically important in black currant production. The causal agent of BCuVClDs is unknown, although suspected as the gooseberry vein banding agent.

In nature, BCuVClDs occurs in black currant, gooseberry, and red currant plants, and the causal agent has been transmitted experimentally to *Ribes*. The most important cvs. of black currant in Great Britain susceptible to graft inoculation differ, based on their tolerance to infection (Adams and Thresh, 1987b). Broad yellow banding potentially accompanied by vein clearing occurs in the midveins of the first emerged leaves. Vein clearing and vein banding are more distinct on leaves that develop later in the season. Vein net patterns and leaf blade malformations and asymmetry characterize symptoms in some cultivars.

The causal agent of BCuVClDs is graft transmitted but cannot be sap inoculated. It is aphid transmitted among gooseberry, black currant, and red currant seedlings by some species (loc. cit. Adams and Thresh, 1987b).

Black Currant Infectious Variegation Disease (BCuInVgDs)

BCuInVgDs, first observed in black currant cvs. in England (Posnette, 1952), was later discovered in other European countries (Kristensen et al., 1962; Putz, 1972). Black currant is the only natural host of the causal agent of BCuInVgDs. Diagnostic symptoms include bright-yellow chrome or pale-yellow mosaic and additionally, during summer, vein patterns plus yellow banding of midveins in leaves of susceptible cvs. Daniels September and Lexton's Nigger. The causal agent of BCuInVgDs is unidentified. The only mode of transmission known is by grafts (Ellenberger, 1962; Kristensen et al., 1962). The seedlings of the Baldwin black currant cv. are useful index plants.

Black Currant Reversion Disease (BCuRvDs)

BCuRvDs was first discovered in the Netherlands and later in England. Widespread in most European countries, it rarely occurs in New Zealand and Australia (loc. cit. Adams and Thresh,

1987c). In Europe, it causes major disease, severely reducing black currant yields (Krczal, 1976; Cropley et al., 1964).

BCuRvDs etiology remains unclear. The disease was first associated with bacteria (Silvere and Romeikis, 1973), then with phytoplasma-like organisms (Zirka et al., 1977), and finally with potato virus Y (Jacob, 1976). However, more of them have been confirmed as the causal agent of BCuRvDs.

Black currant is the primary natural host of BCuRvDs, although infections occur also in other *Ribes* spp., including commercial red currant cvs. All commercial black currant cvs. grown in western Europe are susceptible, although with different degrees of tolerance. Russian cvs. exist that are resistant and some immune to BCuRvDs (loc. cit. Adams and Thresh, 1987c).

Symptoms appear the year following infection. Systemic infection spreads gradually and, by the third or fourth year, causes significant damage to black currant bushes (Adams and Thresh, 1987c). The general appearance of diseased bushes is altered with flat leaves that have smaller basal sinuses plus fewer midveins and marginal serrations; sepals have fewer hairs; and flower buds are nearly glabrous and brightly colored. Descriptions from the former U.S.S.R., Scandinavia, and other areas show that diseased flowers are glabrous and severely malformed, with elongated styles, no stamens, and sepal-like petals (Adams and Thresh, 1987c). Double flowers may be caused by a specific strain of the BCuRvDs.

The vector of the BCuRvDs causal agent is the eriopyhid mite *Cecidophyopsis ribis*, which moves late in the spring and early in the summer, dispersing from older galls into newly formed buds. The mite remains in infected buds, which become surrounded by galls and fail to form leaves or flowers (Thresh, 1965). Black currant reversion can be graft transmitted, and potential transmission of mite vectors must be avoided.

BCuRvDs can be controlled by planting healthy stocks to establish plantations and by spatially isolating plantations from potential sources of infection. Rogueing and destruction of diseased bushes and spraying with acaricides also represent useful control measures. Cvs. known to be resistant to BCuRvDs or harboring vectors should be used to establish new plantations (Keep et al., 1982).

Black Currant Yellows Disease (BCuYsDs)

BCuYsDs was described only in England and only in one nursery from which the diseased plants were spread into a few newly established plantations (Posnette, 1952). BCuYsDs symptoms include stunt of bushes, and lower yields but, due to limited spread, no great losses occurred. The agent of BCuYsDs has not been characterized.

Symptoms develop the year after infection first as pale chlorotic flecks and later as larger olive-green mosaic patches. Infections spread slowly in black currant plantations and no vector is known (Thresh, 1987). Infected bushes are diagnosed by graft-transmission to Amos Black and other black currant cvs.

Virus Diseases of Red Currant (*Ribes* sp.)

Red Currant: Natural Alternative Host for Some Viruses

Aphid-Borne Cucumber Mosaic Virus (CMV)

CMV infection in red currant seldom occurs in nature and is reported only in the Netherlands (van der Meer, 1961) and in England (Thresh, 1967). The virus isolate from red currant was characterized serologically and identified as a strain of CMV (van der Meer, 1987b). CMV is mechanically transmissible into many herbaceous hosts and can be grafted among red currant bushes. Symptoms of disease vary from bright yellow in the central leaf parts of the Maarses Prominent cv. to green mottle and line patterns in leaves of the Jonkheer van Tests cv. CMV can be localized only in lower parts of individual branches. Several aphid species, including some which colonize *Ribes*, are efficient vectors of CMV (Kennedy et al., 1962). Red currant infections spread slowly in nature, resulting from plant resistance or incomplete systemic infection of the plants.

Nematode-Borne Viruses

Red currant is the natural host of several nematode-borne viruses, most of which also infect black currant. These include ArMV, RbRsV, SbLtRsV, TRtV (not known in black currant), and TmRsV.

ArMV is widespread in numerous hosts in nature and cultivated *Ribes* species (gooseberry, black currant, and red currant) are susceptible, although of rare occurrence.

Red currant is susceptible to RbRsV, whose specific strain causes characteristic symptoms of spoon leaf and ringspot in leaves. RbRsV has been discovered in red currant plants in the Netherlands and Germany (loc. cit. van der Meer, 1987c). Bright-yellow mosaic and ringspots occur in leaves of infected red currant in the shock stage, then later accompanied by leaf malformations. Depending on the strain of RbRsV, the leaves of some cvs. display dentations and others become rounded and only slightly dentated. RbRsV-infected bush leaves curl downward or upward and become spoon-shaped. The nematode *Longidorus elongatus* is the vector that transmits RbRsV into red currant roots in nature. RbRsV can be transmitted from red currant into gooseberry or black currant plants by grafting, and from red currant into numerous herbaceous hosts by mechanical inoculations of infective sap.

SbLtRsV is infective in red currant, but seldom occurs naturally. Due to its rare occurrence, it has little influence on yields of infected bushes. SbLtRsV-infected plants usually remain symptomless (Thresh, 1967) but occasionally display bright-yellow symptoms (Kleinhempel, 1968). In nature, SbLtRsV is transmitted by the nematode *Xiphinema diversicaudatum*, the proven vector of SbLtRsV in other crops. It is graft transmitted from currant to currant and by mechanical inoculation of infective sap from red currant to herbaceous hosts. *Chenopodium quinoa* is a suitable indicator plant for detection indexing.

TRtV has been isolated only from scattered individual red currant bushes, and thus is of negligible economic importance (van der Meer, 1987d). Characteristic leaf patterns of light-green mosaic occur in TRtV-infected bushes, a symptom known as red currant leaf pattern. Natural transmission of TRtV is assumed by nematodes of the genus *Trichodorus*, proven vectors of TRtV in other crops. By mechanical sap inoculations, it can be experimentally transmitted from red currant to herbaceous hosts. Only planting material originating from virus-tested mother bushes should be used to establish plantations.

Although in nature TmRsV can infect red currant plants, it seldom occurs, thus seldom causing yield reduction. The disease caused by TmRsV has been described as currant mosaic (Hildenbrand, 1939), American currant mosaic (Hildebrand, 1942), and red currant chlorosis mosaic (Williams and Holdeman, 1970). Natural infectons of red currant by TmRsV were recorded in the U.S., the former U.S.S.R., and former Yugoslavia, transported in by importing American cvs. infected with TmRsV. No reports have been made on the natural spreading of TmRsV in red currant. The nematode *Xiphinema americanum*, well-known as a vector of TmRsV, was discovered in the soils in which red currant was grown in monoculture. Almost certainly, TmRsV was introduced into new red currant growing areas with TmRsV-infected planting material. TmRsV has been transmitted from currant to currant by grafting and by mechanical sap inoculation from red currant to herbaceous hosts, but not vice versa. Only planting material originating from virus-tested, virus-free mother plants should be used to establish new red currant plantations.

Red Currant Interveinal White Mosaic Virus (RCuIvWtMV)

RCuIvWtMV causes a disease of red currant first discovered in a small number of bushes in the Netherlands (van der Meer, 1961) and later in Great Britain (Thresh, 1967). The TIP of Dutch isolates is 45°C; the LIV in tobacco sap is 8 hours at 20°C; and they are serologically related neither to nepoviruses nor alfalfa mosaic virus (AMV) (van der Meer, 1987e). The RCuIvWtMV can be experimentally graft transmitted between currants.

Alfalfa Mosaic Virus (AMV)

The English isolate from red currant causing necrotic local lesions in *Chenopodium quinoa* has been identified as a strain of AMV. The bacilliform AMV has a TIP of 50 to 70°C and retains its infectivity for 4 days at 20°C. Both Dutch and English isolates can be transmitted by sap inoculation to herbaceous hosts.

Diseases of Red Currant Assumed Caused by Viruses

Red Currant Vein Banding Disease (RCuVBdDs)

RCuVBdDs, reported from numerous European countries, is probably widespread wherever red currants are grown. Although the disease etiology has been little studied, its aphid and graft transmissibility suggests a causal agent of viral nature. Its wide-spread distribution makes RCuVBdDs an economically most important red currant disease (van der Meer, 1987f).

The growth of RCuVBdDs-infected red currant seedlings is reduced by as much as 50% (Kleinhempel, 1970) and the growth reduction of 28% of cuttings in the Jonkheer van Tets cv. in the first year (Adams, 1979). RCuVBdDs also reduces the number of cuttings produced in stool beds of red currant (van der Meer, 1980).

Vein banding and vein clearing occur on leaves of most red currant cultivars, frequently only on individual leaf blade parts. Symptom severity depends greatly on environmental factors during the season. The causal agent of RCuVBdDs has been experimentally transmitted by aphids and patch grafting to black currant and gooseberry, but it has not been transmitted mechanically by sap inoculations from diseased currants to herbaceous plants (loc. cit. van der Meer, 1987f).

Vectors of the RCuVBdDs agent include *Aphis schneideri*, *A. triglochinis*, *Cryptomyzus galeopsides*, subsp. *citrinus*, *C. ribis*, *Hyperomyzus lactucae*, and *Nasonovia ribis nigri*. Natural spread of RCuVBdDs has not been studied intensively but has been observed in red currant plantations and in red currant stool beds (van der Meer, 1980). Only vein banding-free planting material should be used to establish new red currant plantations. The elimination by rogueing of infected plants can also be useful in disease control, based on the comparatively slow spread of infection in the field.

Red Currant Yellow Leaf Spot Disease (RCuYLsDs)

RCuYLsDs has been described in several European countries, but no data exist on its effects on plant growth and yield (van der Meer, 1987g). RCuYLsDs etiology is unknown but symptoms and graft tranmissibility suggest a viral agent. Jacob (1976) isolated potato virus Y (PVY) from infected red currant plants, but more results are required for the elucidation of its role in the cause of RCuYLsDs.

Symptoms of disease on red currant leaves occur as small, scattered, light-green and white spots, sometimes light yellow on the leaf blades and leaf margins. Infected plants remain slightly stunted. RCuYLsDs is graft transmitted among currant plants. The causal agent of RCuYLsDs is not transmitted mechanically. Thus, graft inoculation to susceptible cvs. Laxton No. 1 and Fay's Prolific can be used for detection (loc. cit. van der Meer, 1987g).

VIRUS DISEASES OF BLUEBERRY (*VACCINIUM* SP.)

Blueberry Leaf Mottle Virus (BlbLMtV)

Properties of Viral Particles

BlbLMtV is isometric, 28 nm diameter, and sediments as three components with sedimentation coefficients of 53S (Top), 120S (Middle), and 128S (Bottom) (Ramsdell and Stace-Smith, 1981). The viral genome consists of two single-stranded RNAs of 2150 kDa (component M) and 2350 kDa (component B) (Ramsdell and Stace-Smith, 1983). They show that M and B particles contain 37 and 40 to 41% RNA, respectively, and protein subunits of 54 kDa.

The TIP of BlbLMtV is 65°C; the DEP 1×10^{-4}; and the M strain in *Chenopodium quinoa* sap has an LIV of 5 days at 20°C and 14 days at 4°C (Ramsdell and Stace-Smith, 1979). BlbLMtV consists of two strains: MI from blueberry and MY from grapevine. The MI strain does, and the MY strain does not, infect seedlings of cv. Rubel highbush (Ramsdell and Stace-Smith, 1979), while only the MI strain infects a small percentage of *Vitis labrusca* cv. Niagara seedlings (Ramsdell and Stace-Smith, 1980). Otherwise, these two strains are symptomologically indistinguishable in herbaceous hosts. BlbLMtV (MI strain) has strong immunogenic activity. BlbLMtV is distantly related serologically to grapevine Bulgarian latent virus (Ramsdell, 1987).

Geographic Distribution and Economic Importance

BlbLMtV, first observed in 1977 in Rubel cv. blueberry bushes in Michigan, was later discovered in the Jersey cv. Nearly simultaneously, the grapevine Bulgarian latent virus was discovered distantly related serologically to BlbLMtV (Martelli et al., 1977). A virus isolated from one *Vitis labrusca* Concord grapevine in New York was determined as closely related serologically to BlbLMtV, and thus designated a strain of grapevine Bulgarian latent virus (Uyemoto et al., 1977). BlbLMtV causes leaf mottle and stunt of blueberry, and such bushes become unproductive and die in a few years. Economic damage depends on the number of infected plants.

Host Range

Blueberry and grapevine are natural hosts of BlbLMtV. BlbLMtV has been transmitted experimentally, by mechanical sap inoculation into 27 plant species belonging to 7 dicotyledonous families (Ramsdell and Stace-Smith, 1979).

Symptoms

Several symptom types of variable severity depend on cv. susceptibility in BlbLMtV-infected blueberry plants (Figure 197). Mottle plus occasional leaf strap and curl characterize the symptoms in cv. Rubel. Dieback of older stems plus malformed and stunted new growth from crowns occur in bushes infected during several years, which then cease production. Symptoms of disease are milder in the BlbLMtV-infected cv. Jersey with somewhat smaller leaves and mottle of the lower leaves in the crown plus rare stem dieback and growth stagnation is less pronounced. The yields of the diseased Jersey cv. are reduced significantly.

Suitable indicator plants are *Chenopodium quinoa* (chlorotic lesions, mottling, and apical death), *C. amaranticolor* (systemic mottling), and *Nicotiana clevelandii* (pinpoint local lesions in inoculated leaves and also in systemic leaves) (Ramsdell, 1987).

FIGURE 197 Small terminal leaves on stems of Jersey highbush blueberry infected with blueberry leaf mottle virus. (Reproduced with permission from APS.)

Pathogenesis and Control

BlbLMtV-infected blueberry bushes represent permanent sources of infection. Spread of infection appears random in affected fields. Attempts to transmit the virus with the nematode *Xiphinema americanum* were unsuccessful (loc. cit. Ramsdell and Stace-Smith, 1983). Pollen from infected blueberry bushes contained high amounts of BlbLMtV (Ramsdell, 1987), which was transmitted to healthy plants via pollinating honey bees (Boylan-Pett et al., 1991). BlbLMtV is transmitted by mechanical sap inoculation, which allows ease of detection in infected hosts. Only virus-free propagating material should be used for establishment of plantations. Infected plants should be rogued immediately to reduce potential sources of infection.

Blueberry Red Ringspot Caulimovirus (BlbReRsV)

BlbReRsV is isometric, 42 to 46 nm diameter, and sediments as two components (Kim et al., 1981). Particles consist of DNA and a major 44-kDa protein, sometimes also with a faint protein band of 101 kDa (by extrapolation) (Gillett, 1988). BlbReRsV is moderately immunogenic (Gillett and Ramsdell, 1984). Judged by similar symptoms and inclusion bodies, the bluebery red ringspot and cranberry ringspot diseases may be caused by the same virus (Ramsdell et al., 1987).

BlbReRsV is reported only from the U.S., widespread in Arkansas and considered the economically most important disease in New Jersey. Numerous blueberry cvs. susceptible to BlbReRsV include Blueray, Cabot, Earlyblue, and Rubel; Bluecrop appears field-resistant and Jersey appears field-immune (Gillett and Ramsdell, 1988).

In highbush blueberry, BlbReRsV causes characteristic symptoms, such as red spots and rings by mid- and late-summer in older leaves, which gradually spread to younger leaves; the spots may coalesce into variable shapes. Red spots and rings occur also on 1-year-old, as well as older canes. Infected fruits are smaller and yield may be reduced significantly.

Little information is available on the field spread of BlbReRsV; active spread is recorded in New Jersey but not in Michigan. Although suspicions suggest aphids or mealy bugs as possible efficient vectors of BlbReRsV, no positive data on transmission have been confirmed. Detection of disease is by grafting early, prior to bud break, into highly susceptible cvs. (Blueray, Cabot, etc.) to enable symptom development during the first year. The use of healthy plants and timely rogueing and elimination of diseased plants are the primary disease control measures.

Blueberry Shoestring Virus (BlbSsV)

BlbSsV is isometric, 27 nm diameter, sediments at 120S, and contains a single-stranded RNA of 1450 kDa (20% of the MW of the virion) and protein subunits of 30 kDa (80% of the molecular weight of the virion) (Ramsdell, 1987a). BlbSsV is strongly immunogenic (Lesney et al., 1978).

BlbSsV has been reported from the U.S. (New Jersey, Michigan, and Washington), Nova Scotia, and Canada (loc. cit. Ramsdell, 1979). Yields of infected plants are greatly reduced, and the extent of damage is directly correlated with number of diseased plants.

Highbush blueberry (*Vaccinium corymbosum*) and lowbush blueberry (*Vaccinium angustifolium*) are natural hosts of BlbSsV. Susceptible highbush blueberry cvs. include Jersey, Rubel, etc. The Bluecrop and Atlantic cvs. are immune to field infections (Ramsdell, 1987a). Infected leaves in susceptible cvs. become strap-like (shoestring symptom) and malformed (curled or crescent shape) (Figure 198). Red vein banding, red streaks along midribs, and occasionally oak-leaf patterns can be observed on leaves. Elongated red streaks occur on new or 1-year-old canes and on petals. Reddish-purple coloration develops on the surface of immature fruits.

BlbSsV spread to bushes in fields has been recorded along individual rows. The blueberry aphid *Illinoia pepperi* is the vector of BlbSsV in nature (Ramsdell, 1979). Purified BlbSsV has been transmitted mechanically into blueberry seedlings and young vegetatively propagated woody cuttings, but not into herbaceous plants. BlbSsV can be transmitted experimentally from infected to

FIGURE 198 Jersey highbush blueberry with strap leaf (shoestring leaf) caused by blueberry shoestring virus. (Reproduced with permission from APS.)

susceptible gooseberry cvs. by chip budding and whip grafting. Rogueing to eliminate infected plants from the plantation, when combined with chemical control of vectors, can reduce field spread. Replanting with immune bushes is most useful.

Blueberry: Natural Alternative Host for Some Nematode-Borne Viruses

Peach Rosette Mosaic Virus (PhRoMV)

Blueberry is a rare host in which PhRstMV was discovered in an experimental planting in southwestern Michigan and southeastern Ontario (Ramsdell and Gillett, 1981). Symptoms of infection have been observed in cvs. Jersey (with strap-shaped and/or crescent-shaped leaves) and Berkeley (with smaller and spoon-shaped leaves). However, these symptoms are not diagnostic. For detection, the virus is first transmitted mechanically by sap inoculaton from young apical leaves of suspected diseases bushes into *Chenopodium quinoa* and *C. amaranticolor* indicator plants, and then are tested serologically to establish the presence of PhRoMV.

Tobacco Ringspot Virus (TRsV)

TRsV has been reported in blueberry plants displaying necrotic ringspot disease symptoms in numerous states of the U.S. The necrotic ringspot disease reduces productivity in susceptible cvs. such as Pemberton, Stanley, Rubel, Concord, etc. (Ramsdell, 1987a). The severe strain of TRsV causes necrotic ringspot in the cv. Jersey; until recently, considered resistant, it causes severe symptoms and rapid necrosis in herbaceous plants (Ramsdell, 1978).

Infected leaves of the susceptible cv. Pemberton are malformed and slightly thickened, with necrotic spots that fall out, resulting in a shot-hole symptom and stunted cane death. Strap-like leaves develop in TRsV-infected cvs. Concord and Stanley and the internodes are short. Characteristic necrotic ringspots are diagnostic symptoms. However, for verification of its etiology, TRsV is transmitted by sap inoculation into herbaceous hosts and then tested serologically for verifcation.

Tomato Ringspot Virus (TmRsV)

The natural occurrence of TmRsV in blueberry plants was recorded first in Washington (Johnson, 1972), then in Oregon (Converse and Ramsdell, 1982). Chlorotic spots and malformation of leaves occur in infected blueberry (Figure 199). Chracteristic circular, dark necrotic spots also form on the

FIGURE 199 Young leaves of Earliblue highbush blueberry infected with tomato ringspot virus; note strap leaf and mottling. (Courtesy R.H. Converse. Reproduced with permission from APS.)

canes, twigs, and branches. By bud grafting, Johnson (1972) transmitted the virus from infected blueberry to red raspberry cv. Puyallup plants, in which ringspot and leaf pattern symptoms appeared the following spring. (*See* also the descriptions of PhRoMV, TRsV, and TmRsV in their primary hosts.)

Diseases of Blueberry Assumed Caused by Viruses

Blueberry Mosaic Disease (BlbMDs)

BlbMDs has been reported in the eastern U.S. and British Columbia, Canada (Ramsdell and Stretch, 1987). It was first considered a variegation symptom of genetic origin. Varney (1957) suggested the virus-like etiology of BlbMDs, whose agent is not yet isolated or defined. BlbMDs symptoms were observed also in old highbush blueberry cultivars, including Cabot, Earlyblue, Rubel, and others (Ramsdell and Stretch, 1987). Chracteristic symptoms in leaves are mosaic patterns, yellow, yellow-green, and mild to severe mottle. BlbMDs damages most of the bush and also one or two canes. BlbMDs spreads slowly in the field with no known vectors. The causal agent can be graft transmitted into susceptible cvs., but not mechanically with infective plant sap.

Cranberry Ringspot Disease (CbRsDs)

CbRsDs, reported only in New Jersey (Stretch, 1964) and Wisconsin (Boone, 1966), lacks information about its causal agent. Ramsdell et al. (1987), based on symptom similarity and inclusion bodies, suspect that cranberry ringspot and blueberry red ringspot may be caused by the same virus-like agent.

Diseased Searles cv. displays leaf and severe fruit symptoms. Ringspot symptoms are distinct when the leaves become red in the autumn; the rings are dark green and the remainder of the leaf is red. Ringspots on infected fruits differ in size and appearance, and internal rings are distinct with a deep red coloration; these fruits usually become necrotic at the blossom end, sometimes encompassing entire berries, which is most damaging to quality fresh fruit (Boone, 1967). No

Virus Diseases of Small Fruits 467

information is available either on natural or experimental transmission of the causal agent; thus, only diagnostic symptoms can be used for detection.

Other viruses occur on occasion in blueberry, such as peach rosette mosaic (PhRtMV) (see Figure 201).

FIGURE 200 Herbert highbush blueberry with leaf symptoms of mosaic disease. (Courtesy F.L. Caruso. Reproduced with permission from APS.)

FIGURE 201 Jersey highbush blueberry with leaf distortion caused by peach rosette mosaic virus. (Reproduced with permission from APS.)

REFERENCES

Adams, A.N. 1979. The effect of gooseberry vein banding virus on the growth and yield of gooseberry and red currant. *J. Hort. Sci.* 54:23–25.

Adams, A.N. and Posnette, A.F. 1987. Gooseberry vein banding. In: Converse, R.H., Ed. *Virus Diseases of Small Fruits*, USDA, ARS, Agr. Handb. No. 631, p. 129–130.

Adams, A.N. and Thresh, J.M. 1987. Green mottle of black currant. In: Converse, R.H., Ed. *Virus Diseases of Small Fruits*, USDA, ARS, Agr. Handb. No. 631, p. 136–138.

Adams, A.N. and Thresh J.M. 1987a. Other nepoviruses isolated from black currant. In: Converse, R.H., Ed. *Virus Diseases of Small Fruits*, USDA, ARS, Agr. Handb. No. 631, p. 140.

Adams, A.N. and Thresh, J.M. 1987b. Vein clearing and vein net disease of black currant. In: Converse, R.H., Ed. *Virus Diseases of Small Fruits*, USDA, ARS, Agr. Handb. No. 631, p. 137–138.

Adams, A.N. and Thresh, J.M. 1987c. Reversion of black currant. In: Converse, R.H., Ed. *Virus Diseases of Small Fruits*, USDA, ARS, Agr. Handb. No. 631, p. 133–136.

Adams, A.N. and Barbara, D.G. 1988. Multiplication of strawberry crinkle virus in non-vector insects. (Abstr.) *XIVth Int. Symp. Fruit Tree Virus Dis.* and *5th Int. Symp. Small Fruit Virus Dis.*, 12–18 June 1988, Thessaloniki, Greece.

Agrios, G.N. 1988. *Plant Pathology*, 3rd edition, Academic Press, New York.

Babović, M. 1965. Proučavanje virozne kržljavosti lišća jagode u našoj zemliji. *Zbornik radova Poljoprivrednog fakulteta, Zemun* 396:1–21.

Barbara, D.J. 1988. Raspberry bushy dwarf virus (RBDV). In: Smith, I.M. et al. *European Handbook Pl. Dis.*, Blackwell Sci. Publ., p. 97–98.

Barnett, O.W. and Murant, A.F. 1970. Host range, properties and purification of raspberry bushy dwarf virus. *Ann. Appl. Biol.* 65:435–449.

Barnett, O.W. and Murant, A.F. 1971. Differential hosts of some properties of raspberry bushy dwarf virus. *Ann. Phytopathol., Numero Hors Serie*, 129–139.

Basak, W. and Maskiewicz. 1980. Gooseberry mosaic. *Acta Hort.* 95:49–52.

Baumann, G. 1974. Zur Gewinnung von virusgetesteten Vermehrungs-material bei Johannisbeeren. *Der Erwerbsobstbau* 16:26–29.

Bolton, A.T. and Turner L.H. 1962. Note on obtaining virus-free plants of red raspberry through the use of tip cuttings. *Can. J. Pl. Sci.* 42:210–211.

Bolton, A.T. 1967. The inactivation of vein banding and latent C virus in strawberries by heat treatment. *Can. J. Pl. Sci.* 47:375–380.

Boone, D.M. 1966. Ringspot disease of cranberry. *Pl. Dis. Reptr.* 50:543–545.

Boone, D.M. 1967. A cranberry ringspot disease injuring the Searles variety. *Cranberries* 31:14–15.

Boylan-Pett, W. et al. 1991. Honey bee foraging behavior and the transmission of pollen-borne blueberry leaf mottle virus in highbush blueberry. *XVth Int. Symp. Virus and Virus Dis. Temp. Fruit Crops*, 8–13 July 1991, Vienna, Austria.

Cadman, C.H. and Harris, R.V. 1951. Raspberry virus diseases: a survey of recent work. *East Malling Res. Stn. Reptr. 1950*, p. 127–130.

Cadman, C.H. 1952. Studies in *Rubus* virus diseases. II. Three types of vein chlorosis of raspberries. *Ann. Appl. Biol.* 39:61–68.

Cadman, C.H. 1956. Studies on the etiology and mode of spread of raspberry leaf-curl disease. *J. Hort. Sci.* 31:111–118.

Cadman, C.H. 1960. Studies on the relationship between soil-borne viruses of the ringspot type occurring in Britain and continental Europe. *Virology* 11:653–654.

Cadman, C.H. 1961. Raspberry viruses and virus diseases in Britain. *Hort. Res.* 1:47–61.

Chambers, J. 1954. Heat therapy of virus-infected raspberries. *Nature (London)* 173:595–596.

Caruso, F.L. and Ramsdell, D.C. 1995. *Compendium of Blueberry and Cranberry Diseases*, APS Press, St. Paul, MN. 87 pp.

Converse, R.H. 1963. Influence of heat-labile components of the raspberry mosaic complex on growth and yield of red raspberries. *Phytopathology* 53:1251–1254.

Converse, R.H. et al. 1966. Viruses in black raspberry. *Scottish Hort. Res. Inst., Ann. Rept.* 12:51.

Converse, R.H. 1967. Pollen- and seed-borne raspberry viruses. (Abstr.) *Phytopathology* 57:97–98.

Converse, R.H. and Lister, R.M. 1969. The occurrence and some properties of black raspberry latent virus. *Phytopathology* 59:325–333.

Converse, R.H. et al. 1970. Virus diseases observed in cultivated red raspberry in the Pacific North-West. *Pl. Dis. Reptr.* 54:701–703.

Converse, R.H. and Lister, R.M. 1970. Black raspberry virus. In: Frazier, M.W., Ed. *Virus Dis. Small Fruits and Grapevines*, Univ. Ca., Div. Agr. Sci, Berkeley, p. 151–154.

Converse, R.H. 1970. Black raspberry streak. In: Frazier, N.W., Ed. *Virus Dis. Small Fruits and Grapevines*, Univ. Ca., Div. Agr. Sci., Berkeley, p. 155–157.

Converse, R.H. et al. 1971. Search for biological races in *Amphorophora agathonica* Hottes on red raspberries. *Can. J. Pl. Sci.* 51:81–85.

Converse, R.H. 1973. Occurrence and some properties of raspberry bushy dwarf virus in *Rubus* species in the United States. *Phytopathology* 63:780–783.

Converse, R.H. and Ramsdell, D.C. 1982. Occurrence of tomato and tobacco ringspot viruses and of dagger and other nematodes associated with cultivated highbush blueberries in Oregon. *Pl. Dis.* 66:710–712.

Converse, R.H. 1987. Blackberry calico. In: Converse, R.H., Ed. *Virus Diseases of Small Fruits*, USDA, ARS, Agr. Handb. No. 631, p. 245–246.

Cornuet, P. and Morand, J.D. 1960. Infection naturelle des fraisiers (*Fragaria* spp.) par le virus de la mosaïque du tabac. *C.R. Acad. Sci. Paris* 250:1583–1584.

Cropley, R.A. et al. 1964. The effects of black currant yellows virus and a strain of reversion on yield. *Ann. Appl. Biol.* 54:177–182.

Daubeney, H.A. et al. 1975. Effects of tomato ringspot virus on druplet set of red raspberry cultivars. *Can. J. Pl. Sci.* 55:755–759.

Daubeney, H.A. et al. 1982. Effect of raspberry bushy dwarf virus on yield and cane growth in susceptible red raspberry cultivars. *Hort. Sci.* 17:645–657.

Dunez, J. 1988. Strawberry latent ringspot virus. In: Smith, I.M. et al., Eds. *European Handb. Pl. Dis.*, Blackwell Sci. Publ., p. 30.

Dunez, J. 1988a. Raspberry ringspot virus (RRV). In: Smith, I.M. et al., Eds. *European Handb. Pl. Dis.*, Blackwell Sci. Publ., p. 29.

Ellenberger, C.E. 1962. Transmission of variegation of black currants. *East Malling Res. Stn. Rept.* 1961, p. 101.

Ellis, M.A. et al. 1991. *Compendium of Raspberry and Blackberry Diseases and Insects*, APS Press, St. Paul, MN. 100 pp.

Engelbrecht, D.J. and van der Walt, W.J.K. 1974. Host reaction and some properties of a virus causing yellow mosaic in wild bramble (*Rubus* sp.), *Phytophylactica* 6:311–313.

Engelbrecht, D.J. 1987. Bramble yellow mosaic. In: Converse, R.H., Ed. *Virus Diseases of Small Fruits*, USDA, ARS, Agr. Handb. No. 631, p. 243–244.

EPPO 1974. *European Plant Prot. Organ. Bull.* 101.

Frazier, N.W. 1955. Strawberry vein banding virus. *Phytopathology* 45:307–312.

Frazier, N.W. and Posnette, A.L. 1958. Relationships of the strawberry viruses of England and California. *Hilgardia* 27:455–514.

Frazier, N.W. 1960. Differential transmission of four strains of strawberry vein banding virus by four aphid vectors. *Pl. Dis. Reptr.* 44:436–437.

Frazier, N.W. et al. 1961. Yellow-bud virus endemic along California coast. *Pl. Dis. Reptr.* 45:649–651.

Frazier, N.W. 1968. Transmission of strawberry mottle virus by juice and aphids to herbaceous hosts. *Pl. Dis. Reptr.* 52:64–67.

Frazier, N.W. 1970. Strawberry stunt. In: Frazier, N.W., Ed. *Virus Diseases of Small Fruits and Grapevines*, Univ. Ca., Berkeley, p. 23.

Frazier, N.W. and Mellor, F.C. 1970. Strawberry crinkle. In: Frazier, N.W., Ed. *Virus Diseases of Small Fruits and Grapevines*, Univ. Ca., Berkeley, p. 18–29.

Frazier, N.W. and Mellor, F.C. 1970a. Tomato ringspot virus in strawberry. In: Frazier, N.W., Ed. *Virus Diseases of Small Fruits and Grapevines*, Univ. Ca., Berkeley, p. 43–45.

Frazier, N.W. and Stace-Smith, R. 1970. Necrotic shock in strawberry. In: Frazier, N.W., Ed. *Virus Diseases of Small Fruits and Grapevines*, Univ. Ca., Berkeley, p. 48–49.

Frazier, N.W. and Converse, R.H. 1980. Strawberry vein banding virus. *CMI/AAB Descriptions of Plant Viruses*, No. 219.

Freeman, J.A. and Mellor, C. 1962. Influences of latent viruses on vigor, yield and quality of British Sovereign strawberries. *Can. J. Pl. Sci.* 42:602–610.

Freeman, J.A. et al. 1975. Effects of tomato ringspot virus on the growth and yield of red raspberry. *Can. J. Pl. Sci.* 55:749–754.

Fulton, J.P. 1952. A tobacco necrosis virus associated with strawberry plants. *Pl. Dis. Reptr.* 36:313–314.

Gillett, J.M. and Ramsdell, D.C. 1984. Detecting the inclusion forming red ringspot virus with ELISA. *Phytopathology* 74:862 (Abstr.).

Gillett, J.M. and Ramsdell, D.C. 1988. Blueberry red ringspot virus. *CMI/AAB Descriptions of Plant Viruses*, No. 327.

Gillett, J.M. 1988. Physical and Chemical Properties of Blueberry Red Ringspot Virus. M.Sc. thesis, Michigan State Univ., 104 pp.

Gordejchuk, O.G. et al. 1977. Virus diseases of berry crops in the Soviet Far East. I. Identification of some mechanically transmitted viruses detected in Primorye Terriotry. *Zentrablatt für Bakteriologie, Parasitenkunde, Infektionskrankheiten und Hygiene, Abtailung II*, 132:686–707.

Harris, R.V. 1933. The strawberry "yellow-edge" disease. *J. Pom. Hort. Sci.* 11:56–76.

Harris, R.V. and King, M.E. 1942. Studies in strawberry diseases. V. The use of *Fragaria vesca* L. as an indicator of yellow edge and crinkle. *J. Pom. Hort. Sci.* 19:227–242.

Harrison, B.D. 1958. Cucumber mosaic virus in raspberry. *Pl. Pathol.* 7:109–111.

Harrison, B.D. 1958a. Further studies on raspberry ringspot and tomato black ring soil-borne viruses that affect raspberry. *Ann. Appl. Biol.* 46:571–584.

Hildebrand, E.M. 1939. Currant mosaic. *Phytopathology* 29:369–371.

Hildebrand, E.M. 1942. Tomato ringspot on currant. *Am. J. Bot.* 29:362–366.

Horn, W.T. 1922. Strawberry troubles. *Ca. Agr. Exp. Sta. Rept.* 1921 22:122–123.

Horn, N.L. and Carver, R.G. 1962. Effects of three viruses on plant production and yields of strawberries. *Pl. Dis. Reptr.* 46:762–765.

Horn, N.L. and Carver, R.G. 1970. Strawberry chlorotic flecks. In: Frazier, N.W., Ed. *Virus Diseases of Small Fruits and Grapevines*, University of California, Berkeley, p. 11–12.

Houtman, G. 1951. Net verlopen van kruisbessenstruiken (Deterioration of gooseberries). *Fruitteelt* 41:720–721.

Jacob, H. 1976. Investigations on symptomatology, transmission, etiology and host specificity of black current reversion virus. *Acta Hort.* 66:99–104.

Johnson, F. 1972. Tomato ringspot virus on blueberries in Washington (Abstr.). *Phytopathology* 62:1004.

Jones, A.T. and Murant, A.F. 1972. Some properties of a mechanically transmissible virus wide-spread in raspberry (*Rubus idaeus*) in Scotland. *Pl. Pathol.* 21:166–170.

Jones, A.T. et al. 1974. Association of different kinds of bacilliform particles with vein chlorosis and mosaic disease of raspberry (*Rubus idaeus*). *Ann. Appl. Biol.* 77:283–288.

Jones, A.T. and Murant, A.F. 1975. Etiology of a mosaic disease of Glen Clova red raspberry. *Hort. Res.* 14:89–95.

Jones, A.T. 1976. The effect of resistance to *Amphorophora rubi* in raspberry (*Rubus idaeus*) on the spread of aphid-borne viruses. *Ann. Appl. Biol.* 82:503–510.

Jones, A.T. et al. 1977. Raspberry vein chlorosis virus. *CMI/AAB Descriptions of Plant Viruses*, No. 174.

Jones, A.T. and Roberts, I.M. 1977. Ultrastructural changes and small bacilliform particles associated with infection by *Rubus* yellows net virus. *Ann. Appl. Biol.* 84:305–310.

Jones, A.T. and Wood, G.A. 1978. The occurrence of cherry leaf roll virus in red raspberry in New Zealand. *Pl. Dis. Reptr.* 62:835–838.

Jones, A.T. 1979. Further studies on the effect of resistance to *Anphorophora idaei* in raspberry (*Rubus idaeus*) on the spread of aphid-borne viruses. *Ann. Appl. Biol.* 92:119–123.

Jones, A.T. and Jennings, D.J. 1980. Genetic control of the reactions of raspberry to black raspberry necrosis, raspberry leaf mottle and raspberry leaf spot viruses. *Ann. Appl. Biol.* 96:59–65.

Jones, A.T. 1980. Some effects of latent virus infection in red raspberry. *Acta Hort.* 95:63–70.

Jones, A.T. 1981. Recent research on virus diseases of red raspberry—the viruses, their effects and control. *Scottish Hort. Res. Inst. Assn. Bull.* 19:15–25.

Jones, A.T. 1985. Wineberry latent virus. *CMI/AAB Descriptions of Plant Viruses*, No. 304.

Jones, A.T. 1987. Raspberry leaf mottle and raspberry leaf spot. In: Converse, R.H., Ed. *Virus Diseases of Small Fruits*, USDA, ARS, Agr. Handb. No. 631, p. 183–187.

Jones, A.T. et al. 1982. Association of raspberry bushy dwarf virus with raspberry yellows disease; reaction of *Rubus* species and cultivars, and the inheritance of resistance. *Ann. Appl. Biol.* 100:135–137.

Jones, A.T. et al. 1987. Raspberry vein chlorosis. In: Converse, R.H., Ed. *Virus Diseases of Small Fruits*, USDA, ARS, Agr. Handb. No. 631, p. 194–197.

Jones, A.T. 1987a. Wineberry latent virus. In: Converse, R.H., Ed. *Virus Diseases of Small Fruits*, USDA, ARS, Agr. Handb. No. 631, p. 239–241.

Jones, A.T. 1988. Raspberry leaf mottle (RLM) disease (raspberry leaf spot (RLS) diseases. In: Smith, I.M. et al. Eds. *European Handb. Pl. Dis.*, Blackwell Sci. Publ., p. 112–113.

Jones, A.T. 1988a. Black raspberry necrosis virus. *CMI/AAB Descriptions of Plant Viruses*, No. 333.

Jordović, M. 1963. Investigation of the economically important virus diseases of raspberries in Yugoslavia. *Arhiv za poljoprivredne nauke* 16:3–27.

Karl, E. and Kleinhempel, H. 1969. Versuche zur Blatlaus- und Propfübertragung verschiedener Herkünfte des Adernbänderungs virus der Stachel-und Johannisbeere. *Acta Phytopathol. Acad. Sci. Hung.* 4:19–28.

Keep, E. et al. 1982. Progress in the integration of characters in gall mite resistant black currants. *J. Hort. Sci.* 57:189–196.

Kegler, H. and Kleinhempel, H. 1977. Beerenobstarten. In: Klinkowski, M. et al., Eds. *Pflanzliche Virologie*, Bd. 3, 3. Aufl., 316–374, Akademic-Verlag, Berlin.

Kennedy, J.S. et al. 1962. *A Conspectus of Aphids As Vectors of Plant Viruses*, Commonwealth Institute of Entomology (London), 114 p.

Kim, K.S. et al. 1981. Virions and ultrastructural changes associated with blueberry red ringspot disease. *Phytopathology* 71:673–678.

Kleinhempel, H. 1968. Zur Analyse von Viren in Johannis-und Stachelbeerbesten der Deutschen Demokratischen Republik. *Archiv für Gartenbau* 16:151–159.

Kleinhempel, H. 1970. Verbreitung und Schadwrikung von Virosen an Johannis-und Stachelbeere. *Archiv für Gartenbau* 18:319–325.

Kleinhempel, H. 1972. Diagnose und Bekämpbung von Virusinfektionen an Johannis-und Stachelbeere. *Archiv für Gartenbau* 19:234–236.

Krczal, H. 1976. Investigations on the effects of reversion disease on crop and growth of black currant. *Acta Hort.* 66:91–97.

Krczal, H. 1980. Transmission of the strawberry mild yellow edge and strawberry crinkle virus by the strawberry aphid *Chaetosiphon fragarifolii*. *Acta Hort.* 95:23–30.

Krczal, H. 1988. Strawberry crinkle virus. In: Smith, I.M. et al., Eds. *European Handb. Pl. Dis.*, Blackwell Sci. Publ., p. 78–79.

Krczal, H. 1988a. Strawberry mild yellow edge virus (SMYEV). In: Smith, I.M. et al., Eds. *European Handb. Pl. Dis.*, Blackwell Sci. Publ., p. 24.

Kristensen, H.R. et al. 1962. New attacks of virus diseases, fungi and pests, 1960. *Tidsskrift for Planteavl.* 65:611–614.

Leone, J. et al. 1991. Attempts to purify strawberry viruses by non-conventional separation methods. *XVth Int. Symp. Virus and Virus Dis. of Temp. Fruit Crops*, 8–13 July, Vienna, Austria.

Lesney, M.S. et al. 1978. Etiology of blueberry shoestring disease and some properties of the causal virus. *Phytopathology* 68:295–300.

Lister, R.W. 1958. Soil-borne virus diseases in strawberry. *Pl. Pathol.* 7:92–94.

Lister, R.W. 1960. Soil-borne virus diseases. *Sci. Hort.* 14:90–96.

Lister, R.W. 1963. Studies on Viruses and Virus Diseases Affecting Strawberry in Scotland. Ph.D. thesis, St. Andrews, Scotland.

Lister, R.W. 1964. Strawberry latent ringspot: a new nematode-borne virus. *Ann. Appl. Biol.* 54:167–175.

Lister, R.W. 1970. Nematode-borne viruses as pathogens in strawberry. In: Frazier, N.W., Ed. *Virus Diseases of Small Fruits and Grapevines*, University of California, Berkeley, p. 32–45.

Lister, R.W. and Converse, R.H. 1972. Black raspberry latent virus. *CMI/AAB Descriptions of Plant Viruses*, No. 106.

Liu, S.C.Y. 1958. Biochemical relations of certian strawberry viruses and their mechanical transmission. *Dissertation Abstr.* 19:4.

Maas, J.L. 1984. *Compendium of Strawberry Diseases*, APS Press, St. Paul, MN. 138 pp.

Martelli, G.P. et al. 1977. Some properties of grapevine Bulgarian latent virus. *Ann. Appl. Biol.* 85:51–58.

Martin, R.R. and Converse, R.H. 1982. Purification, some properties and serology of strawberry mild yellow-edge virus. *Acta Hort.* 129:75.

Martin, R.R. et al. 1988. Molecular cloning of the RNA associated with strawberry mild yellow edge virus. (Abstr.) *XIVth Int. Symp. on Fruit Tree Virus Diseases and Vth Int. Sympt. on Small Fruit Virus Dis.*, 12–18 June, Thessaloniki, Greece.

Mayo, M.A. et al. 1971. New evidence on the structure of nepoviruses. *J. Gen. Virol.* 12:175–178.

McGrew, J.R. and Scott, D.H. 1964. The effect of strawberry latent A virus on growth and fruiting of seven varieties of strawberry. *Pl. Dis. Rept.* 48:929–932.

McGrew, J.R. 1970. Strawberry latent C. In: Frazier, N.W., Ed. *Virus Diseases of Small Fruits and Grapevines*, Univ. Ca., Berkeley, p. 16–18.

Mellor, F.C. and Forbes, R.R. 1960. Studies of virus diseases of strawberries in British Columbia. III. Transmission of strawberry viruses by aphids. *Can. J. Bot.* 38:343–352.

Mellor, F.C. and Frazier, N.W. 1970. Strawberry mild yellow edge virus. In: Frazier, N.W., Ed. *Virus Diseases of Small Fruits and Grapevines*, Univ. Ca., Berkeley, p. 14–16.

Mellor, F.C. and Frazier, N.W. 1970a. Strawberry mottle. In: Frazier, N.W., Ed. *Virus Diseases of Small Fruits and Grapevines*, Univ. Ca., Berkeley, p. 4–8.

Miller, P.N. 1951. Studies on mechanical transmission of certain strawberry viruses and their mechanical transmission. *Diss. Abstr.* 19 (4).

Miller, P.W. and Belkengren, R.O. 1962. Obtaining virus-free strawberries by excising and culturing of apical meristem of plants infected with the yellow-edge, crinkle and vein-banding viruses and certain other virus complexes. *Phytopathology* 52:743–744.

Miller, P.W. and Belkengren, R.O. 1963. Elimination of yellow edge, crinkle and vein banding viruses and certain other virus complexes from strawberries by excision and culturing of apical meristem. *Pl. Dis. Reptr.* 47:298–300.

Miller, P.W. and Frazier, N.W. 1970. Strawberry vein banding. In: Frazier, N.W., Ed. *Virus Diseases of Small Fruits and Grapevines*, University of California, Berkeley, p. 8–10.

Morris, T.L. et al. 1980. Isolation of caulimovirus from strawberry tissue infected with strawberry vein banding virus. *Phytopathology* 70:156–160.

Murant, A.F. 1970. Tomato black ring virus. *CMI/AAB Descriptions of Plant Viruses*, No. 38.

Murant, A.F. 1974. Strawberry latent ringspot virus. *CMI/AAB Descriptions of Plant Viruses*, No. 126.

Murant, A.F. et al. 1976. Recent research on 52V virus in raspberry. *Acta Hort.* 66:39–46.

Murant, A.F. 1976. Raspberry bushy dwarf virus. *CMI/AAB Descriptions of Plant Viruses*, No. 165.

Murant, A.F. and Jones, A.T. 1976. Comparison of isolates of raspberry bushy dwarf virus from red and black raspberries. *Acta Hort.* 66:47–52.

Murant, A.F. and Roberts, I.M. 1980. Particles of raspberry vein chlorosis virus in the aphid vector, *Aphis idaei*. *Acta Hort.* 95:31–35.

Murant, A.F. 1981. Nepoviruses. In: Kurstak, E., Ed. *Handb. of Pl. Virus Infections and Comparative Diagnosis*, Elsevier, Amsterdam, p. 198–238.

Murant, A.F. 1987. Raspberry ringspot and associated diseases of Rubus caused by raspberry ringspot and tomato black ring viruses. In: Converse, R.H., Ed. *Virus Diseases of Small Fruits*, USDA, ARS, Agr. Handb. No. 631, p. 211–220.

Murant, A.F. 1987a. Raspberry yellow dwarf and associated diseases of Rubus caused by arabis mosaic and strawberry latent ringspot viruses. In: Converse, R.H., Ed. *Virus Diseases of Small Fruits*, USDA, ARS, Agr. Handb. No. 631, p. 204–211.

Murant, A.F. 1987b. Raspberry bushy dwarf. In: Converse, R.H., Ed. *Virus Diseases of Small Fruits*, USDA, ARS, Agr. Handb. No. 631, p. 229–234.

Natsuaki, T. et al. 1991. The genome of raspberry bushy dwarf virus. *XVth Int. Symp. on Virus and Virus Dis. of Temp. Fruit Crops*, 8–13 July, Vienna, Austria.

Polák, J. and Bezpalcová. 1988. Strawberry mottle virus studies in Czechoslovakia. *XIVth Int. Symp. on Fruit Tree Virus Dis.* and *XVth Int. Symp. on Small Fruit Virus Dis.*, 12–28 June, Thessaloniki, Greece (Abstr.).

Polák, J. and Jokeš, M. 1991. Intracytoplasmic inclusions in the leaves of *Chenopodium quinoa* Willd. infected by the strawberry mottle agent. *XVth Int. Symp. on Virus and Virus Dis. Temp. Fruit Crops*, 8–13 July 1991, Vienna, Austria.

Posnette, A.P. 1952. New virus diseases of *Ribes*. *East Malling Res. Stn. Rept.* 1951, p. 133–135.

Posnette, A.P. and Cropley, R. 1958. Heat treatment for the inactivation of strawberry viruses. *J. Hort. Sci.* 33:282–288.

Prentice, I.W. 1952. Resolution of strawberry complexes. V. Experiments with viruses 4 and 5. *Ann. Appl. Biol.* 39:487–494.

Putz, C. and Stocky, G. 1970. Premières observations sur une souche de strawberry latent ringspot virus transmise par *Xiphinema coxi* Tarjan et associée à une maladie du framboiser en Alsace. *Ann. Phytopathol.* 2:329–347.

Putz, C. and Stocky, G. 1971. Identifcation due virus de la mosaïque de l'Arabis chez le framboisier et le cassis en France. *Ann. Phytopathol.* 3:503–507.

Putz, C. and Meignoz, R. 1972. Electron microscopy of virus like particles found in mosaic diseased raspberries in France. *Phytopathology* 62:1477–1478.

Putz, C. 1972. Les viruses des petits fruits. II. Les viroses du cassis et des groseilliers. *Ann. Phytopathol.* 4:59–75.

Ramsdell, D.C. 1978. A strain of tobacco ringspot virus associated with a decline disease of Jersey highbush blueberry. *Pl. Dis. Reptr.* 62:1047–1051.

Ramsdell, D.C. and Stace-Smith, R. 1979. Blueberry leaf mottle, a new disease of highbush blueberry. *Acta Hort.* 95:37–45.

Ramsdell, D.C. 1979. Blueberry shoestring virus. *CMI/AAB Descriptions of Plant Viruses*, No. 204.

Ramsdell, D.C. and Stace-Smith, R. 1980. A serological and physicochemical comparison between blueberry leaf mottle virus and grapevine Bulgarian latent virus. In: *Proc. 7th Meeting Int. Conf. Study of Viruses and Virus-Like Diseases of Grapevine*, Niagara Falls, Canada, p. 119–129.

Ramsdell, D.C. and Stace-Smith, R. 1981. Physical and chemical properties of the particles and ribonucleic acid of blueberry leaf mottle virus. *Phytopathology* 71:468–472.

Ramsdell, D.C. and Gillett, J.M. 1981. Peach rosette mosaic virus in highbush blueberry. *Pl. Dis. Reptr.* 65:757–758.

Ramsdell, D.C. and Stace-Smith, R. 1983. Blueberry leaf mottle virus. *CMI/AAB Descriptions of Plant Viruses*, No. 267.

Ramsdell, D.C. 1987. Blueberry leaf mottle. In: Converse, R.H., Ed. *Virus Diseases of Small Fruits*, USDA, ARS, Agr. Handb, No. 631, p. 112–114.

Ramsdell, D.C. et al. 1987. Red ringspot of blueberry. In: Converse, R.H., Ed. *Virus Diseases of Small Fruits*, USDA, ARS, Agr. Handb., No. 631, p. 121–123.

Ramsdell, D.C. 1987a. Blueberry shoestring. In: Converse, R.H., Ed. *Virus Diseases of Small Fruits*, USDA, ARS, Agr. Handb., No. 631, p. 103–105.

Ramsdell, D.C. 1987b. Necrotic ringspot of blueberry. In: Converse, R.H., Ed. *Virus Diseases of Small Fruits*, USDA, ARS, Agr. Handb., No. 631, p. 114–116.

Ramsdell, D.C. and Stretch, A.W. 1987. Blueberry mosaic. In: Converse, R.H., Ed. *Virus Diseases of Small Fruits*, USDA, ARS, Agr. Handb., No. 631, p. 119–120.

Schmelzer, K. 1969. Das latente Erdbeerringflecken-Virus aus Euonymus, Robinia und Aesculus. *Phytopathol. Z.* 66:1–24.

Scotto la Massese et al. 1973. Analyse d'un phénomène de dégénérescence du Pêcher dans la vallée de l'Eyrieux. *C.R. Acad. Agric. Fr.* 59:327–339.

Silvere, A.P. and Romeikis, M.A. 1973. Formation and release of mesosome-like microbodies in the sporogenesis of the endophytic currant reversion Bacillus. *Esti NSV Teaduste Akad. Toimetiseds Biol.* 22:274–277.

Smith, K.M. and Markham, R. 1944. Two new viruses affecting tobacco and other plants. *Phytopathology* 34:324–329.

Smith, H.E. and Moore, D.J. 1952. Dodder transmission of strawberry viruses. (Abstr.) *Phytopathology* 42:20.

Smith, K.M. 1957. *A Textbook of Plant Virus Diseases*, J. and A. Churchill, London.

Spiegel, S. and Martin, R.R. 1991. Detection of the strawberry mild yellow-edge disease in strawberry plantlets grown *in vitro*. (Abstr.) *XVth Int. Symp. on Virus and Virus Diseases of Temperate Fruit Crops*, 8–13 July, Vienna, Austria.

Stace-Smith, R. 1955. Studies on *Rubus* virus diseases in British Columbia, I. *Rubus* yellow-net. *Can. J. Bot.* 33:267–274.

Stace-Smith, R. 1955a. Studies on *Rubus* virus diseases in British Columbia. II. Black raspberry necrosis. *Can. J. Bot.* 33:314–332.

Stace-Smith, R. 1956. Studies on *Rubus* virus diseases in British Columbia. III. Separation of components of raspberry mosaic. *Can. J. Bot.* 34:435–442.

Stace-Smith, R. 1958. Studies on *Rubus* virus diseases in British Columbia. V. Thimbleberry ring spot. *Can. J. Bot.* 36:385–388.

Stace-Smith, R. 1961. Studies on *Rubus* virus diseases in British Columbia. VII. Raspberry vein chlorosis. *Can. J. Bot.* 39:559–565.

Stace-Smith, R. 1962. Studies on *Rubus* virus diseases in British Columbia. VIII. Raspberry leaf curl. *Can. J. Bot.* 40:651–657.

Stace-Smith, R. and Frazier, N.W. 1971. Tobacco streak virus isolated from strawberry infected with necrotic shock. *Phytopathology* 61:757.

Stace-Smith, R. and Lo, E. 1973. Morphology of bacilliform particles associated with raspberry vein chlorosis virus. *Can. J. Bot.* 51:1343–1345.

Stace-Smith, R. and Jones, A.T. 1987. *Rubus* yellow net. In: Converse, R.H., Ed. *Virus Diseases of Small Fruits*, USDA, ARS, Agr. Handb., No. 631, p. 175–178.

Stace-Smith, R. and Jones, A.T. 1987a. Black raspberry necrosis. In: Converse, R.H., Ed. *Virus Diseases of Small Fruits*, USDA, ARS, Agr. Handb., No. 631, p. 178–183.

Stace-Smith, R. and Converse, R.H. 1987. Raspberry leaf curl. In: Converse, R.H., Ed. *Virus Diseases of Small Fruits*, USDA, ARS, Agr. Handb., No. 631, p. 187–190.

Stace-Smith, R. 1987. *Rubus* virus diseases of minor or undetermined significance. In: Converse, R.H., Ed. *Virus Diseases of Small Fruits*, USDA, ARS, Agr. Handb., No. 631, p. 248–250.

Stace-Smith, R. 1987a. Thimbleberry ringspot. In: Converse, R.H., Ed. *Virus Diseases of Small Fruits*, USDA, ARS, Agr. Handb., No. 631, p. 192–194.

Stephon Liu, C.Y. 1958. Biochemical reactions of certain strawberry viruses and their mechanical transmission. *Dissertation Abstr.* 19 (4).

Stitt, L.L. and Breakey, E.P. 1952. Evidence that aphid control has suppressed virus diseases of potatoes and strawberries in northwestern Washington. *Mededelingen van de Landbouwhogesschool en de Opzoekings - stations van de Staatte Gent* 17:93–100.

Stretch, A.W. 1964. Cranberry disease investigations—1962. *Proc. Am. Cranberry Growers Assn.* 1961–64:32–33.

Sylvester, E.S. et al. 1974. Serial passage of strawberry crinkle virus in the aphis *Chaetosiphon jacobi*. *Virology* 59:301.

Sylvester, E.S. et al. 1976. Strawberry crinkle virus. *CMI/AAB Descriptions of Plant Viruses*, No. 163.

Taylor, C.E. 1962. Transmission of raspberry ringspot by *Longidorus elongatus* (de Man) (Nematode: Dorylaimidae), *Virology* 17:493–494.

Taylor, C.E. et al. 1965. The effect of tomato black ring virus on the growth and yield of Malling Exploit raspberry. *Hort. Res.* 5:19–24.

Taylor, C.E. and Thomas, P.R. 1968. The association of *Xiphinema diversicaudatum* (Micoletski) with strawberry latent ringspot and arabis mosaic viruses in raspberry plantations. *Ann. Appl. Biol.* 62:147–157.

Taylor, C.E. and Murant, A.F. 1969. Transmission of strains of raspberry ringspot and tomato black ring viruses by *Longidorus elongatus* (de Man). *Ann. Appl. Biol.* 64:43–48.

Thresh, J.M. 1965. The effects of gall mite on the leaves and buds of black currant bushes. *Pl. Pathol.* 14:26–30.

Thresh, J.M. 1966. Virus disease of black currant. *East Malling Res. Stn. Rept.* 1965, p. 158–163.

Thresh, J.M. 1967. Virus diseases of red currant. *East Malling Res. Stn. Rept.* 1966, p. 146–152.

Thresh, J.M. 1987. Black currant yellows. In: Converse, R.H., Ed. *Virus Diseases of Small Fruits*, USDA, ARS, Agr. Handb., No. 631, p. 140–141.

Uyemoto, J.K. et al. 1977. Isolation and identification of a strain of grapevine Bulgarian latent virus in Concord grapevine in New York state. *Pl. Dis. Reptr.* 61:949–953.

Van der Meer, F.A. 1961. Virusziekten bij rode bessen. (Virus disease of red currant). *Fruitteelt* 51:166–168.

Van der Meer, F.A. 1980. De bestrijding van virussen in bessen, frambozen en bramen (Control procedures of viruses in currants, raspberries and blackberries). *Fruitteelt* 70:342–344.

Van der Meer, F.A. 1987. Latent infection of gooseberry with arabis mosaic virus. In: Converse, R.H., Ed. *Virus Diseases of Small Fruits*, USDA, ARS, Agr. Handb., No. 631, p. 131–132.

Van der Meer, F.A. 1987a. Gooseberry deterioration. In: Converse, R.H., Ed. *Virus Diseases of Small Fruits*, USDA, ARS, Agr. Handb., No. 631, p. 131.

Van der Meer, F.A. 1987b. Green mottle of red currant. In: Converse, R.H., Ed. *Virus Diseases of Small Fruits*, USDA, ARS, Agr. Handb., No. 631, p. 145–146.

Van der Meer, F.A. 1987c. Spoon leaf of red currant. In: Converse, R.H., Ed. *Virus Diseases of Small Fruits*, USDA, ARS, Agr. Handb., No. 631, p. 146–150.

Van der Meer, F.A. 1987d. Leaf pattern of red currant. In: Converse, R.H., Ed. *Virus Diseases of Small Fruits*, USDA, ARS, Agr. Handb., No. 631, p. 150–152.

Van der Meer, F.A. 1987e. Interveinal white mosaic. In: Converse, R.H., Ed. *Virus Diseases of Small Fruits*, USDA, ARS, Agr. Handb., No. 631, p. 153–155.

Van der Meer, F.A. 1987f. Red currant vein banding. In: Converse, R.H., Ed. *Virus Diseases of Small Fruits*, USDA, ARS, Agr. Handb., No. 631, p. 143–144.

Van der Meer, F.A. 1987g. Yellow leaf spot of red currant. In: Converse, R.H., Ed. *Virus Diseases of Small Fruits*, USDA, ARS, Agr. Handb. No. 631, p. 152–153.

Varney, E.H. 1957. Mosaic and shoestring virus disease of cultivated blueberry in New Jersey. *Phytopathology* 47:307–309.

Vegetti, C. et al. 1979. Identificazione e prime caratteristiche di un ceppo di SLRV (strawberry latent ringspot virus) isolato da coltivazioni di lampone in Lombardia. *Rivista di Patologia Vegetale* 15:51–63.

Williams, H.E. 1987. Tomato ringspot virus in red currant. In: Converse, R.H., Ed. *Virus Diseases of Small Fruits*, USDA Handb. No. 631, p. 149–150.

Wilcox, R.B. 1923. Eastern Blue-stem of Black Raspberry. U.S. Dept. Agr. Circ. 227, 10 p.

Yoshikawa, N. and Inouye, T. 1988. Strawberry viruses occurring in Japan (Abstr.). *XIVth Int. Symp. on Fruit Tree Virus Dis.* and *XVth Int. Symp. on Small Fruit Virus Dis.*, 12-16 June, Thessaloniki, Greece.

Zeller, S.M. and Weaver, L.E. 1941. Stunt disease of strawberry. *Phytopathology* 31:849–851.

Zirka, T.I. et al. 1977. Possible etiological factors of black currant reversion. *Tegungsberichte Akademie der Landwirtschaftwissenschaften zu Berlin, DDR* 152:101–105.

7 Virus and Virus-like Diseases of Grapevine (*Vitis* spp.)

Grapevine, an ancient food plant in the life of man for several millennia (Younger, 1966), has constantly increased in production because the fresh fruit is quite nutritious for humans and now is increasingly useful as an industrial raw material for developing new food products, juices, and beverages. Sustained efforts to ensure high, reliable yields are continuously threatened with damage to production caused by numerous diseases, which are well summarized in the *Compendium of Grape Diseases* (Pearson and Goheen 1990). It is quite difficult to control viruses that infect grapevines.

Grapevine virus diseases, historically long known as the causal agents that have not been well defined, are assumed to originate from ancient Persia according to Hewitt (1970), who discusses the potential origins of grapevine fanleaf, with specific observations. If accepted, then grapevine fanleaf which stems from some millennia ago, represents probably the oldest grapevine virus disease in the world. About 20 viruses have been identified in grapevine plants. The grapevine serves as the primary natural host for grapevine fanleaf (GFLV), grapevine chrome mosaic (GChMV), and grapevine Bulgarian latent viruses (GBgLtV), and as the alternate natural host for all others, such as arabis mosaic (ArMV), tobacco ringspot (TRsV), tomato ringspot (TmRsV), and tomato bushy stunt (TmBuSnV) viruses.

Grapevine is naturally affected by some 20 presumed virus-caused diseases whose etiology remains unclear; therefore, their detection and identification should receive special attention. The important discovery of closterovirus particles associated with grapevine leafroll disease (GLrDs) should contribute to the further characterization of its etiology (Hu et al., 1990; Gugeril et al., 1990). Closterovirus-like particles have also been isolated from a corky bark diseased grapevine (GCyBkDs) (Namba et al., 1990). Grapevines displaying symptoms of the fleck disease, etiology undetermined, revealed isometric, mechanically nontransmissible particles (Boscia et al., 1990).

Viroid-like RNAs seemingly present in many grapevine cvs. and rootstocks with and/or without apparent associated disease have been discovered recently (Hewitt, 1990). Rezaian et al. (1990) have categorized five different viroids in commercial grapevine cvs., two of which cause the grapevine speckle disease (GSpcDs), while the remaining ones were the hop stunt and citrus exocortis viroids. (More details concerning other viroids can be found in the chapter section on "Grapevine Viroid Diseases.")

ECONOMIC IMPORTANCE

Virus-caused diseases rank as the most economically damaging of any grapevine diseases because, in contrast to most fungal and bacterial diseases, once infected, the canes remain systemically infected for life with no prospect for a cure.

Viruses seriously disrupt the structure and all functions of infected grapevine plants. Damaging effects of viral infections are expressed by various types of symptoms. First, and most important, they reduce grape yield and quality, often also reducing the productive life of grapevine canes. In production of grapevine stock, some viruses prevent rootstock and scion unions (i.e., the cause of incompatibility). The extent of damage depends on the characteristic of individual viruses and their strains, the susceptibility of a grapevine cv., and the mode of virus transmission spread.

The effects of economic damage caused by viruses have been established for several viruses; for example, the court-noué (grapevine fanleaf) virus infection may, under severe conditions of infections, reduce grapevine yields by 50 to 95% (Bovey, 1973). It also reduces grapevine longevity, plus adversely affects its rooting ability and graft union compatibility. It is distributed worldwide.

Legin (1972) documented that grapevine leafroll disease can reduce average grape and sugar yields by up to half, thus reducing the quantity of alcohol by 1 to 2%. Infected stocks do not easily join at the graft union. Leafroll is spread worldwide. Leafroll disease considerably reduced cane and root growth of cv. Mission vines, but had little effect on Bacco 22A hybrid; in addition, it reduced fruit yields of Mission by 66% and sugar content by 30%, whereas respective effects in the Bacco 22A hybrid amounted to 44 and 9%, respectively (Over de Linden and Chamberlain, 1970).

While studying the effects of grapevine stem pitting (legno riccio) on 6-year-old canes of Italian and Banat Riesling, Lehoczky (1972) recorded up to 23% dead canes. Nel and Engelbrecht (1972) reported effects caused by latent viruses in grapevine stocks of some cvs. as incompatibility and reduced vitality.

By the application of new general and specific methods, researchers will most likely soon elucidate the etiology of some of these virus-like diseases and even discover others yet unknown.

Epidemiology and Vectors of Grapevine Viruses

The exclusive vegetative propagation of grapevine contributes greatly to the massive scale of general spread of viruses when unnoticed. Infections of grapevines by virus and virus-like causal agents of disease are systemic, which accounts for their regular transmission during propagation. Such transmission of viruses was intense during long history of grapevine propagation and distribution worldwide when the existence of viruses was still unknown. Interestingly, the probability of viruses as the causal agents of grapevine diseases was recognized only near the end of the 19th and the early 20th centuries (Pierce, 1892). It is now known that the massive-scale build-up of populations of the root aphid *Phylloxera vastatrix* greatly endangered grapevine production by significantly contributing to the role of virus spread during the 19th century. The aphid was as readily preserved and transferred by cuttings used as planting materials as were the viruses. This susceptibility was reduced or eliminated when the cuttings normally used in production were substituted by grapevine stocks obtained through grafting susceptible European grapevines onto resistant American grapevine rootstocks. However, such grafting did not eliminate the dangers of virus transmission because new buds originated from infected mother canes.

Infected American rootstocks of *Vitis riparia*, *V. rupestris*, *V. berlandieri*, and their hybrids, most with visible symptoms of disease and some as latent virus carriers, represent a significant danger in virus spread, which can cause large-scale spread of disease, while grapevine stock producers remain generally unaware.

International exchange of planting material and sales in large quantities contribute significantly to the general worldwide spread of the viruses.

Nematodes responsible for transmitting some 20 grapevine viruses in several countries play an important role in the natural wide and local spread of virus diseases (Bovey, 1973; Bovey et al., 1980). Infected nematodes in the soil that can transmit the viruses and that remain on grapevine root stocks can be transported long distances. Following only one feeding on infected cane roots, most nematodes become infective and retain the virus for many days or several weeks, and are thus capable of infecting many healthy plants during that time (Taylor and Raski, 1964; Raski et al., 1965). Although diseased grapevine plants are removed, the nematodes remain infective in those soils and can transmit viruses to newly sown cane for several years (Dalmasso and Weischer, 1976).

All grapevine viruses have been classified into four groups based on vectors of the respective viruses (Bovey, 1973; Bovey et al., 1980; Šutić, 1977).

The largest group is the nematode-transmitted viruses, well known to damage grapevine production. Based on their nematode transmissibility and polyhedral (isometric) shape, these viruses are classified as nepoviruses (Cadman, 1963), and many are named ringspot viruses based on the symptoms that result from disease. The most important representative of this virus group is the court-noué (fanleaf) disease virus. All viruses in this group are isometric, 20 to 30 nm in diameter, and are naturally transmitted by nematodes belonging to the genera *Xiphinema* and *Longidorus*. Both are easily transmitted mechanically by infective plant sap and detected by serological analyses (Bovey et al., 1980).

Vector nematodes become infective by feeding on the roots or fresh grapevine root residues and can maintain this infectivity for several months. Nematodes cause circular patches of disease in the fields, and they can be transmitted even further during irrigation; they are often introduced into new localities with nematode-infested soil particles on the roots of infected plants. Herbaceous weed plants are also hosts through which several nepoviruses are transported into new, uncontaminated areas. Dias (1976) isolated the TmRsV, which infects grapevine from *Taraxacum officinale* and *Plantago major* plants. Ramsdell and Myers (1976) discovered in *T. officinale* peach rosette mosaic virus, which also infects grapevine. Brückbauer and Rudel (1970) found that *Stellaria media*, an important weed in vine-growing areas, serves as a tolerant host for viruses isolated from grapevine plants.

Well-known nepoviruses infective for grapevine include ArMV, artichoke Italian latent virus (AkILtV), GBgLtV, GChMV, GFLV, RbRsV, PhRoMV, TRsV, TmBRV, TmRsV, and SbLtRsV.

The second group of viruses that infects grapevines includes the soil-borne tobacco necrosis virus (TNV) transmitted in the soil by the zoospores of the fungus *Olpidium brassicae*, but it is of little importance in grapevine growing.

The third group includes the grapevine viruses transmitted by aerial insects, including alfalfa mosaic virus, aphid vectors, which transmit alpha mosaic virus, leafhopper *Halitcus citri*, the vector of sowbane mosaic virus, beet leafhopper *Circulifer tenellus*, and leafminer fly *Liriomyza langei* (Bennett and Costa, 1961; Kado, 1971). These viruses also are of little importance in grapevine production.

The fourth group comprises viruses isolated from naturally infected grapevine plants, having unknown vectors, and including tobacco mosaic (TMV) and tomato bushy stunt (TmBuSnV). Their effects in causing grapevine diseases are not yet determined.

Vectors are unknown for ten virus-like diseases of grapevine whose etiology is unstudied, including grapevine asteroid mosaic, corky bark, leafroll, stem pitting (legno riccio), yellow speckle, and fleck diseases, among others. Grapevine leafroll represents one of the most serious grapevine diseases.

Control

All measures possible should be aimed at prevention of infection in control of grapevine viruses. These measures are based on two main principles: first, plant healthy stocks, then protect it against subsequent infections.

Production of healthy planting materials consists of a series of sanitary selection measures (Šutić and Tadijanović, 1974; Vuittenez, 1977; Dimitrijević, 1977). Selection begins with visual inspection of symptomless plants, this is, those displaying no signs of disease in mother plants as a source of scions and rootstocks. All infected, or potentially infected, grapevines detected during inspection should be eliminated from mother plant plantations and destroyed. Following such procedures, the establishment of mother block plantations can be maintained three to four years or longer.

Regardless of careful inspection, some viruses can remain undetected in visually selected canes; therefore, one should also perform thermotherapy to ensure availability of virus-free material for further vineyard establishment from selected plants. Thermotherapy has successfully eliminated GFLV and allowed plants displaying grapevine leafroll and grapevine fleck disease symptoms to recover (Bovey, 1973).

An additional bioassay for the presence of viruses in visually selected material, and/or treated with thermotherapy, is done with susceptible indicator plants called the biological selection stage.

Herbaceous plants in the genus *Chenopodium* (i.e., *C. amaranticolor* and *C. quinoa*) and those in the genus *Vitis* (usually *V. rupestris* cv. St. George) are suitable indicators to detect GFLV. Those suitable as indicators of GLrDs include susceptible European *V. vinifera* cvs. Mission, Gamay, and Pinot Noir; hybrids Bacco 22A and LN-33 are suitable to detect GCyBkDs.

Although extensive and time-consuming—that is, it requires at least two seasons of field work—sanitary control (indexing) of the selected material by grafting onto woody indicator hosts is a reliable method to detect all virus diseases. The "green-grafting" technique enabling the detection of GLrDs, GCyBkDs, fleck, vein mosaic, and vein necrosis diseases in glass houses is a speedy indexing procedure (Walter et al., 1990). This green-grafting method enables a more rapid diagnosis of disease symptoms than traditional field indexing with woody indicators.

Indexing propagative material that tests virus-free is obtained through this sanitary selection method. Since a possible 20 or more viruses must be checked, sanitary selection is used primarily to detect GFLV and GLrDs, the two most important ones in grapevine production. These virus-free planting materials must continue to be protected from subsequent virus inoculations by various natural vectors. GLrDs does not endanger grapevine in this manner because vector(s) remain unknown.

These selected virus-free materials should be increased only on soils known to be free of nematode vectors as the next step in the initial preventive measure. Soils colonized by nematode vectors of grapevine viruses create a potentially dangerous situation even if grapevines have not grown there for two or three years. Subsequent viral infections cannot automatically be excluded in such soils, whether or not disinfestation was done with nematicides; thus, if at all possible, it is best to establish mother plant plantations exclusively on soils tested free of nematode vectors, that is, use new soils on which grapevine has not been grown for many years.

Nematodes should be controlled regularly in newly established vineyards to destroy focal points of infection and to prevent their spread.

Selecting and growing resistant plants represents the most reliable control method to avoid grapevine virus diseases, but finding satisfactory cvs. that are tolerant or resistant to most viruses and their strains, which differ in virulence, is difficult. Because of these difficulties, the creation of resistant hybrid plants to the most economically damaging viruses and nematode vectors remains the most important task in grapevine breeding. Since there are limited possibilities to apply many of these measures, sanitary selection currently plays a decisive role in control of grapevine viruses.

Grapevine Bulgarian Latent Virus (GBgLtV)

Properties of Viral Particles

GBgLtV is isometric and 28 to 30 nm in diameter (Martelli et al., 1978; Kuzmanovi_, 1992). Purified preparations of GBgLtV contain bicomponent particles: one an empty protein devoid of RNA (T), and the other is nucleoprotein that consists of two subcomponents (B_1 and B_2). These particles have 127S (B_2) (Martelli et al., 1978). The B_1 and B_2 components contain single-stranded RNAs of 2100 kDa and 2200 kDa, which account for 39 and 40% of the particle weights, respectively. The coat protein is a single polypeptide of 54 kDa (Martelli et al., 1978) or 55 kDa (Kuzmanović, 1992).

Based on host or location of origin, the viral isolates differ somewhat in their biophysical properties; the TIP ranged from 65 to 70°C (Bari) to 40-60°C (Kostinbrod), the DEP from 1×10^{-5} to 1×10^{-6} (Bari) and 1×10^{-1} and 5×10^{-2} (Kostinbrod), and the LIV from 15 to 20 days (Bari) to 1 to 3 days (Kostinbrod) (Martelli et al., 1978a). Isolates tested in Belgrade matched most closely the properties of the Bari isolate (Kuzmanović, 1992). Titres of antisera of various isolates ranged from 1/256 to 1/1024 (Martelli et al., 1978a) to 1/128 (Kuzmanović, 1992). These distinct differences in biophysical and immunogenic properties have been researched to further characterize strains of GBgLtV (Ramsdell and Stace-Smith, 1980). For example, the American isolate GBgLtV-NY is closely related, and thus considered a strain of blueberry leaf mottle virus (BlbLMtV).

Geographic Distribution

GBgLtV, discovered in 1973 and symptomless in Rcatzitelli grapevine cvs. from Bulgaria, was later isolated in that country from other cvs. (Martelli et al., 1978a). Uyemoto et al. (1977) described the NY strain from the cv. Concord in New York. GBgLtV was isolated from symptomless hybrids of the *Vitis berlandieri* × *V. riparia* SO4 vine rootstock in Yugoslavia, where it was relatively limited in distribution (Dimitrijeviĉ, 1980; Kuzmanoviĉ, 1992).

Host Range and Symptoms

GBgLtV has been isolated from the *Vitis vinifera* cvs. Rcatzitelli (symptomless), Julski Biser, Cabernet Sauvignon, Cardinal, and Bolgar (reduced growth and fanleaf-like symptoms) (Martelli et al., 1978). Symptoms in infected *V. labrusca* cv. Concord include delayed budbreak, irregular shoot elongation, slightly green leaves, and compact fruit clusters (Uyemoto et al., 1977). In the vine rootstock *V. berlandieri* × *V. riparia* SO4, budbreak is delayed 4 to 5 days and fewer shoots form with shorter and thinner internodes and smaller leaves (Kuzmanoviĉ, 1992).

Bulgarian isolates have been experimentally graft transmitted into rooted cuttings of LN-33 and mechanically transmitted with the aid of purified preparations into seedlings and rooted cuttings of five cvs., including Italia and Mission, which remained symptomless (Martelli et al., 1978a). The Yugoslav isolate was graft transmitted into eight European grapevine cvs., including Afus-Ali, Cardinal, Italian Riesling, and Muscat Hamburg, which also remained symptomless during a 3-year trial (Kuzmanoviĉ, 1992).

GBgLtV was mechanically transmitted by sap inoculation into ten species in four dicotyledonous families (loc. cit. Martelli et al., 1978). Herbaceous hosts of Bulgarian isolates belong primarily to the Chenopodiaceae and Amaranthaceae, which display local lesions and/or systemic infection. The Yugoslavian isolate causes local lesions and systemic symptoms in inoculated *Chenopodium amaranticolor*, *C. murale*, *C. quinoa*, and *Nicotiana tabacum* cv. Samsun; local lesions in *Gomphrena globosa* and latent infections in *N. clevelandii*, *N. rustica*, and *Petunia hybrida* (Kuzmanoviĉ, 1992). GBgLtV isolates from Bulgaria, the U.S., and Yugoslavia in the diagnostic host *C. quinoa* cause local lesions, followed by systemic symptoms.

Pathogenesis

Natural vector spread of GBgLtV is unknown, but it is spread by movement of propagative material from infected mother plants. Based on its properties, GBgLtV is a nepovirus, although its nematode transmission has not been confirmed. The American strain (NY) is transmitted through *Vitis labrusca* (ca. 5%) and *Chenopodium quinoa* (ca. 12%) seed (Uyemoto et al., 1977). Transmission of GBgLtV by infected grapevine sap onto herbaceous hosts is experimentally useful.

Grapevine Chrome Mosaic Nepovirus (GChMV)

Properties of Viral Particles

GChMV is isometric, 30 nm in diameter, and it sediments as two equal-size components (M and B) containing 31 and 40% single-stranded RNA, respectively (Martelli and Quacquarelli, 1972). The TIP of GChMV in bean sap is 60 to 62°C; the DEP is 1×10^{-3} to 1×10^{-4}; and LIV is 1 week at 22°C (Martelli and Quacquarelli, 1970). GChMV is moderately immunogenic (Hollings et al., 1969). Based on morphological and biophysical properties, GChMV resembles tomato black ring virus (TmBRV), but it differs serologically and by host reactions, by tolerance to pH, and by electrophoretic mobility, and therefore, GChMV is distinct from other known nepoviruses (Martelli and Quacquarelli, 1970).

Geographic Distribution and Economic Importance

GChMV disease was first observed in Hungary, where it probably existed for a very long time. Its localization to specific vine-growing districts can be associated with the presence of certain vectors.

Studies of the disease caused by GChMV in Hungary were initiated by Lehoczky (1966) and Sárospataki (1965).

The first isolates of GChMV, as yet incompletely characterized, were obtained from Hungarian vines displaying fanleaf and yellow mosaic symptoms (Martelli, 1965). This disease was also recorded in England (Martelli and Quacquarelli, 1972). GChMV is of limited distribution, thus of little economic importance, and yet is potentially dangerous due to the severity of symptoms in infected canes.

Host Range and Symptoms

Natural infections with GChMV have become established in many different *Vitis vinifera* cvs. (33) and in some rootstocks and hybrids (*V. berlandieri* × *V. riparia* Kober 5BB, *V. vinifera* × *V. rupestris* G1, etc.) (Martelli et al., 1970). GChMV has been experimentally, mechanically transmitted into 14 species in 5 dicotyledonous families, and the solanaceous plants usually remain symptomless (Martelli, 1965; Martelli and Quacquarelli, 1970).

Chrome yellow or whitish-yellow discolorations occur on grapevine leaves in the spring, localized along the veins on individual leaf blades or affecting whole leaves. Symptoms sometimes appear only on a single arm of the vine. Short nodes, double nodes, and bifurcations on the leaves and canes have been observed. Infected canes grow slowly, exhibit progressive decline, remain unfertile, and may die in five or six years.

Mechanically infected herbaceous hosts—including *Chenopodium amaranticolor*, *C. quinoa*, *Gomphrena globosa*, *Cucumis sativus*, *Cucurbita pepo*, *Phaseolus vulgaris*, *Datura stramonium*, *Nicotiana clevelandii*, *N. glutinosa*, *N. tabacum*, and *Petunia hybrida*—react variably with both local lesion and/or systemic symptoms (Martelli and Quacquarelli, 1970). Suitable diagnostic plants are *C. quinoa* (small chlorotic lesions, systemic chlorotic mottle, and apical necrosis), *G. globosa* (chlorotic local lesions becoming red and a light transient mosaic plus vein clearing), *D. stramonium* (inoculated leaves symptomless and systemic yellow zonate spots), and *P. vulgaris* French bean (inoculated leaves symptomless, systemic mosaic, yellow-green rings, and chlorotic specks) (Martelli and Quacquarelli, 1970).

Pathogenesis and Control

GChMV is naturally soil-borne in the field, although a vector has not been proven. GChMV had a patchy distribution in the vineyards where *Xiphinema vuittenesi*, and in one case also *X. index*, nematodes occurred simultaneously (Martelli et al., 1970). Roberts and Brown (1980) report that GChMV was acquired by *X. index* under experimental conditions, but it remains to prove whether they represent natural virus vectors.

GChMV has been experimentally graft transmitted from infected into symptomless vines, which indicates that it could be introduced by natural root grafts or by root/scion grafts into new plantations. GChMV is readily transmitted mechanically from infected grapevine sap into herbaceous hosts, which has furthered investigations. No data are available concerning effective controls of GChMV; thus, when establishing a new grapevine plantation, planting healthy propagative material is essential as is selecting fields in which no obvious known potential virus vectors occur.

Grapevine Fanleaf Nepovirus (GFLV)

Properties of Viral Particles

GFLV is isometric, 30 nm in diameter, and consists equally in purified preparations of three components with sedimentation coefficients of 50S (T), 86S (M), and 120S (B) (Vuittenez and Martelli, 1988). They also show that T contains no RNA, M contains one 1400-kDa RNA_2 molecule, and B contains one 2400-kDa RNA_1 or two RNA_2 molecules. RNA_1 directs the synthesis of a polypeptide involved in the composition of two proteins, one of which (the 58 kDa) is the coat protein (Morris-Krsinich et al., 1983).

In the F13 isolate of GFLV, which induces severe symptoms in *Chenopodium quinoa*, Pink et al. (1988) found three RNA species of 6800 (RNA_1), 3900 (RNA_2), and 1150 (RNA_3) nucleotides, respectively. RNA_3 is satellite (-like); multiplies in *C. quinoa* and depends on RNA_1 and RNA_2, and directs the synthesis of a 39-kDa protein.

The TIP of GFLV in infective sap of herbaceous plants is 60 to 65°C; the DEP is 1×10^{-3} to 1×10^{-4}; and some strains maintain an LIV of 15 to 30 days at 20°C (Hewitt et al., 1970). GFLV has good immunogenicity (Vuittenez and Martelli, 1988).

Strains of GFLV have been described, based on host range and variability of symptoms in grapevine plants such as fanleaf, yellow mosaic, and vein banding (Hewitt et al., 1970). These strains were considered initially as distinct, but serological data proved their close relationships. A strain of GFLV is the potential causal agent of the New Zealand grapevine enation disease (Chamberlain et al., 1970) (Figure 202a). No antigenic differences were found in serological tests of isolates with origins from different geographic regions.

FIGURE 202a Outgrowths (enations) on the underside of a basal leaf from a vine with grape enation. (Courtesy W. Gärtel. Reproduced with permission from APS.)

Geographic Distribution and Economic Importance

GFLV is among the most widespread and frequently occurring of grapevine viruses, generally everywhere that *Vitis vinifera* and hybrid grapevine rootstocks are grown. GFLV, a most damaging virus, adversely affects grapevine growth and development, resulting in reduced fruit yields, quality, and vineyard productivity.

Due to reduced vigor, GFLV-infected grapevine becomes susceptible to adverse climatic factors, resulting in rapid plant death. The damaging effects in propagative material include reduced grafting success and rooting ability. Yield reductions of 50 to 70%, and even 90 to 95%, occur in infected grapevine (Bovey, 1973). Brückbauer (1962) reported 75% rooted plants on a healthy SO4 rootstock as compared with 25 and 40% in moderately and heavily GFLV-infected rootstocks, respectively. A 40% success of grafts occurred with healthy Riesling on a healthy SO4 rootstock, but only 6% success when on GFLV-infected rootstocks.

Host Range

GFLV infects only grapevine plants in nature. All *Vitis vinifera* cvs. are susceptible, with no known tolerant ones. Under conditions of both natural and artificial infections, all American grapevine species imported into Europe are derived hybrids of *V. labrusca*, *V. riparia*, *V. rupestris*, *V. berlandieri*,

FIGURE 202b Grapevine fanleaf virus-infected grape leaf.

V. aestivalis, and *V. candicans* (Vuittenez, 1970). Experimentally infected hosts include over 30 species in 7 families (Vuittenez and Martelli, 1988).

Ornamental Asiatic species *Vitis davidi*, *V. coignetiae*, *V. piasezkii*, *V. ishikari*, and the fruiting *V. amurensis* were experimentally graft infected (Vuittenez, 1957). Plants in the genera *Ampelopsis* and *Parthenocissus* were experimentally infected with GFLV (Dias, 1957/1958).

Vuittenez (1970) was able to experimentally transmit GFLV from grapevine into herbaceous hosts including *Chenopodium amaranticolor*, *C. polyspermum*, *C. quinoa*, *C. globosa*, and *Amaranthus* spp. From some of these infected herbaceous hosts, GFLV was transmitted into such plants as *C. album*, *C. foetidum*, *C. foliosum*, *Amaranthus caudatus*, *A. patulus*, *Cucumis sativus*, *Phaseolus vulgaris*, *Nicotiana clevelandii*, *Impatiens holstani*, and others, which greatly facilitated determination of its known properties.

Symptoms

GFLV is characterized by variable symptoms dependant on virus strain, host reaction, age at which canes become infected, and potential interactions in infection complexes with other viruses, especially other nepoviruses (including ArMV, RbRsV, SbLtRsV, and TmBRV) that may cause similar symptoms. Characteristic symptoms of fanleaf occur on all grapevine organs and are useful for disease diagnosis. Color alterations in leaves include various chlorotic, yellow, or mosaic spots of different shapes. Grapevine yellow mosaic and grapevine vein banding diseases purported as types of GFLD caused by virus strains cause specific discolorations. Color changes of leaves characterize the primary symptom immediately following infection.

Secondary symptoms include alterations in the shape and size of leaves of canes infected earlier in the season. Due to uneven growth, leaf blades have the characteristic irregular, distorted appearance. Irregular petiolar sinuses cause the parsley-like appearance of leaves or the complete absence of petiolar sinuses, which causes the round shapes. Primary leaf veins are swollen and gathered, resembling an open fan. Grapevine leaves in New Zealand also had characteristic enations, possibly caused by a strain of GFLV (Chamberlain et al., 1970). Leaf blades are rough and brittle.

GFLV-infected shoots have shortened internodes and, when severely shortened, the nodes coalesce to form double-nodes, notably between the 9th and 10th internodes. Many shoots emerge

Virus and Virus-like Diseases of Grapevine (*Vitis* spp.)

FIGURE 203 Vein banding symptoms on a Cabernet Sauvignon leaf infected with grapevine fanleaf virus. (Courtesy A.C. Goheen. Reproduced with permission from APS.)

as forks from compacted nodes. Branches are reduced in number and size, have occasional stem pitting, and poor fruit with small berries, all of which directly affects yields (Figure 204a). The root system of GFLV-infected canes is underdeveloped and rootlets of cuttings produced from infected canes are fewer and elongated.

Radially distributed light-refracting threads visible in cross-sections of vessels under microscope, known as intravascular cordons, are formed in the vessels of lignified shoots. These cordons are specific products of cell membranes indicative of infected tissues, and abundant in European grapevines, thus of certain diagnostic importance, especially in disease detection in rootstocks.

Symptoms for disease diagnosis in indicator plants are for *Vitis rupestris* cv. St. George under glasshouse conditions, chlorotic spots, rings, lines, and finally leaf mottle; for *Chenopodium*

FIGURE 204a Fruit from grapevine fanleaf virus-infected plant; healthy (left).

amaranticolor and *C. quinoa*, chlorotic local lesions followed by vein clearing, mottle, and deformation; and for *Gomphrena globosa*, chlorotic/reddish local lesions and twisted systemically infected leaves (Vuittenez and Martelli, 1988).

Pathogenesis

Systemically infected canes are perennial sources of GFLV for further infections. GFLV remains infective for more than 1 year in residual live, infected root parts in the soil. Vegetative multiplication of GFLV-infected cuttings, stocks, or scions plays the most important role in its general continued distribution worldwide. GFLV is naturally transmitted by the nematode *Xiphinema index* in soil (Hewitt et al., 1958) and experimentally by *X. italiae* (Cohn et al., 1970). While feeding on GFLV-infected roots, these nematodes acquire the virus and transmit it in a radial fashion to nearby canes, giving the appearance of a circular spread in vineyards.

Xiphinema index lives preferably on grapevine roots. Either larvae or adult nematodes become infective by feeding on infected roots for 15 minutes; then they are capable of transmitting the GFLV during feeding on healthy plants for about 15 minutes. After feeding on GFLV-infected plants, the nematodes maintain their ability to transmit it for at least 8 weeks, and even up to 8 months under experimental sterile conditions, without food (Taylor and Raski, 1964). GFLV does not pass on to the next larval stage during molting. A few nematodes effectively transmit GFLV. Two to three nematodes detected per liter of soil may represent several thousand individuals in the cane rhizosphere, and only one suffices to transmit the GFLV (Bovey, 1973).

GFLV is transmitted via seed of experimental *Chenopodium amaranticolor*, *C. quinoa*, and soybean hosts. It is present in the pollen of grapevine and experimental herbaceous hosts (loc. cit. Hewitt et al., 1970). No reliable data exist on GFLV transmission through grapevine seed. Its mechanical transmission via sap of infected herbaceous hosts is only of experimental importance. GFLV experimentally graft transmissible is a useful diagnostic tool for indexing grapevine cvs.

Control

All the recommendations listed in the introductory section apply to GFLV. They encompass sanitary selection to ensure virus-free planting stock, and selection of soils that are both uncontaminated by nematode vectors and free of grapevine residues as potential sources of infection. Ecological changes in the grapevine-nematode complex can affect nematode populations. Nematode control by fumigation is economically feasible only in the soils used for grapevine replanting for vineyard establishment. GFLV-free stocks can be obtained by exposing the plants to 38°C for 4 to 6 weeks, combined with rooting of shoot tips under mist (Goheen et al., 1965), culturing excised tips *in vitro* (Ottenwaelter et al., 1973), tip micrografting (Bass and Vuittenez, 1977), or meristem-tip culturing (Barlass et al., 1982).

Grapevine: Natural Alternative Host for Nematode-Borne Viruses

Arabis Mosaic Nepovirus (ArMV)

ArMV was discovered in grapevine with the aid of serological tests comparing it with isolates of GFLV and/or tomato black ring viruses (Bercks and Stellmach, 1964; 1966; Bercks, 1967). All infected plants displayed symptoms of GFLV and yellow mosaic disease. ArMV was identified serologically directly from grapevine sap, and its transmission was followed in experimental *Chenopodium quinoa* plants (Bercks, 1967). It always occurred only in mixed infections; thus, symptomology and economic effects by ArMV alone in grapevine remain unknown. ArMV naturally infects *Vitis vinifera* plants with GFL symptoms and rootstock cvs. with reduced growth. *Xiphinema diversicaudatum*, known as the vector of ArMV in several other hosts, has been associated with grapevine plants infected with ArMV; thus circumstantially, it probably can transmit a grapevine-adapted ArMV strain, although this is not proven (Bovey, 1973).

Artichoke Italian Latent Nepovirus (AkILtV)

AkILtV was first isolated from grapevine plants in Bulgaria, displaying fanleaf-like symptoms (Jankulova et al., 1978). These isolates from grapevine proved closely related serologically to typical AkILtV strains. By mechanical inoculations, it has been transmitted from grapevine into 68 species in 15 families (Jankulova et al., 1978). They report that most infected *Chenopodium* and *Nicotiana* hosts display symptoms characteristic of nepoviruses. Other hosts include *Chenopodium quinoa* as assay host; *Gomphrena globosa* with chlorotic or necrotic ring-like local lesions, systemic ringspots, and deformation of apical leaves, and *Phaseolus vulgaris* with chlorotic or necrotic local lesions and systemic necrosis. The actual effect of AkILtV on grapevine remains unknown.

Peach Rosette Mosaic Nepovirus (PhRoMV)

PhRoMV has been reported in grapevine in Ontario, Canada, and in Italy (loc. cit. Klos, 1976). It is observed most frequently in grapevines grown in soils on which peach plantations infected with PhRoMV had grown previously (Bovey, 1973). Isolates of PhRoMV from both peach and grapevine showed no serological differences (Dias, 1975). Grapevine, the natural host of PhRoMV, transmits it through seed (Bovey et al., 1980). PhRoMV has been transmitted experimentally by graft and mechanical inoculations into various herbaceous plants. Mosaic spots and asymmetry characterize symptoms of infected grapevine leaves, bud development is delayed, clusters have fewer berries that frequently drop early, and stunted canes gradually die. *Xiphinema americanum* and *Criconemoides sp.* are PhRoMV vectors in nature (Klos et al., 1967), but vector transmission of PhRoMV into grapevine plants has not been proven (Bovey, 1973).

Raspberry Ringspot Nepovirus (RbRsV)

RbRsV was discovered in naturally infected grapevine in Germany (Bercks, 1968). It was documented serologically in the canes of the cvs. Riesling, Silvaner, Müller-Thurgau, and S88. The nematode *Longidorus macrosoma*, known as the vector that transmits RbRsV into several hosts, was found in an infected grape vineyard (Weischer, 1968).

Strawberry Latent Ringspot Nepovirus (SbLtRsV)

SbLtRsV was discovered in naturally infected grapevine with the aid of serological tests (Vuittenez et al., 1970). No specific symptoms are associated with SbLtRsV in grapevine, and its economic effects in grapevine are unknown. *Xiphinema diversicaudatum* and *X. coxi* are known vectors of SbLtRsV into several hosts, but that role in grapevine has not been proven.

Tobacco Ringspot Nepovirus (TRsV)

TRsV was discovered in infected grapevine in New York (Gilmer et al., 1970; 1972). Diseased grapevine canes display symptoms mimicking those caused by GFLV, including a delay in budbreak, chlorotic ringspots on lower cane leaves plus deformations, and stunt of terminal (apical) leaves. Mosaic with irregular veins is frequently observed in TRsV-infected leaves, shoots have short internodes, and plants bear few fruits or remain fruitless in the second year after infection (Bovey, 1973). By graft and mechanical inoculations, TRsV has been transmitted into several herbaceous plants and grapevine seedlings (Gilmer et al., 1970). *Xiphinema americanum*, known to transmit several strains of TRsV, was recovered from soil in which TRsV-infected grapevine was grown; however, its role as the vector in transmission of TRsV into grapevine has not been proven.

Tomato Black Ring Nepovirus (TmBRV)

TmBRV was discovered in the *Vitis vinifera* Aramon × *V. riparia* 143 AMG rootstock in Germany in 1962 and 1963 (Stellmach, 1970). Both transmissions into susceptible *Chenopodium* and *Nicotiana* plants and serological analyses have shown that the isolates from grapevine are a strain of TmBRV (Stellmach and Bercks, 1963). TmBRV was also reported in American *V. vinifera* rootstocks and cvs. in several vine-growing districts, frequently as a complex with GFLV and ArMV (Bercks and Stellmach, 1966). TmBRV-infected plants in vineyards grow slowly, and chlorotic spots and edge yellowing appear on leaves. Stellmach and Bercks (1965) recorded 40% of healthy plants. The grapevine strain of TmBRV in Germany is related to the peach strain and to the potato bouquet virus (Stellmach et al., 1966). Isolates of TmBRV are transmitted experimentally from *Vitis* to *Vitis*, plus mechanically into susceptible herbaceous hosts, including *Chenopodium quinoa*, local necrotic rings and spots plus systemic chlorotic spots on young leaves; *C. murale*, local necrotic rings plus systemic vein clearing and occasional apical necrosis; and *Nicotiana clevelandii*, local necrotic rings and systemic chlorotic ring mosaic. The soil vector of TmBRV in grapevine vineyards is the nematode *Longidorus attenuatus* (Stellmach et al., 1966). Stellmach (1970) also recognized the omnipresence of *Stellaria media*, a weed known as the virus carrier of TmBRV in infected grapevine.

Tomato Ringspot Nepovirus (TmRsV)

TmRsV was discovered in naturally infected grapevine exhibiting yellow vein symptoms in California (Hewitt, 1956; Gooding and Hewitt, 1962), and in grapevine with ringspot symptoms in New York (Uyemoto, 1970; Gilmer and Uyemoto, 1972). Isolates from grapevine are serologically related to TmRsV and the peach yellow bud mosaic strain of TmRsV; thus, they are considered strains of TmRsV (Gooding, 1963). Yellow vein is reminiscent of the vein banding disease. It has small chlorotic spots along leaf veins that can spread gradually throughout the leaf blade. Initially, bright yellow spots become white. Fewer berries in each cluster represents the major damage caused by TmRsV; but based on spread, it is of little economic importance, although up to 15 to 22% of canes can become unproductive in some vineyards (Figure 204b). TmRsV can be transmitted experimentally by graft and mechanical inoculations to several herbaceous hosts (Gooding, 1966) and onto etiolated grapevine seedlings (Gooding and Teliz, 1970). *Xiphinema americanum* naturally transmits TmRsV into other herbaceous hosts, but such transmission to grapevine has not been proved.

Grapevine: Natural Alternative Host for the Soil-Borne Fungal Transmissible Tobacco Necrosis Virus

Tobacco necrosis virus (TNV) was discovered with the aid of serological tests in crude extracts from grapevines exhibiting yellow spots and mosaic symptoms in South Africa (Cesati and Van Regenmortel, 1969). The TNV isolate from South Africa reacted strongly with the serotype D and weakly with the serotype A antisera of TNV. *Chenopodium quinoa*, with necrotic local lesions, and *Gom-*

FIGURE 204b Cascade berry clusters: healthy (left) and infected with tomato ringspot virus (right). (Courtesy J.K. Uyemoto. Reproduced with permission from APS.)

phrena globosa, with red local lesions, serve as indicator herbaceous hosts of the grapevine TNV isolates to study their properties. TNV is soil transmitted by the zoospores of the *Olpidium brassicae* fungus (Teakle, 1962), but no proof exists for its transmission of the grapevine TNV isolates.

Grapevine: Natural Alternative Host for Air-Borne Insect Transmissible Viruses

Alfalfa Mosaic Virus (AMV)

AMV, first observed in the grapevine cv. Kerner in Germany (Bercks et al., 1973), did not spread from its initial location in a 3-year study. AMV was found in and isolated from naturally infected grapevine in Czechoslovakia (Novak and Lanzova, 1976). Bovey and Cazelles (1978) discovered AMV in Switzerland in a hybrid seedling obtained from the Grézot and *Vitis berlandieri* × *V. riparia* 5C rootstocks. AMV was transmitted from infected plants with ringspot and line pattern symptoms by mechanical inoculations into *Chenopodium quinoa* with chlorotic local lesions and systemic mottle, *Phaseolus vulgaris*, and *Vigna sinensis* with necrotic local lesions. The presence of AMV in infective plant sap has been confirmed serologically. The virus isolate was transmitted by chip-budding grafts to *Vitis rupestris* cv. St. George, displaying line pattern symptoms. The common aphid vectors of AMV have not been experimentally confirmed for grapevine. AMV can be eliminated by heat therapy at 37 to 38°C during 37 days. (Bovey and Cazelles, 1978). Due to lack of localized spread, AMV is of no threat to grapevine growers.

Sowbane Mosaic Virus (SwMV)

SwMV, isolated from naturally infected grapevine in Germany (Bercks and Querfurth, 1969), was transmitted mechanically to *Chenopodium quinoa* with small ringspots followed by vein clearing and chlorotic mosaic. It was then transmitted by sap from *C. quina* into plants in ten families (Bercks and Querfurth, 1969). *Circulifer tenellus*, *Halticus citri*, and *Liriomyza langei* (Bennett and Costa, 1961) are known vectors of SwMV, but their role in its transmission to grapevine must be experimentally proved.

Grapevine: Natural Alternative Host for Viruses with Unknown Vectors

Tobacco Mosaic Virus (TMV)

TMV, first discovered in infected grapevine in the U.S. (Gilmer and Kelts, 1965; 1968), was then reported in Germany (Bercks, 1967a), Bulgaria (Jankulova, 1970), and Yugoslavia (Dimitrijević, 1972). Gilmer and Kelts (1968) showed the presence of TMV in 29 of 67 seedlings of grape breeding lines and proved its transmission through seed of grapevine. With the approach-graft technique, Dimitrijević (1972) transmitted TMV from naturally infected grapevine to *Chenopodium murale* and vice versa. Although it attracted much attention from researchers, TMV in grapevine has no economic impact on grapevine production.

Tomato Bushy Stunt Virus (TmBuSnV)

TmBuSnV was confirmed by serological tests in infected grapevine cvs. Riesling, Silvaner, S88, and 5C (Bercks, 1967b). TmBuSnV isolates from grapevine are serologically identical or closely related to a strain of TmBuSnV isolated earlier from petunia plants in Italy. Since TmBuSnV usually occurs in complex infections, its effect on grapevine alone remains unknown.

Grapevine Diseases Assumed Caused by Viruses

A number of virus-like diseases have been described in grapevine, of which the causal agents are unidentified. Because they all can be transmitted by grafts and systemic symptoms of disease are similar to those caused by known viruses, they have been labeled virus-like diseases. Graft transmission is an important diagnostic feature for identification of obligate pathogens, which now include phytoplasma-like organisms (grapevine flavescence dorée) and bacteria (grapevine Pierce's disease caused by *Xilella fastidiosa*), in addition to viruses. Symptoms of disease are specific and represent an important component of identification of pathogens; but unfortunately, they cannot be completely diagnostic because two or more causal agents may cause similar symptoms on a given host plant.

Numerous diseases described solely on the basis of symptoms have been later confirmed as caused by the strains of previously described viruses; for example, grapevine fanleaf virus and its strains are now known to cause grapevine vein banding, yellow mosaic, and enation diseases; TmBRV and its strains are now known to cause grapevine yellow mosaic and stunting; and TmRsV isolates are now known to cause grapevine yellow vein disease. Some separately described diseases were discovered in grapevine plants as a complex with other diseases; for example, leafroll/stem pitting/fleck diseases and ajinashika/leafroll/fleck diseases.

Recently, substantial efforts have been made to establish the etiology of this group of disease complexes, but progress is slow. This group includes economically important diseases of grapevine leafroll, corky bark, and stem pitting, all of which require special attention in grapevine production. The most frequently encountered virus-like diseases in grapevine are described in some detail.

Grapevine Asteroid Mosaic Disease (GAdMDs)

GAdMDs, first discovered in California (Hewitt, 1954; Hewitt and Goheen, 1959; Hewitt et al., 1962), was later observed in Italy and South Africa (Refatti, 1970; Nel and Engelbrecht, 1972a). The causal agent(s), and thus its properties, are unknown; but based on graft-transmissibility and symptoms, it is believed viral in nature. It has been graft transmitted to several *Vitis vinifera* cvs., including Carignane, Emperor, and Colombard, and to the *V. rupestris* St. George cv. (Reffati, 1965; 1970). No other mode of transmission is known. Star-shaped spots irregularly distributed on leaf blades characterize symptoms in *V. vinifera* and, when numerous, these spots coalesce and the centers become dark and necrotic. Net-like vein banding sometimes occurs along smaller leaf veins and leaves develop asymmetrically. During the summer, all symptoms are milder (i.e., they are less evident). Affected canes are frequently stunted and usually bear no fruit, or only few. *V. rupestris* cv. St. George is a suitable indicator for GAdMDs, wherein 1 to 2 months after inoculation, creamy yellow bands with slightly feathered margins occur along major veins and develop unevenly and remain smaller.

Grapevine Corky Bark Disease (GCyBkDs)

GCyBkDs, first described as grapevine rough bark in California (Hewitt, 1954), is now corky bark disease (Hewitt et al., 1962). Now reported in South America (Mexico, Brasil), South Africa, Japan, and Europe, its distribution is worldwide (Bovey et al., 1980).

The etiology of GCyBkDs has not been confirmed. No causal agent has been re-inoculated into grapevine and subsequently re-isolated. Most recently, isolates of closterovirus-like particles from corky bark-diseased grapevines have been researched. Conti and Martelli (1988) noted filamentous closterovirus-like particles over 1500-nm long associated with the corky bark-stem pitting complex in California. From petioles of the cv. Semillon with symptoms of corky bark disease, Namba et al. (1990) isolated and purified closterovirus-like particles 13×1400 to 2000 nm, with a coat protein of 24 kDa. A double-stranded RNA of 1×10^{-4} to 1×10^{-6} Da was isolated from the bark phloem of these plants. Namba et al. (1990) correlated GCyBkDs from corky bark-diseased plants in over 75% of ELISA tests. Tanne and Meir (1990) also isolated from various GCyBkDs grapevine tissues, a double-stranded RNA of high molecular weight specific for corky bark-diseased plants. The presence of closterovirus-like particles in grapevines with GCyBkDs symptoms is of particular interest in etiological investigations.

European *Vitis vinifera* and American *V. rupestris* grapevines are natural hosts of GCyBkDs. Some *Vitis* hybrids have been infected experimentally. Attempts to mechanically transmit GCyBkDs causal agents into herbaceous and other plants outside the genus *Vitis* (by inarch grafting) have been unsuccessful (Beukman and Goheen, 1970).

No *Vitis vinifera* cvs. display symptoms of GCyBkDs; thus, the LN-33 indicator should be used to detect it (Beukman and Goheen, 1965). They also showed that the *V. vinifera* Carignan cv. has mild symptoms, and the cvs. Cabernet Franc, Gamay, Mondeuse Palomino, and Petite Sirah have moderately severe symptoms. Since the hybrid LN-33 (Couderc 1613 × Thompson Seedless) has the most severe symptoms, it is used as the indicator for detection of GCyBkDs.

Most European grapevine cvs. react to GCyBkDs with reduced vigour, but have no other symptoms. Characteristic corky bark symptoms occur on both canes and leaves of GCyBkDs-susceptible cvs. Cabernet Franc, Gamay, and others. Spring budbreak is delayed, some emerging spurs die, canes tend to bend downward, and their wood is limber or rubbery. Longitudinal cracks appear at the base of the canes (see Figure 205) and prevent regular wood maturation. Infected leaves are usually smaller and turn red, including the veins, in red-berried cvs., a trait that distinguishes them from the red color caused by GLrDs. Symptoms in LN-33 indicator plants include longitudinal cracks and splits along the shoots, abnormal tissue development (i.e., shoot swellings between internodes) at the rootstock/scion graft joint, reddened leaves, and much reduced growth.

FIGURE 205 Stem pitting resulting from corky bark disease.

Damage by corky bark causes heavy economic losses in susceptible cvs. (Bovey et al., 1980). Yield reductions by GCyBkDs of 76% occurred in the cv. Cardinal in Mexico, plus shortened productive life of the vineyard (Teliz et al., 1982). The main control measure for GCyBkDs is to use healthy planting material to establish new grapevine plantation. Healthy scions result from treatment of buds over 90 days at 38°C, which are either grafted into healthy rootstock or removed from shoot tips and rooted under mist (Goheen, 1977).

Grapevine Fleck Disease (GFkDs)

The causal agent of GFkDs is unknown. Some graft transmission tests suggested its resemblance with the vein clearing mosaic disease (Vuittenez, 1966), but later it was shown as two etiologically different diseases. The GFkDs is frequently associated with leafroll, stem pitting, and corky bark—but not consistently (Martelli and Vuittenez, 1988). Jasuo Terai and Ryn Yano (1980) suggested that grapevine ajinashika disease in Japan is closely associated with fleck and leafroll agents.

Boscia et al. (1990) isolated a grapevine phloem-limited, isometric virus (GPhLdV) in Italy, prepared its antiserum and, using a *Vitis rupestris* indicator, established a very strong association of GPhLdV with fleck disease. Thus, they suggest that GPhLdV is likely involved in the etiology of GFkDs.

GFkDs, first observed in California (Hewitt et al., 1962), has been found in both Europe and South Africa. Although believed widely spread, symptoms in tested plants have been attributed frequently to other virus diseases. The disease is systemic, thus preserved in a latent form in several grapevine cvs. and rootstocks, which may account for its unknown important spread in the past.

Mottled flecks occur either sporadically or in large numbers in leaves, characteristic of susceptible cvs.; severely infected leaves twist and wrinkle. Kober 5 BB and *Vitis rupestris* St. George cvs. express chlorotic translucent vein break on young and medium-sized leaves, and thus are suitable indicators (Hewitt et al., 1972). Because of the possible spread of GFkDs by latent infections, only tested disease-free vine stocks should be used to establish new vineyards. The causal agent of GFkDs is heat stable, so elimination requires lengthy thermotherapy—even more than 200 days (Legin et al., 1979).

Grapevine Leafroll Disease (GLrDs)

GLrDs is distributed worldwide wherever grapevine is grown. Known in the mid-19th century by its unique symptoms (Fabre, 1853), Scheu (1936) experimentally graft transmitted it successfully from infected into healthy grapevine plants, which proved its infectivity.

Although long known as an important disease of grapevine, its etiology is still undefined. Several different viruses have been isolated from GLrDs grapevine plants with leafroll symptoms. From grapevines infected with this disease, Tanne et al. (1978) isolated a potyvirus with particles 13×790-nm, containing a single-stranded RNA of 3500 kDa and a protein coat of 2100 identical protein subunits, each 31 kDa. They also transmitted the infectious agent into solanaceous plants, then retransmitted it back to virus-free cuttings of the Mission grapevine cv. from infected *Nicotiana glutinosa*. Special efforts should be made to research the closterovirus-like particles observed by scientists from several countries in thin sections and/or leaf dip preparations from grapevines displaying leafroll symptoms (Milne et al., 1984; Hu, 1990). Milne et al. (1984) named one grapevine virus A (GVA) and another grapevine virus B (GVB). GVA particles, 11 to 12×800 nm, have a single-stranded RNA 2550 kDa and a coat protein of a single polypeptide 22 kDa. GVA and GVB are not serologically related.

In the former U.S.S.R., Hu et al. (1990) mechanically transmitted a closterovirus of 800-nm model length from a leafroll-diseased grapevine to *Nicotiana occidentalis* with necrotic local lesions and systemic mosaic, yellow, and curl of leaves; *N. benthamiana* with mild systemic chlorosis; and to *Cucurbita maxima, Cucumis sativus, Datura stramonium, Gomphrena globosa,* and *Vinca rosea* with latent infections.

In purified preparations from cv. Merlot grapevines in Italy with severe leafroll symptoms, Faggioli et al. (1990) determined a 1100-nm modal length of closterovirus (clostero type III) particles. In crude extracts from leafroll-diseased grapevines in the former U.S.S.R., Bondarchuk et al. (1990), with the aid of an electron microscope, discovered thread-like particles 17×700 to 2300 nm.

Interestingly, Namba et al. (1979) in Japan found 26 to 28-nm diameter, isometric, virus-like particles associated with the ajinashika disease, a severe form of leafroll and fleck diseases of Koshu grapevine. In Italy, Castellano et al. (1983) found 22 to 24-nm diameter, isometric, virus-like particles in the phloem of grapevine exhibiting leafroll symptoms.

Finally, it is important to note that the various "viruses" discovered in grapevine displaying leafroll symptoms shows that the GLrDs phenomenon remains etiologically unclear and still complex as a disease, probably caused by several different viral agents.

These many investigations show that the damaging effects of GLrDs on fruit yield, fruit quality, graft compatibility, and rooting ability of canes ranks it among the most serious of grapevine virus-like diseases. Over de Linden and Chamberlain (1970) showed that GLrDs reduced the yields of *Vitis vinifera* cv. Mission 66% and sugar content 30%, while the hybrid grapevine Bacco 22A had only 44 and 9% reductions, respectively. Sugar content of infected cv. Gamay fruits was reduced

36% (Dimitrijević, 1970). Martelli and Vuittenez (1988) found reductions of yield of 10 to 70% and of sugar content by 13% Oechsle.

European and American grapevine species and cvs. are the only natural GLrDs hosts. American *Vitis* species and rootstock hybrids are usually symptomless (Vuittenez, 1958), while France has GLrDs which causes symptoms on *Vitis riparia* leaves (Bass and Legin, 1981). American rootstocks with latent infections call for added precaution because these symptomless carriers of masked infections in cultivated grapevine cultivars can be easily introduced inadvertantly into other countries.

Leafroll symptoms vary, depending both on the cvs. and on climatic conditions in individual vine-growing regions. The most severe symptoms occur in black-fruited *Vitis vinifera* cvs. of Gamay, Pinot Noir, and others.

Necrotic spots occur on the lower leaves of black-fruited cultivars from June through July. As disease develops, these spots spread, coalesce, and affected leaf blades become red, with the exception of auxiliary veins. Leaves become swollen, brittle, and bent downward along the margins. At season end, these symptoms are evident on the majority of infected cane leaves. The most severely infected leaf blades drop prematurely. Fruit ripening is irregular and delayed; thus, green berries occur at harvest. The fruits of the typical red-fruited cv. Emperor from GLrDs plants remain pale yellow—a symptom that depicted earlier leafroll as the white Emperor disease. The spots that occur on leaves of white-fruited cvs. are pale chlorotic, which makes disease detection much more difficult than on black-fruited cvs.

Phloem degeneration due to death of sieve tubes, companion cells, and parenchyma cells in the canes, petioles, leaves, cluster stems, and fruit pedicles is a characteristic histological and anatomical symptom in GLrDs grapevines (Goheen, 1970). Degeneration symptoms occur early and may be noticed before leaf symptoms.

Low levels of potassium (Ravaz et al., 1933; Lafon et al., 1955), or potassium and calcium (Goheen and Cook, 1959), in leaves of GLrDs grapevines led to the conclusion that such deficiencies caused symptoms of leafroll. Later, it was proved that the deficiency caused by transport disorders is only one of several symptoms and not the cause of the disease.

Symptoms caused by GLrDs can be mimicked by insects feeding on grapevine leaves: for example, leafhopper (*Empoasca flavescens*), buffalo treehopper (*Ceresa bubalus*), and cottony-cushion scale (*Icerya purchasi*), which also can cause changes in the shape and color of leaves. Since other factors have the potential to influence symptom analysis, positive diagnosis of the GLrDs must be achieved by indexing canes by infectivity assays on indicator plants, such as the European grapevine cvs. Mission and Gamay, the LN-33 hybrid, and others.

Infected mother canes and rootstocks are carriers from which this disease is transmitted by grafting. Therefore, infected grapevine stocks represent the main mode of massive scale dissemination of leafroll disease. Although natural vectors that transmit the causal agent(s) of GLrDs remain unknown, the widespread nature of GLrDs suggests their potential existence. GLrDs has been transmitted experimentally by mealy bugs (loc. cit. Hewitt, 1968), but their possible vector role in nature is still unproven.

The use of disease-free stocks for the establishment of vineyards is the primary recommended measure to control GLrDs. Disease-free stocks can be obtained by the use of thermotherapy at 38°C for 60 to 120 days, removal and rooting of shoot tips under mist (Martelli and Vuittenez, 1988), micrografting techniques (Bass and Legin, 1981), and meristem-tip culturing (Barlass et al., 1982).

Grapevine Stem Pitting Disease (GStmPgDs)

GStmPgDs (legno riccio, rugose wood), first observed in 1959 in Italy (Ciccarone, 1961; Graniti and Ciccarone, 1961), was established as an infectious disease by graft transmission (Graniti and Martelli, 1965). However, the etiology of GStmPgDs remains unclassified.

GFLV has frequently been detected in grapevines suffering from stem pitting disease (Graniti and Martelli, 1970). With the aid of an indicator assay host (5 BB), GLrDs symptoms were reproduced from vines with stem pitting symptoms (Legin, 1972). A form of GStmPgDs first described in Italy (Graniti and Martelli, 1970) may be related or identical to GCyBkDs (Martelli and Conti, 1988), while GVA has also been isolated from GStmPgDs vines in Italy (Conti et al., 1980). Kartuzova et al. (1990) discovered 800-nm long, closterovirus-like particles in vines affected with GStmPgDs. In *Vitis rupestris* cv. St. George infected with GStmPgDs, Azzam et al. (1990) found two double-stranded RNA particles, one with two bands of 5300 kDa and 4400 kDa associated with the disease. These various studies of the etiology of GStmPgDs raise the question of whether it is caused by one or more pathogens, or if it may represent a leafroll–corky bark–stem pitting complex, the etiology of which should be studied as an integral unit.

GStmPgDs has been found recently in several countries; thus, it seems widespread wherever grapevine is grown, and occurs more frequently than believed previously. Abracheva (1981) observed obvious disease symptoms in 86% of 646 different grapevine cvs. forming the Bulgarian collection.

Based on its influence on grapevine growth and yield, GStmPgDs is economically damaging. The extent of damages depends on the scion-rootstock combination, the susceptibility of European cvs., and the virulence of the causal agent (Graniti and Martelli, 1970). When all the most favorable conditions exist, GStmPgDs causes gradual decline, progressive yield loss, infertility and finally plant death.

European and American grapevine species are the only natural hosts of GStmPgDs. Most European cvs. and rootstock hybrids are susceptible; they include Cardinal, Italia, Regina, Primus, Moscatello, Razaki, Muscat Lunel, Italian Riesling, and several others. Numerous American rootstocks are affected and include *Vitis berlandieri* × *V. riparia* 420A, 157-11, 34E, Teleeki 5A, Kobber 5BB; *V. riparia* × *V. rupestris* 3309; *V. berlandieri* × *V. rupestris* 17-37, 140; and *V. rupestris* Du Lot (Graniti and Martelli, 1970). Since the introduction of susceptible and infected American rootstocks into grapevine production which were earlier grown on their own rootstocks to ensure protection against Phylloxera, GStmPgDs has become increasingly important economically as it spreads.

Grapevine plants attacked by GStmPgDs are less vigorous, and budbreak is delayed by 4 to 5 weeks (Bovey, 1973). Characteristic swellings occur above graft joints, and rootstocks usually remain thinner than the scions. Such incompatibility alterations are accompanied by phloem necrosis. The bark on the basal part of the canes (rootstock, scion or both) is rough, abnormally swollen, and spongy. Pits and grooves that form along the woody cylinder beneath the bark match the mirror image of pegs and ridge-like protrusions in the cambial part of the bark.

The causal agent of GStmPgDs is graft transmitted from rootstock into scion, and vice versa. American grapevine spp. and their hybrids LN-33, *Vitis rupestris*, *V. riparia* × *V. berlandieri* Kober 5BB or 157-11, *V. rupestris* × *V. berlandieri* 1103P, which exhibit characteristic symptoms after 1 to 3 years, serve as suitable indicators (Martelli and Conti, 1988). The primary disease control measures are planting of disease-free stocks obtainable by thermotherapy at 35°C for more than 150 days (Legin et al., 1979) and removal of shoot tips and grafting or culturing *in vitro*.

Grapevine Yellow Speckle Disease (GYSpeDs)

The causal agent of GYSpeDs has neither been isolated nor identified, although based on symptoms and graft transmissibility, it probably is virus- or viroid-like. Vein banding-like symptoms which were neither expressed by *Vitis rupestris* St. George nor *Chenopodium quinoa* were observed in infected grapevine with yellow speckle symptoms (Figure 206) (Taylor and Woodham, 1972). Grapevine is the only known natural host of GYSpeDs. Krake and Woodham (1983) suggested that vein banding results from mixed infections of GFLV and the agent of GYSpeDs.

GYSpeDs was first observed, described, and detected by indexing of grapevine clones in Australia (Taylor and Woodham, 1972) that had been introduced from California; these clones were either symptomless or displayed only mild symptoms. Among 100 clones introduced from California, 16, including rootstock 1613, in Australia displayed severe symptoms; but in California, only few yellow speckles were noted on occasion in the same clones. Clones of cvs. Grey Riesling

FIGURE 206 Chlorotic speckled bands caused by yellow speckle disease of Mission. (Courtesy A.C. Goheen. Reproduced with permission from APS.)

and Trousseau displayed severe grapevine banding symptoms in Australia, yet only mild symptoms in California (loc. cit. Taylor and Woodham, 1972). Evidently, climatic, and possibly other conditions, favor symptom development in Australia over those in California.

GYSpeDs was named after the small chrome yellow speckles occurring on grapevine leaves that, when abundant, coalesce and encompass large parts of the leaf. When speckles are distributed along the veins, they resemble vein banding or yellow vein symptoms. The symptoms usually become evident only late in the summer; when severe symptoms appear earlier, then the initial chrome yellow color becomes creamy white in the autumn.

The Esparte cv. is most suitable for indexing suspect-diseased grapevine canes. Since GYSpeDs is graft transmitted, planting disease-free stocks is an indispensable general control measure.

Grapevine Viroid Diseases (GVdDs)

Viroid plant pathogens were recently discovered; the first was potato spindle tuber viroid, in 1971, the type member of one of the three known viroid groups. A viroid is the smallest plant pathogen, consisting of only one circular single-stranded RNA molecule, no coat protein, and comprising 246 to 376 nucleotides (Tsagris et al., 1990). The viroids cause either economically important diseases or latent infections in plant.

Viroids were recently discovered in grapevine in Australia, Europe (Greece, France, Italy, Spain), Israel, and the U.S. as presented at the *10th Meeting of ICVG*, Sept. 3-7, 1990, Volos, Greece. The research results so far have shown that viroids are widespread among all cvs. and rootstocks in all vine-growing areas (Hewitt, 1990; Juarez et al., 1990).

The following five viroids are established in commercial grapevine cvs. in Australia: hop stunt (HSnVd), citrus exocortis (CsExVd), yellow speckle (GYSpeVd), grapevine 1B (G1BVd), and grapevine Australia (GAaVd) viroids (Rezaian et al., 1990). GYSpeVd, G1BVd, and GAaVd share unique structural properties with the apple scar skin viroid as a separate group. GYSpeVd and G1BVd are the causal agents of yellow speckle disease. Low molecular weight RNAs tentatively identified as grapevine yellow speckle viroid (GYSpeVd), grapevine viroid 2 (G2Vd), and HSnVd were discovered in all *Vitis vinifera* cvs. and American rootstock hybrids, but not grapevine seedlings, tested in Italy (Minafra et al., 1990). GYSpeVd and HSnVd were also present in grapevines displaying yellow speckle-like and vein banding symptoms. A survey of leaves of some grapevine cvs. in fields in Greece revealed that most samples were infected latently with the hop stunt viroid (Tsagris et al., 1990). A similar distribution of viroids was observed in grapevine samples from Europe and California (Szychowski et al., 1990). The samples from Europe were dominated by the viroid profile of G1Vd + G3Vd; only cvs. Cot and Merlot showed the profile of a single viroid G3Vd. Some samples from California contained all these three viroids, whereas others had only G1Vd or G3Vd.

In order to more precisely define the relationships among viroids and to provide an orderly nomenclature for synonyms, Semancik and Szychowski (1990) suggested classifying similar viroids into two groups: (1) apparent viroids and (2) enhanced viroids. Apparent viroids are characterized by their ability to be isolated directly from grapevine tissues. This group includes widespread viroids containing about 300 nucleotides and related in sequence homology to the hop stunt viroid (HSnVd). They share common hosts with HSnVd, but their biological properties relating to HSnVd have been little studied. Enhanced viroids are characterized by those that must multiply in alternate hosts before they can be isolated and studied. Enhanced viroids contain 360 to 370 nucleotides and have never been isolated directly from grapevine. One of these viroids, isolated from inoculated *Gynura aurantiaca* or tomato plants, has 371 nucleotides and is closely related to the citrus exocortis viroid (CsExVd), while the other, isolated from inoculated cucumber plants, has been denoted as Australia grapevine viroid (GAaVd). The systematization and classification of viroids, although complex, represents progress, which will continue as new information is discovered through research.

Insufficient information is available concerning viroid–grapevine interactions. Viroids occur as single or mixed infections. Their effect on fruit development and quality is unknown and of great importance to clarify. That viruses replicate undetected in grapevine breeding programs without visible damage has been mentioned as a reason that viroids are transmitted by grapevine during vegetative propagation.

Viroids differ from viruses by being able to multiply at elevated temperatures, and thus they cannot be eliminated from grapevines by thermotherapy (Tsagris et al., 1990). Shoot-tip culture, even in the absence of thermotherapy, represents an efficient method to create viroid-free grapevines (Juarez et al., 1990). Further research on properties of viroids and their interaction with grapevine will lead to a better understanding of this enigmatic group of pathogens and to the health of grapevines in commercial production.

REFERENCES

Abracheva, P. 1981. La sensibilité de certaines variétés de la vigne à la maladie du bois strié (legno riccio). *Phytopathol. Medit.* 20:203–205.

Azzam, O.I. et al. 1990. Investigations on the grapevine rupestris stem pitting disease etiology. *10th Meeting of ICVG*, Sept. 3–7, 1990, Volos, Greece, 31.

Barlass, M. et al. 1982. Regeneration of virus-free grapevines using *in vitro* apical culture. *Ann. App. Biol.* 101:291–295.

Bass, P. and Vuittenez, A. 1977. Amélioration de la thermothérapie des vignes virosées au moyen de la cultuer d'apex sur milieux nutritifs ou par greffage sur vignes de semis cultivées aseptiquement *in vitro. Ann. Phytopathol.* 9:539–540.

Bass, P. and Legin, R. 1981. Thermothérapie et multiplication *in vitro* d'apex de vigne. Application à la séparation ou à l'élimination de diverses maladies de type viral et à l'évaluation des dégâts. *C.R. Acad. Agric. Fr.* 67:922–933.

Bennett, C.W. and Costa, S.A. 1961. Sowbane mosaic caused by a seed transmitted virus. *Phytopathology* 51:546–550.

Bercks, R. and Stellmach, G. 1964. Untersuchungen über Rebenvirosen. Biologische Bundesantalt fur Land- und Forstwirtschaft in Berlin und Braunschweig. *Jaresber.*, 1964, A 62.

Bercks, R. and Stellmach, G. 1966. Nachweis verschiedener Viren in Reisigkranken Reben. *Phytopathol. Z.* 56:288–296.

Bercks, R. 1967. Vorkommen und Nachweis von Viren in Reben. *Weinberg 4. Keller* 14:151–162.

Bercks, R. 1967a. Über den Nachweis des Tabakmosaik-Virus in Reben. *Z. Pflanzenkrankheiten und Pflanzenschutz.* 74:346–349.

Bercks, R. 1967b. Über den Nachweis des Tomatenzwergbusch-Virus (tomato bushy stunt virus) in Reben. *Phytopathol. Z.* 60:273–277.

Bercks, R. 1968. Über den Nachweis des Himberringsflecken-Viren (raspberry ringspot virus) in Reben. *Phytopathol. Z.* 62:169–173.

Bercks, R. and Querfurth, G. 1969. Über den Nachweis des sowbane mosaic virus in Reben. *Phytopathol. Z.* 66:365–373.

Bercks, R. et al. 1973. Über den Nachweis des alfalfa mosaic virus in einer Weinrebe. *Phytopathol. Z.* 76:166–171.

Beukman, E.F. and Goheen, A.C. 1965. Corky bark, a tumor-inducing virus of grapevines. In: *Proc. Int. Conf. Virus Vectors Peren. Hosts Spec. Ref. Vitis,* University of California, Davis, Div. Agr. Sci., p. 164–166.

Beukman, E.F. and Goheen, A.C. 1970. Grape corky bark. In: Frazier, N.W., Ed. *Virus Diseases of Small Fruits and Grapevines,* Univeristy of California, Div. Agr. Sci., Ca. 1970, 207–209.

Bondarchuk, V.V. et al. 1990. Closterovirus-like particles associated with leafroll of grapevine in Moldavia. (Abstr.) *10th Meeting ICVG,* Volos, Greece, Sept. 3–7, p. 60.

Boscia, D. et al. 1990. Association of a phloem-limited nonmechanically transmissible isometric virus with fleck disease. (Abstr.) *10th Meeting ICVG,* Volos, Greece, Sept. 3–7, p. 22.

Bovey, R. 1973. *Maladies à Virus et à Mycoplasmes de la Vigne.* Station fédérale de recherches agronomiques de Changins, p. 76, Suisse.

Bovey, R. 1976. Les viroses en viticulture, leur importance économique. *Second Symp. Int. sur la Sélection Clonale de la Vigne,* 10–11 Sept., A.N.T.A.V.

Bovey, R. and Cazelles, O. 1978. Alfalfa mosaic virus on grapevines. *Proc. 6th Conf. Virus and Virus Dis. Grapevine,* 13–21 September, 1976, Madrid, p. 131–134.

Bovey, R. et al. 1980. Maladies à virus et affections similaires de la vigne. Atlas en couleurs des symptômes. *Virus and Virus-like Diseases of Grapevines,* Payot, Lausanne, p. 181.

Brückbauer, H. 1962. Die Wirtschaftliche Bedeutung der Virus-krankheiten und der Virusverdachtungen Erscheinungen an Reben Dtsch. *Weinbau-Kalender, 1962, 13 Jahrg.,* p. 108–119.

Brückbauer, H. and Rüdel, M. 1970. *Die Viruskrankheiten der Rebe,* Verlag Eugen Ulmar, Stuttgart.

Cadman, C.H. 1963. Biology of soil-borne viruses. *Annu. Rev. Phytopathol.* 1:143–172.

Castellano, A.M. et al. 1983. Virus-like particles and ultrastructural modifications in the phloem of leafroll-affected grapevines. *Vitis* 22:23–39.

Cesati, R.R. and van Regenmortel, M.H.V. 1969. Serological detection of a strain of tobacco necrosis virus in grapevine leaves. *Phytopathol. Z.* 64:362–366.

Chamberlain, E.E. et al. 1970. Virus diseases in New Zealand. *N.Z. J. Agr. Res.* 13:338–358.

Ciccarone, A. 1961. La degenerazione infettiva ed altre alterazioni della vite. *Italia Agr.* 98:901–921.

Cohn, E. et al. 1970. *Xiphinema italiae,* a new vector of grapevine fanleaf virus. *Phytopathology* 60:181–182.

Conti, M. et al. 1980. A closterovirus from a stem-pitting-diseased grapevine. *Phytopathology* 70:394–399.

Conti, M. and Martelli, G.P. 1988. Grapevine corky bark disease. In: Smith, I.M. et al., Eds. *European Handbk. Pl. Dis.,* Blackwell Sci. Publ., p. 109–110.

Dalmasso, A. and Weischer, B. 1976. Les nematodes vecteurs de virus et la sélection sanitaire en France et en Allemagne. *II Symp. Int. sur la Selection de la Vigne,* 10–11 Sept., Dom. de l'Espiguette, Le Grau du Roi.

Dias, H.F. 1957-58. Obtencâo experimental de infecceos da videira. (Experimental production of mixed infections of grapevine fanleaf and yellow mosaic.) *Annis J.N.V.,* 1957-58, p. 17–27.

Dias, H.F. 1975. Peach rosette mosaic virus. *CMI/AAB Descriptions of Plant Viruses,* No. 150.

Dias, H. 1976. Incidence and geographic distribution of tomato ringspot virus in Dechaunac vineyards in the Niagara Peninsula. (Abstr.) *VI. Conf. Int. Sobre Virus y Virosis de la Vid,* p. 6.

Dimitrijević, B. 1970. Pojava uvijenosti lišcá vinove loze (leafroll) u našoj zemlji. *Zaštita bilja* 21:373–378.

Dimitrijević, B. 1972. Quelques observations concernant le viris de la mosáique du tobac chez la vigne atteinte de dégénérescence infectieuse en Yougoslavie. IVe Conference du groupe international d'étude des virus et des maladies à virus de la vigne. Colmar, 16-18 Juin 1970. *Ann. Phytopathol., No. hors série* p. 107–111.

Dimitrijević, B. 1977. Savremene metode suzbijanja viroznih oboljenja vinove loze. *Savetovanje o ekskoriozi i virusnim bolestima vinove loze*, 16-18 Nov., Mostar, p. 117–123.

Dimitrijević, B. 1980. Some properties of the new latent virus from grapevine rootstocks in Yugoslavia. *Proc. 7th Mtg. Int. Conf. Study Viruses and Virus-like Disease of the Grapevine*, Niagara Falls, Can., Sept. 8–12, p. 21–24.

Fabre, E. 1853. Observations sur les maladies regnantes de la vigne. *Bull. Soc. Centrale Agr. Hérault* 40:11–75.

Faggioli, F. et al. 1990. Further characterization and serology of closterovirus type III isolated from grape in Italy, (Abstr.) *10th Meeting ICVG*, Volos, Greece, Sept. 3–7, p. 76.

Gilmer, R.M. and Kelts, L.J. 1965. Isolation of tobacco mosaic virus from grape foliage and roots. (Abstr.) *Phytopathology* 55:1283.

Gilmer, R.M. and Kelts, L.J. 1968. Transmission of tobacco mosaic virus in grape seed. *Phytopathology* 58:277–278.

Gilmer, R.M. et al. 1970. A new grapevine disease induced by tobacco ringspot virus. *Phytopathology* 60:619–627.

Gilmer, R.M. et al. 1972. Infection of grapevine in New York by tobacco ringspot virus. IVe Conférence du Groupe International d'Étude des Virus et des Maladies à Virus de la Vigne, Colmar, 16–18 June 1970. *Ann. Phytopathol. No. hors série*, p. 121–122.

Gilmer, R.M. and Uyemoto, J.K. 1972. Tomato ringspot virus in "Baconoir" grapevines in New York. *Pl. Dis. Reptr.* 56:133–135.

Goheen, A.C. and Cook, J.A. 1959. Leafroll (red leaf or rougeau) and its effects on vine growth, fruit quality and yields. *Ann. J. Enol. Vitic.* 10:173–181.

Goheen, A.C. et al., 1965. Inactivation of grapevine viruses *in vivo*. In: *Proc. Int. Conf. Virus Vectors Peren. Hosts Spec. Ref. Vitis*, University of California, Davis, Div. Agr. Sci. p. 255–265.

Goheen, A.C. 1970. Grape leafroll. In: Frazier, N.W., Ed. *Virus Dis. Small Fruits Grapevines*, Univeristy of California, Davis, Div. Agr. Sci., p. 209–212.

Goheen, A.C. 1977. Virus and virus-like diseases of grape. *Hort. Sci.* 12:465–569.

Gooding, G.V. and Hewitt, W.B. 1962. Grape yellow vein: symptomatology, identification and the association of a mechanically transmissible virus with a disease. *Am. J. Enol. Vitic.* 13:196–203.

Gooding, G.V., Jr. 1963. Purification and serology of a virus associated with the grape yellow-vein disease. *Phytopathology* 53:475–480.

Gooding, G.V., Jr. 1965. The reaction of grapevines and herbaceous plants to grapevine yellow-vein virus. In: *Proc. Int. Conf. Virus Vector Peren. Hosts Spec. Ref. Vitis*, University of California, Davis, Div. Agr. Sci. p. 211–222.

Gooding, G.V., Jr. and Teliz, D. 1970. Grapevine yellow vein. In: Frazier, N.W. Ed. *Virus Diseases of Small Fruits and Grapevines*, University of California, Davis, Div. Agr. Sci., p. 238–241.

Graniti, A. and Ciccarone, A. 1961. Osservazioni su alterazioni virosiche e virus-simili della vite in Puglia. *Notiz. Mal. Piante* 55:99–102.

Graniti, A. and Martelli, G.P. 1965. Further observations on legno riccio (rugose wood) and graft-transmissible stem-pitting of grapevine. In: *Proc. Int. Conf. Virus Vector Peren. Hosts Spec. Ref. Vitis*, University of California, Davis, Div. Agr. Sci. p. 168–179.

Graniti, A. and Martelli, G.P. 1970. Legno riccio. In: Frazier, N.W., Ed. *Virus Diseases of Small Fruits and Grapevines*, Univeristy of California, Berkeley, p. 243–245.

Gugeril, P. et al. 1990. Further characterization of grapevine leafroll disease. (Abstr.) *10th Meeting ICVG* Volos, Greece, Sept. 3–7, p. 6.

Hewitt, W.B. 1954. Some virus and virus-like diseases of grapevines. *Bull. Ca. Dept. Agr.* 43:47–64.

Hewitt, W.B. 1956. Yellow vein, a disease of grapevine caused by a graft-transmissible agent. *Phytopathology* 46:15.

Hewitt, W.B. et al. 1958. Nematode vector of soil-borne fanleaf virus of grapevines. *Phytopathology* 48:586–595.

Hewitt, W.B. and Goheen, A. 1959. Asteroid mosaic of grapevines in California. (Abstr.) *Phytopathology* 49:541.

Hewitt, W.B. et al. 1962. Studies on virus diseases of the grapevine in California. *Vitis* 3:57–83.

Hewitt, W.B. 1968. Viruses and virus diseases of the grapevine. *Rev. Appl. Mycol.* 47:433–455.

Hewitt, W.B. 1970. Virus and virus-like diseases of the grapevine. In: Frazier, N.W. Ed. *Virus Diseases of Small Fruits and Grapevines*, University of California, Davis, Div. Agr. Sci., p. 195–196.

Hewitt, W.B. et al. 1970. Grapevine fanleaf virus. *CMI/AAB Descriptions of Plant Viruses*, No. 28.

Hewitt, W.B. et al. 1972. Grapevine fleck disease, latent in many varieties is transmitted by graft inoculation. IVe Réunion du Groupe International d'Étude des Virus et Maladies à Virus de la Vigne, 16-18 June 1970. *Phytopathol., No. hors série* p. 43–47.

Hewitt, W.B. 1990. Viruses and virus-like diseases of grapevine: an overview. *10th Meeting ICVG*, Volos, Greece Sept. 3–7, p. 3.

Hollings, M. et al. 1969. Celery yellow vein and Hungarian chrome mosaic viruses. *Res. Glasshouse Crops Res. Inst.*, 1968, p. 102–103.

Hu, J.S. et al. 1990. Characterization of grapevine leafroll disease associated closteroviruses. (Abstr.) *10th Meeting ICVG*, Volos, Greece, Sept. 3–7, p. 5.

Hu, J.S. 1990. Mechanical transmission and characterization of a closterovirus from grapevine leafroll diseased grapevine. (Abstr.) *10th Meeting ICVG*, Volos, Greece, Sept. 3–7, p. 63.

Jankulova, M. 1970. Nachweis des Tabakmosaik-Virus an der Rebe in Bulgarien. *Dokl. Akad. Sel.-Khoz. Nauk. Bolg.*, 3:123–130.

Jankulova, M. et al. 1978. Isolation of artichoke Italian latent virus from the grapevine in Bulgaria. *Proc. 6th Conf. Virus and Virus Dis. Grapevine*, 13–21 September, 1976, Madrid, p. 143–147.

Juarez, J. et al. 1990. Shoot tip culture and the recovery of viroid-free grapevines. (Abstr.) *10th Meeting ICVG*, Volos, Greece, Sept. 3–7, p. 40.

Kado, C.I. 1971. Sowbane mosaic virus. *CMI/AAB Descriptions of Plant Viruses*, No. 64.

Kartuzova, V.I. et al. 1990. Investigation of virus diseases of grapevine in Ukraina. (Abstr.) *10th Meeting ICVG*, Volos, Greece, Sept. 3–7, p. 56.

Klos, E.J. et al. 1967. Peach rosette mosaic transmissiona and control studies. *Michigan Agric. Exp. Sta. Quart. Bull.* 49:287–293.

Klos, J. 1976. Rosette mosaic. In: *Virus Diseases and Noninfectious Disorders of Stone Fruits in North America*, U.S. Dept. Agr. Handbk. 347:135–138.

Krake, L.R. and Woodham, R.C. 1983. Grapevine yellow speckle agent implicated in the etiology of vein banding. *Vitis* 22:40–50.

Kuzmanović, S. 1992. Proučavanje bugarskog latentnog virusa (Investigation of grapevine Bulgarian latent virus). Manuscript, 39 pp., Inst. Pl. Prot., Belgrade.

Lafon, J. et al. 1955. *Maladies et Parasites de la Vigne*, Librairie J.B. Baillière et Fils, Paris, 364.

Legin, R. 1972. Expérimentation pour étudie l'effect des principales viroses sur al végétation et la production de la vigne. IVe conférence du groupe intenational d'étude des virus et des maladies à virus. Colmar, 16-18 June 1970. *Ann. Phytopathol., No. hors série*, 49–57.

Legin, R. et al. 1979. Premieurs résultats de guérison par thermothérapie et culture *in vitro* d'une maladie de type cannelure (legno riccio) produite par le graffage du cultivar Servant de *Vitis* vinifera sur le porte greffe *Vitis riparia* × *V. berlandieri* Kober 5 BB. Comparison avec diverses viroses de la vigne. *Phytopathol. Medit.* 18:207–210.

Lehoczky, J. 1965. Research on virus diseases of grapevine in Hungary. In: *Proc. Int. Conf. Virus Vector Per. Hosts Spec. Ref. Vitis*, University of California, Davis, Div. Agr. Sci., p. 311–318.

Lehoczky, J. 1972. Destructive effect of legno riccio (rugose wood) on European grapevine varieties. *Ann. Phytopathol. (No. hors série)*, I.N.R.A. Publ. 72–4:59–65.

Martelli, G.P. 1965. Preliminary report on the purification and serology of a virus associated with Hungarian grapevines showing macroscopic symptoms of fanleaf and yellow mosaic. In: *Proc. Int. Conf. Virus and Vector Per. Hosts Spec. Ref. Vitis*, University of California, Davis, Div. Agr. Sci., p. 402–410.

Martelli, G.P. and Quacquarelli, A. 1970. Hungarian chrome mosaic of grapevine and tomato black ring: two similar but unrelated plant viruses. IVth Intl. Coun. Viruses and Virus Dis. Grapevine. *Ann. Phytopathol. (No. hors serie)*, 1972, INRA, p. 123–133.

Martelli, G.P. et al. 1970. Hungarian chrome mosaic. In: Frazier, N.W., Ed. *Virus Diseases of Small Fruits and Grapevines*, University of California, Davis, Div. Agr. Sci., p. 236–237.

Martelli, G.P. and Quacquarelli, A. 1972. Grapevine chrome mosaic virus. *CMI/AAB Descriptions of Plant Viruses*, No. 103.

Martelli, G.P. et al. 1978. Grapevine Bulgarian latent virus. *CMI/AAB Descriptions of Plant Viruses*, No. 186.

Martelli, G.P. et al. 1978a. A manually transmissible latent virus of the grapevine from Bulgaria. *ICGV*, Madrid, Sept. 13-21, 1976, p. 135–141.

Martelli, G.P. and Vuittenez, A. 1988. Grapevine leafroll disease. In: Smith, I.M. et al. Eds. *European Handbk. Pl. Dis.*, Blackwell Sci. Publ., London.

Martelli, G.P. and Conti, M. 1988. Grapevine stem pitting disease. In: Smith, I.M. et al., Eds. *European Handbk. Pl. Dis.*, Blackwell Sci. Publ., London.

Milne, R.G. et al. 1984. Closterovirus-like particles of two types associated with diseased grapevines. *Phytopathol. Z.* 110:360–368.

Minafra, A. et al. 1990. Viroids in a grapevine collection of southern Italy. *10th Meeting ICVG*, Volos, Greece, Sept. 3–7, p. 42.

Morris-Krsinich, B.A.M. et al. 1983. The synthesis and processing of the nepovirus grapevine fanleaf proteins in rabbit reticulocyte lysate. *Virology* 130:523–526.

Namba, S. et al. 1979. A small spherical virus associated with the Ajinashika disease of Koshu grapevine. *Ann. Phytopathol. Soc. Japan* 45:70–73.

Namba, S. et al. 1990. Purification and properties of closterovirus-like particles isolated from a corky bark diseased grapevine. (Abstr.) *10th Meeting ICVG*, Volos, Greece, Sept. 3–7, p. 7.

Nel, A.C. and Engelbrecht, J.D. 1972. Grapevine virus diseases in the nursery. *Ann. Phytopathol. (No. hors série)*, I.N.R.A. Publ. 72–4:47–74.

Nel, A.C. and Engelbrecht, D.J. 1972a. Grapevine virus diseases in South Africa and the influence of latent viruses in the nursery. IVe Conf. Groupe Intel. d'Étude des Virus et des Maladies à Virus de la Vigne, Colmar, June, 16–18, 1970. *Phytopathol. No. hors série*, p. 67–74.

Novak, J.B. and Lanzova, J. 1976. Identification of alfalfa mosaic virus and tomato bushy stunt virus in hop (*Humulus lupulus* L.) and grapevine (*Vitis vinifera* subsp. *sativa* (D.C./Hegi) plants in Czechoslovakia. *Biol. Plantarum* 18:152–154.

Ottenwaelter, M.M. et al. 1973. Amélioration du rendiment de la thermothérapie sur plants en pots par l'utilisation de la culture sur milieu gélosé stérile. *Vitis* 2:46–48.

Over de Linden, A.J. and Chamberlain, E.E. 1970. Effect of grapevine leafroll virus on vine growth and fruit yield and quality. *N.Z. J. Agr. Res.* 13:689–698.

Pearson, R.C. and Goheen, A.C. 1990. *Compendium of Grape Diseases*, APS Press, St. Paul, MN. 93 pp.

Pierce, N.B. 1892. The California vine disease. *Bull. Div. Veg. Physiol. Pathol.* U.S. Dept. Agr. 2:222.

Pinck, L. et al. 1988. A satellite RNA in grapevine fanleaf virus scion F13. *J. Gen. Virol.* 69:233–239.

Ramsdell, D.C. and Myers, L.R. 1976. Rate of spread of peach rosette mosaic virus in *Vitis labrusca* L. "Concord" grape vineyard, some weed hosts and the seed and pollen-borne nature of the virus. (Abstr.) *VI Conf. Int. Sobre Virus y Virosis de la Vid.*, p. 14.

Ramsdell, D.C. and Stace-Smith, R. 1980. A serological and physico-chemical comparison between bleuberry leaf mottle virus and grapevine Bulgarian latent virus. *Proc. 7th Mtng. Int. Conf. Study Viruses and Viruses-like Dis. Grapevine*, Niagara Falls, Canada, p. 119–129.

Raski, D.J. et al. 1965. Survival of *Xiphinema index* and reservoirs of fanleaf virus in followed vineyard soil. *Nematologica* 11:349–352.

Ravaz, L. et al. 1933. Recherches sur le rougeau de la vigne. *Ann. Agron.* 3:225–231.

Refatti, E. 1965. Grapevine asteroid mosaic. *Proc. Int. Conf. Virus and Vector on Per. Hosts Vitis*, September 6–10, Univ. CA, Div. Agr. Sci. CA, p. 157–164.

Refatti, E. 1970. Asteroid mosaic of grapevine. In: Frazier, N.W., Ed. *Virus Diseases of Small Fruits and Grapevines*, Univeristy of California, Davis, Div. Agr. Sci., p. 212–214.

Rezaian, M.A. et al. 1990. Grapevine viroids. (Abstr.) *10th Meeting ICVG*, Volos, Greece, Sept. 3–7, p. 41.

Roberts, I.M. and Brown. 1980. Detection of six nepoviruses in their nematode vectors by immunosorbent electron icroscopy. *Proc. 15th Int. Symp. Eur. Soc. Nematol.*, Bari, 1980.

Sarospataki, G. 1965. Untersuchungen von Viruskrankheiten der Rebe in Ungarien. *D. Wein-Wiss.* 20:20–37.

Scheu, G. 1936. *Mein Winzerbuch*, Reichsnährstand Verlags-Ges.m.b., Berlin.

Semancik, J.S. and Szychowski, J.A. 1990. Comparative properties of viroids of grapevine origin isolated from grapevines and alternate hosts. *10th Meeting ICVG*, Volos, Greece, Sept. 3–7, p. 37.

Stellmach, G. and Bercks, R. 1963. Untersuchungen an Rebevirosen: Nachweis des Tomatenschwartzringflecken-Virus (tomato black ring virus) in kranken Stäcken der Sorte Aramon × 143 AMG (Amerikanerrebe). *Phytopathol. Z.* 48:200–202.

Stellmach, G. and Bercks, R. 1965. Fortgeführte Untersuchungen zur Charakteriesirung des Tomatenschwarzringflecken-Virus (tomato black ring virus) in kranken Reben der Sorte "Aramon × Riparia 143 A.M.G.". *Phytopathol. Z.* 53:383–390.

Stellmach, G. et al. 1965. Tomato black ring virus on grapevine. In: *Proc. Int. Conf. Virus and Vector Peren. Hosts Spec. Ref. Vitis*, Univeristy of California, Davis, Div. Agr. Sci., p. 166–168.

Stellmach, G. 1970. Tomato black ring in *Vitis*. In: Frazier, N.W. Ed. *Virus Diseases of Small Fruits and Grapevines*, Univeristy of California, Davis, Div. Agr. Sci., p. 234–236.

Szychowski, J.A. et al. 1990. Relationships among grapevine viroids from sources maintained in California and Europe. *10th Meeting ICVG*, Volos, Greece, Sept. 3–7, p. 39.

Šutić, D. and Tadijanovi_, D.J. 1974. Primena zdravstveno-varijetalne selekcije u proizvodnji loznog sadnog materijala. *Poljoprivreda* 9:244–269.

Šutić, D. 1977. O nekim karakteristikama vizoza vinove loze. *Savetovanje o ekskoriozi i virusnim bolestima vinove loze*, 16-18 November, Mostar, p. 97–105.

Tanne, E. et al. 1978. Further studies of a leaf roll-associated virus: transmission from herbaceous plants, purification and characterization. *VIth Meeting ICVG*, Madrid, Sept. 13–21, p. 213–215.

Tanne, E. and Meir, E. 1990. Double-stranded RNA associated with the corky bark disease in grapevines. (Abstr.) *10th Meeting ICGV*, Volos, Greece, Sept. 3–7.

Taylor, C.E. and Raski, D.J. 1964. On the transmission of grapevine fanleaf by *Xiphinema index*. *Nematologica* 10:489–495.

Taylor, R.H. and Woodham, R.C. 1972. Indexing grapevines for viruses in Australia. IVe Conférence du Groupe International d'Étude des Virus et des Maladies à Virus de la Vigne, Colmar, 16–18 June 1970. *Ann. Phytopathol., No. hors série*, p. 85–87.

Teakle, D.S. 1962. Transmission of tobacco necrosis by a fungus *Olpidum brassicae*. *Virology* 18:224–231.

Teliz, D. et al. 1982. Grape corky bark and stem pitting in Mexico. I. Occurrence, natural spread, distribution, effects on yield and evaluation of symptoms in 128 grape cultivars. In: *Proc. 7th Mtg. Int. Coun. Study Viruses and Virus-like Diseases of the Grapevine*, Niagara Falls, Ontario, p. 51–67.

Tsagris, M.G. et al. 1990. Viroids in grapevine cultivars in Greece. *10th Meeting ICVG*, Volos, Greece, Sept. 3–7, p. 77.

Uyemoto, J.K. 1970. Symptomatologically distinct strains of tomato ringspot virus isolated from grape and elderberry. *Phytopathology* 60:1838–1841.

Uyemoto, J.K. et al. 1977. Isolation and identification of a strain of grapevine Bulgarian latent virus in a Concord grapevine in New York State. *Pl. Dis. Reptr.* 61:949–953.

Vuittenez, A. 1957. Inoculation de différentes espèces de *Vitis* par les virus du groupe de la dégénérescence infectieuse. Application au diagnostic de la maladie. *C.R., Quatrième Cogrès Int. de Lutte Contre les Ennemis des Plantes*, Hamburg, September (Publ. in Braunschweig, 1959), 1:361–366.

Vuittenez, A. 1958. Transmission par greffage d'une virose du type "Enroulement foliare" commune dans les vignobles de l'est et du centre-est de la France. *C.R. Acad. Agric. Fr.* 44:313–316.

Vuittenez, A. 1966. Oservations sur une mosaïque de la vigne, probablement indépendante du virus du court-noué. Ann. Epiphyt. A., p. 67–73.

Vuittenez, A. 1970. Fanleaf of grapevine. In: Frazier, N.W., Ed. *Virus Diseases of Small Fruits and Grapevines*, Univeristy of California, Davis, Div. Agr. Sci. p. 217–228.

Vuittenez, A. 1977. Metode borbe protiv viroza vinove loze. *Savetovanje o ekskoriozi i virusnim bolestima vinove loze*, Nov. 16-18, Mostar, p. 153–164.

Vuittenez, A. and Martelli, G.P. 1988. Grapevine fanleaf virus (GFLV). In: Smith, I.M. et al., Eds. *European Handbk. Pl. Dis.*, Blackwell Sci. Publ., p. 27–28.

Walter, B. et al. 1990. The use of a green-grafting technique for the detection of virus-like diseases of the grapevine. *J. Phytopathol.* 128:137–145.

Weischer, B. 1968. Das Vorkommen von Arten der Gattung Xiphinema, Longidorus and Trichodorus (Nematoda) in Rebenanlagen in Deutschland. Weinberg. *Keller* 15:540–542.

Yasuo, T. and Yano. R. 1980. Ajinashika disease of the grapevine cultivar Koshu in Japan. *Proc. 7th Int. Coun. Study Viruses and Virus-like Dis. Grapevine*, Niagara Falls, Canada, Sept. 8-12, p. 15-19.

Younger, W. 1966. *Gods, Men and Wine*, Wine and Food Society and World Publishing Co., Cleveland, OH.

Glossary

Ab 1: The first antibody produced by an animal's B lymphocytes in response to the stimulus provided by an antigen.

Absorbance: Amount of light absorbed by a substance at a particular wavelength.

Absorption spectrum: Graphical representation of absorbance of a substance at different wavelengths. Valuable in obtaining an approximate estimate of the percentage of nucleic acid in a virus from the ratio of absorbance at 254 and 280 nm. *See* Specific absorbance.

Acid phosphatases: Enzymes with an acidic pH optimum that catalyze cleavage of inorganic phosphate; found particularly in lysosomes and secretory vesicles.

Acid proteases: Proteolytic enzymes with an acidic pH optimum generally found in lysosomes. *See* Proteases.

Acquired resistance (syn. **induced resistance** or **acquired immunity**): A noninherited resistance response developed by a normally susceptible host after a predisposing treatment, such as prior infection with a related virus, fungus, bacterium, or certain chemicals. *See* Cross protection.

Acquisition access period: The period of time a vector has feeding access to a source of inoculum.

Acquisition feeding: The feeding of a vector on a source of inoculum in transmission tests.

Acquisition feeding period: The time during which a vector feeds on an infected plant to acquire a virus for subsequent transmission (e.g., to become viruliferous).

Acronym: A word composed of the first letter of the principal words in a compound term. *See* Sigla.

Actinomycin D: Antibiotic that inhibits RNA transcription by interacting with the guanine residues of helical DNA; inhibits the replication of DNA-containing viruses and RNA viruses that require DNA to RNA transcription.

Activator: A DNA-binding protein binds upstream of a gene and activates its transcription.

Acute symptoms: *See* Shock symptoms.

Adenine: One of the purine bases found in DNA, RNA, nucleosides, and nucleotides. In DNA, it pairs with thymine.

Adenosine: A mononucleoside consisting of adenine and D-ribose, which is produced in the hydrolysis of adenosine monophosphate (AMP).

Adenosine diphosphate (ADP): A nucleotide composed of adenine and D-ribose with two phosphate groups attached; ADP and ATP participate in metabolic reactions (both anabolic and catabolic). These molecules, through the process of being phosphorylated or dephosphorylated, transfer energy within cells to drive metabolic processes.

Adenosine 5′-triphosphate (ATP): A compound of one molecule each of adenine and D-ribose with three molecules of phosphoric acid, it is the phosphorylated condition of ADP. It conveys energy needed for metabolic reactions, then loses one phosphate group to become ADP.

Adjuvant: (of serology) substance injected with antigens (usually mixed with) which nonspecifically enhances or modifies the immune response to that antigen. Thus, antibody production or the reaction of well-mediated immunity are more vigorous than when antigen is injected without adjuvant.

Affinity: The strength of the antigen–antibody interaction; strength of bonds formed by this reversible interaction, like other reversible bimolecular equilibrium reactions, determines the rate of association between antibody (i.e., monospecific, such as a monoclonal antibody) and antigen (e.g., a hapten), versus the rate of dissociation. High-affinity antibodies have a higher rate of association with antigen and bind more antigen (e.g., 10^8 to 10^{12} mol^{-1}) than low-affinity antibodies, which have a lower rate of association with antigen and bind less antigen (e.g., 10^5 to 10^7 mol^{-1}).

Agarose: A constituent of agar lacking inhibitors of virus development. Widely used in gel electrophoresis due to more uniform pore size than in agar.

Agarose gel electrophoresis: Technique used for separating proteins or nucleic acids by passage of an electric current through the gel.

Agent of disease: An organism or abiotic factor that causes disease; a pathogen.

Agent of inoculation: That which transports inoculum from its source to or into an infection court, e.g., wind, rain splash, flowing water, insects, mites, humans, tools, and equipment.

Agglutination: A serological test in which viruses or bacteria suspended in a liquid collect into clumps whenever the suspension is treated with antiserum containing antibodies specific against these viruses or bacteria; visible clumping of particulate antigens (e.g., red blood cells, bacteria, etc.) when reacted with a specific antibody, or visible clumping of inert particles to which antibodies are adsorbed (e.g., latex microspheres) when reacted with a specific antigen. Clumping is caused by antigen–antibody bridging between adjacent cells or particles.

Agglutinin: An antibody (or compound) that causes another particular antigen (compound) to clump and settle out of suspension.

Aggregation: Intramolecular interactions that do not involve covalent linkage.

Alate: Winged; a form in the life cycle of certain insects (e.g., aphids). *See* Apterous.

Alternate host: A plant, different from the principal host, on which a pathogen (e.g., a heteroecius rust) must develop to complete its life cycle. Do *not* confuse with secondary host or alternative host.

Alternative host: One of a pathogen's several hosts; alternative hosts are not required for completion of a parasites developmental cycle; often weeds and wild plants that support viral replication; not one from which the virus was first described.

Amber mutation: Point mutation producing a UAG termination codon and thereby premature termination of translation.

Amino acid: Basic building block of proteins that contains amino and carboxyl groups plus a variable side chain that determines the properties of the individual amino acid; 20 amino acids occur commonly in nature.

Amino acid sequence: Linear order of the amino acids in a peptide or protein.

Amino group ($-NH_2$): A chemical group, characteristically basic; binds a proton to form $-NH_3^+$. The *amino terminus* of a polypeptide is the end with a free α-amino group.

Angstrom (Å): A unit of length equal to 1/10 millimicron (mµ) or nanometer (nm) = 1/10,000 micron or micrometer (µm) = 10^{-10} meter to measure wavelengths and dimensions of intracellular structures of microorganisms and viruses.

Anti- (prefix): Against.

Antibiosis: Antagonistic association between two organisms or between one organism and a metabolic product of another organism, to the detriment of one of the organisms. *See* Vector resistance.

Antibiotic: Damaging to life; especially a chemical compound produced by one microorganism, which inhibits growth or kills other living organisms in very small amounts.

Antibody: A new or altered specific protein (modified serum immunoglobulin molecule) produced by an animal's B lymphocytes in response to an antigenic stimulus (a foreign substance); binds specifically with the antigen to render it harmless. An antibody that causes lysis is called a *lysin*; those causing agglutination and precipitation *agglutinins* and *precipitins*. *See* Complement fixation, ELISA, and Serology.

Anticodon: A sequence of three nucleotides complementary to the codon triplet in mRNA; three bases in a tRNA molecule that recognizes a codon in the mRNA molecule.

Antigen (adj. **antigenic**): Foreign (viral) protein, occasionally a complex lipid or polysaccharide and some nucleic acids, that, when injected into living animal tissues, stimulates the production and release of specific antibodies into bloodstream antagonistic to the substance injected and capable of specifically binding to antigen binding sites of antibody molecules. Specific responsiveness is a property of host tissues, not of the injected substance. All immunogens are antigenic, but not all antigens are immunogenic (e.g., haptens). The capacity of an antigen to react specifically with an antibody is referred to as an *antigenic reactivity*. *See* Immunogen.

Antigen-antibody reaction: The specific interaction between an antigen and an antibody that recognizes a specific structural feature of the antigen and binds to it.

Antigen binding capacity: A measurement of the ability of an antibody to bind antigen, based on the effects of dilution of the antibody.

Antigen binding site = Fab.: That area of an antibody molecule that binds to antigen; paratope.

Antigenic: The capacity to induce antibody formation and the ability to react with antibodies *in vivo* or *in vitro*.

Antigenic determinant: The small site or epitope on an antigen molecule to which an antibody is specifically bound, determined by structural complementarity between antibody (Fab) and antigen molecules. An antigen molecule may comprise more than one determinant. Attached haptens add further determinants. *See* Epitope.

Antiserum (pl. antisera): The blood serum (fluid fraction of coagulated blood) of vertebrates that contains antibodies after exposure to specified antigens.
Antiserum titer: The reciprocal of the highest dilution of an antiserum that can react with its homologous antigen (virus). *See* Homologous and Heterologous reaction.
Aphid: A small, sucking insect of the family Aphididae (order Homoptera), capable of transmitting many plant viruses. Injures plants when in large populations.
Apical: At or near the end or tip (apex).
Apical meristem: A mass of undifferentiated cells capable of division at the tip of a root or shoot. These cells differentiate by division, which allows a plant to grow in depth and height.
Apterous: Wingless stage in the life cycle of certain insects (e.g., aphids). *See* Alate.
Arbovirus: A virus replicating in both an arthropod and a vertebrate ("arthropod borne").
Asexual reproduction: Refers to any type of reproduction not involving the union of gametes (karyogamy) and meiosis (independent of sexual processes). *See* Replication.
ATCC: Initials of the American Type Culture Collection located at 10801 University Blvd., Manassas, VA 20110-2209 (Telex ATCCROVE 908-768, Telephone 703-365-2700, fax 703-365-2701); a large collection of bacteria, fungi (including yeasts), viruses, viroids, cells, etc. available on request by payment of a small fee.
ATP: *See* Adenosine 5′-triphosphate.
Attenuate, attenuated: To weaken or decrease in virulence or pathogenicity; DNA sequences that retard or stop RNA transcription.
Attenuated virus strain: A selected strain of a virulent virus that does not cause severe disease symptoms associated with the parent virus but still replicates.
Attenuation: The process of producing an attenuated (weakened) virus strain; reduction in virulence of a pathogen.
Aucuba: Bright yellow mosaic leaf variegation of genetic or virus origin.
Avidity: The stability of the antigen–antibody complex as governed by the affinity of the antibody for the antigen, the valency of the antibody or antigen, and the spatial arrangement of the epitope and respective paratope. An IgM antibody (i.e., ten paratopes potentially interacting with ten epitopes) is capable of forming a more stable antigen–antibody complex than an IgG antibody (two paratopes, for example, with the same epitope affinities as IgM).
Avirulent: Nonpathogenic; virus strain lacking virulence; unable to cause disease.
Bacilliform: Shaped like a short, blunt, thick, cylindrical rod, rounded on the ends; bacillus-shaped.
Bacteriophage: A virus replicating in bacteria only; also called a "phage."
Base analogue: A compound resembling one of the natural bases of RNA or DNA that is incorporated into newly synthesized nucleic acid by substituting for the normal "base," which can result in mutation or growth inhibition. Example: 5-fluorouracil substitutes for uracil.
Base pair: A pair of nucleotides (nitrogen bases) held together by hydrogen bonding and found in double-stranded nucleic acids. DNA contains the pairs A-T and G-C, and RNA contains G-C and A-U. *See* Nucleic acid.
Base ratio: The ratio of adenine (A) and thymine (T) to guanine(G) and cytosine (C) in the deoxyribonucleic acid (DNA) in an organism.
Base sequences: The order in which the purine and pyrimidine molecules A, G, C, U (T) occur along the polynucleotide chain of nucleic acids.
B cells: Lymphocytes that produce antibodies. *See* Immunoglobulin.
Binary symmetry: Combination of cubic and helical symmetry.
Binomial nomenclature: The scientific method of naming organisms.
Bioassay: Determine virus amount by measuring its biological activity (infectivity).
Biotechnology: The use of genetically engineered microorganisms and/or modern techniques and processes with biological systems for industrial production.
Booster dose: A dose of immunogen given several days, weeks, months, or years after the initial immunizing dose to stimulate continued production of antibody.
5-Bromouracil: Mutagenic thymine analog in which the 5-CH_3 group has been replaced by bromine.
Buoyant density: The density at which a virus or other macromolecule neither sinks nor floats when suspended in an aqueous solution of a heavy metal salt, calcium chloride, or a sugar such as sucrose; i.e., is in equilibrium.

°C: Centigrade (formerly Celsius); unit of temperature between boiling and freezing points of water at a standard pressure. *See* Celsius and Fahrenheit.

C-terminus: End of the peptide chain with a free α-carboxyl group (traditionally the right end).

Cap: Sequence of methylated bases joined to the 5′-terminus of a eukaryotic mRNA in the opposite (i.e., 5′ to 5′) orientation and interacting with protein factors involved in the initiation of protein synthesis.

Capsid: The protein coat of a virion composed of protein subunits or capsomeres forming a closed shell or tube surrounding the DNA or RNA; together forms the nucleocapsid.

Capsid polypeptide: Protein-forming part of the capsid structure of a virus particle.

Capsomer(e): A protein subunit; a small protein molecule, or group of molecules, that is the structural and chemical unit (polypeptide chains) of the protein coat (capsid) of a virus.

Carboxyl group: COOH (*See* C-terminus).

Carboxymethylcellulose: A cellulose derivative used to separate proteins by ion exchange chromatography.

Carna-5 RNA (syn. Cucumber mosaic virus RNA 5): A satellite RNA of CMV dependent on the remainder of the CMV genome for its own replication, but not essential for the replication of the CMV particle. *See* Genome and Satellite.

Carrier: Plant or animal harboring an infectious disease agent (e.g., virus), but not showing symptoms; a source of infection to others. *See* Vector.

cDNA: Abbreviation for complementary DNA that is synthesized from a messenger RNA template.

Cellular immunity: Immunity ascribed to various cellular functions other than those that produce antibody.

Cellular response: That part of the immune response which involves the interaction between cells and an antigen.

Central dogma molecular biology: The concept that the basic relationship among DNA, RNA, and protein is "one way"; i.e., that DNA serves as a template for both its own duplication and the synthesis of RNA; RNA, in turn, serves as messenger for protein synthesis; and that information cannot be retrieved from the amino acid sequence of a protein.

Centimorgan: A measurement of recombination frequency; one centimorgan = 0.01 chance that a genetic locus will separate from a marker by recombination in one generation.

Cesium chloride density gradient centrifugation: Method for separating viruses or macromolecules based on their density. Sedimentation ceases when the virus reaches a position in the gradient equal to their own buoyant density. *See* Isopycnic density.

Chemotaxonomy: Taxonomy using chemical characteristics.

Chemotype: Group of chemically differentiated individuals of a species of unknown or of no taxonomic importance.

Chloramphenicol: Antibiotic acts as phenylalanine analog isolated from *Streptomyces venazuelae* which inhibits prokaryotic (as well as mitochondrial and chloroplast) protein synthesis.

Chlorophyll (adj. chlorophyllous): The green, light-sensitive pigments found chiefly in the chloroplasts of leaves and other green parts of higher plants, that absorbs the light energy used in the process called photosynthesis.

Chloroplast: Specialized cytoplasmic organelle (plastid) in plant cells that contains chlorophyll and is the site of photosynthesis.

Chlorosis (adj. chlorotic): Paling, yellowing, or whitening of normally green tissue characterized by the partial to complete destruction of chlorophyll. May be due to a virus, the lack of or unavailability of some element (e.g., iron, manganese, zinc, nitrogen, boron, magnesium), lack of oxygen in a waterlogged soil, alkali injury, or some other factor. *See* Yellowing.

Cistron: The basic unit of a genetic function. Sequence of nucleotides within a certain area of DNA or RNA that codes for a particular protein; a chromosomal segment when any two mutations within the cistron give it a mutant phenotype in the *trans*-configuration but a normal phenotype in the *cis*-configuaration;

Classification: The systematic arrangement of names for organisms into categories on the basis of characteristics. Subdivisions (e.g., family, subfamily, tribe, subtribe, genus, subgenus, etc.) are based on detailed differences in specific properties. The relative order of plant groups is governed by the International Code of Botanical Nomenclature and that of virus groups is determined by the International Committee on Taxonomy of Viruses.

Clone, noun or verb (syn. colony): A population of cells, viruses, or organisms of identical genotype; an aggregate of individual organisms produced asexually (vegetatively) originating from one sexually produced individual (e.g., rooted cuttings) or from a mutation; in virology, a population of recombinant DNA

Glossary

molecules all carrying the same inserted sequence; in microbiology or virology, a colony of microorganisms or viruses containing a specific DNA or RNA fragment inserted into a vector; use of *in vitro* recombination techniques to insert a particular DNA or RNA sequence into a vector.

CMI AAB: Initials of the Commonwealth Mycological Institute/Association for Applied Biologists in the United Kingdom.

Coat protein: The protective protein layer(s) surrounding the viral nucleic acid.

Coding capacity: The amount of protein that a given DNA or RNA sequence can, in theory, encode.

Coding sequence: The process by which nucleotides within a certain area of RNA or DNA determines the sequence of amino acids in the synthesis of a particular protein; noncoding sequences may contain various control sequences.

Codon: Three adjacent nucleotides coding for either an amino acid or a chain termination.

Complementary DNA (cDNA): ssDNA synthesized by reverse transcription from an RNA template; if the cDNA is made double-stranded and cloned, it is labeled a cDNA clone.

Complementary RNA (cRNA): A ssRNA molecule complementary in base sequence to the single strand from which it was transcribed. Most ssRNA viruses use complementary RNA as intermediates in replication.

Complementary strand: A double-stranded nucleic acid molecule complementary in base sequence to the single strand from which it was transcribed.

Complementation: Occurs when one virus is assisted by another (or a strain of the same virus) to replicate; repairing a gene defect by the presence of another, functional copy *in trans*; the process by which one genome provides functions that another genome lacks.

Complementation test: A test to determine whether two virus mutants are defective in the same cistron.

Complement fixation: A sensitive test for antigen–antibody reactions that depends on binding (consumption) of complement by antigen–antibody complexes.

Complement-fixation test (CFT): A sensitive test where the antigen–antibody reaction can be detected and quantified; often used for comparing different or related antigens, e.g., viruses.

Conjugate: Joined; in twos; (of serology) the product of joining two or more dissimilar molecules by covalent bonds. In immunological contexts, one is usually a protein and the other a hapten or a label such as fluorescein, ferritin, or enzyme.

Conservative replication: Opposite of semiconservative replication; whereby no displacement of SS form from the double-stranded genome occurs during replication.

Control, of plant diseases: Prevention, retardation, or alleviation of disease by four principal methods: *Eradication* - destruction (roguing) of infected plants or plant parts or killing of the pathogen or agent on or in the host; *Protection* - application of a chemical or physical barrier to prevent entrance of the pathogen; and *Immunization* - production of genetically resistant or immune plant cultivars, chemotherapy, or other treatment to inactivate or nullify the effect of the pathogen within the plant; *Exclusion* - keeping the pathogen away from a disease-free area through quarantines, embargoes, and disinfection of plants and plant parts; plus a similar fifth method, *Avoidance*, i.e., choosing cultural practices (e.g., sanitation, rotation, choice of planting site and date, propagating and planting only disease-free material, etc.) that avoid disease.

Covalent bonds: Strong bonds formed by the sharing of electron pairs between atoms; also termed *primary bonds*.

cRNA: *See* Complementary RNA.

Cross inoculation: Inoculation with one virus, then simultaneously or later inoculation with a second virus.

Cross protection: A susceptible host infected with an avirulent pathogen (usually a virus), thereby resists infection by a second, usually related virulent pathogen in the same host. Upon challenge infection, the host expresses relatively mild symptoms, lower virus concentration, and/or restricted virus movement.

Cross reactive antigen: Antigen capable of combining with antibody produced in response to a different antigen; may cross react due to sharing of determinants by the two antigens or because the antigenic determinants of each, although not identical, are related closely enough stereochemically to combine with antibody against one of them.

Cross reactive antiserum: Antiserum capable of combining with different antigens that share common epitopes; such antisera can be purified to isolate several specific antibodies.

Cubic symmetry: Form of capsid in which protein subunits are assembled to a compact shell with three axes of symmetry, passing through apices, edges, and faces.

Cultivar (cv.): A cultivated variety; assemblage of closely related plants of common origin within a species that differ from other cultivars in certain minor details (e.g., form, color, flower, or fruit) which, when reproduced, sexually or asexually, retain their distinguishing features.

Culture collection: The microbiological equivalent of a botanical herbarium and a zoological museum; a repository (i.e., the ATCC) of cultures of characterized bacteria, fungi, viruses, viroids, cells, and other organisms.

cu m: (cubic meter) = 1.30794 cu yard = 35.3144 cu ft = 28.3776 bushels = 264.173 U.S. gal = 1056.7 qt (liquid) = 2113.4 pints (liquid) = 61,023 cu in.

cv.: Varieties released for cultivation; *see* Cultivar.

Cycloheximide: Antibiotic isolated from *Streptomyces griseus*; reversibly inhibits eukaryotic (but not prokaryotic) protein biosynthesis.

Cytopathic effect (cpe): Changes in the microscopic appearance or morphology of cultured cells following virus infection, e.g., cell fusion, inclusion bodies.

Da: Dalton.

Dalton: Unit of atomic size equal to the mass of a proton or single hydrogen atom; an expression of molecular weight. *See* Kilodalton.

Damage threshold: The lowest pest population density at which damage occurs.

Deaminase: An enzyme involved in the removal of an amino group from a molecule, in which ammonia is liberated.

Deamination: Removal of an amino group, especially from an amino acid.

Decarboxylation: Removal of a carboxyl group, -COOH.

Defective interfering (DI) particles: Virus particles that lack part of the genome nucleic acid of the standard virus and that often interfere with its replication.

Defective virus: A virus that lacks part of its genome, or some function, and is thus unable to replicate fully.

Degeneracy of the code (of virology): The code is degenerate because 64 codons specify 20 amino acids. Most amino acids are coded for by several codons differing in the third base; thus, changes in nucleic acid sequence do not necessarily result in changes in amino acid sequence. *See* Genetic code, Wobble hypothesis.

Deletion: Loss of a portion of the genetic material. Deletions range in size from a single nucleotide to entire genes (opposite = **insertion**).

Denaturation: The destruction of secondary and tertiary (H-bonds) structure of a protein, nucleic acid, or virus by physical or chemical means.

Denatured protein: A protein that has been altered by treatment with a physical or chemical agent to change its properties.

Density gradient: A gradient of a solute in a solvent used to support macromolecules during fractionation; separation of macromolecular species by centrifugation or electrophoresis.

Density-gradient centrifugation: A method of centrifugation in which particles (or components) of a purified virus separate in layers based on differing buoyant densities in sucrose or cesuim chloride.

Deoxyribonuclease (DNase): An enzyme that degrades DNA.

Deoxyribonucleic acid (DNA): The gene-bearing material of each DNA virus, plant and animal cell. *See* DNA.

Deoxyribose: A five-carbon sugar, one of the components of DNA. It has one oxygen atom less than its parent sugar ribose.

DEP: Dilution end point.

Dependent transmission: Transmission of a virus (by aphids) that occurs only when the vector feeds on a source plant jointly infected by a second virus. The second virus is referred to as a *helper virus*, and the virus not transmissible on its own is called the *dependent virus*.

Determinant: *See* Antigenic determinant.

Diagnosis (pl. **diagnoses**): Identification of the nature and cause of a disease problem; a shortened, Latin version of a taxonomic description of a species or other taxon.

Diagnostic: A distinguishing characteristic important for identification of disease or other condition.

Dicotyledon, dicot (pl. **dicotyledonae**, adj. **dicotyledonous**): A flowering plant having two cotyledons (seed leaves), in contrast to monocotyledons (grasses, cereals). *See* Monocotyledoneae.

Differential centrifugation: Cycles of alternate low- (to clarify) and high-speed (to sediment) centrifugation used to purify a virus.

Differential hosts: Special species, cultivars, inbreds, or isogenic plants varying from susceptible to one virus but not another, that express distinctive symptoms; used to separate a virus from a mixture to facilitate identification. *See* Indicator plant.

Differentially permeable: Referring to membranes that allow certain substances to pass through and that retard or prevent the passage of others.

Diffusion coefficient: A measure of the rate at which a solute moves along a gradient from a higher to a lower concentration.

Dilution: The process of increasing the proportion of solvent or diluent to solute or particulate matter, e.g., bacterial cells.

Dilution end point: The last point at which a virus remains infective in a progressive dilution series; the greatest dilution of an antibody that gives a measurable reaction with an antigen in a serological test.

Dilution, serial: Successive dilution of a specimen, e.g., one containing bacterial cells. A 1:10 dilution equals 1 ml of specimen plus 9 ml of diluent (e.g., sterile water); a 1:100 dilution equals 1 ml of a 1:10 dilution plus 9 ml of diluent, etc.

Dimethyl sulfoxide (DMSO): A solvent used (1) for dissolving both inorganic and organic chemicals; (2) for substances being applied to cells in tissue culture; (3) in the preparation of cells for storage in liquid nitrogen; (4) to strand-separate double-stranded nucleic acids.

DI particle: Defective interfering particle.

Disease complex: A plant disease caused by the interaction of two or more pathogens, often manifested by a greater-than-normal array of symptoms.

Disease cycle: The sequence of events involved in disease development, including the stages of development of the pathogen and the effect of the disease on the host; the chain of events that occurs between the time of infection and the final expression of disease.

Disease gradient: Change in incidence of a disease with increasing distance from the source of infection.

Disease pyramid: A concept describing the four factors needed simultaneously for a plant disease to occur: a susceptible host, presence of a virulent pathogen, a favorable environment for infection, after which over time, the disease can develop.

Disease, plant: Any disturbance of a plant that interferes with its normal growth and development (e.g., structure and function), economic value, or aesthetic quality, and leads to development of symptoms. A continuously, often progressively affected condition in which any part of a plant is abnormal (e.g., structure, function, or economic value) or interferes with the normal activity of the plant's cells or organs. Injury, in contrast, results from a momentary damage. *See* Disorder and Injury.

Disease range: The geographic distribution of a disease. *See* Range.

Disease tolerance: Capacity of a plant to maintain fairly normal vigor, without excessive injury or loss in yield, although infection by a pathogen is established within the plant. *See* Tolerance.

Disease triangle: A concept describing the simultaneous occurrence of a pathogen, a susceptible host, and a favorable environment such that a disease can develop. *See* Disease pyramid.

Dispersal: *See* Dissemination.

Dissemination (dispersal): In relation to plant diseases, the transfer (spread) or transport of infectious material (inoculum) to healthy plant tissue by wind, water, insects, animals, machinery, humans, or other means. *See* Agent of inoculation.

Distribution: Spread of a pathogen to areas outside of its previous geographical range; "geographical distribution" is synonymous with "range."

Disulfide bonds: Chemical bonds between sulfhydryl-containing amino acids that bind polypeptide chains together; a bond which forms when two sulfhydryl (-SH) groups of cysteine side chains of a protein are close together and are oxidized by the same reagent.

DMSO: Dimethyl sulfoxide.

DNA (deoxyribonucleic acid): A molecule occurring in nuclei of plant and animal cells composed of repeating subunits of nucleotides containing deoxyribose (a five-carbon sugar), phosphoric acid, and one of four nitrogenous bases (adenine, cytosine, guanine, and thymine). Every inherited characteristic has its origin in the code of an individual's complement of DNA. The linear chain of deoxyribonucleotides is double-stranded in some DNA viruses (DS-DNA) and single-stranded in others (SS-DNA). DS-DNA is held together by bonds between base pairs of nucleotides (adenosine, guanosine, cytidine, and thymidine); bonding occurs only between A-T and G-C; thus, one can determine the sequence of either strand from that of its partner. Also, in DS-DNA, the two strands are antiparallel in a double helix. Purine bases of one strand form hydrogen bonds with pyrimidine bases of the other strand. *See* Nucleic acid.

DNA-binding protein: A protein that binds directly to DNA; often determined by specific amino acid sequences. One function including maintaining DNA in a single-stranded form for transcription or replication.

DNA cloning: Means of isolating individual fragments from a mixture and multiplying each.

DNA-dependent DNA polymerases (I, II, III alpha, beta, gamma): Enzymes that synthesize DNA from a DNA template.

DNA ligase: An enzyme that catalyzes phosphodiester bond formation between the 5′-phosphate of one oligonucleotide and the 3′-hydroxyl of another involved in DNA synthesis and in linking DNA in genetic manipulaton.

DNA primase: *See* Primase.

DNA-RNA hybrid: Double helix containing one strand of DNA hydrogen-bonded to one strand of RNA.

DNAse: An enzyme that breaks down DNA by hydrolysis.

DNA sequencing: The relative order of nucleotide pairs in a stretch of DNA, a gene, a chromosome, or an entire genome.

Dorsal: Referring to the back or upper surface of an organism. *See* Ventral.

Double antibody sandwich (DAS): A method in enzyme-linked immunosorbent assay (ELISA) in which the reactants are added to the test plate in the order of antibody, virus, antibody-enzyme complex, and enzyme substrate. *See* ELISA and Enzyme-linked immunosorbent assay.

Double-diffusion test: Antigen–antibody precipitation reaction in agar or a similar gel in which antigen and antibody are allowed to diffuse toward one another and react at equivalence in the agar.

ds: Double-stranded (nucleic acid).

Early genes: Viral genes expressed early in the replication cycle; usually involved in the replication of the viral nucleic acid. *See* Late genes.

EBIA: electro-blot immunoassay (*see* Western blot). A technique that combines polyacrylamide gel electrophoresis and ELISA on nitrocellulose or nylon membranes.

Economic injury level: The lowest population density of pests that will cause economic damage.

Economic threshold level: The density of pests at which control measures should be used to prevent an increasing pest population from reaching an economic injury level; pest or pathogen population density or damage level above which the value of crop losses, absent managment efforts, would exceed the cost of management practices (especially the use of pesticides).

Ecosystem: A community of living things and its environment; the interacting, functional system comprised of all living organisms in an area and their nonliving environment.

ED: Effective dose.

ED_{50}: The dose that gives a 50% response.

Electrofocusing: Separating protein molecules by gel or density-gradient electrophoresis where a pH gradient exists. Each protein moves to a pH at its isoelectric point.

Electron microscope serology: *See* Immunosorbent electron microscopy (ISEM).

Electron microscopy: Use of an electron microscope, in which a focused beam of electrons produces a greatly enlarged image of minute objects (such as virus particles). *See* Scanning electron microscope and Transmission electron microscope.

Electrophoresis: The differential migration of charged molecules and macromolecular ions (e.g., fragments of nucleic acid or proteins) in a free solution or through a porous medium in an electric field; the porous supporting medium may be filter paper, cellulose acetate, or a gel. *See* Chemotaxonomy and Chromatography.

Electrophoretic mobility: The relative rate of movement of a charged particle or virus per unit potential gradient where mobility is toward the cathode or the anode, depending on its net positive or negative charge at the pH used.

Electroporation: A method using an electric pulse, whereby nucleic acids or virus particles can be introduced into protoplasts or cells by creating transient pores in the plasma membrane.

ELISA: *See* Enzyme-linked immunosorbent assay.

Empty virus particles: Virions-absent nucleic acid, which can be identified by negative staining and electron microscopy and of lower buoyant density than complete virus particles.

Enation: A small abnormal outgrowth of host tissue or eruption from a plant surface, often from veins (mostly from leaves, petioles, and flowers); usually induced by certain virus infections; literally "small leaf." *See* Hyperplasia.

Encapsidated: Enclosed as if in a capsule.

Encapsidation: The enclosure of a virus' nucleic acid genome within a protein shell.

Endemic: Pertaining to a persistent low and steady level of normal disease occurrence; a disease native to or restricted to a certain country or geographic region; also plant species native to a particular environment or locality.

Endonuclease: An enzyme that cleaves a polypeptide chain internally.

Enhancement: An increased concentration of a virus in joint infections with another virus.

Envelope: Lipoprotein membrane derived from a host cell membrane or synthetized *de novo*, surrounds the capsid or nucleocapsid, and usually a bilayer carrying virus-specified proteins.

Enzyme: A complex, high molecular weight protein produced in living cells by protoplasm that catalyzes a specific biochemical reaction but does not enter into the reaction itself.

Enzyme conjugate: Usually a preparation of an antibody to which an enzyme is linked covalently; the enzyme, by decomposing the appropriate substrate, produces a color reaction; used in ELISA and Western blotting.

Enzyme-linked immunosorbent assay (ELISA): A sensitive serologic procedure, widely used to detect and quantify antibodies with many variations to measure relative virus concentrations, and commonly used to identify plant viruses and microorganisms; where an antibody carries with it, an enzyme that releases a colored compound on reaction with a substrate. *See also* Double antibody sandwich (DAS).

Epidemic: A widespread and rapidly developing outbreak of an infectious disease of humans in a community; used loosely for plants and animals. *See* Epiphytotic and Epizootic.

Epidemic (or epiphytotic) rate: Amount of increase of a disease per unit of time in a population.

Epidemiology: Study of epidemics (epiphytotics). Of plant pathology concerned with disease in plant populations; the factors influencing the initiation, development, and spread of causes of infectious disease.

Epidermal-strip test: An electron microscope technique by which a virus can be quickly examined in crude-sap extracts. The *quick leaf-dip* and *quick dip* techniques are similar procedures.

Epiphytology: The science or study of epiphytotics (epidemics).

Epiphytotic: The sudden, widespread, and destructive development of a disease on many plants, usually over large areas (corresponds to an epidemic of a human disease).

Epitope: An antigenic determinant of defined structure, e.g., an identified oligosaccharide, or a chemical hapten that elicits the formation of antibodies; a grouping of amino acid sequences on a protein, or between adjacent protein subunits.

Equilibrium density gradient centrifugation: Isopycnic gradient.

Equilibrium dialysis: A technique used to measure the affinity of antibody–antigen binding, based on diffusion of unbound antigen across a dialysis membrane.

Equivalence: The ratio of antibody to antigen that gives the maximum precipitation in a quantitative precipitation reaction.

Escape: A condition in which a susceptible plant avoids infectious disease through some character of the plant or its location.

Etiologic agent: The parasite or virus causing a disease in a plant or animal.

Etiology, aetiology: The science of the causes or origins of disease; the study of nature of the causal factor and its relations with the host.

Evolution: The development of a species, genus, or other larger group of plants, animals, or other organisms over a long period of time.

Exclusion: Control of disease by excluding the pathogen or infected plant material from crop production areas (e.g., by quarantines and embargoes).

Exonuclease: An enzyme requiring a free end in order to digest an RNA or DNA molecule; both 5′- or 3′-exonucleases.

°F: Fahrenheit; a unit of temperature. *See* Fahrenheit and °C.

Fab fragment: Fragment obtained by papain hydrolysis of immunoglobulin molecules, about M_r 45,000 with a light chain linked to the N-terminal half of the contiguous heavy chain. Two Fab fragments constitute each 7S antibody molecule. A Fab contains one antigen-combining site and can combine with antigen as a univalent antibody but cannot form precipitates.

Fahrenheit (F): A thermometer scale in which the freezing point of water (ice point) is set at 32° and the boiling point (steam point) is 212°. To convert from Fahrenheit to Celsius: subtract 32 from the Fahrenheit reading, multiply by 5, and divide the product by 9. Example: 132°F − 32 = 100 × 5 = 500; 500 ÷ 9 = 55.5°C. *See also* Celsius.

Filamentous: Composed of long thread-like structures, often used to describe bacterial colonies or virus particles.

Filter, bacterial: A special type of filter through which bacterial cells cannot pass.

Filterable virus: Capable of passing through the pores of a bacterial filter; applies to most viruses.

Fingerprinting: A procedure for characterizing DNA, RNA, or proteins using electrophoretic or chromatographic analysis of specific fragments.

Flexuous, flexuose: Having turns, bends, or windings alternately in opposite directions; capable of bending.

Flocculate (n: **flocculation**): To aggregate (clump together) into a loose fluffy mass; (of serology) precipitation reaction of antibody and antigen in which the precipitate appears as flakes of insoluble protein.

Fluorescence microscope: A compound light microscope arranged to emit radiation of specific wavelengths, such as UV, to the spectrum, which then fluoresces.

Fluorescent antibody: An antibody conjugated with a fluorescent dye, e.g., fluorescein isothiocyanate, which then can be used in a fluorescent microscope to detect viral antigen in cells. *See* Enzyme conjugate.

Foundation seed: Seed stocks increased from breeder seed, and handled to closely maintain the genetic identity and purity of a cultivar.

Freeze-drying or **lyophilization:** Preservation of living microorganisms (or concentration of macromolecules with little or no loss of activity), etc., by removing water under a high vacuum while tissue remains in a frozen state.

Freund's adjuvant: A substance containing an emulsifier (lanolin) and mineral oil that is mixed with an antigen then injected into muscles of an animal to produce antiserum. Adjuvant allows slow release of antigen-enhancing immune responses. *Freund's complete adjuvant* includes killed mycobacteria; *Freund's incomplete adjuvant* contains no mycobacteria.

g: Gram = 1000 mg = 0.035274 oz. = 0.0222046 lb. = 0.001 kg (1 lb = 453.59 g).

G and C content: The total guanine (G) and cytosine (C) content of a nucleic acid, usually refers to double-stranded DNA. The G and C content is a good measure of such physical properties as melting temperature; it also affects the banding density in isopycnic gradients.

Gamma globulin: A fraction of serum proteins containing antibodies; an obsolete term for immunoglobulin G (IgG). *See* IgG.

Gel: A jelly-like colloidal mass; coagulated colloid.

Gel chromatography: A molecular sieving (purification) procedure that separates viruses or proteins of different sizes when passed through the pores of gel beads such as agarose.

Gel diffusion: *See* Immunodiffusion.

Gel double-diffusion: A serological test in which the antibody and antigen reactants diffuse toward each other in gel and react to form a visible precipitation line. *See also* Radial immunodiffusion and immunodiffusion.

Gel electrodiffusion: Electrophoresis of macromolecules in a matrix of agarose, polyacrylamide, or similar gel.

Gel filtration: A type of column chromatography that separates molecules on the basis of size.

Gene: The base triplets of the DNA molecule in a chromosome that determines one or more hereditary characters. The smallest functioning unit on a chromosome; bearer of an inherited factor; an ordered sequence of nucleotides that specifies the manufacture of a single type of protein (or for some genes, certain RNAs). *See* Allele.

Major gene: Many genes that have large observable effects on the phenotype.

Minor gene: One gene that has a specific effect on the phenotype.

Gene cloning: The isolation and multiplication of an individual gene sequence by its insertion, usually into a bacterium, where it can multiply.

Gene expression: Expression of genetic material as specific traits; transcription of mRNA from the DNA sequence of a gene; mRNA then translates into a protein = gene product.

Gene therapy: Insertion of more desirable DNA into cells to correct a given genetic defect.

Gene-for-gene hypothesis: The concept that corresponding genes for resistance and virulence exist in the host and pathogen, respectively. *See* Vertical resistance.

Genetic code: The sequence of nitrogen bases in a DNA molecule that codes for an amino acid or protein. More broadly, the sequence of all events from translation of chromosomal DNA to the final synthesis of an enzyme. Three nucleotides (a codon) each specify a single amino acid. The code is degenerate as 64 codons specify 20 amino acids, i.e., many amino acids are determined by more than one triplet. *See* Start and Stop codon.

Genetic engineering or **genetic manipulation:** Nucleotide insertion into or removal from an organism so a cell can produce more or different chemicals, or perform new functions. *See* Transformation of DNA.

Genetic map: A graphic representation of the linear arrangement of genes on a chromosome or genome by nucleotide sequence analysis.

Genetic marker: A mutation in a gene that allows its phenotypic identification.

Genetic variability: The ability of an organism to change inherited characteristics from one generation to next.

Genetics: The science or study of patterns of inheritance of specific traits.

Genome: The complete genetic complement of an organism or virus; size is usually denoted in gene pairs; (of viruses) the nucleic acid component, either DNA or RNA, which may consist of one (*monopartite*), two (*bipartite*), three (*tripartite*), or more (*multipartite*) molecular species of RNA. *See* Multicomponent virus.

Genomic library: Several clones of overlapping DNA fragments representing a genome.

Genotype: The set of genes that influences phenotype; individuals sharing a specific genetic makeup; the genetic constitution, expressed and latent, of an organism.

Germplasm: Material capable of transmitting heritable characteristics sexually or asexually; the total genetic variability available within an organism; a pool of germ cells or seed.

Glycoprotein (adj. **glycosylated**): A conjugated protein with at least one carbohydrate covalently attached to an amino acid.

gp: Gene product.

Gradient of infection: *See* Disease gradient.

Graft, grafting: Transfer of aerial parts of one plant (e.g., buds or twigs—the scion) into close cambial contact with the root or trunk (the rootstock) of a different plant; a method of plant propagation. Also, the joining of cut surfaces of two plants so as to form a living union.

Graftage: Method of inserting buds, twigs, or shoots in other stems or roots for fusion of tissues.

Graft indexing: A procedure to determine the presence or absence of a virus, mycoplasma, spiroplasma, etc. in a plant. The plant is grafted to a healthy plant known to show symptoms if infected. Used to detect disease agents not readily transmitted mechanically.

Green islands (in virology)**:** Nonchlorotic regions in a leaf showing mosaic or ringspot symptoms.

Group-specific antigen: An antigen specific to a group of viruses. *See* Type-specific antigen.

Guanine: One of the purine bases found in DNA and RNA.

Guanosine: A nucleoside of guanine and ribose. *See* Nucleoside and Nucleotide.

ha: *See* Hectare.

Hapten: A partial antigen; substance that can combine with antibody but can initiate an immune response only if bound to a carrier before injection into an animal. Most hapten molecules ($M_r < 1000$) carry only one or two antigenic determinants, but some macromolecules, e.g., pseumococcal polysaccharides, are haptenic. Haptenic groups can be conjugated to carriers *in vitro* and then regarded as capable of reacting with antibody but requiring the carrier molecule to become immunogenic *in vivo*.

HAT medium: A cell culture medium (containing hypoxanthine, aminopterin and thymidine) used in monoclonal antibody production to select hybridomas from unfused myeloma cells.

Hectare (ha): A land area in the metric system = 2.471 acres = 395.367 rods = 10,000 m^2 = 0.01 km^2 = 0.0039 mi^2.

Helenin, helenine: An antiviral antibiotic from *Penicillium funiculosum*; considered to be a RNA of viral origin.

Helical symmetry (of virology)**:** A form of capsid structure in many RNA viruses in which the protein subunits interact with the nucleic acid to form a helix. The only axis of symmetry is the length axis of the particle. All rod-shaped plant viruses have coat protein subunits arranged in helical symmetry.

Helix (adj. **helical**)**:** Coiled or spiral in shape with a repeating pattern; often used in reference to the double spiral of the DNA molecule.

Helper virus: A virus required for replication of a defective or satellite virus or a satellite RNA. *See* Dependent transmission.

Heterologous: Derived from a different type or species.

Heterologous reaction: Different but similar; for example, a serological reaction between an antiserum and an antigen closely resembling but not identical to the antigen that caused the production of antibody. *See* Homologous reaction.

Histopathology: Microscopic study of diseased tissues.

Homologous, homology: In serology, the specific relationship between an antigen and the antibody induced by its interaction with animal immune systems; likeness in structure; degree of relatedness between two nucleic acids or amino acid sequences; considered evidence of evolutionary relatedness.

Homologous antiserum: A serum containing antibodies raised against a specific antigen and that will react with that antigen.

Homologous reaction: A serological reaction in which an antiserum reacts positively with the antigen used for its preparation.

Hopperburn: Marginal yellowing, scorching, and curling of leaves (e.g., alfalfa, dahlia, potato) due to the feeding of certain leafhopper species.

Host: A living organism (e.g., a plant) harboring or invaded by a parasite and from which the parasite obtains part or all of its nourishment; (in virology) an organism or cell culture in which a given virus can replicate. *See* Suscept.

Host indexing: (1) A procedure to determine if a plant is a carrier of a virus, mycoplasma, or spiroplasma; (2) a procedure in which material is transferred from one plant to another that then will develop characteristic symptoms if infected by the pathogen in question.

Host range: The complete range of plant species susceptible to a virus.

Hybridoma: A hybrid animal cell or cell line produced from the fusion of a spleen cell (lymphocyte) and a tumor (immortal mouse myeloma) cell able to multiply *in vitro* to produce monoclonal antibodies.

Hyperchromicity: Increase in UV absorbance resulting from the denaturation of a macromolecule; provides an indication of the amount of base-pairing in a nucleic acid.

Hyperimmune serum: Serum from an animal that has received two or more injections of a foreign antigen to produce a reagent for use in serology.

Hyperimmunization: A condition resulting from immunization, usually with extensive booster doses, designed to stimulate the production of a high-affinity antibody relatively late in the immune response.

Hyperplasia (adj. **hyperplastic**): A plant overgrowth (gall, enation, tumor, witches' broom) due to *increased cell division*; excessive, abnormal, usually pathological multiplication of the cells of a tissue or organ. *See also* Hypertrophy.

Hypersensitive (in virology): An extreme reaction to a virus, e.g., the production of local lesions or the necrotic response of a leaf to a plant virus.

Hypertrophy (adj. **hypertrophic**): A plant overgrowth (gall or tumor) due to abnormal cell enlargement; excessive, abnormal, usually pathologically induced enlargement of individual cells in a tissue or organ. *See also* Hyperplasia.

Hypoplasia (adj. **hypoplastic**): Underdevelopment (malformation) of plant tissue due to decreased cell division. *See* Hyperplasia, Hypertrophy, and Hypotrophy.

Hypotrophy: Underdevelopment of plant tissue due to reduced cell enlargement.

Hypovirulence: Reduced virulence of a pathogenic strain due to the presence of transmissible double-stranded RNA (DS-RNA).

Icosahedral (n. **icosahedron**): a regular polyhedron with 20 equilateral triangular faces or sides, 30 edges, 12 apices, and two-, three-, and fivefold axes of symmetry; the symmetry forms the basis for the arrangement of the protein subunits of isometric virus particles. One of the five Platonic solids.

ID: Infective dose. Number of units of a pathogen required to infect a host.

ID$_{50}$: Median infective dose.

-idae (suffix): Used in virology and zoology to be added to the stem (of a type genus name) to form a family name.

Identification: The study of the characters of an organism to determine its name.

Idiotope: An antigenically distinct region associated with an immunoglobulin's paratope or hypervariable region.

Idiotype (in serology): The antigenic signature associated with an immunoglobulin's paratope or hypervariable region; a specimen identified by an author as typical of his/her species.

Ig: Immunoglobulin.

IgE: Immunoglobulin class that fixes to most cells and is responsible for anaphylactic and allergic sensitivity.

IgG: The major immunoglobulin class in the serum of humans; found in most species from amphibians upward, but not in fish. Human IgG molecular weight = 150,000, $S_{20,w}$ = 7S, fixes complement and crosses the human placenta.
IgM: Pentameric immunoglobulin; first class of antibody produced to most antigens during an immune response.
Immobilized DNA or RNA: Nucleic acid linked to nitrocellulose or activated paper. *See* Northern and Southern blotting.
Immortalization: Growth of cells in culture long after normally expected to cease; cloning of cDNA in a bacterial plasmid or production of monoclonal antibodies in a hybridoma.
Immune (n. **immunity**)**:** Not affected by or responsive to disease; exempt from infection due to its inherent properties (e.g., tough outer wall, hairiness, nature of natural openings, waxy coating, thick cuticle, etc.). *See* Acquired resistance and Resistance.
Immune response: The ability of an animal to produce antibodies to foreign antigens, such as proteins or carbohydrate, introduced by either infection with a pathogenic agent or artificial injection. The antigen ability to induce this response is referred to as *immunogenicity*. The substances capable of inducing the response are called *immunogens*.
Immune serum: The liquid portion of blood containing one or more specific protein antibodies.
Immunity: The natural or acquired state of being immune, exempt from infection.
Immunization: The process of increasing or of giving resistance to a living organism; (in serology) injection of an antigen into an animal in order to induce antibody production (the immune response).
Immunoblot (Western blot): Reaction of labeled antibodies with proteins adsorbed to nitrocellulose paper.
Immunochemistry: The identification of the sites of antigens in cells using antibodies to which a reporter molecule, e.g., ferritin, gold, or a fluorescent dye, is attached; a study of the mechanism of antigen–antibody binding.
Immunocytological methods: Procedures used to study cytopathological disorders in ultrathin sections of virus-diseased tissues, using labeled antibodies to diagnose the virus.
Immunodiffusion: A serological procedure in which the antigen–antibody reaction occurs by allowing the reactants to diffuse toward each other in a gel matrix. *See* Gel double-diffusion and Radial immunodiffusion.
Immunoelectrophoresis: A technique involving separation of proteins in a gel, using an electric field, followed by a precipitation reaction in the gel with antibodies to the separated proteins.
Immunogen: A substance (virus) that elicits an immune response when injected into an animal; must normally be foreign to the animal of a molecular weight greater than 1000 and a protein or polysaccharide. This may take the form of antibody production, together with the development of cell-mediated immunity, or of specific immunological tolerance. *See* Antigen.
Immunogenic: Causing immunity; a protein's (antigen) ability to induce antibody production.
Immunoglobulin classes: Subfamily of proteins produced by B cells based on large differences in H-chain amino acid sequence: isotypes IgA, IgD, IgE, IgG, and IgM.
Immunoglobulin G (IgG): Serum globular glycoprotein with least mobility to the positive electrode during electrophoresis, constituting a distinct class of antibodies. *See* IgG.
Immunoglobulin subclasses: Subpopulations of an Ig class based on more subtle structural or antigenic differences in the H chains than are class differences (e.g., IgG1, IgG2, IgG3, IgG4).
Immunology: A study of acquired immunity in animals against infectious disease.
Immunosorbent electron microscopy (ISEM) (syn. **electron microscope serology** and **immunoelectron microscopy**)**:** Techniques to identify viruses involving the visualization of the antibody–antigen reaction in the electron microscope (EM) to detect viruses, especially their locations *in situ*. These include:
Trapping: The EM grid first coated with antiserum, referred to as an antibody-coated grid (ACG), which then attracts virus particles from a virus preparation placed on it.
Decoration: Virus particles are attached to the EM grid and antiserum is then added. Homologous antibodies react with the particles to coat or "decorate" them.
Inactivation: The loss of the ability of a virus to initiate infection.
Inapparent infection: An infection that does not give obvious symptoms.
Inarch (adj.)**:** A type of plant graft often used to detect plant viruses, phytoplasmas, and viroids with an indicator plant(s). Cut diseased stem (scion) upward and healthy stem (stock) downward. One third of each stem of similar diameter cut diagonally; both bound at this point with grafting tape.

Incidence, of disease: Number of plants affected within a population; disease incidence should be distinguished from disease severity.
Incipient: Early in the development of a disease or condition.
Inclusion body: Subcellular structures (any matrix or array of virus particles, abnormal proteinaceous bodies, or areas of abnormal staining) in the cytoplasm or nucleus of virus-infected plant cells.
  1. Virus-coded intracellular body containing protein and viruses.
  2. Any array of viruses or assembly of abnormal protein crystals detectable by light microscopy.

Incomplete antibody: An antibody that binds antigen but does not precipitate or agglutinate the antigen.
Incomplete virus: *See* Defective virus.
Indicator: A plant that reacts to certain pathogens or environmental factors, producing specific symptoms and used to detect and identify these factors; a substance that changes color as conditions change, e.g., pH indicators reflect changes in acidity or alkalinity. *See* Differential hosts.
Indicator cell: A cell that reacts in a characteristic manner to infection with a specific virus.
Indicator plant or host: A plant that responds specifically to certain viruses, other pathogens, or environmental factors with specific characteristic symptoms; used for identification of the specific pathogen(s) or environmental factor(s); also used in diagnosis of pathogens.
Induced resistance: *See* Acquired resistance.
Induction: The activation of a latent virus infection; an increased rate of synthesis of an enzyme, caused by a small molecule, generally the substrate or a closely related compound.
Infected, (of an organism): Successfully attacked by a pathogen.
Infection (v. **infect**): Process or act of a pathogen entering (invasion, penetration) and establishing an infection in a host plant; after entry, multiply (viruses replicate), and persist in a carrier.
Infection court: Site of a host plant where infection can occur: root, shoot, leaf, flower, or fruit.
Infectious: Capable of infection and spreading disease (a pathogen) from plant to plant.
Infectious disease: A disease caused by a pathogen that can multiply and spread from a diseased to a healthy plant.
Infectious particle: A virion containing the complete viral genome capable of infecting a susceptible cell.
Infectious unit: The smallest number of virus particles that, in theory, can cause infection.
Infective (n. **infectivity**): (Of a pathogen) able to infect a living organism; (of a vector, medium, etc.) having the ability to transmit a pathogen.
Infectivity assay: A bioassay using mechanical sap transmission to quantitatively determine the number of infectious virus particles.
Initiation: The start of synthesis of a polypeptide or nucleic acid chain; or an infection.
Initiation codon: *See* Start codon.
Injury: Momentary (transitory) damage by a causal agent, e.g., insect feeding, action of a chemical, physical, or electrical agent, or an adverse environmental factor. *See* Disease, plant.
Inoculate (n. **inoculation**): (1) Artifical introduction of a pathogen at the site of infection of a host (the infection court) to induce disease or into a culture medium. Loosely, and incorrectly, used to describe the transfer of living cells of a microorganism to any place where they will grow and develop; (2) to treat seeds of leguminous plants with bacteria to induce nitrogen fixation in the roots.
Inoculation access period: The time that vectors are caged on test plants.
Inoculation feeding period (syn. **test feeding period**): The length of time a vector feeds on a test plant during pathogen transmission experiments.
Inoculation threshold period: The minimum feeding period an inoculative vector needs on a susceptible test plant to transmit a disease agent.
Inoculative insects: Infective insects that will transmit during a given test access period.
Inoculum (pl. **inocula**): The pathogen or its parts (e.g., fungus spores, mycelium, bacterial cells, nematodes, virus particles, etc.) used for inoculating to produce disease.
Inoculum potential: The number of independent infections and amount of tissue invaded per infection that may occur in a population of susceptible hosts at any time or place.
Insect: Member of the class Hexapoda (phylum Arthropoda). In the adult stage, true insects have six walking legs, wings, and three body divisions.
Insect vector: An insect that transmits a disease-inducing organism or agent. Plant viruses are transmitted by insects in three ways: (1) mechanically or stylet-borne (also termed nonpersistent or nonpropagative); (2) semipersistent or circulative indicating an ability to transmit over longer periods of time, but the virus

does not replicate in the insect; and (3) persistent or propagative for viruses that replicate in the vector as well as in the host plant.
Insertion: *See* Deletion.
Integrated pest management (IPM), integrated control: The use of all available methods (biological, chemical, cultural, genetic, legal, and physical) to control diseases and pests for best results with least cost and damage to the environment.
Integration (in virology): Insertion of viral DNA into the host genome, usually involving a virus-coded enzyme, integrase.
Interference (in virology): Effect of the replication of one virus on another.
Interferon: Proteins released by animal cells in response to viral infection that protect other cells from infection.
Intermolecular recombination: Recombination due to the reassortment of species of nucleic acid between viruses whose genomes are segmented.
Intramuscular: A route commonly used for injection of purified viruses usually prepared with Freund's adjuvant (*See* Freund's adjuvant).
Intraperitoneal: A route commonly used for injection (e.g.,antigens and viruses) into an animal via the peritoneum.
Intravenous: Describing a route sometimes used for the injection of antigens into the venous blood system of an animal.
Introns: DNA sequences that interrupt the protein-coding sequences of a gene; introns are transcribed into mRNA but eliminated from the message before translation to protein; region of DNA transcribed initially to RNA, then cut out during the formation of functional RNA.
Invasion (v. **invade**): Growth of a pathogen into a plant and its establishment.
Invasion court: The place on the host where the pathogen enters.
IPM: *See* Integrated pest management.
ISEM: *See* Immunosorbent electron microscopy.
Isoelectric focusing (electrofocusing): A separation technique where mixtures of proteins and/or viruses are resolved into their components by subjecting them to an electric field in a supporting gel or stabilized solution in which a pH gradient is established.
Isoelectric point: The pH at which a virus particle or protein molecule has a zero net charge; the point of highest probability of crystallizing or precipitating out of solution.
Isolate: (Verb) to separate a culturable microorganism from an infected suscept and grow it in the absence of other organisms; (noun) a sample (e.g., a virus) from a defined source.
Isolation: Prevention of crossing among plant populations due to distance or geographic barriers; securing an organism or virus into pure culture, or the culture itself.
kb: *See* Kilobase; refers to single-stranded nucleic acid.
kbp: Kilobase pairs; refers to double-stranded nucleic acid.
kDa: Kilodalton; molecular weight designated per 1000 units.
kg: Kilogram = 1000 g = 35.273957 oz. av = 2.20462 lb.
Kilobase (pairs) (kb, kbp): 1000 nucleotides in a polynucleotide chain; a measure of the size of a nucleic acid molecule.
Kinase: Enzymes catalyzing phosphorylation reactions. *See* Polynucleotide kinase and Protein kinase.
Koch's postulates: Rules by Robert Koch to prove the pathogenicity of a microorganism modified for plant viruses are: (1) isolation of the virus from a diseased host; (2) cultivation of the virus in experimental hosts or cells; (3) purification or filterability following inoculation of the filtered or purified virus of the pathogen; (4) production of a comparable disease in the original or related host species following inoculation with the processed virus; and (5) re-isolation of the virus.
l (Liter): = 2.1134 pt = 1.0567 liquid qt. (U.S.) = 0.9081 dry qt. (U.S.) = 0.264178 gal. (U.S.) = 1000 ml or cc = 33.8147 fl. oz. = 61.025 in.^{3} = 0.0353 ft.^{3} = 0.028378 bushel.
Late genes: Genes in a viral nucleic acid that are expressed late in the virus replication cycle; often those coding for capsid proteins. *See* Early genes.
Latency: Stage of an infectious disease, other than the incubation period, where no symptoms are expressed in the host.
Latent: Present but not manifested or visible.
Latent infection, latency: Infection in a plant without visual symptoms.

Latent period (phase): Period after a vector acquires a virus before it can transmit, as in persistent or propagative transmission; (in virology) the time between the apparent disappearance of the infecting virus and the appearance of newly synthesized virus.
Latent virus: A virus that does not induce symptom development in its host; often used to describe inapparent virus infection.
Leafhopper: Active insects (order Homoptera, family Cicadellidae) with sucking mouth parts; a vector of pathogens, especially persistent viruses; may cause direct plant injury during feeding, even in the absence of virus.
Lesion: Well-marked, localized area of diseased or disordered tissue; a wound.
Life cycle, life history: Cyclical stages in the growth of an organism (plant, animal, or pathogen) between the appearance and reappearance of the same stage; not so obvious for virus replication.
Linear DNA: One of several forms of DNA. The two linear strands of double-stranded DNA may be free, bound to a specific protein, or closed by a hairpin loop.
LIV: *See* Longevity *in vitro*.
Local lesion: A small, localized, chlorotic or necrotic spot produced on a leaf upon mechanical inoculation with a virus (e.g., a hypersensitive reaction).
Longevity end point or **longevity** *in vitro* **(LIV):** The time a virus in plant sap diluted 1/10 in a buffer retains its infectivity; usually determined at 0° or 20°C.
Lyophilization, lyophilize (syn. **freeze-drying**): A technique for water removal under high vacuum while tissue is frozen. Used for long-term preservation of antisera, viruses, and other types of pathogens. *See also* Freeze-drying.
Macroscopic: Visible to the unaided eye.
Macromolecules: Molecules with molecular weights of a few thousand to hundreds of million daltons.
Masked: Pertains to symptoms absent under certain environmental conditions but apparent under other conditions; a condition whereby a virus-infected host exhibits no disease symptoms due to unfavorable environment for disease expression and/or development.
Masked virus: A virus carried by a plant that shows no symptoms.
Mealybug: Small, oval insects (family Pseudococcidae, superfamily Coccoidea) with sucking mouth parts and a cottony, scalelike covering. Unlike scale insects, mealybugs possess functional legs and reproduce by producing eggs or living young. Mealybugs transmit a number of plant-infecting viruses, especially of cocoa (*Theobroma*).
Mechanical injury: Injury of a plant part by abrasion, mutilation, or wounding.
Mechanical transmission (or **inoculation**): Artificial introduction of inoculum in presence of mild abrasive (e.g., a virus) to create a wound (infection court) by hand manipulation also in the fields when a virus is transmitted from one plant to another by machinery, leaves rubbing, or root contact.
Median: In the middle.
Median effective dose (ED_{50}): The amount of a substance required to produce a response in 50% of the hosts (subjects) to which it is given.
Median effective time (ET_{50}): The amount of time for a substance to produce a response in 50% of the hosts (subjects) to which it has been given.
Median infective dose (ID_{50}): The dose of a chemical, pathogen, etc. that, on average, will infect 50% of the hosts (individuals) to which it is administered.
Median inoculation access period (IAP_{50}): Inoculation access period during which 50% of the inoculative insects will transmit. *See* Inoculative insects and Inoculation access period.
Median lethal dose (LD_{50}): The dose of a substance that is fatal to 50% of the test population.
Median lethal time (LT_{50}): The period of time required for 50% of a group of organisms to die following a specific dose of an injurious agent, e.g., virus, radiation, drug, etc.
Median survival time (ST_{50}): The period of time at which half the subjects have died following the administration of an injurious agent, e.g., virus, radiation, drug, etc.
-mer: Suffix derived from Greek *meros*, part; e.g., dimer, trimer, pentamer.
Messenger RNA (mRNA): A chain of ribonucleotides that codes for a specific protein. Messenger RNA by definition + sense moves from nucleus to ribosomes, where protein is synthesized in the cytoplasm.
Microbial control: Use of microorganisms and viruses as biological control agents for pests and diseases.
Microbiology: The science or study of microorganisms, e.g., bacteria, actinomycetes, fungi, algae, viruses, and protozoa.

Micrometer (μm): A unit of length = one-thousandth of a millimeter (0.001 mm or 1 micron [μm])]; 0.000001 meter (10^{-6} m) or 0.00003937 of an inch; also a disc or slide of glass ruled with lines forming a metric scale for measuring objects under a microscope in microns.

Micron (μm): One millionth of a meter (m) = 0.001 mm = 0.00003937 of an inch. Same as micrometer.

Microscopic: Too small to be seen except with the aid of a microscope.

Millimeter (mm): See mm.

Millimho (mmho): A measure of electrical conductivity, 1 mmho = 0.001 mho. The mho is the reciprocal of an ohm.

Millimicron (mμm): One-thousandth of a micron or 10 Ä; a nanometer (nm).

Mite: Minute (1/64- to 1/32-inches long) animals divisions; (order *Acarina*, families Tetranchidae and Eriophyidae) without evident body; six-legged as larvae and eight-legged as adults. Red spider mites commonly cover leaf surfaces with fine webbing. Eriophyid mites transmit a number of plant-infecting viruses.

Mitomycin C: Antibiotic; selectively inhibits DNA replication by cross-linking the single strands.

ml (milliliter): = 1 cm^3 (approximate) = 0.001 liter = 0.061 $in.^3$ = 0.03815 fl oz.

mm (Millimeter): = 0.1 cm = 0.01 decimeter = 0.001 m = 1000 μm = 0.03937 in. (about 1/24 in.).

MNPV: Bundle virion. Abbreviation for the subtype of nucleus polyhedrosis viruses in which the majority of enveloped virions contain more than one nucleocapsid.

Modal length: The length that occurs most frequently in a population of virus particles; a criterion used when classifying rod-shaped viruses.

Mode infection (in virology): Inoculation of cells or an organism with a solution that lacks virus particles; a control in virus infection experiments to ascertain any possible side effects of materials in the inoculum other than infectious particles.

Molecular biology: A field of biology concerned with the interaction of biochemistry and genetics in the life of an organism. Molecular techniques in use include DNA-DNA hybridization, DNA fingerprinting, PCR (polymerase chain reaction), RFLP (restriction fragment length polymorphisms), RAPD (random amplified polymorphic DNA), and DNA sequencing; techniques based on extracted DNA.

Molecular radius (Mr): The radius of the space occupied by a molecule and considered a more correct term for macromolecules than molecular weight, as it allows for hydration.

Molecular weight (mw): Sum of the atomic weights of constituent atoms in a molecule. *See* Dalton.

Molecule: A unit of matter; the smallest portion of an element or a compound composed of one or more atoms that retains the chemical identity with the substance in mass. A molecule usually consists of the union of two or more atoms; some organic molecules contains hundreds to a million or more atoms.

Monoclonal antibody: A homogeneous antibody population specific to a single antigenic epitope. Produced by fusing specific antibody-producing cells (in a single clone of lymphocytes) with immortalized tissue culture cells; synthesized and secreted by clonal populations of hybrid cells (hybridoma) prepared by the fusion of individual B lymphocyte cells from an immunized animal (usually a mouse or rat) with individual cells from a lymphocyte tumor (e.g., myeloma).

Monocotyledoneae, monocots (pl. **monocotyledonae**, adj. **monocotyledonous**): The subclass of flowering plants, including the grasses and cereals, having one cotyledon (seed leaf) in the embryo; characterized by parallel-veined leaves and fibrous roots. *See* Dicotyledon.

Monolayer cells: Animal cells grown in culture while attached to a solid surface, in contrast to cells grown in *suspension culture*.

Monomer: A single polypeptide chain. Basic components of Ig molecules are four monomers: two H chains and two L chains.

Morphological subunit: The structural subunit of a virus particle viewed by electron microscopy; often clusters of protein subunits (capsomeres) especially in isometric particles.

Morphology: Study or science of the form, structure, and development of organisms.

Mosaic: Disease symptom characterized by a clearly differentiated patchy mottling of the foliage or by sharp variegated patterns of dark and light green to yellow; disarrangement or unequal development of the chlorophyll content; symptomatic of many viral infections.

Mottle, mottling: Disease symptom of alternate light/dark irregular patterns more subtley differentiated than mosaic; often symptomatic of viral diseases. *See* Mosaic.

mRNA: Messenger ribonucleic acid.

Multicistronic messenger RNA: An mRNA that contains the coding sequences for two or more proteins. *See* Operon.

Multicomponent virus: A virus whose genome, required for infection, consists of two or more parts, each separately encapsidated; differs from a multipartite genome. *See* Genome.

Multipartite genome: A viral genome split between two or more nucleic acid molecules that may be encapsidated in the same particle; if in separate particles, they are multicomponent.

Multipartite virus: Multicomponent virus.

Multiplicity of infection (m.o.i.): Number of infectious virus particles added to a known number of cells required to cause infection.

MW: Molecular weight.

Mycoplasma viruses: Phage-like viruses in mycoplasmas (prokaryotes without cell walls).

Mycovirus: A virus that replicates in fungal cells.

Naked virus: A virus without a lipoprotein envelope (*see also* Viroid).

Nanometer (nm): A unit of length equal to one billionth (10^{-9}) of a meter (m) or one millimicron = 10 Å.

Necrosis (pl. **necroses**, adj. **necrotic**): Localized or general death and disintegration of plant cells or plant parts, usually resulting in tissue turning brown or black due to oxidation of phenolics; commonly a symptom of fungus, nematode, virus, or bacterial infection; a symptom of disease or injury.

Negative contrast staining: Staining procedure used to prepare virus particles for examination in an electron microscope.

Negative-sense strand (negative strand): Nucleic acid complementary to the plus strand.

Negative-strand virus: Virus whose genome is negative-sense RNA. Five families of negative-strand viruses are recognized.

Nepovirus (NEPO): Member of the tobacco ringspot virus group; usually transmitted by nematodes; *NE*matode-transmitted *PO*lyhedral-shaped viruses.

NETU: A term for *NE*matode-transmitted, *TU*bular-(rod-)shaped viruses.

Neutralization: Inactivation of infectious virus by reaction with its specific antibody, thereby blocking sites on the virus that normally adsorb to susceptible cells.

Neutralizing antibody: An antibody that inhibits virus infectivity.

nm (nanometer): = 10^{-9} meter; a millimicron (mμm).

Nomenclature: The system by which names are assigned to organisms and microorganisms. The name is governed by the International Code of Botanical Nomenclature. The Code is comprised of principles, rules (termed articles), and recommendations. The names of viruses are governed by the International Committee on Taxonomy of Viruses.

Nonpathogenic: Incapable of causing disease.

Nonpermissive cells: Cells in which a specific virus will not infect and replicate.

Nonpersistent transmission (syn. **stylet-borne transmission**): Type of insect transmission in which the virus is acquired by the vector after short (minutes) acquisition feeding times and is transmitted during short inoculation feeding periods. The vector (e.g., aphid) remains viruliferous for only a short period (e.g., a few minutes to perhaps 4 hours) unless it recharges by again feeding on an infected plant. *See* Persistent and Semipersistent transmission.

Nonproducer cells: Cells carrying all or part of a viral genome but not producing virus particles; they are usually transformed by the virus.

Nonsense codon: *See* Stop codon.

Nonstructural protein: Protein encoded by a viral genome but not involved in the structure of the virus particle. The protein is usually functional (enzymatic) during replication.

Normal length (of virus particles): The most common length, i.e., the intrinsic natural length of rod-shaped virus particles.

Northern blotting: A procedure analogous to Southern blotting (*see* Southern blotting) but involving the transfer of RNA to nitrocellulose or activated paper sheets.

N-termini: The portion of polypeptide chain ending with NH_2, often surface located in the protein subunit of a viral capsid; a site known to induce a specific antibody response.

Nuclease: An enzyme that can hydrolyze the internucleotide linkages in a nucleic acid.

Nucleic acid: A compound of high molecular weight of pentose (ribose or deoxyribose), phosphate, and nitrogen bases (purines and pyrimidines) joined in a long chain of repeating units (nucleotide complex); present in all living cells; constitutes the infectious parts of plant viruses; all either DNA or RNA.

Nucleocapsid: The nucleic acid (RNA or DNA) of a virus particle or virion enclosed by a protein capsid.

Nucleoprotein: Referring to viruses or another complex consisting of nucleic acid and protein.
Nucleoside: The combination of a sugar (ribose or deoxyribose) joined to a base molecule. *See* Nucleic acid.
Nucleotide(s): The building blocks of DNA and RNA, consisting of a phosphate group joined to a five-carbon sugar (ribose or deoxyribose) and joined to a base, a phosphate ester of a nucleoside. Thousands of nucleotides link to form a DNA or RNA.
Nucleotide phosphohydrolase: An enzyme that removes a phosphate group from the triphosphate end of nucleotides.
Nucleotide sequence: *See* Sequence.
Obligate parasite: A parasite that can grow in nature only on or in living tissue; not on an artificial medium.
Obligate pathogen: Viruses that replicate naturally only in living host tissue.
Occlusion body: Large virus-coded protein crystal containing viruses, referred to in the more general term "inclusion body."
Okazaki fragments: Short pieces of DNA, with attached RNA primers, produced during DNA synthesis. The primers are later replaced by DNA and the fragments are joined together.
Oligogenic: Applied to a character determined by only a few genes as contrasted with many genes—polygenic.
Oligonucleotide: A short-chain nucleic acid molecule.
Oligonucleotide probe: A short DNA probe for which hybridization is sensitive to a single base-mismatch.
Oligopeptide: A short chain of amino acids.
Open reading frame (ORF): A set of codons for amino acids (usually encoding a protein) uninterrupted by stop codons.
Open-circular DNA: One of three forms that double-stranded DNA can take. This circular DNA has one or both strands not covalently closed.
Operator: Region in the chromosome contiguous with an operon and controlling its functioning.
Operon: A cluster (two or more) of functionally related genes in the chromosome, the function of which function is subject to a common control mechanism; regulated and transcribed as a unit.
ORF: *See* Reading frame.
Ouchterlony gel diffusion test: Serological test in which antibody and antigen are placed in separate wells in agar and allowed to diffuse toward each other. A positive reaction is a visible band of precipitate between the antibody and antigen wells. Serological relationships of viruses can be determined by the bands interacting with each other.
Oudin tube: Simple diffusion in agar. Antigen in solution is placed over agar containing antibody in a test tube. Antigen diffuses into agar and forms a precipitin band at equivalence.
oz av (ounce avoirdupois): = 437.5 grains = 0.911458 troy oz. = 28.349527 g = 0.0625 lb.
oz troy (Troy ounce): = 480 grains = 1.097 oz. av = 31.10348 g.
PAGE: Polyacrylamide gel electrophosesis (*see* Polyacrylamide gel electrophoresis).
Partite: Suffix meaning "consisting of parts"; e.g., bipartite, tripartite.
Parts per billion (ppb): A method of expressing the concentration of chemicals in a substrate or diluent. One ppb = 1 inch in nearly 16,000 miles = 1 drop in 20,000 gal = 1 oz. in 753 million gal = 2 crystals of sugar in 1000 lb.
Parts per million (ppm): milligrams per liter; 1 ppm = 1 mg l^{-1}.
Passage: Experimental infection of a host with a parasite and later re-isolated; methods to increase virulence of parasites or to separate a specific virus from a mixture.
Passive hemagglutination: A serological test to detect virus-specific antigens by coating red blood cells with viral antigen. If viral antibody is present, red blood cells agglutinate.
Path-, patho- (prefix): Suffering, hence disease.
Pathogen, pathogene: An organism or agent (e.g., fungus, bacterium, nematode, virus, or viroid) capable of causing disease in a particular host (*suscept*). Most pathogens are parasites, but a few exceptions exist.
Pathogenesis: The sequence of processes in disease development from the time of infection to the final reaction in the host; production and development of disease.
Pathogenesis related (PR) proteins: Proteins that accumulate within cells in plant tissues in reaction to hypersensitive responses to certain chemicals or to fungal or viral infections.
Pathogenic: Ability of a pathogen to cause disease; the state of being pathogenic.
Pathogenicity: Causing, or capable of producing, disease.
Pathology, plant: Science or study of diseases, their nature and effects on plants, causes, and their control.

Pathotoxin: A toxin or poison produced by a pathogen and/or its host within the infected plant that functions in the production of disease, not itself the initial causal agent; vivotoxin. Tobacco mosaic virus was first labeled *contagium vivum fluidum* (Beijerinck, 1898) before viruses were named and known to be infectious.
Pathovar (pv): Synonym for subspecies, strain, pathotype of infectious microorganism or virus.
PBS: Phosphate-buffered saline.
Peck, U.S.: = 0.25 bushel = 2 gal = 8 qt =16 pt = 32 cups = 8.80958 liter = 537.605 in^3.
PEG: Polyethylene glycol.
Peplomer: Glycoprotein projection (spike, knob) located on the envelope.
Permissive cells: Cells in which infection results in production of infectious virus.
Persistent (n. **persistence**): Of circulatory viruses that remain infectious within their living vector for long periods without lysis, and are transmitted via salivary fluids.
Persistent transmission: Type of insect or nematode transmission in which a pathogen (e.g., a virus or mycoplasma) is acquired by a vector only after a long (minutes, hours, or days rather than seconds) acquisition feeding period. A latent period usually follows the acquisition feed, before the vector can transmit the pathogen. The vector often remains inoculative its full life. The pathogen sometimes multiplies in the vector. Nematode vectors transmit immediately after acquisition; they lose the capacity to transmit when they molt. Nonmolting dagger and stubby-root nematodes retain the capacity to transmit longer than do needle nematodes. *See* Nonpersistent transmission, Semipersistent transmission, and Propagative virus.
Pesticide: [*Note*: no virucides are yet known]. Any chemical or physical agent that destroys, prevents, mitigates, repels, or attracts pests (e.g., acaricide or miticide, bactericide, fungicide, herbicide, insecticide, nematicide, rodenticide, etc.).
Phenotypic mixing (or **genomic masking**): Individual progeny from a mixed viral infection encapsidated with structural proteins derived from both viruses; or the genome of one virus encapsidated in that of the other, which is called *genomic masking* or *transcapsidation*.
Phosphodiester bond: Linkage between nucleotides of polypeptide chains by covalent bonding of phosphoric acid with the 3′-hydroxyl group of one ribose or deoxyribose sugar and the 5′-hydroxyl group of the next ribose or deoxyribose ring. *See* Nucleic acid.
Phospholipase: An enzyme that catalyzes the hydrolysis of a phospholipid and is used in the study of cytoplasmic membranes.
Phosphoprotein: A protein with one or more amino acids phosphorylated by a protein kinase.
Photoreactivation: The enzymic repair of DNA damaged by UV light. The enzymes involved are activated by exposure to long-wavelength light.
Physiogenic (physiological) disease: A disease (or disorder) produced by some unfavorable genetic, physical, or environmental factor (e.g., excess or deficiency of light, water, soil nutrients, chemical, physical, or other injury, etc.) that often mimicks virus disease symptoms.
Phytopathogenic: Capable of causing disease in plants.
Phytopathology: Plant pathology; science or study of plant diseases.
Phytoplasma: A mycoplasma-like organism that infects only plants.
Pint, U.S.: = 16 fl. oz. = 32 tablespoons = 2 cups = 0.125 gal. = 473.167 ml = 1.04 lb. water = 28.875 in^3 = 0.473167 liter = 0.01671 ft^3.
Pinwheel inclusions: Shape of some virus-induced, cytoplasmic inclusion bodies seen in cross- and thin-section in an electron microscope, composed of sheets of virus-coded protein and found in cells of plants infected with potyviruses for which they are diagnostic.
Plant disease: *See* Disease, plant.
Plant-disease interaction: The concurrent infection of a host by two or more pathogens in which symptoms are distinctive or other effects are greater than the sum of the effects of each pathogen acting alone; an example of synergism.
Planthopper: Small, leaping, homopterous insects (family Delphacidae) with piercing-sucking mouth parts known to transmit plant-infecting viruses and mycoplasma-like organisms.
Plant pathology: The science or study of plant disease; also phytopathology.
Polyacrylamide gel electrophoresis (PAGE): Electrophoresis in a gel of polyacrylamide made by cross-linking acrylamide with N,N′-methylene-*bis*-acrylamide; for separation of nucleic acid or protein molecules according to molecular size to estimate molecular weight.

Polyclonal antibody: Antibodies produced at different sites and by different B lymphocytes against more than one epitope of an antigen; rabbits preferred. *See* Monoclonal antibody.

Polyhedron: Large, many-sided or rounded occlusion body.

Polymerase: An enzyme that forms RNA or DNA by adding ribonucleotide or deoxyribonucleotide triphosphates.

Polymerase chain reaction (PCR): The selective amplification of DNA by repeated cycles of: (1) heat denaturation of DNA; (2) annealing of two oligonucleotide primers that flank the DNA segment to be amplified; and (3) the extension of the annealed primers with the heat-stable DNA polymerase; to produce probes for virus diagnosis and amplify low copy number sequences; to selectively amplify sequences of interest in the presence of irrelevant DNA(s).

Polynucleotide kinase: An enzyme that phosphorylates the 5′-OH terminal ends of RNA and DNA nucleotide chains.

Polynucleotide ligase: A generic term for enzymes that catalyze the linking or repair of DNA or RNA strands. *See* DNA ligase and RNA ligase.

Polypeptide: A chain of amino acids linked together by peptide bonds, either by synthesis or partial hydrolysis of a protein.

Polyploid virus: A virus that contains a variable number of genomes, depending on such factors as host cell and cultural conditions, found in viruses of various groups.

Polyprotein: A large precursor protein later cleaved into two or more functional proteins.

Positive-sense strand (positive strand): For RNA, the strand that functions as the messenger (mRNA); and for DNA, the strand with the same sequence as the mRNA.

Post-transcriptional cleavage: Cleavage of polyprotein or RNA into functional units, usually monocistronic mRNAs. *See* Subgenomic RNA.

Post-transcriptional processing: Alterations in the structure of a mRNA after its transcription from either DNA or RNA.

Precipitation (syn. **precipitin**) **reaction:** In immunology, a visible fine precipitate when adequate quantities of soluble antibodies (precipitins) and soluble antigens form an insoluble lattice; detected in test tubes as a sediment or in agar gels as a white line appearing where equivalent amounts of antigen and antibody interact. Also termed immunoprecipitation.

Precipitin: The reaction in which an antibody causes precipitation of soluble antigens.

Primary host: (1) The virus host of most economic importance and/or on which the virus was first observed. (2) Plant on which the sexual forms of an aphid mate and lay eggs to overwinter. *See* Secondary host.

Primary symptom (syn. **local symptom**): Symptom(s) produced immediately after infection at the site of entry, in contrast to a secondary symptom, which follows more complete invasion (colonization) of the host. *See* Secondary symptom.

Primase: An enzyme that synthesizes the RNA primers for DNA synthesis on a DNA template. *See* Okazaki fragments.

Primer: Small fragment of nucleic acid with a free 3′-hydroxyl group necessary for the initiation of DNA and, sometimes, RNA synthesis. *See* Primase.

Prion: A "proteinaceous infectious particle," apparently absent nucleic acid, that causes certain neurological diseases of vertebrates.

Probe: A sequence of DNA or RNA used to detect complementary ones by hybridization.

Procapsid: A viral capsid lacking nucleic acid, possibly a stage in virion formation.

Propagative virus: A virus that multiplies (replicates) in its insect vector.

Proteases: Enzymes that digest proteins by cleaving polypeptide chains into fragments; involved in activating and degrading proteins; for peptide mapping or structural studies.

Protectant (of pesticides): An agent, usually a chemical, that tends to prevent or inhibit infection by a pathogen.

Protection: Placement of a chemical or physical barrier between the pathogen and the host that prevents infection.

Protein: Complex, high molecular weight, organic, nitrogenous substance (polymer compound), built up of amino acids joined by peptide bonds.

Protein kinase: Enzyme that catalyzes the phosphorylation of proteins, usually in the presence of cyclic AMP or cyclic GMP.

Protein subunit: A small protein molecule that is the structural and chemical unit of the protein coat of a virus; one or several subunits, depending on shape of the virus, comprise a capsomere.

Proteinase: An enzyme that hydrolyzes proteins to polypeptides.

Proteolytic cleavage: The splitting or digesting of polyproteins into simpler structural compounds.

Provirus: A viral genome inserted into a host genome or a plasmid; DNA passively replicated by host and transmitted to the next generation.

Pseudotype: *See* Phenotypic mixing.

psi: Pounds per square inch.

Purification: To separate molecules, cellular components, or viruses from cellular parts.

Purine: A heterocyclic compound containing fused pyrimidine and imidazole rings. The purine bases of nucleic acid are adenine and guanine. *See* Nucleic acid.

pv: Pathovar (*see* Pathovar).

Pyrimidine: A heterocyclic organic compound containing nitrogen atoms at positions 1 and 3; thymadine, uracil, and cytosine. *See* Nucleic acid.

Quarantine: A period of detention for animals, plants, or people coming from a place where a disease is known to exist; place where people or animals are kept for inspection.

R-loop mapping: A technique where single-stranded RNA is annealed to the complementary strand of partially denatured DNA. The formation of the RNA:DNA hybrid displaces the opposite DNA strand as a loop visible in an electron microscope. The DNA segment complementary to the RNA can then be mapped.

Radial immunodiffusion: A serological test in which liquid antigen (or antibody) is placed in wells cut in agar gel containing the other reactant previously added to the warm gel prior to solidification and allowed to diffuse out into the gel. The resulting antigen–antibody complex forms a halo or ring of precipitate around the well.

Radioimmunoassay: Sensitive, versatile protocols similar to those of enzyme-linked immunosorbent assay (ELISA), with solid-phase binding of either antibodies or antigens, but substituting radiolabeled antibodies or antigens for enzyme-IgG conjugation, then record the radiation (gamma or beta) counts. *See* Radio-immunoprecipitation.

Radioimmunoprecipitation: A serological test in which a reactant, usually antibody, is labeled, usually with ^{125}I. The amount of isotope precipitated in the antigen–antibody complex can be used to quantify immunoprecipitation. The antibody in the immunoprecipitate can also be measured using radiolabeled protein A.

Rate zonal gradient: A sample of macromolecules centrifuged through a gradient of an inert material (e.g., sucrose or glycerol), and the constituents separate as bands based on their sedimentation rates.

Reactivation: The activation of a virus from a latent stage. *See* Rescue.

Reading frame: A sequence of codons in RNA or DNA beginning with the initiation codon AUG. *See* Open reading frame.

Readthrough: Reading (translation) of mRNA through a stop codon. *See* Suppressor tRNA.

Reassortment: Production of a hybrid virus that contains parts derived from the genomes of two viruses in a mixed infection. Also termed pseudorecombination.

Recognition factors: Specific molecules or structures in or on the host (or pathogen) that can be recognized by the pathogen (or host).

Recombinant (n. **recombination**): A new strain of a virus resulting from the breakage and renewal of covalent links in a nucleic acid; genes replaced with different alleles; the exchange of genetic material from two or more virus particles into recombinant progeny virus during a mixed infection. *See* Reassortment.

Recombinant DNA: A new hybrid DNA molecule produced by enzymatic insertion of one piece into another from a different source *in vitro*. *See* Genetic engineering.

Reflective mulch: Plastic, aluminum, or straw layer placed on the soil surface around crop plants to control air-borne insect vectors.

Re-inoculation: Inoculated with the same virus a second or third time at successive intervals to ensure adequate test exposure to a virus when testing for host range or susceptibility.

Reiterated sequence: A nucleotide sequence that occurs many times in a nucleic acid.

Repeated sequence: A nucleotide sequence occurring more than once in a DNA or RNA sequence in the same or opposite orientation.

Replicase: The enzyme involved in the replication of viral genomic nucleic acids.

Replication (v. **replicate**): Duplication of the genomic DNA or RNA molecule of a virus; repetition of an experiment at the same time and place (one of several identical experiments, procedures, or samples).

Replicative form (RF): The intracellular structual form of viral nucleic acid active in replication; usually the double-stranded form in viruses with single-stranded genomes.
Replicative intermediate (R±): Nucleic acid form produced during viral replication.
Replicon: A portion of DNA able to replicate from a single origin. Since most viruses have one origin of replication, the entire genome is a replicon.
Repressor: A regulatory protein that binds to a specific DNA sequence upstream of the transcription initiation site and that prevents RNA polymerase from starting RNA synthesis.
Rescue: The activation of a defective virus by recombination or by complementaton of the defective functions.
Resistance: The inherent ability of an organism (host plant) to overcome or retard, completely or partially, the infective activity of a pathogen or other damaging factor; (of a pest) able to withstand exposure to certain pesticides. *Acquired resistance* is a noninherited resistance response in a normally susceptible host following a predisposing treatment. Various types of resistance include continuous, discontinuous, field, and horizontal; degrees of resistance include hypersensitivity, susceptibility, tolerance, and immunity.
Resistant: Possessing qualities that hinder the development of a particular pathogen. A plant may be slightly, moderately, or highly resistant.
Restriction endonuclease map: The cutting sites for restriction endonucleases marked on a linear or circular representation of a DS-DNA molecule.
Restriction endonucleases: Enzymes, usually bacterial, that hydrolyze DNA at highly specific sites. In recombinant DNA experiments, restriction enzymes are used to cut DS-DNA from a given organism into characteristic size by breaking internal bonds before it is recombined with a vector. Used extensively in recombinant DNA technology; type I enzymes which bind to the recognition site but usually cut DNA randomly; type II enzymes bind at the recognition site and cut either at or near that site.
Restriction enzyme: Restriction endonuclease.
Restriction fragment: A polynucleotide fragment produced by cutting DNA with a restriction endonuclease.
Reverse passive hemagglutination: A serological test in which red blood cells coated with a virus-specific antibody are used to test for an antigen. If virus antigen is present, the red blood cells agglutinate. *See* Passive haemagglutination.
Reverse transcriptase: An enzyme coded by certain viruses that makes a DNA copy of a primed RNA molecule. The enzyme can also synthesize DNA from a DNA template.
Reverse transcription: Copying of the genetic information from RNA into DNA.
RF: *See* Replicative form.
Ribonucleases (RNAses): Enzymes that hydrolyze phosphodiester bonds of RNA; used to characterize and sequence RNA.
Ribonucleic acid (RNA): *See* nucleic acid and RNA.
Ribonucleoside: A purine or pyrimidine base covalently bound to a ribose sugar molecule. *See* Nucleic acid.
Ribonucleotide: A ribonucleoside with one or more phosphate groups esterified to the 5′ position of the sugar. *See* Nucleic acid.
Ribose: A five-carbon sugar; a component of RNA.
Ribosome: A subcellular ribonucleoprotein particle, made up of two subparticles and several proteins, responsible for translating mRNA into protein.
Ribosome binding site: Nucleotides in mRNA to which ribosomes bind, 3 to 9 bases long, which precedes the translation start codon by 3 to 12 bases. In eukaryotes, the ribosomes are thought to bind to the 5′ end of the mRNA.
RNA: Ribonucleic acid. A nucleic acid found in the nucleus and cytoplasm involved in control of cellular activities and protein synthesis; the genetic material of many plant viruses. Nucleotides with a ribose sugar and a base: A, G, C, and U.
RNA ligase: An enzyme that can join RNA molecules and requires ATP.
RNA polymerase: An enzyme synthesizing RNA from a DNA or RNA template. Specific functions include DNA-dependent RNA polymerase and RNA-dependent RNA polymerase.
RNA processing: Modification of RNA after transcribing. *See* Post-transcriptional processing.
RNA replicase: RNA-dependent RNA polymerase.
RNA segment: A distinct piece of RNA; frequently refers to genomic segments of segmented genome viruses.
RNA transcriptase: RNA- or DNA-dependent RNA polymerase.
RNAases: *See* Ribonuclease.
RNA-dependent DNA polymerase: *See* Reverse transcriptase.

Rocket immunoelectrophoresis: An immunological technique to determine a single constituent in a protein mixture in a number of samples. Diluted samples are applied side by side to circular wells in an agarose gel which contains an antiserum to the protein of interest. Rocket-shaped precipitates that form upon electrophoresis identify the protein.

Rolling circle: A replication template is a single-stranded circular molecule found in the DNA replication of geminiviruses, phages, and RNA replication of viroids.

rRNA: Ribosomal RNA.

Rugose, rugous: Coarsely wrinkled surface; covered with coarse, netlike lines; roughened. Descriptive of certain virus diseases characterized by warty, roughened, or severely crinkled leaves or other plant parts.

Rugose mosaic: Striking mosaic accompanied by deformation such as leaf crinkling, curling, or roughening of the leaf surface.

Rugulose: Finely wrinkled.

Russet: Brownish, roughened surface areas (epidermis) of leaves, fruit, stems, and tubers resulting from abnormal cork formation; may result from disease, insects, spray or other mechanical injury.

S: Sedimentation coefficient, coefficient of sedimentation velocity = 10^{-13} cm s^{-1}. *See* Svedberg units.

S: Svedberg unit (*see* Svedberg units).

S$_{20w}$: Sedimentation coefficient expressed in Svedberg units and adjusted to 20°C in water.

S-adenosyl-L-homocysteine (ADoHcy, SAH): An inhibitor of methylation, e.g., nucleic acids, and an analog of S-adenosyl-L-methionine.

S-adenosyl-L-methionine (ADO, SAM): An intracellular source of activated methyl groups, including those used for RNA and DNA methylation; stimulates transcription in some viruses; required by class I restriction endonucleases for initial binding.

SAH: S-adenosyl-L-homocysteine.

SAM: S-adenosyl-L-methionine.

Sanitation (v. sanitize): Destruction (removal) of infected and infested plants or plant parts; elimination of disease inoculum and insect vectors; decontamination of tools, equipment, containers, workspace, hands, etc.; cultural methods of disease control that reduce inoculum.

Saprogenesis (adj. saprogenic): Part of the life cycle of a pathogen when not directly associated with a living host; not applicable to plant viruses. *See* Pathogenesis.

Satellite RNA: A small RNA packaged in protein shells of an unrelated helper virus, on which the satellite RNA depends for its replication, but with little or no sequence similarity to the genome of the helper virus.

Satellite virus: A small defective virus associated with tobacco necrosis virus (TNV) that depends on the "helper" TNV genome to replicate it. *See* Carna 5-RNA.

Schlieren optics: An optical system that detects gradients of concentration of a solute in a solvent using a phase plate. A differentiated curve of the gradient can be observed.

Scientific method: Problem solving consisting of stating the problem, establishing one or more hypotheses as solutions, testing hypotheses by experimentation/observation with controls, and accepting or rejecting the hypotheses.

Scion: Detached bud, twig, or shoot inserted by grafting into a stock (rootstock).

Scion-stock interaction: Effect of rootstock on scion and vice versa, wherein a scion on one kind of rootstock grows differently than on its own roots or on a different rootstock.

Scorch: Sudden browning and death (necrosis) of large, indefinite areas on a leaf, fruit, or stem from infection, lack or excess of some element, chemical injury, or unfavorable weather conditions.

Screening test: Test of response of plant cultivars or types to virus infection; by diagnosticians to detect or eliminate a virus. Screening tests can be morphological, biochemical, or serological.

SDI: *See* Serological differentiation index.

SDS: *See* Sodium dodecyl sulfate.

SDS-polyacrylamide gel electrophoresis: Electrophoresis in polyacrylamide gels of proteins denatured with SDS, which gives proteins equal charge per unit molecular weight, thus separates proteins by molecular weight.

Secondary cycle: Of plant disease: any cycle initiated by inocula generated within the season.

Secondary host: A virus host usually of lesser economic importance, but one that adequately supports virus replication in nature. The host plant on which the asexual aphid form occurs; also to indicate alternative hosts for pathogens. Do *not* confuse with *alternate host*, being required for completion of the life cycle of a fungus. *See* Alternative host, Primary host.

Secondary infection: Infection resulting from the spread of infectious material (inoculum) after a primary infection (the first infection by a pathogen after a resting period) or from subsequent infections without an intervening inactive period.

Secondary inoculum: Inoculum from primary infections or other secondary infections.

Secondary symptom: Symptoms of virus infections subsequent to primary symptoms.

Sedimentation coefficient: Influenced both by shape and molecular weight of a particle, determined by its rate of settling in a solvent; the rate of sedimentation of a macromolecule, such as a virus, per unit centrifugal field is measured in *Svedberg units* (S) = 10^{-13} cm s^{-1}.

Sedimentation rate: The rate at which a macromolecule (virus) moves in a standard gravitational field (the sedimentation coefficient is 10^{-13} cm s^{-1}); influenced by its mass, shape, hydration, and partial specific volume, plus the temperature and viscosity of the solvent.

Seed-borne disease: Disease in which inoculum is borne in or on seed.

Seed-borne inoculum: Propagules of a pathogen capable of causing disease of the seed itself; seedling or plant derived from the seed.

Seed-borne microflora: Bacteria, fungi, nematodes, mycoplasma, or, incorrectly, viruses associated with seeds.

Seed-borne pathogen: An infectious agent associated with seeds and having the potential of causing a disease of a seedling or plant.

Seed contamination: Similar to infestation.

Seed infection: Colonization of seed tissues by pathogens with or without conspicuous symptoms.

Seed infestation: Pathogen propagules carried on the outside of seeds or associated with seed lots.

Seed transmission: Passage of inocula from an infected/infested seed to its seedling plant.

Seed treatment: A chemical or nonchemical used to control pathogen propagules associated with true seeds and seedpieces.

Semiconservative replication: DNA-dependent DNA replication in which the synthesis uses both strands of dsDNA as templates, resulting in progeny, each comprising one parent strand and one new strand. *See* Okazaki fragments.

Semipersistent transmission: Transmission by a virus that circulates for a short time but does not multiply in its arthropod vector; intermediate between *nonpersistent* and *persistent* transmission. The acquisition feed by the vector is short, with no latent period; it may transmit the virus for hours to days.

Sequence: The order of nucleotides in RNA or DNA or of amino acids in a polypeptide.

Sequencing: The determination of a sequence of a nucleic acid or protein.

Sequential epitope: An antigenic site composed of five to seven sequential peptides in an unfolded random coil.

Serological differentiation index (SDI): A measure of the degree of serological cross reactivity of two antigens; the number of twofold dilution steps separating the homologous and heterologous titers.

Serology: A method using the specificity of the antigen–antibody reaction to detect and identify antigenic (protein) substances and/or organisms that carry them; the study, detection, and identification of antigens, antibodies, and their reactions.

Serum: The watery amber portion of blood remaining after coagulation of red blood cells; used in serological tests for viruses. Serum contains no fibrinogen, and thus is commonly used in immunological procedures because no clots form when other materials are added.

Serum neutralization: The inhibition of the infectivity of a virus by antiserum.

Shock (syn. **acute**) **symptoms:** The severe, often necrotic symptoms produced on new growth following infection with some viruses.

Shot-hole: Disease symptom in which small, roundish to irregular dead fragments drop out of leaves, making them appear as if riddled by shot.

Sigla: A name formed from letters or other characters taken from words in compound terms (e.g., virus group names). *See* Acronym.

Signal peptide: Short segments, usually of 15 to 30 amino acids, at the N-terminus of a secreted or exported protein, enabling it to pass through the membrane of a cell or organelle, usually removed by a protease, thus not present in mature proteins.

Signal sequence: Signal peptide or the nucleotide sequence encoding a signal peptide.

Silent gene (message): Gene not expressed; ribosome binding site is blocked/unavailable.

"Silent" infection: An infection with no detectable symptoms; synonym: latent infection.

Single diffusion: A test in which antigen diffuses into agar containing antiserum, resulting in a ring of antigen-antibody precipitation. *See* Radial immunodiffusion.

Single radial diffusion test: Radial immunodiffusion.

Single-stranded (ss): Nucleic acid molecules (DNA or RNA) as a single polypeptide chain.

Sodium dodecyl sulfate (SDS): Synonym: sodium lauryl sulfate (SLS), an anionic detergent capable of disruption of virus particles and cells, inhibiting nucleases during nucleic acid extractions, and denaturing proteins for gel electrophoresis.

Southern blotting: A technique in which DNA fragments, separated by gel electrophoresis, are immobilized on a nitrocellulose or nylon-based membrane. The immobilized DNA is available for hybridization with a labeled ssDNA or RNA probe.

Specific absorbance (A): Absorbance per unit mass of a substance, usually measured at the wavelength where absorbance is maximal. For 0.1% solutions (i.e., 1 mg ml^{-1}) the A_{260nm} is 25 for RNA and single-stranded DNA, and 20 for double-stranded DNA. The corresponding A_{280nm} value for proteins is approximately 1, but the actual value depends on their aromatic amino acid content. *See* Absorbance.

Specific infectivity: The infectivity of a virus expressed as percent number or per unit weight of particles; can be measured by plaque, pock, or local lesion assays.

Specificity (of serology): A term defining selective reactivity between substances, e.g., of an antigen with its corresponding antibody or primed lymphocyte.

Spiroplasma viruses: Phage-like viruses isolated from helical mycoplasms (spiroplasmas). Isolate SV-C3 from *Spiroplasma citri* is polyhedral with a short tail. SV-CE virus, also from *S. citri*, has a polyhedral head and a long tail (morphotype B).

ss: Single-stranded (nucleic acid).

Starch lesion: Local accumulation of starch in a virus-infected leaf; demonstrated by decolorizing the leaf and staining with iodine; useful to assay for virus concentration. *See* Local lesion.

Start codon: The trinucleotide in a mRNA at which ribosomes start the process of translation; sets the reading frame for the translation; usually AUG that encodes methionine.

Stop codon: The trinucleotide sequence at which protein synthesis stops for lack of a tRNA molecule to insert an amino acid into the polypeptide chain; the three stop codons are UAP (ochre), UAG (amber), and UGA (opal).

Strain: An isolate of one virus related closely to another determined by a combination of characteristics including serological or biophysical properties, amino acid sequences or nucleotide mean homology minor differences, or host range and differential host reactions.

Structural unit: Protein subunit, the basic building block of the capsid.

Stunted: Plant reduced in size and vigor due to unfavorable environmental conditions caused by a range of pathogens or abiotic agents.

Stylet: Relatively long, pointed, stiff, slender, hollow, feeding organ in the mouth portion of plant-parasitic nematodes and some insects (e.g., aphids) for piercing and withdrawing nutrients from plant cells. Nematodes have three types:
1. *Stomatostyle*: A spear apparently developed through evolution by the fusion or sclerotization of the stomal wall; anterior aperture is ventrally oblique; commonly in species of *Secernentea*.
2. *Odontostyle*: A spear evolved from a large tooth and derived prior to each molt from a single cell in the esophagus; anterior aperature is broadly oblique; found in some members of the *Adenophorea*.
3. *Onchiostyle*: Stiff, slender, curved, grooved, tooth-like found in the adenophorean genus *Trichodorus*; derived evolutionarily from an onchium; originating in the stomal wall.

Stylet-borne: A virus borne on the stylet of its insect vector; a noncirculative virus. *See* Nonpersistent transmission.

Subcutaneous: Under the epidermis or skin.

Subcutaneous injection: Injection of antigens between the skin and underlying tissues used in the preparation of antisera.

Subcuticular: Beneath a cuticle; between the cuticle and the upper epidermal wall.

Subcuticular layer, subcuticle (of nematodes): The hypodermis.

Subgenomic RNA: RNA less than the genomic length in infected cells, sometimes encapsidated. If encapsidated, not involved in natural infection. Each species of subgenomic RNA has a different cistron at the 5' end, opening it for translation.

Sucrose gradient, sucrose density gradient centrifugation: Particles (e.g., viruses) of different sedimentation coefficients are separated during centrifugation in a gradient of sucrose concentrations, increasing from top to bottom, constructed in a centrifuge tube with the solution containing the test particles placed on top. On centrifugation in a swing-out rotor, the various particle species sediment as bands at different rates. *See* Isopycnic centrifugation and Rate zonal gradient.

Superinfection: Inoculation of a host with a second virus, usually a strain of the first infecting virus; result (1) interference between the two viruses, the first virus preventing replication of the second (*see* Cross protection); (2) the two viruses replicating independently (*see* Synergism); (3) recombination occurring between the two viruses; and (4) phenotypic mixing.

Suppressor tRNA: tRNA molecule containing an altered anticodon that either reads a mutated codon in the same sense as the original or introduces a different amino acid that does not abolish protein function.

Suscept: Any living organism susceptible to a disease, pathogen, or toxin; an abbreviated term for "susceptible plant" or "susceptible species." *See* Host.

Susceptibility: The inability of a plant to resist the effect of a pathogen.

Susceptible: Not immune; lacking resistance or with little resistance; prone to infection. *See* Tolerance, Tolerant.

Suspect: Any plant potentially subject to infection and colonization by a pathogen.

Svedberg equation: A formula for estimating the molecular weight of a macromolecule; $s = (1/w^2r)\ dr/dt$; r = distance in cm of sedimenting boundary from the center of rotation, t = time in seconds, w = angular velocity radians per second; standard defined as 20°C in water (s_{20},w).

Svedberg units (S): The units used to measure the rate of sedimentation of a macromolecule, subcellular component, or virus to determine its *sedimentation coefficient*.

Sv$_p$: Subviral particle; devoid of outer capsomeres, the core containing the nucleic acid; refers to isometric viruses.

Swing-out rotor: A centrifuge rotor with buckets free to swing out horizontally when the rotor is spinning. The gravitational force is thus along the long axis of the tubes. *See* Sucrose gradient.

Symmetry: Structural principle of viral capsid or nucleocapsid.

Symmetrical: Of equal morphology about a designated axis.

Symptom: Indication of disease; visible effects produced in or on a plant by the presence of a pathogen.

Symptomatology, symptomology: The study of symptoms of disease caused by pathogens for the purpose of diagnosis; class of symptoms diagnostic for a particular disease.

Symptomless carrier: A plant infected by a pathogen (i.e., a virus) with no visible effects.

Symptomless virus: A virus present in a plant that produces no visible symptoms.

Systemic infection: Infection resulting from the spread of virus from the site of infection to all or most of the cells of the host.

Tailing: The addition of stretches of nucleotides to the 3′- terminus of ds nucleic acid to facilitate cloning into a vector.

Taxon (pl. **taxa**): A taxonomic group of any rank such as family, genus, species, etc.

Taxonomy: Science of systematically describing, naming, and classifiying organisms (viruses) based on natural relationships.

tbs: Tablespoonful = 3 tsp = 14.8 ml or cc = 1/2 fl. oz. = 0.902 in.^{3} = 0.063 cup.

TDP (thermal death point): *See* Thermal death (or inactivation) point.

TEM: *See* Transmission electron microscope.

Temperature-sensitive mutant: A mutant not capable of infection at as high a temperature as its wild-type strain.

Template: A nucleic acid molecule from which a complementary nucleic acid molecule is being synthesized.

Terminal codon: *See* Stop codon.

Terminal redundancy: Identical sequences at each end of a linear polynucleotide molecule.

Terminal repetition: An identical nucleotide sequence at each end of a nucleic acid molecule (*see* Terminal redundancy) or identical but inverted.

Terminal transferase, deoxynucleotidyl transferase: An enzyme that adds deoxynucleotide triphosphates (dNTP) to the 3′ hydroxyl groups of the DNA fragment; used in 3′ end labeling of DNA and in tailing DNA fragments with complementary dNPSs to facilitate cloning.

Test plant: A plant species, variety, or cultivar used to diagnose, or evaluate, the infectivity of a pathogen. *See* Indicator plant.

Thermal death (or **inactivation**) **point:** The lowest temperature at which death or inactivation of an organism occurs (after heating) for usually 10 minutes.

Thermotherapy: The curing of a host (or propagative part, e.g., a seed, bulb, corm, tuber) or cell line of infection by a pathogen by heat (rarely cold) treatment.

Thrips: Small slender insects (order Thysanoptera), usually with four long, narrow wings. Thrips have rasping-sucking mouth parts and a ten-segmented abdomen. They influence the development of disease in plants by acting as vectors of pathogens like fungi, bacteria, and the tomato spotted wilt virus.

Thymidine kinase: An enzyme catalyzing the phosphorylation of thymidine to thymidic acid and induced in cells infected with DNA viruses.

Thymine: Pyrimidine nitrogenous base found in DNA. *See* Nucleic acid.

TIP: Thermal inactivation point; the temperature (varying from 45 to 95°C) at which virus particles lose their infectivity or an enzyme its activity. *See* Thermal death point.

Titer or titre (in serological reactions): Antibody: a relative measure of the amount of antibody in an antiserum per unit volume of original serum. The antiserum is serially diluted, antigen is added, and the antibody titer is the reciprocal of the highest serum dilution to produce a discernible antigen–antibody reaction. Virus: the concentration of virus present in a preparation as measured by bioassay or a serological assay.

Titration: The measurement of titer.

Tolerance (adj. **tolerant**): Ability of a plant to endure virus infection without serious damage or yield loss; (in serology) active state of unresponsiveness to a given immunogen; immune responses to other immunogens are normal. *See* Susceptible, Resistant, Resistance, and Vector resistance.

Ton (long, U.S.): = 2,240 lb. = 2,722.22 troy lb. = 1.120 short tons = 1016 metric tons = 1016.04 kg.

Ton (metric): = 2,204.6 lb. = 1.1023 short tons = 0.984 long ton = 1000 kg.

Ton (short, U.S.): = 2000 lb. av = 2430.56 troy lb. = 0.892857 long ton = 0.907185 metric ton = 907.18486 kg = 32,000 oz.

Transcript: The RNA molecule produced by transcription. The primary transcript is often processed or modified into mature functional RNA, e.g., mRNA, rRNA, or tRNA.

Transcription: Synthesis of RNA with one strand of DNA as template; catalyzed by the enzyme DNA-dependent RNA polymerase; to copy a gene into RNA. *See* Reverse transcription.

Transencapsidation: Encapsidation of nucleic acid of one virus with the protein of another strain during simultaneous replication of two strains. *See* Phenotypic mixing.

Transfection: Virus infection of cells after inoculation with viral nucleic acid.

Transfer RNA (tRNA): The RNA that moves amino acids to the ribosome in order prescribed by the messenger RNA (mRNA). Each tRNA has a specific trinucleotide sequence that interacts with a complementary sequence in mRNA. Also called soluble RNA (sRNA) and acceptor RNA (aRNA).

Transformation of DNA: A technique wherein cell properties can be modified by insertion and expression of foreign DNA. *See* Genetic engineering.

Transgressive segregation: Genetic mechanism transferring characters (to progeny) not expressed by parents.

Translation (adj. **translational**): Synthesis of proteins on ribosomes as directed by successive triplets of nucleotides moved along mRNA.

Transmission: The transfer of virus from an infected plant to a healthy plant. Transmission by physical contact, vector, or pollen are termed *horizontal*; *vertical* transmission is from mother (seed) to progeny.

Transmission electron microscope (TEM): An electron microscope in which an electric beam passes through a thin specimen (e.g., a preparation of virus particles negatively stained or shadowed), or a thin section of tissue mounted on a grid. The image of the specimen is magnified with electrostatic lenses, observed on a fluorescent screen, or photographed.

Transport protein: A protein involved in the transport of small molecules around the cell.

Transstadial: Describing a pathogen (e.g., a virus) that is retained through the molt of its insect vector.

Treehopper: Small, jumping, homopterous insects of the family Membracidae with piercing-sucking mouth parts; known to transmit several plant-infecting viruses.

Triangulation number: Number of identical equilateral triangles composing a face of a given icosahedron, e.g., T = 4, 9, or 16.

Tris: Tris(hydroxymethyl) methylamine; common biological buffer with pH range of 7.0 to 9.2.

Triton 100: A nonionic detergent (iso-octyl phenoxypolyethoxy ethanol) for disruption of cells, stabilizing proteins, solubilizing aqueous samples in scintillation fluids, etc.

Triturate: To grind, as with a mortar and pestle.

tRNA: Transfer ribonucleic acid.
Trypsin: A proteolytic enzyme that catalyzes the hydrolysis of peptide bonds on the carboxyl side of arginine, lysine, and aminoethyl cysteine residues.
Tryptic peptide: A peptide formed from a protein by the action of trypsin. It has arginine, lysine, or aminoethyl cysteine at the C-terminus.
tsp: Teaspoonful = 5 ml = 0.17 fl. oz.
Tuber indexing: Propagation of a plant from a tuber or a tuber assess presence of a tuber-borne disease.
Type (in nomenclature): Specimen(s) on which the name of a taxonomic group (taxon) is based; an object that serves as the basis for the name of a taxonomic category (e.g., a specimen is the type for a species, a species for a genus, a genus for a family, etc.).
Type-specific antigen: An antigen specific to a certain type of virus. *See* Group-specific antigen.
Ultracentrifuge: A high-speed (up to 70,000 rpm and 500,000 g) centrifuge used to separate cellular components and for determining the particle size of viruses and proteins.
Ultracentrifugation: Processing in a high-speed centrifuge capable of sedimenting small particles. *See* Rate zonal gradient.
Ultrafiltration: A method for the removal of all but the very smallest particles (e.g., viruses) from a fluid medium.
Ultramicroscopic: Too small to be seen with a light microscope.
Uncoating: The removal of the outer layers of a virus particle on infection, thus releasing the viral nucleic acid. Ribosomes may be involved in uncoating the particles for some plant viruses; a process, called cotranslational disassembly, gives rise to complexes of virus particles and ribosomes, structures that have been termed "striposomes."
Uracil: One of two pyrimidine nitrogenous (cytosine) bases found in RNA. *See* Nucleic acid.
Uranyl acetate: A negative stain to aid visualizing viruses in an electron microscope; often used as a 1% solution, natural pH 4.0.
URF: Unidentified reading frame deduced from a DNA sequence but in which no protein or genetic function is known. *See* Open reading frame.
Urea: Low molecular weight compound used to denature proteins and nucleic acids.
Uridine: The nucleoside of uracil and ribose. *See* Nucleic acid.
Uridine 5′-triphosphate (UTP): A pyrimidine nucleotide, one of the four major constituents of RNA. *See* Nucleic acid.
UV: Ultraviolet (radiation).
Valence (of antibody): The number of epitopes with which one antibody can combine. The valence of most antibodies is 2. IgM has up to 10; IgA has different valences, depending on the degree of polymerization.
Valid (of taxonomic epithets): Names published in accordance with the Code. Such names may be legitimate or illegitimate. *See* Nomenclature.
Variability: The ability of an organism to change its characteristics from one generation to another due to adaptation to the environment or a mutation. *See* Genetic variability.
Variant: An organism showing some variation from the parent culture.
Vector: A living organism (e.g., insect, mite, bird, higher animal, nematode, parasitic plant, human, etc.) able to carry and transmit a pathogen (virus, bacterium, fungus, nematode), thus disseminate disease; (in genetic engineering) a vector *or* cloning vehicle is a self-replicating DNA molecule, such as a plasmid or virus, used to introduce a fragment of foreign DNA into a host cell (as the receptor for DNA molecules). *See* Transmission.
Vector efficiency: Usually expressed as percentage, indicating the proportion of vectors among a population tested.
Vector propensity: Probability that a vector with opportunity to acquire a pathogen in nature will transmit it; i.e. it probes a noninfected plant.
Vector resistance: Resistance of a host plant to the vector of a virus. Three basic types of vector resistance are:
Antibiosis: Resistance where the growth and multiplication of the vector on the host is inhibited.
Nonpreference: Resistance because a vector prefers not to feed on a host; the vector may probe-feed a few times after landing, then move on, which may reduce the transmission of a persistent virus, but increase the spread of a nonpersistent virus.
Tolerance: Ability of a host plant to withstand a vector's attack without damage; does not control virus spread. *See* Resistance and Tolerance.

Vein banding: Symptom of a virus disease in which regions along the leaf veins remain darker green than the tissue between the veins.

Vein clearing: Symptom of a virus or other pathogenic disease characterized by the disappearance of green color adjacent to the veins of young leaves.

Ventral: Front or lower surface, as opposed to dorsal.

Vertical resistance: Resistance with complete protection of a host only against specific strains of a pathogen. *See* Gene-for-gene hypothesis.

Vertical transmission or **genetic transmission:** Passage of a viral genome from host generation to the next, either as a provirus or associated with the gamete genome, i.e., transovarial transmission in insect vector, or seed transmission of plant viruses.

Virion: Complete infectious virus particle (nucleic acid plus protein shell or capsid).

Viroid: An infectious pathogen smaller than a virus; unencapsidated, free-floating, circular, single-stranded, 247 to 374 nucleotides, ribonucleic acid (RNA) that can infect certain plant cells, directs its own replication from host metabolites, and causes disease. Replication in the nucleus occurs by rolling, circle-producing, oligomeric forms processed to unit length. The nucleic acid cannot code for proteins. Heat stability for viroids range from narrow to wide.

Viroplasm: An amorphous, usually proteinaceous, cytoplasmic inclusion body absent membrane, associated with virus infection in plant cells infected with wound tumor, cauliflower mosaic, and other viruses.

Virosis (pl. **viroses**): Any virus disease of plants.

Virucidal: Describing a virus-inactivating agent, e.g., formalin and chlorine.

Virucide: A chemical or physical agent capable of inactivating or permanently suppressing virus replication.

Virulence: The degree of pathogenicity of a virus; capacity to produce disease.

Virulent: Highly pathogenic, the capacity to cause a severe disease; able to produce the symptoms typical of the disease in a susceptible host.

Viruliferous: Containing or carrying a virus; term applied particularly to virus-laden insect and nematode vectors capable of transmitting the virus.

Virus: A submicroscopic, filterable agent that causes disease and multiplies (actually replicates) only in living cells; intracellular, obligate parasite consisting of a core of one or more infectious nucleic acids (either RNA or DNA) encapsidated by protein. Viruses contain no energy-producing enzymes, no functional ribosomes or cellular organelles; all are supplied by the cell in which viruses replicate. *See* Helper virus and Latent virus.

Virus assembly: The coming together of the constituent parts of a virus particle and the formation of that particle. The process varies from autoassembly of protein subunits around nucleic acid, to a more complicated assembly of many phages.

Virus cryptogram: A descriptive code summarizing the main properties of a virus; seldom used.

Virus inactivator: A chemical that inactivates a virus, causing loss of infectivity.

Virus inhibitor: A chemical that prevents virus transmission without inactivating the virus; inhibition may be temporary and sometimes avoided by dilution; acts on host, not on virus.

Virus particle: The morphological form of a virus. In some viruses, it consists of a genome surrounded by a protein capsid; other virus particles have additional structures such as envelopes, tails, etc.

Virus replication: The process of forming progeny virus from input virus. It involves the expression and replication of the viral genomic nucleic acid and the assembly of progeny virus particles.

Virus-like particle: A structure resembling a virus but pathogenicity is unproven.

Virusoid: The very small circular RNA component of some isometric RNA viruses.

Virustatic: Able to inhibit viral replication, which can be resumed once the virustatic agent is removed. Actinomycin D and 5-fluorodeoxyuridase inhibit DNA viruses. Thiouracil, 5-fluorouracil, and puromycin inhibit RNA viruses. *See* Base analogue.

von Magnus phenomenon: The increase in the proportion of defective virus particles produced during the repeated passage of a virus at a high multiplicity of infection.

VPg: A genome-linked virion protein.

Western blotting: The transfer of proteins, separated on a polyacrylamide gel, to an immobilizing matrix, often nitrocellulose. Proteins are frequently transferred using electrophoresis, called *electroblotting* (*See* EBIA).

Witches' broom: Disease symptom with an abnormal, massed, brush-like development of many weak shoots or roots of mainly woody plants, arising at or close to the same point or resulting from the proliferation of buds; caused by mites, viruses, fungi (especially rusts and Taphrinales), bacteria, false mistletoes, nematodes, etc.

X bodies: Inclusion bodies in tobacco mosaic virus (TMV)-infected plant cells.

Yd (Yard): = 36 in. = 3 ft = 0.181818 rod = 0.9144 m.

Yellowing (distinct from chlorosis)**:** To make or render tissue yellow that once was green due to the underdevelopment or partial distruction of chlorophyll. *See* Chlorosis.

Yellows: A plant disease characterized by yellowing and stunting of the host plant usually caused by xylem-limited fastidious bacteria or elemental deficiencies; yellows of the cabbage family (Brassicaceae) is caused by *Fusarium oxysporum* f. sp. *conglutinans*; peach yellows is caused by a virus.

Zonal centrifugation: Centrifugation in a cylindrical or bowl-shaped rotor where the solutions are in a large central cavity instead of individual tubes or buckets; usually for large quantity gradient centrifugation. The macromolecules (e.g., viruses) separate as concentric rings (zones) in the gradient.

Zonate: Marked with concentric zones (bands or lines) of different colors and/or textures; target-like; any leaf or stem development appearing in concentric rings.

Zoochoric: Dependent on animals for dissemination or dispersal.

+ sense: Plus-sense or positive-sense. In RNA viruses, the strand that functions as a messenger (mRNA).

− sense: Minus-sense or negative-sense; nucleic acid complementary to the + strand.

INDEX

INTRODUCTION

We followed A.A. Brunt et al., 1996, *Viruses of Plants*; descriptions and lists from the VIDE database, CAB International, University Press, Cambridge for use of virus names. We changed many acronyms as pointed out in the Preface and Introduction. For example, in current literature "T" can symbolize tobacco, tomato, or turnip, so we retained "**T**" to symbolize tobacco. Then to specify other hosts we added a lower case letter, e.g., Tm for tomato, Tu for turnip, etc. Confusion also occurs when designating symptoms — i.e., "M" currently indicates mosaic, mild, mottling, and mottle. We retained "**M**" for mosaic, then used **Mt** for mottle, **Md** for mild, and **Mg** for mottling. We retained both letters and numbers to designate certain viruses — i.e., potato virus Y (**PVY**) — which are placed after the **V** for viruses; these are not easily confused with the same designation which occurs before the V. A question mark (**?**) means the virus classification indicated is most likely correct.

While making changes we observed the standard protocol:
- First position is the symbol for the name of the host plant in which the virus or disease was first observed, for example, tobacco mosaic virus (**TMV**).
- Second position, name and symbol, may designate geographical locations where the virus (disease) was first observed, i.e., wheat American striate mosaic virus (**WAStMV**).
- Second position usually (but third position if geography was used) is reserved for symbols which refer to symptoms. For example, **D** = dwarf, **Dh** = death, **Dk** = dieback, **Dl** = dapple, **Dn** = deformation, **Dp** = drop, **Ds** = disease, etc.
- The last symbol refers to etiology, or causal agent of disease; for example, **V** = virus, **Vd** = viroid, and **Ds** = disease, which should be used until the etiology has been proved.

Therefore, in order to assist the reader to correlate the acronyms we used with those commonly found in print (acronyms in parentheses) the "Standard Acronym Appendix 2" can be referred to in Matthews, R.E.F., 1991, *Plant Virology*, Academic Press, pages 685–694.

Old names (if changed) and abbreviations (if changed) are given in parentheses. The abbreviation "syn." indicates synonymous with the noted virus or disease. **Boldface** page numbers indicate the key description of the virus or disease.

A

AAaLtv, *see* Alfalfa Australian latent nepovirus
Abaca mosaic virus, AbMV, 293
AbMV, *see* Abaca mosaic virus
Abutilon mosaic geminivirus III, AnMV (AbMV), 303
AcSbcVd, *see* Avocado sunblotch viroid
AdBdFlDs, *see* Almond bud failure disease
AdCaDs, *see* Almond Calico disease
AdLnPtDs, *see* Almond line pattern disease
AdMDs, *see* Almond mosaic disease
AdNRsDs, *see* Almond necrotic ringspot disease
AgMV, *see* Agropyron mosaic potyvirus

Agropyron mosaic potyvirus, AgMV, 35, 36, 54, 85, **90**
Agropyron repens, *see* Couch grass
AkClDV, *see* Artichoke curly dwarf potexvirus
AkILtV, *see* Artichoke Italian latent nepovirus
AkLtV, *see* Artichoke latent nepovirus
AkMDs, *see* Artichoke mosaic disease
AkMtCrV, *see* Artichoke mottled crinkle tombusvirus
AkVBaV, *see* Artichoke vein banding nepovirus
AkYRsV, *see* Artichoke yellow ringspot nepovirus
Alfalfa, 99–106
Alfalfa Australian latent nepovirus, AAaLtV

(LALV), **104–105**
ALtV, *see* Alfalfa latent carlavirus
Alfalfa latent carlavirus, ALtV (ALV), **104**
Alfalfa mosaic alfamovirus, AMV, **99-104,** 105, 106, 113, 114, 116, 118, 125, 128–129, 145, 146, 149, 174, 183, 193, 198, 203, 206, 207, 208, 211, 214, 257, 258, 260, 274, 275, 303, 304, 329, 399, 461, 479, 489
Alfalfa transient streak sobemovirus, ATtSkV (LTSV), **105**
AmBlDV, *see* Arrhenatherum blue dwarf Fijivirus
Allium spp, *see* Lily Vegetables (onion, leek, shallot, garlic)
Almond, 388–389
Almond calico disease, AdCaDs, 388
  bud failure disease, AdBdFlDs, 389
  line pattern disease, AdLnPtDs (see PsNcRsV), 388
  mosaic disease, AdMDs, **389**, 390
  necrotic ringspot disease, AdNcRsDs, 388
Alsike Clover, 114
AMV, *see* Alfalfa mosaic alfamovirus
AnMV, *see* Abutilon mosaic geminivirus
Anthoxanthum mosaic potyvirus, AxMV (AnMV), **96**, 106
Anthoxanthum odoratum, *see* Sweet vernal grass
ApCLsV, *see* Apple chlorotic leafspot closterovirus
ApDlVd, *see* Apple Dapple viroid
ApGnCrDs, *see* Apple green crinkle disease
Apium graveolens, *see* Celery
Apple, 330–341
Apple chlorotic leaf spot closterovirus, ApCLsV (ACLV), 323, 326, 328, 329, 330, **332–333**, 341–342, 346–347, 355, 356, 358-359, 362, 364, 370, 371, 379, 381, 387, 389
  Dapple viroid, ApDlVd (Dapple apple viroid, DAVd), 330, **337–338**
  green crinkle disease, ApGnCrDs (AGCD), 321, **335**
  latent virus, ApLtV, syn. ApCLsV
  leaf pucker disease, ApLPkDs (ALPD), 330, **334–335**, syn. ApCLsV
  mosaic ilarvirus, ApMV, 229, 309, 321–324, 326, 328, **330–332**, 334, 356, 358, 364, 370, 379, 388, 391, 451, 452, *see* PuNcRsV
  platycarpa scaly bark disease, ApPcSyBkDs (APSBD), **340**
  ringspot disease, ApRsDs (ArsD), **336**
  rough skin disease, ApRgSknDs (ARSD), **336**
  scar skin viroid, ApScSknVd (ASSVd), 330, **337–338**, 495
  spy epinasty and decline disease, ApSpyEyDeDs, 330, 339, 345–346, *see* PrVYsDs; ApStmPgDs
  star crack disease, ApStrCkDs (ASCD), **336**
  stem grooving capillovirus, ApStmGgV (ASGV), 326, 330, **338–339**, 345, 346
  stem pitting virus, ApStmPgV (ASPD), 330, **339–340**
  Tulare mosaic virus, ApTrMV (Tulare apple mosaic virus, TAMV), **334**, 391
ApLPkDs, *see* Apple leaf pucker disease
ApMV, *see* Apple mosaic ilarvirus
ApPcSyBkDs, *see* Apple platycarpa scaly bark disease
ApRgSknDs, *see* Apple rough skin disease
ApRsDs, *see* Apple ringspot disease
ApScSknVd, *see* Apple scar skin viroid
ApSpyEyDeDs, *see* Apple spy epinasty and decline disease
ApStmGgV, *see* Apple stem grooving capillovirus
ApStrCkDs, *see* Apple star crack disease
ApStmPgV, *see* Apple stem pitting virus
ApTrMV, *see* Apple Tulare mosaic virus
Apricot, 370–374
Apricot ring pox disease, AtRPDs (AtRPD), **373–374**
  asteroid spot disease, AtAdsDs, **374**
  chlorotic leaf mottle disease, AtCLMtDs, **374**
  Moore park mottle disease, AtMooreMtDs, **374**
  mosaic disease, AtMDs (see PhMDs), **374**
  pucker leaf disease, AtPkLDs, 374
Arabis mosaic nepovirus, ArMV, 116, 162, 167, 173, 207, 214, 306, 309, 325, 367, 377, 409, 440, 441, 442, 447, 457, 459, 461, 477, 479, 484, 486, 487
Arachis hypogaea, *see* Peanut

ArMV, *see* Arabis mosaic nepovirus
Arrhenatherum blue dwarf Fijivirus, AmBlDV (ABDV), **90–91**
Arrhenatherum elatius, *see* False oatgrass
Artichoke, 208–211
Artichoke curly dwarf potexvirus, AkClDV (AkCDV), **210**
  Italian latent nepovirus, AkILtV (AkILV), **208–209**, 216, 479, 487
  latent nepovirus, AkLtV (AkLV), **210**
  mosaic disease, AkMDs (AkMD), **210**
  mottled crinkle tombusvirus, AkMtCrV (AMCV), **210**
  vein banding nepovirus, AkVBaV (AVBV), **209**
  yellow ringspot nepovirus, AkYRsV (AYRSV), 186, 198, 206, **209–210**
AtAdsDs, *see* Apricot asteroid spot disease
AtCLMtDs, *see* Apricot chlorotic leaf mottle disease
AtMDs, *see* Apricot mosaic disease
AtMooreMtDS, *see* Moore park mottle disease
AtPkLDs, *see* Apricot pucker leaf disease
AtRPDs, *see* Apricot ring pox disease
ATtSkMV, *see* Alfalfa transient streak mosaic virus
Avena Sativa, *see* Oat
Avocado, 392–394
Avocado sunblotch viroid, AcSbcVd (ASBVd), **392–394**
AxMV, *see* Anthoxanthum mosaic potyvirus

B

BaBtMV, *see* Banana bract mosaic potyvirus
BaByTpV, *see* Banana bunchy top virus
Banana, 394–396
Banana bract mosaic potyvirus, BaBtMV (BBMV), **394–395**
  bunchy top nanavirus, BaByTpV (BBTV), 325, **394**
  streak badnavirus, BaSkV, 297
Barley, 38–48
Barley stripe mosaic hordeivirus, BySpMV (BSMV), 36, **39–42**, 54; *see* WCSkV
  yellow dwarf luteovirus, ByYDV (BYDV), 22, 23, 36, 37, **42–45**, 53, 54, 65–66, 70, 85, 86, 89, 90, 92, 94, 96, 97, 157, 297
  yellow mosaic potyvirus, ByYMV (BaYMV), **45–46**
  yellow striate mosaic rhabdovirus, ByYStMV (BYSMV), *see* wheat chlorotic streak rhabdovirus, 35, 37, **46–47**, 54, 70
BaSkV, *see* Banana streak badnavirus
BbCaDs, *see* Blackberry calico disease
BBMdMV, *see* Broad bean mild mosaic virus
BBMtV, *see* Broad bean mottle bromovirus
BBNV, *see* Broad bean necrosis furovirus
BBSV, *see* Broad bean stain comovirus
BBTMV, *see* Broad bean true mosaic comovirus
BBWV, *see* Broad bean wilt fabavirus
BCmMV, *see* Bean common mosaic potyvirus
BcNcYsV, *see* Broccoli necrotic yellows rhabdovirus
BCuInVgDs, *see* Black currant infectious variegation disease
BCuRvDs, *see* Black currant reversion disease
BCuVClVNtDs, *see* Black currant vein clearing and vein net disease
BCuYsDs, *see* Black currant yellows disease
Bean, 174–186
Bean common mosaic potyvirus, BCmMV (BCMV), 117, 125, **174–176**, 203, 237
  golden mosaic geminivirus III, BGMV 134, **180–181**, 182
  leafroll luteovirus, BLrV (BLRV), 116, 117, 118, 174, **179–180**, 193, 198, 203
  mild mosaic carmovirus, BMdMV (BMMV), **181–182**
  pod mottle comovirus, BPoMtV (BPMV), **182**, 303, 304
  rugose mosaic comovirus, BRgMV (BRMV), **182–183**
  summer death geminivirus, BSmDhV (BSDV), **183**
  summer mosaic virus, BsmMV, **203**
  southern mosaic sobemovirus, BSuMV (SBMV) **180**, 203, 303
  yellow mosaic potyvirus, BYMV, 98, 106, 110–114, 116, 117, 118, 125, 174, **176–179**, 194, 198, 203, 211, 240, 279, 299, 303, 304
BeCpMV, *see* Blackeye cowpea mosaic potyvirus
Beet, 276–292

Beet curly top geminivirus II, BtCuTpV
   (BCTV), 148, 167, 186, 208, 218, 257,
   275, **276–279**
   leaf curl rhabdovirus, BtLCuV (BLCV), **290**
   mild yellowing luteovirus, BtMdYgV
      (BMYV), 218, 276, **290–291**
   mosaic potyvirus, BtMV, 193, 218, 276,
      279–280
   necrotic yellow vein furovirus, BtNtYVV
      (BNYVV), 218, 276, **280–286**
   pseudo yellows closterovirus, BtPdYsV
      (BPYV), 167, 216, 229, **291**
   western yellows luteovirus, BtWsYsV
      (BWYV), 157, 211, 214, 218, 276,
      286–287, 290, 299, 303
   yellow net luteovirus, BtYNtV (BYNV), **291**
   yellow stunt closterovirus, BtYSnV (BYSV),
      216, **291–292**
   yellows closterovirus, BtYsV (BYV), 218,
      229, 276, **288–290**, 291
Beta vulgaris, see Beet
BgMtV, *see* Blackgram mottle carmovirus
BGMV, *see* Bean golden mosaic geminivirus III
Bidens mottle potyvirus, BiMtV (BiMoV),
   211, 216
BiMtV, *see* Bidens mottle potyvirus
Birdsfoot trefoil, 117
Blackberry calico disease, BbCaDs (BCD),
   456–457
Black currant, 458–460
Black currant infectious variegation disease,
   BCuInVgDs (BDIVD), **459**
   reversion disease, BCuRvDs (BCRD),
      459–460
   vein clearing and vein net disease, BCuV
      ClVNtDs (?), **459**
   yellows disease, BCuYsDs (BCYD), **460**
Black raspberry, 452–457
Black raspberry latent ilarvirus, BRbLtV
   (BRLV), **454**
   necrosis virus, BRbNV (BRNV), 325, 444,
      452–454
   streak disease, BRbSkDs, **455**
Blackeye cowpea mosaic potyvirus, BeCpMV
   (BlCMV), **202**
Blackgram mottle carmovirus, BgMtV
   (BMoV), **203**
BlbLMtV, *see* Blueberry leaf mottle virus
BlbMDs, *see* Blueberry mosaic disease

BlbReRsV, *see* Blueberry red ringspot
   caulimovirus
BlbSsV, *see* Blueberry shoestring sobemovirus
BLrV, *see* Bean leafroll luteovirus
Blueberry, 462–467
Blueberry leaf mottle nepovirus, BlbLMtV
   (BbLMV), **462–464**, 480
   mosaic disease, BlbMDs (BBMD), **466**, 467
   red ringspot caulimovirus, BlbReRsV
      (BRRV), **464**
   shoestring sobemovirus, BlbSsV (BSSV),
      464–465
BMdMV, *see* Bean mild mosaic carmovirus
BmYMV, *see* Bramble yellow mosaic potyvirus
BPoMtV, *see* Bean pod mottle comovirus
Bramble yellow mosaic potyvirus, BmYMV
   (BrYMV), **455–456**
Brassica oleracea, see Cabbage, Cauliflower,
   and Brussels sprouts
Brassica rapa, see Turnip
BRbLtV, *see* Black raspberry latent ilarvirus
BRbNV, *see* Black raspberry necrosis virus
BRbSkDs, *see* Black raspberry streak disease
BRgMV, *see* Bean rugose mosaic comovirus
BrMV, *see* Brome (grass) mosaic bromovirus
Broad Bean, 194–198
Broad bean mild mosaic virus, BBMdMV
   (BBMMV), 193, **198**
   mottle bromovirus, BBMtV (BBMV),
      194–196, 206
   necrosis furovirus, BBNV, **197**
   stain comovirus, BBSV, 114, 117, 193,
      197–198
   true mosaic comovirus, BBTMV, 193, **196**
   wilt fabavirus, BBWV, 125, 134, 148, 194,
      196–197, 208, 216, 219
Broccoli necrotic yellows rhabdovirus,
   BcNcYsV (BNYV), **153**
Brome (Grass), 85–86
Brome (grass) mosaic bromovirus, BrMV
   (BMV), 24, 38, 54, 71, **85–86**, 90, 92, 96
Bromus inermis, see Brome grass
Brussels Sprouts, 150–153
BSmMV, *see* Bean summer mosaic virus
BSmDhV, *see* Bean summer death geminivirus
BSuMV, *see* Bean southern mosaic
   sobemovirus
BtClTpV, *see* Beet curly top geminivirus II
BtLCuV, *see* Beet leaf curl rhabdovirus

Index

BtMdYgV, *see* Beet mild yellowing luteovirus
BtMV, *see* Beet mosaic potyvirus
BtNtYVV, *see* Necrotic yellow vein furovirus
BtWsYsV, *see* Beet western yellows luteovirus
BtPdYsV, *see* Beet pseudo yellows closterovirus
BtYNtV, *see* Beet yellow nut luteovirus
BtYSnV, *see* Beet yellow stunt closterovirus
BtYsV, *see* Beet yellows closterovirus
BYMV, *see* Bean yellow mosaic potyvirus
BySpMV, *see* Barley stripe mosaic hordeivirus
ByYDV, *see* Barley yellow dwarf luteovirus
ByYMV, *see* Barley yellow mosaic potyvirus
ByYStV, *see* Barley yellow striate mosaic rhabdovirus

C

Cabbage, 150–153
Cabbage black ringspot virus, CgBRsV (TuMV), 125, **150–151**, *see* TuMV
CaIRsv, *see* Carnation Italian ringspot tombusvirus
CaLtV, *see* Carnation latent carlavirus
Capsicum annuum, *see* Pepper
Carica papaya, *see* Papaya
Carnation Italian ringspot tombusvirus, CaIRsV (CIRSV), **381**
  latent carlavirus, CaLtV (CLV), 257
  ringspot dianthovirus, CaRsV, 325, 340, 346, 359
Carrot, 203–206
Carrot latent rhabdovirus, CoLtV (CaLV), **205**
  mosaic potyvirus, CoMV (CtV), 125, **205**
  mottle togavirus, CoMtV (CMoV), **203–204**, 205, 208
  red leaf luteovirus, CoReLV (CaRLV), **204–205**
  thin leaf potyvirus, CoTnLV (CTLV), **205**
  yellow leaf closterovirus, CoYLV (CYLV), **206**
CaRsV, *see* Carnation ringspot dianthovirus
Castanea sativa, *see* Chestnut
Cauliflower, 150–153
Cauliflower mosaic caulimovirus, CfMV (CaMV), 125, 150, **151–153**, 437
CbRsDs, *see* Cranberry ringspot disease
Celery, 206–207

Celery latent potyvirus, CeLtV (CeLV), **207**
  mosaic potyvirus, CeMV, 125, 203, **206**, 208
  ringspot virus, CeRsV (CeRsV), **207**
  yellow net virus, CeYNtV (CeYNV), **207**
  yellow spot luteovirus, CeYSpoV, **207**
  yellows virus, CeYsV (CeYSV), **207**
CeLtV, *see* Celery latent potyvirus
CeMV, *see* Celery mosaic potyvirus
Cereal chlorotic mottle rhabdovirus, CrCMtV (CeCMV), 23, 25, **48**, 54
  northern mosaic cytorhabdovirus, CrNnMV (Northern cereal mosaic virus, NCMV), 48, 53
  tillering disease Fijivirus, CrTlDsV (CTDV), 13, **47–48**, 53, 54, 90, 97
CeRsV, *see* Celery ringspot virus
CeYNtV, *see* Celery yellow net virus
CeYSpoV, *see* Celery yellow spot luteovirus
CfMV, *see* Cauliflower mosaic caulimovirus
CgBRsV, *see* Cabbage black ringspot virus
CGnMtMV, *see* Cucumber green mottle mosaic tobamovirus
ChBCnDs, *see* Cherry black canker disease
ChEpV, *see* Cherry Epirus virus
Cherry, 374–388
Cherry black canker disease, ChBCnDs (CBCD), 385
  epirus virus, ChEpV (Epirus cherry virus, EpCV), **378**
  European rasp leaf nepovirus, ChERpLV, 377, 378, 382
  green ring mottle disease, ChGnRMtDs (CGRMtD), **382**, 387
  leaf mottle disease, ChLMtDs (CLMtD); 325, **385**
  leafroll nepovirus, ChLrV (CLRV), 324, 326, 327, 348, **374–376**, 377–378, 387, 389–391, 409, 447
  little cherry disease, ChLlChDs (CLCD); **384**
  necrotic ringspot disease, ChNcRsDs (CNRsD), 322, 383, **386**
  rasp leaf nepovirus, ChRpLV (CRLV), 325, 340, **376–378**
  rusty mottle disease, ChRyMtDs (CryMtD), 321, 382–384
  spur cherry disease, ChSprChDs (CSCD), 385
  twisted leaf disease, ChTdLDs (CTLD), 385
  Utah Dixie rusty mottle disease,

ChUDxRyMtDs, 374
ChERpLV, *see* Cherry European rasp leaf nepovirus
Chestnut, 392
Chestnut mosaic virus, CtMV (ChMV), **392**
  line pattern mosaic disease, CtLnPtMDs (ChLPMD), **392**
ChGnRMtDs, *see* Cherry green ring mottle disease
Chicorum endivia, *see* Endive
Chicory yellow mottle nepovirus, CcyYMtV (ChYMV), 208
ChLlChDs, *see* Cherry little cherry disease
ChLMtDs, *see* Cherry leaf mottle disease
ChLrV, *see* Cherry leafroll nepovirus
ChNcRsDs, *see* Cherry necrotic ringspot disease
ChRpLV, *see* Cherry rasp leaf nepovirus
ChRyMtDs, *see* Cherry rusty mottle disease
Chrysanthemum stunt viroid, CmSnVd, 400
ChSprChDs, *see* Cherry spur cherry disease
ChTdLDs, *see* Cherry twisted leaf disease
ChUDxRuMtDs, *see* Cherry Utah Dixie rusty mottle disease
CicRsV, *see* Crimson clover ringspot virus
Citrus, 396–407
Citrus cachexia-xylorporosis viroid, CsCxDs (CCaVd), **403**
  concave gum disease, CsCgDs (CCGD), **404**
  crinkly leaf ilarvirus, CsCrLV (CCLV), 399
  cristacortis disease, CsCcsDs (CCD), **404**
  exocortis viroid, CsExVd (CEVd), **400–401**, 402, 477, 495–496
  gummy bark disease, CsGmBkDs (CGBD), **405**
  impietrature disease, CsImDs (CID), **405**
  leafroll virus, CsLrV, 399
  leaf rugose ilarvirus, CsLRgsV (CiLRV), **396–397**, 399
  psorosis disease, CsPsDs (CpsD), **405–406**
  ringspot disease, CsRsDs (CrsD), **406**
  tristeza closterovirus, CsTzV (CTV), 321, 325, 329, **397–399**
  variegation ilarvirus, CsVgV (CVV), **399**, 400
  vein enation luteovirus, CsVEnV (CVED), **406–407**
  woody gall virus, WdGlV, 406–407
Citrus spp, *see* Citrus

CkMdMV, *see* Cocksfoot mild mosaic sobemovirus
CkMtV, *see* Cocksfoot mottle sobemovirus
CkSkV, *see* Cocksfoot streak potyvirus
Clover yellow mosaic potexvirus, ClYMV, **112**, 334
  yellow vein potyvirus, ClYVV, 98, 110, **112–113**, 303
  subterranean mottle sobemovirus, ClSrMtV (SuCMtV), 115
ClSrMtV, *see* Clover subterranean mottle sobemovirus
CLsV, *see* Cucumber leaf spot carmovirus
ClYMV, *see* Clover yellow mosaic potexvirus
ClYVV, *see* Clover yellow vein potyvirus
CmSnVd, *see* Chrysanthemum stunt viroid
CMV, *see* Cucumber mosaic cucumovirus
CNV, *see* Cucumber necrosis tombusvirus
Cocksfoot grass, 86–90
Cocksfoot mild mosaic sobemovirus, CkMdMV (CMMV), 54, 85, **86**, 92, 93, 96 (see TtMtV)
  mottle sobemovirus, CkMtV (CoMV), 54, 85, **87–88**
  streak potyvirus, CkSkV (CSV), 85, 86, **88–89**, 92, 96
Coconut cadang-cadang viroid, CcCcgVd (CCCVd), **409–410**
CoLtV, *see* Carrot latent rhabdovirus
CoMV, *see* Carrot mosaic potyvirus
CoMtV, *see* Carrot mottle togavirus
CoReLV, *see* Carrot red leaf luteovirus
Corn stunt spiroplasma, (CSS), *see* Maize stunt spiroplasma
Corylus avelana, *see* Hazelnut
CoTnLV, *see* Carrot thin leaf potyvirus
Couch Grass, 90
Courgette, *see* Melon
COWPEA, 199–203
Cowpea aphid-borne mosaic potyvirus, CpApBnMV (CABMV), **199**
  chlorotic mottle bromovirus, CpCMtV (CCMV), **199–200**, 240, 303
  mild mottle carlavirus, CpMdMtV (CPMMV), **202**, 239–240
  mosaic comovirus, CpMV (CPMV), **200–201**
  mottle carmovirus, CpMtV (CPMoV), **202**
  severe mosaic comovirus, CpSeMV

(CPSMV), **201–202**
CoYLV, *see* Carrot yellow leaf closterovirus
CpApBnMV, *see* Cowpea aphid-borne mosaic potyvirus
CpCMtV, *see* Cowpea chlorotic mottle bromovirus
CpMdMtV, *see* Cowpea mild mottle carlavirus
CpMtV, *see* Cowpea mottle carmovirus
CpMV, *see* Cowpea mosaic comovirus
CpSeMV, *see* Cowpea severe mosaic comovirus
Cranberry ringspot disease, CbRsDs (CbRsD), **466**
CrCMtV, *see* Cereal chlorotic mottle rhabdovirus
Crimson Clover, 114–115
Crimson clover ringspot virus, CicRsV, 115
CrNnMV, *see* Cereal northern mosaic cytorhabdovirus
CrnSnSpp, *see* Corn stunt spiroplasma
CrTlDsV, *see* Cereal tillering disease Fijivirus
CsCcsDs, *see* Citrus cristacortis disease
CsCgDs, *see* Citrus concave gum disease
CsCrLV, *see* Citrus crinkly leaf ilarvirus
CsCxVd, *see* Citrus cachexia-xyloporosis viroid
CsExVd, *see* Citrus exocortis viroid
CsGmBkDs, *see* Citrus gummy bark disease
CsImDs, *see* Citrus impietrature disease
CsPsDs, *see* Citrus psorosis disease
CsLRgV, *see* Citrus leaf rugose ilarvirus
CsRsDs, *see* Citrus ringspot disease
CsTzV, *see* Citrus tristeza closterovirus
CsVEnDs, *see* Citrus vein enation disease
CsWdGlDs, *see* Citrus woody gall disease
CsVgV, *see* Citrus variegation ilarvirus
CtLnPtMDs, *see* Chestnut line pattern mosaic disease
CtMV, *see* Chestnut mosaic virus
Cucumber, 162–173
Cucumber green mottle mosaic tobamovirus, CGnMtMV (CGMMV), 162, **166**, 172, 473
  leaf spot carmovirus, CLsV (CLSV), **162–163**
  mosaic cucumovirus, CMV, 24, 54, 71, 105–106, 113, 114, 116, 117, 125, 129–130, 131, 146, 147, 149, 157, 162, **163–165**, 172, 174, 184, 194, 198, 203, 206, 207, 208, 211, 215, 218, 229, 238–239, 257, 258, 260, 275, 292, 299, 303, 304, 329, 359, 367, 381, 389, 396, 409, 444, 457, 458, 460
  necrosis tombus virus, CNV (CuNV), **165–166**, 167
  pale fruit viroid, CpaFtVd, 307, 308
Cucumis melo, *see* Melon
Cucurbita maxima, *see* Melon
Cucurbita pepo, *see* Melon
Cucumis sativus, *see* Cucumber
Cydonia oblonga, *see* Quince
Cymbidium ringspot tombus virus, CyRsV (CyRSV), **116**
Cynara scolymus, *see* Artichoke
CyRsV, *see* Cymbidium ringspot.

D

Dactylis glomerata, *see* Cocksfoot grass
Daucus carota, *see* Carrot
Dapple apple viroid, *see* Apple Dapple viroid, ApDlVd

E

Endive, 211
Eggplant, 148–150
Eggplant mosaic tymovirus, EMV, **148–149**
  mottled dwarf rhabdovirus, EmtDV (EMDV), 125, **149**
EMV, *see* Eggplant mosaic tymovirus
European wheat striate mosaic tenuivirus, *see* Wheat European striate mosaic tenuivirus

F

False oatgrass, 90–91
FeLSkV, *see* Festuca leaf streak rhabdovirus
Fescuegrass, 91–92
FeNV, *see* Festuca necrosis closterovirus
Festuca leaf streak rhabdovirus, FeLSkV (FLSV), 91, **92**
  necrosis closterovirus, FeNV (FNV), **91**, 96
Festuca spp, *see* Fescuegrass
Fig, 407–408
Ficus carica, *see* Fig

Fig mosaic potyvirus, FMV (FMD), **407–408**
Fiji disease Fijivirus, FjDsV (FDV), 14, 71, **293–295**, *see* Sugarcane
FMV, *see* Fig mosaic potyvirus
Foxtail Millet, 92
Foxtail mosaic potexvirus, FxMV (FoMV), **92–93**
Fragaria chiloensis, *see* Strawberry
FxMV, *see* Foxtail mosaic potexvirus

G

Garlic latent carlavirus, GcLtV (GLV), 158, **162**
  mosaic potyvirus, GcMV (GMV), 158, **161–162**
GAaVd, *see* Grapevine Australian viroid
GAdMDs, *see* Grapevine asteroid mosaic disease
GBgLtV, *see* Grapevine Bulgarian latent nepovirus
GbMDs, *see* Gooseberry mosaic disease
GbVBaDs, *see* Gooseberry vein-banding disease
GlBVd, *see* Grapevine 1B viroid
GcLtV, *see* Garlic latent carlavirus
GChMV, *see* Grapevine chrome mosaic nepovirus
GCyBkV, *see* Grapevine corky bark ? closterovirus
GFkV, *see* Grapevine fleck virus
GFLV, *see* Grapevine fanleaf nepovirus
GLrV, *see* Grapevine leafroll ? closterovirus
Glycine max, *see* Soybean
Gooseberry, 457–458
Gooseberry mosaic disease, GbMDs (GMD), **458**
  vein-banding disease, GbVBaDs (GVBD), **458**
Grapevine, 477–496
Grapevine asteroid mosaic disease, GAdMDs (GAMD), 479, **490**, 495
  Australian viroid, GAaVd (Australian grape vine viroid, AGVd), 307, **495–496**
  1B viroid, GlBVd (GVd1B), **495**, 496
  Bulgarian latent nepovirus, GBgLtV (GBLV), 477, 479, **480–481**
  2 viroid, G2Vd, **495**
  3 viroid, G3Vd, **496**
  chrome mosaic nepovirus, GChMV (GCMV), 477, 479, **481–482**
  corky bark ? closterovirus, GCyBkDs (GCBD), 477, 479, 480, **490–491**
  fanleaf nepovirus, GFLV, 477, 478, 479, 480, **482–486**, 487, 490, 493, 494
  flavescence doreé, 489 (phytoplasma)
  fleck virus, GFkV (GfkD), 477, 479, 490, **491–492**
  leafroll ? closterovirus, GLrV (GLRD), 477, 478, 479, 480, 490, **492–493**
  Pierce's disease, 489 (bacterium)
  speckle disease, *see* GYSpeVd
  stem pitting (legno riccio) (rugose wood) closterovirus, GStmPgV, 477, 479, 490, **493–494**
  vein mosaic disease, GVMDs, 479, 490
  vein necrosis disease, GVNDs, 479
  virus A, GVA, 492
  virus B, GVB, 492
  yellow speckle viroid, GYSpeVd (GYVd), 479, **494–495**, *see* GYSpeVd
  yellow speckle viroid, GYSpeVd, 477, 494, 495
Groundnut, *see* Peanut
Groundnut chlorotic spot ? potexvirus, GtCSpoV, 230
  crinkle carlavirus, GtCrV, 230, 240
  eyespot potyvirus, GtEsV, 230
  rosette assistor luteovirus, GtRoAsV (GRAV), 230, 235
  rosette disease, GtRoDs, 230, 235
  rosette umbravirus, GtRoV(GRV), 230, 235
  veinal chlorosis virus, GtVCV, 230
  yellow mosaic virus, GtYMV (see BGYMV), 230
  yellow mottle tymovirus, GtYMtV (PYMV), 230
GStmPgV, *see* Grapevine stem pitting closterovirus
GtCSpoV, *see* Groundnut Chorotic spot ? potexvirus
GtCV, *see* Groundnut crinkle carlavirus
GtEsV, *see* Groundnut eyespot potyvirus
GtRoAsV, *see* Groundnut rosette asistor luteovirus
GtRoDs, *see* Groundnut rosette disease
GtRoV, *see* Groundnut rosette umbravirus

GtVCV, *see* Groundnut veinal chlorosis virus
GtYMV, *see* Groundnut yellow mosaic virus
GtYMtV, *see* Groundnut yellow mottle tymovirus
GugMV, *see* Guinea grass mosiac potyvirus
Guinea grass mosaic potyvirus, GugMV (GGMV), **93–94**
G2Vd, *see* Grapevine 2 viroid
G3Vd, *see* Grapevine 3 viroid
GVMDs, *see* Grapevine vein mosaic disease
GVNDs, *see* Grapevine vein necrosis disease
GYSpeVd, *see* Grapevine yellow speckle viroid

H

HALtV, *see* Hop American latent carlavirus
Hazelnut, 391–392
Hazelnut dieback disease, HtDkDs (HDD), **392**,
  line pattern disease, HtLnPtDs (HLPD), **391**,
  mosaic disease, HtMDs (HMD), **391**
HCDsV, *see* Hop chlorotic disease virus
Helianthus annuus, *see* Sunflower
Henbane mosaic potyvirus, HnMV (HMV), 263
HLtV, *see* Hop latent carlavirus
HMV, *see* Hop mosaic carlavirus
HMCsV, *see* Hop mosaic chlorosis virus
HNMV, *see* Henbane mosaic potyvirus
Hop, 305–310
Hop American latent carlavirus, HALtV (AHLV), 306, **307**
  chlorotic disease virus, HCDsV, **308**
  latent carlavirus, HLtV (HvLV), **306–307**
  mosaic carlavirus, HMV (HpMV), **306**, 307
  mosaic chlorosis virus, HMCsV, **308**
  stunt viroid, HSnVd (HSVd), 167, 306, **307–308**, 477, 495–496
  yellow net virus, HYNtV, 309
Hordeum vulgare, *see* Barley
Horseradish mosaic virus, HrMV, 151
HrMV, *see* Horseradish mosaic virus
HrMV, *see* Horseradish mosaic virus
HSnVd, *see* Hop stunt viroid
HtDkDs, *see* Hazelnut dieback disease
HtLnPtDs, *see* Hazelnut line pattern disease
HtMDs, *see* Hazelnut mosaic disease
Humulus lupulus, *see* Hop

HYNtV, *see* Hop yellow net virus

I

Indian peanut clump furovirus, PtIdCpV, *see* Peanut Indian clump

J

Johnsongrass, 97
Johnsongrass mosaic potyvirus, JgMV (JGMV), 2, 4, 9, 21, 22, 67, 71, **98–99**, 293
Juglans regia, *see* Walnut

L

Lactuca sativa, *see* Lettuce
Leek, 160, 162
LBiVV, *see* Lettuce big vein varicosavirus
Leek yellow stripe potyvirus, LkYSpV (LYSV), 157, **160–161**
Legume yellows geminivirus, LgYsV, 157
Lettuce, 211–216
Lettuce big vein varicosavirus, LBiVV (LBVV), 211, **212–213**
  infectious yellows closterovirus, LInYsV, 213, 214
  mosaic potyvirus, LMV, **211–212**, 218
  necrosis virus, LNV, **213–214**
  necrotic yellows cytorhabadovirus, LNcYsV (LNYV), **212**
LgYsV, *see* Legume yellows geminivirus
Lily Vegetables, 157–162
LInYsV, *see* Lettuce infectious yellows closterovirus
LkYSpV, *see* Leek yellow stripe potyvirus
LMV, *see* Lettuce mosaic potyvirus
LNV, *see* Lettuce necrosis virus
LNcYsV, necrotic yellows rhabdovirus
LoEnV, *see* Lolium enation Fijivirus
Lolium enation Fijivirus, LoEnV, *see* Ryegrass enation virus, RgEnV
  mottle virus, LoMtV (LoMV), 54, **94**, *see* Ryegrass mottle virus, RgMtV
Lolium multiflorum, *see* Ryegrass
Lolium perenne, *see* Ryegrass
LoMtV, *see* Lolium mottle virus

Lotus corniculatus, see Birdsfoot trefoil
Lucerne Australian latent nepovirus, AAaLtV, **104**, *see* Alfalfa Australian latent nepovirus
    transient streak sobemovirus, AtTSkV, **105**, *see* Alfalfa transient streak virus
Lycopersicon lycopersicum, see Tomato

M

Maize, 1–24
Maize bushy stunt phytoplasma, MBuSnPy (MBSM), 11
    chlorotic dwarf waikavirus, MCDV, **10–13**, 70
    chlorotic mottle machlomovirus, MCMtV (MCMV), **19**, 70
    dwarf mosaic potyvirus, MDMV, **1–10**, 11, 12, 21, 22, 24, 66-70, 86, 90, 92, 93, 94, 97, 98, 293, 296
    leaf fleck virus, MLFkV (MLFV), **21**
    mosaic nucleorhabdovirus, MMV, **15–16**, 68, 70
    rayado fino marafivirus, MRaFnV (MRFV), **20**
    rough dwarf Fijivirus, MRgDV (MRDV), **13–15**, 48, 53, 90, 96
    streak geminivirus I, MSkV (MSV), **17–18**, 53, 70
    stripe tenuivirus, MSpV (MStpV), **20**, 70
    stunt spiroplasma, MSnSpa, 11, 12
    vein enation virus, MVEnV (MVEV), 70
    white line mosaic virus, MWtLnMV (MWLMV), **19**
Malva yellows luteovirus, MaYsV (MvYV), 157, 290
Malus sylvestris, see Apple
MaYsV, *see* Malva yellows luteovirus
MBuSnPy, *see* Maize bushy stunt phytoplasma
MbYMV, *see* Mung bean yellow mosaic geminivirus III
MCDV, *see* Maize chlorotic dwarf waikavirus
MCMtV, *see* Maize chlorotic mottle machlomovirus
MDMV, *see* Maize dwarf mosaic potyvirus
Medicago sativa, see Alfalfa
Melilotus latent rhabdovirus, MsLtV (MLV), **118** (*see* StcNcMV)

Melon, 167–173
Melon necrotic spot carmovirus, MnNcsV (MNSV), 162, **167–168**
MMV, *see* Maize mosaic rhabdovirus
MnNcsV, *see* Melon necrotic spot carmovirus
Molinia, 93
Molinia coerulea, see Molinia
Molinia streak virus, MoSkV, 93
MSnSpa, *see* Maize stunt spiroplasma
MoSkV, *see* Molinia streak virus
MraFnV, *see* Maize rayado fino marafivirus
MRgDV, *see* Maize rough dwarf Fijivirus
MSkV, *see* Maize streak geminivirus I
MSnSpa, *see* Maize stunt spiroplasma
MSpV, *see* Maize stripe tenuivirus
MsLtV, *see* Melilotus latent rhabdovirus
Mung bean yellow mosaic geminivirus III, MbYMV (MYMV), **203**
Musa paradisiaca, see Banana
MVEnV, *see* Maize vein enation virus
MWtLnMV, *see* Maize white line mosaic virus

N

Nicotiana tabacum, see Tobacco
Northern cereal mosaic rhabdovirus, *see* cereal northern mosaic virus

O

Oat, 48–53
Oat blue dwarf marafivirus, OtBlDV (OBDV), **48–49**, 54
    golden stripe furovirus, OtGSpV (OGSV), **52** (see OtSoSpV)
    mosaic bymovirus, OtMV (OMV), 48, **51–52**, 70
    necrotic mottle rymovirus, OtNcMtV (ONMV), 48, **52**
    pseudo-rosette virus, OtPdRoV (OPRV), 48, **51**, 54, 66, 70, 90, 93, 94
    soil-borne stripe virus, OtSoSpV (OsbSpV), **52** (see OtGSpV)
    sterile dwarf Fijivirus, OtSrDV (OSDV), **50**, 54, 90, 91, 94, 96, 97
    striate mosaic rhabdovirus, OtStMV (OStMV), 25, **51–52**
Olea europea, see Olive

Olive, 408–409
Olive latent ringspot nepovirus, OlLtRsV (OLRSV), **408–409**
    latent ? sobemovirus 1, OlLtV1 (OLV1), **409**
    latent ? ourmiavirus 2, OlLtV2 (OLV2), **409**
OlLtRsV, see Olive latent ringspot nepovirus
OlLtV1, see Olive latent ? sobemovirus 1
OlLtV2, see Olive latent ? ourmiavirus 2
Onion, 158
Onion yellow dwarf potyvirus, OYDV, 125, **158–59**, 160, 161
Onobrychus spp, see Sanfoin 118
OtBlDV, see Oat blue dwarf marafivirus
OtGSpV, see Oat golden stripe furovirus
OtMV, see Oat mosaic bymovirus
OtNcMtV, see Oat necrotic mottle rymovirus
OtPdRoV, see Oat pseudo-rosette virus
OtSoSpV, see Oat soil-borne stripe virus
OtSrDV, see Oat sterile dwarf Fijivirus
OtStMV, see Oat striate mosaic rhabdovirus
OuMnV, see Ourmia melon virus
Ourmia melon virus, OuMnV (OuMV), **378**
OYDV, see Onion yellow dwarf potyvirus

P

PAcMV, see Potato aucuba potexvirus
PAdMV, see Petunia asteriod mosaic tombusvirus
Palm, 409–410
PaMV, see Papaya mosaic potexvirus
Pangola stunt Fijivirus, PoSnV (PaSV), 13
Panic grasses, 93–94
PAnLtV, see Potato Andean latent virus
Panicum (Panic grass) mosaic sobemovirus, PgMV (PMV), 71
Panicum spp., see Panic grasses
Papaya, 410–413
Papaya mosaic potexvirus, PaMV (PMV), **410**, 413
    ringspot potyvirus, PaRsV (PapMV), 167, 174, 325, 330, **410–413**
Parsley, 208
Parsley latent rhabdovirus, PyLtV (PLV), **208**
    ? potexvirus 5, PyV5 (PV5), **208**
Parsnip, 207–208
Parsnip mosaic potyvirus, PnpMV (ParV), **207**

    yellow fleck sequivirus, PnpYFkV (PYFV), **208**
PaRsV, see Papaya ringspot potyvirus
Passionfruit woodiness potyvirus, PfWdV (PWV), 240
Pastinaca sativa, see Parsnip
Pea, 186–194
Pea early browning tobravirus, PeElBrV (PEBV), 106, 113, 114, 116, 174, **186–187**, 198
    enation mosaic enamovirus, PeEnMV (PEMV), 106, 107, 113, 114, 117, **187–189**, 198
    leafroll virus, PeLrV, 106, 116, 117, see Bean leafroll virus
    green mottle comovirus, PeGnMtV (PGMV), **184**
    mild mosaic comovirus, PeMdMV (PMMV), **192**, 271
    mosaic (common) potyvirus, PeCmMV (*see* Bean yellow mosaic virus), 106, 113, 114, 177, 186, 303
    rhabdovirus, PeRbV, 192
    seedborne mosaic potyvirus, PeSdbMV (PSbMV), 106, **189–190**, 198
    streak carlavirus, PeSkV (PeSV), syn. WclMV, **190–192**, 303
    stunt virus, PeSnV, 192, 193
    wilt virus, syn. WclMV 303
Peach, 360–369
Peach asteroid spot disease, PhAdsDs (PhASD), 359, **368**
    bark and wood grooving disease, PhBkWdGgDs (PhBWGD), 369
    enation ? nepovirus, PhEnV (PhEV), **361**
    latent mosaic viroid, PhLtMVd (PhLMVd), 325, **362**
    line pattern and leaf curl V, PhLnPtLElV, 361–362
    leaf necrosis disease, PhLNDs (PhIND), 359
    mosaic disease, PhMDs (PhMD), 325, 327–328, **367–368**, 374
    oil blotch disease, PhOlBhDs (PhOBD), 369
    pseudo stunt disease, PhPdSnDs (PhPsSD), 359, 500
    rosette mosaic nepovirus, PhRoMV (PRMV), 325, **360–361**, 465, 467, 479, 487

seedling chlorosis disease, PhSgCsDs (PhSCD), 369
stem pitting disease, PhStmPgDs, 366, 367, see TmRsV
stubby twig disease, PhSbTwDs (PhSTD), 369
wart disease, PhWrDs (PhWD), 368
yellow bud mosaic disease, PhYBdMDs, 366, 367, see TmRsV
yellow mosaic disease, PhYMDs (PhYMD), 369

Peanut, 229–240
Peanut bud necrosis virus, PtBdNV (PBNV) (and Tomato spotted wilt virus, TmSpoWV), **229–232**
chlorotic streak caulimovirus, PtCSkV (PClSV), 230, **240**
clump pecluvirus, PtCpV (PCV), 71, 230, **232–234**, 296
green mosaic poptyvirus, PtGnMV, 230
Indian clump virus, PtIdCpV, 232-234, see peanut clump pecluvirus
mottle potyvirus, PtMtV (PeMoV), 184, 186, **236**, 237, 303, 305
rosette assistor virus (see groundnut rosette assistor virus)
rosette disease (see groundnut rosette disease)
rosette virus (see groundnut rosette virus)
stripe potyvirus, PtSpV, **237**
stunt cucumovirus PtSnV (PSV), 106, 117, 185, 186, 194, 230, **238**
yellow spot tospovirus, PtYSpoV, 230, **241**

Pear, 341–346
Pear bark measles, PrBkMsDs, 345
Pear bark necrosis disease, PrBkNDs (PBND), 345
bark split disease, PrBkSplDs (PBSD), 345
blister canker viroid, PrBsCnVd (PBCVd), 326, 328, 345
bud drop disease, PrBdDpDs **342**
freckle pit disease, PrFrPiDs (PFPD), **345**
necrotic spot disease, PrNcsDs (PNSD), 343
ring mosaic disease, PrRMDs (PRD), **342**, see ApCRsV
ring pattern mosaic disease, PrRPtMDs (PRPMD), **342**, see ApCRsV
rough bark disease, PrRgBkDs (PRBD) 345

stony pit disease, PrStnPiDs (PSPD), 321, 326, 328, 330, 339, 341, **343–345**, 345, 347
vein yellows disease, PrVYsDs, 341

PeCmMV, see Pea mosaic (common) potyvirus
PeElBrV, see Pea early browning tobravirus
PeEnMV, see Pea enation mosaic enamovirus
PeGnMtV, see Pea green mottle comovirus
PeMdMV, see Pea mild mosaic comovirus
Pepper, 126–134
Pepper mild mottle tobamovirus, PpMdMtV (PMMV), **126**
golden mosaic geminivirus, PpGMV, 134, 135
mottle potyvirus, PpMtV (PepMoV), **126–127**
veinal mottle potyvirus, PpVMtV (PVMV), **127–128**
PeRbV, see Pea rhabdovirus
Persea americana, see Avocado
PeSnV, see Pea stunt virus
PeSdBMV, see Pea seedborne mosaic potyvirus
PeSkV, see Pea streak carlavirus
Petroselinum crispum, see Parsley
Petunia asteroid mosaic tombus virus, PnAdMV (PeAMV), 381
PfWdV, see Passionfruit woodiness potyvirus
PgSnV, see Pangola stunt Fijivirus
Phaseolus vulgaris, see Bean
Phleum mottle virus, PmMtV (PhMoV), 54, 93, **96–97** (strain CkMdMV)
PhAdsDs, see Peach asteroid spot disease
PhStmPgDs, see Peach stem pitting disease
PhBkWdGgDs, see Peach bark and wood growing disease
PhEnV, see Peach enation ? nepovirus
Phleum pratense, see Timothygrass
PhLNDs, see Peach leaf necrosis disease
PhLtMVd, see Peach latent mosaic viroid
PhMDs, see Peach mosaic disease
PhOlBhDs, see Peach oil blotch disease
PhPdSnDs, see Peach pseudo stunt disease
PhRoMV, see Peach rosette mosaic nepovirus
PhSbTwDs, see Peach stubby twig disease
PhSgCsDs, see Peach seedling chlorosis disease
PhWrDs, see Peach wart disease

PhYBdMDs, *see* Peach yellow bud mosaic disease
PhYMDs, *see* Peach yellow mosaic disease
PiRbV, *see* Pisum rhabdovirus
Pisum rhabdo virus, PiRbV (PRV), **230**
Pisum sativum, *see* Pea
PlALnPtV, *see* Plum American line pattern ilavirus
PlBkSplDs, *see* Plum bark spit disease
PlELnPtV, *see* plum European line pattern ilarvirus
PlFtCrDs, *see* Plum fruit crinkle disease
PlMtLDs, *see* Plum mottle leaf disease
PlOMDs, *see* Plum ochre mosaic disease
PlPdPDs, *see* Plum pseudopox disease
PlPV, *see* Plum pox potyvirus
PLrV, *see* Potato leafroll luteovirus
Plum, 348–360
Plum American line pattern ilarvirus, PlALnPtV (APLPV), **357–358**
　European line pattern virus, PlELnPtV (EPLPV), **358**
　fruit crinkle disease, PlFtCrDs (PFCD), 359
　bark split disease, PlBkSplDs (PBSD), 358, *see* ApCLsV
　mottle leaf disease, PlMtLDs (PMLD), 359
　ochre mosaic disease, PlOMDs (POMD), 325, 359
　pox potyvirus, PlPV (PPV), 321–330, 347, **348–354**, 355, 359, 362, 364–365, 370–373, 388, 391
　pseudopox disease, PlPdPDs (PpsPD), 358, *see* ApCLsV
PMpTpV, *see* Potato mop top furovirus
PmMtV, *see* Phleum mottle virus
PnAdMV, *see* Petunia asteriod mosaic tombus virus
PnMV, *see* Panicum mosaic sobemovirus
PnpMV, *see* Parsnip mosaic potyvirus
PnpYFkV, *see* Parsnip yellow fleck sequivirus
Potato, 240–259
Potato Andean latent virus, PAnLtV, 149
　aucuba mosaic potexvirus, PAcMV (PAMV), 241, **250–251**
　leafroll luteovirus, PLrV (PLRV), 229, **241–244**, 249
　mop-top furovirus, PMpTpV (PMTV), 241, **253–255**

　spindle tuber viroid, PSlTbVd (PSTVd), 244, **252–253**, 400
　A potyvirus, PVA, 241, **244–245**, 250, 255
　M carlavirus, PVM, **245–246**
　S carlavirus, PVS, 241, **246–247**
　X potexvirus, PVX, 130, 147–148, 229, 244, 245, 249, **255–256**, 260, 275
　yellow dwarf rhabdovirus, PYDV, 241, **251–252**
　Y potyvirus, PVY, 125, 130, 147–148, 229, 241, 245, 246, 276, **247–250**, 260, 275, 279, 298
PpGMV, *see* Pepper golden mosaic geminivirus
PpMdMtV, *see* Pepper mild mottle tobamovirus
PpMtV, *see* Pepper Mottle potyvirus
PpVMtV, *see* Pepper veinal mottle potyvirus
PrBdDpDs, *see* Pear bud drop disease
PrBkNDs, *see* Pear bark necrosis disease
PrBsCnVd, *see* Pear blister canker viroid
PrBkSplDs, *see* Pear bark split disease
PrFrPiDs, *see* Pear freckle pit disease
PrNcsDs, *see* Pear necrotic spot disease
PrRgBkDs, *see* Pear rough bark disease
PrRMDs, *see* Pear ring mosaic disease
PrRPtDs, *see* Pear ring pattern mosaic disease
PrStnPiDs, *see* Pear stony pit disease
Prune dwarf ilarvirus, PuDV (PDV), 321–324, 326–328, 330–331, 347, **354–355**, 356, 357, 359, 362, 363, 367, 370–371, 377–379, 386–387, 388
Prunus necrotic ringspot ilarvirus, PsNcRsV (PNRSV), 229, 309, 321–324, 326–328, 330, 331, 347, 354, 355, **356–357**, 358, 359, 361–364, 367, 370, 371, 375, 377, 379, 381, 386–389
Prunus armeniaca, *see* Apricot
Prunus avium, *see* Sweet cherry
Prunus cerasus, *see* Sour cherry
Prunus communis, *see* Almond
Prunus persica, *see* Peach
Prunus spp., *see* Plum
PrVYsDs, *see* Pear vein yellow disease
PsNcRsV, *see* Prunus necrotic ringspot ilarvirus
PtBdNV, *see* Peanut bud necrosis virus
PtCpV, *see* Peanut clump furovirus
PtCSkV, *see* Peanut chlorotic streak caulimovirus
PtGnMV, *see* Peanut green mosaic potyvirus

PtIdCpV, *see* Peanut Indian clump virus
PtMtV, *see* Peanut mottle potyvirus
PtSnV, *see* Peanut stunt cucumovirus
PtSlVd, *see* Potato spindle tuber viroid
PtSpV, *see* Peanut stripe potyvirus
PtYSpoV, *see* Peanut yellow spot tospovirus
PuDV, *see* Prune dwarf ilarvirus
PVA, *see* Potato A potyvirus
PVM, *see* Potato M carlavirus
PVS, *see* Potato S carlavirus
PVX, *see* Potato X potexvirus
PVY, *see* Potato Y potyvirus
PYDV, *see* Potato yellow dwarf ilarvirus
PyLtV, *see* Parsley latent rhabdovirus
Pyrus communis, *see* Pear
PyV5, *see* Potato Parsley potexvirus 5

Q

QFtDnDs, *see* Quince fruit deformation disease
QStyRsDs, *see* Quince sooty ringspot disease
Quince, 346–347
Quince fruit deformation disease, QFtDnDs (QFDD), 347, *see* PrStnPiDs
    ring pattern disease, *see* ApCLsV, 346
    sooty ringspot disease, QStyRsDs (QSRsD), 339, 342, **346–347**
    stunt disease, *see* ApCLsV, 346

R

Radish, 153–155
Radish mosaic comovirus, RaMV, 125, 150, **153–154**
    yellow edge alphacryptovirus, RaYEgV (RYEV), **154–155**
RaMV, *see* Radish mosaic comovirus
Raphanus sativus, *see* Radish
Raspberry, 443–452
Raspberry bushy dwarf ilarvirus, RbBuDV (RBDV), 326, 328, 433, **449–450**, 454
    leaf curl ? luteovirus, RbLcDs (RLCD), 321, 451, 454
    leaf mottle disease, RbLMtDs (RLMD), 325, 450–451, 455
    leaf spot disease, RbLsDs (RLSD), **450–451**, 452
    ringspot nepovirus, RbRsV (RRSV), 321, 325, 326, 359, 377, 433, 440, 441, **445–447**, 451, 457–459, 461, 479, 484, 487
    vein chlorosis ? nucleorhabdovirus, RbVCsV (RVCV), 325, 433, **443–444**
RaYEaV, *see* Radish yellow edge alphacryptovirus
RbBuDV, *see* Raspberry bushy dwarf ilarvirus
RbLcDs, *see* Raspberry leaf curl luteovirus
RbLMtDs, *see* Raspberry leaf mottle disease
RbLsDs, *see* Raspberry leaf spot disease
RbRsV, *see* Raspberry ringspot nepovirus
RBSkDV, *see* Rice black-streaked dwarf Fijivirus
RbVCsV, *see* Raspberry vein chlorosis ? nucleorhabdovirus
RClCtRsV, *see* Red clover chlorotic ringspot virus
RClMtV, *see* Red clover mottle comovirus
RClNcMV, *see* Red clover necrotic mosaic dianthovirus
RClVMV, *see* Red clover vein mosaic carlavirus
RCuIvWtMV, *see* Red currant interveinal white mosaic virus
RCuVBdDs, *see* Red currant vein banding disease
RCuYLsDs, *see* Red currant yellow leafspot disease
Red clover, 106–114
Red clover chlorotic ringspot virus, RClCtRsV (RCCRsV), **110–111**
    mottle comovirus, RClMtV (RCMV), **106–108**, 194
    necrotic mosaic dianthovirus, RClNcMV (RCNMV), **108–109**, 118
    vein mosiac carlavirus, RClVMV (RCVMV), **109–110**, 117, 194, 195
Red currant, 460–462
Red currant interveinal white mosaic virus, RCuIvWtMV (RCIWMV), **461**
    vein banding disease, RCuVBdDs (RCVBD), **462**
    yellow leaf spot disease, RCuYLsDs (RCYLSD), 462
RDV, *see* Rice dwarf phytoreovirus
RgBC, *see* Ryegrass bacilliform nucleorhabdovirus
RgCcV, *see* Ryegrass cryptic cryptovirus
RgEnV, *see* Ryegrass enation virus

Index

RGlDV, *see* Rice gall dwarf phytoreovirus
RgMV, *see* Ryegrass mosaic rymovirus
RGrSnV, *see* Rice grassy stunt tenuivirus
RgMtV, *see* Ryegrass mottle sobemovirus
RHBV, *see* Rice hoja blanca tenuivirus
Ribes sp., *see* Gooseberry, Black currant, Red currant
Rice, 54–66
Rice black-streaked dwarf Fijivirus, RBSkDV (RBSDV), 13, 14, **55–57**
    dwarf phytoreovirus, RDV, **54–55**, 56
    gall dwarf phytoreovirus, RG1DV (RGDV), 54, **57–58**
    grassy stunt tenuivirus, RGrSnV (RGSV), **62–63**
    hoja blanca tenuivirus, RHBV, 54, **61–62**
    necrosis mosaic bymovirus, RNMV, **63–64**
    ragged stunt oryzavirus, RRdSnV (RRSV), 54, **59**, 60
    stripe tenuivirus, RSpV (RSV), **59–61**, 70, 71
    transitory yellowing nucleorhabdovirus, RTyYgV (RTYV), **64**, 65
    tungro waikavirus, RTgV (RTBV), **64–65**
    yellow mottle sobemovirus, RYMtV (RYMV), **65**
RoMV, *see* Rose mosaic virus
Rose mosaic virus, RoMV, 356
RNMV, *see* Rice necrotic mosaic bymovirus
RRdSnV, *see* Rice ragged stunt oyzavirus
RSpV, *see* Rice stripe tenuivirus
RsYNtV, *see* Rubus yellow net badnavirus
RTgV, *see* Rice tungro waikavirus
RTyYgV, *see* Rice transitory yellowing nucleorhabdovirus
Rubus yellow net ? badnavirus, RsYNtV (RYNV), **444**, 452, 455
Rubus Idaeus/occidentalis, *see* Raspberry
Rye, 53–54
Ryegrass, 94–96
Ryegrass bacilliform nucleorhabdovirus, RgBcV (RyBV), **95**
    cryptic cryptovirus, RgCcV (RGCV), **95–96**
    enation virus, RgEnV, 94, *see* Lolium enation virus
    mosaic rymovirus, RgMV (RGMV), 66, 85, 89, 92, **94–95**
    mottle ? sobemovirus, RgMtV, **94**
RYMtV, *see* Rice yellow mottle sobemovirus

S

Saccharum officinarum, *see* Sugarcane
SaDV, *see* Satsuma dwarf nepovirus
Sanfoin, 118
Satsuma dwarf nepovirus, SaDV (SDV), **401–403**
SChUV, *see* Sour cherry yellow virus
SbCFkDs, *see* Strawberry chlorotic fleck disease
SbCrV, *see* Strawberry crinkle cytorhabdovirus
SbLtCV, *see* Strawberry latent C rhabdovirus
SbLtRsV, *see* Strawberry latent ringspot nepovirus
SbMdYEgV, *see* Strawberry mild yellow edge luteovirus
SbMtV, *see* Strawberry mottle virus
SbSnDs, *see* Strawberry stunt disease
SbVBdV, *see* Strawberry vein banding caulimovirus
ScBcV, *see* Sugarcane bacilliform badnavirus
ScCSkV, *see* Sugarcane chlorotic streak virus
ScMdMV, *see* Sugarcane mild mosaic virus
ScMV, *see* Sugarcane mosaic potyvirus
ScRamuSkDs, *see* Sugarcane Ramu streak disease
ScRamuSnDs, *see* Sugarcane Ramu stunt disease
ScShDs, *see* Sugarcane serch disease
SChGnRMtDs, *see* Sour cherry green ring mottle disease
ScSkV, *see* Sugarcane streak geminivirus
ScStMDs, *see* Sugarcane striate mosaic disease
ScYLV, *see* Sugarcane yellow leaf virus
Secale cereale, *see* Rye
Setaria spp., *see* Foxtail millet
SfMV, *see* Sunflower mosaic potyvirus
SfRgMDs, *see* Sunflower rugose mosaic disease
SfYBhV, *see* Sunflower yellow blotch virus
SfYgMDs, *see* Sunflower yellowing mosaic disease
Shallot latent carlavirus, ShLtV (SLV), 125, 157, **159–160**, 162
ShLtV, *see* Shallot latent carlavirus
SltV, *see* Spinach latent ilarvirus
SlYVV, *see* Sowthistle yellow vein nucleorhabdovirus

Soil-borne wheat mosaic furovirus, *see* Wheat soil-borne mosaic furovirus, WsoMV
Solanum melongena, *see* Eggplant
Solanum tuberosum, *see* Potato
Sorghum, 67–71, *see also* Forage Sorghum, 92
Sorhum bicolor, *see* Sorghum
Sorghum chloroticspot furovirus, SrCsV (SCSV), **68**
Sorghum halepense, *see* Johnsongrass
Sorghum mosaic potyvirus, SrMV, 2, 4, 22, **67–68**, 92, 98, 293, 296
    stunt mosaic rhabdovirus, SrSnMV (SSMV), **68**
    yellow banding virus, SrYBaV (SYBV), **68**
Sour Cherry, 386–388
Sour cherry green ring mottle disease, SChGnRMtDs, 367, **387–388**
    yellow virus, SChYV, 378
Southern bean mosaic sobemovirus, *see* Bean southern mosaic virus, BSuMV (SBMV)
Sowbane mosaic sobemovirus, SwMV (SoMV), 341, 359, 479, 489, *see* ApLtV
Sowthistle yellow vein nucleorhabdovirus, SlYVV (SwYVV), 216, 217
Soybean, 299–305
Soybean chlorotic mottle caulimovirus, SyCMtV (SbCMV), **302–303**
    dwarf luteovirus, SyDV (SbDV), 157, **302**, 303
    mosaic potyvirus, SyMV (SbMV), **300–302**
    yellow mosaic potyvirus, *see* BYMV
Spinach, 216–219
Spinacia oleracea, *see* Spinach
Spinach latent ilarvirus, SLtV (SPLV), **217–218**
SqLcV, *see* Squash leaf curl geminivirus
Squash, *see* Melon
Squash leaf curl geminivirus, SqLcV, **172**, 173
SqMV, *see* Squash mosaic comovirus
Squash mosaic comovirus, SqMV (SMV), 125, 162, **169–170**
    yellow leaf curl geminivirus, SqYLcV, 134, 173
SrCsV, *see* Sorghum chlorotic spot furovirus
SrMV, *see* Sorghum mosaic potyvirus
SrNcMtV, *see* Sorghum necrotic mottle virus
SrSnMV, *see* Sorghum stunt mosaic rhabdovirus
SrYBaV, *see* Sorghum yellow banding virus

StcLtV, *see* Sweet clover latent (?) nucleorhabdovirus
StcNcMV, *see* Sweet clover necrotic mosaic dianthovirus
Strawberry, 434–443
Strawberry chlorotic fleck disease, SbCFkDs (SCFD), **422**
    crinkle cytorhabdovirus, SbCrV (SCrV), 325, 433, **434–435**, 437
    latent C rhabdovirus, SbLtCV (SLCD), **422**
    latent ringspot (?) nepovirus, SbLtRsV (SLRSV), 207, 325, 327, 359, 362, 366, 373, 377, 409, **438–440**, 447, 457–459, 461, 479, 484, 487
    mild yellow edge luteovirus, SbMdYEgV (SMYEV), 325, 433, 434, **435–436**, 437
    mottle virus, SbMtV (SmtV), 433, 434, 435, **436–437**
    stunt disease, SbSnDs (SSnD), **442–443**
    vein banding (?) caulimovirus, SbVBdV (SVBV), 325, 434, **437–438**, 443
    yellows virus, syn. yellow edge
Subterranean Clover, 115
Subterranean clover mottle sobemovirus, *see* Clover subterranean mottle sobemovirus
Sugarbeet (see Beet)
Sugarcane, 292–297
Sugarcane bacilliform badnavirus, ScBcV (ScBV), 295, 297
    chlorotic streak virus, ScCSkV, 71, **296**
    Fiji disease Fijivirus, *see* Fiji disease Fijivirus, FjDsV, 71, **293–295**
    mild mosaic virus, ScMdMV, **297**
    mosaic potyvirus, ScMV (SCMV), 1, 4, 7, 8, 9, 21, 24, 66, 71, 98, 229, **292–293**, 330
    Ramu streak disease, ScRamuSkDs, 297
    Ramu stunt disease, ScRamuSnDs, 297
    streak geminivirus I, ScSkV (SCSV), **295**
    striate mosaic disease, ScStMDs (ScStMD), **296**
    serch disease, ScShDs (ScSD), **295**
    yellow leaf virus, ScYLV, 297
Sunflower, 297–299
Sunflower mosaic (?) potyvirus, SfMV, 297–298
    rugose mosaic disease, SfRgMDs, 298
    yellow blotch virus, SfYBhV, 240
    yellowing mosaic disease, SfYgMDs, 298
Sweet cherry, *see* Cherry
Sweet vernal grass, 96

Index

Sweetclover latent nucleorhabdovirus, StcLtV,
  see Melilotus latent virus, MsLtV, **118**
  necrotic mosaic dianthovirus, StcNcMV
    (SCNMV), **118**
SyCMtV, see Soybean chlorotic mottle
  caulimovirus
SyDV, see Soybean dwarf luteovirus
SyMV, see Soybean mosaic potyvirus

T

TBRV, see Tobacco black ring nepovirus
TbRsDs, see Thimbleberry ringspot disease
TEtV, see Tobacco etch potyvirus
Thimbleberry ringspot disease, TbRsDs (TrsD),
  457
Timothygrass, 96–97
Timothy mottle virus, TtMtV, **96–97**
  black ring nepovirus, TBRV, **194**
TLCuV, see Tobacco leaf curl geminivirus
TmAsV, see Tomato aspermy cucumovirus
TmBRV, see Tomato black ring nepovirus
TmBuSnV, see Tomato bushy stunt
  tombusvirus
TmGMV, see Tomato golden mosaic
  geminivirus III
TmMV, see Tomato mosaic tobamovirus
TmPuV, see Tomato Peru potyvirus
TmRsV, see Tomato ringspot nepovirus
TmSpoWV, see Tomato spotted wilt tospovirus
TmTpNV, see Tomato top necrosis nepovirus
TmYDV, see Tomato yellow dwarf geminivirus
TmYLCuV, see Tomato yellow leaf curl
  geminivirus III
TMV, see Tobacco mosaic tobamovirus
TmYNtV, see Tomato yellow net virus
TmYTpV, see Tomato yellow top luteovirus
TNV, see Tobacco necrosis necrovirus
TNcDV, see Tobacco necrotic dwarf luteovirus
Tobacco, 259–276
Tobacco etch potyvirus, TEtV (TEV), 260,
  263–265
  leaf curl geminivirus, TLCuV (TLCV), 131,
    148, 260, **266–267**, 303
  mosaic tobamovirus, TMV, 125, 126,
    131-134, 140, 146, 147, 149, 150, 156,
    229, 257, 259, **260–263**, 269, 275, 276,
    329, 341, 346, 359, 381, 441, 479,
    489

necrosis necrovirus, TNV, 162, 167, 173,
  194, 215, 229, 257, 260, **271–272**, 303,
  306, 310, 341, 359, 441, 479, 488–489
necrotic dwarf luteovirus, TNcDV (TNDV),
  260, **266**
rattle tobravirus, TRtV (TRV), 215, 229, 257-
  258, 260, **267–269**, 292, 299, 461
ringspot nepovirus, TRsV (TRSV), 117, 167,
  168, 203, 216, 219, 260, **269–271**, 275,
  299, 303, 305, 341, 413, 455, 456, 461,
  465, 477, 479, 487
streak ilarvirus, TSkV (TSV), 118, 131, 148,
  184-186, 194, 203, 211, 216, 217, 229,
  240, 257, 260, **273–274**, 299, 303, 304,
  399, 441, 443
stunt varicosavirus, TSnV (TStV), 260,
  272–273
vein distortion luteovirus, TVDiV, 274
vein mottling potyvirus, TVMtV (TVMV),
  260, **265**
yellow dwarf geminivirus I, TYDV, 260, **266**
yellow net luteovirus, TYNtV, 274
Tomato, 134–148
Tomato aspermy cucumovirus, TmAsV (TAV),
  125, **134–136**, 207
  black ring nepovirus, TmBRV (TBRV), 106,
    125, **136–137**, 162, 167, 173, 185, 203,
    207, 208, 216, 257, 292, 325, 327, 365,
    389, 441, 447, 451, 459, 479, 484,
    487–488, 490
  bushy stunt tombusvirus, TmBuSnV
    (TBSV), 125, 134, **137–139**, 211, 219,
    341, 381, 477, 479, 489
  golden mosaic geminivirus III, TmGMV
    (TGMV), **139**
  mosaic tobamovirus, TmMV (ToMV),
    139–141
  necrosis necrovirus, TmNV, 185
  Peru potyvirus, TmPuV, 144
  ringspot nepovirus, TmRsV(ToRSV), 117,
    141–142, 275, 325, 327, 328, 341, 347,
    359, 366–367, 373, 381, 389, 413, 441,
    447–449, 457, 459, 461, 465–466, 477,
    479, 487–490
  spotted wilt tospovirus, TmSpoWV (TSWV),
    125, 134, **142–144**, 149, 150, 162, 174,
    186, 194, 203, 215, 216, 229–232, 257,
    258, 260, 275–276, 277, 299, 303, 329,
    413

streak virus, TmSkV, 455
top necrosis (?) nepovirus- TmTpNV (TTNV), 144
yellow dwarf geminivirus, TmYDV (TYDV), 144, 183, see Bean summer death geminivirus
yellow leaf curl geminivirus III, TmYLCuV (TYLCV), 134, 135, 144–145
yellow net virus, TmYNtV (TYNV), 144
yellow top luteovirus, TmYTpV (TYTV), 144
TmNV, see Tomato necrosis Necrovirus
TmSkV, see Tomato streak virus
Trifolium hybridum, see Alsike Clover
Trifolium incarnatum, see Crimson Clover
Trifolium pratense, see Red Clover
Trifolium repens, see White Clover
Trifolium subterraneum, see Subterranean Clover
Triticum aestivum, see Wheat
TRsV, see Tobacco ringspot nepovirus
TRtV, see Tobacco rattle tobravirus
TSkV, see Tobacco streak ilarvirus
TSnV, see tobacco stunt varicosavirus
TtMtV, see Timothy mottle virus
TuCrV, see Turnip crinkle carmovirus
TuMV, see Turnip mosaic potyvirus
Turnip, 155–157
Turnip crinkle carmovirus, TuCrV (TCV), 150, 155
mosaic potyvirus, TuMV, **150–151**, 152, 155, 216, 299
rosette mosaic sobemovirus, TuRoMV (TRoMV), 150, **155–156**
yellow mosaic tymovirus, TuYMV (TYMV), 150, **156–157**
yellows luteovirus, TuYsV (TYV) 150, **157**, 290
TuRoMV, see Turnip rosette mosaic sobemovirus
TuYMV, see Turnip yellow mosaic tymovirus
TuYsV, see Turnip yellows luteovirus
TVDiV, see Tobacco vein distortion luteovirus
TVMtV, see Tobacco vein mottling potyvirus
TYDV, see Tobacco yellow dwarf geminivirus I
TYNtV, see Tobacco yellow net luteovirus

V

Vaccinium sp., see Blueberry
Vetch, 117
Vetch spotted crinkle disease, see Pea enation mosaic enamovirus, 117
vein banding chlorosis disease, see Broad bean stain comovirus, 117
yellow mosaic disease, see Bean yellow mosaic potyvirus, 117
Vicia faba, see Broad bean
Vicia spp., see Vetch
Vigna spp., see Cowpea
Vitis spp., see Grapevine

W

Walnut, 389–391
Walnut line pattern disease, WtLnPtDs (WtLPD), **391**
mosaic disease WtMDs (WtMD), **391**
WAStMV, see Wheat American striate mosaic rhabdovirus
Watercress mosaic virus, WcMV, 151
Watermelon, see Melon
Watermelon mosaic potyvirus 1, WmMV1 (WMV1), 410–413, see Papaya ringspot virus
mosaic potyvirus 2, WmMV2 (WMV2), 98, 125, 162, 167, **168–169**, 172, 186, 194, 413
WClRMtDs, see White clover ring mottle disease
WCMV, see Watercress mosaic virus
WbLtV, see Wineberry latent potexvirus
WcCcV2, see White clover betacryptovirus 2
WClEnDs, see White clover enation disease
WClMV, see White clover mosaic potexvirus
WCSkV, see Wheat chlorotic streak rhabdovirus
WDV, see Wheat dwarf geminivirus I
WEStMV, see Wheat European striate tenuivirus
Wheat, 24–38
Wheat American striate mosaic rhabdovirus, WAStMV (AWSMV), 23, **25–26**, 53
chlorotic streak rhabdovirus, WCSkV, 35, 37, 85

dwarf geminivirus I, WDDs (WDV), **34–35**, 53, 86
European striate mosaic (?) tenuivirus, WEStMV, 24, **26–27**, 86, 96, 97
soil-borne mosaic furovirus, WSoMV (WSBMV) (Soil-borne wheat mosaic virus, SBWMV), **27–28**, 29, 48, 53
spindle streak mosaic bymovirus, WSlSkMV (WSSMV), **32–34**
streak mosaic rymovirus, WSkMV, 23, 24, **28–32**, 36, 48, 49, 54, 93, 94
striate mosaic virus, WStMV, 16, **24–25**, 48
winter mosaic tenuivirus, WWnMV (WWMV), **34**, 48, 53
yellow leaf closterovirus, WYLV, White Clover, 115–117
White clover betacryptovirus 2, WcCcV2 (WCCV2), 115, **116**
 enation disease, WClEnDs (WCED), 116
 mosaic potexvirus, WClMV (WCMV), 113, 114, **115**, 303
 ring mottle disease, WclRMtDs, 116, *see Arabis mosaic virus*

Wineberry latent potexvirus, WbLtV (WLV), 456
Winter wheat mosaic tenuivirus (see Wheat winter mosaic tenuivirus)
WSkMV, *see* Wheat streak mosaic rumovirus
WSlSkV, *see* Wheat spindle streak mosaic bymovirus
WmMV1, *see* Watermelon mosaic potyvirus 1
WmMV2, *see* Watermelon mosaic potyvirus 2
WSoMV, *see* Wheat soil-borne mosaic furovirus
WStMV, *see* Wheat striate mosaic virus
WWnMV, *see* Wheat winter mosaic tenuivirus
WtLnPtDs, *see* Walnut line pattern disease
WtMDs, *see* Walnut mosaic disease

Z

Zea Mays, *see* Maize
Zucchini yellow fleck potyvirus, ZYFkV (ZYFV), 162, 167, **172**
 yellow mosaic potyvirus, ZYMV, 125, 162, 167, **170–172**
ZYFkV, *see* Zucchini yellow fleck potyvirus
ZYMV, *see* Zucchini yellow mosaic potyvirus